MULTIMEDIA IMAGE and VIDEO PROCESSING

IMAGE PROCESSING SERIES

Series Editor: Phillip A. Laplante

Published Titles

Image and Video Compression for Multimedia Engineering
Yun Q. Shi and Huiyang Sun

Forthcoming Titles

Adaptive Image Processing: A Computational Intelligence Perspective
Ling Guan, Hau-San Wong, and Stuart William Perry

Shape Analysis and Classification: Theory and Practice
Luciano da Fontoura Costa and Roberto Marcondes Cesar, Jr.

MULTIMEDIA IMAGE and VIDEO PROCESSING

Edited by
Ling Guan
Sun-Yuan Kung
Jan Larsen

CRC Press
Boca Raton London New York Washington, D.C.

Library of Congress Cataloging-in-Publication Data

Multimedia image and video processing / edited by Ling Guan, Sun-Yuan Kung, Jan Larsen.
p. cm.
Includes bibliographical references and index.
ISBN 0-8493-3492-6 (alk.)
1. Multimedia systems. 2. Image processing—Digital techniques. I. Guan, Ling. II. Kung, S.Y. (Sun Yuan) III. Larsen, Jan.

QA76.575 2000
006.4'2—dc21 00-030341

Direct all inquiries to CRC Press LLC, 2000 N.W. Corporate Blvd., Boca Raton, Florida 33431.

International Standard Book Number 0-8493-3492-6
Library of Congress Card Number 00-030341
Printed in the United States of America 1 2 3 4 5 6 7 8 9 0
Printed on acid-free paper

Contents

Preface

Multimedia is one of the most important aspects of the information era. Although there are books dealing with various aspects of multimedia, a book comprehensively covering system, processing, and application aspects of image and video data in a multimedia environment is urgently needed. Contributed by experts in the field, this book serves this purpose.

Our goal is to provide in a single volume an introduction to a variety of topics in image and video processing for multimedia. An edited compilation is an ideal format for treating a broad spectrum of topics because it provides the opportunity for each topic to be written by an expert in that field.

The topic of the book is processing images and videos in a multimedia environment. It covers the following subjects arranged in two parts: (1) standards and fundamentals: standards, multimedia architecture for image processing, multimedia-related image processing techniques, and intelligent multimedia processing; (2) methodologies, techniques, and applications: image and video coding, image and video storage and retrieval, digital video transmission, video conferencing, watermarking, distance education, video on demand, and telemedicine.

The book begins with the existing standards for multimedia, discussing their impacts to multimedia image and video processing, and pointing out possible directions for new standards.

The design of multimedia architectures is based on the standards. It deals with the way visual data is being processed and transmitted at a more practical level. Current and new architectures, and their pros and cons, are presented and discussed in Chapters 2 to 4.

Chapters 5 to 8 focus on conventional and intelligent image processing techniques relevant to multimedia, including preprocessing, segmentation, and feature extraction techniques utilized in coding, storage, and retrieval and transmission, media fusion, and graphical interface.

Compression and coding of video and images are among the focusing issues in multimedia. New developments in transform- and motion-based algorithms in the compressed domain, content- and object-based algorithms, and rate–distortion-based encoding are presented in Chapters 9 to 12.

Chapters 13 to 15 tackle content-based image and video retrieval. They cover video modeling and retrieval, retrieval in the transform domain, indexing, parsing, and real-time aspects of retrieval.

The last chapters of the book (Chapters 16 to 19) present new results in multimedia application areas, including transcoding for multipoint video conferencing, distance education, watermarking techniques for multimedia processing, and telemedicine.

Each chapter has been organized so that it can be covered in 1 to 2 weeks when this book is used as a principal reference or text in a senior or graduate course at a university.

It is generally assumed that the reader has prior exposure to the fundamentals of image and video processing. The chapters have been written with an emphasis on a tutorial presentation so that the reader interested in pursuing a particular topic further will be able to obtain a solid introduction to the topic through the appropriate chapter in this book. While the topics covered are related, each chapter can be read and used independently of the others.

This book is primarily a result of the collective efforts of the chapter authors. We are very grateful for their enthusiastic support, timely response, and willingness to incorporate suggestions from us, from other contributing authors, and from a number of our colleagues who served as reviewers.

Ling Guan

Sun-Yuan Kung

Jan Larsen

Contributors

Tülay Adali University of Maryland, Baltimore, Maryland

Horst Bunke Institute für Informatik und Angewandte Mathematik, Universität Bern, Switzerland

Frank M. Candocia University of Florida, Gainesville, Florida

Chang Wen Chen University of Missouri, Columbia, Missouri

Tsuhan Chen Carnegie Mellon University, Pittsburgh, Pennsylvania

Tat-Seng Chua National University of Singapore, Kentridge, Singapore

Sachin G. Deshpande University of Washington, Seattle, Washington

Li Fan University of Missouri, Columbia, Missouri

Ling Guan University of Sydney, Sydney, Australia

Lars Kai Hansen Technical University of Denmark, Lyngby, Denmark

N. Herodotou University of Toronto, Toronto, Ontario, Canada

Yu Hen Hu University of Wisconsin-Madison, Madison, Wisconsin

Shih-Kun Huang Institute of Information Science, Academia Sinica, Taiwan, China

Thomas S. Huang Beckman Institute, University of Illinois at Urbana-Champaign, Urbana, Illinois

Jenq-Neng Hwang University of Washington, Seattle, Washington

Yi Kang Beckman Institute, University of Illinois at Urbana-Champaign, Urbana, Illinois

Aggelos K. Katsaggelos Northwestern University, Evanston, Illinois

S.W. Kim Korea Advanced Institute of Science and Technology, Taejon, Korea

Surin Kittitornkun University of Wisconsin-Madison, Madison, Wisconsin

Ut-Va Koc Lucent Technologies Bell Labs, Murray Hill, New Jersey

Thomas Kolenda Technical University of Denmark, Lyngby, Denmark

Sun-Yuan Kung Princeton University, Princeton, New Jersey

Jan Larsen Technical University of Denmark, Lyngby, Denmark

Jose A. Lay University of Sydney, Sydney, Australia

Hong-Yuan Mark Liao Institute of Information Science, Academia Sinica, Taipei, Taiwan

K.J. Ray Liu University of Maryland, College Park, Maryland

Chun-Shien Lu Institute of Information Science, Academia Sinica, Taipei, Taiwan

Gerry Melnikov Northwestern University, Evanston, Illinois

Jörn Ostermann AT&T Labs — Research, Red Bank, New Jersey

K.N. Plataniotis University of Toronto, Toronto, Ontario, Canada

Jose C. Principe University of Florida, Gainesville, Florida

K.R. Rao University of Texas at Arlington, Arlington, Texas

Kim Shearer Curtin University of Technology, Perth, Australia

Ming-Ting Sun University of Washington, Seattle, Washington

S. Suthaharan Tennessee State University, Nashville, Tennessee

Chwen-Jye Sze Institute of Information Science, Academia Sinica, Taiwan, China

A.N. Venetsanopoulos University of Toronto, Toronto, Ontario, Canada

Svetha Venkatesh Curtin University of Technology, Perth, Australia

Yue Wang Catholic University of America, Washington, D.C.

H.R. Wu Monash University, Clayton, Victoria, Australia

Tzong-Der Wu University of Washington, Seattle, Washington

Yi Zhang National University of Singapore, Kent Ridge, Singapore

Chapter 1

Emerging Standards for Multimedia Applications

Tsuhan Chen

1.1 Introduction

Due to the rapid growth of multimedia communication, multimedia standards have received much attention during the last decade. This is illustrated by the extremely active development in several international standards including H.263, H.263 Version 2 (informally known as H.263+), H.26L, H.323, MPEG-4, and MPEG-7. H.263 Version 2, developed to enhance an earlier video coding standard H.263 in terms of coding efficiency, error resilience, and functionalities, was finalized in early 1997. H.26L is an ongoing standard activity searching for advanced coding techniques that can be fundamentally different from H.263. MPEG-4, with its emphasis on content-based interactivity, universal access, and compression performance, was finalized with Version 1 in late 1998 and with Version 2 1 year later. The MPEG-7 activity, which has begun since the first call for proposals in late 1998, is developing a standardized description of multimedia materials, including images, video, text, and audio, in order to facilitate search and retrieval of multimedia content. By examining the development of these standards in this chapter, we will see the trend of video technologies progressing from pixel-based compression techniques to high-level image understanding. At the end of the chapter, we will also introduce H.323, an ITU-T standard designed for multimedia communication over networks that do not guarantee quality of service (QoS), and hence very suitable for Internet applications.

The chapter is outlined as follows. In Section 1.2, we introduce the basic concepts of standards activities. In Section 1.3, we review the fundamentals of video coding. In Section 1.4, we study recent video and multimedia standards, including H.263, H.26L, MPEG-4, and MPEG-7. In Section 1.5, we briefly introduce standards for multimedia communication, focusing on ITU-T H.323. We conclude the chapter with a brief discussion on the trend of multimedia standards (Section 1.6).

1.2 Standards

Standards are essential for communication. Without a common language that both the transmitter and the receiver understand, communication is impossible. In multimedia communication systems the language is often defined as a standardized *bitstream syntax*. Adoption of standards by equipment manufacturers and service providers increases the customer base and hence results in higher volume and lower cost. In addition, it offers consumers more freedom of choice among manufacturers, and therefore is welcomed by the consumers.

For transmission of video or multimedia content, standards play an even more important role. Not only do the transmitter and the receiver need to speak the same language, but the language also has to be efficient (i.e., provide high compression of the content), due to the relatively large amount of bits required to transmit uncompressed video and multimedia data.

Note, however, that standards do not specify the whole communication process. Although it defines the bitstream syntax and hence the decoding process, a standard usually leaves the encoding processing open to the vendors. This is the *standardize-the-minimum* philosophy widely adopted by most video and multimedia standards. The reason is to leave room for competition among different vendors on the encoding technologies, and to allow future technologies to be incorporated into the standards, as they become mature. The consequence is that a standard does not guarantee the quality of a video encoder, but it ensures that any standard-compliant decoder can properly receive and decode the bitstream produced by any encoder.

Existing standards may be classified into two groups. The first group comprises those that are decided upon by a mutual agreement between a small number of companies. These standards can become very popular in the marketplace, thereby leading other companies to also accept them. So, they are often referred to as the *de facto* standards. The second set of standards is called the *voluntary* standards. These standards are defined by volunteers in open committees. These standards are agreed upon based on the consensus of all the committee members. These standards need to stay ahead of the development of technologies, in order to avoid any disagreement between those companies that have already developed their own proprietary techniques.

For multimedia communication, there are several organizations responsible for the definition of voluntary standards. One is the International Telecommunications Union–Telecommunication Standardization Sector (ITU-T), originally known as the International Telephone and Telegraph Consultative Committee (CCITT). Another one is the International Standardization Organization (ISO). Along with the Internet Engineering Task Force (IETF), which defines multimedia delivery for the Internet, these three organizations form the core of standards activities for modern multimedia communication.

Both ITU-T and ISO have defined different standards for video coding. These standards are summarized in Table 1.1. The major differences between these standards lie in the operating bit rates and the applications for which they are targeted. Note, however, that each standard allows for operating at a wide range of bit rates; hence each can be used for all the applications in principle. All these video-related standards follow a similar framework in terms of the coding algorithms; however, there are differences in the ranges of parameters and some specific coding modes.

1.3 Fundamentals of Video Coding

In this section, we review the fundamentals of video coding. Figure 1.1 shows the general data structure of digital video. A video sequence is composed of pictures updated at a certain rate, sometimes with a number of pictures grouped together (group of pictures [GOP]). Each picture is composed of several groups of blocks (GOBs), sometimes called the slices. Each GOB contains a number of macroblocks (MBs), and each MB is composed of four luminance

Table 1.1 Video Coding Standards Developed by Various Organizations

Organization	Standard	Typical Bit Rate	Typical Applications
ITU-T	H.261	$p \times 64$ kbits/s, $p = 1 \ldots 30$	ISDN Video Phone
ISO	IS 11172-2 MPEG-1 Video	1.2 Mbits/s	CD-ROM
ISO	IS 13818-2 MPEG-2 Video[a]	4–80 Mbits/s	SDTV, HDTV
ITU-T	H.263	64 kbits/s or below	PSTN Video Phone
ISO	IS 14496-2 MPEG-4 Video	24–1024 kbits/s	A variety of applications
ITU-T	H.26L	<64 kbits/s	A variety of applications

[a]ITU-T also actively participated in the development of MPEG-2 Video. In fact, ITU-T H.262 refers to the same standard and uses the same text as IS 13818-2.

blocks, 8×8 pixels each, which represent the intensity variation, and two chrominance blocks (C_B and C_R), which represent the color information.

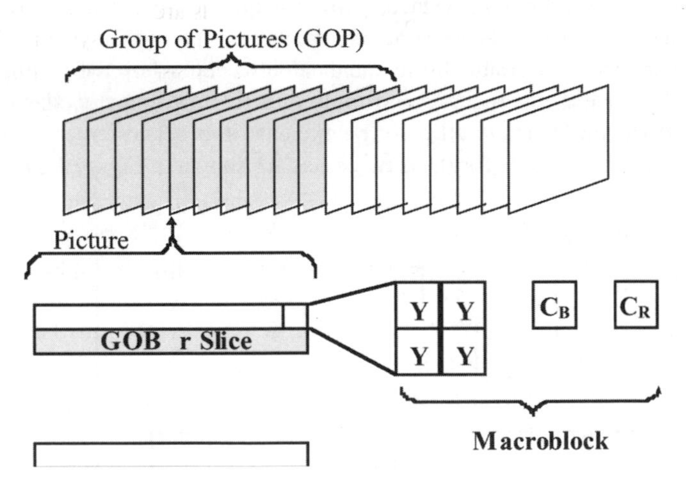

FIGURE 1.1
Data structure of digital video.

The coding algorithm widely used in most video coding standards is a combination of the *discrete cosine transform (DCT)* and *motion compensation.* DCT is applied to each block to transform the pixel values into DCT coefficients in order to remove the spatial redundancy. The DCT coefficients are then quantized and zigzag scanned to provide a sequence of symbols, with each symbol representing a number of zero coefficients followed by one nonzero coefficient. These symbols are then converted into bits by entropy coding (e.g., variable-length coding [VLC]). On the other hand, temporal redundancy is removed by motion compensation (MC). The encoder estimates the motion by matching each macroblock in the current picture with the reference picture (usually the previous picture) to find the *motion vector* that specifies the best matching area. The residue is then coded and transmitted with the motion vectors. We now discuss these techniques in detail.

1.3.1 Transform Coding

Transform coding has been widely used to remove redundancy between data samples. In transform coding, a set of data samples is first linearly transformed into a set of *transform coefficients*. These coefficients are then quantized and coded. A proper linear transform should decorrelate the input samples, and hence remove the redundancy. Another way to look at this is that a properly chosen transform can concentrate the energy of input samples into a small number of transform coefficients, so that resulting coefficients are easier to code than the original samples.

The most commonly used transform for video coding is the DCT [1, 2]. In terms of both objective coding gain and subjective quality, the DCT performs very well for typical image data. The DCT operation can be expressed in terms of matrix multiplication by:

$$\mathbf{Z} = \mathbf{C}^T \mathbf{X} \mathbf{C}$$

where **X** represents the original image block and **Z** represents the resulting DCT coefficients. The elements of **C**, for an 8×8 image block, are defined as

$$C_{mn} = k_n \cos\left[\frac{(2m+1)n\pi}{16}\right] \quad \text{where } k_n = \begin{cases} 1/(2\sqrt{2}) & \text{when } n = 0 \\ 1/2 & \text{otherwise} \end{cases}$$

After the transform, the DCT coefficients in **Z** are quantized. Quantization implies loss of information and is the primary source of actual compression in the system. The quantization step size depends on the available bit rate and can also depend on the coding modes. Except for the intra-DC coefficients that are uniformly quantized with a step size of 8, an enlarged "dead zone" is used to quantize all other coefficients in order to remove noise around zero. Typical input–output relations for these two cases are shown in Figure 1.2.

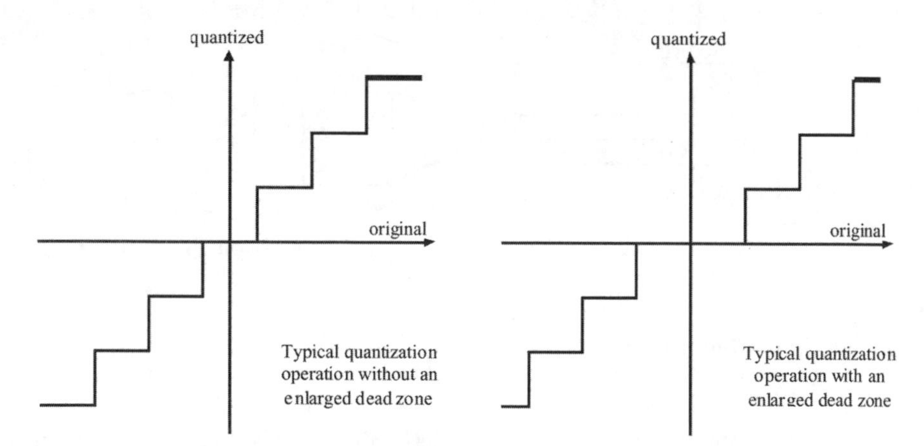

FIGURE 1.2
Quantization with and without the "dead zone."

The quantized 8×8 DCT coefficients are then converted into a one-dimensional (1D) array for entropy coding by an ordered scanning operation. Figure 1.3 shows the zigzag scan order used in most standards for this conversion. For typical video data, most of the energy concentrates in the low-frequency coefficients (the first few coefficients in the scan order) and the high-frequency coefficients are usually very small and often quantized to zero. Therefore, the scan order in Figure 1.3 can create long runs of zero-valued coefficients, which is important for efficient entropy coding, as we discuss in the next paragraph.

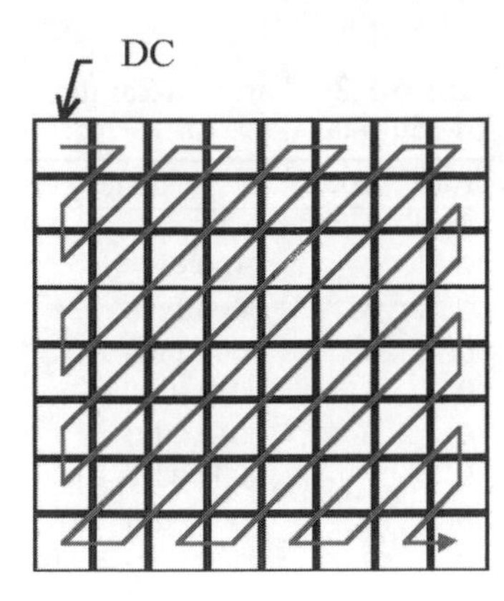

FIGURE 1.3
Scan order of the DCT coefficients.

The resulting 1D array is then decomposed into segments, with each segment containing either a number of consecutive zeros followed by a nonzero coefficient or a nonzero coefficient without any preceding zeros. Let an *event* represent the pair *(run, level)*, where "run" represents the number of zeros and "level" represents the magnitude of the nonzero coefficient. This coding process is sometimes called "run-length coding." Then, a table is built to represent each event by a specific codeword (i.e., a sequence of bits). Events that occur more often are represented by shorter codewords, and less frequent events are represented by longer codewords. This entropy coding process is therefore called VLC or Huffman coding. Table 1.2 shows part of a sample VLC table. In this table, the last bit "s" of each codeword denotes the sign of the level, "0" for positive and "'1" for negative. It can be seen that more likely events (i.e., short runs and low levels), are represented with short codewords, and vice versa.

At the decoder, all the above steps are reversed one by one. Note that all the steps can be exactly reversed except for the quantization step, which is where loss of information arises. This is known as "lossy" compression.

1.3.2 Motion Compensation

The transform coding described in the previous section removes spatial redundancy within each frame of video content. It is therefore referred to as *intra* coding. However, for video material, *inter* coding is also very useful. Typical video material contains a large amount of redundancy along the temporal axis. Video frames that are close in time usually have a large amount of similarity. Therefore, transmitting the difference between frames is more efficient than transmitting the original frames. This is similar to the concept of differential coding and predictive coding. The previous frame is used as an estimate of the current frame, and the residual, the difference between the estimate and the true value, is coded. When the estimate is good, it is more efficient to code the residual than the original frame.

Consider the fact that typical video material is a camera's view of moving objects. Therefore, it is possible to improve the prediction result by first estimating the motion of each region in the scene. More specifically, the encoder can estimate the motion (i.e., displacement) of each block between the previous frame and the current frame. This is often achieved by matching each block (actually, macroblock) in the current frame with the previous frame to find the best matching area,[1] as illustrated in Figure 1.4. This area is then offset accordingly to form the estimate of the corresponding block in the current frame. Now, the residue has much less energy than the original signal and therefore is much easier to code to within a given average error.

Table 1.2 Part of a Sample VLC Table

Run	Level	Code
0	1	11s
0	2	0100 s
0	3	0010 1s
0	4	0000 110s
0	5	0010 0110 s
0	6	0010 0001 s
0	7	0000 0010 10s
0	8	0000 0001 1101 s
0	9	0000 0001 1000 s
0	10	0000 0001 0011 s
0	11	0000 0001 0000 s
0	12	0000 0000 1101 0s
0	13	0000 0000 1100 1s
0	14	0000 0000 1100 0s
0	15	0000 0000 1011 1s
1	1	011s
1	2	0001 10s
1	3	0010 0101 s
1	4	0000 0011 00s
1	5	0000 0001 1011 s
1	6	0000 0000 1011 0s
1	7	0000 0000 1010 1s
2	1	0101 s
2	2	0000 100s
2	3	0000 0010 11s
2	4	0000 0001 0100 s
2	5	0000 0000 1010 0s
3	1	0011 1s
3	2	0010 0100 s
3	3	0000 0001 1100 s
3	4	0000 0000 1001 1s
...	...	...

This process is called motion compensation (MC), or more precisely, motion-compensated prediction [3, 4]. The residue is then coded using the same process as that of intra coding.

Pictures that are coded without any reference to previously coded pictures are called *intra* pictures, or simply I pictures (or I frames). Pictures that are coded using a previous picture as a reference for prediction are called *inter* or *predicted* pictures, or simply P pictures (or P frames). However, note that a P picture may also contain some intra-coded macroblocks. The reason is as follows. For a certain macroblock, it may be impossible to find a good enough matching area in the reference picture to be used for prediction. In this case, direct intra coding of such a macroblock is more efficient. This situation happens often when there is occlusion or intense motion in the scene.

[1] Note, however, that the standard does not specify how motion estimation should be done. Motion estimation can be a very computationally intensive process and is the source of much of the variation in the quality produced by different encoders.

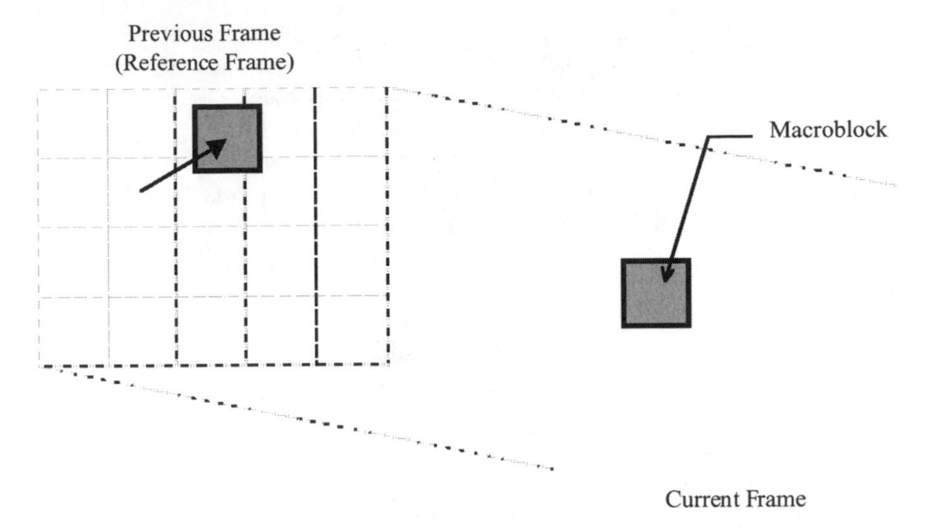

FIGURE 1.4
Motion compensation.

During motion compensation, in addition to bits used for coding the DCT coefficients of the residue, extra bits are required to carry information about the motion vectors. Efficient coding of motion vectors is therefore also an important part of video coding. Because motion vectors of neighboring blocks tend to be similar, differential coding of the horizontal and vertical components of motion vectors is used. That is, instead of coding motion vectors directly, the previous motion vector or multiple neighboring motion vectors are used as a prediction for the current motion vector. The difference, in both the horizontal and vertical components, is then coded using a VLC table, part of which is shown in Table 1.3. Note two things in

Table 1.3 Part of a VLC Table for Coding Motion Vectors

MVD	Code
...	...
−7 & 25	0000 0111
−6 & 26	0000 1001
−5 & 27	0000 1011
−4 & 28	0000 111
−3 & 29	0001 1
−2 & 30	0011
−1	011
0	1
1	010
2 & −30	0010
3 & −29	0001 0
4 & −28	0000 110
5 & −27	0000 1010
6 & −26	0000 1000
7 & −25	0000 0110
...	...

this table. First, short codewords are used to represent small differences, because these are more likely events. Second, one codeword can represent up to two possible values for motion vector difference. Because the allowed range of both the horizontal component and the vertical component of motion vectors is restricted to the range of -15 to $+15$, only one will yield a motion vector with the allowable range. Note that the ± 15 range for motion vector values may not be adequate for high-resolution video with large amounts of motion; some standards provide a way to extend this range as either a basic or optional feature of their design.

1.3.3 Summary

Video coding can be summarized into the block diagram in Figure 1.5. The left-hand side of the figure shows the encoder and the right-hand side shows the decoder. At the encoder, the input picture is compared with the previously decoded frame with motion compensation. The difference signal is DCT transformed and quantized, and then entropy coded and transmitted. At the decoder, the decoded DCT coefficients are inverse DCT transformed and then added to the previously decoded picture with loop-filtered motion compensation.

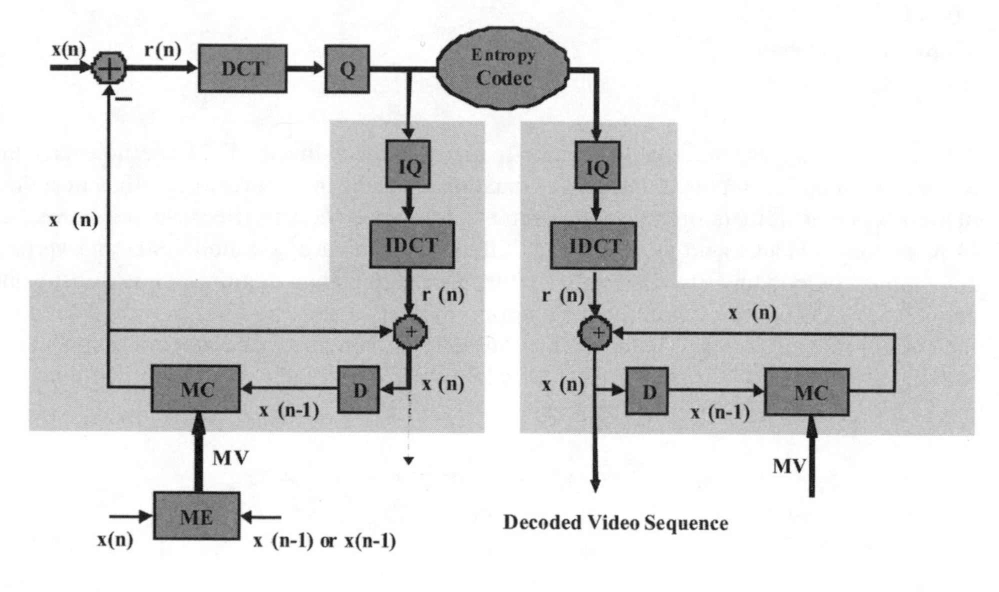

x(n): Current frame — x (n): Prediction for x(n) — x (n): Reconstructed Frame

r(n): Residue, x(n)-x (n) — r (n): Decoded residue — DCT: Discrete Cosine Transform

MC : Motion Compensation — ME: Motion Estimation — Q: Quantizer

D: Delay — IDCT: Inverse DCT — IQ: Inverse Quantizer

FIGURE 1.5
Block diagram of video coding.

1.4 Emerging Video and Multimedia Standards

Most early video coding standards, including H.261, MPEG-1, and MPEG-2, use the same hybrid DCT-MC framework as described in the previous sections, and they have very specific functionalities and targeted applications. The new generation of video coding standards, however, contains many optional modes and supports a larger variety of functionalities. We now introduce the new functionalities provided in these new standards, including H.263, H.26L, MPEG-4, and MPEG-7.

1.4.1 H.263

The H.263 design project started in 1993, and the standard was approved at a meeting of ITU-T SG 15 in November 1995 (and published in March 1996) [5]. Although the original goal of this endeavor was to design a video coding standard suitable for applications with bit rates around 20 kbits/s (the so-called very-low-bit-rate applications), it became apparent that H.263 could provide a significant improvement over H.261 at any bit rate. In essence, H.263 combines the features of H.261 with several new methods, including the half-pixel motion compensation first found in MPEG-1 and other techniques. Compared to an earlier standard H.261, H.263 can provide 50% or more savings in the bit rate needed to represent video at a given level of perceptual quality at very low bit rates. In terms of signal-to-noise ratio (SNR), H.263 can provide about a 3-dB gain over H.261 at these very low rates. In fact, H.263 provides superior coding efficiency to that of H.261 at all bit rates (although not nearly as dramatic an improvement when operating above 64 kbits/s). H.263 can also provide a significant bit rate savings when compared to MPEG-1 at higher rates (perhaps 30% at around 1 Mbit/s).

H.263 represents today's state of the art for standardized video coding. Essentially any bit rate, picture resolution, and frame rate for progressive-scanned video content can be efficiently coded with H.263. H.263 is structured around a "baseline" mode of operation, which defines the fundamental features supported by all decoders, plus a number of optional enhanced modes of operation for use in customized or higher performance applications. Because of its high performance, H.263 was chosen as the basis of the MPEG-4 video design, and its baseline mode is supported in MPEG-4 without alteration. Many of its optional features are now also found in some form in MPEG-4.

In addition to the baseline mode, H.263 includes a number of optional enhancement features to serve a variety of applications. The original version of H.263 had about four such optional modes. The latest version of H.263, known informally as H.263+ or H.263 Version 2, extends the number of negotiable options to 16 [5]. These enhancements provide either improved quality or additional capabilities to broaden the range of applications. Among the new negotiable coding options specified by H.263 Version 2, five of them are intended to improve the coding efficiency. These are the advanced intra coding mode, alternate inter VLC mode, modified quantization mode, deblocking filter mode, and improved PB-frame mode. Three optional modes are especially designed to address the needs of mobile video and other unreliable transport environments. They are the slice structured mode, reference picture selection mode, and independent segment decoding mode. The temporal, SNR, and spatial scalability modes support layered bitstream scalability, similar to those provided by MPEG-2.

There are two other enhancement modes in H.263 Version 2: the reference picture resampling mode and reduced-resolution update mode. The former allows a previously coded picture to be resampled, or warped, before it is used as a reference picture.

Another feature of H.263 Version 2 is the use of supplemental information, which may be included in the bitstream to signal enhanced display capabilities or to provide tagging information for external use. One use of the supplemental enhancement information is to specify the chroma key for representing transparent and semitransparent pixels [6].

Each optional mode is useful in some applications, but few manufacturers would want to implement all of the options. Therefore, H.263 Version 2 contains an informative specification of three levels of preferred mode combinations to be supported. Each level contains a number of options to be supported by an equipment manufacturer. Such information is not a normative part of the standard. It is intended only to provide manufacturers some guidelines as to which modes are more likely to be widely adopted across a full spectrum of terminals and networks.

Three levels of preferred modes are described in H.263 Version 2, and each level supports the optional modes specified in lower levels. In addition to the level structure is a discussion indicating that because the advanced prediction mode was the most beneficial of the original H.263 modes, its implementation is encouraged not only for its performance but for its backward compatibility with the original H.263.

The first level is composed of

- The advanced intra coding mode
- The deblocking filter mode
- Full-frame freeze by supplementary enhancement information
- The modified quantization mode

Level 2 supports, in addition to modes supported in Level 1

- The unrestricted motion vector mode
- The slice structured mode
- The simplest resolution-switching form of the reference picture resampling mode

In addition to these modes, Level 3 further supports

- The advanced prediction mode
- The improved PB-frames mode
- The independent segment decoding mode
- The alternative inter VLC mode

1.4.2 H.26L

H.26L is an effort to seek efficient video coding algorithms that can be fundamentally different from the MC-DCT framework used in H.261 and H.263. When finalized, it will be a video coding standard that provides better quality and more functionalities than existing standards. The first call for proposals for H.26L was issued in January 1998. According to the call for proposals, H.26L is aimed at very-low-bit-rate, real-time, low-end-to-end delay coding for a variety of source materials. It is expected to have low complexity, permitting software implementation, enhanced error robustness (especially for mobile networks), and adaptable rate control mechanisms. The applications targeted by H.26L include real-time conversational services, Internet video applications, sign language and lip-reading communication, video storage and retrieval services (e.g., VOD), video store and forward services (e.g., video mail), and multipoint communication over heterogeneous networks. The schedule for H.26L activities is shown in Table 1.4.

Table 1.4 Schedule for H.26L

Jan 1998	Call for proposals
Nov 1998	Evaluation of the proposals
Jan 1999	1st test model of H.26L (TML1)
Nov 1999	Final major feature adoptions
Aug 2001	Determination
May 2002	Decision

1.4.3 MPEG-4

MPEG-4 [7] was originally created as a standard for very low bit rate coding of limited-complexity audiovisual material. The scope was later extended to supporting new functionalities such as content-based interactivity, universal access, and high-compression coding of general material for a wide bit-rate range. It also emphasizes flexibility and extensibility. The concept of content-based coding of MPEG-4 is shown in Figure 1.6. Each input picture is decomposed into a number of arbitrarily shaped regions called video object planes (VOPs). Each VOP is then coded with a coding algorithm that is similar to H.263. The shape of each VOP is encoded using context-based arithmetic coding.

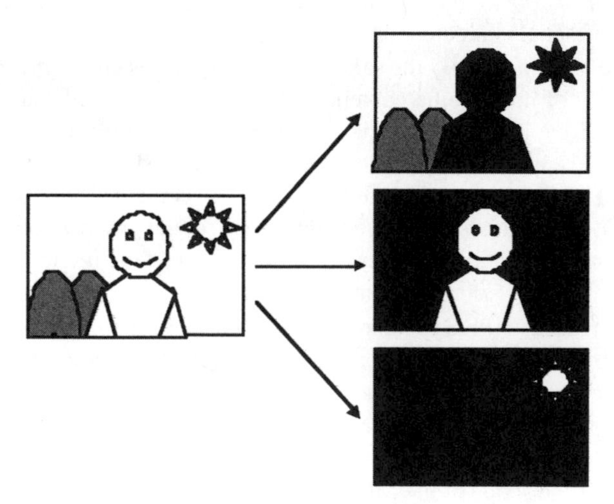

FIGURE 1.6
Object-layer-based video coding in MPEG-4.

Comparing MPEG-4 video coding with earlier standards, the major difference lies in the representation and compression of the shape information. In addition, one activity that distinguishes MPEG-4 from the conventional video coding standards is the synthetic and natural hybrid coding (SNHC). The target technologies studied by the SNHC subgroup include face animation, coding and representation of 2D dynamic mesh, wavelet-based static texture coding, view-dependent scalability, and 3D geometry compression. These functionalities used to be considered only by the computer graphics community. MPEG-4 SNHC successfully brings these tools into the scope of a video standard, and hence bridges computer graphics and image processing.

1.4.4 MPEG-7

MPEG-7 is targeted to produce a standardized description of multimedia material including images, text, graphics, 3D models, audio, speech, analog/digital video, and composition information. The standardized description will enable fast and efficient search and retrieval of multimedia content and advance the search mechanism from a text-based approach to a content-based approach. Currently, feature extraction and the search engine design are considered to be outside of the standard. Nevertheless, when MPEG-7 is finalized and widely adopted, efficient implementation for feature extraction and search mechanism will be very important. The applications of MPEG-7 can be categorized into pull and push scenarios. For the pull scenario, MPEG-7 technologies can be used for information retrieval from a database or from the Internet. For the push scenario, MPEG-7 can provide the filtering mechanism applied to multimedia content broadcast from an information provider.

As pointed out earlier in this chapter, instead of trying to extract relevant features, manually or automatically, from original or compressed video, a better approach for content retrieval should be to design a new standard in which such features, often referred to as meta-data, are already available. MPEG-7, an ongoing effort by the Moving Picture Experts Group, is working exactly toward this goal (i.e., the standardization of meta-data for multimedia content indexing and retrieval).

MPEG-7 is an activity triggered by the growth of digital audiovisual information. The group strives to define a "multimedia content description interface" to standardize the description of various types of multimedia content, including still pictures, graphics, 3D models, audio, speech, video, and composition information. It may also deal with special cases such as facial expressions and personal characteristics.

The goal of MPEG-7 is exactly the same as the focus of this chapter (i.e., to enable efficient search and retrieval of multimedia content). Once finalized, it will transform the text-based search and retrieval (e.g., keywords), as is done by most of the multimedia databases nowadays, into a content-based approach (e.g., using color, motion, or shape information). MPEG-7 can also be thought of as a solution to describing multimedia content. If one looks at PDF (portable document format) as a standard language to describe text and graphic documents, then MPEG-7 will be a standard description for all types of multimedia data, including audio, images, and video.

Compared with earlier MPEG standards, MPEG-7 possesses some essential differences. For example, MPEG-1, 2, and 4 all focus on the representation of audiovisual data, but MPEG-7 will focus on representing the meta-data (information about data). MPEG-7, however, may utilize the results of previous MPEG standards (e.g., the shape information in MPEG-4 or the motion vector field in MPEG-1 and 2).

Figure 1.7 shows the scope of the MPEG-7 standard. Note that feature extraction is outside the scope of MPEG-7, as is the search engine. This is owing to one approach constantly taken by most of the standard activities (i.e., "to standardize the minimum"). Therefore, the analysis (feature extraction) should not be standardized, so that after MPEG-7 is finalized, various analysis tools can be further improved over time. This also leaves room for competition among vendors and researchers. This is similar to MPEG-1 not specifying motion estimation and MPEG-4 not specifying segmentation algorithms. Likewise, the query process (the search engine) should not be standardized. This allows the design of search engines and query languages to adapt to different application domains, and also leaves room for further improvement and competition. Summarizing, MPEG-7 takes the approach of standardizing only what is necessary so that the description for the same content may adapt to different users and different application domains.

We now explain a few concepts of MPEG-7. One goal of MPEG-7 is to provide a standardized method of describing features of multimedia data. For images and video, colors or

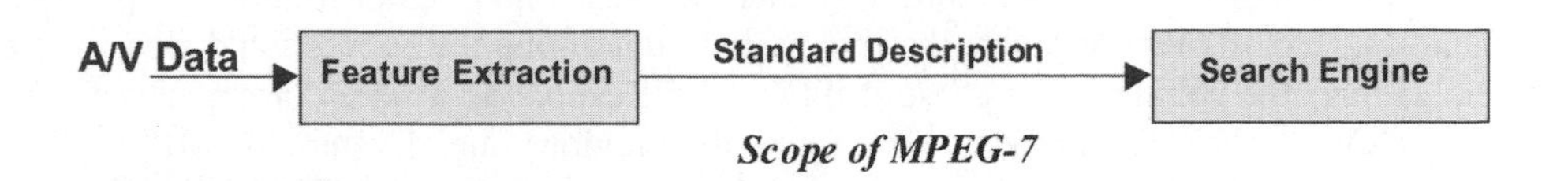

FIGURE 1.7
The scope of MPEG-7.

motion are example features that are desirable in many applications. MPEG-7 will define a certain set of descriptors to describe these features. For example, the color histogram can be a very suitable descriptor for color characteristics of an image, and motion vectors (commonly available in compressed video bitstreams) form a useful descriptor for motion characteristics of a video clip. MPEG-7 also uses the concept of description scheme (DS), which means a framework that defines the descriptors and their relationships. Hence, the descriptors are the basis of a description scheme. Description then implies an instantiation of a description scheme. MPEG-7 not only wants to standardize the description, but it also wants the description to be efficient. Therefore, MPEG-7 also considers compression techniques to turn descriptions into coded descriptions. Compression reduces the amount of data that need to be stored or processed. Finally, MPEG-7 will define a description definition language (DDL) that can be used to define, modify, or combine descriptors and description schemes. Summarizing, MPEG-7 will standardize a set of descriptors and DSs, a DDL, and methods for coding the descriptions. Figure 1.8 illustrates the relationship between these concepts in MPEG-7.

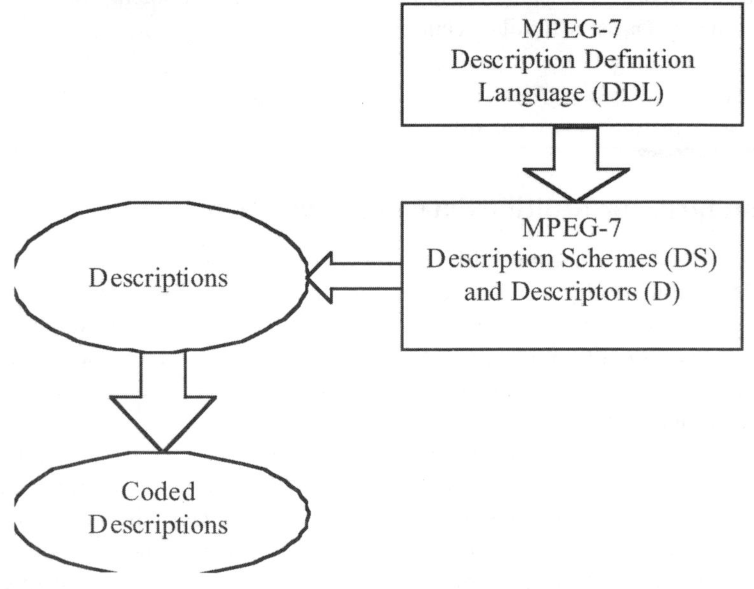

FIGURE 1.8
Relationship between elements in MPEG-7.

The process to define MPEG-7 is similar to that of the previous MPEG standards. Since 1996, the group has been working on defining and refining the requirements of MPEG-7 (i.e., what MPEG-7 should provide). The MPEG-7 process includes a competitive phase followed by a collaborative phase. During the competitive phase, a call for proposals is issued and participants respond by both submitting written proposals and demonstrating the proposed techniques. Experts then evaluate the proposals to determine the strength and weakness of each. During the collaborative phase, MPEG-7 will evolve as a series of experimentation models (XMs), where each model outperforms the previous one. Eventually, MPEG-7 will evolve into an international standard. Table 1.5 shows the timetable for MPEG-7 development. At the time of this writing, the group is going through the definition process of the first XM.

Table 1.5 Timetable of MPEG-7

Call for test material	Mar 1998
Call for proposals	Oct 1998
Proposals due	Feb 1999
First experiment model (XM)	Mar 1999
Working draft (WD)	Dec 1999
Committee draft (CD)	Oct 2000
Final committee draft (FCD)	Feb 2001
Draft international standard (DIS)	July 2001
International standard (IS)	Sep 2001

Once finalized, MPEG-7 will have a large variety of applications, such as digital libraries, multimedia directory services, broadcast media selection, and multimedia authoring. Here are some examples. With MPEG-7, the user can draw a few lines on a screen to retrieve a set of images containing similar graphics. The user can also describe movements and relations between a number of objects to retrieve a list of video clips containing these objects with the described temporal and spatial relations. Also, for a given content, the user can describe actions and then get a list of similar scenarios.

1.5 Standards for Multimedia Communication

In addition to video coding, multimedia communication also involves audio coding, control and signaling, and the multiplexing of audio, video, data, and control signals. ITU-T specifies a number of system standards for multimedia communication, as shown in Table 1.6 [8]. Due to the different characteristics of various network infrastructures, different standards are needed. Each system standard contains specifications about video coding, audio coding, control and signaling, and multiplexing.

For multimedia communication over the Internet, the most suitable system standard in Table 1.6 is H.323. H.323 [9] is designed to specify multimedia communication systems on networks that do not guarantee QoS, such as ethernet, fast ethernet, FDDI, and token ring networks. Similar to other system standards, H.323 is an umbrella standard that covers several other standards. An H.323-compliant multimedia terminal has a structure as shown in Figure 1.9. For audio coding, it specifies G.711 as the mandatory audio codec, and includes G.722, G.723.1, G.728, and G.729 as optional choices. For video coding, it specifies H.261 as the mandatory coding algorithm and includes H.263 as an alternative. H.225.0 defines the multiplexing of audio, video, data, and control signals, synchronization, and the packetization mechanism. H.245 is used to specify control messages, including call setup and capability exchange. In addition, T.120 is chosen for data applications. As in Figure 1.9, a receive path

Table 1.6 ITU-T Multimedia Communication Standards

Network	System	Video	Audio	Mux	Control
PSTN	H.324	H.261/263	G.723.1	H.223	H.245
N-ISDN	H.320	H.261	G.7xx	H.221	H.242
B-ISDN/ATM	H.321	H.261	G.7xx	H.221	Q.2931
	H.310	H.261/H.262	G.7xx, MPEG	H.222.0/H.222.1	H.245
QoS LAN	H.322	H.261	G.7xx	H.221	H.242
Non-QoS LAN	H.323	H.261	G.7xx	H.225.0	H.245

*Note:*G.7xx represents G.711, G.722, and G.728.

delay is used to synchronize audio and video (e.g., for lip synchronization) and to control jitters.

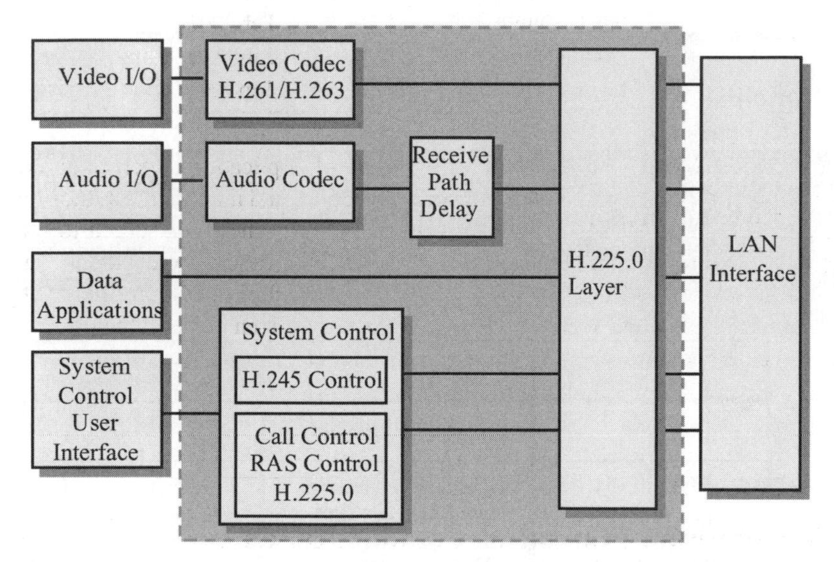

FIGURE 1.9
H.323 terminal equipment.

In addition to terminal definition, H.323 also specifies other components for multimedia communication over non-QoS networks. These include the gateways and gatekeepers. As shown in Figure 1.10, the responsibility of a gateway is to provide interoperability between H.323 terminals and other types of terminals, such as H.320, H.324, H.322, H.321, and H.310. A gateway provides the translation of call signaling, control messages, and multiplexing mechanisms between the H.323 terminals and other types of terminals. It also needs to support transcoding when necessary. For example, for the audio codec on an H.324 terminal to interoperate with the audio codec on an H.323 terminal, transcoding between G.723.1 and G.711 is needed. On the other hand, a gatekeeper serves as a network administrator to provide the address translation service (e.g., translation between telephone numbers and IP addresses) and to control access to the network by H.323 terminals or gateways. Terminals have to get permission from the gatekeeper to place or accept a call. The gatekeeper also controls the bandwidth for each call.

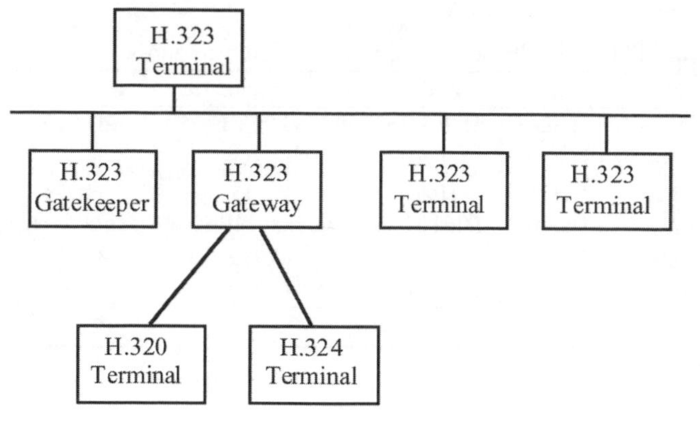

FIGURE 1.10
Interoperability of H.323.

1.6 Conclusion

In this chapter, we described several emerging video coding and multimedia communication standards, including H.263, H.26L, MPEG-4, MPEG-7, and H.323. Reviewing the development of video coding, as shown in Figure 1.11, we can see that the progress of video coding and multimedia standards is tied to the progress in modeling of the information source. The finer the model, the better we can compress the signals, and with more content accessibility to

MODELS	CODED INFORMATION	EXAMPLES
Pixels	Color of pixels	PCM
Statistically dependent pixels	Prediction error or transform coeffs	Predictive Coding Transform Coding
Moving blocks	Motion vectors and prediction error	Block-based coding H.261/263, MPEG-1/2
Moving regions	Shapes, motion, and colors of regions	Region-based coding H.263+, MPEG-4
Moving objects	Shapes, motion, and colors of objects	Model-based coding MPEG-4
Facial models	Action units	MPEG-4
A/V objects	Descriptive languages	MPEG-7

FIGURE 1.11
Trend of video coding standards.

the user. At the same time, the price to pay includes higher complexity and less error resilience. The complexity manifests itself not only in the higher computation power that is required, but also in higher flexibility. For example, whereas H.261 is a well-defined and self-contained compression algorithm, MPEG-4 and MPEG-7 are toolboxes of a large number of different algorithms.

References

[1] Ahmed, N., Natarajan, T., and Rao, K.R., "Discrete cosine transform," *IEEE Trans. on Computers,* C-23, pp. 90–93, 1974.

[2] Rao, K.R., and Yip, P., *Discrete Cosine Transform,* Academic Press, New York, 1990.

[3] Netravali, A.N., and Robbins, J.D., "Motion-compensated television coding: Part I," *Bell Systems Technical Journal,* 58(3), pp. 631–670, March 1979.

[4] Netravali, A.N., and Haskell, B.G., *Digital Pictures,* 2nd ed., Plenum Press, New York, 1995.

[5] ITU-T Recommendation H.263: "Video coding for low bit rate communication," Version 1, Nov. 1995; Version 2, Jan. 1998.

[6] Chen, T., Swain, C.T., and Haskell, B.G., "Coding of sub-regions for content-based scalable video," *IEEE Trans. on Circuits and Systems for Video Technology,* 7(1), pp. 256–260, February 1997.

[7] Sikora, T., "MPEG digital video coding standards," *IEEE Signal Processing Magazine,* pp. 82–100, Sept. 1997.

[8] Schaphorst, R., *Videoconferencing and Videotelephony: Technology and Standards,* Artech House, Boston, 1996.

[9] Thom, G.A., "H.323: The multimedia communications standard for local area networks," *IEEE Communication Magazine* (Special Issue on Multimedia Modem), pp. 52–56, December 1996.

Chapter 2

An Efficient Algorithm and Architecture for Real-Time Perspective Image Warping

Yi Kang and Thomas S. Huang

2.1 Introduction

Multimedia applications are among the most important embedded applications. HDTV, 3D graphics, and video games are a few examples. These applications usually require real-time processing. The perspective transform used for image warping in MPEG-4 is one of the most demanding algorithms among real-time multimedia applications. An algorithm is proposed here for a real-time implementation of MPEG-4 sprite warping; however, it can be useful in general computer graphics applications as well.

MPEG-4 is a new standard for digital audio–video compression currently being developed by the ISO (International Standardization Organization) and the IEC (International Electrotechnical Commission). It will attempt to provide greater compression, error robustness, interactiveness, support of hybrid natural and synthetic scenes, and scalability. MPEG-4 will require more computational power than existing compression standards, and novel architectures will probably be necessary for high-complexity MPEG-4 systems. Whereas current video compression standards transmit the entire frame in a single bitstream, MPEG-4 will separately encode a number of irregularly shaped objects in the frame. The objects in the frame can then be encoded with different spatial or temporal resolutions [1].

By studying the MPEG-4 functions, we find that there are two critical parts for real-time implementation: one is motion estimation in the encoder and the other is sprite warping in the decoder. The algorithm for motion estimation in MPEG-4 is similar to those in previous standards. There has already been plenty of work on algorithms and architectures for real-time motion estimation. However, there have been few discussions on real-time sprite warping. We therefore focus on algorithm and architecture development for sprite warping.

Real-time sprite warping involves implementing a perspective transform, a bilinear interpolation, and high-bandwidth memory accesses. It is both computationally expensive and memory intensive. This poses a serious challenge for designing real-time MPEG-4 architectures. With the goal of real time and cost-effectiveness in mind, we first optimize our algorithm to reduce the computation burden of the perspective transform by proposing the constant denominator algorithm. This algorithm dramatically reduces divisions and multiplications in the perspective transform by an order of magnitude. Based on the proposed algorithm, we designed an architecture which implements the real-time sprite warping. To make our architecture feasible for implementation under current technologies, we address the design of the data path as well as the memory system according to the real-time requirement of computations and memory accesses in the sprite warping. Other related issues for implementation of real-time sprite warping are also discussed.

2.2 A Fast Algorithm for Perspective Transform

The perspective transform is widely used in image and video processing, but it is computationally expensive. The most expensive part is its huge number of divisions. It is well known that a division unit has the highest cost and the longest latency among all basic data path units. The number of divisions in the perspective transform would make its real-time implementation formidable without any fast algorithm. This motivates us to explore a new algorithm for real-time perspective transform. The constant denominator method reduces the number of required division operations to $O(N)$ while maintaining high accuracy. It also has fewer multiplications and divisions.

2.2.1 Perspective Transform

Perspective transforms are geometric transformations used to project scenes onto view planes along lines which converge to a point. The perspective transform which maps two-dimensional images onto a two-dimensional view plane is defined by

$$x' = \frac{ax + by + c}{gx + hy + 1} \tag{2.1}$$

$$y' = \frac{dx + ey + f}{gx + hy + 1} \tag{2.2}$$

where (x, y) is a coordinate in the reference image, (x', y') is the corresponding coordinate in the transformed image, and a, b, c, d, e, f, g, and h are the transform parameters.

The perspective transform has many applications in computer-aided design, scientific visualization, entertainment, advertising, image processing, and video processing [3]. One new application for the perspective transform is MPEG-4. In MPEG-4 one of the additional functionalities proposed to support is sprite coding [7]. A sprite is a reference image used to generate different views of an object. The reference image is transmitted once, and future images are produced by warping the sprite with the perspective transform. Because the transform parameters a, b, c, d, e, f, g, and h are rational numbers, they are not encoded directly. Instead, the image is encoded using four (x', y') pairs, since the transform parameters can be determined from the reference and warped coordinates of four reference points using the following system of equations:

$$\begin{bmatrix} x'_1 \\ x'_2 \\ x'_3 \\ x'_4 \\ y'_1 \\ y'_2 \\ y'_3 \\ y'_4 \end{bmatrix} = \begin{bmatrix} x_1 & y_1 & 1 & 0 & 0 & 0 & -x_1x'_1 & -y_1x'_1 \\ x_2 & y_2 & 1 & 0 & 0 & 0 & -x_2x'_2 & -y_2x'_2 \\ x_3 & y_3 & 1 & 0 & 0 & 0 & -x_3x'_3 & -y_3x'_3 \\ x_4 & y_4 & 1 & 0 & 0 & 0 & -x_4x'_4 & -y_4x'_4 \\ 0 & 0 & 0 & x_1 & y_1 & 1 & -x_1y'_1 & -y_1y'_1 \\ 0 & 0 & 0 & x_2 & y_2 & 1 & -x_2y'_2 & -y_2y'_2 \\ 0 & 0 & 0 & x_3 & y_3 & 1 & -x_3y'_3 & -y_3y'_3 \\ 0 & 0 & 0 & x_4 & y_4 & 1 & -x_4y'_4 & -y_4y'_4 \end{bmatrix} \begin{bmatrix} a \\ b \\ c \\ d \\ e \\ f \\ g \\ h \end{bmatrix} \tag{2.3}$$

High compression is therefore possible using sprite coding, especially for background sprites and synthetic objects. After the original image is transmitted, the new view on the right can be described using four points.

The warped image can be transmitted using fewer reference points. If three reference points are transmitted, the affine transform is used for estimation. The affine transform is equivalent to the perspective transform, with g and h equal to zero. Only two reference points are required using an isotropic transformation, where $g = h = 0, d = -b$, and $e = a$. If only one reference point is used, the transformation becomes simple translation, where $g = h = 0$, $a = e = 1$, and $b = d = 0$. These simpler approximations provide less complexity, but generally provide a less accurate estimate of the warped image.

To prevent holes or overlap in the warped sprite, backward perspective mapping is used. Each point (x', y') in the warped sprite is obtained from point (x, y) in the reference image. The backward perspective mapping can be obtained from the adjoint and determinant of the forward transform matrix [10]:

$$x = \frac{(hf - e)x' + (b - hc)y' + (ec - bf)}{(eg - dh)x' + (ah - bg)y' + (db - ae)} = \frac{a'x' + b'y' + c'}{g'x' + h'y' + i'} \tag{2.4}$$

$$y = \frac{(d - fg)x' + (cg - a)y' + (af - dc)}{(eg - dh)x' + (ah - bg)y' + (db - ae)} = \frac{d'x' + e'y' + f'}{g'x' + h'y' + i'} \tag{2.5}$$

Though x' and y' are integers, x and y generally are not. Bilinear interpolation is used to approximate the pixel value at point (x, y) from the four nearest integer points.

The perspective transform is computationally expensive. Computation of x and y using equations (2.4) and (2.5) requires one division, eight multiplications, and nine additions per pixel. The division is especially expensive. Since the transform parameters are not integers, floating point computations are typically used. For real-time hardware implementations using high-resolution images, direct computation of the transform is too slow. An approximation method must be used.

2.2.2 Existing Approximation Methods

The perspective transform can be approximated using polynomials to avoid the expensive divisions needed to compute the rational functions in equations (2.4) and (2.5). Linear approximation is the simplest and most widely used approximation technique. However, it usually results in large errors due to the simplicity of the approximation [2, 4]. To achieve greater accuracy, more complex methods such as quadratic approximation, cubic approximation, biquadratic approximation, and bicubic approximation have been proposed [6, 10]. Additional methods to reduce aliasing and simplify resampling have also been developed, such as the two-pass separable algorithm [10].

The Chebyshev approximation is a well-known method in numerical computation that also has been used to approximate the perspective transform [2]. Its main advantage over other methods is that its error is evenly distributed [8]. The result thus visually appears closer to the ideal result. The formula for the Chebyshev approximation is

$$f(x) \approx \sum_{k=0}^{N-1} c_k T_k(x) - 0.5c_0 \tag{2.6}$$

where c_j's are the coefficients computed as

$$c_j = \frac{2}{N} \sum_{k=1}^{N} f(x_k) T_j(x_k) , \tag{2.7}$$

$T_j(x)$ is the jth base function for the approximation, $f(x)$ is the target function to approximate, and N is the order of the approximation. $N = 2$ for the quadratic Chebyshev approximation; $N = 3$ for the cubic Chebyshev approximation.

Biquadratic and bicubic Chebyshev methods have also been proposed to approximate the perspective transform [2]. These methods first calculate the Chebyshev control points, then use transfinite interpolation to approximate the rational functions using polynomials.

All of the above approximation methods require more multiplications and additions than direct computation of the original rational functions. For complex approximations such as the Chebyshev methods, the additional multiplications and additions offset the benefit of avoiding division. Simpler approximations such as linear approximation require fewer additional operations, but often achieve poor quality. These methods also require an initialization procedure to compute the approximation coefficients on every scan line. This increases the hardware overhead.

In the following section, a new method to perform the perspective transform is proposed. This new method does not increase the number of multiplications and additions, has a simple initialization procedure, and decreases the number of divisions from $O(N^2)$ to $O(N)$.

2.2.3 Constant Denominator Method

Equations (2.4) and (2.5) both contain the same denominator: $g'x' + h'y' + i'$. Setting the denominator equal to a constant value defines a line in the $x'y'$ plane.

$$k = g'x' + h'y' + i' \tag{2.8}$$

Furthermore, lines defined by different values of k are all parallel and all have slope equal to $-g'/h'$. The constant k for the line with y' intercept equal to q can be calculated as

$$k_q = h'q + i' \tag{2.9}$$

By calculating the perspective transform along lines of constant denominator, the number of divisions is reduced from one per pixel to one per constant denominator line.

The constant denominator method begins by calculating $(d - fg)$, $(cg - a)$, $(af - dc)$, $(hf - e)$, $(b - hc)$, $(ec - bf)$, $(eg - dh)$, $(ah - bg)$, and $(db - ae)$. These coefficients need only be calculated once per frame. Next, $(eg - dh)$ and $(ah - bg)$ are used to calculate the slope m of the constant denominator lines. There are four possible cases: $m < -1$, $-1 \leq m \leq 0$, $0 < m \leq 1$, and $1 < m$. The case determines whether the constant denominator lines are scanned in the horizontal or vertical direction.

Figure 2.1 illustrates a case where $0 < m < 1$. The lines all have slope $m = -g'/h'$ and represent constant values of $g'x' + h'y' + i'$. The pixels are shaded to indicate which constant denominator line they approximately fall on. The pixels for the initial line are determined by starting at the origin and applying Bresenham's Algorithm. Bresenham's Algorithm requires only incremental integer calculations [3]. The result is the table in Figure 2.1, which lists the corresponding vertical position for every horizontal position on the constant denominator line that passes through the origin. By storing the table as the difference of subsequent entries, the number of bits required to store the table is the larger of the width or height of the image.

After the position of the constant denominator line has been determined, the actual warping is performed. The reciprocal of the denominator is first calculated for the constant denominator line which crosses the origin:

$$r = \frac{1}{k_0} = \frac{1}{h' * 0 + i'} = \frac{1}{i'} \tag{2.10}$$

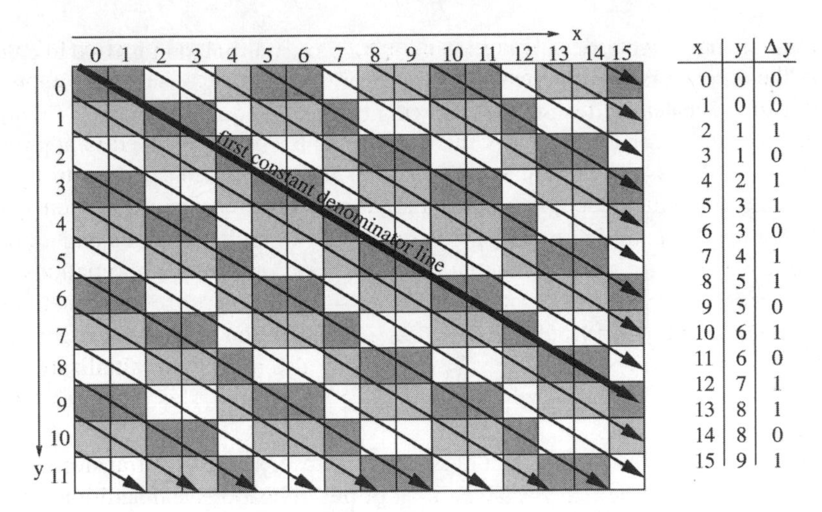

x	y	Δy
0	0	0
1	0	0
2	1	1
3	1	0
4	2	1
5	3	1
6	3	0
7	4	1
8	5	1
9	5	0
10	6	1
11	6	0
12	7	1
13	8	1
14	8	0
15	9	1

FIGURE 2.1
Lines of constant denominator with $0 < \text{slope} < 1$.

This is the only division required for the first constant denominator line. This reciprocal is then multiplied by d', e', f', a', b', and c' to obtain the coefficients in equations (2.11) and (2.12).

$$x = ra'x' + rb'y' + rc' \tag{2.11}$$

$$y = rd'x' + re'y' + rf' \tag{2.12}$$

The horizontal position x' is incremented from 0 to $M - 1$, where M is the width of the image. For each value of x', $\Delta y'$ is obtained from the line table. The current value of the x and y coordinates, x_n and y_n, are calculated from the previous values of the x and y coordinates, x_{n-1} and y_{n-1}, using the following equations. If $\Delta y' = 0$,

$$x_n = x_{n-1} + ra' \tag{2.13}$$

$$y_n = y_{n-1} + rd' \tag{2.14}$$

If $\Delta y' = 1$,

$$x_n = x_{n-1} + [ra' + rb'] \tag{2.15}$$

$$y_n = y_{n-1} + [rd' + re'] \tag{2.16}$$

Only two additions are required to calculate x_n and y_n for each pixel on the constant denominator line. No multiplications or divisions are required per pixel.

The next constant denominator line is warped by calculating r for point $(x', y') = (0, 1)$ using the following equation:

$$r = \frac{1}{k_1} = \frac{1}{h' * 1 + i'} = \frac{1}{h' + k_0} \tag{2.17}$$

One addition and one division are required to calculate r. The line table is used to trace the new line, and equations (2.13)–(2.16) are used to warp the pixels on the new line. Every constant denominator line below the original line is warped, followed by the constant denominator lines above the original line.

Because x_n and y_n are generally not integers, bilinear interpolation is used to calculate the value of the warped pixel using the four pixels nearest to (x_n, y_n) in the original sprite. The warped pixel P is calculated using the following three equations, as shown in Figure 2.2:

$$P_{01} = P_0 + (P_1 - P_0) * dx \tag{2.18}$$

$$P_{23} = P_2 + (P_3 - P_2) * dx \tag{2.19}$$

$$P = P_{01} + (P_{23} - P_{01}) * dy \tag{2.20}$$

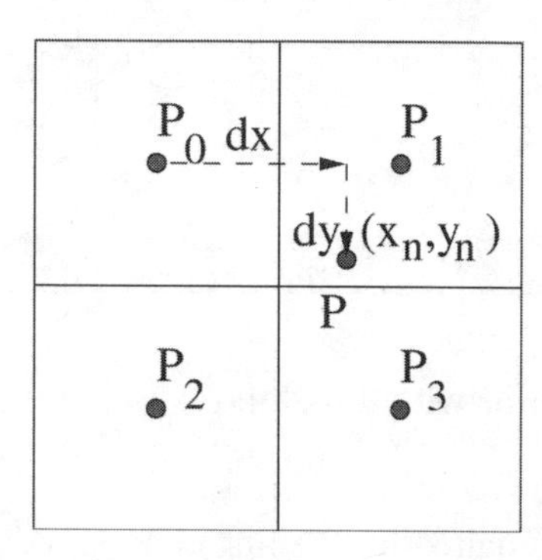

FIGURE 2.2
Bilinear interpolation.

As shown above, the constant denominator method reduces the number of divisions required to calculate (x, y) from one per pixel, using equations (2.4) and (2.5) directly, to one per constant denominator line. For an image M pixels wide and N pixels high, the number of divisions is reduced from MN using the direct method to, at most, $M + N - 1$. The number of multiplications needed to calculate (x, y) is reduced from $8MN$ to $8(M + N - 1) + 17$. The drastic reduction in divisions and multiplications makes the constant denominator method suitable for real-time sprite decoding.

In addition, the constant denominator method can be used to calculate the backward affine transform when only three reference points are transmitted. In this case, $r = 1$ for every point in the plane. No divisions and only 14 multiplications per frame are therefore required for the affine transform.

2.2.4 Simulation Results

To compare the visual quality of the warping approximations, five methods were implemented in C++: direct warping, constant denominator, quadratic, quadratic Chebyshev, and cubic Chebyshev. The methods were then used to warp the checkerboard image, which is a standard test image for computer graphics. The checkerboard image is useful because the perspective transform should preserve straight lines. The parameters are set to $a = 1.2, b = 0$, $c = -100$, $d = 0$, $e = 1.2$, $f = -20$, $g = -.0082$, and $h = 0$. The simulation shows that straight lines in the original image are curved greatly by the quadratic and quadratic Chebyshev methods. They are curved slightly by the cubic Chebyshev method. The constant denominator method preserves the straight lines.

To generate test data for a wide range of cases, simulations were conducted varying g and h over $\{-.1, -.01, -.001, -.0001, 0, .0001, .001, .01, .1\}$. Parameters a and e were set to 1, and the remaining parameters were set to 0. An error image was calculated for each method using the direct warping image as a reference, and the mean squared error (MSE) was computed from each error image. The mean, median, and maximum values of mean squared error for each method are shown in Table 2.1. A histogram of the MSE for the four methods is shown in Figure 2.3. The MSE is plotted on a logarithmic scale, and all MSEs less than 1 are plotted at 1. One third of the simulations for the constant denominator method had MSEs below 1. The largest error occurred for the case where $g = 0.01$ and $h = -0.1$. The other three methods were significantly less accurate than the constant denominator method.

Error in the constant denominator method occurs because the pixels do not fall exactly on constant denominator lines. Each pixel can lie a maximum of one-half pixel off the actual constant denominator line if we treat each pixel as a square. An additional source of error is from sprite resampling via the bilinear interpolation. Most of the error in Table 2.1 for the constant denominator method is due to position computation because the direct warped image with resampling is used as the error reference.

Table 2.1 Checkerboard Mean Squared Error Table

Method	Mean	Median	Max
Constant denominator	73	20	428
Quadratic	2,831	693	15,888
Quadratic Chebyshev	2,118	457	14,313
Cubic Chebyshev	1,822	392	14,116

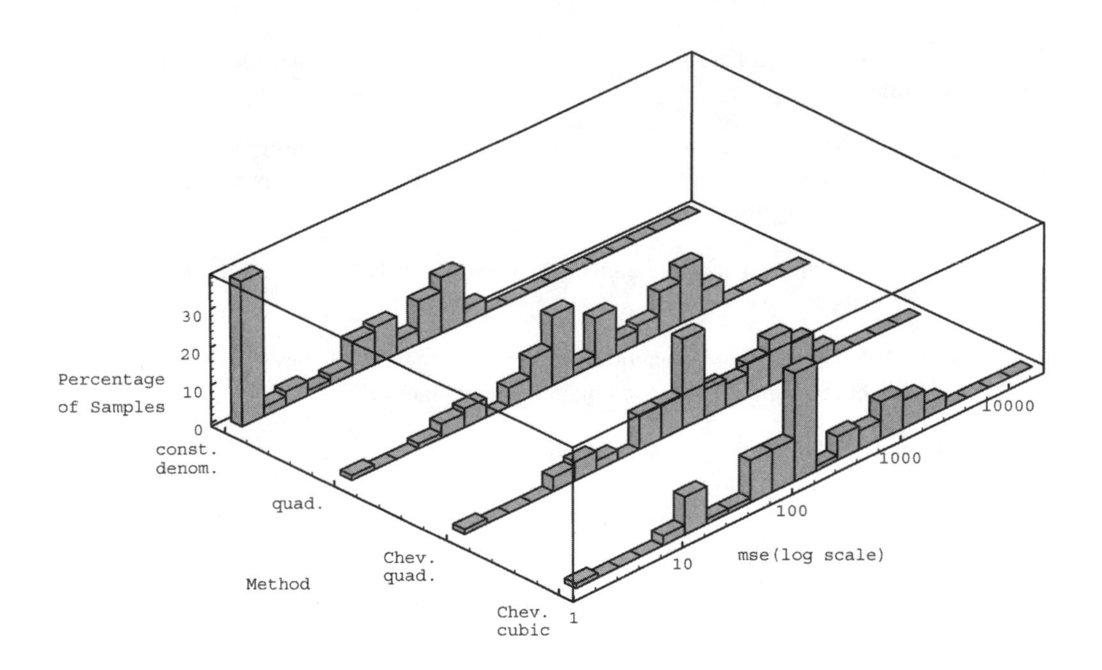

FIGURE 2.3
Checkerboard mean squared error histogram.

The constant denominator method was also tested on natural images. Simulation was done for the $a = 1$, $b = 0$, $c = 0$, $d = 0$, $e = 1$, $f = 0$, $g = -0.1$, and $h = 0.002$ case using a coastguard image. The MSE for the constant denominator method was 0.00043. The error is so small that it can hardly be picked up by the eyes. Table 2.2 shows a performance comparison between the various approximation methods as g and h are varied between -0.1 and 0.1 for the coastguard image.

Table 2.2 Coastguard Mean Squared Error Table

Method	Mean	Median	Max
Constant denominator	73	20	428
Quadratic	2,831	693	15,888
Quadratic Chebyshev	2,118	457	14,313
Cubic Chebyshev	1,822	392	14,116

2.2.5 Sprite Warping Algorithm

We designed an algorithm to perform sprite warping using the perspective transform as specified in MPEG-4. The sprite warping algorithm performs the following tasks:

- Step 1: Compute the eight perspective transform parameters a, b, c, d, e, f, g, and h from the reference coordinates.
- Step 2: Compute the nine backward transform coefficients $(d-fg)$, $(cg-a)$, $(af-dc)$, $(hf-e)$, $(b-hc)$, $(ec-bf)$, $(eg-dh)$, $(ah-bg)$, and $(db-ae)$.
- Step 3: Use Bresenham's Algorithm to calculate the line table for the first constant denominator line.
- Step 4: Compute the constant r in equation (2.17) using restoring division [5]. Then compute the coefficients in equations (2.11) and (2.12). This step is performed once per constant denominator line.
- Step 5: Perform the backward transform for every pixel along the constant denominator line described above.
- Step 6: Fetch the four neighboring pixels from memory for every warped pixel and perform bilinear interpolation to obtain the new pixel value.

Step 1 entails solving the system of equations given in equation (2.3). Using LU decomposition, the eight sprite warping parameters can be calculated using 36 divisions, 196 multiplications, and 196 additions. Steps 2 through 5 use the constant denominator method to perform the perspective transform. The computation of the backward transform coefficients in step 2 requires 14 multiplications and nine additions. Calculating the line table in step 3 requires three multiplications, one division, and either M or N additions, depending on the slope of the line. These three steps are performed once per frame. Step 4 requires one division, eight multiplications, and three additions for every constant denominator line. Step 5 requires two additions for every pixel. After the warped coordinate has been computed, the bilinear interpolation in step 6 requires three multiplications and six additions for every pixel.

For gray-scale sprites M pixels wide and N pixels high and with horizontal scanning, the entire sprite warping process requires at most $M+N+36$ divisions, $3MN+8M+8N+205$ multiplications, and $8MN+4M+3N+202$ additions. Color sprites require additional operations. For YUV images with 4:2:0 format, sprite warping requires at most a total of $1.5M+1.5N+35$ divisions, $4.5MN+12M+12N+200$ multiplications, and $11.5MN+6M+4.5N+199$ additions.

The computation burden can be reduced by using fixed point instead of floating point operations wherever possible. Steps 1, 2, and 4 are best suited for floating point operations. However, since steps 1 and 2 are performed once per frame, and step 4 is performed once per constant denominator line, they consume only a small fraction of the computational power. Step 3 is also performed once per frame. The additions in step 3 can be performed in fixed point.

Most of the computations are performed in steps 5 and 6, since these steps are performed on each pixel. In step 5, a floating point coefficient is multiplied by the integer coordinate x' or y'. Therefore, instead of using true floating point, the coefficients can be represented in block floating point format. Fixed point operations can then be used for step 5. After (x, y) is calculated for each pixel, it is translated to a long fixed point number. Thus, only fixed point computation is required for the bilinear interpolation in step 6.

By using fixed point operations for steps 5 and 6, the number of floating point multiplications is reduced to at most $12M + 12N + 196$ and the number of floating point additions becomes $4.5M + 4.5N + 199$. The number of floating point divisions remains $1.5M + 1.5N + 35$. Almost all of the operations are now fixed point. $4.5MN$ fixed point multiplications and $11.5MN + 1.5M$ fixed point additions at most are required for steps 3, 5, and 6. Table 2.3 lists the number of operations required for various full-screen sprites.

Table 2.3 Number of Operations per Second Required for 30 Frames per Second

Sprite Size	QCIF	CIF	ITU-R 601
Sprite width	176	352	720
Sprite height	144	288	576
Float. divide	15,000	30,000	59,000
Float. multiply	120,000	240,000	470,000
Float. add	49,000	92,000	180,000
Fixed multiply	3.4 million	14 million	56 million
Fixed add	8.8 million	35 million	140 million

2.3 Architecture for Sprite Warping

An MPEG-4 sprite warping architecture is described which uses the constant denominator method. The architecture exploits the spatial locality of pixel accesses and pipelines an arithmetic logic unit (ALU) with an interpolation unit to perform high-speed sprite warping. Several other implementation issues (e.g., boundary clipping and error accumulation) are also discussed.

2.3.1 Implementation Issues

One issue inherent to the perspective transform is aliasing. Subsampling the sprite can cause aliasing artifacts for perspective scaling. However, sprite warping is intended for video applications where aliasing is less of a problem due to the motion blur. To address aliasing in the constant denominator method, techniques such as adaptive supersampling could be used. Supersampling would be performed when consecutive accesses to the sprite memory are widely separated.

Boundary clipping can also be a concern. Sprite warping can attempt to access reference pixels beyond the boundaries of the reference sprite. If the simple point clipping method is used, four comparisons per pixel are required. Instead, a hybrid point–line clipping method can be used with the constant denominator method. For each constant denominator line, the endpoints are first checked to see if they fall within the boundaries of the reference sprite. If both endpoints are in the reference sprite, the line is warped. If only one endpoint is outside the boundary, warping begins with this endpoint using point clipping. Once a point within the boundary is warped, clipping is turned off, because the remaining points on the line are within the sprite. If both endpoints lie outside the reference sprite, point clipping is used beginning with one of the endpoints. Once a point inside the reference sprite is reached, warping switches to the other endpoint. Point clipping is used until the next point with the sprite is reached, when point clipping is turned off. Using this method, comparisons are only required when the reference pixel is out of bounds. Because memory accesses and interpolations are not required for the out-of-bound pixels, and clipping computations are not required for in-bound pixels, the clipping procedure does not slow the algorithm.

Error accumulation in the fixed-point, iterative calculation of equations (2.13)–(2.16) must also be considered. Sufficient precision of the fractional part of x_n and y_n must be used to prevent error from accumulating to 1. The number of bits k required for the fractional part depends on the height N and width M of the warped sprite according to the following inequality:

$$k \geq \log_2(MAX[M, N]) \tag{2.21}$$

The integral part of x_n and y_n must contain enough bits to avoid overflow. Because (x_n, y_n) is a coordinate in the reference plane, they theoretically have infinite range. Practically, the number of integral bits j is chosen according to the size of the reference sprite plus additional bits to prevent overflow. If a is the number of overflow bits and the reference sprite is $P \times Q$ pixels, then

$$j \geq \log_2(MAX[P, Q]) + a \tag{2.22}$$

For example, if the reference and warped sprite are both 720×576 pixels and four overflow bits are used, then $a = 4$, $k = 10$, $j = 10$, and 24 total bits are required for calculating x_n and y_n.

2.3.2 Memory Bandwidth Reduction

Memory bandwidth is a concern for high-resolution sprites. Warped pixels are interpolated from the four nearest pixels in the original sprite. Warping a sprite can therefore require four reads and one write for every pixel in the sprite. An ITU-R 601 sprite requires 89 MB/s of memory bandwidth at 30 frames per second.

Figure 2.4 illustrates the memory access pattern for sprite warping using the constant denominator method. It shows lines of slope $-g/h$ in the original sprite which correspond to the lines of constant denominator in the warped sprite. While the memory access lines in the original sprite are parallel to each other, they are not evenly spaced, and memory accesses on different lines do not have the same spacing. Points in the warped sprite can also map to points outside the original sprite.

The total memory access time required to warp a sprite can be reduced by either decreasing the time required for each memory access or decreasing the number of accesses. Unlike scan-line algorithms which enjoy the advantage of block memory access in consecutive addresses, the constant denominator method must contend with diagonal memory access patterns. However, spatial locality inherent in diagonal access can be exploited. Figure 2.4 shows the use of spatial locality to reduce the time per access. The original sprite is divided into rectangular pages, which correspond to pages in the sprite memory. Consecutive accesses on a line will frequently lie on the same page. Fast page mode can therefore be used to retrieve the data quickly.

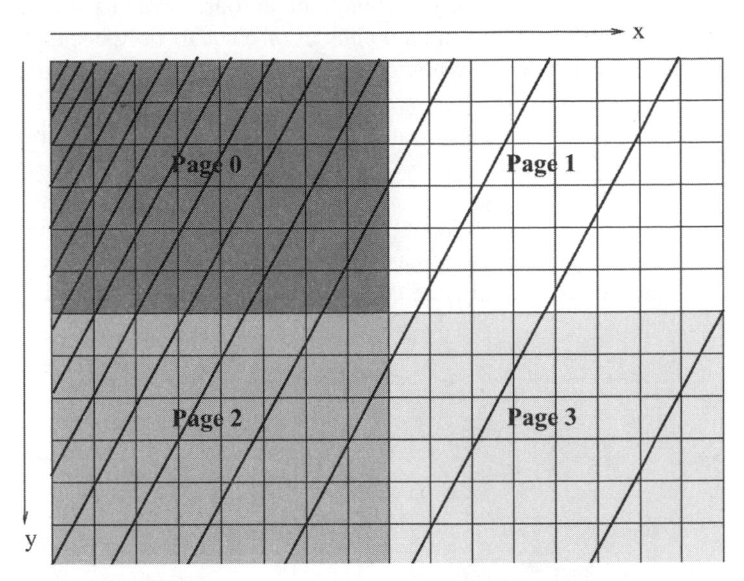

FIGURE 2.4
Example memory access pattern.

A cache can be used to reduce the number of accesses per pixel. Consecutive accesses on a constant denominator line often reference common pixels in the original sprite. Consecutive constant denominator lines frequently use many of the same pixels. By retaining pixel values in a cache, accesses to main memory can be avoided.

Cache effectiveness is dependent on the spacing between memory accesses. In the upper left area of the example in Figure 2.4, memory access lines are closely spaced. Pixels on the upper left will be accessed many times, and a cache will save memory accesses. In the lower right area, however, memory access lines are widely spaced. Pixels are not shared between consecutive lines, and a cache will not be as effective. However, because the lines are widely spaced, most of the pixels in the lower right area are not accessed from memory. Many accesses will instead occur outside the boundaries of the sprite memory and will be resolved by boundary clipping instead of being retrieved from memory or the cache. The worst-case memory access situation therefore does not occur for widely spaced lines. The cache should be designed for line spacings small enough such that most of the reference sprite pixels are read four times.

A very small cache which holds only four pixels will reduce the number of memory reads per sprite. By keeping the four pixels used to interpolate the previous point in the cache, the worst-case number of memory reads per sprite will be reduced from four times the number of warped pixels to three times the number of warped pixels. The worst case occurs when pixels on diagonal lines are accessed. If consecutive accesses on the lines are widely spaced, then the cache will be of no use. However, many pixels on the diagonal lines will not be accessed and the total number of accesses to sprite memory will be small. This is therefore not the worst case. Instead, the worst case occurs when consecutive pixels on the diagonal lines are accessed. One pixel in the cache can be reused; the three remaining pixels must be read in from memory.

A larger cache will further reduce the memory bandwidth required. Figure 2.5 illustrates the use of a cache with a three-line capacity. The cache is three-way set associative to remove conflict cache misses. For lines with slope greater than 1 or less than -1, as in the figure, there is one set for every y coordinate in the sprite, and pixels are tagged with the x coordinate. Shallower lines have one set for every x coordinate and are tagged with the y coordinate. The three-line cache reduces the worst-case number of reads to one per pixel. For an ITU-R 601 sprite, a three-line cache requires approximately 17 Kbit.

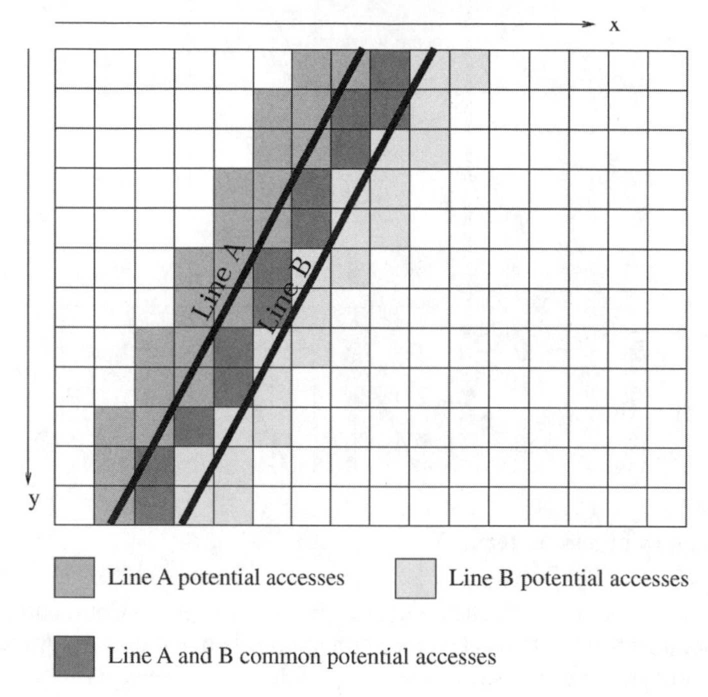

FIGURE 2.5
Cache operation example.

2.3.3 Architecture

The data path for a sprite warping architecture is shown in Figure 2.6. It contains two processors: an ALU to perform steps 1 through 5 in the sprite warping process and an interpolation unit to perform step 6. Since steps 5 and 6 are the two steps executed per pixel, they are assigned to different processors.

The ALU performs integer addition and multiplication. It reads reference coordinates from the coordinate buffer and calculates the perspective transform coefficients, using the small scratch memory for intermediate storage. The nine backward transform coefficients are then stored in the floating point coefficient buffer. The ALU uses Bresenham's Algorithm to compute the incremental line table for the first constant denominator lines. The line table is stored in the Bresenham shift register, which is simply a line of serially connected, 1-bit flip-flops. For each line, the ALU computes the six coefficients in equations (2.11) and (2.12). For each pixel, the coordinates of the corresponding pixel in the original frame are calculated and partitioned into an integer part (x_I, y_I) and a fractional part (dx, dy). The integer part is output to the pixel cache while the fractional part is passed to the interpolation unit.

The pixel cache outputs pixels P_0, P_1, P_2, and P_3. These are the four pixels with coordinates (x_I, y_I), (x_{I+1}, y_I), (x_I, y_{I+1}), and (x_{I+1}, y_{I+1}), which are shown in Figure 2.2. If the pixels

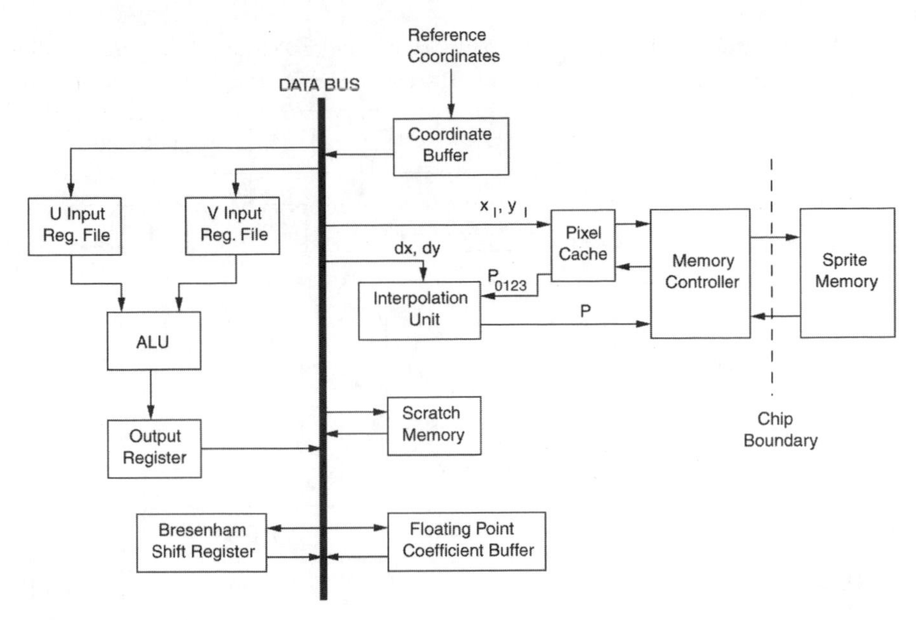

FIGURE 2.6
Sprite warping architecture.

are not in the cache, they are retrieved from memory. The pixels are transmitted serially to the interpolation unit.

The interpolation unit is based on a design commonly used for half-pixel motion compensation [9]. It is shown in detail in Figure 2.7. The unit reads a new pixel whenever the cache signals that the value P_{0123} is ready. It receives dx and dy from the ALU and outputs a bilinearly interpolated pixel after reading every fourth pixel. P_0 and P_1 are first linearly interpolated using dx to compute P_{01}. P_2 and P_3 are then interpolated using dx to compute P_{23}. Finally, the vertical fraction dy is used to linearly interpolate P_{01} and P_{23} and obtain the bilinearly interpolated pixel P. P is then output to the sprite memory.

If the four interpolation pixels are not in the cache, memory access time is critical. Table 2.4 lists the memory requirements for warping sprites with various resolutions. The memory size listed is for a single sprite buffer. Since the warped sprite and original sprite are stored in separate areas of sprite memory, two sprite buffers are required. To provide additional memory bandwidth, the sprite buffers can be stored on separate memory chips. If a single warping unit is used to warp k sprites, $k + 1$ sprite buffers are required.

Table 2.4 Memory Requirements for Sprite Warping

Sprite Format	QCIF	CIF	ITU-R 601
Memory size	297 Kbits	1,188 Kbits	4,860 Kbits
Pixel reads/frame	152,000	608,000	2,488,000
Pixel writes/frame	38,000	152,000	622,080
Time/read at 30 fps	219 ns	54 ns	13 ns

Table 2.4 lists the worst-case number of pixel reads and writes required to warp a sprite. The table also lists the average time per read that must be met if the sprite is to be warped at 30 frames per second. It assumes that the warped sprite and original sprite are contained in

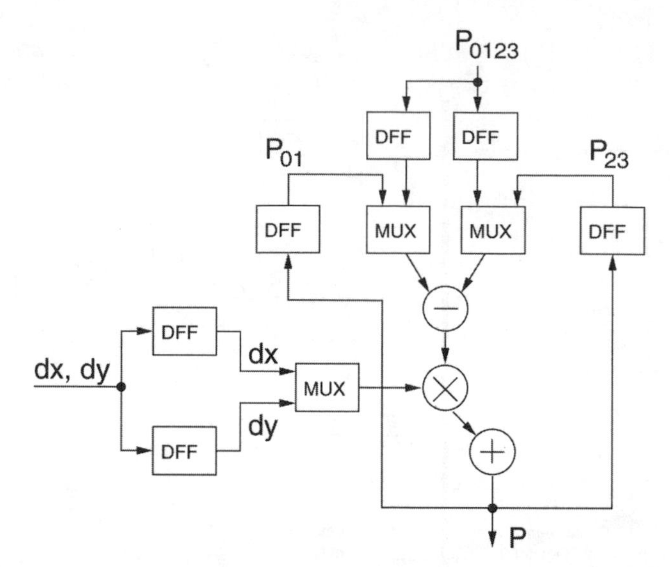

FIGURE 2.7
Interpolation unit.

separate memories. The times were obtained using

$$t_{\text{read}} = \left[\left(\frac{\text{number of frames}}{\text{second}} \right) \left(\frac{\text{number of pixel reads}}{\text{frame}} \right) \right]^{-1} \tag{2.23}$$

If the pixels are currently in the cache, they will be transmitted to the interpolation unit quickly, and the computation time in the interpolation unit becomes critical. A new pixel cannot be read in until the previous pixel has been linearly interpolated. Assuming the linear interpolation time, $t_{\text{interpolate}}$, is less than $t_{\text{page hit}}$ (the time to read from memory on a page hit), then the average read time is determined by

$$t_{\text{read}} = ct_{\text{interpolate}} + (1 - c)[pt_{\text{page hit}} + (1 - p)t_{\text{page miss}}] \tag{2.24}$$

where c is the cache hit ratio, p is the page hit ratio, and $t_{\text{page miss}}$ is the time to read from memory on a page miss.

If no cache is used, equation (2.24) reduces to

$$t_{\text{read}} = pt_{\text{page hit}} + (1 - p)t_{\text{page miss}} \tag{2.25}$$

Assuming DRAM access times of 20 ns on a page hit and 85 ns on a page miss, the 219-ns cycle time listed in Table 2.4 for QCIF sprites can be easily obtained without a cache. CIF sprites can also be warped without a cache, because the 73-ns cycle time can be met for $p > .5$, which is a very low page hit ratio. For both sprite sizes, the interpolation unit can be designed to match the memory access time.

With a four-pixel cache, the average read time equation becomes

$$t_{\text{read}} = \frac{1}{4}t_{\text{interpolate}} + \frac{3}{4}[pt_{\text{page hit}} + (1 - p)t_{\text{page miss}}] \tag{2.26}$$

where $c = \frac{1}{4}$ is the cache hit ratio for the worst case. The four-pixel cache can be used to warp sprite sizes larger than CIF. It cannot warp ITU-R 601 sprites, because they require the short 13-ns cycle from Table 2.4. Instead, the three-line cache is used, where $c = \frac{3}{4}$. For $t_{\text{page hit}} = 20$ ns and $t_{\text{page miss}} = 85$ ns, equation (2.24) can then be rewritten as

$$t_{\text{read}} = \frac{3}{4} t_{\text{interpolate}} + \frac{1}{4}[p t_{\text{page hit}} + (1 - p) t_{\text{page miss}}] = (28p - 17.3) \quad \text{nanoseconds} \tag{2.27}$$

which simplifies to

$$t_{\text{interpolate}} < (21.67p - 11) \quad \text{nanoseconds} \tag{2.28}$$

This equation is satisfied by realistic interpolation times and page hit ratios. For example, an 8.5-ns interpolation time and a 0.9 page hit ratio, or a 6.3-ns linear interpolation time and a 0.8 page hit ratio, can be used for real-time warping of ITU-R 601 sprites with 0.35μ m or better VLSI technology.

2.4 Conclusion

We have presented a new fast algorithm for computing the perspective transform. The constant denominator method reduces the number of divisions required from $O(N^2)$ to $O(N)$ and also dramatically reduces multiplications in the computation. The speed of the constant denominator method does not sacrifice the accuracy of the algorithm. Indeed, it has more than 35 times less error compared with other approximation methods. The algorithm primarily targets real-time implementation of sprite warping. However, it is generally for speeding up the perspective transform. Based on this algorithm, an architecture was proposed for the implementation of sprite warping for MPEG-4. Our architecture is feasible under current VLSI technology. We also analyzed the real-time requirement of the architecture and addressed several other implementation issues.

References

[1] CCITT. MPEG-4 video verification model version 11.0. ISO-IEC JTC1/SC29/WG11 MPEG98/N2172, Mar. 1998.

[2] Demirer, M., and Grimsdale, R.L. Approximation techniques for high performance texture mapping. *Computer & Graphics 20,* 4 (1996).

[3] Hearn, D., and Baker, M.P. *Computer Graphics,* 2 ed., Prentice-Hall, Englewood Cliffs, NJ, 1994.

[4] Heckbert, P., and Moreton, H.P. Interpolation for polygon texture mapping and shading. In *State of the Art in Computer Graphics Visualization and Modeling,* D.F. Rogers and R.A. Earnshaw, Eds., Springer-Verlag, Berlin, 1991.

[5] Hennessy, J.L., and Patterson, D.A. *Computer Architecture: A Quantitative Approach,* 2 ed., Morgan Kaufmann Publishers, 1996.

[6] Kirk, D., and Vorrhies, D. The rendering architecture of the dn10000vs. *Computer Graphics 24* (1990).

[7] Lee, M.C., Chen, W., Lin, C.B., Gu, C., Markoc, T., Zabinsky, S.I., and Szeliski, R. A layered video object coding system using sprite and affine motion model. *IEEE Transactions on Circuits and Systems for Video Technology 7,* 1 (Feb. 1997).

[8] Press, W.H., Flannery, B.P., Teukolsky, S.A., and Vetterling, W.T. *Numerical Recipes in C,* 2 ed., Cambridge University Press, London, 1994.

[9] Sun, M.T. Algorithms and VLSI architectures for motion estimation. In *VLSI Implementations for Communications,* P. Pirsh, Ed., Elsevier Science Publishers, New York, 1993.

[10] Wolberg, G. *Digital Image Warping.* IEEE Computer Society Press, 1990.

Chapter 3

Application-Specific Multimedia Processor Architecture

Yu Hen Hu and Surin Kittitornkun

3.1 Introduction

Multimedia signal processing concerns the concurrent processing of signals generated from multiple sources, containing multiple formats and multiple modalities. A key enabling technology for multimedia signal processing is the availability of low-cost, high-performance signal processing hardware including programmable digital signal processors (PDSPs), application-specific integrated circuits (ASICs), reconfigurable processors, and many other variations.

The purposes of this chapter are (1) to survey the micro-architecture of modern multimedia signal processors, and (2) to investigate the design methodology of dedicated ASIC implementation of multimedia signal processing algorithms.

3.1.1 Requirements of Multimedia Signal Processing (MSP) Hardware

Real-Time Processing

With *real-time* processing, the results (output) of a signal processing algorithm must be computed within a fixed, finite duration after the corresponding input signal arrives. In other words, each computation has a *deadline*. The real-time requirement is a consequence of the *interactive* nature of multimedia applications. The amount of computations per unit time, also known as the *throughput* rate, required to achieve real-time processing varies widely for different types of signals. If the required throughput rate cannot be met by the signal processing hardware, the quality of service (QoS) will be compromised. Real-time processing of higher dimensional signals, such as image, video, or 3D visualization, requires an ultra-high throughput rate.

Concurrent, Multithread Processing

A unique feature of MSP hardware is the need to support concurrent processing of multiple signal streams. Often more than one type of signal (e.g., video and sound) must be processed concurrently as separate task threads in order to meet deadlines of individual signals. Synchronization requirements also impose additional constraints.

Low-Power Processing

Multimedia signal processing devices must support mobile computing to facilitate ominous accessibility. Low-power processing is the key to wireless mobile computing. Technologies (TTL vs. CMOS, power supply voltages) are the dominating factor for power consumption. However, architecture and algorithm also play a significant role in system-wide power consumption reduction.

3.1.2 Strategies: Matching Micro-Architecture and Algorithm

To achieve the performance goal (real-time processing) under the given constraint (low power consumption), we must seek a close match between the multimedia signal processing algorithm formulation and the micro-architecture that implements such an algorithm. On the one hand, micro-architecture must be *specialized* in order to custom fit to the given algorithm. On the other hand, alternative algorithm formulations must be explored to exploit its inherent *parallelism* so as to take advantage of the power of parallel micro-architecture.

Specialization

Specialized hardware can be customized to execute the algorithm in the most efficient fashion. It is suitable for low-cost, embedded applications where large-volume manufacturing reduces the average design cost. Hardware specialization can be accomplished at different levels of granularity. Special function units such as an array multiplier or multiply-and-accumulator (MAC) have been used in programmable DSPs. Other examples include a bit reversal unit for fast Fourier transform and so forth.

Another approach of specialization is to use a special type of arithmetic algorithm. For example, CORDIC arithmetic unit is an efficient alternative when elementary functions such as trigonometric, exponential, or logarithmic functions are to be implemented. Another example is the so-called distributed arithmetic, where Boolean logic functions of arithmetic operations are replaced with table-lookup operations using read-only memory.

At a subsystem level, specialized hardware has also been developed to realize operations that are awkward to be realized with conventional word-based micro-architecture. For example, the variable-length entropy-coding unit is often realized as a specialized subsystem.

Specialized hardware consisting of multiple function units to exploit parallelism is also needed to handle computation-intensive tasks such as motion estimation, discrete cosine transform, and so forth. At the system level, specialized hardware has also been developed to serve large-volume, low-cost, and embedded consumer applications, such as the MPEG decoder chip.

Parallelism

Parallelism is the key to achieving a high throughput rate with low power consumption. To reduce power consumption, power supply voltage must be reduced. Lower power supply voltage implies lower switching speed. As such, to meet the real-time processing throughput constraint, more function units must be activated together, taking advantage of the potential parallelism in the algorithm.

Many MSP algorithms can be formulated as nested iterative loops. For this family of algorithms, they can be *mapped algebraically* into regular, locally interconnected pipelined processing arrays such as the systolic array. Examples include discrete cosine transform, full search motion estimation, discrete wavelet transform, and discrete Fourier transform.

In addition to the systolic array, parallelism can be exploited in different formats. A vector-based parallel architecture is capable of performing vector operations efficiently. A specific vector-parallel architecture is known as the subword parallelism. It appears as the multimedia extension (MMX) instructions in general-purpose microprocessors.

Some algorithms do not have a regular structure such as nested iterative loops. However, since MSP applications often deal with indefinite streams of signals, it is also possible to develop pipelined special-purpose hardware to exploit the parallelism. Examples include fast discrete cosine transform (DCT) algorithms.

For programmable DSP processors, instruction-level parallelism (ILP) has dominated modern superscalar microprocessor architecture. A competing ILP approach is known as the very long instruction word (VLIW) architecture. The main difference between ILP and VLIW is that ILP architecture relies on a hardware-based instruction issuing unit to exploit the potential parallelism inherent in the instruction stream during the *run time,* whereas the VLIW micro-architecture relies heavily on a compiler to exploit ILP during the *compile time.*

3.2 Systolic Array Structure Micro-Architecture

3.2.1 Systolic Array Design Methodology

Systolic array [1, 2] is an unconventional computer micro-architecture first proposed by H.T. Kung [3]. It features a regular array of identical, simple processing elements operated in a pipelined fashion. It can be visualized that data samples and intermediate results are processed in a systolic array in a manner analogous to how the blood is pumped by the heart — a phenomenon called systole circulation — which is how this architecture received its name.

A systolic array exhibits characteristics of parallelism (pipelining), regularity, and local communication. If an algorithm can be described as a nested "do" loop with simple loop body, specifically known as a *regular iterative algorithm,* then it can be *mapped algebraically* onto a systolic array structure.

A number of multimedia signal processing algorithms can be implemented using systolic arrays. Examples include two-dimensional DCT (2D DCT), video block motion estimation, and many others. To illustrate systolic array design methodology, consider the convolution of a finite length sequence $\{h(n); 0 \leq n \leq M-1\}$ with an infinite sequence $\{x(n); n = 0, 1, \ldots$

$$y_n = \sum_{k=0}^{\min(n,M-1)} h(k)x(n-k) \qquad n = 0, 1, \ldots \tag{3.1}$$

This algorithm is usually implemented with a two-level nested do loop:

```
Algorithm 1:

For n = 0, 1, 2,...
        y(n)  = 0
        For k = 0 to min(n,M-1),
                y(n)  = y(n)+h(k)*x(n-k)
        end
end
```

It can be implemented using a systolic array containing M processing elements as depicted in Figure 3.1. In Figure 3.1, the narrow rectangular box represents delay, and the square

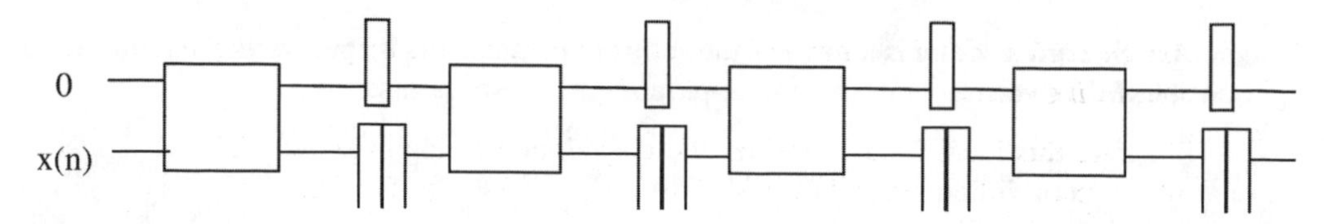

FIGURE 3.1
Convolution systolic array.

box represents a processing element (PE). Moreover, every PE is identical and performs its computation in a pipelined fashion. The details of a PE are shown in Figure 3.2. In this

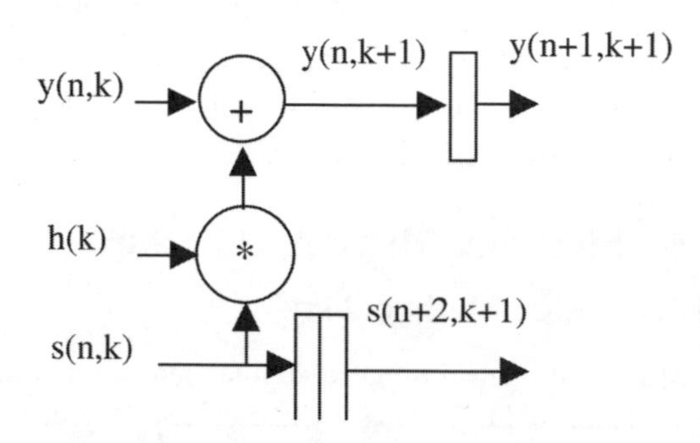

FIGURE 3.2
A processing element of the convolution systolic array.

figure, the circle represents arithmetic operations. The above implementation corresponds to the following algorithm formulation:

```
Algorithm 2:

s(n,0)  = x(n);     g(n,0)  = 0;  n = 0, 1, 2,...
g(n,k+1)  =g(n,k)+h(k)*s(n,k); n = 0, 1, 2,...; k = 0 to M-1,
s(n,k+1)  = s(n,k);            n = 0, 1, 2,...; k = 0 to M-1,
g(n+1,k+1)  = g(n,k+1);        n = 0, 1, 2,...; k = 0 to M-1,
s(n+2,k+1)  = s(n,k+1);        n = 0, 1, 2,...; k = 0 to M-1,
y(n)  = g(n+M,M)  ;            n = 0, 1, 2,...
```

In the above formulation, n is the time index and k is the processing element index. It can be verified manually that such a systolic architecture yields correct convolution results at the sampling rate of $x(n)$.

Given an algorithm represented as a nested do loop, a systolic array structure can be obtained by the following three-step procedure:

1. *Deduce a localized dependence graph of the computation algorithm.* Each node of the dependence graph represents computation of the innermost loop body of an algorithm represented in a regular nested loop format. Each arc represents an inter-iteration dependence relation. A more detailed introduction to the dependence graph will be given later in this chapter.

2. *Project each node and each arc of the dependence graph along the direction of a projection vector.* The resulting geometry gives the configuration of the systolic array.

3. *Assign each node of the dependence graph to a* schedule *by projecting them along a scheduling vector.*

To illustrate this idea, let us consider the convolution example above. The *dependence graph* of the convolution algorithm is shown in Figure 3.3. In this figure, the input $x(n)$ is from the bottom. It will propagate its value (unaltered) along the northeast direction. Each of the coefficients $\{h(k)\}$ will propagate toward the east. The partial sum of $y(n)$ is computed at each node and propagated toward the north. If we *project* this dependence graph along the [1 0] direction, with a schedule vector [1 1], we obtain the systolic array structure shown on the right-hand side of the figure. To be more specific, each node at coordinate (n, k) in the dependence graph is mapped to processing element k in the systolic array. The coefficient $h(k)$ is stored in each PE. The projection of the dependence vector [1 1] associated with the propagation of $x(n)$ is mapped to a physical communication link with two delays (labeled by 2D in the right-hand portion of the figure). The dependence vector [0 1] is mapped to the upward communication link in the systolic array with one delay. Figure 3.1 is identical to the right side of Figure 3.3 except more details are given.

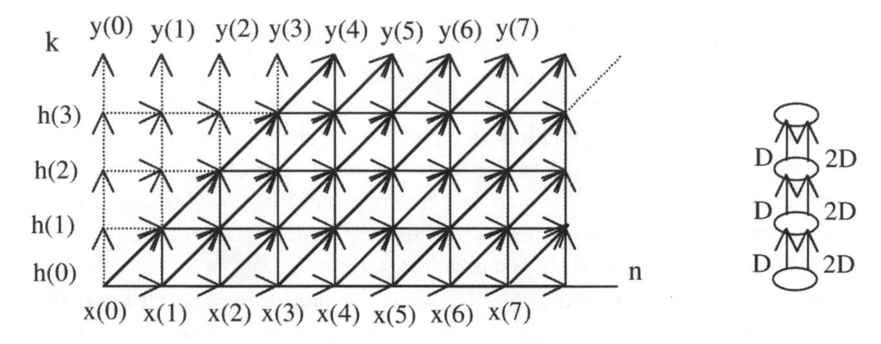

FIGURE 3.3
Dependence graph of convolution (left) and systolic array projection (right).

The systolic design methodology of mapping a dependence graph into a lower dimensional systolic array is intimately related to the loop transformation methods developed in parallel program compilers. A detailed description of loop transform can be found in [4].

3.2.2 Array Structures for Motion Estimation

Block motion estimation in video coding standards such as MPEG-1, 2, and 4, and H.261 and H.263 is perhaps one of the most computation-intensive multimedia operations. Hence it is also the most implemented algorithm.

We will briefly explain block-based motion estimation using Figure 3.4. A basic assumption of motion estimation is that there is high *temporal correlation* between successive frames in video streams; hence, the content of one frame can be *predicted* quite well using the contents of adjacent frames. By exploiting this *temporal redundancy,* one need not transmit the predictable portion of the *current frame* as long as these *reference frame(s)* have been successfully transmitted and decoded. Often, it is found that the effectiveness of this scheme can be greatly enhanced if the basic unit for comparison is reduced from the entire frame to a much smaller "block." Often the size of a block is 16×16 or 8×8 (in the unit of pixels). This is illustrated on the right-hand side of Figure 3.4. Let us now focus on the "current block" that has a dotted pattern in the current frame. In the reference frame, we identify a *search area* that surrounds a block having the same coordinates as the current block. The hypothesis is that within this search area, there is an area equal to the size of the current block which best *matches*

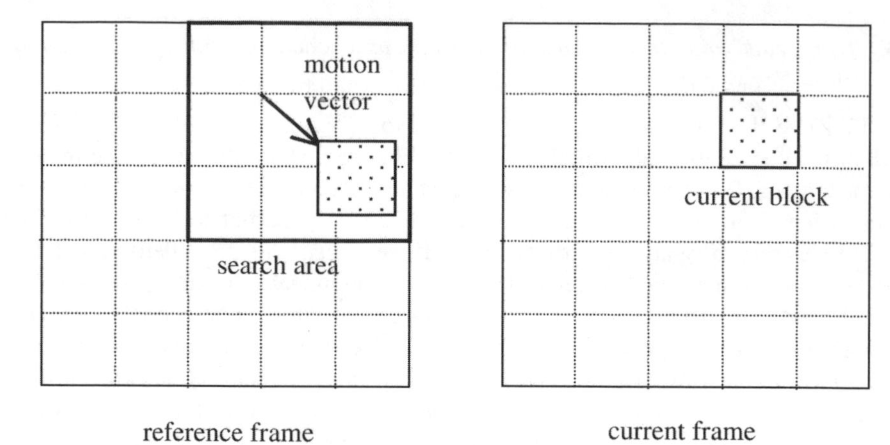

FIGURE 3.4
Block motion estimation.

(is similar to) the current block. Then, instead of transmitting all the pixels in the current block of the current frame, all we need is to specify the displacement between the current block location and the best matched blocking area on the reference frame. Then we cut-and-paste this area from the reference frame to the locations of the current block on a *reconstructed* current frame at the receiving end. Since the reference frame has been transmitted, the current block at the current frame can be reconstructed this way *without transmitting any bit in addition to the displacement values,* provided the match is perfect.

The displacement we specified above is called the *motion vector.* It inherits this name from the motion estimation task in computer vision researches. However, there, the motion estimation is performed on individual pixels, and the objective is to identify object motion in sequential image frames. Since each pixel within the search area can be the origin of a matching block, its coordinates become a candidate for a motion vector. If every pixel within the search area is tested in order to find the best matching block, it is called a *full-search* block-matching method. Obviously, a full search block-matching algorithm offers the best match. But the computation cost is also extremely high. On the other hand, the matching operations can be written in a regular six-level nested do loop algorithm. Thus, numerous systolic array or other dedicated array architectures have been proposed. We note that there are also many fast block-matching algorithms proposed to skip pixels in the search area in order to reduce computation without significantly compromising matching quality. Unfortunately, most of these fast search algorithms are too complicated for a systolic array implementation. In this section, we will survey systolic array structures for the implementation of only the full-search block-matching motion estimation algorithm. First, we review some notations and formulas of this algorithm.

FBMA (Full-Search Block-Matching Algorithm)

Assume a *current* video frame is divided into $N_h \times N_v$ blocks in the horizontal and vertical directions, respectively, with each block containing $N \times N$ pixels. The most popular similarity criterion is the mean absolute difference (MAD), defined as

$$\mathrm{MAD}(m, n) = \frac{1}{N^2} \sum_{i=0}^{N-1} \sum_{j=0}^{N-1} |x(i, j) - y(i + m, j + n)| \tag{3.2}$$

where $x(i, j)$ and $y(i+m, j+n)$ are the pixels of current frame and previous frame, respectively. The motion vector (MV) corresponding to the minimum MAD within the search area is given by

$$\mathrm{MV} = \arg\{\min \mathrm{MAD}(m,n)\} \qquad -p \leq m, n \leq p \,, \tag{3.3}$$

where p is the search range parameter. We focus on the situation where the search area is a region in the reference frame consisting of $(2p+1)^2$ pixels.

In the FBMA, MAD distortions between the current block and all $(2p+1)^2$ candidate blocks are to be computed. The displacement that yields the minimum MAD among these $(2p+1)^2$ positions is chosen as the motion vector corresponding to the present block. For the entire video frame, this highly regular FBMA can be described as a six-level nested do loop algorithm, as shown below.

```
Algorithm 3: Six-level nested do loop of full-search
block-matching motion estimation

Do h=0 to Nh-1
Do v=0 to Nv-1
        MV(h,v)=(0,0)

     Dmin(h,v)=∞

     Do m=-p to p (-1)
     Do n=-p to p (-1)
         MAD(m,n)=0
         Do i=hN to hN+N-1
         Do j=vN to vN+N-1
             MAD(m,n)= MAD(m,n)+|x(i,j)-y(i+m,j+n)|
         End do j
         End do i
         If Dmin(h,v) > MAD(m,n)
             Dmin(h,v)=MAD(m,n)
             MV(h,v)=(m,n)
         End if
     End do n
     End do m
End do v
End do h
```

The frame rate for a particular resolution standard (e.g., MPEG-2, H.261) can be used as a performance metric. Assuming that time to compute an MV of one block of $N \times N$ pixels is T_{block}, then the time to compute the whole video frame is

$$T_{\text{frame}} = N_h N_v T_{\text{block}} \,, \tag{3.4}$$

and the frame rate F_{frame} is determined by

$$F_{\text{frame}} = \frac{1}{T_{\text{frame}}} \,. \tag{3.5}$$

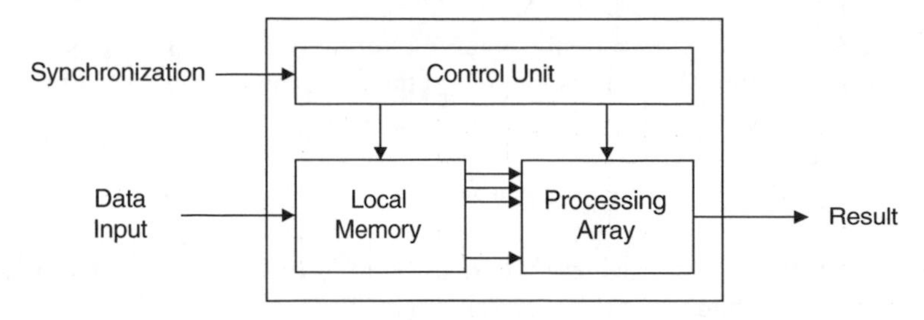

FIGURE 3.5
MEP block diagram.

Motion Estimation Subsystem Architecture

A generic block diagram of a motion estimation subsystem consists of a processing array, local (on-chip) memory, and a control unit as shown in Figure 3.5.

The control unit provides the necessary clock timing signals and flags to indicate the beginning and completion in processing the current block. The local memory unit not only acts as an on-chip cache but also facilitates data reordering. The size of the local memory depends on the specific systolic mapping performed. Based on the geometry of the processing array (in conjunction with local memory), existing motion estimation array structures can be roughly classified into four categories:

- 2D array
- linear array
- tree-type structure (TTS)
- hybrid

We will briefly survey each of these array structures.

2D Array Micro-Architecture

The AB2 architecture [5] shown in Figure 3.6 and its sibling AS2 (not shown) were among the first motion estimation array structures. Subsequently, AB2 has been modified [6] to scan the search area data sequentially in raster scan order using shift registers. This reduces the need for a large number of input–output (I/O) pins. However, the overall processing element utilization is rather inefficient. An improved AB2-based architecture is presented by [7]. The movement of search area data is carefully studied so that it can exploit a spiral pattern of data movement. On average, this processor array is able to compute two MADs in every cycle. However, it requires a PE that is twice as complicated. This can reduce the computation latency at the expense of more complicated PE architecture. These earlier array structures are often derived in an ad hoc manner without employing a formal systolic array mapping strategy.

A modular semisystolic array derived by performing the systolic mapping of a six-level nested do loop algorithm on an array is presented in [8]. First, we transform the three pairs of indices (v, h), (m, n), (i, j) of the six-level nested do loop in Algorithm 3 to a three-level nested do loop with indices (b, l, k), where b, l, and k represent block, search vector, and pixel, respectively, of the entire frame. A systolic multiprojection technique [1] is then used to project the 3D dependence graph (DG) into a linear array. Next, exploiting the fact that the neighboring search area shares many reference frame pixels, this linear array is further folded into a spiral 2D array as shown in Figure 3.7. In this configuration, the search area pixel y

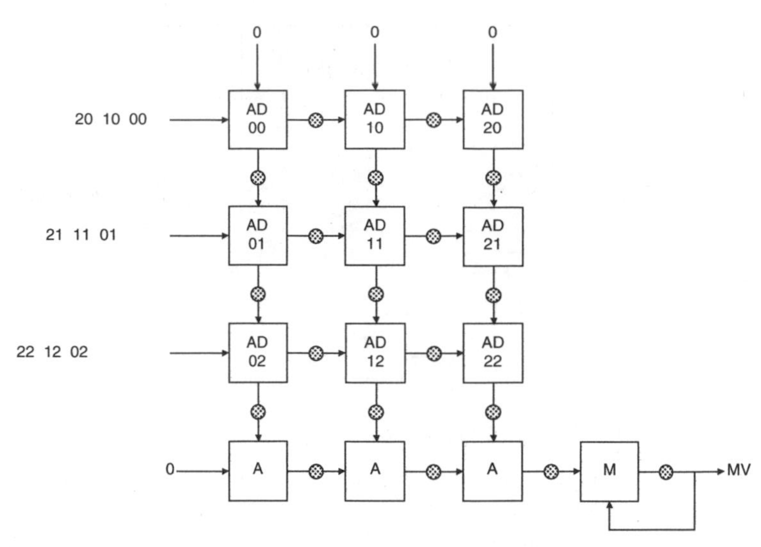

FIGURE 3.6
AB2 architecture [5]. AD: absolute difference, A: addition, M: memory.

is broadcast to each processing element in the same column, and the current frame pixel x is propagated along the spiral interconnection links. The constraint of $N = 2p$ is imposed to achieve a low I/O pin count. A simple PE is composed of only two eight-bit adders and a comparator, as shown in Figure 3.7.

In [9] the six-level nested do loop is transformed into a two-level nested do loop, which is then mapped into a linear array and then folded into a 2D spiral array. The resulting design has better scalability to variable block sizes and search ranges and does not need data broadcasting. In [10], another 2D array structure is proposed. It uses multiprojection directly to transform the dependence graph corresponding to the six-level nested do loop into a 2D fully pipelined systolic array. Two levels of on-chip caches are required to handle the data movements. Furthermore, it has been shown that the previous motion estimation array architecture [6] is a special case of this 2D array structure. In the architectures proposed in [11] and [12], attention is paid to data movement before and after the motion estimation operations. Data broadcasting is used to yield a semisystolic array [11]. Two sets of shift register arrays are used to switch back and forth between two consecutive current blocks to ensure 100% PE utilization (Figure 3.8).

Linear Array Architecture

A linear array configuration uses fewer processing elements but has a lower data throughput rate. It is suitable for applications with a lower frame rate and lower resolution such as videoconferencing and/or videophone. The AB1 [5] depicted in Figure 3.9 is an example of linear array architecture.

The performance of a linear array architecture can be enhanced using data broadcasting to reduce the pipelining latency in a systolic array where data are propagated only to its nearest neighboring PE. In [13], it is suggested to broadcast either the current block pixels or the search area pixels so that PEs that need these data can be computed earlier. Obviously, when the array size grows, long global interconnection buses will be needed to facilitate data broadcasting. This may increase the critical path delay and hence slow down the applicable clock frequency.

A hybrid SIMD (single instruction, multiple data) systolic array, consisting of four columns of 16 PEs, has been proposed by [14]. It is essentially the collection of four independent 16 ×

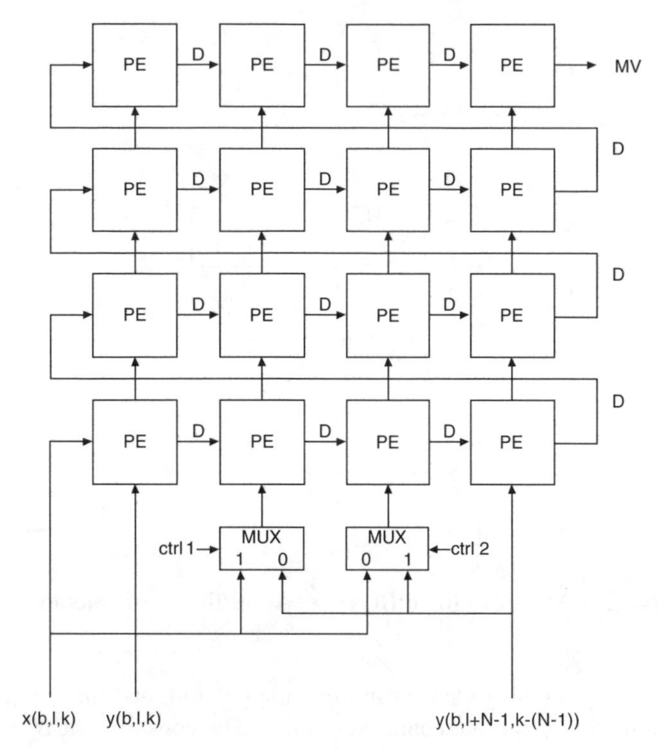

FIGURE 3.7
2D array with spiral interconnection ($N = 4$ and $p = 2$). PE: processing element, D: delay, ctrl: control line, MUX: multiplexer.

1 linear arrays; hence, it should be considered as a variant of linear array architecture. More recently, a linear array structure was reported in [15]. It is based on slicing and tiling of a 4D DG onto a single 2D plane in order to make the projection easier. Global buses are needed to broadcast search area data. Additional input buffers are required to reorder the input sequence into a format suitable for the processing array. On the other hand, modules can be linearly cascaded for better parallelism or to handle bigger block size as well as a larger search range.

Tree-Type Structure (TTS) Architecture

TTS is suitable for not only FBMA but also irregular block-matching algorithms such as the three-step hierarchical search. Since each tree level shown in Figure 3.10 can be viewed as a parallel pipeline stage, the latency is shorter. Nevertheless, the computation time is still comparable to those of 1D or 2D array architectures. The problem associated with TTS is the memory bandwidth bottleneck due to the limited number of input pins. This can be alleviated by a method called 1/M-cut subtree, as proposed in [16], to seek a balance between memory bandwidth and hardware complexity.

Hybrid Architecture

Several hybrid architectures proposed in the literature are now briefly reviewed.

In [17], two types (type 1 and type 2) of hybrid architectures are proposed. In these architectures, search area data y are injected into a 2D array with tree adders in a meander-like pattern. The type-1 architecture is similar to the AB2 array [5] shown in Figure 3.6. It imposes

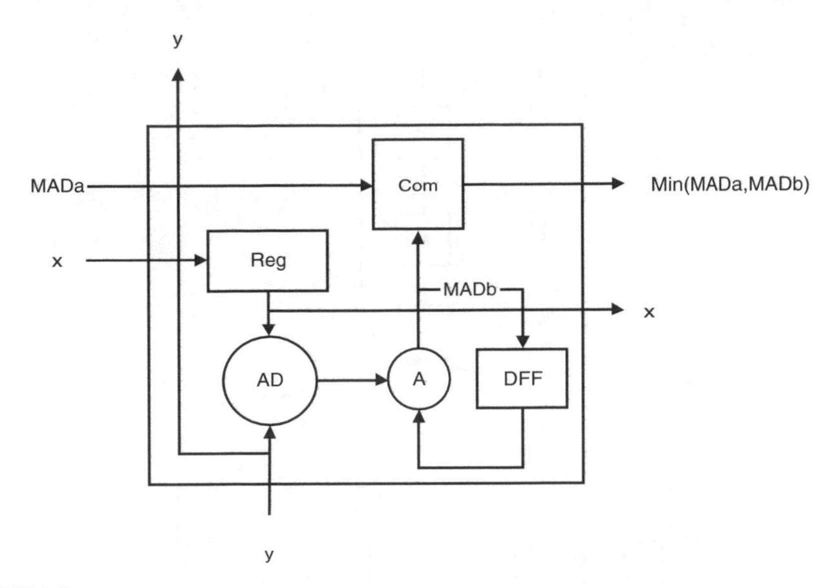

FIGURE 3.8
Diagram of an individual processing element. Reg: register, Com: compare, AD: absolute difference, A: addition, DFF: D flip-flop.

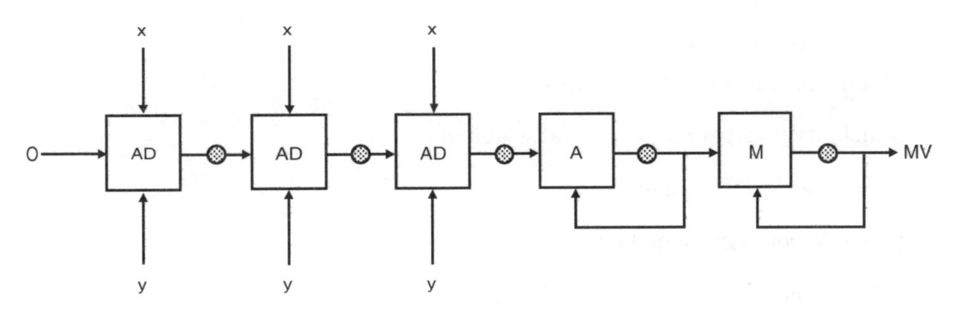

FIGURE 3.9
AB1 architecture [5]. AD: absolute difference, M: memory, -o-: delay.

the constraint that $N = 2p + 1$. The type-2 architecture is analogous to the AS2 array in [5]. These array architectures have registers on both the top and bottom of the processing array to support meander-like movement of search area data.

In [17], a hybrid TTS/linear structure has been suggested. This architecture consists of a parallel tree adder to accumulate all the partial sums calculated by a linear array of PEs. To achieve the same throughput as a 2D array, clock frequency must be increased n times from the 2D array, where n is the degree of time-sharing. A register ring is added to accumulate SAD after a tree adder, as reported in [18, 19]. Another hybrid architecture [20] utilizes a linear array of N 1/2-cut subtrees with systolic accumulation instead of a single 1/32-cut subtree, as shown in [16].

Performance Comparison

We use the following features to compare different motion estimation array architectures:

- Area and complexity
- Number of I/O ports and memory bandwidth

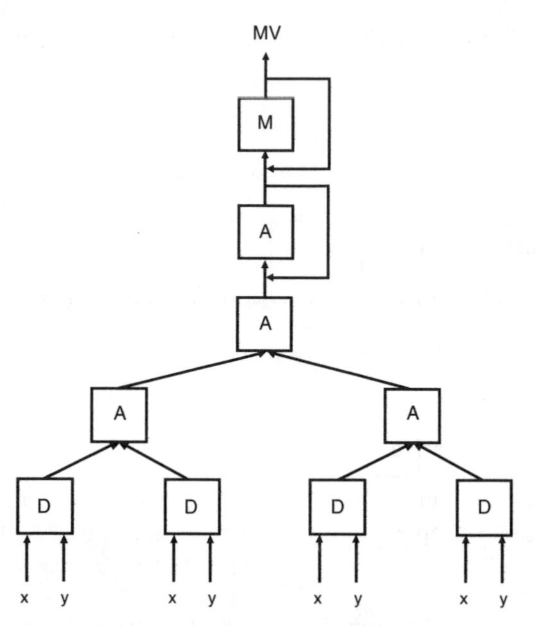

FIGURE 3.10
Tree-type structure [16]. D: absolute difference, A: addition, M: memory.

- Throughput rate of motion vectors
- Scalability to larger block size and search range
- Operating clock frequency
- Dynamic power consumption
- PE utilization

Area and complexity can be represented by the number of PEs, the micro-architecture of an individual PE, and the number of on-chip memory units such as latches, pipeline registers, shift registers, etc. Motion vector computation throughput rate can be determined by block computation time. The memory bandwidth is proportional to the number of I/O ports required by the processing array. I/O ports include current block, search area data, and motion vector output ports. A multiple-chip solution provides the ability to support a bigger block size and search range.

With today's technology, a single-chip solution or subsystem solution is more practical and cost-efficient. A few architectures can truly scale well but require a large number of fan-outs as a result of broadcasting. Block-level PE utilization is taken into consideration rather than the frame level. Power consumption becomes more and more important to support mobile communication technology. The block size of $N = 16$ and search range of $p = 8$ are used as common building blocks. In Tables 3.1 and 3.2, the performance parameters are formulated as functions of N and p.

For simulated or fabricated layouts, important parameters such as maximum operating frequency, die size, transistor count, and power consumption can be used to evaluate the performance of each architecture in Table 3.2. For example, the bigger the die size, the more likely lower yield becomes, leading to the higher list price. Within a certain amount of broadcasting, the higher the transistor count, the more power is consumed. Otherwise, power consumed by the inherent capacitance and inductance of long and wide interconnection may become more apparent. This can affect the battery time of a digital video camcorder and/or multimedia mobile terminal.

3.3 Dedicated Micro-Architecture

3.3.1 Design Methodologies for Dedicated Micro-Architecture

A dedicated micro-architecture is a hardware implementation specifically for a given algorithm. It achieves highest performance through both specialization and parallelism.

Implementation of Nonrecursive Algorithms

Any computing algorithm can be represented by a directed graph where each node represents a task and each directed arc represents the production and consumption of data. In its most primitive form, such a graph is called a *data flow graph*. Let us consider an algorithm with the following formulation.

```
Algorithm 4:

tmp0= c4*(-x(3)+x(4));
y(3)  = ic6*(x(3)  + tmp0);
y(7)  = ic2*(-x(3)  + tmp0);
```

It can be translated into a data flow diagram as shown in Figure 3.11. In this algorithm, three additions and three multiplication operations are performed. There are two input data samples, `x(3)` and `x(4)`, and two output data samples, `y(3)` and `y(7)`. `c4`, `ic2`, and `ic6` are precomputed constant *coefficients* which are stored in memory and will be available whenever needed. To *implement* this algorithm, one must have appropriate hardware devices to perform addition and multiplication operations. Moreover, each device will be assigned to perform a specific task according to a *schedule*. The collection of task assignment and schedule for each of the hardware devices then constitutes an *implementation* of the algorithm.

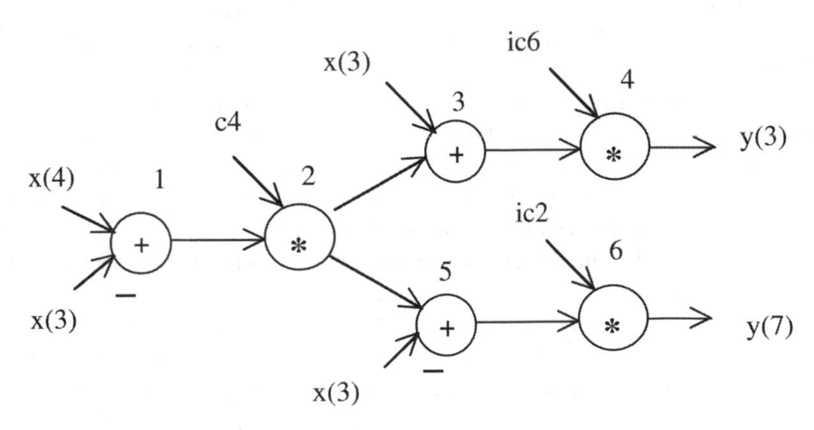

FIGURE 3.11
An example of a data flow diagram.

Table 3.1 Architecture Comparison for Search Range $p = N/2 = 8$

Architecture	Search Range	PE	Computation Time (cycles)	I/O Ports (8 bits)	Memory Units (8 bits)
Komarek and Pirsch [5]					
AS1	$-p/+p$	$2p+1$	$N(N+2p)(2p+1)$	3	$10p+6$
AB1	$-p/+p$	N	$N(N+2p)(2p+1)$	$2N+1$	$2N+1$
AS2	$-p/+p$	$N(2p+1)$	$N(N+2p)$	$N(N+2P)$	$3(N+P)(3N+2)+1$
AB2	$-p/+p$	N^2	$(N+2p)(2p+1)$	$2N+1$	$2N^2+N+1$
Vos and Stegherr [17] (2D)					
2D array (type 1)	$-p/+p$	N^2	N^2	4	$7N^2+2Np$
Linear array	$-p/+p$	N	$N(2p+1)^2$	4	$3N^2+2Np$
Yang et al. [13]	$-p/+p-1$	N	$2p(N^2+2p)$	4	$4N$
Hsieh and Lin [6]	$-p/+p$	N^2	$(N+2p)^2+5$	3	$3N^2+(N-1)(2p-1)$
Jehng et al. [16]	$-p/+p$	$N^2/16$	$32(2p+1)^2$	4	$N^2/16+1$
Wu and Yeh [14]	$-p/+p$	$4N$	$2N(2N+p)$	4	N^2
Nam et al. [18] & Nam and Lee [19]	$-p/+p-1$	N	$(2p)^2N+N+\log_2 N$	4	$8N+1$
Chang et al. [15]	$-p+1/+p$	$2N$	$(2p)^2N$	6	$9N+4p$
Yeo and Hu [8]	$-p/+p-1$	N^2	N^2	4	$2N^2$
Pan et al. [7]	$-p+1/+p-1$	$2N^2$	$(N+2p)(p+3)$	$N+3$	$2N^2+4N+1$
Chen et al. [20]	$-p/+p$	$2N^2+2$	$(2p+2N/M)(2p+1)$	3	$2N^2+2N+2$
Lee and Lu [11]	$-p/+p-1$	N^2	$(2p)^2$	4	$5N^2+2(N-1)(N+2p)$
You and Lee [12]	$-p/+p-1$	kv	$(2pN)^2/kv$	10	$(N+2p)^2$
Chen and Kung [10]	$-p/+p$	N^2	N^2	3	$2N^2+(N+2p)^2$
STi3220 [21]	$-p/+p-1$	N^2	N^2+46	5	$2N^2$
Kittitornkun and Hu [9]	$-p/+p$	$(2p+1)^2$	N^2	4	$3(2p+1)^2+N^2$

Note: The number of PE corresponds to the number of arithmetic units that perform absolute difference (AD), addition (A), and comparison (M).

Table 3.2 Parameter Comparison of Fabricated or Simulated Layouts

Architecture	Techno. (μm)	Max Freq (MHz)	I/O Pads	Die size (mm^2)	Transistor Count	Power Consum. (W)
Yang et al. [13]	1.2	25	116	3.15×3.13	52,000	Na.
Hsieh and Lin [6]	1.0	120	Na.	Na.	Na.	Na.
Wu and Yeh [14]	0.8	23	65	5.40×4.33	86,000	Na.
Chang et al. [15]	0.8	Na.	100	6.44×5.26	102,000	Na.
Vos and Schobinger [22]	0.6	72	Na.	228	1,050,000	Na.
Nam and Lee [19]	0.8	50	Na.	Na.	Na.	Na.
Chen et al. [20]	0.8	30	97	12.0×4.3	Na.	Na.
Lee and Lu [11]	0.8	100	84	9.5×7.2	310,000	1.95 @ 50 MHz
Sti3220 [21]	Na.	20	144	Na.	Na.	2.4 @ 20 MHz

Na.: not available.

Assume that four hardware devices, two adders and two multipliers, are available. The delay for an addition is one time unit, whereas for a multiplication it is two time units. Furthermore, assume that after the execution of each task, the result will be stored in a temporary storage element (e.g., a register) before it is used as the input by a subsequent task. A possible implementation of Algorithm 4 is illustrated in Table 3.3.

Table 3.3 Implementation # 1 of Algorithm 4

devices\time units	1	2	3	4	5	6
adder #1	1			3		
adder #2				5		
multiplier #1		2	2		4	4
multiplier #2					6	6

In this table, each column represents one time unit, and each row represents a particular device. The numerical number in each shaded box corresponds to the particular task in the data flow graph. Blanked cells indicate that the corresponding device is left idle. Note that task 2 cannot be commenced before task 1 is completed. This relationship is known as *data dependence*. Also note that in time unit 4, tasks 3 and 5 are executed in both adders in parallel. This is also the case in time units 5 to 6 where tasks 4 and 6 are executed in the two multipliers in parallel. Thus, with a sufficient number of hardware devices, it is possible to exploit parallelism to expedite the computation.

Suppose now that only one adder and one multiplier are available; then an implementation will take longer to execute. An example is given in Table 3.4. Note that the total execution time is increased from 6 to 8 time units. However, only half the hardware is needed.

Let us consider yet another possible implementation of Algorithm 4 when there is a stream of data samples to be processed by the hardware.

Table 3.4 Implementation # 2 of Algorithm 4

devices\time units	1	2	3	4	5	6	7	8
adder	1			3		5		
multiplier		2	2		4	4	6	6

```
Algorithm 5:

for i = 1 to . . .,
tmp0(i)= c4*(x(3,i)+x(4,i));
y(3,i)  = ic6*(x(3,i) + tmp0(i));
y(7,i)  = ic2*(x(3,i) + tmp0(i));
end
```

Algorithm 5 contains an infinite loop of the same loop body as Algorithm 4. Since the output of loop i `(tmp0(i),y(3,i), y(7,i))` does not depend on the output of other iterations, the corresponding DG of Algorithm 5 will contain infinitely many copies of the DG of a single iteration shown in Figure 3.11. Since the DGs of different iteration index `i` are independent, we need to focus on the realization of the DG of a single iteration. Then we may duplicate the implementation of one iteration to realize other iterations. In particular, if the input data samples `x(3,i)` and `x(4,i)` are sampled sequentially as i increases, multiple iterations of the this algorithm can be implemented using two adders and three multipliers (Table 3.5).

Table 3.5 Multi-Iteration Implementation # 1 of Algorithm 5

devices\time units	1	2	3	4	5	6	7	8	9	10
adder #1	1		1'							
adder #2				3	5	3'	5'			
multiplier #1		2	2	2'	2'					
multiplier #2					4	4	4'	4'		
multiplier #3						5	5	5'	5'	

Note: Cells with the same texture or shade belong to tasks of the same iteration.

In this implementation, each type of box shading corresponds to a particular iteration index `i`. This implementation differs from the previous two implementations in several ways: (1) Multiple iterations are realized on the same set of hardware devices. (2) Each adder or multiplier performs the same task or tasks in every iteration. In other words, each task is assigned to a hardware device *statically,* and the schedule is *periodic*. Also, note that execution of tasks of successive iterations overlap. Thus, we have an overlap schedule. (3) While each iteration will take seven time units in total to compute, every successive iteration can be initiated every two time units. Hence, the throughput rate of this implementation is two time units per iteration. The average duration between the initiation of successive iterations is known as the *initiation interval*.

Comparing these three implementations, clearly there are trade-offs between the amount of resource utilized (number of hardware devices, for example) and the performance (the total delay, in this case) achieved. In general, this can be formulated as one of two constrained optimization problems:

- Resource-constrained synthesis problem — Given the maximum amount of resources, derive an implementation of an algorithm A such that its performance is maximized.
- Performance-constrained synthesis problem — Given the desired performance objective, derive an implementation of an algorithm A such that the total cost of hardware resources is minimized.

The resource-constrained synthesis problem has an advantage in that it guarantees a solution as long as the available hardware resource is able to implement every required task in algorithm A. On the other hand, given the desired performance objective, an implementation may not exist regardless of how many hardware resources are used. For example, if the performance objective is to compute the output `y(3)` and `y(7)` within four time units after input data `x(3)` and `x(4)` are made available, then it is impossible to derive an implementation to achieve this goal.

Implementation of Recursive Algorithms

Let us consider the following example:

```
Algorithm 6:

for i = 1 to . . .
        y(i) = a*y(i-1) + x(i)
end
```

This is a *recursive* algorithm since the execution of the present iteration depends on the output from the execution of a previous iteration. The data flow graph of this recursive algorithm is shown in Figure 3.12. The dependence relations are labeled with horizontal arrows. The

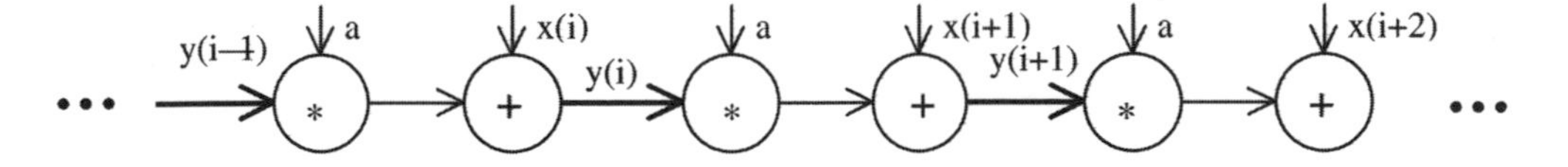

FIGURE 3.12
Data flow graph of Algorithm 6.

thick arrows indicate *inter-iteration dependence relations*. Hence, the execution of the ith iteration will have to wait for the completion of the $(i-1)$th iteration. The data flow graph can be conveniently expressed as an *iterative computation dependence graph (ICDG)* that contains only one iteration, but label the inter-iteration dependence arc with a *dependence distance* d, which is a positive integer. This is illustrated in Figure 3.13. We note that for a nonrecursive algorithm, even if it has an infinite number of iterations (e.g., Algorithm 4), its complete data flow graph contains separate copies of the DG of each iteration. These DGs have no inter-iteration dependence arc linking them.

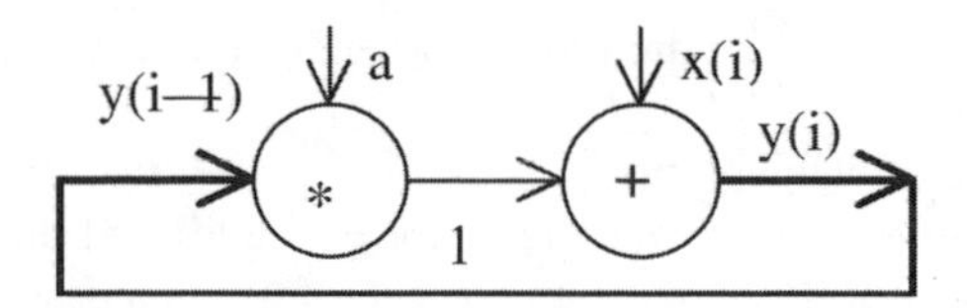

FIGURE 3.13
ICDG of Algorithm 6.

A challenge in the implementation of a recursive algorithm is that one must consider the inter-iteration dependence relations. Many design theories have been developed toward this goal [4], [23]–[25]. The focus of study has been on the feasibility of performance-constrained synthesis. Given a desired throughput rate (initiation interval), one wants to derive an implementation that can achieve the desired performance using the minimum number of hardware modules.

Suppose that multiplication takes two clock cycles and addition takes one clock cycle. It is easy to see that `y(i)` cannot be computed until three clock cycles after `y(i)` is computed. In other words, the minimum initiation interval is $(2+1) = 3$ clock cycles. In a more complicated ICDG that contains more than one tightly coupled cycle, the minimum initiation interval can be found according to the formula

$$I_{\min} = \operatorname*{Max}_{k} \frac{\sum_{i} \tau_i(k)}{\sum_{j} \Delta_j(k)}$$

where $\tau_i(k)$ is the computation time of the ith node of the kth cycle in the ICDG and $\Delta_j(k)$ is the jth inter-iteration dependence distance in the kth cycle. Let us now consider the example in Figure 3.14. There are two cycles in the ICDG in this figure. The initiation interval can be

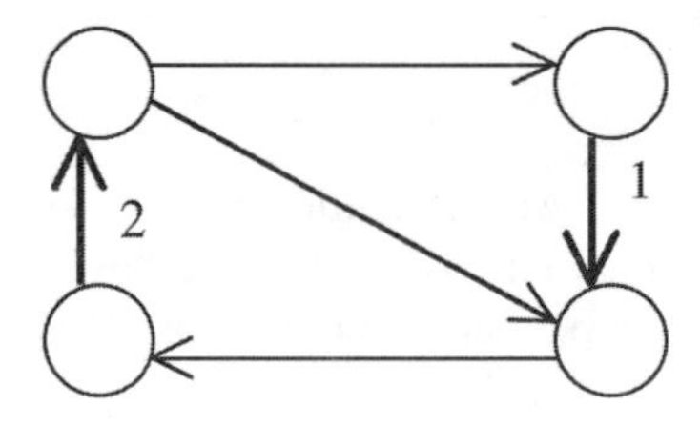

FIGURE 3.14
An ICDG containing two cycles.

calculated as follows:

$$I_{\min} = \max\{(3+1+2+2)/(1+2), (3+2+2)/2\} = \max\{8/3, 7/2\} = 3.5$$

If the desired initiation interval is larger than the minimum initiation interval, one may consider any efficient implementation of a single iteration of the ICDG, and then simply duplicate that implementation to realize computations of different iterations. For example, in the case of Algorithm 6, one may use a single adder and a multiplier module to implement the algorithm if, say, the desired throughput rate is one data sample per four clock cycles. The corresponding implementation is quite straightforward (Table 3.6).

Here we assume that `x(i)` is available at every fourth clock cycle: 4, 8, 12, ... Thus the addition operation can take place only at these clock cycles. The shaded boxes in the adder row of Table 3.6 are also labeled with the corresponding `y(i)` computed at the end of that

Table 3.6 Implementation of Algorithm 6

clock cycles	1	2	3	4	5	6	7	8	9	10	11
multiply											
add				y(1)				y(2)			

Note: Initiation interval = four clock cycles.

clock cycle. The multiplication can then be performed in the immediate next two clock cycles. However, the addition must wait until `x(i)` is ready.

Suppose now the desired throughput rate is increased to one sample per two clock cycles, which is smaller than the minimum initiation interval of three clock cycles. What should we do? The solution is to use an algorithm transformation technique known as the *look-ahead transformation.* In essence, the look-ahead transformation is to substitute the iteration expression of one iteration into the next so as to reduce the minimum initiation interval at the expense of more computations per iteration. For example, Algorithm 6, after applying the look-ahead transformation once, can be represented as:

```
Algorithm 7:

for i = 1 to . . .
        y(i) = a^2*y(i-2) + a*x(i-1) + x(i)
end
```

The corresponding ICDG is displayed in Figure 3.15. The new minimum initiation interval

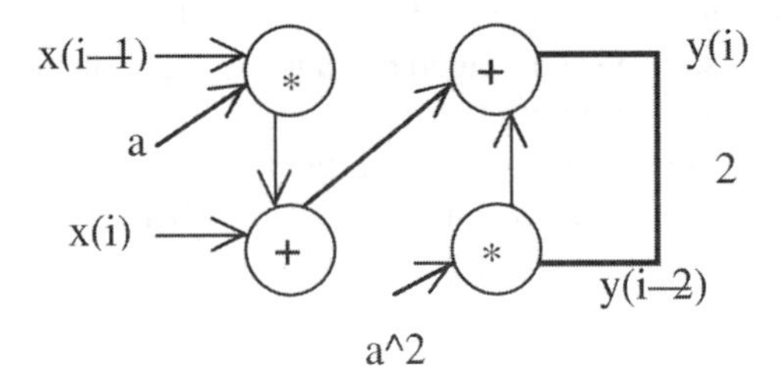

FIGURE 3.15
ICDG of Algorithm 7.

now becomes: $(2 + 1)/2 = 1.5 < 2$ clock cycles, as desired. Next, the question is how to implement this transformed algorithm with dedicated hardware modules. To address this question, another algorithm transformation technique called *loop unrolling* is very useful. Specifically, we consider splitting the sequence {`y(i)`} into two subsequences $\{ye(i)\}$ and $\{yo(i)\}$ such that

$$ye(i) = y(2i) \quad \text{and} \quad yo(i) = y(2i + 1).$$

Then the iterations in Algorithm 7 can be divided into two subiterations with $ye(i)$ and $yo(i)$:

```
Algorithm 8:

for i = 1 to . . .
```

```
            ye(i) = a^2*ye(i-1) + a*x(2i-1) + x(2i)
            yo(i) = a^2*yo(i-1) + a*x(2i) + x(2i+1)
    end
```

To implement Algorithm 8, we denote a new sequence

$$u(i) = x(i) + a * x(i - 1)$$

Then one can see that Algorithm 8 corresponds to two independent subloops:

$$ye(i) = a^2 * ye(i - 1) + u(2i)$$
$$yo(i) = a^2 * yo(i - 1) + u(2i + 1)$$

Each of these subloops will compute at a rate twice as slow as $u(i)$ is computed. Since $x(i)$ is sampled at a rate of one sample per two clock cycles, $ye(i)$ and $yo(i)$ each will be computed at a rate of one sample every four clock cycles. Hence, on average, the effective throughput rate is one sample of $y(i)$ every two clock cycles. A possible implementation is shown in Table 3.7.

Table 3.7 Implementation of the Loop-Unrolled ICDG of Algorithm 8

clock cycles	1	2	3	4	5	6	7	8	9	10	11
x(i) input	x(1)		x(2)		x(3)		x(4)		x(5)		x(6)
Multiplier #1	a*x(1)		A*x(2)		a*x(3)		a*x(4)		a*x(5)		
adder	u(1)		u(2)	ye(1)	u(3)	yo(1)	u(4)	ye(2)	u(5)	yo(2)	u(6)
Multiplier #2	a^2*ye(0)		a^2*yo(0)		a^2*ye(1)		a^2*yo(1)		a^2*ye(2)		

In this implementation, $u(i)$ is computed using the adder and multiplier #1. For example, $u(3)$ is computed after $a * x(2)$ is computed and $x(3)$ is available during the fifth clock cycle. The two subloops share a common multiplier #2 and the same adder that is used to compute $u(i)$. Note that $a^2 * ye(i)$ or $a^2 * yo(i)$ is computed right after $ye(i)$ or $yo(i)$ is computed in the adder. Also note that there are four clock cycles between when $ye(1)$ and $ye(2)$ are computed. This is also the case between $yo(1)$ and $yo(2)$.

In the rest of this section, we survey a few multimedia algorithms and the corresponding implementations.

3.3.2 Feed-Forward Direct Synthesis: Fast Discrete Cosine Transform (DCT)

Dedicated Micro-Architecture for 1D Eight-Point DCT

An N-point DCT is defined as:

$$y(k) = c(k) \sum_{n=0}^{N-1} \cos \frac{2\pi k(2n+1)}{4N} x(n) \tag{3.6}$$

where $c(0) = 1/\sqrt{N}$ and $c(k) = \sqrt{(2/N)}$, $1 \leq k \leq N - 1$. The inverse DCT can be rewritten as:

$$x(n) = \sum_{k=0}^{N-1} \cos \frac{2\pi k(2n+1)}{4N} c(k) y(k) \tag{3.7}$$

For the case of $N = 8$, the DCT can be written as a matrix vector product [26]

$$\mathbf{y} = \mathbf{C}_8\mathbf{x} \tag{3.8}$$

The 8×8 matrix $\mathbf{C}_8$ can be factored into the product of three matrices:

$$\mathbf{C}_8 = P_8 K_8 B \tag{3.9}$$

where P_8 is a permutation matrix, and K_8 is a block diagonal matrix

$$K_8 = \frac{1}{2}\begin{bmatrix} G_1 & & & \\ & G_1 & & \\ & & G_2 & \\ & & & G_4 \end{bmatrix} \tag{3.10}$$

with $G_1 = \cos(\pi/4)$, $G_2 = \begin{bmatrix} \cos(3\pi/8) & \cos(\pi/8) \\ -\cos(\pi/8) & \cos(3\pi/8) \end{bmatrix}$, and

$$G_4 = \begin{bmatrix} \cos(5\pi/16) & \cos(9\pi/16) & \cos(3\pi/16) & \cos(\pi/16) \\ -\cos(\pi/16) & \cos(5\pi/16) & \cos(9\pi/16) & \cos(3\pi/16) \\ -\cos(3\pi/16) & -\cos(\pi/16) & \cos(5\pi/16) & \cos(9\pi/16) \\ -\cos(9\pi/16) & -\cos(3\pi/16) & -\cos(\pi/16) & \cos(5\pi/16) \end{bmatrix} \tag{3.11}$$

is an anticirculant matrix. Finally, B can be further factored into the product of three matrices consisting of 0, 1, and -1 as its entries: $B = B_1 B_2 B_3$. Based on this factorization, Feig and Winograd [26] proposed an efficient eight-point DCT algorithm that requires 13 multiplication operations and 29 additions. An implementation of this algorithm in Matlab™ m-file format is listed below.

```
    Algorithm 9: Fast DCT Algorithm

function y=fdct(x0);
% implementation of fast DCT algorithm by Feig and Winograd
% IEEE Trans. SP, vol. 40, No. 9, pp. 2174-93, 1992.
% (c) copyright 1998, 1999 by Yu Hen Hu
%
% Note that the array index is changed from 0:7 to 1:8

% These are constants which can be stored as parameters.
C1 = 1/cos(pi/16);    C2=1/cos(pi/8);      C3 =1/cos(3*pi/16);
C4 = cos(pi/4);
C5 = 1/cos(5*pi/16); C6 = 1/cos(3*pi/8); C7 =1/cos(7*pi/16);

% Multiply by B3
A1 = x0(1) + x0(8);       A5 = x0(1) - x0(8);
A2 = x0(2) + x0(7);       A6 = x0(2) - x0(7);
A3 = x0(3) + x0(6);       A7 = x0(3) - x0(6);
A4 = x0(4) + x0(5);       A8 = x0(4) - x0(5);

% Multiply by B2
A9  = A1 + A4;    A10 = A2 + A3;
A11 = A1 - A4;    A12 = A2 - A3;
```

```
% Multiply by B1
A13 = A9 + A10;    A14 = A9 - A10;
% multiply by (1/2) G1
M1 = (1/2)*C4*A13;      % y(1)
M2 = (1/2)*C4*A14;      % y(5)
% multiply by (1/2) G2
A15 = -A12 + A11;  M3 = cos(pi/4)*A15;
A20 = A12 + M3;    A21 = -A12 + M3;
M6 = (1/4)*C6*A20;     % y(3)
M7 = (1/4)*C2*A21;     % y(7)
% Now multiply by (1/2)G4
% multiply by H_42
A16 = A8 - A5;    A17 = -A7 + A5;
A18 = A8 + A6;    A19 = -A17 + A18;
% Multiply by 1, G1, G2
M4 = C4*A16;           M5 = C4*A19;
A22 = A17 + M5;        A23 = -A17 + M5;
M8 = (1/2)*C6*A22;     M9 = (1/2)*C2*A23;
% Multiply by H_41, then by D^-1, and then 1/2 this is G4
% then multiply by (1/2) to make it (1/2) G4
A24 = - A7 + M4;       A25 = A7 + M4;
A26 = A24 - M8;        A27 = A25 + M9;
A28 = -A24 - M8;       A29 = -A25 + M9;
M10 = -(1/4)*C5*A26;   % y(2)
M11 = -(1/4)*C1*A27;   % y(4)
M12 = (1/4)*C3*A28;    % y(8)
M13 = -(1/4)*C7*A29;   % y(6)

y(1) = M1; y(2) = M10; y(3) = M6; y(4) = M11;
y(5) = M2; y(6) = M13; y(7) = M7; y(8) = M12;
```

To support high-throughput real-time image and video coding, a DCT algorithm must be executed at a speed that matches the I/O data rate. For example, in HDTV applications, videos are processed at a rate of 30 frames per second, with each frame 2048×4096 pixels. At a 4:1:1 ratio, there can be as many as

$$30 \times (6/4) \times 2048 \times 4096 \times (2 \times 8)/64$$
$$= 45 \times 2^{11+12+1+3-6} = 94,371,840 \approx 94.5 \text{ million 8-point DCT operations}$$

to be performed within one second. Hence, dedicated micro-architecture will be needed in such an application.

The DG shown in Figure 3.16 completely describes the algorithm and dictates the ordering of each operation that needs to be performed. In this figure, the inputs $x(0)$ to $x(7)$ are made available at the left end and the results $y(0)$ to $y(7)$ are computed and made available at the right end. Each shaded square box represents a multiplication operation, and each shaded circle represents an addition. The open circles do not correspond to any arithmetic operations, but are used to depict the routing of data during computation. Since the direction of dependence is always from left to right, it is omitted in Figure 3.16 in the interests of clarity. From Figure 3.16, it can be identified that the critical path is from any of the input nodes to M5, and from there to any of the four output nodes $y(1)$, $y(3)$, $y(5)$, and $y(7)$. The total delay is five additions and three multiplications.

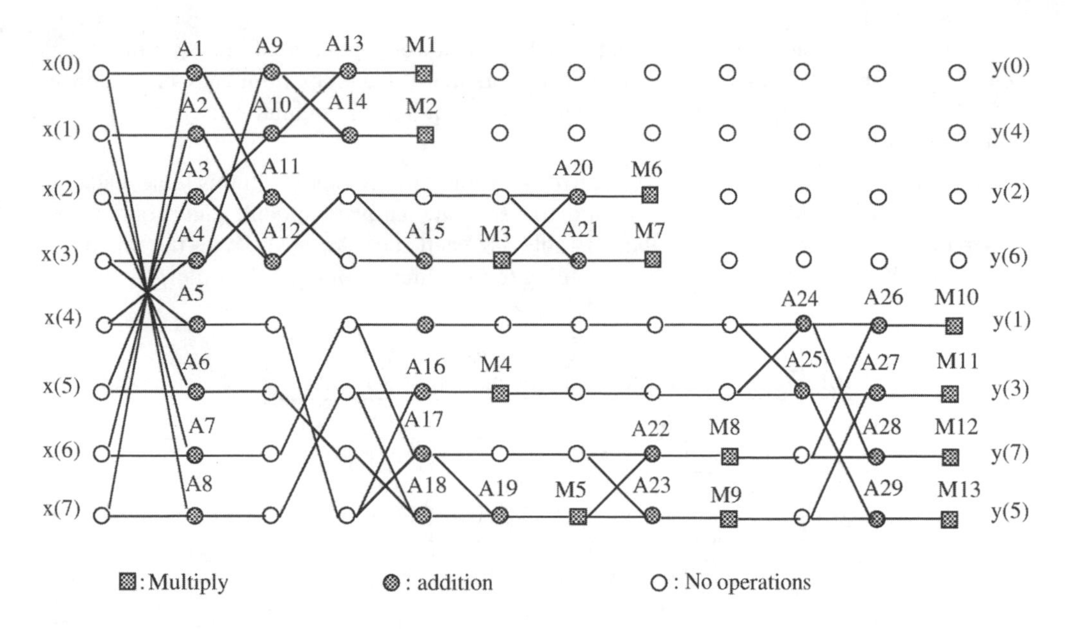

FIGURE 3.16
Dependence graph of the fast DCT algorithm.

Once a dependence graph is derived, one may directly *map* the DG into a dedicated hardware implementation by (1) designating a hardware module to realize each computation node in the DG, and (2) interconnecting these hardware modules according to the directed arcs in the DG.

Two types of hardware modules will be used here: an adder module, which takes one clock cycle to perform an addition, and a multiplier module, which takes two clock cycles to compute a multiplication. The mapping of the DG into a hardware module is a *binding process* where each node of the DG is mapped onto one hardware module which can implement the function to be performed on that node. A single hardware module may be used to implement one or more nodes on the DG. As in the previous section, we assume the output of each hardware module will be held in a register.

a. Performance-Constrained Micro-Architecture Synthesis

Suppose that one may use as many hardware modules as needed. Then, from a theoretical point of view, one may always derive an implementation to achieve the desired throughput rate. This is because successive eight-point DCT operations are independent of each other. For each new arriving eight-point data sample, one can always assign a new set of hardware modules and initiate the computation immediately. Hence the minimum initiation interval can be made as small as possible. The only limiting factor would be the speed to redirect data samples into appropriate hardware modules.

Next, suppose that in addition to the throughput rate, the latency (time between arrival of data samples and when they are computed) is also bounded. The minimum latency, given that a sufficient number of hardware modules are available, is equal to the time delay along the critical path, which includes five addition operations and three multiplication operations. Thus, the minimum latency is $5 \times 1 + 3 \times 2 = 11$ clock cycles. The maximum latency is equal to the total computing time, with every operation executed sequentially. Thus, the upper bound of latency is $29 \times 1 + 13 \times 2 = 55$ clock cycles.

Table 3.8 shows an implementation that achieves a throughput rate of one 8-point DCT per clock cycle and a latency of 11 clock cycles. Note that if the clock frequency is greater than 95 MHz, then this implementation can deliver the required throughput rate for HDTV main profile performance.

The implementation is expressed in a *warped* format to save space. In this table, each item Ai $(1 \leq i \leq 29)$ or Mj $(1 \leq j \leq 13)$ refers to a separate hardware module and should take up a separate raw in the implementation. In Table 3.8, each entry Ai or Mj gives the schedule of the particular hardware module corresponding to the same set of eight data samples.

Table 3.8 A Dedicated Implementation of 8-Point DCT

A1	A9	A13	M1	M1						
A2	A10	A14	M2	M2						
A3	A11	A15	M3	M3	A20	M6	M6			
A4	A12				A21	M7	M7			
A5	A16	M4	M4	A24				A26	M10	M10
A6	A17			A25				A27	M11	M11
A7	A18	A19	M5	M5	A22	M8	M8	A28	M12	M12
A8					A23	M9	M9	A29	M13	M13

Note: Throughput = 1 DCT/clock cycle, latency = 11 clock cycles.
The implementation is shown in a compact format.

In this implementation, 29 full adders and 13 pipelined multipliers are used. By *pipelined multiplier,* we require each multiplication to be accomplished in two successive stages, with each stage taking one clock cycle. A buffer between these two stages will store the intermediate result. This way, while stage 2 is completing the second half of the multiplication of the present iteration, stage 1 can start computing the first half of the multiplication of data from the next iteration. Thus, with two-stage pipelined operation, such a multiplier can achieve a throughput rate of one multiplication per clock cycle.

On the other hand, if one type of multiplier module which cannot be broken into two pipelined stages is used, then two multipliers must be used to realize each multiplication operation in Table 3.6 in an interleaved fashion. This is illustrated in Figure 3.17. The odd number of the data set will use multiplier #1 while the even number of the data set will use multiplier #2. As such, on average, two multiplication operations can be performed in two clock cycles. This translates into an effective throughput rate of one multiplication per clock cycle. However, the total number of multiplier modules needed will increase to $2 \times 13 = 26$.

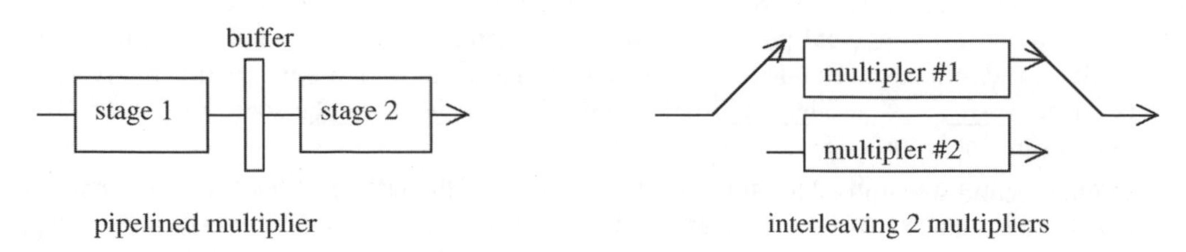

FIGURE 3.17
Illustration of the difference between pipelined and interleaved multiplier implementation.

Let us consider relaxing the performance constraints by lowering the throughput rate to one 8-point DCT per two clock cycles and allowing longer latency. One possible implementation, in a compact format, is shown in Table 3.9.

Table 3.9 Eight-Point DCT Implementation

<table>
<tr><td>A1</td><td>A9</td><td></td><td colspan="2">M1</td><td></td><td></td><td></td><td></td><td></td><td></td><td></td></tr>
<tr><td>A2</td><td>A10</td><td>A13</td><td>A14</td><td colspan="2">M2</td><td></td><td></td><td></td><td></td><td></td><td></td></tr>
<tr><td>A3</td><td>A11</td><td></td><td colspan="2">M3</td><td></td><td colspan="2">M6</td><td></td><td></td><td></td><td></td></tr>
<tr><td>A4</td><td>A12</td><td>A15</td><td></td><td></td><td>A20</td><td>A21</td><td colspan="2">M7</td><td></td><td></td><td></td></tr>
<tr><td>A5</td><td>A16</td><td colspan="2">M4</td><td>A24</td><td>A22</td><td></td><td></td><td></td><td colspan="2">M10</td><td></td></tr>
<tr><td>A6</td><td>A17</td><td></td><td></td><td>A25</td><td>A23</td><td></td><td></td><td>A26</td><td>A27</td><td colspan="2">M11</td></tr>
<tr><td>A7</td><td>A18</td><td></td><td colspan="2">M5</td><td></td><td colspan="2">M8</td><td></td><td colspan="2">M12</td><td></td></tr>
<tr><td></td><td>A8</td><td>A19</td><td></td><td></td><td></td><td colspan="2">M9</td><td>A28</td><td>A29</td><td colspan="2">M13</td></tr>
</table>

Note: Throughput rate: 1 DCT per 2 clock cycles; latency: 12 clock cycles; 15 adder modules and 13 multipliers are used.

In this implementation, we use only regular multiplier modules. If we use two-stage pipelined multiplier modules, the number of multipliers can further be reduced to seven. In order to minimize the number of adder modules, we choose to execute A26 and A27 (as well as A28 and A29) sequentially. This change accounts for the additional clock cycle of latency.

b. Resource-Constrained Micro-Architecture Synthesis

In a resource-constrained synthesis problem, the number of hardware modules is given. The objective is to maximize the performance (throughput rate) under this resource constraint. To illustrate, let us consider the situation where only one adder module and one multiplier module is available. In Table 3.10, the first row gives the clock-by-clock schedule for the adder module,

Table 3.10 Implementation of 8-Point DCT with 1 Adder and 1 Multiplier

<table>
<tr><td>5</td><td>6</td><td>7</td><td>8</td><td>16</td><td>17</td><td>18</td><td>19</td><td>24</td><td>25</td><td>22</td><td>23</td><td>1</td><td>2</td><td>3</td><td>4</td><td>26</td><td>27</td><td>28</td><td>29</td><td>9</td><td>10</td><td>11</td><td>12</td><td>15</td><td>13</td><td>14</td><td>20</td><td>21</td></tr>
<tr><td colspan="2">M2</td><td colspan="2">M6</td><td colspan="2">M7</td><td colspan="2">M4</td><td colspan="2">M5</td><td></td><td></td><td colspan="2">M8</td><td colspan="2">M9</td><td></td><td colspan="2">M10</td><td colspan="2">M11</td><td colspan="2">M12</td><td colspan="2">M13</td><td colspan="2">M3</td><td colspan="2">M1</td></tr>
</table>

and the second row gives the schedule for the multiplier module. The shaded area (M2, M6, M7) indicates that those multiplication operations belong to the previous data set. Thus, this is an overlapped schedule. The initiation interval is 29 clock cycles — the minimum that can be achieved with only one adder module. The execution of the adder and the multiplier are completely overlapped. Hence, we can conclude that this is one of the optimal solutions that maximize the throughput rate (1 DCT in 29 clock cycles), given the resource constraint (one adder and one multiplier module).

c. Practical Implementation Considerations

In the above synthesis examples, the complexity of inter-module communication paths (buses) is not taken into account, nor do we factor in the amount of temporary storage elements (registers) needed to facilitate such realization.

Furthermore, in practical hardware synthesis, not all modules have the same word length. Due to the addition and multiplication operations, the dynamic range (number of significant digits) will increase. The adder at a later stage of computing will need more bits. Therefore, before commencing a hardware synthesis, it is crucial to study the numerical property of this fast DCT algorithm and determine its quantization noise level to ensure that it meets the requirements of the standard.

Generalization to 2D Scaled DCT

In image and video coding standards such as JPEG and MPEG, a 2D DCT is to be performed on an 8×8 image pixel block $\mathbf{X}$:

$$\mathbf{Y} = \mathbf{C}_8 \mathbf{X} \mathbf{C}_8^T \tag{3.12}$$

This corresponds to a consecutive matrix–matrix product. An array structure can be developed to realize this operation using a systolic array. However, it would require many multipliers. In [26], a different approach is taken. First, we note that the above formulation can be converted into a matrix–vector product between a 64×64 matrix formed by the kroenecker product of the DCT matrix, $\mathbf{C}_8 \otimes \mathbf{C}_8$, and a 64×1 vector $\underline{\mathbf{X}}$ formed by concatenating columns of the $\mathbf{X}$ matrix. The result is a 64×1 vector $\underline{\mathbf{Y}}$ that gives each column of the $\mathbf{Y}$ matrix:

$$\underline{\mathbf{Y}} = (\mathbf{C}_8 \otimes \mathbf{C}_8)\, \underline{\mathbf{X}} \tag{3.13}$$

The $\mathbf{C}_8$ matrix can be factorized, in this case, into the product as follows:

$$\mathbf{C}_8 = P_8 D_8 R_{8,1} M_8 R_{8,2} \tag{3.14}$$

where P_8 is the same permutation matrix as in the 1D eight-point DCT algorithm. D_8 is an 8×8 diagonal matrix; $R_{8,1}$ is a matrix containing elements of 0, 1, and -1; and $R_{8,2}$ is the product of three matrices, each of which contains 0, 1, and -1 elements only.

$$M_8 = \begin{bmatrix} 1 &&&&&&& \\ & 1 &&&&&& \\ && 1 &&&&& \\ &&& \cos(\pi/8) &&&& \\ &&&& 1 &&& \\ &&&&& \cos(\pi/8) && \\ &&&&&& \cos(3\pi/16) & \cos(\pi/16) \\ &&&&&& -\cos(\pi/16) & \cos(3\pi/16) \end{bmatrix} \tag{3.15}$$

For the kroenecker product $\mathbf{C}_8 \otimes \mathbf{C}_8$, the factorization becomes

$$\begin{aligned} \mathbf{C}_8 \otimes \mathbf{C}_8 &= \left(P_8 D_8 R_{8,1} M_8 R_{8,2}\right) \otimes \left(P_8 D_8 R_{8,1} M_8 R_{8,2}\right) \\ &= \left[\left(P_8 D_8\right) \otimes \left(P_8 D_8\right)\right] \bullet \left[\left(R_{8,1} M_8 R_{8,2}\right) \otimes \left(R_{8,1} M_8 R_{8,2}\right)\right] \\ &= \left(P_8 \otimes P_8\right) \bullet \left(D_8 \otimes D_8\right) \bullet \left(R_{8,1} \otimes R_{8,1}\right) \bullet \left(M_8 \otimes M_8\right) \bullet \left(R_{8,2} \otimes R_{8,2}\right) \end{aligned} \tag{3.16}$$

Hence a fast 2D DCT algorithm can be developed accordingly. The hardware implementation approach will be similar to that of 1D DCT. However, the complexity will be significantly greater.

One advantage of the factorization expression in (3.13) is that a scaled DCT can be performed. Scaled DCT is very useful for JPEG image coding and MPEG intra-frame coding standards. In these standards, the DCT coefficients will be multiplied element by element to a *quantization matrix* to deemphasize visually unimportant frequency components before applying scalar quantization. Thus, for each block, there will be 64 additional multiplication operations performed before quantization can be applied. In effect, this quantization matrix can be formulated as a 64×64 diagonal matrix $\mathbf{W}$ such that the scaled DCT coefficient vector

$$\underline{\Psi} = \mathbf{W}\underline{\mathbf{Y}} = \mathbf{W} \bullet \left(P_8 \otimes P_8\right) \bullet \left(D_8 \otimes D_8\right) \bullet \left(R_{8,1} \otimes R_{8,1}\right) \bullet \left(M_8 \otimes M_8\right) \bullet \left(R_{8,2} \otimes R_{8,2}\right)\underline{\mathbf{X}} \tag{3.17}$$

A complicated flow chart of the above algorithm is given in the appendix of [26]. Due to space limitations, it is not included here. The basic ideas of designing a dedicated micro-architecture for 2D scaled DCT will be similar to 1D DCT.

3.3.3 Feedback Direct Synthesis: Huffman Coding

In this section, we turn our attention to the dedicated micro-architecture implementation of a different class of recursive multimedia algorithms, known as the Huffman entropy coding algorithm.

Huffman coding encodes symbols with variable-length binary streams *without* a separator symbol. It is based on the probability of symbol appearances in the vocabulary. Often the encoding table is designed off line. During encoding, each symbol is presented to the encoder and a variable-length bitstream is generated accordingly. This is essentially a table-lookup procedure. The decoding procedure is more complicated: For each bit received, the decoder must decide whether it is the end of a specific code or it is in the middle of a code. In other words, the decoder must be realized as a sequential machine. Due to the variable-length feature, the number of cycles to decode a codeword varies. The throughput in this case is 1 bit per clock cycle. Let us consider a Huffman decoding example. Assume the coding table is as in Table 3.11. Then we may derive the Mealy model state diagram, as shown in Figure 3.18.

Table 3.11 A Huffman Coding Table

Symbol	Codeword
A	0
B	10
C	1100
D	1101
E	1110
F	1111

Usually the total number of states is the total number of symbols minus 1, and the longest cycle in the state diagram equals the longest codewords. In practical applications, such as in JPEG or MPEG, there are a large number of symbols and long codewords. For example, in the JPEG AC Huffman table, there are 162 symbols, and many codewords are as long as 16 bits.

Implementation of Finite State Machine

A general structure of implementing finite state machine is shown in Figure 3.19. The state variables are implemented with flip-flops. The combinational circuits can be realized

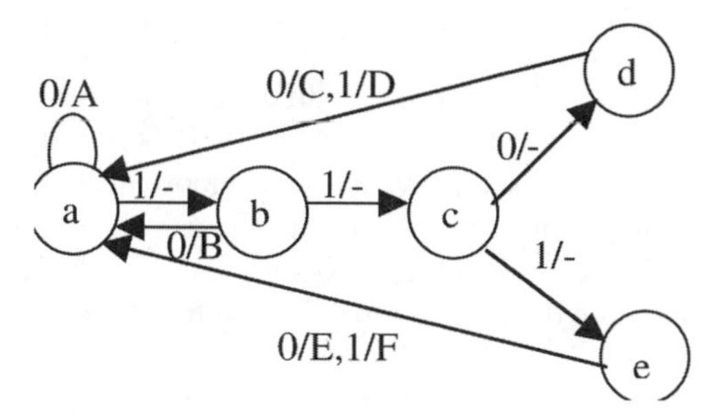

FIGURE 3.18
State diagram of the Huffman coding algorithm.

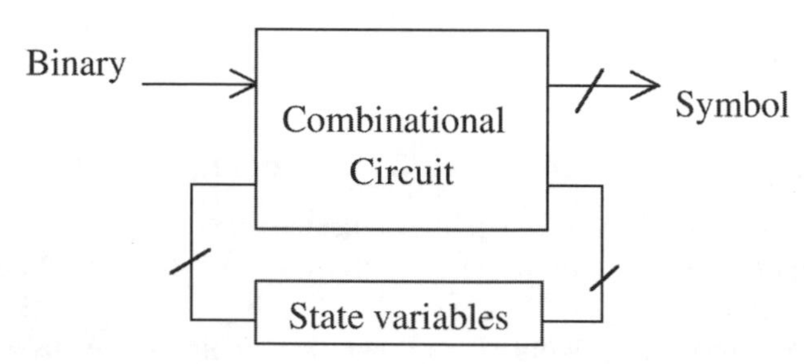

FIGURE 3.19
Finite state machine implementation of Huffman decoding algorithm.

with read-only memory (ROM), programmable logic array (PLA), or dedicated logic gates. The design issues include: (1) how high the clock rate can go, and (2) how complicated the combinational circuit design will be.

In the above example, there are five states (a, b, c, d, and e), which require at least three state variables to represent. There are seven output symbols (A, B, C, D, E, F, and ——) to be encoded in an additional 3 bits. Thus, there are at least six outputs of the combinational circuit. In other words, the combinational circuit consists of six Boolean functions sharing the same set of four Boolean variables (3 state variables + 1 bit input). If a ROM is used, it will have a size of 16 words with each word containing 6 bits. Let us consider yet another example of the JPEG AC Huffman table. The JPEG AC Huffman code contains 161 symbols and has a codeword length smaller than or equal to 16 bits. Since the Huffman tree has 161 nodes, it requires at least eight state variables ($2^8 = 256 > 161$). Output symbol encoding will also require 8 bits. If a ROM is used to realize the combinational circuit, then it will have a size of $2^9 \times (8 + 8) = 512 \times 16 = 8\text{K}$ bits.

The above implementation using a finite state machine ensures a *constant input rate* in that it consumes 1 bit each clock cycle. The number of symbols produced at the output varies. However, on average, the number of clock cycles needed to produce a symbol is roughly equal to the average codeword length L_{avg}. Asymptotically, L_{avg} is a good approximation of the *entropy* of the underlying symbol probability distribution. If the input throughput rate is to be increased, we may scan more than 1 bit at each clock cycle provided the input data rate is at least twice the decoder's internal clock rate. This will not increase the number of states, but it will make the state transition more complicated. For example, if each time 2 bits of input data are scanned, the corresponding state diagram will be as in Figure 3.20.

The size of the state table doubles for each additional bit being scanned in a clock cycle. If a ROM is used to realize the state table, the number of addresses will double accordingly. Moreover, since there can be more than one symbol in the output during each clock cycle, the

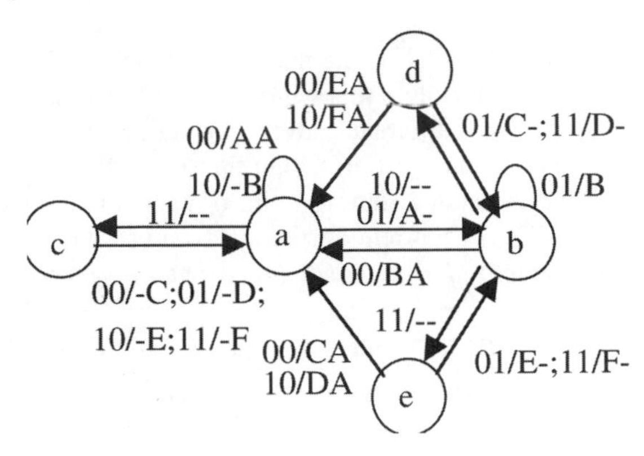

FIGURE 3.20
State diagram decoding 2 bits at a time.

word length of the ROM will also be increased accordingly. Hence it is a trade-off between hardware complexity and throughput rate.

Lei et al. [27] have proposed a *constant output rate* Huffman decoding method using FSM realization. This is accomplished by scanning L bits of input at a time, with M being the maximum codeword length. Each time, exactly one codeword is decoded. The remaining bits, which are not part of the decoded symbols, then will be realigned and decoded again. Let us consider the following bitstream 0011001001110010 0. During decoding, the decoder scans the first 4 bits (0011) and determines that the first symbol is $A(0)$. Next, it shifts by 1 (since A is encoded by 1 bit) and decodes the second bit as A again. Next, after shifting another bit, its window contains 1100, which is decoded as C. The next iteration, it will shift 4 bits instead of 1 bit because the entire 1100 is used. Therefore, during each clock cycle, one symbol is decoded. However, the rate at which the input data stream is consumed depends on the composition of the given sequence. This process is depicted in Figure 3.21. Each double arrow line segment indicates the 4 bits being scanned in a clock cycle. $0 : A$ indicates that the left-most bit 0 is being decoded to yield the symbol A. Of course, one can be more opportunistic by allowing more than one symbol to be decoded in each L-bit window and thereby increase the decoding rate, at the expense of additional hardware complexity.

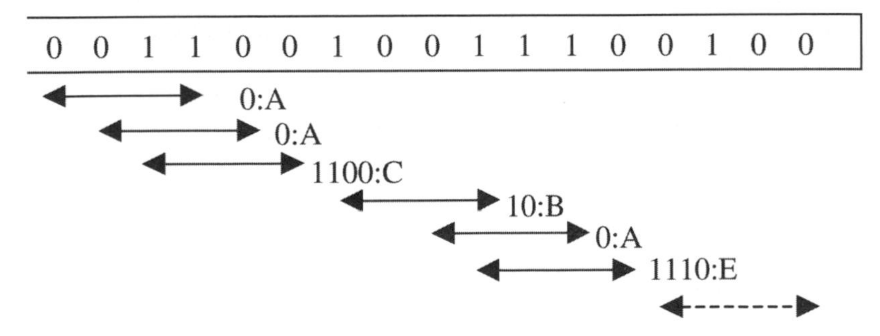

FIGURE 3.21
Illustration of constant symbol decoding rate Huffman decoder.

Concurrent VLC Decoding [28]

One way to further increase the coding speed is to exploit parallelism by decoding different segments of a bitstream concurrently. Successive M-bit segments will be overlapped by an

L-bit window, where $M >> L$ and L is the maximum codeword length. Therefore, there must be a split of two codewords within this window. In other words, in the successive M-bitstreams, each can have at most L different starting bit positions within that L-bit window. By comparing the potential starting bit position within this L-bit window of two M-bitstreams, we can uniquely determine the actual starting point of each stream and therefore *decouple* the successive streams to allow concurrent decoding. To illustrate, consider the bitstream in Figure 3.22 and the partition into $M = 10$ bitstreams with an $L = 4$ bits overlapping window:

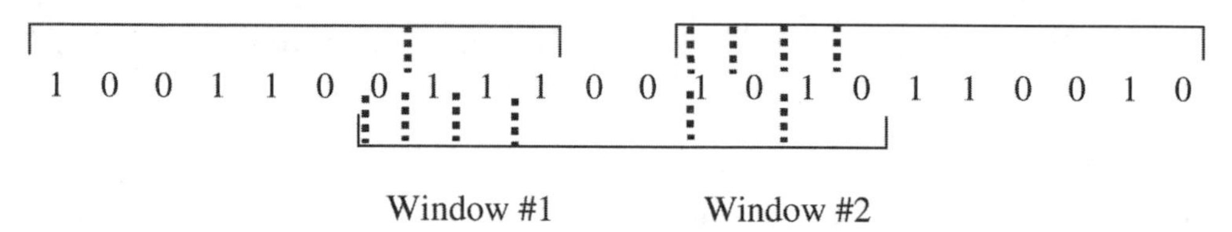

FIGURE 3.22
Concurrent VLC decoding.

In this figure, the dashed lines within each window indicate the legitimate codeword splitting positions. The upper dashed lines are identified from the upper stream segments and the lower dashed lines are from the lower stream segments. If the upper and lower splitting points overlap, it will be accepted as a likely codeword splitting point. To elaborate, let us consider window #1, which is the trailing window of the first upper stream segment. We note that if the splitting point is at the position to the left of the window, then the previous 4 bits (0110) do not correspond to any 4-bit symbols. They do contain the codeword B (10) as the last 2 bits. But then the first 2 bits (01) must be part of a 4-bit codeword. In fact, from the Huffman table, they must be part of the codeword 1101. Unfortunately, the 2 bits to the left of the stream (0110) are 10 (the first 2 bits from the left). Hence, we conclude that such a split is not valid. In other words, for each potential split position, we must trace back to the remainder of the bitstream segment to validate if there is a legitimate split. In practical implementation, for each stream segment, and each potential codeword splitting position in the leading window, a Huffman decoding will be performed. If the decoder encounters an illegitimate Huffman code along the way, the splitting point is deemed infeasible and the next potential splitting point will be tested. If a splitting point in the leading window is consistent up to a codeword that partially falls within the trailing window, the corresponding split position at the trailing window will be recorded together with the splitting point in the leading window of the same segment. The legitimate splitting points in the same window of the successive stream segments then will be regarded as true codeword splitting points. After these points are determined, concurrent decoding of each stream segment will commence.

3.4 Concluding Remarks

In this chapter, we surveyed implementation strategies for application-specific multimedia signal processors. Using the application of video coding as an example, we illustrated how each design style is applied to synthesize dedicated realization under different constraints. Current research efforts have been focused on low-power implementation and reconfigurable architecture. With these new research efforts, there will be more alternatives for designers to choose.

References

[1] Kung, S.Y., *VLSI Array Processors*. 1988, Englewood Cliffs, NJ, Prentice-Hall.

[2] Kung, S.Y., On supercomputing with systolic/wavefront array processors. *Proc. IEEE,* 1984. 72: p. 1054–1066.

[3] Kung, H.T., Why systolic array. *IEEE Computers,* 1982. 15: p. 37–46.

[4] Parhi, K.K., Algorithm transformation techniques for concurrent processors. *Proc. IEEE,* 1989. 77: p. 1879–1895.

[5] Komarek, T., and P. Pirsch, Array architectures for block matching algorithms. *IEEE Trans. on Circuits & Syst.,* 1989. 36: p. 1301–1308.

[6] Hsieh, C.H., and T. P. Lin, VLSI architecture for block motion estimation algorithm. *IEEE Trans. on Video Technol.,* 1992. 2(2): p. 169–175.

[7] Pan, S.B., S.S. Chae, and R.H. Park, VLSI architectures for block matching algorithm. *IEEE Trans. on Circuits Syst. Video Technol.,* 1996. 6(1): p. 67–73.

[8] Yeo, H., and Y.H. Hu, A novel modular systolic array architecture for full-search block matching motion estimation. *IEEE Trans. on Circuits Syst. Video Technol.,* 1995. 5(5): p. 407–416.

[9] Kittitornkun, S., and Y.H. Hu, Systolic full-search block matching motion estimation array structure. *IEEE Trans. on Circuits Syst. Video Technol.* (submitted), 1998.

[10] Chen, Y.-K., and S.Y. Kung, A systolic methodology with applications to full-search block matching architectures. *J. of VLSI Signal Processing,* 1998. 19(1): p. 51–77.

[11] Lee, C.Y., and M.C. Lu, An efficient VLSI architecture for full-search block matching algorithms. *J. of VLSI Signal Processing,* 1997. 15: p. 275–282.

[12] You, J., and S.U. Lee, High throughput, scalable VLSI architecture for block matching motion estimation. *J. of VLSI Signal Processing,* 1998. 19(1): p. 39–50.

[13] Yang, K.M., M.T. Sun, and A.L. Wu, A family of VLSI designs for the motion compensation block-matching algorithm. *IEEE Trans. on Circuits Syst.,* 1989. 26(10): p. 1317–1325.

[14] Wu, C.-M., and D.-K. Yeh, A VLSI motion estimator video image compression. *IEEE Trans. on Consumer Elec.,* 1993. 39(4): p. 837–846.

[15] Chang, S., J.-H. Hwang, and C.-W. Jen, Scalable array architecture design for full search block matching. *IEEE Trans. on Circuits Syst. Video Technol.,* 1995. 5(4): p. 332–343.

[16] Jehng, Y.-S., L.-G. Chen, and T.-D. Chiueh, An efficient and simple VLSI tree architecture for motion estimation algorithms. *IEEE Trans. on Signal Processing,* 1993. 40(2): p. 889–900.

[17] Vos, L.D., and M. Stegherr, Parametrizable VLSI architecture for the full-search block matching algorithms. *IEEE Trans. on Circuits Syst.,* 1989. 26(10): p. 1309–1316.

[18] Nam, S.H., J.S. Baek, and M.K. Lee, Flexible VLSI architecture of full search motion estimation for video applications. *IEEE Trans. on Consumer Elec.,* 1994. 40(2): p. 177–184.

[19] Nam, S.H., and M.K. Lee, Flexibility of motion estimator for video image compression. *IEEE Trans. on Circuits Syst.,* 1996. 43(6): p. 467–470.

[20] Chen, M.-J., L.-G. Chen, K.-N. Cheng, and M.C. Chen, Efficient hybrid tree/linear array architectures for block-matching motion estimation algorithms. *IEEE Proc.-Vis. Image Signal Processing,* 1996. 143(4): p. 217–222.

[21] SGS Thomson Microelectronics, STi3220 data sheet, 1994: `http://www.st.com`.

[22] Vos, L.D., and M. Schobinger, VLSI architecture for a flexible block matching processor. *IEEE Trans. on Circuits Syst. Video Technol.,* 1995. 5(5): p. 417–428.

[23] Wang, D.J., and Y.H. Hu, Fully static multiprocessor array realization of real time recurrence DSP applications. *IEEE Trans. on Signal Processing,* 1994. 42(5): p. 1288–1292.

[24] Wang, D.J., and Y.H. Hu, Rate optimal scheduling of recursive DSP algorithms by unfolding. *IEEE Trans. on Circuits Syst.,* 1994. 41(10): p. 672–675.

[25] Wang, D.J., and Y.H. Hu, Multiprocessor implementation of real time DSP algorithms. *IEEE Trans. on VLSI Syst.,* 1995. 3(3): p. 393–403.

[26] Feig, E., and S. Winograd, Fast algorithms for the discrete cosine transform. *Trans. on Signal Processing,* 1992. 40(9): p. 2174–2191.

[27] Lei, S.-M., M.-T. Sun, and K.-H. Tzou, Design and hardware architecture of high-order conditional entropy coding for images. *IEEE Trans. on Video Technol.,* 1992. 2(2): p. 176–186.

[28] Lin, H.D., and D.G. Messerschmitt, Designing high-throughput VLC decoder. Part II — parallel decoding method. *IEEE Trans. on Video Technol.,* 1992. 2: p. 197–206.

Chapter 4

Superresolution of Images with Learned Multiple Reconstruction Kernels

Frank M. Candocia and Jose C. Principe

4.1 Introduction

Superresolution is the term given to the signal processing operation that achieves a resolution higher than the one afforded by the physical sensor. This term is prevalent within the radar community and involves the ability to distinguish objects separated in space by less than the resolution afforded by radar. In the domain of optical images the problem is akin to that of perfect reconstruction[1, 2]. As such, this chapter will address the issue of image magnification (also referred to as interpolation, zooming, enlargement, etc.) from a finite set of samples. An example where magnification can aid multimedia applications is video teleconferencing, where the bit rate constraints limit video throughput. Such restrictions typically result in the transmission of a highly compressed and size-limited video sequence of low visual quality. In this context, the task of superresolution thus becomes one of restoring lost information to the compressed sequence of images so as to result in their magnification as well as providing a sharper and/or less degraded image. Among the areas in which multimedia can benefit from superresolution, the focus herein is on the image processing resulting in the superresolution of still images. The benefits afforded by the proposed architecture will be examined and several issues related to the methodology will be discussed.

Commonly, image magnification is accomplished through convolution of the image samples with a single kernel — such as the bilinear, bicubic [3], or cubic B-spline kernels [4] — and any postprocessing or subsequent image enhancement would typically be performed in an ad hoc fashion. The mitigation of artifacts, due to either aliasing or other phenomena, by this type of linear filtering is very limited. More recently, magnification techniques based on image domain knowledge have been the subject of research. For example, directional methods [5, 6] examine an image's local edge content and interpolate in the low-frequency direction (along the edge) rather than in the high-frequency direction (across the edge). Multiple kernel methods typically select between a few ad hoc interpolation kernels [7]. Orthogonal transform methods focus on the use of the discrete cosine transform (DCT) [8, 9] and the wavelet transform [10]. Variational methods formulate the interpolation problem as the constrained minimization of a functional [11, 12]. An extended literature survey discussing these methods at great length has been provided by Candocia [1].

The approach presented herein is novel and addresses the ill-posed nature of superresolution by assuming that similar (correlated) neighborhoods remain similar across scales, and that this a priori structure can be *learned locally* from available image samples across scales. Such local information extraction has been prominent in image compression schemes for quite some time, as evidenced by JPEG- [13] and PCA- [14] based approaches, which typically compress the set of nonoverlapping subblocks of an image. Recent compression approaches also exploit the interblock correlation between subblocks [15, 16]. The goal is to divide the set of subblocks into a finite number of disjoint sets that can individually be represented more efficiently than the original set. Our approach is similar in spirit in that we exploit interblock correlation for mapping similar overlapping neighborhoods to their high-resolution counterparts. However, no one before us has proposed using this information to create constraints that can superresolve images. We further show that a very simple local architecture can learn this structure effectively. Moreover, our approach is shown to be equivalent to a convolution with a family of kernels established from available images and "tuned" to their local characteristics, which represents an extension to conventional sampling theory concepts.

The chapter is divided into sections as follows. Section 4.2 conceptually introduces the superresolution that is discussed herein. Comments and observations are made and the methodology from which the local architecture arises is also described. Section 4.3 presents the image acquisition model used for synthesizing our low-resolution images. Section 4.4 describes single and multikernel-based approaches to magnification. Section 4.5 details the local architecture implementing the superresolution methodology. In Section 4.6, several results are presented illustrating the architecture's capability. Section 4.7 discusses several issues regarding the methodology and Section 4.8 provides our conclusions.

4.2 An Approach to Superresolution

The superresolution approach presented here addresses the reconstruction of an image (from a finite set of samples) beyond the limit imposed by the Shannon sampling theory [17, 18]. For the sake of simplicity, our development uses one-dimensional (1D) signals, but the extensions to two dimensions (2D) should be clear.

Let $x(t)$, where $-\infty < t < \infty$, be a continuous signal with maximum frequency content Ω_c rad/s. Thus, our analysis is based on band-limited signals. We can represent $x(t)$ as a linear combination of a set of basis functions as

$$x(t) = \sum_{n} x[n]k(t, n) \tag{4.1}$$

where the linear weighting is given by the samples in $x[n]$ and $k(t, n)$ represents our set of basis functions. Here $x[n] \equiv x(nT_s)$, for integers n satisfying $-\infty < n < \infty$, and sampling period T_s. The equation describing the perfect reconstruction of a signal in sampling theory is

$$x(t) = \sum_{n=-\infty}^{\infty} x(nT_s) \sin c\left(\frac{t}{T_s} - n\right) \tag{4.2}$$

where, by definition, $\sin c(t) \equiv \frac{\sin(\pi t)}{\pi t}$. We see that our basis functions are given by $k(t, n) = \sin c(\frac{t}{T_s} - n)$ and the basis functions in this infinite set are orthogonal [19]; that is,

$$\int_{-\infty}^{\infty} \sin c\left(\frac{t}{T_s} - n\right) \sin c\left(\frac{t}{T_s} - m\right) dt = T_s \delta[n - m] .$$

The perfect reconstruction can be obtained if the sampling period satisfies $T_s < \frac{T_c}{2}$ where $T_c = \frac{2\pi}{\Omega_c}$. For time signals, $\frac{T_c}{2}$ is the critical sampling rate, also called the Nyquist rate.

Therefore, every instance of $x(t)$ can be exactly resolved with an infinite set of samples provided the density of samples is high enough. The sampling period T_s provides the limit to which our signals can be perfectly reconstructed (resolved) from an orthogonal set of linear projections and an infinite number of samples. Notice that the $\sin c$ bases are the universal set of linear projections capable of perfectly reconstructing band-limited signals in time (space). This set of bases is universal in that all appropriately sampled infinite extent band-limited signals can be reconstructed with them, irrespective of their content.

The case of finite extent data is more realistic, in particular for images. For finite extent data, equation (4.2) can be expressed as

$$\hat{x}(t) = \sum_{n=-\infty}^{\infty} [x(nT_s)w[n]] \sin c\left(\frac{t}{T_s} - n\right) \tag{4.3}$$

where $w[n]$ describes samples of our window function

$$w[n] = \begin{cases} 1; & 0 < n < N-1 \\ 0; & \text{otherwise} \end{cases}$$

and N is the extent of our window. Notice that the hat superscript in $\hat{x}$ has been used purposely to denote the approximation afforded by a finite number of data samples. The finite data set reduces the resolvability of the signal. We can see this by examining equation (4.3) in the frequency domain. The continuous Fourier transform of equation (4.3) yields

$$\hat{X}(\Omega) = \frac{T_s}{\Omega_s}[X(\omega) \circledast W(\omega)] \prod\left(\frac{\Omega}{\Omega_s}\right)$$

where $\circledast$ is the periodic convolution operator, $\omega = \Omega T_s$ is the angular frequency in radians, $X(\omega)$ is the DTFT of $x(nT_s)$ and is periodic of period 2π in ω and $\frac{2\pi}{T_s}$ in Ω (similarly for $W(\Omega)$), and $\prod(\Omega) = \{1 \ |\Omega| < \frac{1}{2}; 0 \quad \text{otherwise}\}$. The effect of the windowing function $w[n]$ in equation (4.3) is to smear (distort) the true frequency spectrum $X(\Omega)$, which results in a decreased ability to properly resolve the signal $x(t)$. To illustrate this, consider a down-sampled image of Lena in Figure 4.1a. This image is 128×128 samples in size. It is interpolated to 256×256 using equation (4.3) and is illustrated in Figure 4.1b. A visual comparison of the interpolated image with the original 256×256 image in Figure 4.1c demonstrates that the $\sin c$ basis set limits the reconstruction performance. This is standard digital signal processing knowledge: interpolating a digital representation does not improve frequency resolution. In order to improve the frequency resolution, more samples (either longer windows or a higher sampling frequency) of the original signal are necessary [18]. In image processing only a higher sampling frequency will do the job since the window (the scene) is prespecified.

The objective of superresolution is to reconstruct $x(t)$ more faithfully than the resolution afforded by equation (4.3), that is, the resolution afforded from a finite set of observed data obtained at a sampling rate T_s.

4.2.1 Comments and Observations

The conditions necessary for perfect signal reconstruction are: (1) there must be no noise associated with the collected samples (e.g., no quantization error), (2) the sampling rate must be higher than the Nyquist sampling rate of the signal, and (3) the signal must be of infinite extent. We can immediately say that in image processing, perfect signal reconstruction is impossible because an image has finite extent. In the practical sampling of optical images, the issue of quantization error is usually not critical. The standard use of an 8-bit dynamic range usually

(a)

(b) (c)

FIGURE 4.1
Illustrating the resolution limit imposed by interpolating with the $\sin c$ **bases of the Shannon sampling theory. Artifacts are clearly visible in the** $\sin c$ **interpolated image of (b). (a) Lena 128 × 128 image [obtained from (c)]; (b) image interpolated to 256 × 256; (c) original (desired) Lena 256 × 256 image.**

yields highly acceptable and pleasing images. The issue of sampling frequency is much more critical. The information of a natural scene has typically very high spatial frequency content. The sharp contrast we perceive in order to delineate objects (object boundaries) as well as the textural character of those objects are just two of the attributes inherent to the high-frequency content of optical images. As such, the sampling frequency used in collecting optical images is generally not large enough to fully describe a "continuous image" in the sense of the Shannon theory. An interesting attribute of optical images is their highly structured nature. This structure appears locally and can be used to characterize objects in these images; that is, portions of objects can be described as smooth, edgy, etc. Information such as this is not considered in the sampling theory.

Let us now make a few observations. Equation (4.2) specifies a set of basis functions which are linearly weighted by the collected samples of the signal $x(t)$. If the samples are collected at a sampling rate T_s that does not meet the critical sampling rate, the set of $\sin c$ bases cannot be linearly weighted according to equation (4.2) to perfectly reconstruct our data. *However, this does not preclude the existence of other sets of basis functions that could be linearly combined by samples collected at a rate below the critical sampling rate and still yield the signal's perfect reconstruction.* In fact, the perfect reconstruction of a signal according to the Shannon sampling theory only establishes *sufficient* conditions for the perfect reconstruction from samples [17]. If some other knowledge about the signal corresponding to the observed samples is available, then this can be used to develop bases for superresolving a signal.

The problem is that when these bases are no longer universal, they become signal dependent. As a simple example, let's consider the set of piecewise constant time functions where each constant segment in the signals has duration T seconds (e.g., signals quantized in time). An illustration of such a function is provided in Figure 4.2a. Note that this signal has infinite frequency content due to its staircase nature.

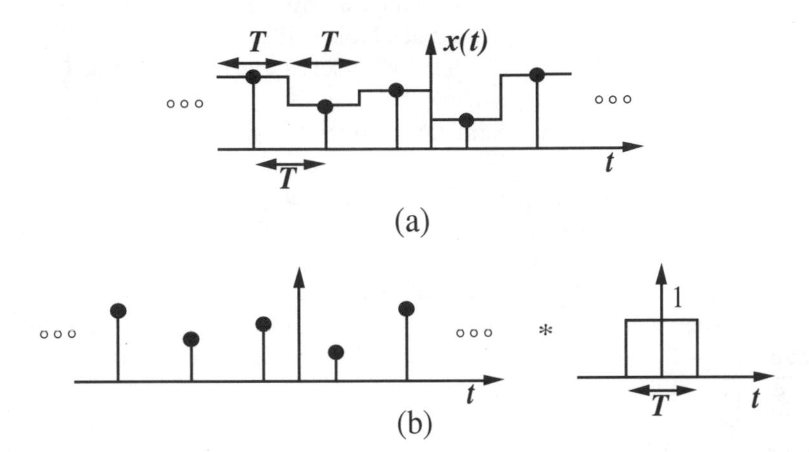

FIGURE 4.2
Simple example illustrating how perfect reconstruction is possible when a priori knowledge of the relation between signal samples and its corresponding continuous function is known. (a) Piecewise constant continuous function grossly undersampled according to the Nyquist criterion; (b) recovering the continuous function in (a) from its samples requires a simple convolution of the samples with a zero-order-hold kernel.

If our observed samples were obtained by sampling this function every T seconds, then clearly the convolution of a zero-order-hold kernel with the observed samples would be optimal for recovering the piecewise constant function. That is, it would yield perfect reconstruction even though the function in question was grossly undersampled according to the Shannon sampling theory. This convolution is pictured in Figure 4.2b.

The set of piecewise constant signals is not typically encountered in practice, so the basis set resulting from the zero-order-hold kernel is of limited use. However, this very simple example illustrates that superresolution could be based on a priori knowledge about the signal given the observed samples — irrespective of the frequency content of the signal. Therefore, *superresolution is directly associated with methods to acquire extra information about the signal of interest and derive from it appropriate bases.*

Recently, optimal reconstruction of signals sampled below their Nyquist rate was proved possible by modeling the signal statistics [20]. Ruderman and Bialek derived the optimal filter (which happens to be linear) for reconstructing a signal $x(t)$, which is assumed band limited, Gaussian, zero mean, and stationary. Their results also show that the signal statistics play no role in perfect reconstruction when the Shannon sampling conditions are met.

The great lesson from this work is that a statistical description can be used to superresolve a signal from a collected set of samples, irrespective of the relation between sampling frequency and maximum frequency content. However, the analytic result is valid only for stationary Gaussian signals. In practice, real-world signals are typically nonstationary and have very complex statistics. The analytical intractability of determining the optimal filters for complicated density functions, as commented by the authors, limits the practical use of this method.

However, statistical information about signals can also be obtained with adaptive algorithms. This is the avenue explored in this work.

4.2.2 Finding Bases for Image Representation

Superresolution of images will be possible if the interpolation system uses more efficiently the information contained in the available image samples. This requires projections onto data-specific sets of bases instead of the ones established by the sampling theorem. Naturally, learning or adaptive system theories play a crucial role in this methodology of designing data-specific projections. The manner in which these models are realized must be consistent with the information character of images and how this relates to the superresolution problem.

The universality of the perfect reconstruction theories is an amazing result, but the price paid is a strict limitation on the resulting resolution. The practical problem is to do the best we can with the available samples in order to superresolve the images. To yield better reconstructed images, we must find alternative sets of projections from which to reconstruct our images. It has been shown that perfect reconstruction can be achieved if a priori signal knowledge is available, but in practice this knowledge is absent. *So a central problem is how to capture statistical knowledge about the domain and effectively use it to design basis functions.* In determining the set of projections to use, we must either make assumptions regarding our data or learn this a priori knowledge from the available data using nonparametric models. In our work we are concerned with the latter.

We herein propose a novel technique for image superresolution by working across scales. From the original image we create a low-resolution version through a down-sampling operation on the original image. The high-resolution image becomes the desired response to a learning system that receives the low-resolution image as input. At this point we have two options to learn a statistical model: either we model the global image statistics or we seek local statistical models. The highly structured and localized nature of images calls for the development of local statistical models. In fact, local models arise naturally from the various structures in images resulting from the objects in the imaged scene. The local models can be practically implemented by learning the relation between low- and high-resolution versions of an image. Two particular traits typical to images can be used in this modeling:

- There is much similarity in local structure throughout an image.
- This structure is maintained across scales.

We now discuss the implications of the existence of these traits in images.

Similarity in Local Structure

The first trait can be exemplified by considering an image of a face. Local neighborhoods in the person's cheeks and forehead are generally indistinguishable when viewed independently. We have assumed that the effects of lighting and other "special" attributes (scars, moles, birthmarks, etc.) are absent in this comparison. An easy method to test this observation is to locate these similar image portions in an image and randomly swap them to form a new image. If the new image resembles the original one then our observation is correct. Similarly, all neighborhoods exhibiting a particular characteristic can be treated in practically the same manner. These neighborhoods can be considered generated by the same statistical process. It has been shown that the targets for the neighborhoods can be interpreted as the mean of each statistical process — one for each model used [1].

Examples of this first trait abound in images. It has recently been exploited to increase compression gains. The standard schemes for lossy image compression are based around the highly redundant information generally present in small image blocks that could be described in a more efficient and compact manner. In the case of compression via principal component analysis (PCA), a single representation is established for all of the blocks in the image. The recent compression approaches exploit the first image trait by grouping the small blocks of an image into clusters that are most similar (correlated) with one another. In this way, each cluster can be described with an efficient representation of its own — separate from those corresponding to other clusters. Overall, this results in a more efficient image representation than that afforded by the approaches of the standard compression schemes.

Across-Scale Similarity

If a strong similarity exists between homologous and highly correlated regions of low- and high-resolution images, then it is foreseeable that a simple transformation can associate these neighborhoods across scales. Experimental analysis has shown that such similar information exists locally across scales — a similarity we term *scale interdependence.* In testing for the existence of such scale interdependence, experiments were performed. The experiments report on the percentage of homologous neighborhoods that were similarly clustered from the low- and high-resolution counterpart images that were analyzed. If a high percentage of these neighborhoods is found, then a strong scale interdependence among neighborhoods is said to exist. A detailed description and analysis of this simple experiment follows.

In brief, the experiment considers a low- and high-resolution version of an image, $x_l[n_1, n_2]$ and $x_h[n_1, n_2]$, respectively. The homologous structural neighborhoods of $x_l[n_1, n_2]$ and $x_h[n_1, n_2]$ are then clustered using vector quantization (VQ) [21] to form K disjoint groups. A confusion matrix is constructed in which the most likely ordering for the K groups is sought. Finally, a measure of across-scale similarity is obtained from the percentage of neighborhoods similarly clustered for the most likely ordering obtained. The definition of a homologous neighborhood will be stated shortly. Also, the definition of a structural neighborhood, as well as why these neighborhoods were chosen, will be provided in Section 4.2.3. For now, just note that a structural neighborhood is an affine mapped version of an image neighborhood.

The $H_1 \times H_2$ neighborhoods in the $N_1 \times N_2$ image $x_l[n_1, n_2]$ form the set of neighborhoods

$$X = \{x_l\,[m_1 : m_1 + H_1 - 1, m_2 : m_2 + H_2 - 1]\}\,\big|_{m_1=0,\ldots,N_1-H_1, m_2=0,\ldots,N_2-H_2}\,.$$

The homologous neighborhoods in the high-resolution image are defined as the $G_1 H_1 \times G_2 H_2$ neighborhoods in the $(M_1 = G_1 N_1) \times (M_2 = G_2 N_2)$ image $x_h[n_1, n_2]$, which forms the set

$$\begin{aligned} D = \{x_h\,[&G_1 m_1 : G_1 m_1 + G_1 H_1 - 1, \\ &G_2 m_2 : G_2 m_2 + G_2 H_2 - 1]\}\,\big|_{m_1=0,\ldots,N_1-H_1, m_2=0,\ldots,N_2-H_2}\,. \end{aligned}$$

Recall that $x_l[n_1, n_2]$ is simulated from $x_h[n_1, n_2]$ through decimation by a factor of $G_1 \times G_2$. The manner in which we have simulated the across-scale neighborhoods yields regions of support that encompass the same physical region of the scene — with the low-resolution neighborhood having fewer samples to describe this region. The corresponding image acquisition model is to be presented.

Now the neighborhoods in X are clustered to form K disjoint groups $X_1, \ldots, X_K$ and the neighborhoods in D are separately clustered to form K disjoint groups $D_1, \ldots, D_K$. If the homologous neighborhoods in X and D form similar clusters in their respective images, then the information content of the low- and high-resolution images must be similar in some sense. To determine how well clustered information from the same image relates across scales, we form a confusion matrix as shown in Figure 4.3.

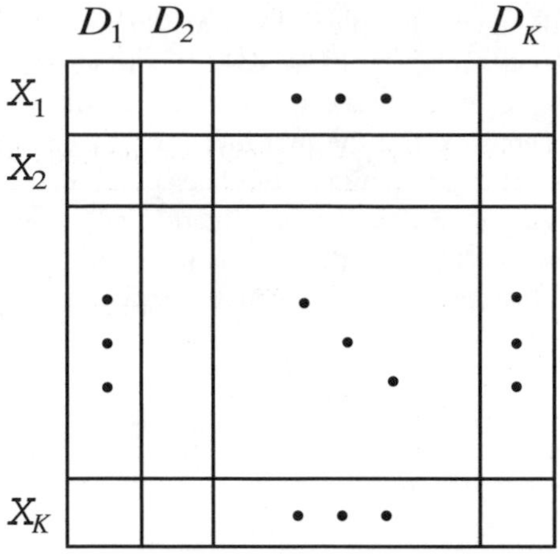

FIGURE 4.3
Confusion matrix for clustered homologous neighborhoods within their low- and high-resolution images. The X_j are the disjoint sets of clustered neighborhoods in the low-resolution image and the D_k are the disjoint sets of clustered neighborhoods in the high-resolution image, where $j, k = 1, \ldots, K$.

The entry in location (j, k) of the matrix is the number of neighborhoods assigned to cluster X_j and D_k, $j, k = 1, \ldots, K$. The interdependence across scales is determined as the maximum number of homologous neighborhoods common to the clusters formed. Since the ordering of the true clusters or "classes" between X_j and D_k is not known, we can't just examine the contents of the confusion matrix's diagonal. Instead, we must search for the most likely ordering. This in turn yields a number that reveals a measure of how similar information in the low- and high-resolution images was clustered. This number is easily found with the following simple algorithm:

Step 1: Initialize $N = 0$.

Step 2: Find the largest number L in the confusion matrix and save its row and column coordinates (r, c).

Step 3: Perform $N \leftarrow N + L$.

Step 4: Remove row r and column c from the confusion matrix to form a new confusion matrix with one less row and column.

Step 5: If the confusion matrix has no more rows and columns: STOP else Go to step 2.

The variable N represents the number of homologous neighborhoods common to similar clusters from the low- and high-resolution images. The percentage of such clustered neighborhoods is $P = \frac{N}{(N_1-H_1+1)(N_2-H_2+1)}$ since there are a total of $(N_1 - H_1 + 1)(N_2 - H_2 + 1)$ homologous neighborhoods to cluster in each image.

In Figure 4.4 this percentage is plotted as a function of the number of clusters. The high-resolution image clustered is 256×256 and is pictured in Figure 4.1c (Lena). Two low-resolution counterpart images have been used. One was obtained through $(G_1 = 2) \times (G_2 = 2)$ decimation of the high-resolution image (Figure 4.1a). The other image uses a DCT compressed version of the low-resolution image. The plots also report on two different neighborhood sizes tested: $H_1 \times H_2 = 3 \times 3$ and $H_1 \times H_2 = 5 \times 5$.

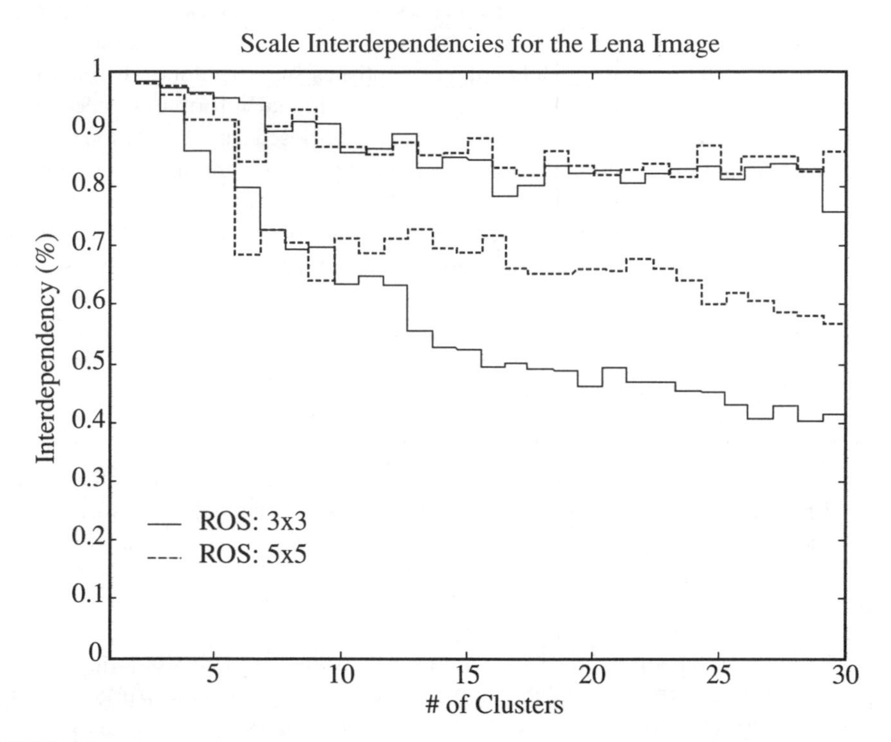

FIGURE 4.4
Scale interdependency plot for the Lena image from Figure 4.1. Two low-resolution images are used in determining the interdependency: a DCT compressed image (lower two curves) and a noncompressed one (top two curves). The high-resolution image was 256 × 256 and the corresponding low-resolution images were 128 × 128. Two regions of support (ROSs) are reported for computing the interdependency: 3 × 3 and 5 × 5. A very high interdependency among homologous neighborhoods is seen between the noncompressed low-resolution image and its high-resolution counterpart even when considering $K = 30$ clusters.

Figure 4.4 illustrates that there is a very strong interdependence of homologous neighborhoods across image scales when the low-resolution (noncompressed) image of Figure 4.1a is used — even as the number of clusters increases toward $K = 30$. This is illustrated by the top two curves in the plot. The interdependency decreases when the compressed low-resolution image is used. This is shown by the bottom two curves on the plot. Note that the case of $K = 1$ always yields an interdependency of 1. This is because for $K = 1$, no clustering is actually being performed. That is, all neighborhoods are assumed to belong to the same cluster. As such, the "disjoint" sets (or single set in this case) have all the homologous neighborhoods in common.

The interdependency generally decreases as the number of clusters increases. This is intuitively expected because an increase in the number of clusters results in the clustering of information increasingly specific to a particular image and scale. Because the frequency content between the low- and high-resolution counterpart images differs, the greater specialization of information within an image is expected to result in less interdependency among them.

4.2.3 Description of the Methodology

As already alluded to, the superresolution problem is one of determining an appropriate mapping, which is applied to a set of collected samples in order to yield a "better" reconstructed image. The manner in which this mapping is determined describes the resulting superresolution process. The methodology presented herein accomplishes superresolution by exploiting the aforementioned image traits in order to extract the additional information necessary (i.e., beyond the collected samples) to obtain a solution to the ill-posed superresolution problem [2]. Rather than assuming smoothness or relying on other typical constraints, we employ the fact that a given class of images contains *similar information locally and that this similarity holds across scales.* So the fundamental problem is to devise a superresolution scheme that will be able to determine similarity of local information and capture similarities across scales in an automated fashion.

Such a superresolution approach necessitates establishing

- which neighborhoods of an image are similar in local structure
- how these neighborhoods relate across scale.

To answer the question of which neighborhoods, the image space of local neighborhoods will be partitioned. As already alluded to, this is accomplished via a VQ algorithm — for which many are available. To determine how they relate, each Voronoi cell resulting from the VQ will be linked to a linear associative memory (LAM) trained to find the best mapping between the low-resolution neighborhoods in that cluster and their homologous high-resolution neighborhoods, hence capturing the information across scales. In other words, the assumption we make is that the information embodied in the codebook vectors and LAMs describes the relation (mapping) between a low-resolution neighborhood and its high-resolution counterpart. As such, our approach does not require the assumptions typically needed to obtain a reasonable solution to the ill-posed superresolution problem.

The LAMs can be viewed as reconstruction kernels that relate the image information across scales. We choose an adaptive scheme to design the kernels because we know how to design optimal mappers given a representative set of training images. We further expect that, if the local regions are small enough, the information will generalize across images. When a new image is presented, the kernel that best reconstructs each local region is selected automatically and the reconstruction will appear at the output.

One can expect that this methodology will yield better reconstruction than methods based on the sampling theory. However, unlike the universal character of the sampling theory, this superresolution method is specific to the character of images. That is, bases obtained for one class of images may perform poorly when reconstructing another class. Because of this, establishing the appropriate models with which to compare our data is important to the successful superresolution of an image.

4.3 Image Acquisition Model

The image acquisition process is modeled in this section. We use this model to synthesize a low-resolution counterpart to the original image. With this model, the regions of support of our low- and high-resolution neighborhoods are homologous (i.e., they encompass the same physical region of the imaged scene). The model herein was used to obtain the 128×128 image of Figure 4.1a, which was sin c interpolated in that figure. In the superresolution architecture, the low-resolution synthesis creates an input from which the information across scales can be modeled (from the pair of images).

Let the function $x_a(t, t_1, t_2)$ represent a continuous, time-varying image impinging on a sensor plane. The spatial plane is referenced by the t_1, t_2 coordinate axes and time is referenced by the variable t. The imaging sensor plane is assumed to be a grid of $N_1 \times N_2$ rectangular sensor elements. These elements serve to sample the spatial plane within the camera's field of view. Each of these elements is said to have physical dimensions $p_1 \times p_2$. The output of each element is proportional to the amount of light that impinges on each sensor during a given time interval. The output of each sensor, given by $x_l[n_1, n_2]$ where $n_1 = 0, 1, \ldots, N_1 - 1$ and $n_2 = 0, 1, \ldots, N_2 - 1$, can then be expressed as

$$x_l[n_1, n_2] = \int_0^1 \int_{p_1 n_1}^{p_1(n_1+1)} \int_{p_2 n_2}^{p_2(n_2+1)} x(t, t_1, t_2)\, dt_2\, dt_1\, dt$$

where the integration over time is one time unit in duration. The subscript l is used to denote a low-resolution image.

To obtain a higher resolution image, a finer sensor grid encompassing the same field of view used in obtaining $x_l[n_1, n_2]$ would have to be employed. Let the resolution in each spatial dimension be increased — by a factor of G_1 and G_2 in their respective spatial dimensions. The physical size of the sensor elements now becomes $\frac{p_1}{G_1} \times \frac{p_2}{G_2}$ units of area. The high-resolution image is then given by $x_h[m_1, m_2]$, where $m_1 = 0, 1, \ldots, M_1 - 1$ and $m_2 = 0, 1, \ldots, M_2 - 1$, and $M_i = G_i N_i (i = 1, 2)$. The output for each of the $M_1 \times M_2$ sensor elements for the high-resolution image can be described by

$$x_h[m_1, m_2] = \int_0^{G_1 G_2} \int_{p_1 m_1/G_1}^{p_1(m_1+1)/G_1} \int_{p_2 m_2/G_2}^{p_2(m_2+1)/G_2} x(t, t_1, t_2)\, dt_2\, dt_1\, dt$$

Notice that the integration limits over time have been extended from one time unit to $G_1 G_2$ time units in order to maintain the average intensity value for each pixel in the image.

The superresolution process is to estimate the high-resolution image $x_h[m_1, m_2]$ from the low-resolution image $x_l[n_1, n_2]$. One can notice that the process of acquiring $x_l[n_1, n_2]$ from $x_h[m_1, m_2]$ is given by

$$x_l[n_1, n_2] = \frac{1}{G_1 G_2} \sum_{m_1 = G_1 n_1}^{G_1(n_1+1)-1} \sum_{m_2 = G_2 n_2}^{G_2(n_2+1)-1} x_h[m_1, m_2] \tag{4.4}$$

The decimation model in the above equation produces a low-resolution image by averaging the pixels of $G_1 \times G_2$ nonoverlapping pixel neighborhoods in the high-resolution image.

4.4 Relating Kernel-Based Approaches

This section introduces kernel-based formalisms for the magnification of images. Conventional approaches to magnification utilize a single kernel and interpolate between samples for increasing the sample density of an image. The superresolution methodology presented herein is related to the use of a family of kernels. Each kernel is tailored to specific information of an image across scales.

4.4.1 Single Kernel

A magnified image can be obtained by expanding the samples of a low-resolution image $x_l[n_1, n_2]$ and convolving with a sampled interpolation kernel [22]. For an expansion rate of $G_1 \times G_2$, where G_1, G_2 are whole numbers greater than 1, the expanded image is given by

$$x_e[n_1, n_2] = \begin{cases} x_l\left[\frac{n_1}{G_1}, \frac{n_2}{G_2}\right] & \begin{matrix} n_1 = 0, \pm G_1, \pm 2G_1, \ldots \\ n_2 = 0, \pm G_2, \pm 2G_2, \ldots \end{matrix} \\ 0 & \text{otherwise} \end{cases} \tag{4.5}$$

and the corresponding interpolation kernel, obtained by sampling a continuous kernel, is denoted $k[n_1, n_2]$. The interpolated image $\hat{x}_h[n_1, n_2]$ that estimates the true image $x_h[n_1, n_2]$ is

$$\hat{x}_h[n_1, n_2] = x_e[n_1, n_2] * *k[n_1, n_2] \tag{4.6}$$

where $**$ denotes 2D convolution. This form of interpolation is a linear filtering that processes the image similarly throughout (i.e., it uses the same linear combination of image samples in determining interpolated points — as does the Shannon sampling theory).

4.4.2 Family of Kernels

Reconstruction with a single kernel is a simple operation since the same function is applied over and over again to every sample. This is not so when we have at our disposal many kernels. Two fundamental questions must be answered to reconstruct signals with a family of kernels: how to choose one member of the family and how to design it. We will formalize these issues next.

The kernel family approach is a scheme in which the kernel used depends on the local characteristics of the image [23]. This is formulated as

$$\hat{x}_h[n_1, n_2] = x_e[n_1, n_2] * *k_{c,l}[n_1, n_2] \tag{4.7}$$

The subscripts c and l, which are functions of image location, select a kernel based on the local image characteristics about the point of interest. The family of kernels is given by $\{k_{c,l}[n_1, n_2] : c = 1, \ldots, C; l = 1, \ldots, L\}$. C represents the number of established local image characteristics (features) from which to compare local neighborhood information and L is the number of kernels created per feature. In summary, equation (4.7) describes a convolution with a shift-varying kernel. It is a generalization of equation (4.6) and defaults to the standard convolution of equation (4.6) when $C, L = 1$.

4.5 Description of the Superresolution Architecture

Figure 4.5 illustrates the proposed architecture for superresolving images using a family of kernels. As we proceed, the relation between the architecture and equation (4.7) will be elucidated. The purpose of data clustering is to partition the low-resolution image neighborhoods into a finite number of clusters where the neighborhoods within each cluster are similar in some sense. Once the clusters are established, a set of kernels can be developed that optimally transforms each clustered neighborhood into its corresponding high-resolution neighborhood. The subsections that follow discuss how the kernel family, implemented here as LAMs (see Figure 4.5), is established and then used for optical image superresolution.

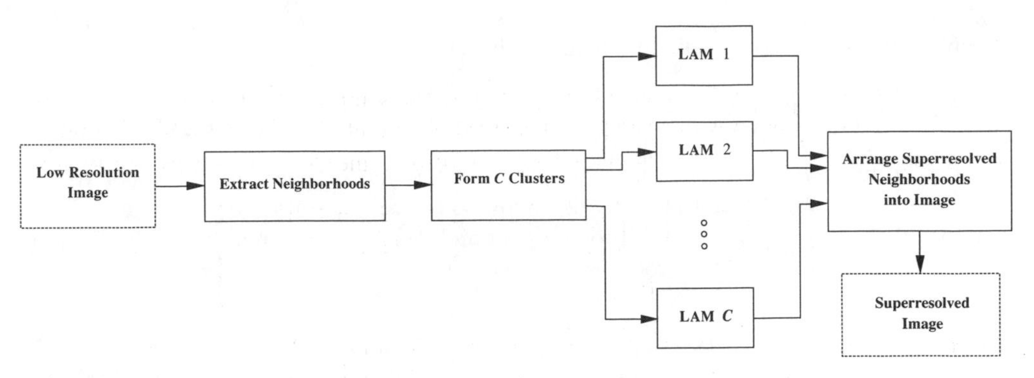

FIGURE 4.5
Superresolution architecture for the kernel family approach. This paradigm performs the equivalent operation of a convolution with a family of kernels.

4.5.1 The Training Data

Ideally, the low- and high-resolution data sets used to train the LAMs of Figure 4.5 would each encompass the same scene and have been physically obtained by hardware with different, but known, resolution settings. Such data collection is not common. Instead, the low-resolution counterparts of the given images are obtained via decimation using the image acquisition model discussed earlier. Once established, the training of the superresolution architecture proceeds as described in Figure 4.6. Note that the decimation is represented by the $\downarrow G_1 \times G_2$ block in the figure. The data preprocessing and high-resolution construction sections of this figure will now be explained.

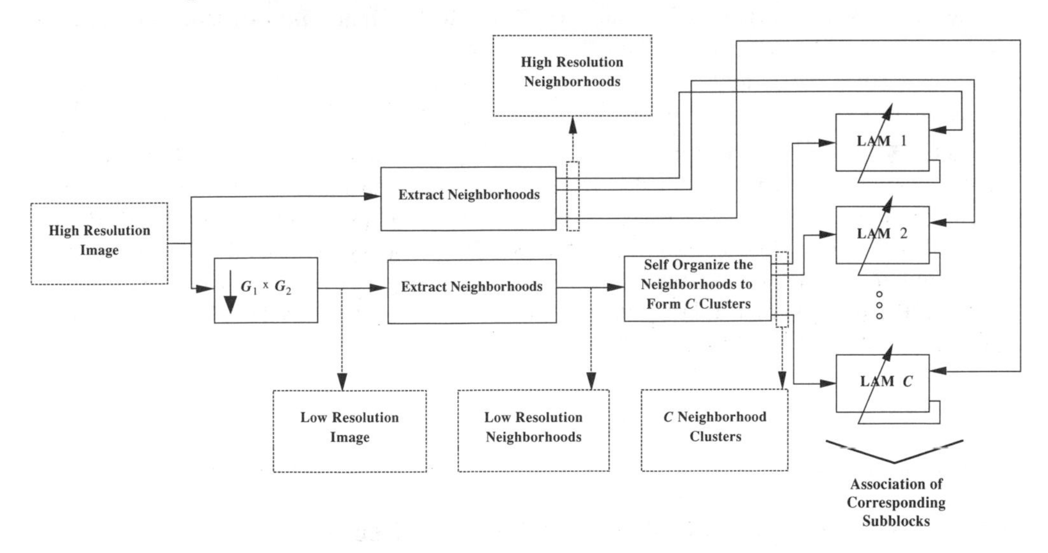

FIGURE 4.6
Training architecture for the superresolution of images via the kernel family approach.

4.5.2 Clustering of Data

The neighborhoods considered consist of all the *overlapping* $H_1 \times H_2$ neighborhoods of the low-resolution image $x_l[n_1, n_2]$. The set of these $N = (N_1 - H_1 + 1)(N_2 - H_2 + 1)$

neighborhoods in the low-resolution image is given by

$$X = \left\{ x_l \left[m_1 : m_1 + H_1 - 1, m_2 : m_2 + H_2 - 1 \right] \right\} \Big|_{m_1 = 0, \ldots, N_1 - H_1, m_2 = 0, \ldots, N_2 - H_2} \tag{4.8}$$

and can be represented by the matrix $\mathbf{X} \in \Re^{H_1 H_2 \times N}$ whose columns are the set of vectors $\{\mathbf{x}_r\}_{r=1}^{N}$ where $\mathbf{x}_r$ is a "vectorized" 2D neighborhood. Each low-resolution neighborhood is paired with its $(2G_1 - 1) \times (2G_2 - 1)$ homologous high-resolution neighborhood. Specifically, these high-resolution neighborhoods are described by

$$S = \left\{ \begin{matrix} x_h \left[G_1 m_1 + \phi_1 + 1 : G_1(m_1 + 2) + \phi_1 - 1, \right. \\ \left. G_2 m_2 + \phi_2 + 1 : G_2(m_2 + 2) + \phi_2 - 1 \right] \end{matrix} \right\} \Bigg|_{m_1 = 0, \ldots, N_1 - H_1, m_2 = 0, \ldots, N_2 - H_2} \tag{4.9}$$

where $\phi_i = \frac{G_i(H_i - 3)}{2}$, and $i = 1, 2$.

Notice that the set of neighborhoods to be clustered here is different from the set used for arriving at the across-scale similarity measure. The previous set of neighborhoods resulted from the *nonoverlapping* neighborhoods in the low- and high-resolution counterpart images. The set now consists of *overlapping* neighborhoods. The reason for the overlap is to obtain multiple estimates of a high-resolution sample. In this way, the final high-resolution sample can be estimated more reliably.

The neighborhoods in S can be represented by a matrix $\mathbf{S} \in \Re^{(2G_1 - 1)(2G_2 - 1) \times N}$ similar to the representation used in $\mathbf{X}$. These low- and high-resolution neighborhoods are depicted in Figure 4.7, where the shaded circles represent a low-resolution neighborhood. For the case of $G_1 = G_2 = 2$ in Figure 4.7a, the shaded circles are used to construct the crossed circles about the center of the low-resolution neighborhood. Note that if we elect not to construct the center pixel, we will be interpolating locally about the observed image samples. If we elect to construct the center pixel (along with the other crossed circles), we are allowing for the ability to change a "noisy" observed sample. Figure 4.7b similarly illustrates this for the case of $G_1 = G_2 = 3$.

In establishing our family of kernels, we have chosen to associate the *structure* between the neighborhoods in $\mathbf{X}$ and $\mathbf{S}$, *not* the observed samples themselves. The structure of a neighborhood is defined as the neighborhood with its mean subtracted out; each neighborhood thus becomes a vector whose component mean is zero. This kind of preprocessing allows us to categorize neighborhoods sharing a particular characteristic (i.e., they could be smooth, edgy at a particular orientation, etc.) as belonging to the same class regardless of the average intensity of the neighborhood. The structure $\mathbf{p}_r$ of neighborhood $\mathbf{x}_r$ is obtained through multiplication with the square matrix $\mathbf{Z} \in \Re^{H_1 H_2 \times H_1 H_2}$ (i.e., $\mathbf{p}_r = \mathbf{Z}\mathbf{x}_r$ for a single neighborhood or $\mathbf{P} = \mathbf{Z}\mathbf{X}$ for all the input neighborhoods), where

$$\mathbf{Z} = \frac{1}{H_1 H_2} \begin{pmatrix} H_1 H_2 - 1 & -1 & \cdots & -1 \\ -1 & H_1 H_2 - 1 & & \vdots \\ \vdots & & \ddots & -1 \\ -1 & \cdots & -1 & H_1 H_2 - 1 \end{pmatrix}. \tag{4.10}$$

The desired exemplars associated with $\mathbf{P}$ are contained in matrix $\mathbf{D}$. Each column in $\mathbf{D}$ is obtained by subtracting the mean of $\mathbf{x}_r$ from its corresponding neighborhood $\mathbf{s}_r$ in $\mathbf{S}$. This is done to compensate for the low-resolution neighborhood mean, which has been subtracted from $\mathbf{x}_r$ and must be added back after the high-resolution neighborhood structure is created. Specifically, $\mathbf{D} = \mathbf{S} - \mathbf{A}\mathbf{X}$, where $\mathbf{A} \in \Re^{(2G_1 - 1)(2G_2 - 1) \times H_1 H_2}$ is a constant matrix with elements $\frac{1}{H_1 H_2}$.

The clusters are formed by performing a VQ on the space of structural neighborhoods in $\mathbf{P}$. This clustering is based on the interblock correlation among the neighborhoods in $\mathbf{P}$ [1]. The

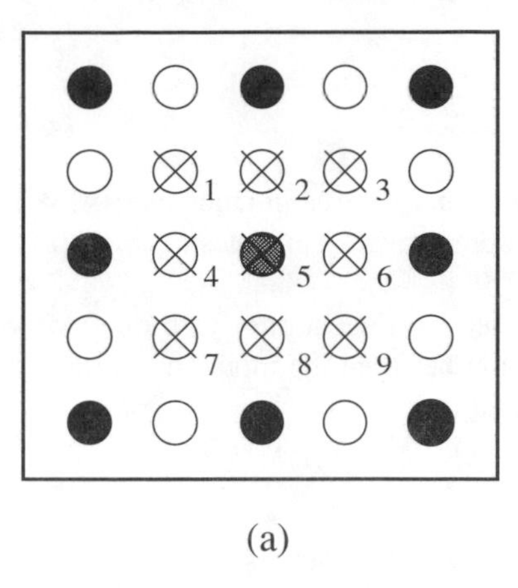

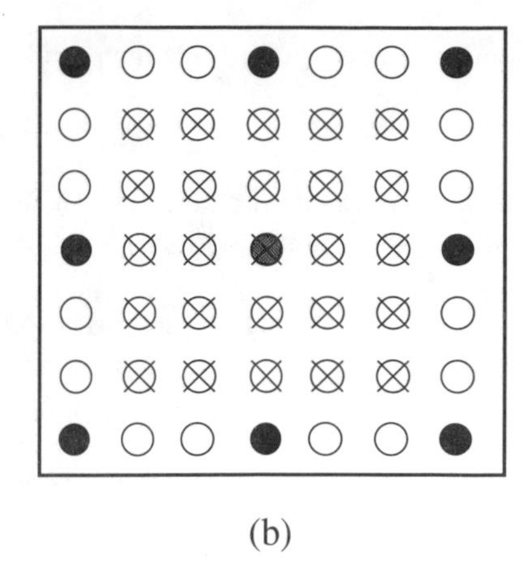

(a) (b)

FIGURE 4.7
Local image neighborhoods and the pixels they superresolve. Each circle represents a 2D high-resolution image pixel. The shaded circles are the low-resolution image pixels obtained via decimation of the high-resolution image. The gray pixel is the center of the low-resolution neighborhood. Each $H_1 \times H_2$ low-resolution neighborhood constructs a $(2G_1 - 1) \times (2G_2 - 1)$ high-resolution neighborhood about the low-resolution neighborhood's center — these are depicted by the crossed circles. The numbers are a convention used to distinguish between constructed pixels in this neighborhood. (a) Decimation factor $G_1 = G_2 = 2$; (b) decimation factor $G_1 = G_2 = 3$.

VQ is accomplished using Kohonen's self-organizing map [24] for reasons discussed later. The VQ operation results in a set of C feature vectors $\{\mathbf{f}_c\}_{c=1}^{C}$, where usually $C << N$. The C clusters $K_c, c = 1, 2, \ldots, C$, formed by our neighborhood, and feature vectors are given by

$$K_c = \left\{\mathbf{p}_r : \left\|\mathbf{p}_r - \mathbf{f}_c\right\|_2 < \left\|\mathbf{p}_r - \mathbf{f}_b\right\|_2; b = 1, 2, \ldots, C; b \neq c; r = 1, 2, \ldots, N\right\} \tag{4.11}$$

4.5.3 Neighborhood Association

The superresolution methodology herein is piecewise local in nature — inherent to the fact that we consider neighborhoods. A mapping is required to produce high-resolution image samples from the low-resolution ones that are available. This mapping could be linear (or affine) or nonlinear. A description of how these mappings have been implemented within the methodology now follows.

The input–output relationship of a LAM [21] is an affine transformation described by

$$\mathbf{y}_r = \mathbf{W}\mathbf{p}_r + \mathbf{b} \tag{4.12}$$

where $\mathbf{W}$ is a weight matrix that specifies the network connectivity of the LAM, $\mathbf{b}$ is a bias vector, and $\mathbf{p}_r$ is the input vector (neighborhood structure). Note that $\mathbf{y}_r$ contains a vector representation of a superresolved 2D neighborhood structure. The neighborhoods in $\mathbf{P}$ and $\mathbf{D}$ are associated in the least square sense to determine the values of the $\mathbf{W}$ and $\mathbf{b}$ parameters. These parameters can be obtained recursively via the least mean squares (LMS) algorithm update equation [21]

$$\mathbf{W}(n+1) = \mathbf{W}(n) + \mu(\mathbf{D} - \mathbf{Y})\mathbf{P}^T \tag{4.13}$$

where T denotes matrix transposition and μ is the learning rate. They can equivalently be obtained in closed form via the pseudo-inverse [21]

$$\mathbf{W} = \mathbf{D}\mathbf{P}^T(\mathbf{P}\mathbf{P}^T)^{-1} . \tag{4.14}$$

We have assumed in equations (4.13) and (4.14) that $\mathbf{W}$ is actually the augmented matrix $[\mathbf{W}|\mathbf{b}]$ and $\mathbf{P}$ is the augmented matrix $[\mathbf{P}^T|\mathbf{v}]^T$, where $\mathbf{v}$ is a column vector of ones of appropriate dimensions.

Nonlinear associative memories (NLAMs) can be used as a substitute for the LAMs of Figures 4.5 and 4.6. The parameterized nonlinear relation between the input and output can be achieved using a multilayer perceptron (MLP) [21] and is given by

$$\mathbf{y}_r = \alpha\left(\{\mathbf{W}_k\}, \{\mathbf{b}_k\}, \mathbf{p}_r\right) \tag{4.15}$$

where, in general, $\alpha(\cdot)$ is a nonlinear function of a set of weight matrices, bias vectors, and the neighborhood structure, and k describes a layer in the MLP feed-forward configuration. The NLAM parameters are readily obtained with back-propagation learning [21]. It is well established as a supervised training method for neural networks. This method generalizes the LMS training algorithm for linear networks to MLPs. The on-line weight update is similar to that of the LMS algorithm. At each time step the weight matrix $\mathbf{W}_k$, for layer k of the MLP, is updated as follows

$$\mathbf{W}_k(n+1) = \mathbf{W}_k(n) + \mu\mathbf{g}_k(n)\mathbf{y}_{r,k-1}^T \tag{4.16}$$

where $\mathbf{g}_k(n)$ is the local gradient and $\mathbf{y}_{r,k-1}$ is the postneural activity of the previous layer (hence the $k-1$) due to input vector $\mathbf{x}_r$. Please note here that the subscript k describes the weight layer of a feed-forward NLAM with several layers. We could describe the kth layer of the cth NLAM by $\mathbf{W}_{c,k}$. The postneural activity of an NLAM at a given layer is recursively defined as $\mathbf{y}_{r,k} = \mathbf{W}_k(n)\mathbf{y}_{r,k-1}$. This is because the output of one layer serves as the input to the next layer in the feed-forward configuration. Note that the postneural activity for layer 0 at time step n is defined as just the input vector at that time step (i.e., $\mathbf{y}_{r,0} \equiv \mathbf{x}_r$).

If layer k is the output layer, then

$$\mathbf{g}_k(n) = \left(\mathbf{d}_r - \mathbf{y}_{r,k}\right) \bullet \varphi'\left(\mathbf{y}_{r,k}\right) \tag{4.17}$$

where $\bullet$ represents the element-by-element multiplication of two matrices (or vectors) and $\varphi'(\cdot)$ is the first derivative of $\varphi(\cdot)$, a differentiable squashing function. If layer k is other than the output layer, then

$$\mathbf{g}_k(n) = \left(\mathbf{W}_{k+1}^T(n)\mathbf{g}_{k+1}(n)\right) \bullet \varphi'\left(\mathbf{y}_{r,k}\right) . \tag{4.18}$$

There are C NLAMs to be trained. Each corresponds to a particular cluster of the input data. NLAM c associates the neighborhoods $\mathbf{x}_r \in K_c$ with its corresponding samples in $\mathbf{d}_r$.

4.5.4 Superresolving Images

The construction of a high-resolution image, as depicted in Figure 4.5, results from transforming the neighborhood structure of the low-resolution input image with the parameters obtained in the training phase. The mean of the neighborhood is subsequently added back to the transformation. When LAMs are used, the superresolution of a low-resolution neighborhood $\mathbf{x}_r$ can be expressed as

$$\hat{\mathbf{s}}_r = \mathbf{W}_c\mathbf{Z}\mathbf{x}_r + \mathbf{b}_c + \mathbf{A}\mathbf{x}_r \quad \text{for} \quad (\mathbf{p}_r = \mathbf{Z}\mathbf{x}_r) \in K_c \tag{4.19}$$

where $\mathbf{W}_c$ and $\mathbf{b}_c$ are the weight matrix and bias vector, respectively, associated with the cth LAM. As discussed before, there is a direct relation between equation (4.19) and equation (4.7). Equation (4.19) constructs the high-resolution neighborhoods' structure $\hat{\mathbf{s}}_r$. The subscript r refers to the neighborhood being constructed. The constructed pixels that overlap are averaged and the high-resolution image is thus constructed. Averaging several high-resolution samples improves the reliability of the final high-resolution sample. Equation (4.19) can be equivalently expressed as

$$\hat{x}_h[n_1, n_2] = \sum_{l=1}^{L} \left[x_e[n_1, n_2] ** \left(k_{c,l}[n_1, n_2] + a[n_1, n_2]\right)\right] \cdot b[n_1, n_2] \tag{4.20}$$

where $L = (2G_1 - 1)(2G_2 - 1)$; x_e is the expanded low-resolution image; the kernel was created with the values $\mathbf{W}_c\mathbf{Z}(l, :)$ and $\mathbf{b}_c(l)$ (i.e., row l of $\mathbf{W}_c\mathbf{Z}$ and $\mathbf{b}_c$); a is a constant kernel with the same extent as $k_{c,l}$, that averages a low-resolution neighborhood (its impulse response samples equal $\frac{1}{H_1 H_2}$); and $b[n_1, n_2] = b_1[n_1]b_2[n_2]$ is responsible for averaging multiple estimates of superresolved samples. Specifically,

$$b_i[n_i] = \begin{cases} 1 & n_i \bmod G_i = 0 \\ \frac{1}{2} & \text{otherwise} \end{cases} \tag{4.21}$$

for $i = 1, 2$. Notice that the index l refers to a specific convolution pass that is constructing the corresponding enumerated crossed circle associated with each low-resolution neighborhood in that pass. Please refer to Figure 4.7a for the case of $G_1 = G_2 = 2$.

The NLAM case differs only by the presence of the nested nonlinearities. The construction, for an M layer MLP topology, is expressed as

$$\hat{\mathbf{s}}_r = \varphi\left(\mathbf{W}_{c,M}\varphi\left(\ldots\varphi\left(\mathbf{W}_{c,l}\mathbf{Z}\mathbf{x}_r + \mathbf{b}_{c,l}\right)\right) + \mathbf{b}_{c,M}\right) + \mathbf{A}\mathbf{x}_r \quad \text{for} \quad (\mathbf{p}_r = \mathbf{Z}\mathbf{x}_r) \in K_c \tag{4.22}$$

where $\mathbf{W}_{c,k}$ and $\mathbf{b}_{c,k}$ are the weight matrix and bias vector, respectively, at layer k of the cth NLAM, and φ denotes the squashing function at each layer of the feed-forward structure.

4.6 Results

The results illustrated in this section make use of the Peppers image for training and the Lena image for testing. The Lena image has already been illustrated; the Peppers image can be found in several references (e.g., [1, 23]). The LAM-based results were compared against several kernel-based interpolation results including the subpixel edge localization and interpolation (SEL) technique [6], which fits an ideal step edge through those image regions where an edge is deemed to exist and otherwise uses a bilinear interpolation. The parameters for the SEL technique were the same as those reported in [6].

Table 4.1 reports on the peak signal-to-noise ratio (PSNR) resulting from kernel-based interpolation of the Lena and Peppers 128×128 images by a factor of 2 in each dimension. The PSNR is defined as $\text{PSNR} \equiv -10 \log_{10}(e_{rms}^2)$ where

$$e_{rms}^2 = \frac{1}{M_1 M_2} \sum_{m_1=0}^{M_1-1} \sum_{m_2=0}^{M_2-1} \left(x_h[m_1, m_2] - \hat{x}_h[m_1, m_2]\right)^2 \tag{4.23}$$

and x_h and $\hat{x}_h$ take values in [0, 1].

Table 4.1 PSNR for Magnified Images

	Zero Order	Bilinear	Bicubic	Cubic B-Spline	SEL	Train[a]	Test[a]	Train[b]	Test[b]
Lena	27.00	27.26	27.45	27.43	27.48	32.63	31.78	32.23	31.71
Peppers	27.48	27.51	27.74	27.74	27.79	34.58	32.90	34.03	32.74

[a] Used 30 LAMs.

[b] Used 30 NLAMs.

Note: The interpolation factor was 2 in each image axis from the listed 128 × 128 images. The training and test cases of the kernel family approach utilized 30 features and a 3 × 3 region of support (ROS). In the test cases, the parameters obtained in training to reconstruct Lena were used for the Peppers and vice versa.

The plot in Figure 4.8 illustrates the PSNR when superresolving the Lena 128 × 128 image with varying numbers of LAMs by a factor of 2 in each dimension. The system parameters (feature vectors, weights, and biases) were *trained using the Peppers 256 × 256 image (i.e., a different image).* The solid and dashed lines in Figure 4.8 denote training and test set reconstruction performance, respectively, using regions of support (ROSs) 3 × 3 and 5 × 5. In general, the PSNR of the training set increased as the number of LAMs increased. This is intuitively expected because an increase in the number of LAMs yields a greater specialization to particular image features, hence a more accurate image reconstruction. The feature set extracted using a 5 × 5 ROS yields more macroscopic image characteristics than does a 3 × 3 ROS. This results in greater specialization of the characteristics particular to the image of interest and generally to a more faithful image reconstruction on the training set.

In the test set, however, the larger ROS tended to show a drop in PSNR performance as the degree of specialization to image features increased. This general trend was encountered in all the tests we have run. It suggests that the similarity between features, as the system specializes more (uses more LAMs), tends to occur at a more microscopic level. It can also be observed that the kernel family approach yielded higher PSNR than those methods listed in Table 4.1.

A visual comparison of the results, utilizing the common approaches and the kernel family approach for the Lena image, can be observed in Figure 4.9. The training and testing images shown in each of these figures were created using 30 LAMs and an ROS of 3 × 3. They correspond to those points in Figure 4.8 marked by a circle. In Figure 4.10 we see the 30 features extracted from the Peppers 128 × 128 image that were used in reconstructing the test image of Figure 4.9. Notice how "regular" and edgy these features are. The features extracted are image dependent, and we would expect a different set to result from texture images, for example. The number below each feature signifies the maximum gray-level difference between the largest and smallest value present in each feature. Therefore, the feature with the "1" below it can be considered a constant feature, that is, one that contains practically no structure. This feature corresponds to those image portions that are very smooth. The features have not been scaled here; instead, the constant feature (which is the zero vector) is represented by gray (128 in an 8-bit scale). Positive feature values become lighter and negative feature values are represented by a proportionately darker shade of gray.

The superresolved training and testing images of Figure 4.9 were of similar quality. The kernel family superresolved images appear crisper than those obtained with the other approaches presented here. In Figure 4.11 we have shown the magnitude spectra of the reconstructed images corresponding to Figure 4.9. The spectra here are for the full reconstructed image, not just the portion shown in the figure. It is evident from viewing these spectra that the LAM-based approach is recovering information above half the sampling frequency and reproducing better the high-frequency information characteristic of the original images. The SEL approach is

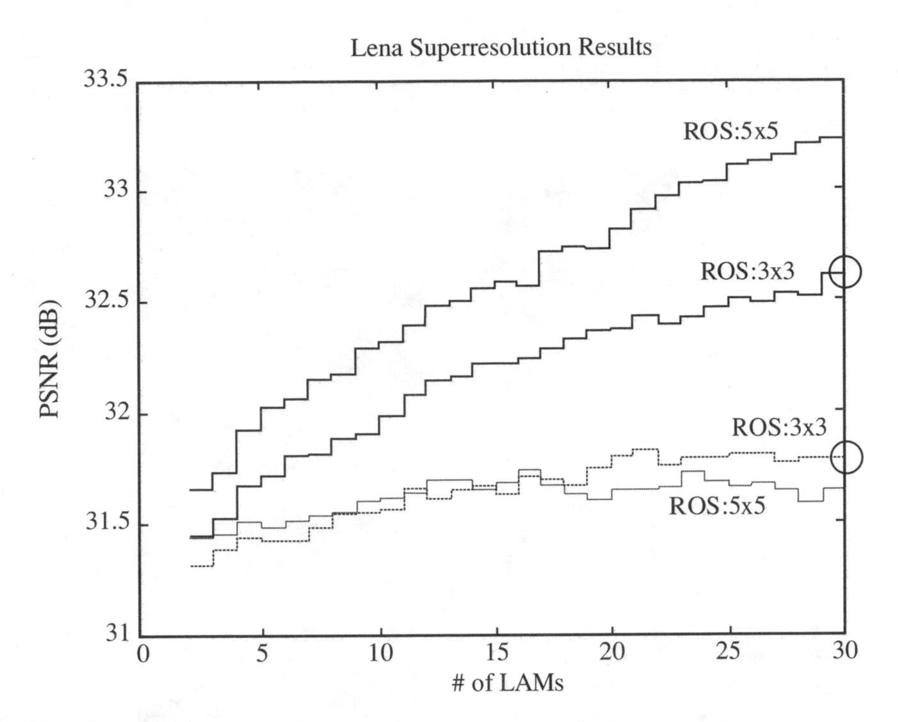

FIGURE 4.8
Training and testing superresolution results for the Lena image considering two different regions of support (ROSs). The solid lines correspond to training set results and the dashed lines are test set results. The curves related to the training data result from superresolving the Lena 128 × 128 image with the systems (features and LAMs) trained to reconstruct the Lena 256 × 256 image from the Lena 128 × 128 image. The curves related to testing result from superresolving the Lena 128 × 128 image with the systems trained to reconstruct the Peppers 256 × 256 image from the Peppers 128 × 128 image. Superresolved images corresponding to the two circled points are shown in Figure 4.9.

also able to reproduce high-frequency information. This is because, as mentioned earlier, the SEL approach fits an ideal step edge through those image regions it deems are edges. The low PSNR of the SEL approach can be attributed to its performance in smoothly varying image regions. This is because, in these regions, the SEL approach uses the bilinear kernel for its interpolation. In summary, the superresolved images of this work generally appear more crisp than those obtained with the other approaches presented here. Edges seemed to be preserved well with our approach, and the higher PSNR obtained with our methodology is evidence of the accuracy of reconstruction in smoothly varying image regions relative to the other approaches reported herein.

To test our hypothesis that the system captures well redundancy across scales, we illustrate in Figure 4.12 the superresolution of the Lena 128 × 128 image using two successive $G_1 = G_2 = 2$ reconstruction stages with the *same codebook and LAMs*. In other words, the resulting "test" image of Figure 4.9 is fed through the system of Figure 4.5 twice with the same parameters used in the first superresolution stage for a total superresolution factor of 4 in each dimension. Notice that the LAM reconstructed image is crisper than the other expanded images. This also supports our claim regarding the similarity of image neighborhoods across scales — which we exploit for superresolution. Figure 4.13 illustrates a case of what can happen when inappropriate sets of bases are used for the image reconstruction. In this figure, the Lena 128

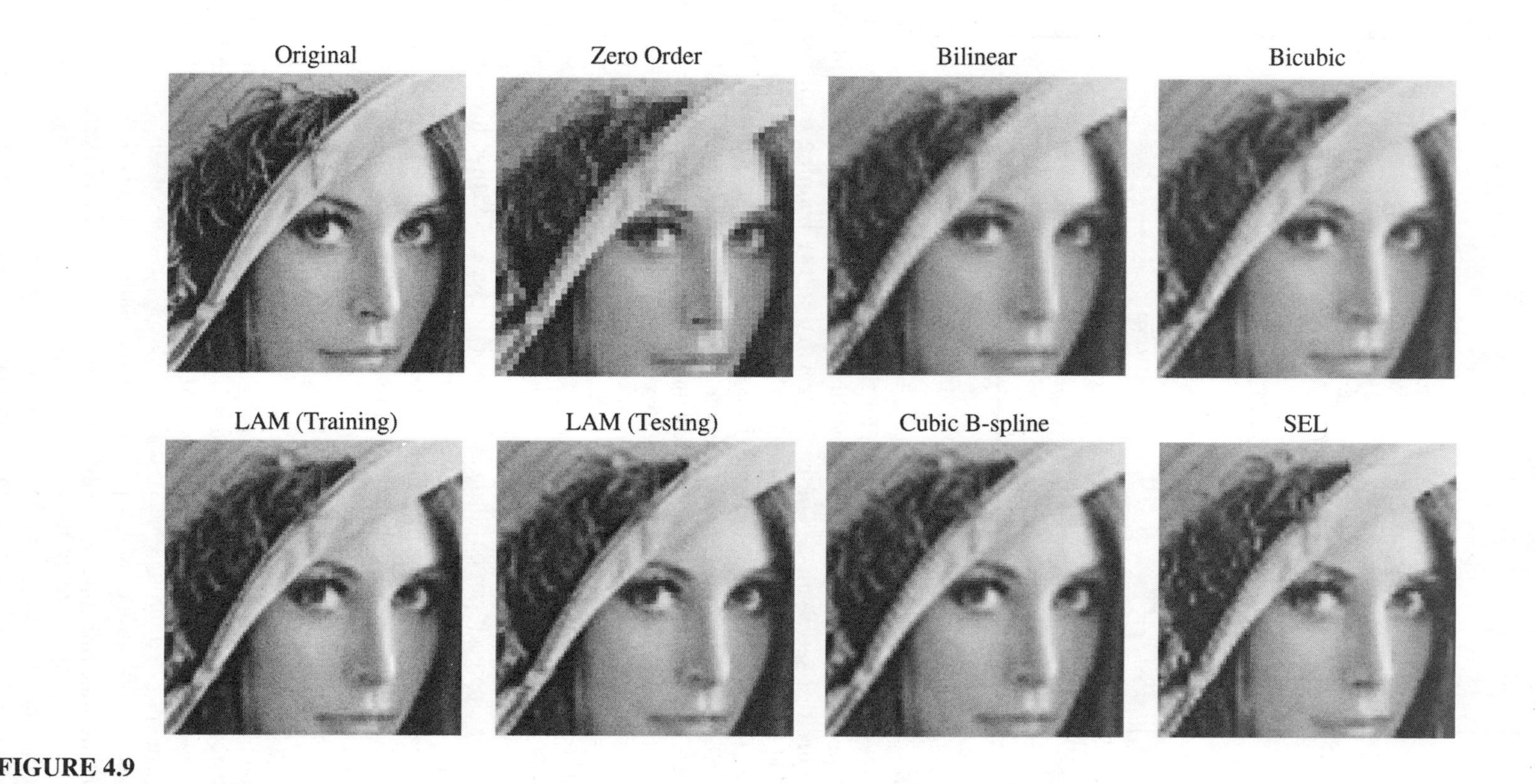

FIGURE 4.9
Visual comparison of the reconstruction results for the Lena image. The 128×128 image was reconstructed to 256×256. A zoomed section (using nearest neighbor replication) of the reconstructed results is displayed. The "training" reconstruction utilized the 30 features and corresponding LAMs obtained in training to reconstruct the Lena 256×256 image from the Lena 128×128 image with an ROS of 3×3. The "testing" reconstruction utilized the 30 features and corresponding LAMs obtained in training to reconstruct the Peppers 256×256 image from the Peppers 128×128 image with an ROS of 3×3.

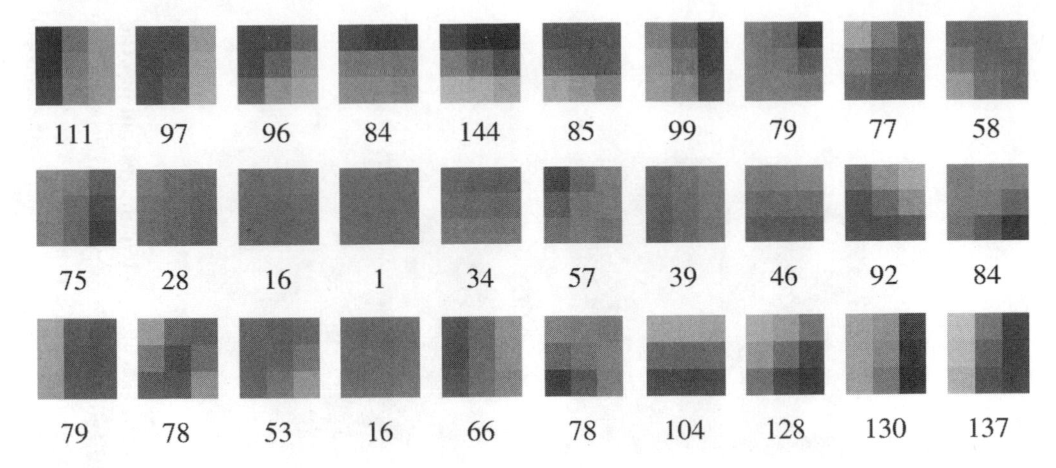

FIGURE 4.10
Features extracted from the Peppers 128 $\times$ 128 image. They were used in superresolving the Lena 128 $\times$ 128 image to a size of 256 $\times$ 256; these results are given in Figure 4.9. Notice the largely edgy nature of these features. The features have not been scaled here; instead, the constant feature is represented by gray (128 in an 8-bit scale). The number below each feature represents the maximum 8-bit gray-level difference between the largest and smallest value of that feature.

$\times$ 128 image is being reconstructed to 256 $\times$ 256. The desired image is given in Figure 4.13a. The image reconstructed from the 30 features and LAMs obtained in training to reconstruct the Peppers 256 $\times$ 256 image from the Peppers 128 $\times$ 128 image is given in Figure 4.13b, and the image reconstructed from the 30 features and LAMs obtained in training to reconstruct the Pentagon 256 $\times$ 256 image from the Pentagon 128 $\times$ 128 image is given in Figure 4.13c. The Pentagon images have not been pictured here. The systems of the Peppers are appropriate for the reconstruction of Lena. However, the systems of the Pentagon are not as appropriate. This is seen particularly by the reconstruction performance about the right portion of the forehead and hat in Figure 4.13c. Incorporating the correct a priori information into the reconstruction process can be beneficial to superresolution, but introducing the wrong information can have the opposite effect. In Figure 4.13d we compensate for the lack of proper bases by incorporating the appropriate bases obtained from the Peppers image used in reconstructing Figure 4.13b. The appropriate bases were simply "appended" to the inappropriate set from the Pentagon that was used in this example. In this manner, we did not have to retrain a system from scratch in order to produce an appropriately reconstructed image. Available bases for reconstruction can simply be incorporated into an existing system to produce adequately reconstructed images. This is possible because of the hard partitioning scheme our procedure is implementing.

Figure 4.14 illustrates results when NLAMs are used in place of the LAMs. Again, we superresolve the Lena 128 $\times$ 128 image using 30 LAMs and an ROS of 3 $\times$ 3. The NLAMs used a single hidden layer and had approximately the same number of free parameters as did the LAMs. The LAM and NLAM results are very similar (both visually and in PSNR). In this and several other tests we have run with superresolution factors of 2 and 3, the added complexity and flexibility afforded by the NLAMs seems unwarranted. This makes sense since many nonlinear mappings are reasonably well approximated locally by linear models.

Figure 4.15 illustrates the superresolution of a DCT-compressed Lena image from 256 $\times$ 256 to 512 $\times$ 512. The parameters used in our local architecture were those trained to superresolve the compressed Peppers 256 $\times$ 256 image to the original peppers 512 $\times$ 512 image. The system used 15 LAMs and an ROS of 5 $\times$ 5. The compressed images were obtained by inverse

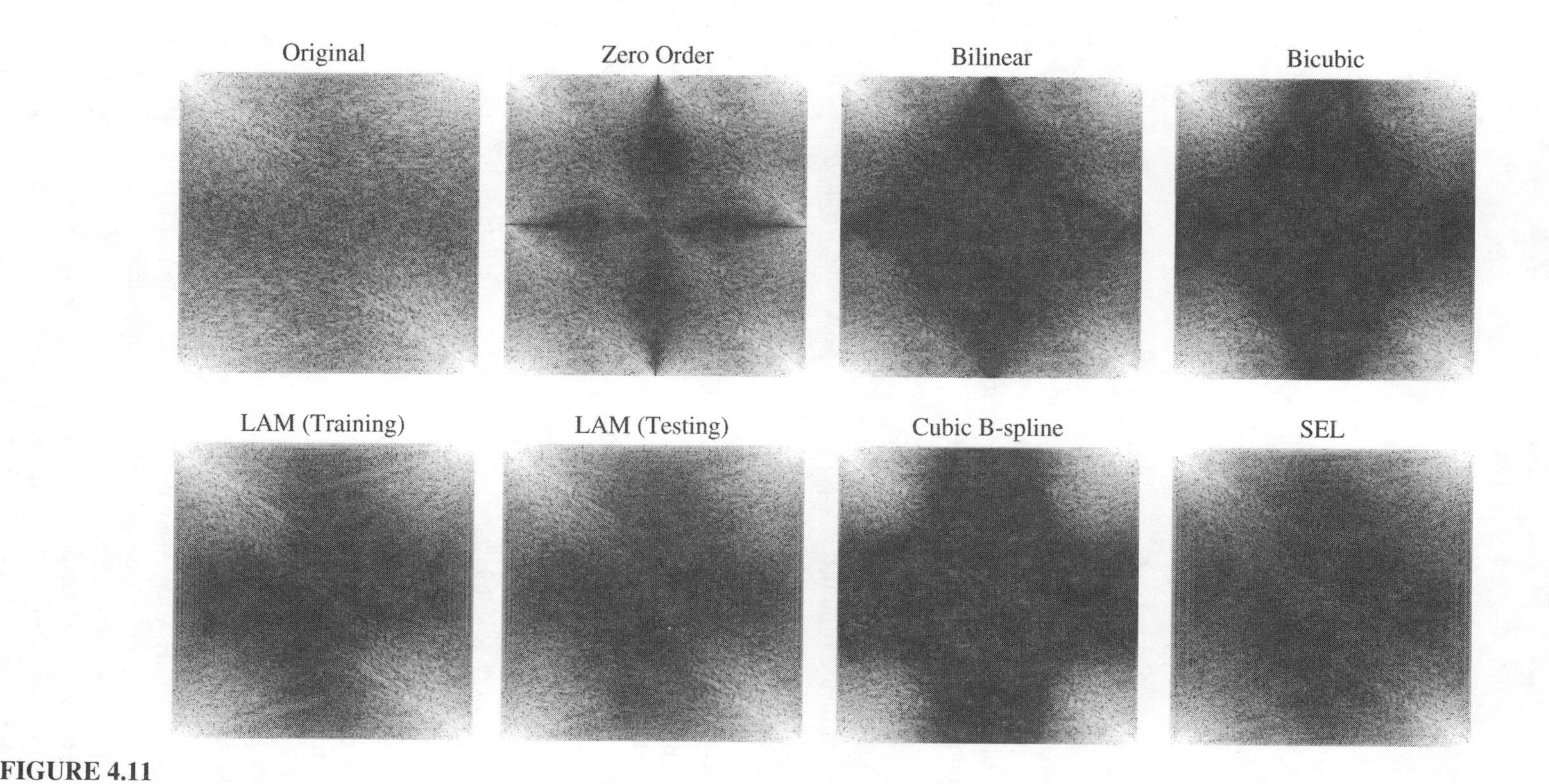

FIGURE 4.11
Magnitude spectra of the reconstructed Lena images in Figure 4.9. The spectra here are for the corresponding full reconstructed images, not just the zoomed sections pictured in Figure 4.9. The spectra $F[n_1, n_2]$ of each image have been enhanced via the log scaling $\frac{2}{\log_{10}(|F[0,0]|)} \log_{10}(|F[n_1, n_2]|)$. The LAM and SEL approaches are better able to produce higher frequency information relative to the other methods compared.

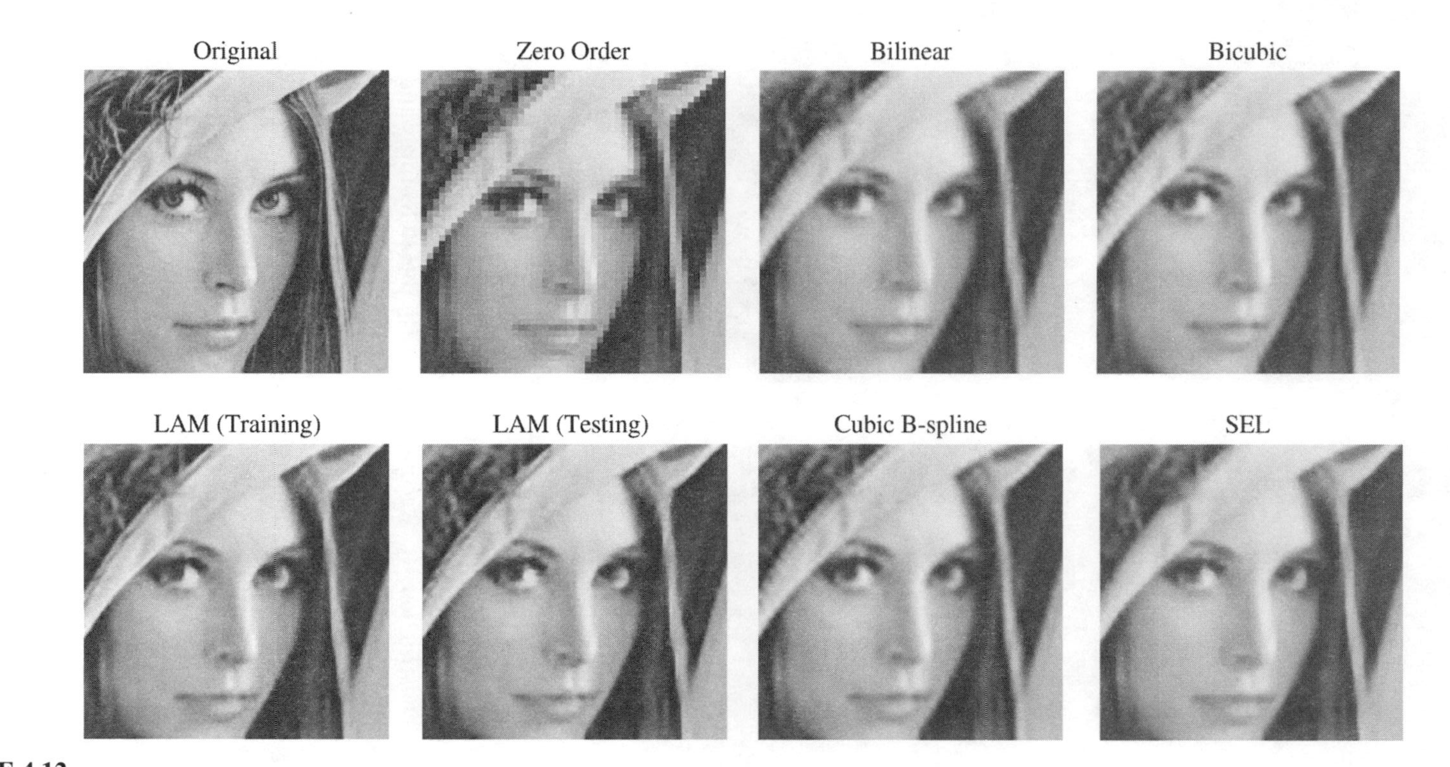

FIGURE 4.12

Example of reconstruction of Lena 128×128 image by a factor of 16 (a factor of 4 along each image axis). The reconstruction was accomplished using two successive stages of reconstruction, each by a factor of 4. The same features and LAMs were used in each stage. The training image was reconstructed using the features and LAMs trained to reconstruct the Lena 256×256 image from the Lena 128×128 image with an ROS of 3×3. The test image was reconstructed using the features and LAMs trained to reconstruct the Peppers 256×256 image from the Peppers 128×128 image with an ROS of 3×3.

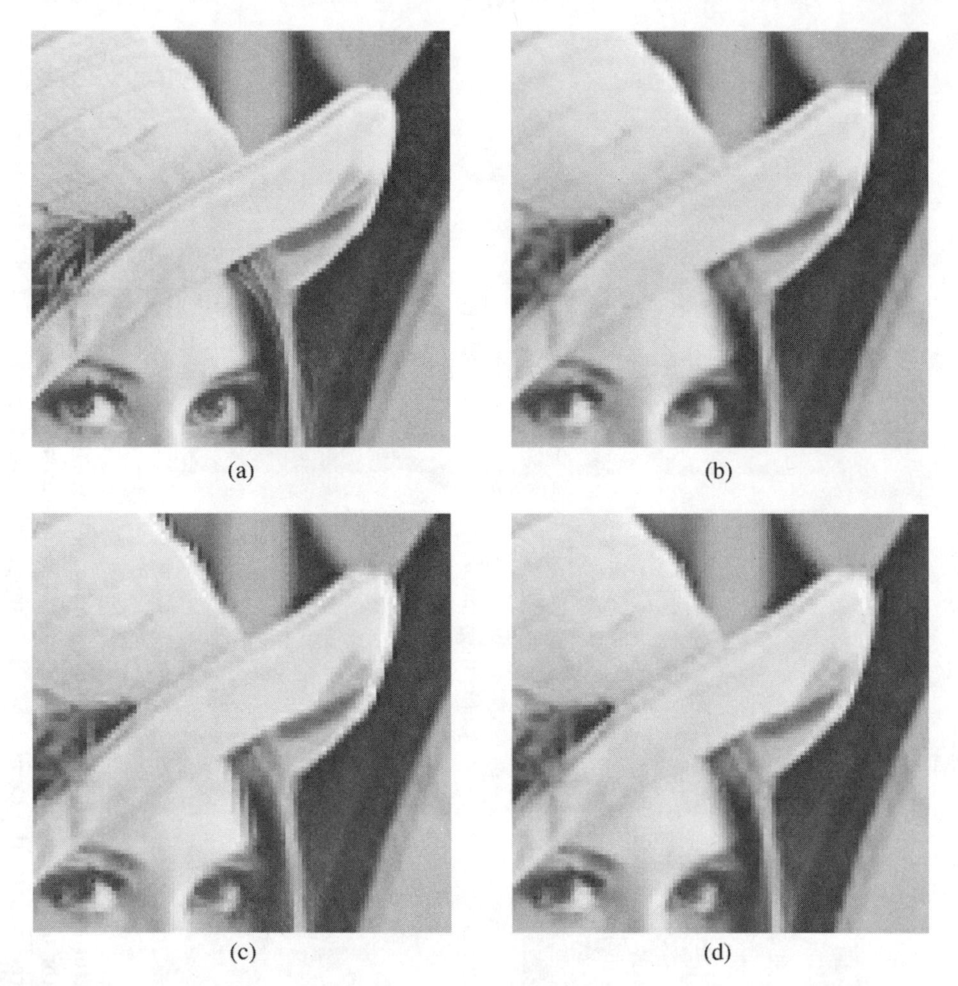

FIGURE 4.13

Compensating for the effects of reconstruction with "inappropriate" bases. The results displayed show a portion of the Lena 256 × 256 image reconstructed from the Lena 128 × 128 image. (a) Original. (b) Reconstructed with the 30 features and LAMs used in reconstructing the Peppers 256 × 256 image from the Peppers 128 × 128 image; this yielded a good reconstruction. (c) Reconstructed with the 30 features and LAMs used in reconstructing the Pentagon 256 × 256 image from the Pentagon 128 × 128 image. The inappropriate reconstruction is most noticeable in the right portion of the forehead and on portions of the hat. (d) Reconstructed with 60 features and LAMs: 30 from the Pentagon image used in (c) and 30 from the Peppers image used in (b). We did not have to retrain our system in establishing an appropriate set of bases. We simply "append" the appropriate features and LAMs of the Peppers image to the existing set from the Pentagon image to reconstruct an adequate image.

transforming each nonoverlapping 8 × 8 subblock of the original 256 × 256 images with only the 3 × 3 low-frequency DCT coefficients — the others were set to zero. The compression results in the loss of information within the borders of each subblock and the introduction of edge artifacts along the borders of the compressed subblocks. Our system was able to substantially suppress the blocking artifacts in the superresolved image with no explicit prior

FIGURE 4.14
Comparing the reconstruction of Lena using LAMs and NLAMs. The overall PSNR performance of the LAM- and NLAM-based results were very similar. Here the Lena 128 $\times$ 128 image is reconstructed by a factor of two in each image axis. The 30 features and LAMs (NLAMs, respectively) used in the reconstruction were those obtained from training to reconstruct the Peppers 256 $\times$ 256 image from the Peppers 128 $\times$ 128 image with an ROS of 3 $\times$ 3.

knowledge of the existence or location of artifacts. This suppression of artifacts is obviously not possible with any of the kernel-based interpolation approaches.

Recall that the scale interdependence between the compressed image and its uncompressed counterpart is significantly reduced relative to using a noncompressed low-resolution image. This reduction in scale interdependence reduces the reliability of the multiple estimates obtained for a high-resolution sample. This limits the extent to which our superresolution can produce a sharp image. However, the averaging of multiple estimates by considering overlapping neighborhoods in the superresolution architecture is responsible for filtering out the effects of blockiness. We can notice from Figures 4.9 and 4.12 that if strong scale interdependencies exist, then our multiple estimates of an image sample are relatively reliable and their averaging does not result in discernible low-pass filtering. This results in crisper images compared to the kernel-based techniques.

4.7 Issues and Notes

Although the preliminary results are very promising, there are many issues requiring further analysis. Noteworthy issues pertaining to the superresolution process herein are:

- The feature vectors and LAMs are established in a manner that is not driven directly by the error rate of superresolution. This is potentially suboptimal. However, because the function defining our input space partition (the clustering stage) is not differentiable, this issue is not easily addressed. We have tested our approach using the hierarchical mixture of experts [25], which trains to minimize the error rate [affine experts (LAMs)

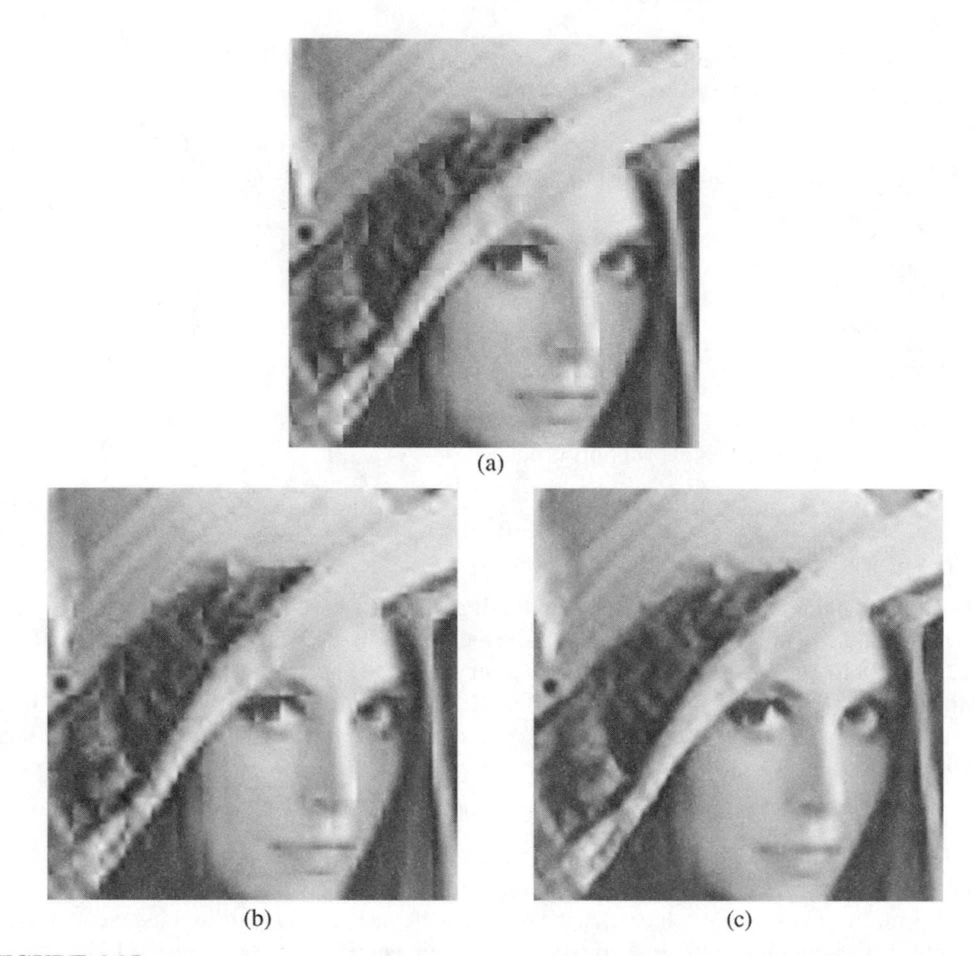

FIGURE 4.15
Results of superresolution on a compressed image with visible blocking artifacts. (a) 128 $\times$ 128 portion of compressed image to magnify. It is shown here as a 256 $\times$ 256 image by using zero-order hold interpolation. (b) Cubic B-spline interpolated result. (c) Superresolution using 15 LAMs and an ROS of 5 $\times$ 5. The architecture was trained to reconstruct the original Peppers 512 $\times$ 512 image with its 256 $\times$ 256 down-sampled and DCT-compressed image.

and affine transformations for the gating structure were used], and our method trained faster and consistently produced higher PSNRs in the reconstructed images [1].

- The topological mapping property of Kohonen's self-organizing map (SOM) was not used for the results presented here. We used the SOM because of its efficient training approach and its tendency for full codebook utilization. We have performed the clustering with the Neural Gas algorithm [26] and have not noticed performance differences [1]. The incorporation of the topological information of the SOM to improve the superresolution is the subject of future research.

- Nonlinear associative memories showed no improvement with respect to LAM performance for the parameters utilized in these experiments. Since the neighborhood sizes and the superresolution factors were small, a linear mapper seems to capture well the redundancy across scales. However, for larger superresolution factors the mapping will tend to be more and more nonlinear, so NLAMs may yield a performance advantage.

- The superresolution approach herein also allows for noninteger (rational) magnification factors. The size of the images associated (as well as what local samples are to be constructed) determines this factor for the feature vectors and LAMs established. Thus different feature and LAM sets must be established for different magnification factors.

- The low-resolution neighborhood size used is a trade-off between the amount of local support considered and how much information is to be constructed. As a rule of thumb, we suggest setting $H_i \geq 2G_i - 1$ but keeping H_i $(i = 1, 2)$ reasonably small. The number of free parameters is determined by the low-resolution support specified by H_i. Note that as G_i increases, there is more missing information to construct; hence, more low-resolution sample support is needed.

- The results presented here use a single image for training — the Peppers. However, multiple images can easily be (and have been [23]) used for training the system parameters in Figure 4.5. Our experiments have revealed that there is much similar local structure among images, which might not be apparent when images are casually viewed.

- The number of input vectors should be much larger than the dimensionality of the input space for proper LAM training. This results in the solution of an overdetermined problem rather than an underdetermined one.

- The system in Figure 4.5 lends itself to the incorporation of new or additional features (and LAMs) and *does not require retraining* of the existing parameters. This allows for quick amending of the bases used for reconstruction.

- The methodology presented for superresolution is general and has been used in the superresolution of synthetic aperture radar (SAR) imagery [1, 27]. Due to the nature of these signals, the processing accounts for local information in the frequency domain — which necessarily implies the learning of nonlocal basis functions upon which our collected samples are projected. This is in contrast to the local processing performed in the spatial domain of the optical images in this chapter.

4.8 Conclusions

A local architecture has been presented for the superresolution of optical images. The procedure was shown to be equivalent to convolution of the image with a family of kernels developed from a training image. The ill-posed superresolution problem was addressed by determining locally the optimal least-squares projections across scales for image neighborhoods of similar character. The similarity between neighborhoods was characterized by their interblock correlation. The key assumption of our approach was that this similarity of neighborhoods in the low-resolution image also held across scales — an assumption that we've noticed experimentally to be very reasonable. The use of LAMs for the local transformation is interesting in that the relation between correlated neighborhoods' structure across scales seems reasonably modeled by an affine mapping. This simplifies the training and eases the need for establishing more complicated nonlinear transformations.

Several interesting traits were demonstrated which favor the use of this architecture. These include: the real-time implementation of the architecture due to its highly parallel nature, the incorporation of new bases into the reconstruction without having to retrain the system, and the inherent ability to regulate errors made in the reconstruction through smoothing. This last trait results from considering overlapping blocks from which multiple sample estimates can be averaged if they are not reliable — this reduces the possibility of introducing artifacts into the image. This bodes well when superresolving images exhibit "blockiness" due to compression. Finally, the need for an analysis that mathematically supports the assumptions we've observed to be reasonable is warranted and has been left for future research.

References

[1] F.M. Candocia, "A Unified Superresolution Approach for Optical and Synthetic Aperture Radar Images," Ph.D. dissertation, University of Florida, Gainesville, 1998.

[2] A.M. Tekalp, *Digital Video Processing,* Ch. 17, Upper Saddle River, NJ: Prentice-Hall, 1995.

[3] A.N. Netravali and B.G. Haskell, *Digital Pictures: Representation, Compression and Standards,* 2nd ed., New York: Plenum Press, 1995.

[4] M. Unser, A. Aldroubi, and M. Eden, "Fast B-Spline Transforms for Continuous Image Representation and Interpolation," *IEEE Trans. Pattern Anal. Mach. Int.,* vol. 13, no. 3, pp. 277–285, 1991.

[5] S.D. Bayrakeri and R.M. Mersereau, "A New Method for Directional Image Interpolation," *Proc. Int. Conf. Acoustics, Speech, Sig. Process.,* vol. 4, pp. 2383–2386, 1995.

[6] K. Jensen and D. Anastassiou, "Subpixel Edge Localization and the Interpolation of Still Images," *IEEE Trans. Image Process.,* vol. 4, no. 3, pp. 285–295, 1995.

[7] A.M. Darwish and M.S. Bedair, "An Adaptive Resampling Algorithm for Image Zooming," *Proc. SPIE,* vol. 2666, pp. 131–144, 1996.

[8] S.A. Martucci, "Image Resizing in the Discrete Cosine Transform Domain," *Proc. Int. Conf. Image Process.,* vol. 2, pp. 244–247, 1995.

[9] E. Shinbori and M. Takagi, "High Quality Image Magnification Applying the Gerchberg-Papoulis Iterative Algorithm with DCT," *Systems and Computers in Japan,* vol. 25, no. 6, pp. 80–90, 1994.

[10] S.G. Chang, Z. Cvetkovic, and M. Vetterli, "Resolution Enhancement of Images Using Wavelet Transform Extrema Extrapolation," *Proc. Int. Conf. Acoustics, Speech, Sig. Process.,* vol. 4, pp. 2379–2382, 1995.

[11] N.B. Karayiannis and A.N. Venetsanopoulos, "Image Interpolation Based on Variational Principles," *Signal Process.,* vol. 25, pp. 259–288, 1991.

[12] R.R. Schultz and R.L. Stevenson, "A Bayesian Approach to Image Expansion for Improved Definition," *IEEE Trans. Image Process.,* vol. 3, no. 3, pp. 233–242, 1994.

[13] G.K. Wallace, "The JPEG Still Image Compression Standard," *Commun. ACM,* vol. 34, no. 4, pp. 30–44, 1991.

[14] R.D. Dony and S. Haykin, "Neural Network Approaches to Image Compression," *Proc. IEEE,* vol. 83, no. 2, pp. 288–303, 1995.

[15] R.D. Dony and S. Haykin, "Optimally Integrated Adaptive Learning," *IEEE Trans. Image Proc.,* vol. 4, no. 10, pp. 1358–1370, 1995.

[16] N. Kambhatla and T. Leen, "Dimension Reduction by Local Principal Component Analysis," *Neural Computation,* vol. 9, pp. 1493–1516, 1997.

[17] R.J. Marks, *Introduction to Shannon Sampling and Interpolation Theory,* New York: Springer-Verlag, 1991.

[18] A.V. Oppenheim and R.W. Schafer, *Discrete-Time Signal Processing,* Englewood Cliffs, NJ: Prentice-Hall, 1989.

[19] P.M. Woodward, *Probability and Information Theory, with Applications to Radar,* 2nd ed., Oxford, NY: Pergamon Press, 1964.

[20] D.L. Ruderman and W. Bialek, "Seeing Beyond the Nyquist Limit," *Neural Computation,* vol. 4, pp. 682–690, 1992.

[21] S. Haykin, *Neural Networks: A Comprehensive Foundation,* New York: Macmillan, 1994.

[22] R.W. Schafer and L.R. Rabiner, "A Digital Signal Processing Approach to Signal Interpolation," *Proc. IEEE,* vol. 61, no. 6, pp. 692–702, 1973.

[23] F.M. Candocia and J.C. Principe, "A Neural Implementation of Interpolation with a Family of Kernels," *Proc. Int. Conf. Neural Networks,* vol. 3, pp. 1506–1510, 1997.

[24] T. Kohonen, "The Self-Organizing Map," *Proc. IEEE,* vol. 78, pp. 1464–1480, 1990.

[25] M.I. Jordan and R.A. Jacobs, "Hierarchical Mixtures of Experts and the EM Algorithm," *Neural Computation,* vol. 6, pp. 181-214, 1994.

[26] T.M. Martinez, S.G. Berkovich, and K.J. Schulten, "'Neural Gas' Network for Vector Quantization and Its Applications to Time-Series Prediction," *IEEE Trans. Neural Networks,* vol. 4, no. 4, pp. 558–569, 1993.

[27] F.M. Candocia and J.C. Principe, "A Method Using Multiple Models to Superresolve SAR Imagery," *Proc. SPIE: Algorithms for Synthetic Aperture Radar Imagery V,* April 1998.

Chapter 5

Image Processing Techniques for Multimedia Processing

N. Herodotou, K.N. Plataniotis, and A.N. Venetsanopoulos

5.1 Introduction

Multimedia data processing refers to a combined processing of multiple data streams of various types. Recent advances in hardware, software, and digital signal processing allow for the integration of different data streams which may include voice, digital video, graphics, and text within a single platform. A simple example may be the simultaneous use of audio, video, and closed-caption data for content-based searching and browsing of multimedia databases or the merging of vector graphics, text, and digital video. This rapid development is the driving force behind the convergence of the computing, telecommunications, broadcast, and entertainment technologies. The field is developing rapidly and emerging multimedia applications, such as intelligent visual search engines, multimedia databases, Internet/mobile audiovisual communication, and desktop video conferencing will all have a profound impact on modern professional life, health care, education, and entertainment.

The full development and consumer acceptance of multimedia will create a host of new products and services including new business opportunities for innovative companies. However, in order for these possibilities to be realized, a number of technological problems must be considered. Some of these include, but are not limited to, the following:

1. Novel methods to process multimedia signals in order to meet quality of service requirements must be developed. In the majority of multimedia applications, the devices used to capture and display information vary considerably. Data acquired by optical, electro-optical, or electronic means are likely to be degraded by the sensing environment. For example, a typical photograph may have excessive film grain noise, suffer from various types of blurring (motion or focus blur), or have unnatural shifts in hue, saturation, or brightness. Noise introduced by the recording media degrades the quality of the resulting images. It is anticipated that the use of digital processing techniques, such as filtering and signal enhancement, will improve the performance of the system.

2. Efficient compression and coding of multimedia signals, in particular, visual signals with an emphasis on negotiable quality of service contracts, must be considered. Rich

data types such as digital images and video signals have enormous storage and bandwidth requirements. Techniques that allow images to be stored and transmitted in more compact formats are of great importance. Multimedia applications are putting higher demands on both the achieved image quality and compression ratios.

Quality is the primary consideration in applications such as DVD drives, interactive HDTV, and digital libraries. Existing techniques achieve compression ratios from 10:1 to 15:1, while maintaining reasonable image quality. However, higher compression ratios can reduce the high cost of storage and transmission and also lead to the advent of new applications (i.e., future display terminals with photo-quality resolution, or the simultaneous broadcast of a larger number of visual programs).

3. Innovative techniques for indexing and searching multimedia data must be developed. Multimedia information is difficult to handle in terms of both its size and the scarcity of tools available for navigation and retrieval. A key problem is the effective representation of this data in an environment in which users from different backgrounds can retrieve and handle information without specialized training. Unlike alphanumeric data, multimedia information does not have any semantic structure. Thus, conventional information management systems cannot be directly used to manage multimedia data. Content-based approaches seem to be a natural choice where audio information along with visual indices of color, shape, and motion are more appropriate descriptions. A set of effective quality measures are also necessary in order to measure the success of different techniques and algorithms.

In each of these areas, a great deal of progress has been made in the past few years, driven in part by the availability of increased computing power and the introduction of new standards for multimedia services. For example, the emergence of the MPEG-7 multimedia standard demands an increased level of intelligence that will allow the efficient processing of raw information; recognition of dominant features; extraction of objects of interest; and the interpretation and interaction of multimedia data. Thus, effective multimedia signal processing techniques can offer promising solutions in all of the aforementioned areas.

This chapter focuses on the intelligent processing of visual information within the research domain of multimedia signal processing using color image processing techniques in conjunction with fuzzy concepts. More specifically, the framework presented includes filtering, segmentation, and meta-data concepts using adaptive techniques for a number of application areas. The organization of the chapter is as follows. Section 5.2 reviews some of the key issues of color imaging with an emphasis on the models needed to support the efficient representation of color information among various devices in a multimedia system. Section 5.3 focuses on the problem of color image filtering for the improvement and enhancement of image quality. New filtering schemes are introduced to meet the challenging high quality of standards necessary in the multimedia era. Color image processing applications demand digital filters that are suitable for complex nonlinear problems, have a reduced complexity, are numerically robust, and are computationally attractive. In Section 5.4, the problem of color image segmentation is addressed for the purposes of audiovisual coding in object-based compression schemes. The segmented regions can be used to form a nonuniform mesh structure which allows for a more accurate motion estimation and compensation in contrast to the conventional block-based methods. Section 5.5 explores the application of color segmentation and fuzzy analysis for the automatic localization of the facial region in an image or video sequence. The extraction process can be utilized for a more efficient coding or for indexing and retrieval in multimedia databases. Lastly, some open technical issues and promising application trends are suggested in the concluding section.

5.2 Color in Multimedia Processing

Color is a key feature used to understand and recollect the contents within a scene. It is found to be a highly reliable attribute for image retrieval because it is generally invariant to translation, rotation, and scale changes [1]. Several color coordinate systems have come into existence for establishing a numerical description of color. The representation of color is based on the classical three-color theory whereby any color can be reproduced by mixing an appropriate set of three primary colors [2]. In this way, the numerical representation of a particular color can be specified by its three component vectors within the 3D color coordinate system. The set of all colors form a vector space called the *color space* or *color model.*

Color information is commonly represented in the widely used RGB (red, green, blue) Cartesian coordinate system. This basis is hardware oriented and is suitable for acquisition or display devices but not particularly applicable in describing the perception of colors. In this coordinate space, the RGB primaries are additive in that the individual contributions of each primary are added to form the overall result. The YIQ (Y is the luminance and I and Q are the chrominance components) and CMYK color models are also hardware-based systems and are utilized for different application purposes. The former is used in color television broadcasting and is a recoding of the RGB components for transmission efficiency and downward compatibility with the earlier monochrome TV standards. The CMYK color space, on the other hand, is important in dealing with printing devices where subtractive primaries are relevant. Colors are specified in this latter model by what is removed or subtracted from white light, rather than by what is added to black.

The need to formulate a simple yet accurate perceptual color distance prompted the development of a perceptually uniform color space [3]. The Commission Internationale de l'Eclairage (CIE) standardized the perceptually uniform L*u*v* and L*a*b* coordinate systems, which are derived by a nonlinear transformation of the RGB values. These color models define a uniform metric space representation of color so that a perceptual color difference is represented by the Euclidean distance. The L*a*b* cube-root color coordinate system was essentially developed to provide a quantitative expression for the Munsell system of color classification [4]. The following transformation equations can be used to convert a set of RGB vector values to the L*a*b* space

$$\begin{bmatrix} X \\ Y \\ Z \end{bmatrix} = \begin{bmatrix} 0.490 & 0.310 & 0.200 \\ 0.177 & 0.813 & 0.011 \\ 0.000 & 0.010 & 0.990 \end{bmatrix} \begin{bmatrix} R \\ G \\ B \end{bmatrix} \tag{5.1}$$

$$L^* = 25 \left(\frac{100Y}{Y_0} \right)^{\frac{1}{3}} - 16 \tag{5.2}$$

$$a^* = 500 \left[\left(\frac{X}{X_0} \right)^{\frac{1}{3}} - \left(\frac{Y}{Y_0} \right)^{\frac{1}{3}} \right] \tag{5.3}$$

$$b^* = 200 \left[\left(\frac{Y}{Y_0} \right)^{\frac{1}{3}} - \left(\frac{Z}{Z_0} \right)^{\frac{1}{3}} \right] \tag{5.4}$$

where the constraint $1 \leq 100Y \leq 100$ must be satisfied, which is indeed the case for most practical purposes [5]. The intermediate values $[XYZ]^T$ are the CIE XYZ tristimulus values, and the $[X_0Y_0Z_0]^T$ triplet is the reference white. In equations (5.2)–(5.4), L^* is correlated with brightness, a^* with the red-green content, and b^* with the yellow-blue content within the image. A similar set of nonlinear expressions can be found for the L*u*v* coordinate system. The computational complexity of the cube-root expressions above, however, may render the perceptually uniform spaces unsuitable for real-time applications. Comprehensive descriptions of the numerous color coordinate systems can be found in [5, 6, 7] along with their appropriate transformation equations.

The HSV (hue, saturation, value) and the TekHVC (hue, value, chroma) color models belong to a group of hue-oriented color coordinate systems that correspond more closely to the human perception of color. These user-oriented color spaces are based on the intuitive appeal of the artist's tint, shade, and tone. The proprietary TekHVC model was developed by Tektronix as a modification of the CIE L*u*v* perceptually uniform color space described earlier. The HSV coordinate system, originally proposed by Smith [8], is cylindrical and is conveniently represented by the hexcone model shown in Figure 5.1.

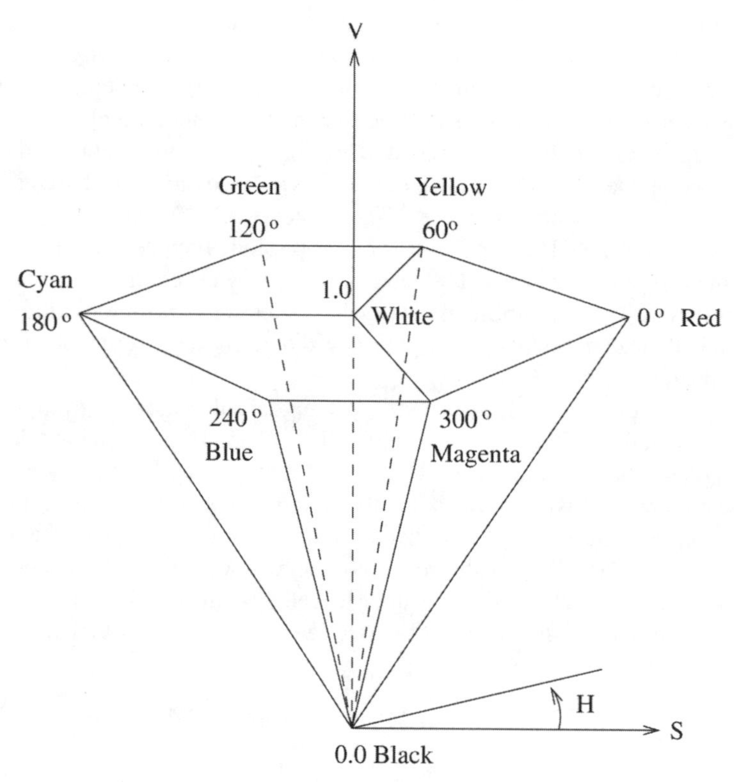

FIGURE 5.1
HSV hexcone color model.

The hue (H) is measured by the angle around the vertical axis and has a range of values between 0 and 360° beginning with red at 0°. It gives us a measure of the spectral composition of a color. The saturation (S) is a ratio that ranges from 0 (i.e., on the V axis), extending radially outward to a maximum value of 1 on the triangular sides of the hexcone. This component refers to the proportion of pure light of the dominant wavelength and indicates how far a color is from a gray of equal brightness. The value (V) also ranges between 0 and 1 and is a measure of the relative brightness. At the origin, V=0 and this point corresponds to black. At this particular value, both H and S are undefined and meaningless. As we traverse upward along the V axis we perceive different shades of gray until the endpoint is reached (where V=1 and S=0), which is considered to be white. At any point along the V axis the saturation component is zero and the hue is undefined. This singularity occurs whenever R=G=B. The set of equations below can be used to transform a point in the RGB coordinate system to the appropriate value in the

HSV space:

$$H_1 = \cos^{-1}\left\{\frac{\frac{1}{2}[(R-G)+(R-B)]}{\sqrt{(R-G)^2+(R-B)(G-B)}}\right\} \tag{5.5}$$

$$H = H_1, \qquad \text{if } B \leq G \tag{5.6}$$

$$H = 360^\circ - H_1, \quad \text{if } B > G \tag{5.7}$$

$$S = \frac{\text{Max (R,G,B)} - \text{Min (R,G,B)}}{\text{Max (R,G,B)}} \tag{5.8}$$

$$V = \frac{\text{Max (R,G,B)}}{255} \tag{5.9}$$

In the expressions above, the Max and Min operators select the maximum and minimum values of the operand, respectively, and R, G, and B range between 0 and 255. A fast algorithm used here to convert the set of RGB values to the HSV color space is provided in [6].

5.3 Color Image Filtering

Filtering of multichannel images has received increased attention due to its importance in processing color images. Numerous filtering techniques have been proposed to date for multichannel image processing. Nonlinear filters applied to images are required to preserve edges and details and remove impulsive and Gaussian noise. On the other hand, vector processing of multichannel images constitutes one of the most effective methods for filtering and edge detection [9, 10]. Nonlinear filters based on order statistics (OS) have been extensively used in the past to smooth and restore images corrupted by noise. Recently, a number of multichannel filters which utilize correlation among multivariate vectors using distance measures have been proposed for image filtering. Among them are the vector median filter (VMF) [11], the vector directional filter (VDF) [12], the fuzzy vector filter (FVF) [13, 14, 15], and different versions of the weighted mean filter [16, 17].

Apart from nonlinear multichannel filters based on order statistics, a number of fuzzy operators have been developed lately for image processing [18, 19]. Local correlation in the data is utilized by applying the fuzzy rules directly on the pixels that lie within the operational window. The output of the fuzzy processing depends on the fuzzy rule and the defuzzification process, which combines the effects of the different rules into an output value. However, there is no optimal way to determine the number and type of fuzzy rules required for the fuzzy image operation. Usually, a large number of rules are necessary and the designer has to compromise between quality and number of rules, because for even a moderate processing window a large number of rules are required [20].

The large number of filters available poses some difficulties to the practitioner, since most of them are designed to perform well in a specific application and their performance deteriorates rapidly under different operation scenarios. Thus, a nonlinear adaptive filter that performs equally well in a wide variety of applications is of great importance. Our goal is to devise a simple, computationally efficient and reliable filter structure, which will deliver acceptable results without making any assumption about signal or noise characteristics. Fuzzy operators are utilized to assist us in this task. Consequently, a second objective is to examine aggregation operators, analyze their properties, and justify their applicability to the design of multichannel filters.

5.3.1 Fuzzy Multichannel Filters

The Filtering Structure

Let $y(x) : Z^l \rightarrow Z^m$ represent a multichannel image and let $W \subset Z^l$ be a window of finite size n (filter length). The noisy image pixels inside the window W are denoted as $\mathbf{x}_j,\ j = 1, 2, \ldots, n$. The general form of the filter class is given as a fuzzy weighted average of the input vectors inside the window W. The uncorrupted multichannel signal is estimated by determining the center of gravity of the cluster of vectors inside the processing window. Therefore, the filter's output at the window center is:

$$\hat{\mathbf{y}} = \sum_{j=1}^{n} \xi_j \mathbf{x}_j \ , \tag{5.10}$$

$$\hat{\mathbf{y}} = \frac{\sum_{j=1}^{n} w_j \mathbf{x}_j}{\sum_{j=1}^{n} w_j} \ . \tag{5.11}$$

where $\xi = \frac{w_j}{\sum_{j=1}^{n} w_j}$.

The weights of the filter are determined adaptively using transformations of a distance criterion at each image position. These weighting coefficients are transformations of the sum of distances between the center of the window (pixel under consideration) and all samples inside the filter window. The transformation has the meaning of membership function with respect to the specific window component. Thus, the fuzzy weights provide the degree to which an input vector contributes to the output, making the filter structure data dependent. From such a viewpoint, a fuzzy clustering approach is introduced to determine the cluster center considering the ambiguity of the multichannel signal. The filter structure proposed here combines distance concepts with data-dependent filters and fuzzy membership functions. Through the normalization procedure, two constraints necessary to ensure that the output is an unbiased estimator are satisfied, namely:

- Each weight is a positive number, $\xi_j \geq 0$.
- The summation of all the weights is equal to one, $\sum_{j=1}^{n} \xi_j = 1$.

In multichannel filtering it is desirable to perform smoothing on all vectors that are from the same region as the vector at the window center. At edges and lines the filter must only smooth pixels at the same side of the edge as the vector at the window center. The proposed algorithm assigns to a given point inside the window some membership function defined on the set of vectors and then uses these membership values to calculate the final output. The fuzzy weights represent the confidence that the vectors under consideration come from the same region. It is therefore reasonable to make the weights proportional to the difference, in terms of a distance measure, between a given vector and its neighbors inside the operational window. In this way, whenever the current pixel is close to an area with high detail, the vectors with the relatively large distance values will be assigned smaller weights and will contribute less to the final filter estimate. Thus, edge or line detection operations prior to filtering can be avoided, with considerable savings in terms of computational effort.

The filtering structure presented can be considered as an *R-ordering*-based multichannel filter because distances inside the operational window are used. However, unlike any *R-ordering*-based filter, the distances are not used to rank the vectors. Rather, they are used to weight the vectors such that negligible weights are assigned to outliers. The structure in (5.10) has the familiar form of an adaptive filter, where the value of the noisy vector at the window center is replaced by a weighted average value of all the points inside the operational window. It can also be viewed as a generalization of existing linear or nonlinear averaging filters. Specifically, if the weighting coefficients are fixed, a linear shift invariant finite impulse response filter is devised. Such a filter smoothes the signal but at the same time blurs signal boundaries (e.g., image edges). In order to alleviate the problem, adaptive methodologies have been introduced, namely filter structures with adaptively determined coefficients [21, 22]. However, a priori knowledge about the signal and the desired response is required. Then the coefficients of the adaptive filter can be optimized for a specific noise distribution with respect to a specific error criterion. However, such information is not available in realistic signal processing applications. Learning schemes based on training signals are iterative processes with heavy computational requirements. Their real-time implementation is usually not feasible. Other adaptive filters are based on different forms of the *Wiener* filter with variable coefficients. These filters, however, are based on the assumption that the input signal and the available desired response are stationary ergodic processes. This is not true for many practical applications. Other approaches use local statistics on part of the signal to adaptively calculate the weights [9, 23, 24]. In these designs noise statistics are often assumed ergodic in order to justify the use of the sample mean and sample noise covariance in the calculations, although it is known that assumption does not always hold. In summary, these filters are more perplexing than useful for engineers faced with real image processing problems. On the contrary, the nonlinear scheme proposed here is simple. It is adaptive but its coefficients are not calculated using complex iterative procedures.

In the last 5 years weighted mean filters with adaptively determined coefficients have been proposed for robust multichannel estimation. In [13], a filter structure which uses a sigmoidal fuzzy transformation to adaptively calculate data-dependent weights was proposed. The measure suggested to calculate distances among the vectors under consideration was the angle between the vectors. In [14] ordered weights based on the same distance criterion as above were used to generate the final filter output. Similarly, a multichannel filter that uses the inverse of the *Euclidean* distance to weight the vectors in the final output was proposed in [16]. This filter extends to multichannel signals the methodology introduced in [17] for univariate input signals. However, weights based on multichannel distance measures can be constructed in more than one way because there is no unique way to define the distance between two multichannel signals. Depending on the distance criteria used and the transformations applied to them, a number of different adaptive filters can be devised. Although it is not clear how to select the appropriate distance-based weight, it is known from experimental results that its form is of paramount importance for the performance of the filter. This work addresses the problem of the selection of the appropriate weight form. Fuzzy connectives are utilized to provide weight transformations that can be considered as a generalization of the transforms already in use.

Before we introduce our methodology to construct a generalized weight function, we will discuss common distance measures and their corresponding fuzzy transformations.

Distances and Fuzzy Weights

The most crucial step in the filter's design is the development of the membership functions. Despite past efforts, a unified form of fuzzy membership functions has not yet been derived [25]. In most cases, it is assumed that somehow they are available. Here, the weights ξ_j in (5.10) are determined using fuzzy membership functions based on selected distance criteria. The fuzzy transformation is not unique. The different fuzzy functions must meet a number of desirable characteristics but mainly are required to have a smooth finite output over the entire input range. Several candidate functions can meet the above specification. According to [25], the most commonly used shapes for membership functions are triangular, trapezoidal, piecewise linear, and Gaussian-like functions. These functions are chosen by the designer arbitrarily, based on experience, problem specifications, and computational constraints imposed by the design. Because the choice of the membership function form is very much problem dependent, the only applicable a priori rule is that designers must confine themselves to those functions that are continuous and monotonic [26].

We devote our attention to fuzzy transformations that are suitable for two important distance measures extensively used for nonlinear filter design.

The objective in the design is to select an appropriate fuzzy transformation, so that the pixel with the minimum distance will be assigned the maximum weight.

The first criterion used to judge similarity (distance) between two vectors is the so-called *vector angle criterion*. This criterion considers the angle between two vectors as their distance. The distance associated with the noisy vector x_i inside the processing window of length n can be defined as:

$$a_i = \sum_{j=1}^{n} A\left(\mathbf{x}_i, \mathbf{x}_j\right) \tag{5.12}$$

with

$$A\left(\mathbf{x}_i, \mathbf{x}_j\right) = \cos^{-1}\left(\frac{\mathbf{x}_i^T \mathbf{x}_j}{|\mathbf{x}_i||\mathbf{x}_j|}\right) \tag{5.13}$$

This similarity measure was introduced to measure distances between color vectors [12]. Because in the RGB color space, color is defined as relative values in the trichromatic channel and not as a triplet of absolute intensity values, it was argued in [12] that the distance measure must respond to relative intensity differences (chromaticity) and not absolute intensity differences (luminance). Thus, the orientation difference between two color vectors was selected as their distance measure, because it correlates well with their spectral ratio difference.

A number of different shapes can be used to generate a membership function based on the vector angle criterion. However, in the neural network and fuzzy systems literature [25], a sigmoidal transformation is usually associated with inner product type distances. Therefore, if the sum of angles is selected as the similarity measure, a sigmoidal membership function should be utilized.

The fuzzy weight w_i has the following form:

$$w_{1i} = \frac{\beta}{(1 + \exp(a_i))^r} \tag{5.14}$$

where β and r are parameters to be determined. The value of r is used to adjust the weighting effect of the membership function, and β is a weight scale threshold. Since, by definition, the vector angle distance criterion delivers a positive number in the interval $[0, n\pi]$ [12], the output of the fuzzy transformation introduced above produces a membership value in the interval $[\frac{\beta}{(1+\exp((n\pi))^r}, \frac{\beta}{2^r}]$. However, even for a moderate size window, such as a 3×3 or 5×5 window, the lower limit of the above interval should safely be considered zero. As an example, for a modest 3×3 window and with $r = 1$ and $\beta = 2$, the corresponding interval is $[1.4 \times 10^{-12}, 1]$ and for a 5×5 window the interval becomes $[1.5 \times 10^{-35}, 1]$. Therefore, we can consider the above membership function as having values in the interval [0, 1]. It can easily be seen through simple calculations that the above transformation satisfies the design objectives.

The generalized *Minkowski* norm (L_p metric) can also be used to measure the distances between two multichannel vectors [14]. The L_p is defined as:

$$d_p(i, j) = \left(\sum_{k=1}^{m} \left|\left(x_i^k - x_j^k\right)\right|^p\right)^{\frac{1}{p}} \tag{5.15}$$

where m is the dimension of the vector x_i. Using this norm the scalar distance measure

$$d_p(i) = \sum_{j=1}^{n} d_p(i, j) \tag{5.16}$$

is associated with the noisy vector x_i inside a filter window of length n. For such a distance an appropriate membership function is the exponential (Gaussian-like) form:

$$w_{2i} = \exp\left[-\frac{d_p(i)^r}{\beta}\right], \tag{5.17}$$

where r is a positive constant and β is a distance threshold. The actual values of the parameters vary with the application. The above parameters correspond to the denominational and exponential fuzzy generators controlling the amount of *fuzziness* in the fuzzy weight. It is obvious that since the distance measure is always a positive number, the output of this fuzzy membership function lies in the interval [0, 1]. The fuzzy transformation is such that the higher the distance value, the lower the fuzzy weight becomes. It can easily be seen that the membership function is one (maximum value) when the distance value is zero and becomes zero (minimum value) when the distance value is infinite.

5.3.2 The Membership Functions

Both membership functions can be used to derive the fuzzy weights introduced in the filter structure of (5.10). However, the shape and the parameters of the functions were chosen intuitively based on our experience and the distance criterion selected. More recently, membership functions have been designed using optimization procedures [25]. The general idea is to tune the shape and the parameters of the membership function using a training signal. The form of the fuzzy membership function is usually fixed ahead of time. Then a set of available training pairs (input, membership values) is used to tune the parameters of the assumed membership function. The most commonly used procedure exploits the mean squared error (MSE) criterion. In addition, since most of the used shapes are nonlinear, iterative schemes (e.g., back-propagation) are used in the calculations [26]. However, in an application such as image processing, in order for the membership function to be tuned adaptively, the original image or an image with properties similar to those of the original must be available. Unfortunately, this is seldom the case in real-time image processing applications, where the uncorrupted original image or knowledge about the noise characteristics is not available. Therefore, alternative ways to obtain the "best" fuzzy transformation must be explored.

To this end, an approach is introduced here in which instead of "training" one membership function, a bank of candidate membership functions are determined in parallel using different distance measures. Then, a generalized nonlinear operator is used to determine the final optimized membership function, which is employed to calculate the fuzzy weights. This method of generating the overall function is closely related to the essence of computations with fuzzy logic. By choosing the appropriate operator, the generalized membership function can meet any specific objective requested by the design. As an example, if a minimum operator is selected, the designer pays more attention to the objectives that are satisfied poorly by the elemental functions and selects the overall value based on the worst of the properties. On the contrary, when using a maximum operator the positive properties of the alternative membership functions are emphasized. Finally, a mean-like operator provides a trade-off among different, possibly incompatible, objectives.

Using the previous setting, the problem of determining the overall function is transformed into a decision-making problem where the designer has to choose among a set of alternatives after considering several criteria. We discuss here only discrete solution spaces since distinct membership function alternatives are available. As in any decision problem, where satisfaction of an objective is required, two steps can be defined, namely, (1) the determination of the efficient solution, and (2) the determination of an optimal compromise solution.

The compromise solution can be defined as the one preferred by the designer to all other solutions, taking into consideration the objective and all the constraints imposed by the design. The designer can specify the nonlinear operator used to combine elemental functions in advance and use this operator to single out the final value from the set of available different solutions. This is the approach followed here. An aggregator (fuzzy connective), whose shape is defined a priori, will be used to combine the different elemental functions in order to produce the final weights at each position.

In fuzzy decision making, connectives or aggregators are defined as mappings from $[0, 1]^{\phi} \rightarrow [0, 1]$ and are often requested to be monotonic with respect to each argument. The subclass of aggregation operators which are continuous, neutral, and monotonic is called the class of *CNM* operators [27]. An averaging operator is a member of the class of compensative CNM operators but different from *min* or *max* operators. Averaging operators M can be characterized under several natural properties, such as monotonicity and neutrality [28]. It is widely accepted that an averaging operator verifies the following properties:

$M : [0, 1]^{\phi} \rightarrow [0, 1]$

(i) Idempotency: $\forall \alpha, M(\alpha, \alpha, \ldots, \alpha) = \alpha$

(ii) Neutrality: the order of arguments is unimportant

(iii) M is nondecreasing in each place

The above implies that the averaging operator lies between *min* and *max*. However, aggregation operators are in general nonassociative or decomposable since associativity may conflict with idempotence [29]. An example of averaging operators is the arithmetic mean, the geometric mean, the harmonic mean, or the root-power mean. The problem of choosing operators for logical combination of criteria is a difficult one. Experiments in decision making indicate that aggregation among criteria is neither a conjunctive or disjunctive type of operation. Thus, *compensatory connectives* which mix both conjunctive and disjunctive behavior were introduced in [30].

In this work a *compensative operator,* first introduced in [31], is utilized to generate the final membership function. Following the results in [31], the operator is defined as the weighted mean of a (*logical AND*) and a (*logical OR*) operator:

$$A \bigodot_{\gamma} B = \left(A \bigcap B\right)^{1-\gamma} \cdot \left(A \bigcup B\right)^{\gamma} \tag{5.18}$$

where A, B are sets defined on the same space and represented by their membership functions. Different *t-norms* and *t-conorms* can be used to express a conjunctive or a disjunctive attitude. If the product of membership functions is utilized to determine intersection (*logical AND*) and the possibilistic sum for union (*logical OR*), the form of the operator for several sets is as follows [30]:

$$w_{ci} = \prod_{j=1}^{\phi} w_{ji}^{(1-\gamma)} \left(1 - \prod_{j=1}^{\phi} (1 - w_{ji})\right)^{\gamma} \tag{5.19}$$

where w_{ci} is the overall membership function for the sample at pixel i, w_{ji} is the jth elemental membership value, and $\gamma \in [0, 1]$. The weighting parameter γ is interpreted as the *grade of compensation,* taking values in the range of [0, 1] [31]. In this work a constant value of 0.5 is used for γ.

The product and the possibilistic sum are not the only operators that can be used in (5.18). A simple and useful *t-norm* function is the *min* operator. In this chapter, we also use this *t-norm* to represent intersection. Subsequently, the *max* operator is the corresponding *t-conorm* [25]. In such a case, the compensative operator of (5.18) has the following form:

$$w_{ci} = \left(\min_{j=1}^{\phi} w_{ji}\right)^{(1-\gamma)} \left(\max_{j=1}^{\phi} w_{ji}\right)^{\gamma} \tag{5.20}$$

The form of the compensative operator is not unique. A number of other mathematical models can be used to represent the *AND* aggregation. An alternative operator, which combines the averaging properties of the arithmetic mean (member of the averaging operator class) with a *logical AND* operator (conjunctive operator) was proposed also in [30].

$$w_{ci} = \gamma \min_{j=1}^{\phi} w_{ji} + (1-\gamma)\left(\phi^{-1}\sum_{j=1}^{m} w_{ji}\right) \tag{5.21}$$

where w_{ci} is the overall membership function for the sample at pixel i and the parameter $\gamma \in [0, 1]$ is interpreted as the grade of compensation. In this equation the *min t-norm* stands for the *logical AND*. Alternatively, the product of membership functions can be used instead of the *min* operator in the above equation. The arithmetic mean is used to prevent higher elemental weights with extreme values from dominating the final outcome. The operator is computationally simple and possesses a number of desirable characteristics.

Compensatory operators are intuitively appealing but are based on ad hoc definitions and properties, such as monotonicity, neutrality, or idempotency, that cannot always be verified. However, despite these drawbacks, these methods are still appealing in that they can express compensatory effects or interactions between design objectives. For this reason, we utilize them in the next subsection to construct the overall fuzzy weights in our adaptive filter designs.

5.3.3 A Combined Fuzzy Directional and Fuzzy Median Filter

In our adaptive filter, we intend to assign higher weights to those samples that are more centrally located (inside the filter window). However, as we have seen in Section 5.3.2 for multichannel data, the concept of vector ordering has more than one interpretation and the vector median inside the processing window can be defined in more than one way. Therefore, the determination of the most centrally positioned vector heavily depends on the distance measure used. Each distance measure described in Section 5.3.2 selects a different most centrally located vector. Since multichannel ordering has no natural basis, it is anticipated that we should expect better filtering results combining ranking criteria which utilize different distances.

Let us assume that the adaptive multichannel filter of (5.10) must be used and the weights $w_i \forall i$ inside the operational window must be assigned. Consider the design objective: *The x_i is centrally located as measured with the angle criterion and x_i is centrally located using the Minkowski distance.* We intend to establish a fuzzy membership function for this statement. The first step is to realize that this statement is a composition between two design objectives, which can be realized using elemental membership functions, such as the ones discussed in the previous section. Then, utilizing the compensative operator, the overall function can be obtained. At this point, we must clarify the effect of the compensatory operator in our filter. In the above design objective, the same degree of attractiveness can be reached by having a less centrally located vector according to the Euclidean distance, but more central using the angle criterion and vice versa. That is, the higher value of "with the angle criterion" compensates for the lower value of membership in "using the Minkowski distance."

For the specific case of two elemental membership functions and equal exponents, the compensative operator defined in (5.18) has the form of a weighted membership product. Thus, depending on the *t-norm* or *t-conorm* used, the overall fuzzy function can be defined as:

$$w_{ci}^{a} = (w_{1i}w_{2i})^{0.5} \tag{5.22}$$

where w_{ci}^{a} is the overall membership function for the sample at pixel i, or

$$w_{ci}^{a} = (w_{1i}w_{2i})^{0.5}\left(1 - \left(\left(1 - w_{1j}\right)\left(1 - w_{2j}\right)\right)\right)^{0.5} \tag{5.23}$$

It can easily be seen from (5.21) that using the *min* and *max* operators and for equal powers the operator in (5.18) actually has the form of the geometric mean, a member of the averaging operators family.

The alternative operator introduced in (5.21) has, for this specific case, the following form:

$$w_{ci}^{b} = 0.5\min_{j=1}^{2} w_{ji} + 0.25\sum_{j=1}^{2} w_{ji} \tag{5.24}$$

or

$$w_{ci}^{b} = 0.5(w_{1i} * w_{2i}) + 0.25\sum_{j=1}^{2} w_{ji} \tag{5.25}$$

In general, additional weighting factors which will absorb possible scale differences in the definition of the elemental membership functions must be used. However, since the two elemental functions used here take values in the interval [0, 1], no such weighting factor is required.

The averaging operator defined in (5.22), and the two compensative operators defined in (5.23) and (5.24), can be used to define the fuzzy weights in (5.10) provided that the elemental fuzzy transforms of (5.14) and (5.17) have been used to construct the elemental weights. However, in order for our results to be meaningful, the nonlinear operator applied must satisfy some properties that will guarantee that its application will not alter in any manner the elemental decisions about the weights. In the literature, there are a number of properties that all the aggregation or compensative operators must satisfy. In this subsection we will examine whether the operators we intend to use to calculate the adaptive weights satisfy these properties [28].

The requisite properties are listed below:

1. *Convexity:*

    - The mean operator in (5.22) is convex.
      Proof:

$$w_{ci}{}^{a} = \left(\min_{k=1,2} w_{ki} \max_{k=1,2} w_{ki}\right)^{0.5} \tag{5.26}$$

$$\min_{k} w_{ki} \leq w_{ci}{}^{a} \leq \max_{k} w_{ki} \tag{5.27}$$

- The operator introduced in (5.24) is convex.
Proof:

$$w^b_{cj} = 0.75\min_k w_{ki} + 0.25\max_k w_{ki} \tag{5.28}$$

Then, we can conclude that:

$$\min_k w_{ki} \leq w^b_{ci} \leq \max_k w_{ki} \tag{5.29}$$

2. *Monotonicity:* The property of monotonicity guarantees that the stronger piece of evidence (larger elemental membership value) generates a stronger support in the final membership function.

- The operator introduced in (5.22) is monotonous.
Proof:

$$w^{a*}_{ci} \geq w^a_{ci} \tag{5.30}$$

where $w^{a*}_{ci} = (w_{1i}w_{ki})^{0.5}$, $w^a_{ci} = (w_{1i}w_{ji})^{0.5}$, and $\forall w_{ki} \geq w_{ji}$.

- The operator introduced in (5.24) is monotonous.
Proof:
For w_{1i} and $\forall w_{ki} \geq w_{ji}$, $\min(w_{1i}, w_{ki}) \geq \min(w_{1i}, w_{ji})$, so using (5.24),

$$w^{b*}_{ci} \geq w^b_{ci} \tag{5.31}$$

3. *Idempotence:* This property guarantees that the outcome of the overall function generates the same value with each elemental value if all functions report the same result.

- The operator introduced in (5.22) is idempotent.
Proof:

$$w^a_{ci} = (ww)^{0.5} = w \tag{5.32}$$

- The operator introduced in (5.24) is idempotent.
Proof:

$$w^b_{ci} = 0.5w + 0.25(w + w) = w \tag{5.33}$$

It can easily be seen from (5.23) that this operator is not idempotent. However, the operator is symmetric and satisfies the monotonicity requirement, namely,

$$w^{a*}_{ci} \geq w^a_{ci} \tag{5.34}$$

where

$$w^{a*}_{ci} = (w_{1i}w_{ki})^{0.5}\,(1 - ((1 - w_{1i})\,(1 - w_{ki})))^{0.5} \tag{5.35}$$

and

$$w^a_{ci} = \left(w_{1i}w_{ji}\right)^{0.5}\left(1 - ((1 - w_{1i})\left(1 - w_{ji}\right))\right)^{0.5} \tag{5.36}$$

If $\forall w_{ki} \geq w_{ji}$, then

$$(1 - w_{ki}) \leq \left(1 - w_{ji}\right) \tag{5.37}$$

$$(1 - ((1 - w_{1i})\,(1 - w_{ki}))) \geq \left(1 - \left((1 - w_{1i})\left(1 - w_{ji}\right)\right)\right) \tag{5.38}$$

Combining (5.34)–(5.38), we can conclude that the operator defined in (5.23) satisfies the monotonicity requirement.

In addition, it is not hard to see that the operators introduced here are symmetric (neutral). This property guarantees that the order of presentation for the elemental functions does not affect the overall membership value.

In summary, we have proven that the compensatory operators we intend to use for the fuzzy weights calculations in (5.10) correspond to an aggregation class which satisfies a number of natural properties, such as neutrality and monotonicity.

The decision to utilize a fuzzy aggregator to construct the overall weight is not arbitrary. On the contrary, it is anticipated that the operator will help us to accomplish the design objective. The introduction of a combination of different distances in the weight determination procedure is expected to enhance the filter performance. Each one of the above defined operators can generate a final membership function, which is sensitive to relative changes in the elemental membership values and helps us to accomplish our objective. A fuzzy filter, which utilizes this form of membership function for its fuzzy weights, constitutes a fuzzy generalization of a combined VMF and VDF.

It must be emphasized that through this design the problem of determining the appropriate membership function is transformed into the problem of combining a collection of possible functions. This constitutes a problem of considerably reduced complexity, since admissible membership functions may be known from physical considerations or design specifications. The proposed adaptive design is a *scalable* one. The designer controls the complexity of the final membership function by determining the number and form of the individual membership functions. Depending on the problem specification and the computational constraints, the designer can select the appropriate number of elemental functions to be used in the final weighting function. The shape of the membership function (e.g., sigmoidal or exponential) is not the only parameter that differentiates between possible elemental fuzzy transformations. The designer may decide to use the same form for the elemental functions and assign different parameter values to them (e.g., different r or β). Then, an overall membership function can be devised using an appropriate combination of the individual functions. The computational efficiency of the proposed filter depends not only on the form of the membership function selected or the operator used for aggregation, but on both of them.

This parallel, adaptive on-line determination of the membership function allows for a fast design without time-consuming iterative processes. The filter's output is calculated in one pass without any recursion. Thus, our filter does not depend on a "good" initial estimate. On the contrary, it is well known that iterative learning filters starting from certain initial value are likely to be trapped in local optima with profound consequences to the filter's performance. Furthermore, in our design there is no requirement for the training signal needed to assist learning in iterative adaptive designs. The final fuzzy membership function is determined without any suboptimal local noise or signal statistic evaluation since such approaches usually lead to biased solutions. Thus, our adaptive multichannel filters can be used in real-time image applications, in contrast to other "trainable" multichannel filters, which are based on unrealistic assumptions about the availability of training sequences.

5.3.4 Application to Color Images

The performance of the new filters introduced here is evaluated below (see also Table 5.1). The evaluation is carried out using a color test image and their performance is measured against popular vector processing filters, such as the VMF, the basic vector directional filter (BVDF), the generalized vector directional filter (GVDF), the arithmetic mean filter (AMF), and the hybrid filters of [33]. Since our objective is not to develop all the different adaptive filters based on fuzzy transformations of the distance but to demonstrate the improvement introduced in terms of performance using a fuzzy aggregator or compensator, we construct five different filters based on the distance criteria and elemental transforms described previously. The following notation is used for convenience.

Table 5.1 Filters Compared

Notation	Filter	Ref.
BVDF	Basic vector directional filter	[12, 32]
GVDF	Generalized vector directional filter	[12, 32]
AMF	Arithmetic mean filter	[32]
VMF	Vector median filter	[11]
HF	Hybrid directional filter	[32, 33]
AHF	Adaptive hybrid directional filter	[32, 33]
FVF1	Fuzzy vector directional filter with weights determined through $w_{1j} = \frac{2}{1+\exp(a_j)}$	[15, 32]
FVF2	Fuzzy vector directional filter with weights determined through $w_{2j} = \exp[-d_p(j)^{0.5}]$	[15]
FVF3	Fuzzy vector directional filter with weights determined through w_{1j}, w_{2j}, (5.13)	[15]
FVF4	Fuzzy vector directional filter with weights determined through w_{1j}, w_{2j}, (5.13)	[15]
FVF5	Fuzzy vector directional filter with weights determined through w_{1j}, w_{2j}, (5.15)	[15]

The filters are applied to the widely used 512×480 RGB color image Lena. The test image has been contaminated using various noise source models in order to assess the performance of the filters under different scenarios. The test image is contaminated with correlated Gaussian noise, and a percentage of the image samples are replaced by outliers, which have very high or low signal values with equal probability (see Table 5.2).

Table 5.2 Noise Distributions

Number	Noise Model
1	Gaussian ($\sigma = 30$)
2	Impulsive (4%)
3	Gaussian ($\sigma = 15$), impulsive (2%)
4	Gaussian ($\sigma = 30$), impulsive (4%)

The normalized mean squared error (NMSE) has been used as a quantitative measure for evaluation purposes. It is computed as:

$$\text{NMSE} = \frac{\sum_{i=0}^{N1} \sum_{j=0}^{N2} \|(y(i, j) - \hat{y}(i, j)\|^2}{\sum_{i=0}^{N1} \sum_{j=0}^{N2} \|(y(i, j)\|^2}, \tag{5.39}$$

where $N1$ and $N2$ are the image dimensions, and $y(i, j)$ and $\hat{y}(i, j)$ denote the original image vector and the estimation at pixel (i, j), respectively. Table 5.3 summarizes the results obtained for the Lena test image for a 3×3 processing window. The results obtained using a 5×5 filter window are given in Table 5.4. The GVDF uses the appropriate gray-scale operator at the *magnitude processing* module to obtain the best possible result. It must be emphasized that these modules are noise dependent. The designer must know a priori the actual noise characteristics. This is hardly the case in a real-time image processing situation. In contrast, the FVF family does not require any information about noise characteristics. However, despite the fact that the GVDF utilizes more information, we select the best filter from the GVDF family for the comparisons below [12].

Table 5.3 NMSE ($\times 10^{-2}$) for the Lena Image (3 $\times$ 3 Window)

Filter	Noise Model			
	1	2	3	4
None	4.2083	5.1694	3.6600	9.0724
BVDF	2.8962	0.3448	0.4630	1.1354
GVDF	1.4600	0.3000	0.6334	1.9820
AMF	0.6963	0.8186	0.6160	1.298
HF	1.3192	0.2182	0.5158	1.6912
AHF	1.0585	0.2017	0.4636	1.4355
FVF1	0.735	0.2481	0.401	1.039
FVF2	0.9812	0.1663*	0.3826	1.1744
FVF3	0.6940*	0.2161	0.3310	0.9130*
FVF4	0.7335	0.1908	0.3234*	0.9445
FVF5	0.7201	0.244	0.3511	0.9903

* Best filter performance in the corresponding row.

Table 5.4 NMSE ($\times 10^{-2}$) for the RGB Lena Image (5$\times$5 Window)

Filter	Noise Model			
	1	2	3	4
None	4.2083	5.1694	3.6600	9.0724
BVDF	2.800	0.7318	0.6850	1.3557
GVDF	1.0800	0.5400	0.4590	1.1044
AMF	0.5977*	0.6656	0.572	0.8896
HF	0.7700	0.3841	0.4890	1.1417
AHF	0.6762	0.3772	0.4367	0.7528
FVF1	0.7549	0.3087	0.4076	0.9550
FVF2	0.6718	0.3040	0.4031	0.7491
FVF3	0.6178	0.3042	0.3813*	0.7224
FVF4	0.6584	0.2984*	0.3817	0.7444
FVF5	0.6239	0.3069	0.387	0.7074*

* Best filter performance in the corresponding row.

From the results listed in the tables, it can be easily seen that our adaptive design with the generalized membership function provides consistently good results in every type of noise situation. The different fuzzy filters attenuate both impulsive and correlated Gaussian noise with or without outliers present in the test image. It must be noted that if no assumption about the noise characteristics is made, the fuzzy filter with the generalized membership weights provides results better than the results obtained by any other filter under consideration. Results also indicate that our fuzzy techniques are less sensitive to the window length, compared to the GVDF or the VMF. As an example, it can be seen that our adaptive fuzzy filters do not suffer from VMF's inefficiency in a nonimpulsive noise scenario and small filtering window.

Finally, considering the number of computations, the computationally intensive part of the fuzzy algorithm is the distance calculation part. However, this step is common in all multichannel algorithms considered here. More than that, the different elemental membership functions can be calculated in parallel, thus reducing the execution time and making our filters suitable for real-time implementation with digital signal processors. The adaptation procedure used to evaluate the generalized membership function does not introduce any additional computational cost. To the best of our knowledge, the adaptation mechanism introduced in this work is the only one capable of providing this form of parallel processing capability.

In conclusion, our adaptive design is simple, scalable, does not increase the numerical complexity of the fuzzy algorithm, and delivers excellent results for complicated multichannel signals, such as real color images. Moreover, as can easily be seen from the attached images, the new filters preserve the chromaticity component, which is very important in the visual perception of color images.

5.4 Color Image Segmentation

Image compression is essential in numerous multimedia applications due to the enormous bandwidth and storage requirements, as mentioned previously. Conventional coding standards such as H.261 and MPEG-1 and -2 fail to adequately model object motion within the scene and also suffer from the familiar blocking artifacts. Furthermore, these schemes deal with video exclusively at the frame level, thereby preventing the manipulation of individual objects within the bitstream. Recently, however, greater attention has been paid to a newer generation of coding schemes that are *object based* [34, 35]. These methods rely on the techniques of image analysis and computer graphics to represent the image signals using their structural features such as contours and regions. In this latter approach, the input video sequence must first be segmented into an appropriate set of arbitrarily shaped regions [36]. Thus, the success of any object-based method depends largely on this segmentation process. This not only improves the coding efficiency, but it can also support various content-based functionalities.

In this section, we focus our attention on the color segmentation problem. A fast color segmentation algorithm is presented that employs the perceptual HSV color space model to partition an image into arbitrarily shaped regions. This is carried out by employing a recursive 1D histogram thresholding procedure. The proposed technique is robust, suitable for real-time implementation (i.e., due to the 1D histogram approach), and very intuitive in describing the color/intensity content of a region.

The hue component of the HSV color model can be effectively employed to segment the color content within a scene. However, the hue attribute is ineffective and unreliable when the saturation or value components are low. Therefore, we partition the image into the following three primary regions so that an appropriate segmentation scheme can be applied within each region: (1) an achromatic, (2) a chromatic, and (3) a transitional area. The achromatic regions are characterized by low values of saturation and value and consist of the black, white, and gray areas within the scene. Threshold values of $S \leq 10\%$ and $V \leq 20\%$ were used to define the achromatic sector of the HSV space. A similar saturation threshold was selected in [37] to partition the achromatic sector of the HVC space without enforcing an intensity restriction. The intensity information, however, is important [38], and erroneous results may be obtained if this latter restriction is not imposed [39]. The value component (i.e., the brightness) is used to segment the achromatic regions of the image. The chromatic region (region 2), on the other hand, is described by high values of saturation and value where the hue has great discriminating power and can be effectively used to segment the chromatic parts of the image. Threshold values of S $\geq$ 20% and V > 20% were selected in defining this second region. Finally, the third region separates the chromatic and achromatic areas and is referred to as the transition region. Thresholds of 10% < S < 20% and V > 20% were chosen for this latter region. Slices of this solid correspond to annular rings in the HSV model. The hue component in this transition region is once again unreliable. Pixel values within this region have very little chroma and, thus, are better characterized by the value component. This partitioning of the HSV hexcone model into the three primary regions is summarized in Table 5.5. A simple two-region model has also been proposed for segmentation purposes in the similar HSI space [40]. In this scheme, the original image is split into only two regions (chromatic and achromatic) by using the average value of the peaks found in the saturation histogram as a threshold value. There are two problems associated with this approach: (1) threshold values may be over- or underestimated due to the averaging process, which may result in an incorrect partition of the chromatic and achromatic regions, and (2) no intensity information is taken into account, which may lead to erroneous results due to the low intensity value pixels.

Table 5.5 Partitioning of the HSV Hexcone Model

Region	Bounding Thresholds		Segmentation Cue
Achromatic	S $\leq$ 10%	V $\leq$ 20%	Value
Transitional	10% <S < 20%	V > 20%	Value
Chromatic	S $\geq$ 20%	V > 20%	Hue

Once the image has been partitioned into the three primary regions above, then a histogram thresholding procedure is carried out within each region using the appropriate cue.

5.4.1 Histogram Thresholding

Segmentation within the achromatic region is performed by using the histogram of the value component. The value histogram is first formed and smoothened by the scale–space filtering approach [41]. The largest peak is then selected and the valleys are subsequently found on either side of this peak. Pixel values within the two valley points are classified as a uniform area. A set of binary operations which include median filtering and region removal are used to remove isolated pixels and small regions (i.e., less than a predefined threshold), respectively. This process is repeated recursively until all the pixels within the achromatic region are segmented into significant areas of uniformity (i.e., no more regions can be further extracted from the histogram after the small region removal step).

The procedure just described is also carried out using the value histogram of the pixels within the transitional region. Areas within this region appear to have some chroma component and, therefore, are kept disjoint from the achromatic region.

Finally, the chromatic region is segmented by using the hue histogram of the chroma pixels, as defined in Table 5.5. However, we have found that subdividing the chromatic area further into subregions yields an improvement in the segmentation results. This division is carried out at the valleys (i.e., between peaks) of the smoothened saturation histogram of the chromatic region. In effect, this partitions the chromatic areas into varying levels of saturation for improved results (i.e., two areas with the same hue but different saturation values are not grouped together). Segmentation is performed within each of these chromatic subregions by using the histogram of the hue component (i.e., as done with the value component above).

5.4.2 Postprocessing and Region Merging

The recursive histogram procedure described in the previous subsection is applied to each of the three primary regions, until no areas of uniformity can be further extracted. However, a number of pixels will still remain unclassified as a result of this process (i.e., due to small region removal, median filtering, etc.). These pixels are subsequently combined into the best matching region (within a spatially local window) from the set of regions obtained in the initial histogram extraction process as follows. The image is progressively scanned (in a raster scan fashion) and a 3×3 window is formed for each unclassified pixel that borders at least one pixel (in the 8th connected nearest neighbor sense) from an initially segmented area. The L_2 norm is computed for each pixel in the window, with respect to the central pixel (i.e., the unclassified pixel). The smallest value is taken and compared to a predefined threshold. If it is less than the threshold value, then the central pixel is incorporated into the area where the corresponding pixel (i.e., the one with the smallest L_2 norm) belongs. If it exceeds the threshold value, then the central pixel is left unclassified. Pixel sites are revisited through a number of iterations until all the unclassified pixels are grouped to an appropriate region. When no groupings are made within a particular iteration, then the threshold values are increased so that the process converges. The selection of the initial threshold value is quite small and is gradually relaxed (i.e., increased) until all pixels are classified. This process is very fast because there are usually a small number of pixels (typically at the borders of regions) requiring few iterations.

Once all of the pixels have been classified, a series of binary morphological operations are used to refine the extracted regions [42]. A binary morphological opening operation is first used to remove small spurs and thin channels, followed by a binary morphological closing operation to fill in small holes and gaps.

At this stage, the segmentation of the image into a set of refined, uniform regions is complete. However, an oversegmented region may result if the threshold for small region removal is set too low. Region merging is used to overcome this situation by joining bordering regions with a similar average hue value. Adjacent regions are merged if the Euclidean distance of the average RGB values of two regions is less than a set threshold. Region merging is performed in the RGB space due to the lack of an appropriate distance metric in the HSV color space. Regions can be merged so that the smallest region is of some minimum size, or a specific number of regions is obtained. Here, we select a fixed threshold based on experimental values to reduce the computational complexity. Setting an appropriate threshold can also reduce the regions so that they coincide with semantically meaningful objects.

5.4.3 Experimental Results

The performance of the proposed segmentation scheme was tested with a number of different video sequences, and the results of the Carphone and Claire sequences are displayed below. In Figure 5.2a and b, the results of the Carphone QCIF (176×144) sequence are shown. Part a illustrates frame 80 of the original image, whereas part b shows the segmentation results after region merging in which adjacent areas are joined and the number of regions is reduced. Small regions were removed if their perimeter was less than 30 pixels. Setting a smaller threshold generates many additional smaller regions, which may result in an oversegmented image. At the same time, however, this may capture some additional detail present within the scene. The postprocessing operations included a 5×5 binary median filter and a circular morphological structuring element. In Figure 5.2c and d, the results of the Claire CIF (360×288) sequence are displayed. Once again, an arbitrary set of regions is effectively extracted in this less detailed scene.

The effectiveness of this segmentation scheme, and its potential for a more suitable content-based representation, is encouraging for future object-based video coding environments. This

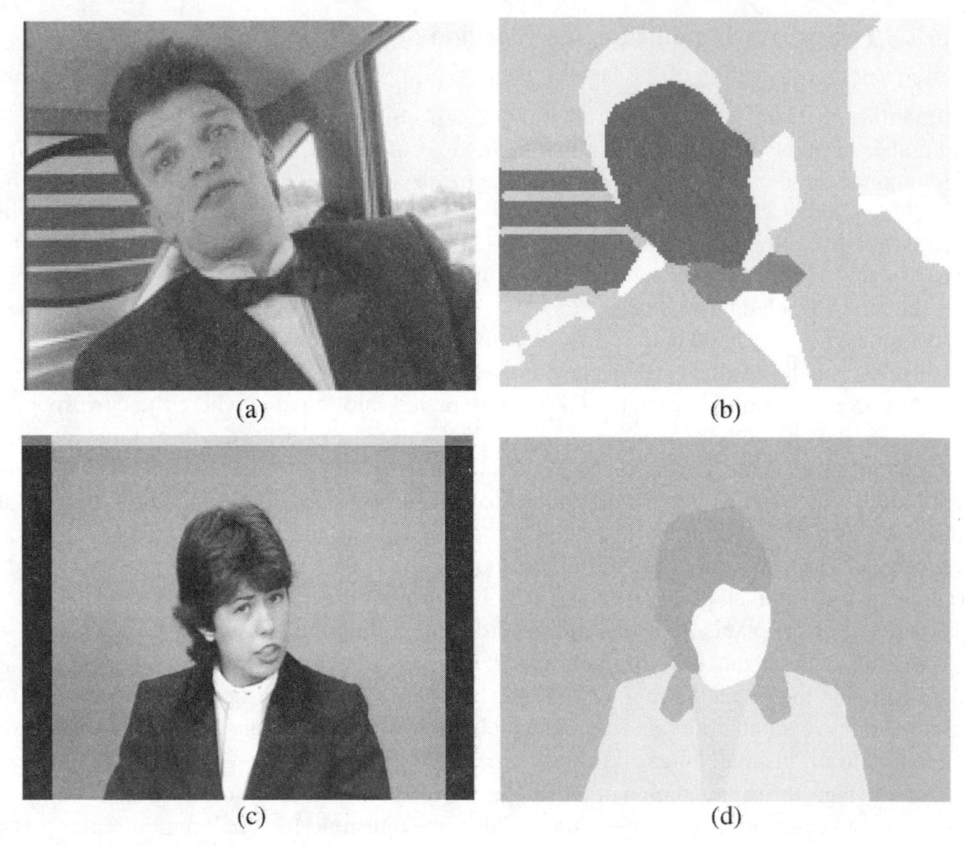

FIGURE 5.2
(a) Original frame 80 of the Carphone sequence. (b) Final segmentation of (a) after region merging. (c) Original frame 100 of the Claire sequence. (d) Final segmentation of (c) after region merging.

approach is being further enhanced to incorporate motion information so that regions can be associated with semantically meaningful objects.

5.5 Facial Image Segmentation

The recognition of human faces is currently an active area of research in computer vision [43]–[46]. The task of recognizing human faces is essentially a two-step process: (1) the detection and automatic location of the human face, and (2) the automatic identification of the face based on the extracted features. Most of the research to date has been directed toward the identification phase, with less emphasis being placed on the initial localization stage. However, the first step is critical to the success of the second and the overall recognition system. Thus, the importance of obtaining an accurate localization of the face is clear and vital in numerous multimedia applications including human recognition for security purposes, human–computer interfaces, and more recently, for video coding, multimedia databases, and video on demand. Nevertheless, determining the location of a face of unknown size in a scene with a complex or moving background still remains a difficult problem that is relatively unexplored.

Several techniques based on shape and motion information have been proposed recently for the automatic location of the facial region [47]–[49]. The former two are related to video coding applications, whereas the latter is part of a facial recognition system. The shape-based approach in [47] models the contours of the face as an ellipse. The location of the facial region is determined by performing an ellipse fitting task to a thresholded binary edge image. In [48], a generic 3D face model is adapted to the extracted facial outline from a videophone-type scene for the case where only one person is talking against a stationary background. In this application, a hierarchical localization scheme is utilized to isolate the facial area. The technique is based on the shape of the extracted head-and-shoulders silhouette, which is obtained using the thresholded frame differences. Finally, in [49], a motion detection algorithm is used to segment the facial area from a complex background. The proposed method locates the facial region by assuming that the object having the greatest motion in the video sequence is the face to be detected. This assumption, however, may limit the success of the approach in applications with nonstationary backgrounds (e.g., mobile videophones) and/or other moving objects in the scene. The authors also acknowledge potential problems caused by noise or other objects moving in the background and also suggest a modification in their technique to better handle the case of tilted or turned faces.

5.5.1 Extraction of Skin-Tone Regions

The identification and tracking of the facial region is determined by utilizing a priori knowledge of the skin tone distributions in the HSV color space outlined earlier. It has been found that skin-colored clusters form within a rather well-defined region in chromaticity space [50], and also within the HSV hexcone model [51], for a variety of different skin types. In the HSV space in particular, the skin distribution was found to lie predominantly within the limited hue range between 0° and 50° (red–yellow), and in certain cases between 340° and 360° (magenta–red) for darker skin types [39]. The saturation component suggests that skin colors are somewhat saturated, but not deeply saturated, with varying levels of intensity.

The hue component is the most significant feature in defining the characteristics of the skin clusters. However, as mentioned earlier, the hue can be unreliable when: (1) the level of brightness (i.e., value) in the scene is low, or (2) the regions under consideration have low saturation values [39]. The first condition can occur in areas of the image where there are shadows, or generally under low lighting levels. In the second case, low values of saturation are found in the achromatic regions of a scene. Thus, we must define appropriate thresholds for the value and saturation components where the hue attribute is reliable. We have defined the following polyhedron with appropriate threshold values that correspond to the skin-colored clusters with well-defined saturation and value components, based on a large sample set [39]:

$$T_{\text{hue1}} = 340^\circ \leq \text{H} \leq T_{\text{hue2}} = 360^\circ \tag{5.40}$$

$$T_{\text{hue3}} = 0^\circ \leq \text{H} \leq T_{\text{hue4}} = 50^\circ \tag{5.41}$$

$$\text{S} \geq T_{\text{sat1}} = 20\% \tag{5.42}$$

$$\text{V} \geq T_{\text{val}} = 35\% \tag{5.43}$$

The extent of the above hue range is purposely designed to be quite wide so that a variety of different skin types can be modeled. As a result of this, however, other objects in the scene with *skin-like* colors may also be extracted. Nevertheless, these objects can be separated by analyzing the hue histogram of the extracted pixels. The valleys between the peaks are used to identify the various objects that possess different hue ranges (i.e., facial region and different colored objects). Scale–space filtering [52] is used to smoothen the histogram and obtain the meaningful peaks and valleys. This process is carried out by convolving the original hue histogram, $f_h(x)$, with a Gaussian function $g(x, \tau)$ of zero mean and standard deviation τ as follows:

$$F_h(x, \tau) = f_h(x) * g(x, \tau) = \int_{-\infty}^{\infty} f_h(u) \frac{1}{\sqrt{2\pi}\,\tau} \exp\left[\frac{-(x-u)^2}{2\tau^2}\right] du \tag{5.44}$$

where $F_h(x, \tau)$ represents the smooth histogram. The peaks and valleys are determined by examining the first and second derivatives of F_h above. In the remote case that another object matches the skin color of the facial area (i.e., separation is not possible by the scale–space filter), the shape analysis module that follows provides the necessary discriminatory functionality.

A series of postprocessing operations which include median filtering and region filling/removal is subsequently used to refine the regions obtained from the initial extraction stage.

5.5.2 Postprocessing

Median filtering is the first of two postprocessing operations that are performed after the initial color extraction stage. The median operation is introduced in order to smoothen the segmented object silhouettes and also eliminate any isolated misclassified pixels that may appear as impulsive-type noise. Square filter windows of size 5×5 and 7×7 provide a good balance between adequate noise suppression and sufficient detail preservation. This operation is computationally inexpensive because it is carried out on the bilevel images (i.e., object silhouettes).

The result of the median operation is successful in removing any misclassified *noise-like* pixels; however, small isolated regions and small holes within object areas may remain after this step. Thus, we follow the application of median filtering by region filling and removal. This second postprocessing operation fills in small holes within objects which may occur due to color differences (e.g., eyes and mouth of the facial skin region), extreme shadows, or any unusual lighting effects (specular reflection). At the same time, any erroneous small regions are also eliminated as candidate object areas.

We have found that the hue attribute is reliable when the saturation component is greater than 20% and meaningless when it is less than 10% [39]. Similar results have also been confirmed in the HVC color model [37]. Saturation values between 0 and 10% correspond to the achromatic areas within a scene, whereas those greater than 20% correspond to the chromatic ones. The range between 10 and 20% represents a sort of transition region from the achromatic to the chromatic areas. We have observed that, in certain cases, the addition of a select number of pixels within this 10 to 20% range can improve the results of the initial extraction process. In particular, the initial segmentation may not capture smaller areas of the face when the saturation component is decreased due to the lighting conditions. Thus, pixels within this transition region are selected accordingly [39] and merged with the initially extracted objects. A pixel within the transitional region is added to a particular object if its distance is within a threshold of the closest object. A reasonable selection can be made if the threshold is set to a factor between 1.0 and 1.5 of the distance from the centroid of the object to its most distant point. The results from this step are once again refined by the two postprocessing operations described earlier.

At this point, one or more of the extracted objects corresponds to the facial regions. In certain video sequences, however, we have found gaps or holes around the eyes of the segmented facial area. This occurs in sequences where the forehead is covered by hair and, as a result, the eyes fail to be included in the segmentation. We utilize two morphological operators to overcome this problem and at the same time smoothen the facial contours. A morphological closing operation is first used to fill in small holes and gaps, followed by a morphological opening operation to remove small spurs and thin channels [42]. Both of these operations maintain the original shapes and sizes of the objects. A compact structuring element such as a circle or square without holes can be used to implement these operations and also help to smoothen the object contours. Furthermore, these binary morphological operations can be implemented by low-complexity *hit-or-miss* transformations [42].

The morphological stage is the final step prior to analysis of the extracted objects. The results at this point contain one or more objects that correspond to the facial areas within the scene. The block diagram in Figure 5.3 summarizes the proposed face localization procedure. The shape and color analysis unit, described next, provides the mechanism to correctly identify the facial regions.

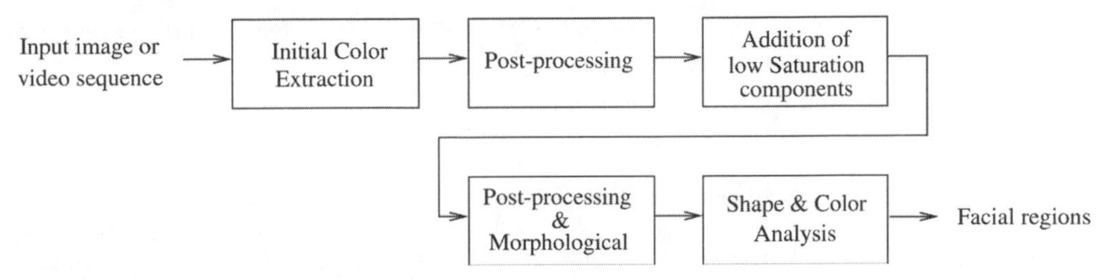

FIGURE 5.3
Overall scheme to extract the facial regions within a scene.

5.5.3 Shape and Color Analysis

The input to the shape and color analysis module may contain objects other than the facial areas. Thus, the function of this module is to identify the actual facial regions from the set of candidate objects. To achieve this, a number of expected facial characteristics such as shape, color, symmetry, and location are used in the selection process. Fuzzy membership functions are constructed in order to quantify the expected values of each characteristic. Thus, the value of a particular membership function gives us an indication of the *goodness of fit* of the object under consideration with the corresponding feature. An overall goodness of fit value can finally be derived for each object by combining the measures obtained from the individual primitives.

In our segmentation and localization scheme we utilize a set of features that are suitable for our application purposes. In facial image databases (employees, models, etc.) or videophone-type sequences (video archives of newscasts, interviews, etc.), the scene consists of predominantly upright faces that are contained within the image (i.e., not typically at the edges of the image). Thus, we utilize features such as the location of the face, its orientation from the vertical axis, and its aspect ratio to assist with the recognition task. These features can be determined in a simple and fast manner, as opposed to measurements based on facial features such as the eyes, nose, and mouth, which may be difficult to compute (i.e., in certain images the features may be small or occluded). More specifically, we consider the following four primitives in our face localization system:

1. *Deviation from the average hue value of the different skin-type categories.* The average hue value for different skin types varies among humans and depends on the race, gender, and age of the person. However, the average hue of different skin types falls within a more restricted range than the wider one defined by equations (5.40) and (5.41) [39]. The deviation of an object's expected hue value from this restricted range gives us an indication of its similarity to skin tone colors.

2. *Face aspect ratio.* Given the geometry and shape of the human face, it is reasonable to expect that the ratio of height to width falls within a specific range. If the dimensions

of a segmented object fit the commonly accepted dimensions of the human face then it can be classified as a facial area.

3. *Vertical orientation.* The location of an object in a scene depends largely on the viewing angle of the camera and the acquisition devices. For the intended applications it is assumed that only reasonable rotations of the head are allowed in the image plane. This corresponds to a small deviation of the facial symmetry axis from the vertical direction.

4. *Relative position of the facial region in the image plane.* By similar reasoning to (3) above, it is probable that the face will not be located right at the edges of the image but, rather, within a central window of the image.

5.5.4 Fuzzy Membership Functions

A number of membership function models can be constructed and empirically evaluated. A trapezoidal function model is utilized here for each primitive in order to keep the complexity of the overall scheme to a minimum. This type of membership function attains the maximum value only over a limited range of input values. Symmetric or asymmetrical trapezoidal shapes can be obtained depending on the selected parameter values. The membership function can assume any value in the interval [0, 1], including both of the extreme values. A value of 0 in the function above indicates that the event is impossible. On the contrary, the maximum membership value of 1 represents total certainty. The intermediate values are used to quantify variable degrees of uncertainty. The estimates for the four membership functions are obtained by a collection of physical measurements of each primitive from a database of facial images and sequences [39].

The hue characteristics of the facial region (for different skin-type categories) were used to form the first membership function. This function is built using the discrete universe of discourse $[-20°, 50°]$ (i.e., $-20° = 340°$). The lower bound of the average hue observed in the image database is approximately 8° (African-American distribution), whereas the upper bound average value is around 30° (Asian distribution) [39]. A range is formed using these values, where an object is accepted as a skin tone color with probability 1 if its average hue value falls within these bounds. Thus, the membership function associated with the first primitive is defined as follows:

$$\mu(x) = \begin{cases} \dfrac{(x+20)}{28}, & \text{if } -20° \leq x \leq 8° \\ 1, & \text{if } 8° \leq x \leq 30° \\ \dfrac{(50-x)}{20}, & \text{if } 30° \leq x \leq 50° \end{cases} \tag{5.45}$$

Experimentation with a wide variety of facial images has led us to the conclusion that the aspect ratio (height/width) of the human face has a nominal value of approximately 1.5. This finding confirms previous results reported in the open literature [49]. However, in certain images we must also compensate for the inclusion of the neck area, which has similar skin tone characteristics to the facial region. This has the effect of slightly increasing the aspect ratio. Using this information along with the observed aspect ratios from our database, we can tune the parameters of the trapezoidal function for this second primitive. The final form of the function is given by

$$\mu(x) = \begin{cases} \dfrac{(x-0.75)}{0.5} & , \text{ if } 0.75 \le x \le 1.25 \\ 1 & , \text{ if } 1.25 \le x \le 1.75 \\ \dfrac{(2.25-x)}{0.5} & , \text{ if } 1.75 \le x \le 2.25 \\ 0 & , \text{otherwise} \end{cases} \tag{5.46}$$

The vertical orientation of the face in the image is the third primitive used in our shape recognition system. As mentioned previously, the orientation of the facial area (i.e., deviation of the facial symmetry axis from the vertical axis) is more likely to be aligned toward the vertical due to the type of applications considered. A reasonable threshold selection of 30° can be made for valid head rotations also observed within our database. Thus, a membership value of 1 is returned if the orientation angle is less than this threshold. The membership function for this primitive is defined as follows:

$$\mu(x) = \begin{cases} 1 & , \text{ if } 0^\circ \le x \le 30^\circ \\ \dfrac{(90-x)}{60} & , \text{ if } 30^\circ \le x \le 90^\circ \end{cases} \tag{5.47}$$

The last primitive used in our knowledge-based system refers to the relative position of the face in the image. Due to the nature of the applications considered, we would like to assign a smaller weighting to objects that appear closer to the edges and corners of the images. For this purpose, we construct two membership functions. The first one returns a confidence value for the location of the segmented object with respect to the x axis. Similarly, the second one quantifies our knowledge about the location of the object with respect to the y axis. The following membership function has been defined for the position of a candidate object with respect to either the x or y axis:

$$\mu(x) = \begin{cases} \dfrac{(x-(d))}{\frac{d}{2}} & , \text{ if } d \le x \le \dfrac{3d}{2} \\ 1 & , \text{ if } \dfrac{3d}{2} \le x \le \dfrac{5d}{2} \\ \dfrac{((3d)-x)}{\frac{d}{2}} & , \text{ if } \dfrac{5d}{2} \le x \le 3d \\ 0 & , \text{otherwise} \end{cases} \tag{5.48}$$

The membership function for the x axis is determined by letting $d = \frac{D_x}{4}$, where D_x represents the horizontal dimensions of the image (i.e., in the x direction). In a similar way, the y axis membership function is found by letting $d = \frac{D_y}{4}$, where D_y represents the vertical dimensions of the image (i.e., in the y direction).

The individual membership functions expressed above must be appropriately combined to form an overall decision. To this end, we utilize the fuzzy aggregators used in Section 5.3.2 to form the overall function used in the filter. In particular, the compensative operator (i.e., overall fuzzy membership function), which assumes the form of a weighted product as follows

$$\mu_c = \left(\left(\min_{j=1}^{m} \mu_j \right) \left(\max_{j=1}^{m} \mu_j \right) \right)^{0.5} \tag{5.49}$$

was selected because it provides a good compromise of conjunctive and disjunctive behavior. The aggregation operator defined in (5.49) is used to form the final decision based on the designed primitives.

5.5.5 Meta-Data Features

Multimedia databases are composed of a number of different media types, such as images and video that are binary by nature, and hence are unstructured. An appropriate set of interpretations must be derived for these media objects in order to allow for content-based functionalities which include storage and retrieval. These interpretations, or *meta-data*, are generated by applying a set of feature-extracting functions on the contained media objects [53]. These functions are media dependent (i.e., audio, video, images) and are unique even within each media type (i.e., satellite images, facial images). The following four steps are necessary in extracting the features from image object types: (1) object locator design, (2) feature selection, (3) classifier design, and (4) classifier training. The function of the object locator is to isolate the individual objects of interest within the image through a suitable segmentation algorithm. In the second step, specific features are selected to identify the different types of objects that might occur within the images of interest. The classifier design stage is then used to establish a mathematical basis for distinguishing the different objects based on the designed features. Finally, the last step is used to train and update the classifier module by adjusting various parameters. In the previous sections we have designed the object locator to automatically isolate and track the facial area within a facial image database or a videophone-type sequence. Now, we propose the use of a set of features that may be used in constructing a meta-data feature vector for the classifier design and training stages.

Having determined the facial regions within the image, we can construct an n-dimensional feature vector, $\mathbf{f} = (f_1, f_2, \ldots, f_n)$, that may be used for content-based storage and retrieval purposes. We present several features that may be incorporated within a more detailed meta-data feature vector. More specifically, we propose the use of hair and skin color and face location and size as a preliminary set.

Hair color is a significant human characteristic that can be effectively employed in user queries to retrieve particular facial images. We have determined a scheme to categorize black, gray/white, brown, and blonde hair colors within the HSV space. First, the H, S, and V component histograms of the hair regions are formed and smoothened using the scale–space filter defined earlier. The peak values from each histogram are subsequently determined and used to form the appropriate classification. The following regions were suitably found from our large sample set for the various categories of hair color:

(1) *Black* $\quad \mathrm{V}_p < 15\%$

(2) *Gray* $\quad \mathrm{S}_p < 20\% \cap \mathrm{V}_p > 50\%$

(3) *Brown* $\quad \mathrm{S}_p \geq 20\% \cap 15 \leq \mathrm{V}_p < 40\%$

(4) *Blonde* $\quad 20° < \mathrm{H}_p < 50° \cap \mathrm{S}_p \geq 20\% \cap \mathrm{V}_p \geq 40\%$

where H_p, S_p, and V_p denote the peaks of the corresponding histograms. Thus, dark or black hair is characterized by low-intensity values and gray or white hair by low saturation and high-intensity values. On the other hand, brown or blonde hair colors are typically well saturated but differ in their intensity values. The expected value component of dark brown hair lies at approximately $\mathrm{V}_p \approx 20\%$, lighter brown at around $\mathrm{V}_p \approx 35\%$, and blonde hair at higher values, $\mathrm{V}_p \geq 40\%$. Therefore, we can use this information to appropriately categorize the facial regions extracted earlier. We use a suitably sized template above each facial area for the classification process as shown in Figure 5.4. The template consists of regions $R_1 + R_2 + R_3$. This provides a fast yet good approximation to the overall description.

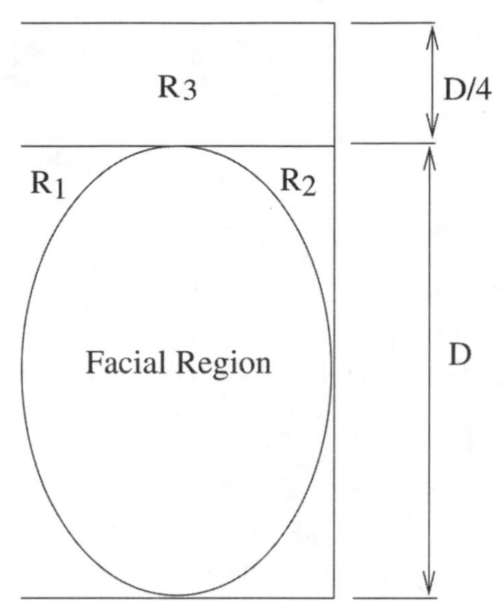

FIGURE 5.4
Template for hair color classification = $R_1 + R_2 + R_3$.

The next feature we propose to use is the average hue value of the facial area. We have found that darker skin types tend to shift toward 0° (i.e., average hue = 8° for the darker skin-type sample set), whereas lighter colored skin types move toward 30° [39]. In certain cases, however, lighter skin types with a reddish appearance may also have a slightly reduced average hue value (i.e., 15°). Nevertheless, the hue sector can be partitioned to discriminate between lighter and darker skin types as follows: (1) darker colored skin, $H < 15°$, and (2) lighter, $H \geq 15°$. This can give us a reasonable approximation; however, we believe that the saturation and value components can improve upon these results.

Finally, the location and size of each facial area (i.e., centroid location and size relative to the image) can provide very useful information in a retrieval system. These combined features can give us an indication of whether the face is a portrait shot or if perhaps the body is included. In addition to this, it can also provide information about the spatial relationships of a particular facial region with other objects or faces within the scene. Further work is being done in this latter area.

5.5.6 Experimental Results

The scheme outlined in Figure 5.3 was used to locate and track the facial region in a number of still images and video sequences. The results from three videophone-type sequences (i.e., newscast or interview-type sequences) are presented below: (1) Carphone, (2) Miss America, and (3) Akiyo.

The segmentation results in Figure 5.5 illustrate the robustness of the technique to the various cases of object/background motion, lighting, and scale variations. A parameter selection of $\tau = 2$ was made in the Gaussian function of equation (5.44) in order to smoothen the histograms. This provided adequate smoothing and was found to be appropriate for the skin tone distribution models [39]. A similar value [37] has also been suggested in the HVC space. The shape and color analysis module was used to identify the facial regions from the set of

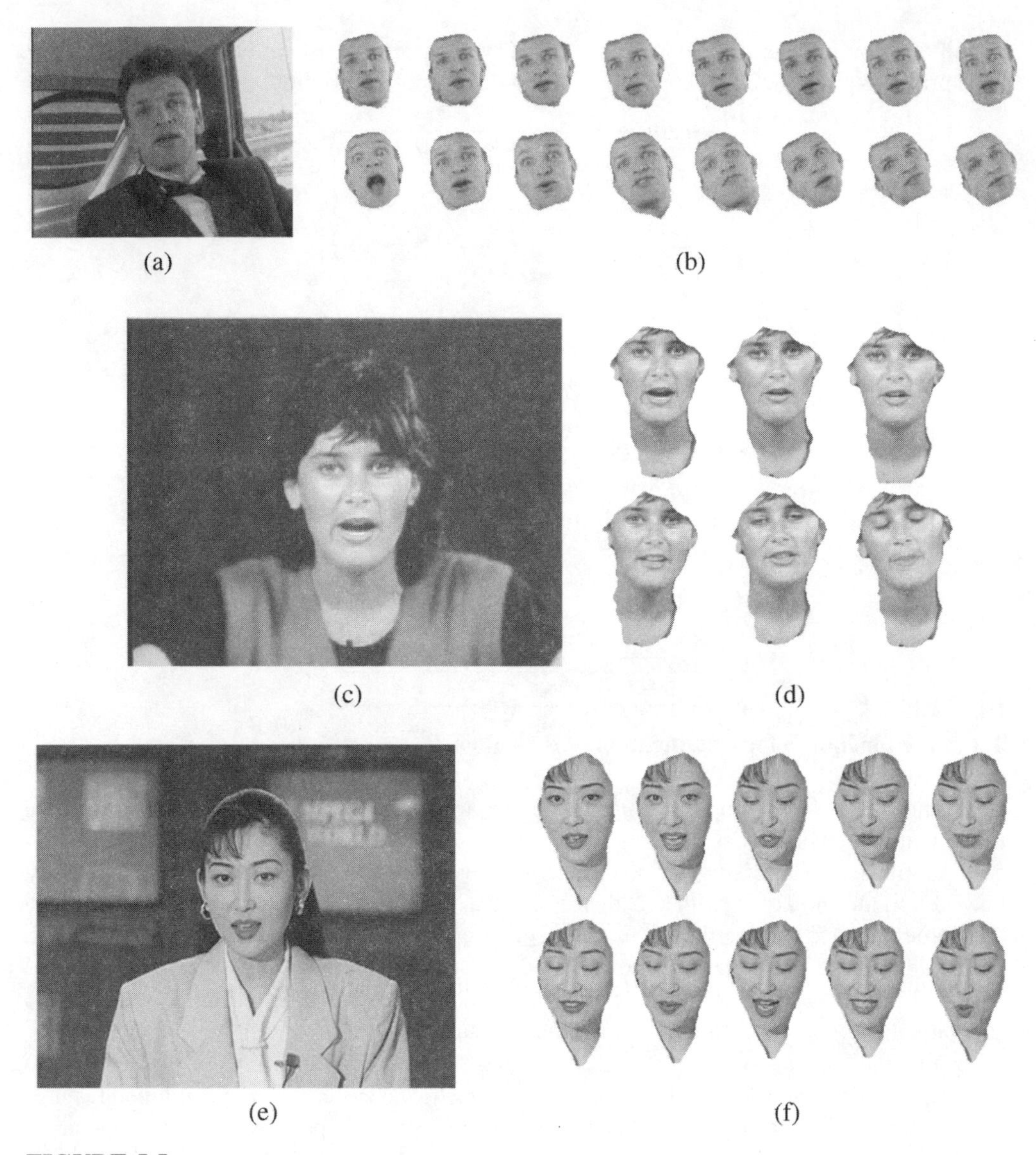

FIGURE 5.5
Location and tracking of the facial region for the following video sequences: (a) Carphone, (b) Carphone frames 20–85, (c) Miss America, (d) Miss America frames 20–120, (e) Akiyo, and (f) Akiyo frames 20–110.

candidate objects. An object was classified as a facial region if its overall membership function, μ_c, exceeded a predefined threshold of 0.75. In the QCIF Carphone sequence of Figure 5.5a, only one candidate region was extracted by the localization procedure in Figure 5.3, which indeed corresponded to the facial area. In Figure 5.5c, a similar procedure was followed with the CIF Miss America sequence. In this case, three objects of significant size were extracted, and the results of these are summarized in Table 5.6.

Only the first object was selected, based on the aggregation of the membership function values. The objects O_2 and O_3 were rejected because they scored poorly in their mean hue value and location and had reduced membership values in the orientation primitive. Finally, in

Table 5.6 Miss America (Width × Height =360×288): Shape and Color Analysis

Attributes	Objects		
	O_1	O_2	O_3
Centroid location			
x	177	245	244
μ_1	1	0	0
y	188	120	269
μ_2	1	1	0.02
Orientation			
θ°	4.92	47.74	44
μ_3	1	0.7	0.77
Object ratio			
r	1.61	1.16	1.32
μ_4	1	0.82	1
Mean hue			
H_m ($^\circ$)	20	-6	-5
μ_5	1	0.5	0.54
Aggregation	1.0	0.0	0.0

Figure 5.5e, the facial region was successfully identified and tracked for the Akiyo sequence. Two candidate objects were extracted in this case and, once again, the face was correctly selected based on the aggregation values.

Once the facial region is identified, the proposed meta-data features can be computed according to the methodology provided in the previous section. The feature values for each of the image sequences are summarized in Table 5.7. The average hue value of the facial area (i.e., skin) is in all three cases greater than 20°, which puts them in the lighter skin category, as expected. Next, we observe the S_p and V_p values of the hair region obtained from our constructed template. According to our classification scheme, the tabulated values indicate that the facial image in the Carphone sequence has brown hair, whereas the other two have black hair. These *fuzzy* descriptions are appropriate representations of the images shown in Figure 5.5. Finally, the last two features give us an indication of the location and size of the face within the scene. In all cases, the facial region is relatively close to the center of the image (location is with respect to the top left corner) and is of significant size (i.e., a closeup).

5.6 Conclusions

The tremendous advances in both software and hardware have brought about the integration of multiple media types within a unified framework. This has allowed the merging of video, audio, text, and graphics with enormous possibilities for new applications. This integration is at the forefront in the convergence of the computer, telecommunications, and broadcast industries. The realization of these new technologies and applications, however, demands a new way of processing audiovisual information. We have shifted from pixel-based models (pulse code modulation) to statistically dependent pixel models (transform coding) to the

Table 5.7 Proposed Meta-Data Feature Values

Meta-Data	Carphone	Miss America	Akiyo
Skin color			
H (°)	24	20	23
Hair color			
S (%)	29	38	16
V (%)	30	15	12
Description	Brown	Black	Black
Centroid location			
Horizontal (%)	40	53	51
Vertical (%)	42	60	47
Image height/face height	1.7	1.5	1.8

current audiovisual object-based approaches (MPEG-7).

In this chapter we have focused on several aspects of the intelligent processing of visual information using color imaging techniques and fuzzy concepts. We have applied this methodology to three problem areas, namely: (1) color image filtering, (2) color segmentation, and (3) automatic face localization and meta-data generation.

Digital images and video signals suffer from several degradations and artifacts. These may include sensor noise and lens aberrations from commercial camcorders or artifacts from the digitization process from analog sources. In some cases this may be acceptable, but for high-resolution multimedia applications they become objectionable. In Section 5.3 we presented a new class of filters based on a fuzzy multichannel filtering structure. Our new adaptive design was computationally efficient, scalable, nonrecursive, and did not require a training signal. The application of our fuzzy filters to a number of noise-contaminated color images indicated a performance that improved upon the conventional vector processing filters.

One of the challenges in the representation of visual information is to decompose a video sequence into its elementary parts. Temporal segmentation refers to finding shot boundaries, whereas spatial segmentation corresponds to the extraction of visual objects in each frame. In Section 5.4 we have addressed the spatial segmentation problem using the visual cue of color. The proposed color segmentation scheme utilized the perceptual HSV color model to effectively partition a sequence into a set of arbitrarily shaped regions. The method was found to be robust and of relatively low computational complexity due to the 1D histogram procedure and the binary nature of the postprocessing operations involved. Thresholds for merging regions and removing small areas were used to avoid the problems associated with oversegmented results. The effectiveness of this technique and its potential for a suitable content-based representation is encouraging for future object-based video coding environments. The incorporation of motion information into the segmentation problem holds more promising results. Automatic segmentation methods are challenged in their quest to extract semantically meaningful objects as opposed to simple regions. The combination of spatial and temporal segmentation with object tracking algorithms remains an active area of research. These solutions will enable true object-based compression schemes and effective indexing and retrieval of the content.

The automatic extraction of facial images in digital pictures is vital in numerous multimedia applications, including multimedia databases, video on demand, human–computer interfaces, and video coding. In Section 5.5, a novel technique was introduced to locate and track the facial area in videophone-type sequences. The proposed method essentially consisted of two components: (1) a color processing unit and (2) a knowledge-based shape and color analysis module. The color processing component utilized the HSV color space, while the shape module employed a number of fuzzy membership functions to correctly identify the facial region. The suggested approach was robust with regard to different skin types and various types of object or background motion within the scene. Having determined the facial regions within an image, we then constructed a meta-data feature vector that could be used for content-based storage and retrieval purposes. Meta-data features such as hair and skin color and face location and size were utilized as a preliminary set. The results of our findings were encouraging in extracting vital information from facial images. Efforts for content-based video description are an active research topic. It is highly desirable to index multimedia data using visual features such as color, texture, and shape; sound features such as audio and speech; and textual features such as script and closed captioning. It is also of great interest to have the capabilities to browse and search for this content using compressed data since most video data will likely be stored in compressed formats. Another area of interest is in temporal segmentation, where it is important to extract shots, scenes, or objects. Furthermore, higher level descriptions for the direction and magnitude of dominant object motion and the entry and exit instances of objects of interest are highly desirable. These are all future research areas to be investigated and fueled with the upcoming MPEG-7 standard.

In this chapter we have examined the concepts of adaptive fuzzy systems and color processing for several multimedia applications. More specifically, the algorithms and architectures were examined; however, further analysis is warranted to address issues of real-time architectures and realizations, modularity, software portability, and system robustness.

References

[1] A.K. Jain, A. Vailaya, Image retrieval using color and shape, *Pattern Recognition,* 29, 1233, 1996.

[2] T. Young, On the theory of light and colors, *Philosophical Transactions of the Royal Society of London,* 92, 20, 1802.

[3] G. Wyszecki, W.S. Stiles, *Color Science: Concepts and Methods, Quantitative Data and Formulae,* John Wiley and Sons, New York, 1982.

[4] A.H. Munsell, *A Color Notation,* Munsell Color Company, Baltimore, 1939.

[5] C.A. Poynton, *A Technical Introduction to Digital Video,* John Wiley and Sons, New York, 1996.

[6] J. Foley, A. van Dam, S. Feiner, J. Hughes, *Computer Graphics, Principles and Applications, 2nd Edition,* Addison-Wesley, Reading, MA, 1990.

[7] A.K. Jain, *Fundamentals of Digital Image Processing,* Prentice-Hall, Englewood Cliffs, NJ, 1989.

[8] A.R. Smith, Color gamut transform pairs, *SIGGRAPH 78,* 12, 1978.

[9] I. Pitas, A.N. Venetsanopoulos, *Nonlinear Digital Filters,* Kluwer Academic, Boston, MA, 1990.

[10] A.N. Venetsanopoulos, K.N. Plataniotis, Multichannel image processing, *Proceedings of the IEEE Workshop on Nonlinear Signal Processing,* I. Pitas (ed.), 2, 1995.

[11] J. Astola, P. Haavisto, Y. Neuvo, Vector median filter, *Proceedings of IEEE,* 78, 678, 1990.

[12] P.E. Trahanias, A.N. Venetsanopoulos, Vector directional filters: A new class of multichannel image processing filters, *IEEE Transactions on Image Processing,* 2, 528, 1993.

[13] K.N. Plataniotis, D. Androutsos, A.N. Venetsanopoulos, Color image processing using fuzzy vector directional filters, *Proceedings of the IEEE Workshop on Nonlinear Signal Processing,* I. Pitas (ed.), 535, 1995.

[14] D. Androutsos, K.N. Plataniotis, A.N. Venetsanopoulos, Color image processing using fuzzy vector rank filters, *Proceedings of the International Conference on Digital Signal Processing,* 614, 1995.

[15] K.N. Plataniotis, D. Androutsos, A.N. Venetsanopoulos, Multichannel filters for image processing, *Signal Processing: Image Communications,* 9, 143, 1997.

[16] A. Buchowicz, I. Pitas, Multichannel distance filters, *Proceedings of IEEE International Conference on Image Processing, ICIP-94,* II, 575, 1994.

[17] G. Economou, S. Fotopoulos, M. Vemis, A family of nonlinear filters with data dependent coefficients, *IEEE Transactions on Signal Processing,* 43, 318, 1995.

[18] F. Russo, G. Ramponi, Fuzzy operator for sharpening of noisy images, *IEEE Electronics Letters,* 28, 1715, 1992.

[19] X. Yang, P.S. Toh, Adaptive fuzzy multilevel median filter, *IEEE Transactions on Image Processing,* 4, 680, 1995.

[20] F. Russo, G. Ramponi, A new class of fuzzy operators for image processing: Design and implementation, *Proceedings of IEEE Conference on Fuzzy Systems, FUZZ-IEEE'93,* 815, 1995.

[21] I. Pitas, A.N. Venetsanopoulos, Adaptive filters based on order statistics, *IEEE Transactions on Signal Processing,* 39, 518, 1991.

[22] C. Kotropoulos, I. Pitas, Constrained adaptive LMS L-filter, *Signal Processing,* 26, 335, 1992.

[23] X.Z. Sun, A.N. Venetsanopoulos, Adaptive schemes for noise filtering and edge detection by use of local statistics, *IEEE Transactions on Circuits and Systems,* 35, 57, 1988.

[24] K. Tang, J. Astola, Y. Neuvo, Nonlinear multirate image filtering techniques, *IEEE Transactions on Image Processing,* 4, 788, 1995.

[25] J.M. Mendel, Fuzzy logic systems for engineering: A tutorial, *Proceedings of IEEE,* 83, 345, 1995.

[26] W. Pedryz, *Fuzzy Sets Engineering,* CRC Press, Boca Raton, FL, 1995.

[27] M. Grabisch, H.T. Nguyen, E.A. Walker, *Fundamentals of Uncertainty Calculi with Applications to Fuzzy Inference,* Kluwer Academic, Dordrecht, The Netherlands, 1995.

[28] F.S. Roberts, *Measurement Theory with Applications to Decision-Making, Utility and the Social Sciences,* Addison-Wesley, Reading, MA, 1979.

[29] J. Fodor, J. Marichal, M. Roubens, Characterization of the ordered weighted averaging operators, *IEEE Transactions on Fuzzy Systems,* 3, 231, 1995.

[30] H.J. Zimmermann, *Fuzzy Sets, Decision Making and Expert System,* Kluwer Academic, Boston, MA, 1987, 235.

[31] H.J. Zimmermann, P. Zysno, Latent connectives in human decision making, *Fuzzy Sets and Systems,* 4, 37, 1980.

[32] K.N. Plataniotis, D. Androutsos, A.N. Venetsanopoulos, Color image filters: The vector directional approach, *Optical Engineering,* 36, 2375, 1997.

[33] M. Gabbouj, F.A. Cheickh, Vector median-vector directional hybrid filter for color image restoration, *Proceedings of EUSIPCO-96,* 879, 1996.

[34] H.G. Musmann, M. Hotter, J. Ostermann, Object-oriented analysis-synthesis coding of moving objects, *Signal Processing: Image Communication,* 1, 117, 1989.

[35] M. Hotter, Object-oriented analysis-synthesis coding based on moving two-dimensional objects, *Signal Processing: Image Communication,* 2, 409, 1990.

[36] N. Herodotou, A.N. Venetsanopoulos, Temporal prediction of video sequences using an image warping technique based on color segmentation, *ICIAP 97,* 1, 494, 1997.

[37] Y. Gong, M. Sakauchi, Detection of regions matching specified chromatic features, *Computer Vision and Image Understanding,* 61, 263, 1995.

[38] T. Uchiyama, M.A. Arbib, Color image segmentation using competitive learning, *IEEE Transactions on Pattern Analysis and Machine Intelligence,* 16, 1197, 1994.

[39] N. Herodotou, K.N. Plataniotis, A.N. Venetsanopoulos, Automatic location and tracking of the facial region in color video sequences, *Signal Processing: Image Communication,* 14, 359, 1999.

[40] Y.J. Zhang, Y.R. Yao, Y. He, Automatic face segmentation using color cues for coding typical videophone scenes, *SPIE Visual Communications and Image Processing,* 3024, 468, 1997.

[41] A. Witkin, Scale-space filtering, *Proceedings IJCAI-83,* 1019, 1983.

[42] J. Serra, *Image Analysis and Mathematical Morphology,* Academic Press, New York, 1982.

[43] A.L. Yuille, Deformable templates for face recognition, *Journal of Cognitive Neuroscience,* 3, 59, 1991.

[44] R. Brunelli, T. Poggio, Face recognition: Features versus templates, *IEEE Transactions on Pattern Analysis and Machine Intelligence,* 15, 10, 1993.

[45] M. Kirby, L. Sirovich, Application of the Karhunen-Loeve procedure for the characterization of human faces, *IEEE Transactions on Pattern Analysis and Machine Intelligence,* 12, 1, 1990.

[46] O. Nakamura, S. Mathur, T. Minami, Identification of human faces based on isodensity maps, *Pattern Recognition,* 24, 263, 1991.

[47] A. Eleftheriadis, A. Jacquin, Automatic face location detection for model-assisted rate control in H.261-compatible coding of video, *Signal Processing: Image Communication,* 7, 435, 1995.

[48] M.J.T. Reinders, P.J.L van Beek, B. Sankur, J.C.A. van der Lubbe, Facial feature localization and adaptation of a generic face model for model-based coding, *Signal Processing: Image Communication,* 7, 57, 1995.

[49] C.H. Lee, J.S. Kim, K.H. Park, Automatic human face location in a complex background using motion and color information, *Pattern Recognition,* 29, 1877, 1996.

[50] T.C. Chang, T.S. Huang, C. Novak, Facial feature extraction from color images, *Proceedings of the 12th International Conference on Pattern Recognition,* 3, 39, 1994.

[51] N. Herodotou, A.N. Venetsanopoulos, Image segmentation for facial image coding of videophone sequences, *13th International Conference on Digital Signal Processing,* 1, 223, 1997.

[52] M.J. Carlotto, Histogram analysis using a scale-space approach, *IEEE Transactions on Pattern Analysis and Machine Intelligence,* 9, 121, 1987.

[53] B. Prabhakaran, *Multimedia Database Management Systems,* Kluwer Academic, Boston, MA, 1997.

Chapter 6

Intelligent Multimedia Processing

Ling Guan, Sun-Yuan Kung, and Jenq-Neng Hwang

6.1 Introduction

The way we access information, conduct business, communicate, educate, learn, and entertain will be profoundly changed by the rapid development of multimedia technologies [17, 79, 124]. Multimedia technologies also represent a new opportunity for research interactions among a variety of media such as speech, audio, image, video, text, and graphics. As digitization and encoding of images and video have become more affordable, computer and Web database systems are starting to store voluminous image and video data. Consequently, massive amounts of visual information online have become closer to a reality. This promises a quantum elevation of the level of tomorrow's world in entertainment and business. However, as the data acquisition technology advances rapidly, we have now substantially fallen behind in terms of technologies for indexing and retrieval of visual information in large archives.

For example, it would be desirable to have a tool that efficiently searches the Web for a desired picture (or video clip) and/or audio clip by using as a query a shot of multimedia information [119, 135]. Nowadays, some popular queries might look like: *"Find frames with 30% blue on top and 70% green in bottom"* or *"Find the images or clips similar to this drawing."* In contrast to the above similarity-based queries, it has been argued that a so-called "subject-based" query [135] might be more likely to be used — for example, *"Find Reagan speaking to the Congress."* The subject-based query offers a more user-friendly interface, but it also introduces a greater technical challenge, which calls for advances in two distinctive research frontiers [17]:

- *Computer networking technology.* Novel communication and networking technologies are critical for multimedia database systems to support interactive dynamic interfaces. A truly integrated media system must connect with individual users and content-addressable multimedia databases. This will involve both logical connection to support information sharing and physical connection via computer networks and data transfer.

- *Information processing technology.* To advance the technologies of indexing and retrieval of visual information in large archives, multimedia content-based indexing would complement well the text-based search. Online and real-time visual information retrieving and display systems would provide popular services to professionals such as business traders, researchers and librarians as well as general users such as students and housewives. Such systems must successfully combine digital video and audio, text animation, graphics, and knowledge about such information units and their interrelationships in real time.

This chapter addresses mainly emerging issues closely related to the research frontier on *information processing technology.*

Because speech, image, and video are playing increasingly dominant roles in multimedia information processing, content-based retrieval has a broad spectrum of applications. Hence, quick and easy access of large speech, image, and video databases must be incorporated as an integral part of many near-future multimedia applications. Future multimedia technologies will need to handle information with an increasing level of *intelligence* (i.e., automatic extraction, recognition, interpretation, and interactions of multimodal signals). This will lead to what can be called *intelligent multimedia processing (IMP)* technology.

Indeed, the technology frontier of information processing is shifting from coding (MPEG-1 [83], MPEG-2 [84], and MPEG-4 [85]) to automatic recognition — a trend precipitated by a new member of the MPEG family, MPEG-7 [86, 87], which focuses on the "multimedia content description interface." Its research domain will cover techniques for object-based tracking and segmentation, pattern detection and recognition, content-based indexing and retrieval, and fusion of multimodal signals. For these, neural networks (NNs), sometimes in combination with two other branches of computational intelligence (CI), fuzzy system (FS) and evolutionary computation (EC), can offer a very promising horizon.

6.1.1 Neural Networks and Multimedia Processing

The main reason CI is perceived as a critical core technology for IMP hinges on its learning, adaptation, reasoning, and evolution capability [4, 57], which enables machines to be taught to interpret possible variations of the same object or pattern (e.g., scale, orientation, and perspective).

More specifically, to build an IMP system, the emerging synthesis of various techniques is required. Each technique plays a specific role in IMP systems. The main characteristics of NNs are to recognize patterns and to classify input, and to adapt themselves to dynamic environments by learning; but the mapping structure of an NN is a black box. The resulting NN behavior is difficult to understand. An FS, on the other hand, can cope easily with human knowledge and can perform inference, but it does not fundamentally incorporate the learning mechanism. Neuro-fuzzy computing has developed for overcoming their respective disadvantages [46, 137]. In general, the neural network part is used for learning, whereas the fuzzy logic part is used for representing knowledge. The learning is fundamentally performed as a necessary change such as incremental learning, back-propagation, and unsupervised learning schemes. EC can also tune NNs and FSs. Furthermore, EC has been used for the structure optimization of NNs and FSs [46, 137]. However, evolution can be defined as a resultant change, not a necessary change, because EC cannot predict and estimate the effect of the change. To summarize, an IMP system can quickly adapt to a dynamically changing environment by NNs and FSs, and the structure of the system can globally evolve by ECs. The capability concerning adaptation and evolution can construct more advanced IMP systems.

Among the three branches of CI, NNs have been the most popular tool for IMP because

- Neural networks offer unsupervised clustering and/or supervised learning mechanisms for recognition of objects which are deformed or have incomplete information. Therefore, NNs can be "trained" to see or hear, to recognize objects or speech, or to perceive human gestures.

- Neural networks are powerful pattern classifiers that appear to be most powerful and appealing when explicit a priori knowledge of underlying probability distributions is unknown, such that properly trained NN classifiers allow the nonparametric approximation of the associated a posteriori class probabilities [101].

- Neural networks offer a universal approximation capability, which allows accurate approximation of unknown systems based on sparse sets of noisy data. In this context, some neural models have also effectively incorporated statistical signal processing and optimization techniques.
- Temporal neural models, which are specifically designed to deal with temporal signals, further expand the application domain in multimedia processing, particularly audio, speech, and audiovisual integration and interactions.
- A hierarchical network of neural modules will be vital to facilitate search mechanisms used in a voluminous, or Web-wide, database. Typically, in a tree network structure, kernels that are common to all the models form the root of the tree. The leaves of the tree correspond to the individual neural modules, whereas the paths from root to leaf connect the modules to their respective kernels.

Consequently, NNs have recently received increasing attention in many multimedia applications. Here we list just a few examples: (1) human perception: facial expression and emotion categorization [105], human color perception [109], and multimedia data visualization [3, 102]; (2) computer–human communication: face recognition [70], lipreading analysis [19, 20, 65, 95], and human–human and computer–human communication/interaction [89]; and (3) multimodal representation and information retrieval: hyperlinking of multimedia objects [64], queries and searches of multimedia information [75], 3D object representation and motion tracking [118], and image sequence generation and animation [78]. More concrete application examples will be discussed in the subsequent sections.

6.1.2 Focal Technical Issues Addressed in the Chapter

This chapter will focus on vital technical issues in the research frontier on information processing technology, particularly those closely related to IMP. More specifically, this chapter will demonstrate why and how CI, with neural networks in particular, offers as a core technology for: *efficient representations for audiovisual information* (Section 6.2.1); *detection and classification techniques* (Section 6.2.2); *fusion of multimodal signals* (Section 6.2.3); and *multimodal conversion and synchronization* (Section 6.2.4). Here let us first offer some motivations as well as a brief explanation on the key technical points.

Efficient Representations for Audiovisual Information

An efficient representation of the information can facilitate many useful multimedia functionalities, such as object-based indexing and access. To this end, it is vital to have sophisticated preprocessing of the image or video data. For many multimedia applications, preprocessing is usually carried out on the input signals to make the subsequent processing modeling and classification tasks easier (e.g., segmentation of 2D or 3D images and video for content-based coding and representation in the context of the MPEG or JPEG standards). The more sophisticated the representation obtained by preprocessing, the less sophisticated the classifier would need to be. Hence, a synergistic balance (and eventually interaction) between representation and indexing needs to be explored.

An efficient representation of vast amounts of multimedia data can often be achieved by adaptive data clustering or model representation mechanisms, which happen to be the most promising strength of many well-established unsupervised neural networks [e.g., self-organizing feature map (SOFM) and principal component analysis (PCA) neural network]. The evolution from conventional statistical clustering and/or contour and shape modeling to these unsupervised NNs will be highlighted in Section 6.2.1.

Some of these NNs have been incorporated for various feature extraction, moving object tracking, and segmentation applications. Illustrative samples for such preprocessing examples are provided in Section 6.3.1.

Detection and Classification for Audiovisual Databases

As most digital text, audio, and visual archives exist on various servers throughout the world, it becomes increasingly difficult to locate and access the information. It thus necessitates automatic search tools for indexing and access. Detection and classification constitute a very basic tool for most search and indexing mechanisms. Detection of a (deformable) pattern or object has long been an important machine learning and computer vision problem. The task involves finding a specific (but locally deformable) pattern in images (e.g., human faces). What is critically needed are powerful search strategies to identify contents on speech or visual clues, possibly without the benefit of textual information. These will have important commercial applications, including automatic teller machine (ATM), access control, surveillance, and video conferencing systems.

Several *static* supervised NNs (i.e., no feedback connections are used in the network), which are useful for detection and classification, will be covered in Section 6.2.2. Built upon these NNs, many NN content-based image search systems have been developed for various applications. On the horizon are several promising tools which allow users to specify image queries by giving examples, drawing sketches, selecting visual features (e.g., color, texture, shape, and motion), and arranging the spatiotemporal structure of features. Some exemplar NN systems will be presented in Section 6.3.4. They serve to demonstrate the fact that unsupervised and supervised NN models are useful means for developing reliable search mechanisms.

Multimodal Media Fusion: Combine Multiple Sources

Multimedia signal processing is more than simply "putting together" text, audio, images, and video. The correlation between audio and video can be utilized to achieve more efficient coding and recognition. New application systems and thus new research opportunities arise in the area of fusion and interaction among these media.

Humans perform most perception and recognition tasks based on joint processing of the input multimodal data. The biological cognitive machines of humans handle multimodal data through visual, auditory, and sensory mechanisms via some form of adaptive processing (learning/retrieving) algorithms, which remain largely mysterious to us. Motivated by the nature of biological information processing, fusion NN models which combine information from multiple sensor and data sources are being pursued as a universal data processing engine for multimodal signals. Linear fusion networks and nonlinear fusion networks are discussed in Section 6.2.3.

Audio–video interaction can be used for personal authentication and verification. A visual/auditory fusion network for such an application is discussed in Section 6.3.2.

Multimodal Conversion and Synchronization

One of the most interesting interactions among different media is the one between audio and video. In multimodal speech communication, audio–video interaction has a significant role, as evidenced by the McGurk effect [74]. It shows that human perception of speech is bimodal in that acoustic speech can be affected by visual cues from lip movements. For example, one experiment showed that when a person *sees* a speaker saying /ga/, but *hears* the sound /ba/, the person perceives neither /ga/ nor /ba/, but something close to /da/. In video conferencing applications, it is conceivable that the video frame rate is severely limited by the bandwidth and is by far very inadequate for lip synchronization perception. One solution is to warp the acoustic signal to synchronize it with the person's mouth movements, which will be useful for dubbing in a studio and other non-real-time applications.

There is a class of *temporal* neural models (i.e., feedback connections are used to keep track of temporal correlation of signals) that can facilitate the conversion and synchronization processes. Prominent temporal NN models and popular statistical approaches will be reviewed in Section 6.2.4. Verbal communication has been efficiently achieved by combining speech recognition and visual interpretation of lip movements (or even facial expressions or body language). As another example, an NN-based lipreading system via audio and visual integration will be presented in Section 6.3.3. Other potential applications include dubbing of movies, segmentation of video scenes, and human–computer interfaces.

6.1.3 Organization of the Chapter

Section 6.2 reviews some of the key NNs, then highlights their usefulness to IMP applications. Built upon these NN models, exemplar IMP applications will be illustrated in Section 6.3. Some open technical issues and promising application trends will be suggested in Section 6.4.

6.2 Useful Neural Network Approaches to Multimedia Data Representation, Classification, and Fusion

We will discuss in this section a variety of statistical learning techniques adopted by NNs. Through these techniques, machines may be taught to automatically interpret and represent possible variations of the same object or pattern. Some of these NNs (e.g., the self-organization feature map) can be perceived as a natural evolution from traditional statistical clustering and parameter estimation techniques (e.g., vector quantization (VQ) and expectation maximization). These NNs can also be incorporated into traditional pattern recognition techniques (e.g., active contour model) to enhance the performance.

6.2.1 Multimedia Data Representation

From the learning perspective, neural networks are grouped into *unsupervised learning* and *supervised learning* networks. Static features extraction is often inadequate for an adaptive environment where users may require adaptive and dynamic feature extraction tools. Unsupervised neural techniques are very amenable to dynamic feature extraction. The SOFM is one representative of an unsupervised NN, which combines the advantages of statistical data clustering (such as vector quantization and PCA) and local continuity constraint (as imposed in the active contour model search).

Self-Organizing Feature Map (SOFM)

The basic idea of constructing an SOFM is to incorporate into the competitive learning (clustering) rule some degree of sensitivity with respect to the neighborhood or history. This provides a way to avoid totally uncommitted neurons, and it helps enhance certain topological properties which should be preserved in the feature mapping (or data clustering).

Suppose that an input pattern has n features and is represented by a vector $\mathbf{x}$ in an n-dimensional pattern space. The network maps the input patterns to an output space. The output space in this case is assumed to be 1D or 2D arrays of output nodes, which possess a certain topological ordering. The question is how to cluster these data so that the ordered relationship can be preserved. Kohonen proposed to allow the centroids (represented by output nodes of an SOFM) to interact laterally, leading to the *self-organizing feature map* [52, 53], which was originally inspired by a biological model.

The most prominent feature is the concept of excitatory learning within a neighborhood around the winning neuron. The size of the neighborhood slowly decreases with each iteration. A version of the training rule is described below:

1. First, a winning neuron is selected as the one with the shortest Euclidean distance (nearest neighbor),
$$\|\mathbf{x} - \mathbf{w}_i\| ,$$
between its weight vector and the input vector, where $\mathbf{w}_i$ denotes the weight vector corresponding to the ith output neuron.

2. Let i^* denote the index of the winner and let I^* denote a set of indices corresponding to a defined neighborhood of winner i^*. Then the weights associated with the winner and its neighboring neurons are updated by
$$\Delta\mathbf{w}_j = \eta\left(\mathbf{x} - \mathbf{w}_j\right) ,$$
for all the indices $j \in I^*$, and η is a small positive learning rate. The amount of updating may be weighted according to a preassigned "neighborhood function," $\Lambda(j, i^*)$.
$$\Delta\mathbf{w}_j = \eta\Lambda\left(j, i^*\right)\left(\mathbf{x} - \mathbf{w}_j\right) , \tag{6.1}$$
for all j. For example, a *neighborhood function* $\Lambda(j, i^*)$ may be chosen as
$$\Lambda\left(j, i^*\right) = \exp\left(-\left|\mathbf{r}_j - \mathbf{r}_{i^*}\right|^2 / 2\sigma^2\right) \tag{6.2}$$
where $\mathbf{r}_j$ represents the position of the neuron j in the output space. The convergence of the feature map depends on a proper choice of η. One plausible choice is that $\eta = 1/t$, where t denotes the iteration number. The size of neighborhood (or σ) should decrease gradually.

3. The weight update should be immediately succeeded by the normalization of $\mathbf{w}_i$.

In the retrieving phase, all the output neurons calculate the Euclidean distance between the weights and the input vector and the winning neuron is the one with the shortest distance.

By updating all the weights connecting to a neighborhood of the target neurons, the SOFM enables the neighboring neurons to become more responsive to the same input pattern. Consequently, the correlation between neighboring nodes can be enhanced. Once such a correlation is established, the size of a neighborhood can be decreased gradually, based on the desire of having a stronger identity of individual nodes.

Application examples: There are many examples of successful applications of SOFMs. More specifically, the SOFM network was used to evaluate the quality of a saw blade by analyzing its vibration measurements, which ultimately determines the performance of a machining process [7]. The major advantage of SOFMs is their unsupervised learning capability, which makes them ideal for machine health monitoring situations (e.g., novelty detection in medical images can then be performed online or classes can be labeled to give diagnosis [35]). A good system configuration algorithm produces the required performance and reliability with maximum economy. Actual design changes are frequently kept to a minimum to reduce the risk of failure. As a result, it is important to analyze the configurations, components, and materials of past designs so that good aspects may be reused and poor ones changed. A generic method of configuration evaluation based on an SOFM has been successfully reported [88]. The SOFM architecture with activation retention and decay in order to create unique distributed response patterns for different sequences has also been successfully proposed for mapping between arbitrary sequences of binary and real numbers, as well as phonemic representations of English words [45]. By using a selective learnable SOFM, which has the special property of effectively creating spatially organized internal representations and nonlinear relations of various input signals, a practical and generalized method was proposed in which effective nonlinear shape restoration is possible regardless of the existence of distortion models [34]. There are many other examples of successful applications (e.g., [24, 62, 111]).

The Self-Organizing Tree Map (SOTM)

The motivation for the self-organizing tree map (SOTM) [54] — SOFM with a hierarchical structure is different from Kohonen's motivation for the original SOFM — is a nonparametric regression model, but it is an effective tool for accurate clustering/classification leading to segmentation and other image/multimedia processing applications.

The SOFM is a good clustering method, but it has some undesirable properties when an input vector distribution has a prominent maximum. The results of the best-match computations tend to be concentrated on a fraction of nodes in the map. Therefore, the reference vectors lying in zero-density areas may be affected by input vectors from the surrounding nonzero distribution areas. Such phenomena are largely due to the nonparametric regression nature of the SOFM.

In order to overcome the aforementioned problems, tree-structured SOFMs were proposed. A typical example is the SOTM [54]. The main characteristic of the SOTM is that it exhibits better fitting of the input data.

In the SOTM, the relationships between the output nodes are defined adaptively during learning. Unlike the SOFM, which has a user-predefined and fixed number of nodes in the network, the number of nodes is determined automatically by the learning process based on the distribution of the input data. The clustering algorithm starts from an isolated node and coalesces the nearest patterns or groups according to a hierarchy control function from the root node to the leaf nodes to form the tree. The proposed approach has the advantage of K-means, with their ability to accurately locate cluster centers, and the SOFM's topology-preserving property. The SOTM also provides a better and faster approximation of prominently structured density functions.

Using the definitions of the input vector $\mathbf{x}(t)$ and the weight vector $\mathbf{w}_j(t)$, the SOTM algorithm is summarized as follows:

1. Select the winning node j^* with minimum Euclidean distance d_j,

$$d_{j^*}\left(\mathbf{x}, \mathbf{w}_{j^*}\right) = \min_j d_j\left(\mathbf{x}, \mathbf{w}_j\right)$$

2. If $d_{j^*}(\mathbf{x}, \mathbf{w}_{j^*}) \leq H(t)$ where $H(t)$ is the hierarchy control function, which controls the number of levels of the tree and decreases with time, then assign x to the jth cluster and update the weight vector $\mathbf{w}_j$ according to the following learning rule:

$$\mathbf{w}_{j^*}(t+1) = \mathbf{w}_{j^*}(t) + \eta(t)\left[\mathbf{x}(t) - \mathbf{w}_{j^*}(t)\right] \tag{6.3}$$

where $\eta(t) = e^{(-t/T_1)}$ (with T_1 determining the rate of convergence) is the learning rate, which decreases with time and satisfies $0 < \eta(t) < 1$.

Else form a new subnode with $\mathbf{x}$ as the weight vector.

3. Repeat by going back to step 1.

The hierarchy control function $H(t) = e^{(-t/T_2)}$ (with T_2 being a constant which regulates the rate of decrease) controls the number of levels of the tree. It adaptively partitions the input vector space into subspaces.

With the decrease of the hierarchy control function $H(t)$, a subnode forms a new branch. The evolution process progresses recursively until it reaches the leaf node. The entire tree structure preserves topological relations from the root node to the leaf nodes.

The SOTM is much better than the SOFM at preserving the topological relations of the input dataset, as shown in the example. The learning of the tree map in Figure 6.1a is driven by sample vectors uniformly distributed in the English letter "K." The tree mapping starts from the root node and gradually generates its subnodes as $H(t)$ decreases. By properly controlling the rate of decrease $\alpha(t)$, the final representation of the letter "K" is shown in Figure 6.1b. For comparison, the SOFM is also used in this example, as shown in Figure 6.1c. The superiority of the SOTM is apparent.

The other tree-structured SOFM models that share many similarities with the SOTM include the self-generating neural networks [128], the hierarchical SOTM [51], and the self-partitioning neural networks [104].

Application examples: The SOTM and the other tree-structured SOFMs have been used in many image and multimedia applications. Self-generating neural networks have been applied to visual communications [128], the hierarchical SOFM for range image segmentation [51], the self-partitioning neural networks for target detection and recognition [104], and the SOTM for quality cable TV transmission [54], image segmentation, and image/video compression [55].

Principal Component Analysis (PCA)

Principal component analysis (PCA) provides an effective way to find representative components of a large set of multivariate data. The basic learning rules for extracting principal components follow the Hebbian rule and the Oja rule [57, 94]. PCA can be implemented using an unsupervised learning network with traditional Hebbian-type learning. The basic network is one where the neuron is a simple linear unit with output $a(t)$ defined as follows:

$$a(t) = \mathbf{w}(t)^T \mathbf{x}(t) . \tag{6.4}$$

To enhance the correlation between the input $\mathbf{x}(t)$ and the output $a(t)$, it is natural to use a Hebbian-type rule:

$$\mathbf{w}(t+1) = \mathbf{w}(t) + \beta \mathbf{x}(t) a(t) . \tag{6.5}$$

The above Hebbian rule is impractical for PCA, taking into account the finite-word-length effect, since the training weights will eventually overflow (i.e., exceed the limit of dynamic range) before the first component totally dominates and the other components sufficiently diminish. An effective technique to overcome the overflow problem is to keep normalizing the weight vectors after each update. This leads to the Oja learning rule or, simply, the Oja rule:

$$\mathbf{w}(t+1) = \mathbf{w}(t) + \beta \left[\mathbf{x}(t) a(t) - \mathbf{w}(t) a(t)^2 \right] . \tag{6.6}$$

In contrast to the Hebbian rule, the Oja rule is numerically stable.

For the extraction of multiple principal components, a lateral *network structure* was proposed [57]. The structure incorporates lateral connections into the network. The structure, together with an orthogonalization learning rule, helps ensure the preservation of "orthogonality" between multiple principal components. A numerical analysis on their learning rates and convergence properties has also been established.

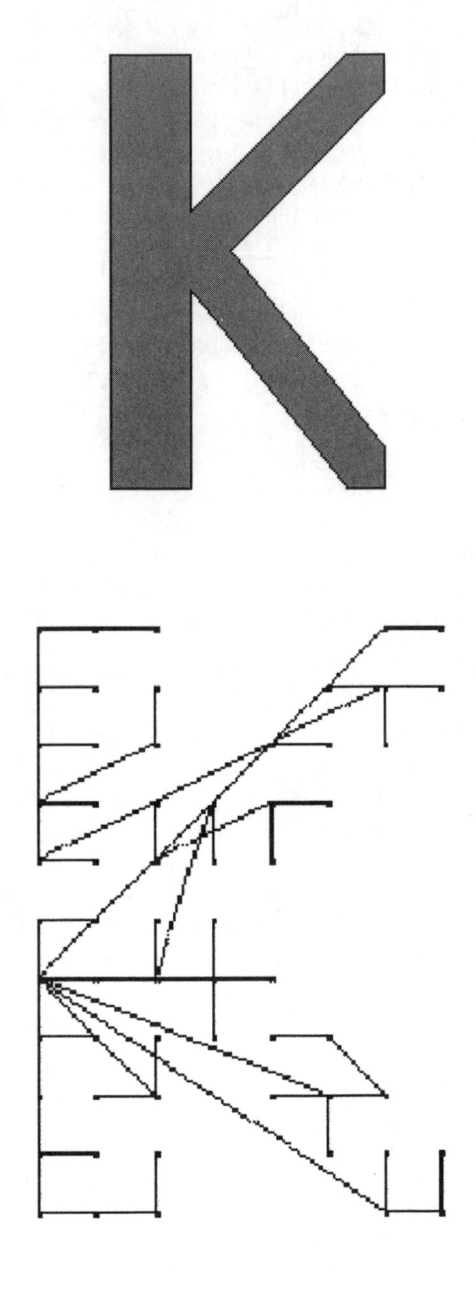

FIGURE 6.1
The SOTM for representation: (a) English letter "K;" (b) the representation of "K" by the SOTM; (c) the representation of "K" by the SOFM. ***(Cont.).***

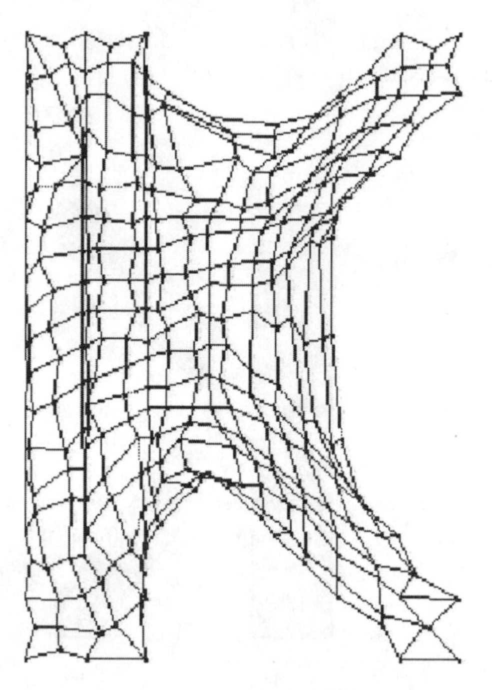

FIGURE 6.1
***(Cont.)* The SOTM for representation: (a) English letter "K;" (b) the representation of "K" by the SOTM; (c) the representation of "K" by the SOFM.**

Application examples: The lipreading system of Bregler and Konig [8], an early attempt in using both audio and visual features, used PCA to guide the snake search (the so-called active shape models [23]) on gray-scale video for the visual front end. There are two ways to perform PCA: (1) contour-based PCA is directly based on the located points from the snake search (form feature vectors using the located points and projected onto a few principal components); (2) area-based PCA is directly based on the gray-level matrix surrounding the lips. Instead of reducing the dimensionality of the visual features, as performed by the contour-based KLT, one can reduce the variation of mouth shapes by summing fewer principal components to form the contours. It was concluded that gray-level matrices contain more information for classifying visemes. Another attempt in PCA-based lip motion modeling is to express the PCA coefficients as a function of a limited set of articulatory parameters which describe the external appearance of the mouth [66]. These articulatory parameters have been directly estimated from the speech waveform based on a bank of (time-delay) NNs. A PCA-based Eigenface technique for a face recognition algorithm was studied in [6]. Its performance was compared with a computationally compatible "Fisherface" method based on tests on the Harvard and Yale Face Databases.

6.2.2 Multimedia Data Detection and Classification

In many application scenarios [e.g., optical character recognition (OCR), texture analysis, face detection] several prior examples of a targeted class or object are available for training, whereas the a priori class probability distribution is unknown. These training examples may be best exploited as valuable teacher information in supervised learning models. In general, detection and classification based on supervised learning models by far outperform those via unsupervised clustering techniques. That is why supervised neural networks are generally adopted for detection and classification applications.

Multilayer Perceptron

Multilayer perceptron (MLP) is one of the most popular NN models. In this model, each neuron performs a linear combination on its inputs. The result is then nonlinearly transformed by a sigmoidal function. In terms of structure, the MLP consists of several layers of *hidden neuron units* between the input and output neuron layers. The most commonly used learning scheme for the MLP is the *back-propagation* algorithm [106]. The weight updating for the hidden layers is performed based on a back-propagated corrective signal from the output layer. It has been shown that the MLP, given its flexible network/neuron dimensions, offers a universal approximation capability. It was demonstrated in [129] that two-layer perceptrons (i.e., networks with one hidden layer only) should be adequate as universal approximators of any nonlinear functions.

Let us assume an L-layer feed-forward neural network (with N_l units at the lth layer). Each unit, say the ith unit at the $(l+1)$th layer, receives the weighted inputs from other units at the lth layer to yield the net input $u_i(l+1)$. The net input value $u_i(l+1)$, along with the external input $\theta_i(l+1)$, will determine the new activation value $a_i(l+1)$ by the *nonlinear activation function* $f_i(l+1)$. From an algorithmic point of view, the processing of this multilayer feed-forward neural network can be divided into two phases: *retrieving* and *learning*.

Retrieving phase: Suppose that the weights of the network are known. In response to the input (test pattern) $\{a_i(0), i = 1, \ldots, N_0\}$, the system dynamics in the retrieving phase of an L-layer MLP network iterate through all the layers to generate the response $\{a_i(L), i = 1, \ldots, N_L\}$ at the output layer.

$$
\begin{aligned}
u_i(l+1) &= \sum_{j=1}^{N_l} w_{ij}(l+1)a_j(l) + \theta_i(l+1) \\
a_i(l+1) &= f_i\left(u_i(l+1)\right) = f_i(l+1)
\end{aligned}
\tag{6.7}
$$

where $1 \le i \le N_{l+1}, 0 \le l \le L-1$, and f_i is nondecreasing and differentiable (e.g., sigmoid function [106]). For simplicity, the external inputs $\{\theta_i(l+1)\}$ are often treated as special modifiable synaptic weights $\{w_{i,0}(l+1)\}$ which have clamped inputs $a_0(l) = 1$.

Learning phase: The learning phase of this L-layer MLP network follows a simple gradient descent approach. Given a pair of input/target training patterns, $\{a_i(0), i = 1, \ldots, N_0\}$, $\{t_j, j = 1, \ldots, N_L\}$, the goal is to iteratively (by presenting a set of training pattern pairs many times) choose a set of $\{w_{ij}(l), \forall l\}$ for all layers so that the squared error function E can be minimized:

$$
E = \frac{1}{2} \sum_{i=1}^{N_L} (t_i - a_i(L))^2 \tag{6.8}
$$

To be more specific, the iterative gradient descent formulation for updating each specific weight $w_{ij}(l)$ given a training pattern pair can be written as

$$
w_{ij}(l) \Longleftarrow w_{ij}(l) - \eta \frac{\partial E}{\partial w_{ij}(l)} \tag{6.9}
$$

where $\frac{\partial E}{\partial w_{ij}(l)}$ can be computed effectively through a numerical chain rule by back-propagating the error signal from the output layer to the input layer.

Other popular learning techniques of MLPs include discriminative learning [49], the support vector machine [36], and learning by evolutionary computation [137].

Due to the popularity of MLPs, it is not possible to exhaust all the numerous IMP applications using them. For example, Sung and Poggio [114] used MLP for face detection and Huang [40] used it as preliminary channels in an overall fusion network. More details about using MLPs for multimodal signal will be discussed in the audiovisual processing section.

RBF and OCON Networks

Another type of feed-forward network is the *radial basis function* (RBF) network. Each neuron in the hidden layer employs an RBF (e.g., a Gaussian kernel) to serve as the activation function. The weighting parameters in the RBF network are the centers, the widths, and the heights of these kernels. The output functions are the linear combination (weighted by the heights of the kernels) of these RBFs. It has been shown that the RBF network has the same universal approximation power as an MLP [98].

The conventional MLP adopts an all-class-in-one-network (ACON) structure, in which all the classes are lumped into one supernetwork. The supernet has the burden of having to simultaneously satisfy all the teachers, so the number of hidden units tends to be large. *Empirical results confirm that the convergence rate of ACON degrades drastically with respect to the network size because the training of hidden units is influenced by (potentially conflicting) signals from different teachers* [57].

In contrast, it is natural for the RBF to adopt another type of network structure — the one-class-in-one-network (OCON) structure — where one subnet is designated to one class only. The difference between these two structures is depicted in Figure 6.2. Each subnet in the OCON network specializes in distinguishing its own class from the others, so the number of hidden units is usually small. In addition, OCON structures have the following features:

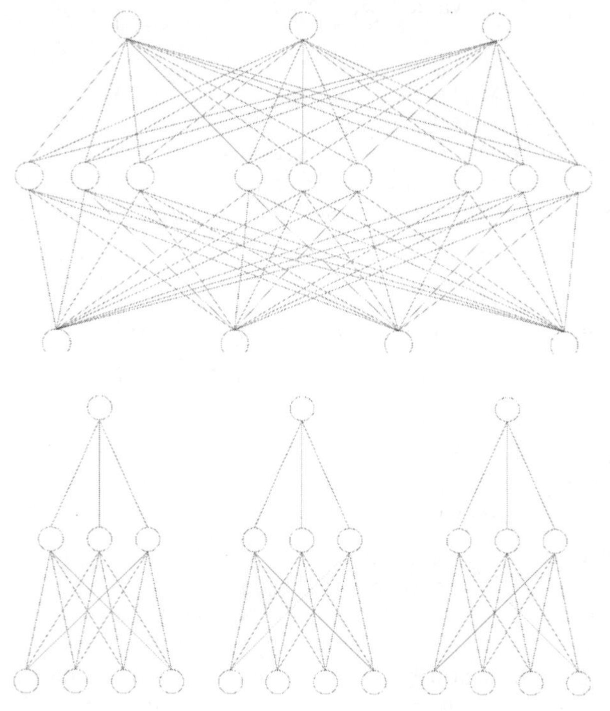

FIGURE 6.2
(a) An ACON structure; (b) an OCON structure.

- Locally, unsupervised learning may be applied to determine the initial weights for individual subnets. The initial clusters can be trained by VQ or K-mean clustering techniques. If the cluster probabilities are desired, the EM algorithm can be applied to achieve maximum likelihood estimation for each *class conditional likelihood density*.
- The OCON structure is suitable for incremental training (i.e., network upgrading through the addition/removal of memberships [57, 58]).
- The OCON network structure supports the notion of distributed processing. It is appealing to smart card biometric systems. An OCON-type classifier can store personal discriminant codes in individual class subnets, so the magnet strip in the card needs to store only the network parameters in the subnet that have been designated to the card holder.

Application examples: In [11], Brunelli and Poggio proposed a special type of RBF network called the "hyperBF" network for successful face recognition applications. In [72], the associated audio information is exploited for video scene classification. Several audio features have been found to be effective in distinguishing audio characteristics of different scene classes. Based on these features, a neural net classifier can successfully separate audio clips from different TV programs.

Decision-Based Neural Network

A decision-based neural network (DBNN) [58] has two variants: one is a hard-decision model and the other is a probabilistic model. A DBNN has a modular OCON network structure: one subnet is designated to represent one object class. For multiclass classification problems, the outputs of the subnets (the *discriminant functions*) will compete with each other, and the subnet with the largest output value will claim the identity of the input pattern.

Decision-Based Learning Rule The learning scheme of the DBNN is decoupled into two phases: *locally unsupervised* and *globally supervised* learning. The purpose is to simplify the difficult estimation problem by dividing it into several localized subproblems and, thereafter, the fine-tuning process would involve minimal resources.

- Locally Unsupervised Learning: VQ or EM Clustering Method

  Several approaches can be used to estimate the number of hidden nodes, or the initial clustering can be determined based on VQ or EM clustering methods.

  - In the hard-decision DBNN, the VQ-type clustering (e.g., K-mean) algorithm can be applied to obtain initial locations of the centroids.
  - For the probabilistic DBNN, called PDBNN, the EM algorithm can be applied to achieve the maximum likelihood estimation for each *class conditional likelihood density*. (Note that once the likelihood densities are available, the posterior probabilities can be easily obtained.)

- Globally Supervised Learning

  Based on this initial condition, the decision-based learning rule can be applied to further fine-tune the decision boundaries. In the second phase of the DBNN learning scheme, the objective of the learning process changes from maximum likelihood estimation to *minimum classification error*. Interclass mutual information is used to fine-tune the decision boundaries (i.e., the *globally supervised* learning). In this phase, DBNN applies the reinforced–antireinforced learning rule [58], or discriminative learning rule [49], to adjust network parameters. Only misclassified patterns are involved in this training phase.

- Reinforced–Antireinforced Learning Rules

  Suppose that the mth training pattern $\mathbf{x}^{(m)}$ is known to belong to class Ω_i, and that the leading challenger is denoted as $j = arg\max_{j \neq i} \phi(\mathbf{x}^{(m)}, \mathbf{w}_j)$. The learning rule is

$$\begin{array}{ll}\text{Reinforced learning:} & \mathbf{w}_i^{(m+1)} = \mathbf{w}_i^{(m)} + \eta \nabla \phi \left(\mathbf{x}^{(m)}, \mathbf{w}_i\right) , \\ \text{Antireinforced learning:} & \mathbf{w}_j^{(m+1)} = \mathbf{w}_j^{(m)} - \eta \nabla \phi \left(\mathbf{x}^{(m)}, \mathbf{w}_j\right) .\end{array}$$

Application examples: DBNN is an efficient neural network for many pattern classification problems, such as OCR and texture classification [57] and face and palm recognition problems [68, 71]. A modular neural network based on DBNN and a model-based neural network have recently been proposed for interactive human–computer vision.

Mixture of Experts

Mixture of experts (MOE) learning [44] has been shown to provide better performance due to its ability to effectively solve a large complicated task by smaller and modularized trainable networks (i.e., experts), whose solutions are dynamically integrated into a coherent one using the trainable gating network. For a given input $\mathbf{x}$, the posterior probability of generating class $\mathbf{y}$ given $\mathbf{x}$ using K experts is computed by

$$P(\mathbf{y}|\mathbf{x}, \phi) = \sum_{i=1}^{K} g_i P(\mathbf{y}|\mathbf{x}, \theta_i) , \tag{6.10}$$

where $\mathbf{y}$ is a binary vector, ϕ is a parameter vector $[\mathbf{v}, \theta_i]$, g_i is the probability for weighting the expert outputs, $\mathbf{v}$ is a vector of the parameters for the gating network, θ_i is a vector of the parameters for the ith expert network ($i = 1, \ldots, K$), and $P(\mathbf{y}|\mathbf{x}, \theta_i)$ is the output of the ith expert network.

The gating network can be a nonlinear neural network or a linear neural network. To obtain the linear gating network output, the softmax function is utilized [10]:

$$g_i = \exp(b_i) / \sum_{j=1}^{K} \exp(b_j) \tag{6.11}$$

where $b_i = \mathbf{v}_i^T \mathbf{x}$, with $\mathbf{v}_i$ denoting the weights of the ith neuron of the gating network.

The learning algorithm for the MOE is based on the maximum likelihood principle to estimate the parameters (i.e., choose parameters for which the probability of the training set given the parameters is the largest). The gradient ascent algorithm can be used to estimate the parameters.

Assume that the training dataset is $\{\mathbf{x}^{(t)}, \mathbf{y}^{(t)}\}$, $t = 1, \ldots, N$. First, we take the logarithm of the product of N densities of $P(\mathbf{y}|\mathbf{x}, \phi)$:

$$l(\mathbf{y}, \mathbf{x}, \phi) = \sum_t \sum_i \log \left[g_i^{(t)} P\left(\mathbf{y}^{(t)}|\mathbf{x}^{(t)}, \theta_i\right)\right] . \tag{6.12}$$

Then, we maximize the log likelihood by gradient ascent. The learning rule for the weight vector v_i in a linear gating network is obtained as follows:

$$\Delta \mathbf{v}_i = \rho \sum_t \left(h_i^{(t)} - g_i^{(t)}\right) \mathbf{x}^{(t)} , \tag{6.13}$$

where ρ is a learning rate and $h_i = g_i P(\mathbf{y}|\mathbf{x}, \theta_i) / \sum_t g_j P(\mathbf{y}|\mathbf{x}, \theta_j)$.

The MOE [44] is a modular architecture in which the outputs of a number of "experts," each performing classification tasks in a particular portion of the input space, are combined in a probabilistic way by a "gating" network which models the probability that each portion of the input space generates the final network output. Each local expert network performs multi-way classification over K classes by using either a K-independent binomial model, with each model representing only one class, or one multinomial model for all classes.

Application example: The MOE model was applied to a time series analysis with well-understood temporal dynamics, and produced significantly better results than single networks. It also discovered the regimes correctly. In addition, it allowed the users to characterize the subprocesses through their variances and avoid overfitting in the training process [76]. A Bayesian framework for inferring the parameters of an MOE model based on ensemble learning by variational free energy minimization was successfully applied to sunspot time series prediction [126]. By integrating pretrained expert networks with constant sensitivity into an MOE configuration, the trained experts are able to divide the input space into specific subregions with minimum ambiguity, which produces better performance in automated cytology screening applications [42]. By applying a likelihood splitting criterion to each expert in the HME, Waterhouse and Robinson [125] first grew the HME tree adaptively during training; then, by considering only the most probable path through the tree, they pruned branches away, either temporarily or permanently, in case of redundancies. This improved HME showed significant speedups and more efficient use of parameters over the standard fixed HME structure for both simulation problems and real-world applications, as in the prediction of parameterized speech over short time segments [127]. The HME architecture has also been applied to text-dependent speaker identification [16].

A Network of Networks

A network of networks (NoN) is a multilevel neural network consisting of nested clusters of neurons capable of hierarchical memory and learning tasks. The architecture has a fractal-like structure, in that each level of organization consists of interconnected arrangements of neural clusters. Individual elements in the model form level zero cluster organization. Local groupings among the elements via certain types of connections produce level one clusters. Other connections link level one clusters to form level two clusters, while the coalescence of level two clusters yields level three clusters, and so on [115]. A typical NoN is schematically depicted in Figure 6.3. The structure of the NoN makes it a natural choice for massive parallel processing and a hierarchical search engine.

Training of the NoN is very flexible. Mean field theory [116] and Hebbian learning algorithms [2] were among the first to be used in the NoN.

Recently, EP was proposed to discover clusters in the NoN in the context of adaptive segmentation/image regularization [131]. First a population of potential processing strategies is generated and allowed to compete under a k-pdf error measure E^k_{pdf}, of data quality which is defined as the following weighted probability density error measure:

$$E^k_{pdf} = \int_0^N w(k) \left(p^m_k(k) - p_k(k)\right)^2 dk \tag{6.14}$$

where the variable k is defined in [131], E is a factor which characterizes the correlation of each item in the dataset with a prescribed neighboring subset, $p_k(k)$ is the probability density function of k within the dataset to be processed, $p^m_k(k)$ characterizes the density function of a

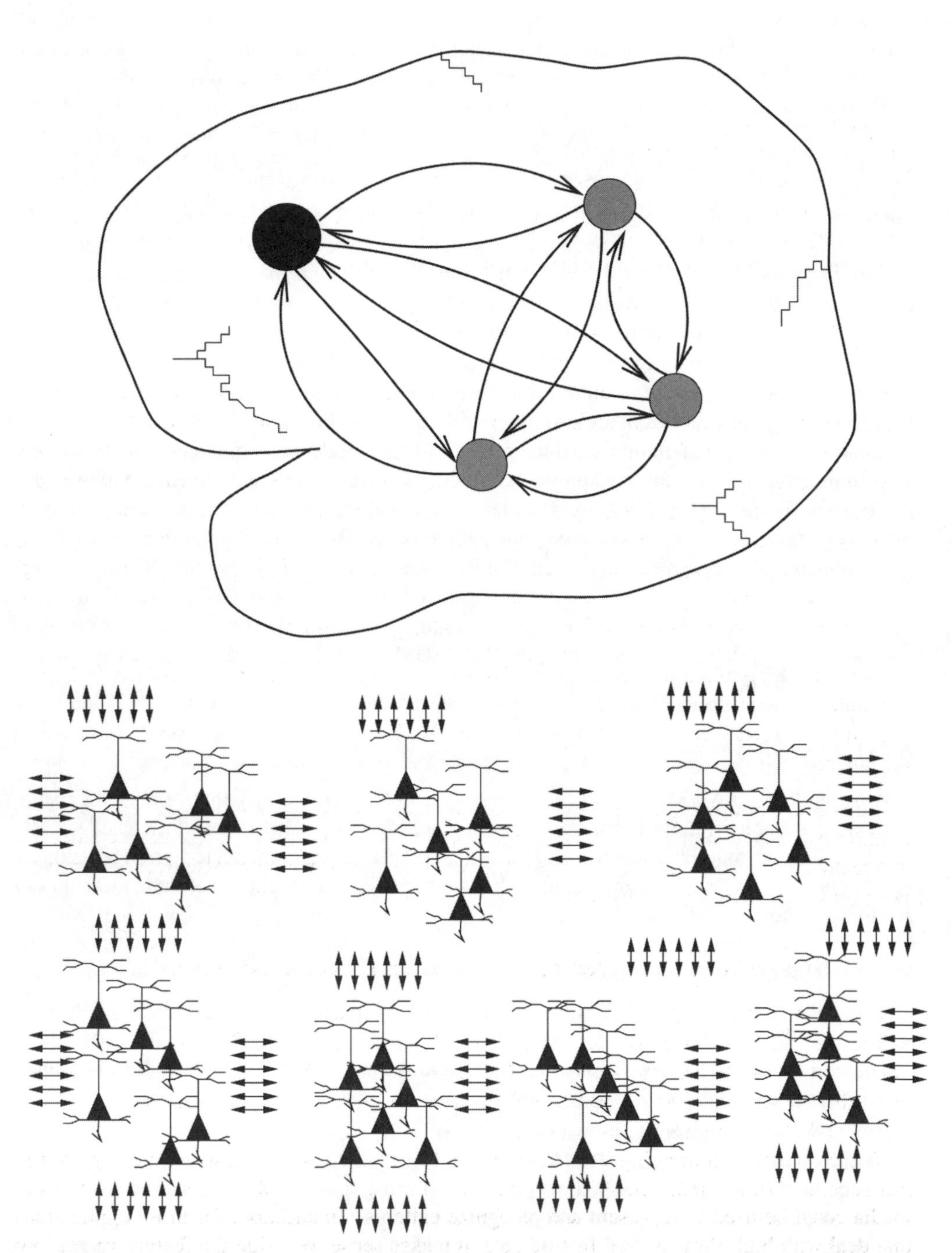

FIGURE 6.3
Schematic representation of a biologically inspired network: (a) the overall network; (b) a simplified connection model within one part of the network in (a) (the black dot at the top left corner, for example), which itself is a three-level NoN.

model dataset with certain desired properties, and $w(k)$ is the weighting coefficient defined as

$$w(k) = \frac{1}{\max\left(p_k^m(k), p_k(k)\right)^2} \tag{6.15}$$

to compensate for the generally smaller contribution of the tail region to the total probability. In the context of image processing, it was shown in [131] that a small k represents a smooth image region, a medium k represents an area with one or two dominant edges, and a large k represents a texture area. Optimization is carried out on (6.14) in order to identify the clusters in terms of the following regularization parameter assignment function $\lambda(\sigma)$, which is a decreasing sigmoid function of the local data standard deviation σ:

$$\lambda(\sigma) = \frac{\lambda_{\max} - \lambda_{\min}}{1 + e^{\beta(\sigma - \alpha)}} + \lambda_{\min} \tag{6.16}$$

where $\lambda_{\min}$ and $\lambda_{\max}$ are the minimum and the maximum regularization parameters used, α represents the offset of the sigmoidal function from the origin, and β controls the steepness of the function. Apart from the assignment of the local regularization parameters, this function indirectly achieves segmentation if we identify image pixels with similar associated λ values as a single cluster. Concatenating them with their respective strategy parameters [29] $\sigma_{\lambda_{\min}}, \sigma_{\lambda_{\max}}, \sigma_\alpha, \sigma_\beta$ into an eight-tuple, we define the following regularization strategy S_p as the pth potential optimizer in the population:

$$S_p = \left(\lambda_{\min,p}, \lambda_{\max,p}, \alpha_p, \beta_p, \sigma_{\lambda_{\min},p}, \sigma_{\lambda_{\max},p}, \sigma_{\alpha,p}, \sigma_{\beta,p}\right) \tag{6.17}$$

We generate a population P consisting of μ instances of S_p in the first generation and apply the mutation operation [29] to each of these μ parents to generate μ descendants. In this and subsequent generations, the potential optimizers undergo a competition process from which the emerged winners are incorporated into the new population in the next generation.

Application example: The first engineering application of the NoN was in signal categorization by Anderson et al. [1]. Guan studied the NoN and proposed a hierarchical adaptive image processing based on it [32]. He later developed a low-level vision model to recursively perform segmentation and edge extraction [33].

6.2.3 Hierarchical Fuzzy Neural Networks as Linear Fusion Networks

In many multimedia applications, it is useful to have a versatile multimedia fusion subsystem, where information from various sensors are laterally combined to yield improved classification. Neural networks offer a natural solution for sensor or media fusion. This is because of their capability for nonlinear and nonparametric estimation in the absence of complete knowledge on the underlying models or sensor noises.

The problem of combining the classification power of several classifiers is of great importance to various applications. First, for several recognition problems, numerous types of media could be used to represent and recognize patterns. In addition, for those applications that deal with high-dimensional feature data, it makes sense to divide the feature vector into several lower-dimensional vectors before integrating them for a final decision (i.e., divide and conquer).

Most of the current information fusion models are based on a linear combination of outputs weighted by some proper confidence parameters. This is largely motivated by the following statistical and computational reasons:

- It can make use of the popular Bayesian formulation.

- It can facilitate adoption of EM training of the confidence parameters.

Channel Fusion

Two channel fusion models were proposed to deal with information from different media sources: class-dependent channel fusion and data-dependent channel fusion.

- The class-dependent channel fusion scheme deploys one PDBNN for each sensor channel. Each PDBNN receives only the patterns from its corresponding sensor. The *class* and *channel* conditional likelihood densities ($p(\mathbf{x}|\omega_i, C_j)$) are estimated. The outputs from different channels are combined in the weighted sum fashion. *The weighting parameters, $P(C_j|\omega_i)$, represent the confidence of the channel C_j producing the correct answer for the object class ω_i.* $P(C_j|\omega_i)$ can be trained by the EM algorithm; after that, its value is fixed during the identification process (recall that the values of the weighting parameters in the HME are functions of the input pattern). Figure 6.4a illustrates the

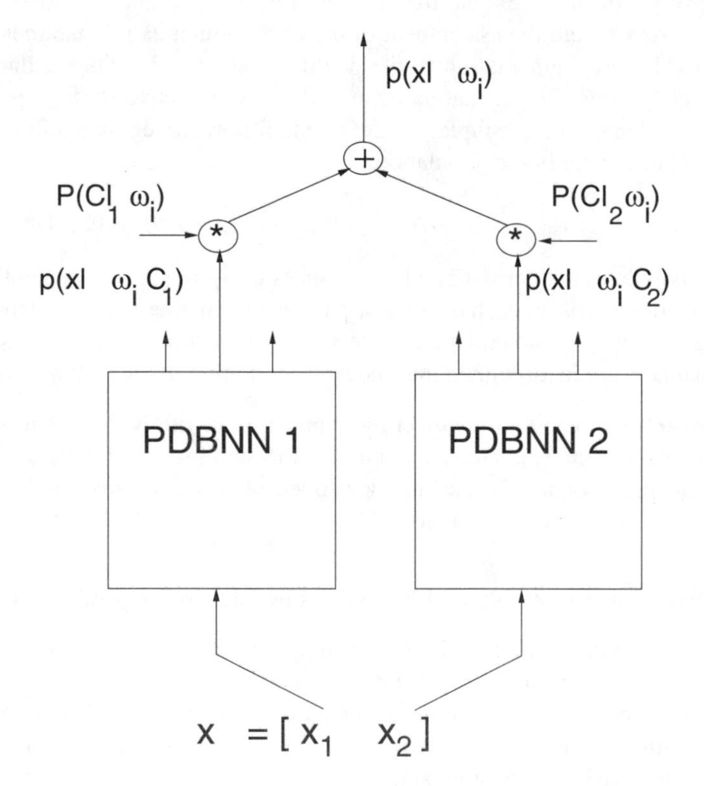

FIGURE 6.4
A media fusion network: linear fusion of probabilistic DBNN classifiers. (a) For the applications where there are several sensor sources, the class-dependent channel fusion scheme can be applied for classification. $P(C_j|\omega_i)$ is a trainable parameter. Its value is fixed during the retrieving phase.

structure of the class-dependent channel fusion scheme. The class-dependent channel fusion scheme considers the data distribution as the mixture of the likelihood densities from various sensor channels. This is a simplified density model. If the feature dimension is very large and the number of training examples is relatively small, the direct estimation approaches can hardly obtain good performance due to the curse of dimen-

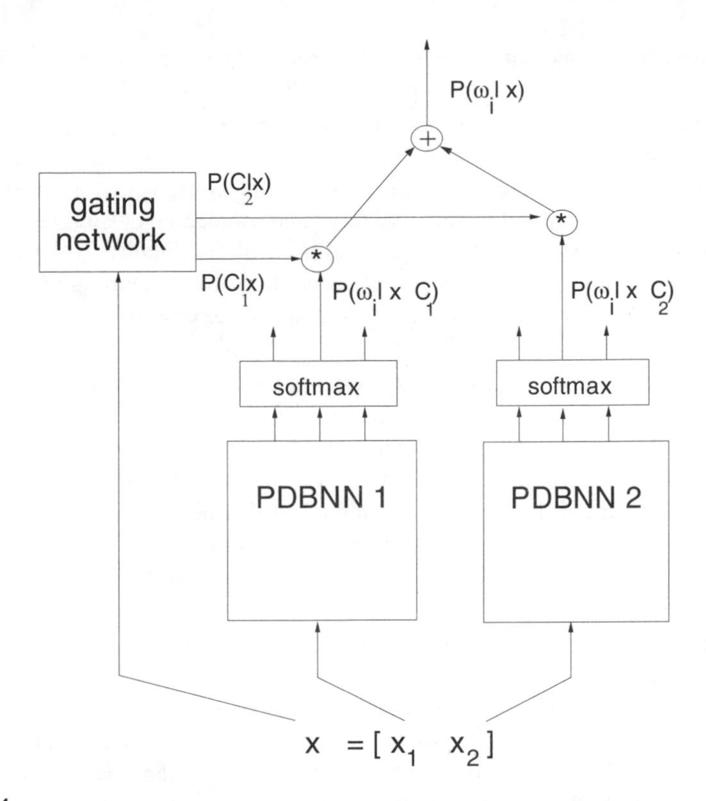

FIGURE 6.4
(Cont.) **A media fusion network: linear fusion of probabilistic DBNN classifiers. (b) Data-dependent channel fusion scheme. In this scheme, the channel weighting parameters are functions of the input pattern x ($P(C_j|\mathbf{x})$).**

sionality. For this kind of problem, since the class-dependent fusion scheme greatly reduces the number of parameters, it could achieve better estimation results.

- Another fusion scheme is the data-dependent channel fusion. Figure 6.4b shows the structure of this scheme. Like the class-dependent fusion method, each sensor channel has a PDBNN classifier. The outputs of the PDBNNs are transformed into the posterior probabilities by the softmax functions [10]. In this fusion scheme, the channel weighting $P(C_j|\mathbf{x})$ is a function of the input pattern $\mathbf{x}$. Therefore, the importance of an individual channel may vary if the input pattern is different.

Fuzzy Systems and Modular Neural Networks

The basic idea behind a fuzzy inference system is to incorporate the human "expert's experience" into system design. The input–output relationship is described by a collection of fuzzy inference rules involving linguistic variables. The typical architecture of a fuzzy system is composed of four components:

- A fuzzifier, which maps crisp numbers into suitable linguistic values
- A fuzzy rule base, which stores the knowledge of the human experts and the empirical observations
- An inference engine, which deduces the desired output by performing approximate reasoning

- A defuzzifier, which extracts a crisp value from a fuzzy set as a representative value.

Fuzzy logic systems, in contrast to neural networks, offer a structural framework with high-level fuzzy rule thinking and reasoning. Fuzzy systems base their decisions on inputs in the form of linguistic variables defined by membership functions, which are formulas used to determine the fuzzy set to which a value belongs and the degree of membership in that set. The variables are then matched with the preconditions of the linguistic rules to calculate the firing strengths of the rules, and the response of each rule is obtained through fuzzy implication. Following a compositional rule of inference, the response of each rule is weighted according to the rule firing strength.

It has recently become popular for a fuzzy system to utilize Gaussian membership functions and a centroid defuzzification scheme to calculate the output. This is in part due to the capability of this combination to approximate any real continuous functions on a compact set to an arbitrary accuracy, provided sufficient fuzzy logic rules are available [56, 123]. In the neural network literature, it has also been established that neural networks with normalized RBFs as the hidden node functions are also universal approximators [98]. Therefore, neural networks, especially those with modular structures, and fuzzy systems are similar in terms of approximation capabilities.

They also bear very sharp structural resemblance. A good example is to compare the fuzzy inference engine and the MOE modular neural networks. Stretching the similarity further, the intersection of fuzzy systems and neural networks actually defines a large family of learning networks. In the following it can be shown that this family of models can be built upon a common mathematical formulation and system architecture. In terms of learning capabilities, neural networks with RBFs as hidden nodes are basically equivalent to fuzzy systems using Gaussian membership function, product inference, and fuzzy rules with singleton consequents. It has been shown that an RBF MOE network and a fuzzy inference system are essentially equivalent as long as the gating network of the MOE generates the fuzzy membership values according to the membership function and the **And** operation in the fuzzy **If-Then** rule.

Bearing the above analysis in mind, Kung et al. [60] demonstrated that a hierarchical fuzzy neural network designed by combining the expert-level partitioning strategies of the MOE and the class-level partitioning of the DBNN offers an attractive processing structure for linear channel fusion. In particular, they proposed to adopt expert-in-class hierarchical structure (ECHS) for class-dependent channel fusion and class-in-expert hierarchical structure (CEHS) for data-dependent channel fusion.

Hierarchical Fuzzy Neural Networks for Class-Dependent Channel Fusion

The architecture of the ECHS is exactly the same as for the class-dependent channel fusion model illustrated in Figure 6.4a. The inner blocks comprise expert-level modules, whereas the outer blocks are on the class level. A typical example of this type of network is the hierarchical DBNN [59], which describes the class discriminant function as a mixture of multiple probabilistic distribution. That is, the discriminant function of the class ω_c in the hierarchical DBNN is a class conditional likelihood density which can be described as follows:

$$p\left(\mathbf{x}(t)|\omega_i\right) = \sum_{k=1}^{K} P\left(C_k|\omega_i\right) p\left(\mathbf{x}(t)|\omega_i, C_k\right) ,$$

where $p(\mathbf{x}(t)|\omega_i, C_k)$ is the discriminant function of subnet i in channel k, and $p(\mathbf{x}(t)|\omega_i)$ is the combined discriminant function for class ω_i. The channel confidence $P(C_k|\omega_i)$ can be learned by the following procedure. Define $\alpha_k = P(C_k|\omega_i)$. At the beginning, assign

$\alpha_k = 1/K, \forall k = 1, \ldots, K$. At step j,

$$h_k^{(j)}(t) = \frac{\alpha_k^{(j)} p(\mathbf{x}(t)|\omega_i, C_k)}{\sum_l \alpha_l^{(j)} p(\mathbf{x}(t)|\omega_i, C_l)}, \qquad \alpha_k^{(j+1)} = \frac{1}{N} \sum_{t=1}^{N} h_k^{(j)}(t) . \tag{6.18}$$

In an ECHS, each expert processes only the local features from its corresponding class. The outputs from different experts are linearly combined. The weighting parameters, $P(C_k|\omega_i)$, represent the confidence of expert E_k producing the correct answer for the object class ω_i. Once they are trained, their values remain constant during the retrieving phase. By definition, $\sum_{k=1}^{K} P(C_k|\omega_i) = 1$, where K is the number of experts in the subnet ω_i. So it has the property of a probability function. Note that, within this expert-level (or rule-level) hierarchy, each hidden node in one class must be used to model a certain local expert with a varying degree of confidence, which reflects its ability to interpret a given input vector. The locally unsupervised and globally supervised schemes described in the previous section can be adopted to train the OCON network.

Hierarchical Fuzzy Neural Networks for Data-Dependent Channel Fusion

The architecture of the CEHS is exactly the same as for the data-dependent channel fusion model illustrated in Figure 6.4b. The inner blocks comprise class modules, whereas the outer blocks are the expert modules. Each expert has its own hierarchical DBNN classifier. The outputs of the hierarchical DBNNs are transformed into the posterior probabilities by softmax functions. In this fusion scheme, the expert weighting $P(E_j|\mathbf{x})$ is a function of input pattern $\mathbf{x}$. Therefore, the importance of an individual expert may vary with different input patterns observed.

The network adopts the posterior probabilities of electing a class given $\mathbf{x}(t)$ (i.e., $P(\omega_i|\mathbf{x}(t), C_k)$), instead of the likelihood of observing $\mathbf{x}(t)$ given a class (i.e., $p(\mathbf{x}(t)|\omega_i, C_k)$), to model the discriminant function of each cluster. For this version of hierarchical fuzzy neural networks, a new confidence $P(C_k|\mathbf{x}(t))$ is assigned, which stands for the confidence on expert k when the input pattern is $\mathbf{x}(t)$. Accordingly, the probability model is modified to become

$$P\left(\omega_i|\mathbf{x}(t)\right) = \sum_{k=1}^{K} P\left(C_k|\mathbf{x}(t)\right) P\left(\omega_i|\mathbf{x}(t), C_k\right) ,$$

where $P(\omega_i|\mathbf{x}(t), C_k) = P(\omega_i|C_k)p(\mathbf{x}(t)|\omega_i, C_k)/p(\mathbf{x}(t)|C_k)$, and the confidence $P(C_k|\mathbf{x})$ can be obtained by the following equation:

$$P\left(C_k|\mathbf{x}(t)\right) = \frac{P(C_k)p(\mathbf{x}|C_k)}{\sum_l P(C_l)p(\mathbf{x}(t)|C_l)} ,$$

where $p(\mathbf{x}(t)|C_k)$ can be computed as $p(\mathbf{x}(t)|C_k) = \sum_i P(\omega_i|C_k)p(\mathbf{x}(t)|\omega_i, C_k)$ and $P(C_k)$ can be learned by equation (6.18) with $p(\mathbf{x}(t)|\omega_i, C_k)$ replaced by $p(\mathbf{x}(t)|C_k)$. The term $P(C_k)$ can be interpreted as "the general confidence" we have in channel k. Unlike in the class-dependent approach, the fusion weights need to be computed for each testing pattern during the retrieving phase. Notice that this data-dependent fusion scheme can be considered a combination of PDBNN and MOE [44].

Application example: The class-dependent channel fusion scheme has been observed to have very good classification performance on vehicle recognition and face recognition problems [67]. The experiment in [67] used six car models from different view angles to create the training and testing database. Approximately 30 images (each 256 × 256 pixels) were taken for each car model from various viewing directions. There were 172 examples in the dataset.

Two classifier channels were built from two different feature extraction methods: one used intensity information and the other edge information. With the fusion of these two channels (with 94% and 85% recognition rates), the recognition rate reached 100%.

The fusion model was compared with a single network classifier. The input vectors of these two networks were formed by concatenating the intensity vector with the edge vector. Therefore, the input vector dimension became $144 \times 2 = 288$. The RBF-typed DBNN was used as the classifier. The experimental result showed that the performance was worse than for the fusion network (about 95.5% recognition rate).

6.2.4 Temporal Models for Multimodal Conversion and Synchronization

The class of neural networks that are most suitable for applications in multimodal conversion and synchronization is the so-called *temporal* neural network. Unlike the feed-forward type of artificial neural network, temporal networks allow bidirectional connections between a pair of neuron units, and sometimes feedback connections from a unit to itself. Let us elaborate further on this difference. From the perspective of connection patterns, neural networks can be grouped into two categories: *feed-forward* networks, in which the associated network graphs have no loops, and *recurrent* networks, where loops occur because of the existence of the feedback connections. Feed-forward networks are *static;* that is, a given input can produce only one set of output values rather than a sequence of data. Thus, they carry no memory. In contrast, many *temporal* neural networks employ some kind of *recurrent* network structure. Such an architectural attribute enables temporal information to be stored in the networks.

A simple extension to the existing feed-forward structure to deal with temporal sequence data is the *partially recurrent network* (sometimes called the *simple recurrent network*). The connection in a simple recurrent network (SRN) is mainly of the feed-forward type, but a carefully chosen set of feedback connections is also included. In most cases the feedback connections are fixed and not trainable. This judicious incorporation of recurrence allows the network to remember cues from the past without appreciably complicating the overall training procedure. The most widely used SRNs are Elman's network and Jordan's network [28, 47]. The time-delay neural network (TDNN) is a further extension to cope with the shift-invariance property required in speech recognition. It is achieved by making time-shifted copies of the hidden units and linking them to the output layer [122]. Several fully recurrent neural network architectures with the corresponding learning algorithms are real-time recurrent learning (RTRL) networks [130] and back-propagation through time (BPTT) networks [39, 106]. The computational requirements of these and several variants are very high. Among all the recurrent networks, BPTT's performance is the best unless online learning is required, in which case the RTRL is required instead. But for many applications involving temporal sequence data, an SRN or a TDNN may suffice and is much less costly than RTRL or BPTT.

Time-Delay Neural Network

Figure 6.5 shows the TDNN architecture [122] for a three-class temporal sequence recognition task. A TDNN is basically a feed-forward multilayer (four layers) neural network with time-delay connections at the hidden layer to capture varying amounts of contexts. The basic unit in each layer computes the weighted sum of its inputs and then passes this sum through a nonlinear sigmoid function to the higher layer. The TDNN classifier shown in Figure 6.5 has an input layer with 12 units, a hidden layer with 8 units, and an output layer with 3 units (one output unit represents one class).

When a TDNN is used for speech recognition, the speech utterance is partitioned frame by frame (e.g., 30 ms frame with 15 ms advance). Each frame is transformed into 12 coefficients, and every three frames with successive time delay 0, 1, and 2 are used as inputs to the 8 time

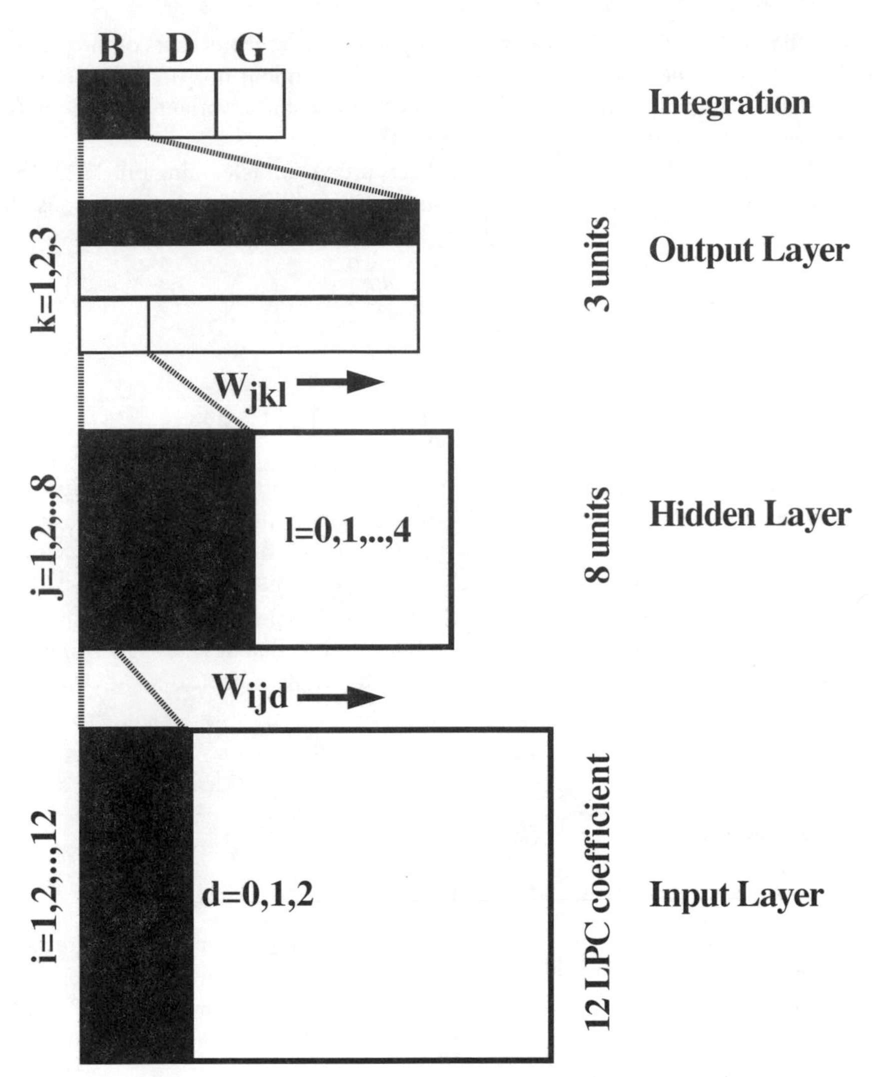

FIGURE 6.5
The architecture of a time-delay neural network (TDNN).

delay hidden units [i.e., each neuron in the first hidden layer now receives input (via 3×12 weighted connections) from the coefficients in the 3-frame window]. The 8-unit hidden layer is delayed 5 times to form a 40-unit layer. At the second hidden layer, each unit looks at all 5 copies of the delayed 8-unit hidden blocks of the first hidden layer. Finally, the output is obtained by integrating the information from the second hidden layer over time. This procedure can be formalized using the following equations:

$$y_c = \frac{1}{T-6} \sum_{t=7}^{T} b_c^{(t)} \tag{6.19}$$

$$b_c^{(t)} = S\left(\sum_{j=0}^{7} \sum_{l=0}^{4} w_{cjl}^{H} S\left[\sum_{i=1}^{12} \sum_{d=0}^{2} w_{ijd}^{I} x_i^{(t-l-d)} + \theta_j^{I} \right] + \theta_c^{H} \right), \tag{6.20}$$

where T is the total number of frames, $\mathbf{x}$ is the input, $\{b_c^{(t)}\}$ are the outputs of the c class at the second layer at different time instances, and $S(\cdot)$ is the sigmoid function. The tapped delay line structure of the input layer implies the adoption of the shift invariant assumption (i.e., the absolute time of a particular event is not important).

Like an MLP, a TDNN is also trained by the back-propagation learning rule [122]. Suppose the input to the TDNN is a vector $\mathbf{x}$; then the updating of the weights, $\mathbf{w}$, can be described by

$$\mathbf{w} \Longleftarrow \mathbf{w} - \eta \frac{\partial E}{\partial \mathbf{w}}$$

where

$$E = E(\{\mathbf{w}\}, \{\mathbf{x}\}) = \frac{1}{2} \sum_{c=1}^{C} (\mathbf{t}_c - \mathbf{y}_c(\mathbf{x}))^2 \ .$$

Therefore, through this training procedure, the local short duration features in speech signal can be formed at the lower layer and more complex longer duration features formed at the higher layer. The learning procedure ensures that each of the units in each layer has its weights adjusted in a way that improves the network's overall performance [41].

After the TDNN has learned its internal representation, it performs recognition by passing input speech over the TDNN neurons and selecting the class that has the highest output value. Section 6.3.3 presents an example employing such a TDNN model to audiovisual synchronization in the lipreading application.

6.3 Neural Networks for IMP Applications

Neural networks have played a very important role in the development of multimedia application systems [17, 43, 92, 124]. Their usefulness ranges from low-level preprocessing to high-level analysis and classification. A complete multimedia system consists of many of the following information processing stages, for which neural processing offers an efficient and unified core technology:

- Visualization, Tracking, and Segmentation
  - Neural networks have been found useful for some visualization applications, such as optimal image display [61] and color constancy and induction [25].
  - Feature-based tracking is crucial to motion analysis and the motion/shape reconstruction problem. Neural networks can be applied to motion tracking schemes for feature- and object-level tracking [18].
  - Segmentation is a very critical task for both image and video processing. Object boundary detection methods can use a hierarchical technique by adopting pyramid representation of images for computation efficiency [14, 73]. Active contour (e.g., snake) could also take advantage of the NN's adaptive learning capability for continuous and fast tracking of the region of interest (ROI) [21]. Both unsupervised and supervised neural networks may be adopted for object boundary detection methods, based on a variety of cues including motion, intensity, edge, color, and texture.

- Detection and Recognition
  - Neural networks can be applied to machine learning and computer vision problems with applications to detection and recognition of a specific object class. Examples are online OCR applications [12, 30], signature verification [5], currency recognition [117], and structure from motion [63].
  - Neural networks can facilitate detection or recognition of high-level features such as human faces in pictures or a certain object shape under inspection.
  - Multimodality recognition and authentication will have useful applications in network security and access control.
- Multimodal Coding, Conversion, and Synchronization
  - Multimodal coding, conversion, and synchronization will remain a challenging research task. Static MLP networks for multimodal facial image coding driven by speech and phonemes were already studied in [82].
  - Temporal NN models (e.g., TDNN) for multimodality synchronization, integrating audio and visual signals for lipreading, will be elaborated in Section 6.3.3.
- Video and Image Content Indexing and Browsing

  It is important for a system to possess the ability to fast access audiovisual objects, manipulate them, and present them in a highly flexible way. For video content selection, the ability to extract and utilize proper information content inherent in video clips may lead to efficient search schemes for many disciplines:

  - object-based and subject-based video indexing and databases
  - video skimming and browsing
  - content-based retrieval

  Again, neural processing presents a promising approach for these tasks.
- Interactive Human–Computer Communications

  Teaching a computer to understand human behavior and imitate human action can have profound impact on successful multimedia systems. The process needs investigation in the following two areas:

  - multimodal human–computer interaction
  - interactive human–computer vision

  In this area, neural networks also offer attractive solutions.

6.3.1 Image Visualization and Segmentation

The task of feature extraction is critical to search schemes, because an efficient representation of the information can facilitate many subsequent multimedia functionalities, such as feature-based or object-based indexing and access. Efficient representation of multimedia data can be achieved by neural clustering mechanisms. The general objectives are (1) to extract the most salient features to make classification tasks easier, and (2) to extract representation of media information needed at various levels of abstraction.

Although perfect segmentation and tracking of 3D video objects may not always be required, it is desirable to have such capability in telemedicine and biomedically related applications. Using the local energy surface as a principal feature, an SOFM can provide sufficient D resolution of surface details of specific objects through the process of 3D segmentation. The technique has been applied to the segmentation and visualization of specimen chromosomes in microscopy images and the CAT images of human brains [93, 103].

6.3.2 Personal Authentication and Recognition

Neural networks have been recognized as an established and mature tool for many pattern classification problems. In particular, they have been successfully applied to face recognition applications. By combining face information with other biometric features such as speech, this feature fusion approach offers improved accuracy as well as some degree of fault tolerance (i.e., it could tolerate temporary failure of one of the bimodal channels).

Face Detection and Recognition

For many visual monitoring and surveillance applications, it is important to determine human eye positions from an image or an image sequence containing a human face. Once the human eye positions are determined, all of the other important facial features, such as positions of nose and mouth, can easily be determined. The basic facial geometry information, such as the distance between two eyes, nose and mouth size, etc., can further be extracted. This geometry information can then be used for a variety of tasks, such as the recognition of a face from a given face database.

There are many successful neural network examples for face detection and recognition. Brunelli and Poggio have adopted an RBF network for face recognition [11]. Pentland et al. [80, 96, 120] used eigenface subspace to determine the classes of face patterns. Eigenface and Fisherface recognition algorithms were studied and compared in [6]. Cox et al. [26] proposed a *mixture–distance* VQ network for face recognition and reached a 95% rate in a large (685 persons) database. In [67, 69], neural networks were successfully applied to the detection of human faces and the location of eyes on the face.

6.3.3 Audio-to-Visual Conversion and Synchronization

There already exist a few application examples that apply temporal neural models to conversion and/or synchronization. Included in this subsection is an example using TDNN for lipreading applications.

Audio and Visual Integration for Lipreading Applications

Although the theory of automatic speech recognition (ASR) is well advanced, it is still not widely adopted in practical applications due to the contamination of the speech signals with background noise in adverse environments such as offices, automobiles, aircraft, and factories. To improve the performance of the speech recognition system, the following approaches can be used: (1) compensate for the noise in the acoustic speech signals prior to or during the recognition process [81], or (2) use multimodal information sources, such as semantic knowledge and visual features, to assist acoustic speech recognition. The latter approach is supported by the evidence that humans rely on other knowledge sources, such as visual information, to help constrain the set of possible interpretations [133].

Due to the maturity of digital video technology, it is now feasible to incorporate visual information in the speech understanding process (lipreading). These new approaches offer effective integration of visually derived information into the state-of-the-art speech recognition systems so as to gain an improved performance in noise without suffering degraded performance on clean speech [108]. Other important evidence to support the use of lipreading in human speech perception is offered by the auditory–visual blend illusion or the McGurk effect [74].

Three mechanisms concerning the means by which the two disparate (audio and visual) streams of information are integrated have been proposed [113]. First, vision is used to direct the attention, which commonly occurs in situations such as crowded rooms where several people are talking at once. Second, visual information provides redundancy to the audio information. Finally, visual information complements the audio information, especially when listening conditions are poor. Most current research efforts concentrate on the third mechanism of integration. A complete audiovisual lipreading system can be decomposed into the following three major components [108]:

1. Audiovisual information preprocessing: explicit feature extraction from audio and visual data
2. Pattern recognition strategy: hidden Markov modeling, pattern matching with dynamic or linear time warping, and various forms of neural networks
3. Integration strategy: decision from audio and visual signal recognition

Audiovisual Information Preprocessing

Audio information processing has been well documented in speech recognition literature [99]. Briefly, digitized speech is commonly sampled at 8 KHz. The sampled speech is pre-emphasized, then partitioned into frames with a fixed time interval (say, 32 ms long) and with some overlap (say, 16 ms). For each frame, an N-dimensional feature vector is extracted (e.g., 12-order LPC cepstral coefficients, 12-order delta cepstral coefficients, 12-order delta–delta coefficients, a log–energy coefficient, a delta–log–energy coefficient, and a delta–delta–log–energy coefficient).

There are two major types of visual features useful for lipreading: contour-based and area-based features. The active contour model [50] is a good example of an approach based on contour-based features, which have been applied to locating object contours in many image analysis problems [21, 22]. PCA of a gray-level image matrix, a typical area-based method, has been successfully used for principal feature extraction in pattern recognition problems [77, 120]. Most early systems used explicit contour feature extraction. Petajan [97] extracted contour features from binary thresholded mouth images. This approach was also used by Goldschen [31]. Deformable template approaches to obtain contour features, such as snake, have been the dominant method for contour feature extraction [8, 37, 100]. Chiou and Hwang made the first attempt in using neural networks to guide the search of the deformable template for lipreading applications [20]. These methods attempt to directly measure physical aspects of the mouth that are invariant to changes in lighting, camera distance, and orientation. Area-based techniques have primarily been based on neural networks [112, 136]. These area-based features are directly derived from the gray-level matrix surrounding the lips and allow the extraction of more detailed information in the vicinity of the mouth, including the cheek and chin. However, purely area-based approaches tend to be very sensitive to changes in position, camera distance, rotation, and the identity of the speaker.

Pattern Recognition Strategies

Most lipreading systems have used similar pattern recognition strategies as adopted in traditional speech recognition approaches, such as dynamic time warping [97] and hidden Markov models [20, 107]. Neural network architectures have also been extensively explored, such as the static feed-forward back-propagation networks used by Yuhas et al. [136], the TDNNs used by Stork et al. [112], the multistage TDNNs used in [27], and the HMM recognizer, which uses neural networks for performing the observation probabilities calculation [9].

The speech data used in Yuhas' experiments were captured from a male speaker under a well-lit condition. It is based on an NTSC video with 30 frames per second. Nine different phonemes were recognized. A reduced subimage (20×25) centered around the mouth was automatically identified for visual features, which were then converted into the corresponding "clean" audio short-term cepstrum magnitude envelope (STSAE) by a feed-forward back-propagation network. The resulting cepstrum were weight averaged, with the noisy cepstrum directly derived from the audio signals. The weighting between the visual converted STSAE and the audio STSAE was determined based on the environment's SNR. Another feed-forward neural network collected the sequence of the combined STSAE as the inputs and performed the recognition of vowels.

The work presented by Stork et al. [112] used a TDNN for recognizing the combined audio and video speech data for five speakers. In their experiments, a video-only (VO) TDNN was used to recognize the visual speech inputs, which were acquired every 10 ms. From the 10-ms visual frame, five features (noise–chin separation, vertical separation of mouth opening, horizontal separations estimated from upper and lower lips, and horizontal separation of mouth opening) were estimated and combined by the VO TDNN to produce the classification posterior probabilities $P(C|V)$, where C represents one of the 10 spoken letters. Similarly, an audio-only (AO) TDNN was used to recognize the audio speech inputs, which again were acquired every 10 ms. From the 10-ms audio frame, 14 mel-scale coefficients (from 0 to 5 KHz) were estimated and used by the AO TDNN to produce the classification posterior probabilities $P(C|A)$. The resulting classification posterior probability $P(C|V, A)$ is approximated as

$$P(C|V, A) \propto P(C|V)P(C|A) .$$

It was shown in [112] that this combined VO and AO TDNN network, a single video–audio (VA) TDNN, receives the concatenated video and audio features (19 dimensions) as inputs, thus illustrating the importance of adopting separate modules for different media types.

The See Me, Hear Me project [27] developed at Carnegie Mellon University extended the idea of using two separate (VO and AO) TDNNs in performing continuous letter recognition encountered in the continuous spelling tasks. The audio features consist of 16 mel-scale Fourier coefficients obtained at a 10-ms frame rate. The visual features were formed from the PCA transform with reduced dimensionality (only 32 out of 24×16 smoothed eigenlips). The two TDNNs were used for recognizing the phoneme (out of 62) and viseme (out of 42), which were then combined statistically for recognition of the continuous letter sequence based on the dynamic time warping algorithm.

The project presented in [8] also combined acoustic and visual features for effective lipreading. Instead of using neural networks as the temporal sequence classifier, this project adopted the HMMs and used an MLP to calculate the observation probabilities $\{P(phoneme|audio, visual)\}$. The system combined the 10-order PCA transform coefficients (and/or the delta features) from the gray-level eigenlip matrix (instead of the PCA from the snake points) from the video data and nine of the acoustic features from audio data [38]. They used a discriminatively trained MLP to compute the observation probabilities (the likelihood of the input speech data given the state of a subword model) needed by the Viterbi algorithm. Theoretically, the MLP provides the posterior probabilities, instead of the likelihood, which can be easily converted to likelihood according to Bayes' rule using the prior probability information. This bimodal hybrid speech recognition system has already been applied to a multispeaker spelling task, and work is in progress to apply it to a speaker-independent spontaneous speech recognition system, the "Berkeley Restaurant Project (BeRP)."

Decision Integration

As discussed in the previous subsection, audio and visual features can be combined into one vector before pattern recognition; then the decision is solely based on the result of the pattern recognizer. In the case of some lipreading systems, which perform independent visual and audio evaluation, some rule is required to combine the two evaluation scores into a single one. Typical examples have included the use of heuristic rules to incorporate knowledge of the relative confusability of phonemes in the evaluation of two modalities [97]; others have used a multiplicative combination of independent evaluation scores for each modality. These postintegration methods possess the advantages of conceptual and implementational simplicity as well as giving the user the flexibility to use just one of the subsystems if desired.

6.3.4 Image and Video Retrieval, Browsing, and Content-Based Indexing

Digital video processing has recently become an important core information processing technology. The MPEG-4 audiovisual coding standards tend to allow content-based interactivity, universal accessibility, and a high degree of flexibility and extensibility. To accommodate voluminous multimedia data, researchers have long suggested the content-based indexing and retrieval paradigm. Content-based intelligent processing is so critical because it encompasses various application domains including video coding, compaction, object-oriented representation of video, content-based retrieval in the digital library, video mosaicing, video composition (a combination of natural and synthetic scenes), and so forth [15].

Subject-Based Retrieval for Image and Video Databases

A neural network–based tagging algorithm has been proposed for subject-based retrieval for image and video databases [135]. Object classification for tagging is performed offline using DBNN. A hierarchical multiresolution approach is used which helps cut down the search space of looking for a feature in an image. The classification is performed in two phases, first using color, and then texture features are applied to refine the classification (both via DBNN). The general indexing scheme and tagging procedure are depicted in Figure 6.6. The system [135] allows the customer to search the image database by supplying the semantic subject. The images are not manipulated directly in the online phase. Each image is classified into a series of predefined subjects offline using color and texture features and neural network techniques. Queries are answered by searching the tag database. Unlike previous approaches, which directly manipulate images online using templates or low-level image parameters, this system tags the images offline, which greatly enhances performance.

Compared to most of the other existing content-based retrieval systems, which only support similarity-based retrieval, this system supports subject-based retrieval by using descriptions of visual objects as search keys. The difference between subject-based and similarity-based retrieval lies in the necessity for identifying visual objects in the images. Therefore, previous low-level models are not suitable for subject-based retrieval. Novel models are needed for subject-based retrieval that could be utilized in film- and TV program-oriented digital video databases. Neural networks provide a natural effective technology for intelligent information processing.

The tagging procedure includes four steps. In the first step, each image is cut into 25 equal size blocks. Each block may contain single or multiple objects. In the second step, color information is employed for an initial classification where each block is classified into one of the following families: black family, gray family, white family, red family, yellow family, green family, cyan family, blue family, or magenta family in the HSV color space. In the next step, texture features are applied to refine the classification using DBNN if the result of color

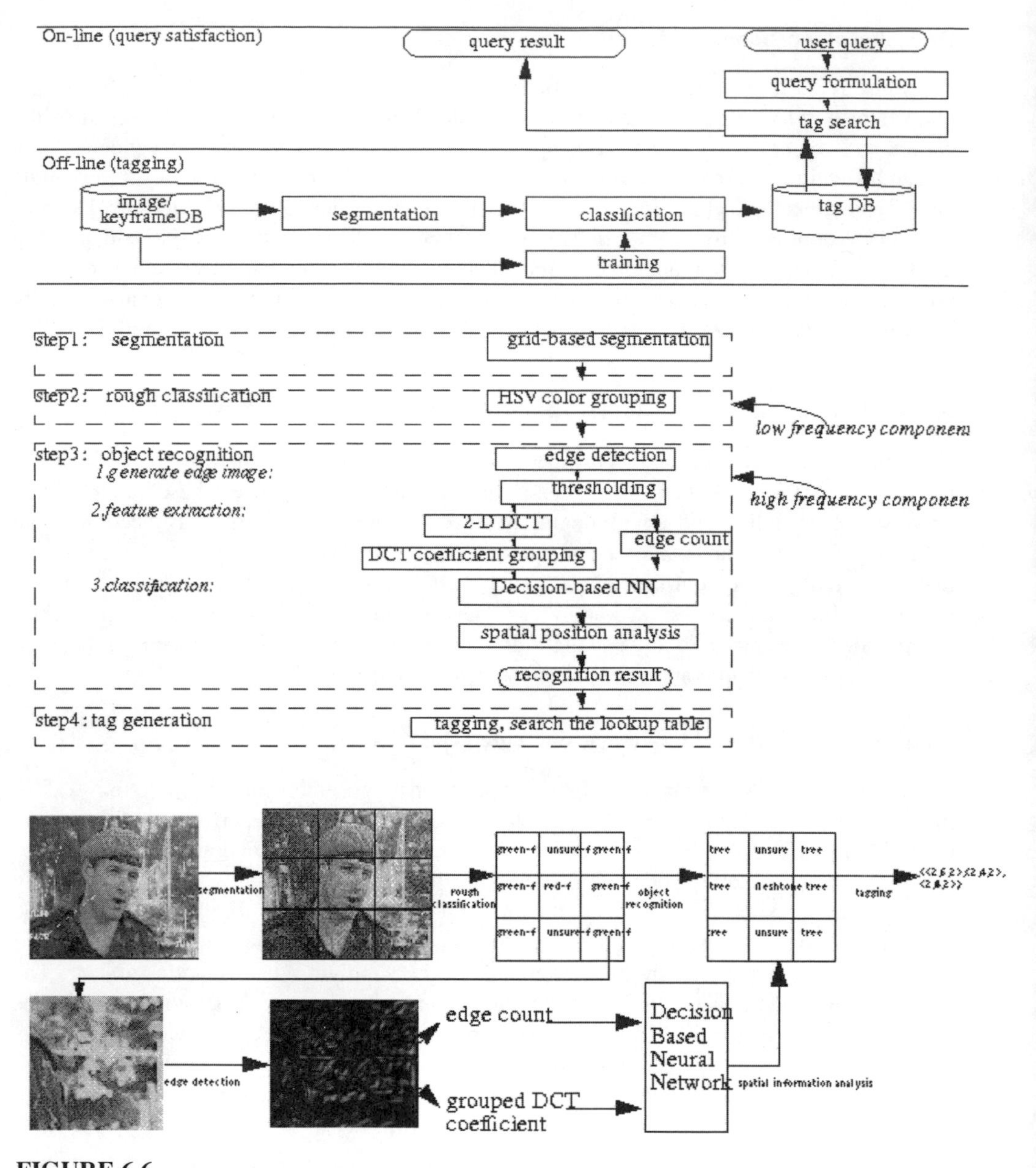

FIGURE 6.6
A subject-based indexing system: (a) visual search methodology; (b) tagging procedure; (c) tagging illustration.

classification is a non-singleton set of subject categories. Each block may be further classified into one of the following categories: sky, foliage, fleshtone, blacktop, white object, ground, light, wood, unknown, and unsure. Finally, an image tag generated from the lookup table using the object recognition results is saved in the tag database. The experimental results of the Web-based implementation shows that this model is very efficient for a large film- or TV program-oriented digital video database.

Transform Domain–Based Retrieval for Digital Image and Video Library (DIVL)

Transform domain–based retrieval offers an attractive alternative to content-based retrieval. With the increasing popularity of the use of compressed images and videos, an intuitive approach for lowering computational complexity and increasing the efficiency of image and video retrieval systems is to perform retrieval directly in the compressed domain. The advantages of this approach are that no extra time is required to calculate features and no extra space is required to store them. Chapter 14 of this book presents a method of using energy histograms of the low-frequency DCT coefficients as features for the retrieval of images and videos compressed in the DCT domain. One of the attractive features of this approach is that the DCT coefficients obtained from coding are representative features of the images, and there is no need to process the images to obtain features as required by most other content-based methods. It is observed that the method is sufficient for performing high-level retrieval on medium-size DIVLs, and it represents a promising solution to efficient retrieval. However, when the size of the DIVLs gets larger (i.e., when the number of images are in the range of millions), any of the current retrieval methods based on matching, including those in the compressed domain, would inevitably slow down considerably. Real-time processing becomes a critical issue. The intuitive solution is to introduce a preprocessing scheme to cut down the amount of matching performed. Neural networks offer attractive solutions to this problem.

One proposal consists of the following four basic steps for the preprocessing stage:

1. Average the corresponding DCT coefficients in all the 8×8 DCT transformed blocks. This operation results in an 8×8 feature matrix representing the image.

2. Cluster the images in a DIVL into categories by the SOFM or the SOTM, to ensure more precise clustering by using the most significant coefficients in the feature matrix, which are normally the low-frequency coefficients.

3. A general regression neural network (GRNN) [110] or a PCA network is then used to identify the coefficients that are most effective to distinguish between the categories.

4. The features selected in step 3 are used to train a classification machine.

When the DIVL receives a query, the classification machine first determines the specific category to which the query belongs, prior to matching.

Further considerations must be taken into account to ensure reliable performance. Averaging the DCT coefficients in a large image may result in too much loss of information. One way to preserve information is to adopt a divide-and-conquer strategy. In particular, each image is partitioned into N subimages, and then the GRNN is used to identify the most effective features for categorizing the subimages at the same geographical location in the various images. Then we can regard the problem as a multisensor fusion problem where we consider those features extracted from the same subimage as arising from the same sensor. Afterward, the modular structured fusion network or the FNN can be applied to this task. In this model, there are N experts, with each of them specializing in representing one particular subimage. Each expert is a DBNN consisting of M neurons, which measures the similarity of that subimage to a particular category.

The matching process will only be performed with those images in a particular category as predicted by the FNN. To minimize the possibility of matching images in a wrong category due to misclassification by the FNN, the ranking of the similarity should be checked. If the differences between the top two or three categories are small, matching should be carried out in all of these categories, instead of the top category only.

A prototype system has been built based on steps 1, 3, and 4 of the above principle and tested on a small database (3000 images). Ten categories were identified in the database. It was observed that a correct matching rate of 95% was achieved if only the top-ranked categories were searched. The rate was further increased to 99.5% if the three top-ranked categories were searched [121].

The aforementioned DIVL architecture is hierarchical and clustered. Such architecture is well adapted for searching, but it would be difficult to encode new information in this hierarchy due to the following facts: (1) the strictly top-down links in the architectures make it hard to merge and split clusters or change the borders of the clusters when new data is entered into the database, and (2) a global training has to be performed to accommodate new information. The hierarchically structured NoN and SOTM offer potential solutions to this problem due to the coexistence of both top-down and lateral links in these networks.

For these two networks, each data cluster (class of images) is represented by a particular subnetwork. It has been shown in both the NoN [2] and the SOTM [54] that the clusters are not isolated from one another, but are sparsely connected. Therefore, the structure of the networks dynamically changes according to the availability of new information. Split and merge, or change of borders, is executed smoothly and continuously. In addition, because of the modularized architecture, retraining is restricted to some limited subarchitecture of the network (e.g., the cluster directly affected and a number of surrounding clusters).

Face-Based Video Indexing and Browsing

A video indexing and browsing scheme based on human faces has been proposed by S.H. Lin et al. [69]. The scheme is implemented by applying face detection and recognition techniques. In many video applications, browsing through a large amount of video material to find the relevant clips is an extremely important task. The video database indexed by human faces provides users with the facility to efficiently acquire video clips featuring the person of interest. For example, a film-study student may conveniently extract the clips of his/her favorite actor/actress from a movie archive to study a performance, and a TV news reporter may quickly find in a news database the clips containing images of some politician in order to edit the evening news.

The scheme contains three steps. The first step of the face-based video browser is to segment the video sequence by applying a scene change detection algorithm. Scene change detection gives an indication of when a new shot starts and ends. Each segment created by scene change detection can be considered as a story unit of this sequence. After video sequence segmentation, a probabilistic DBNN face detector [69] is invoked to find the segments (shots) that most possibly contain human faces. From every video shot, we take its representative frame and feed it into a face detector. Those representative frames from which the detector gives high face detection confidence scores are annotated and serve as the indices for browsing.

This scheme can also be very helpful to algorithms for constructing hierarchies of video shots for video browsing purposes. One such algorithm [134], for example, proposes using global color and luminance information as similarity measures to cluster video shots in an attempt to build video shot hierarchies. Their similarity metrics enable very fast processing of videos. However, in their demonstration, some shots featuring the same anchorman fail to be grouped together due to insufficient image content understanding. For this type of application, we believe that the existence of similar objects, and human objects in particular, should provide a good similarity measure. As reported in [13], this scheme successfully classifies these shots to the same group.

6.3.5 Interactive Human–Computer Vision

The importance of interaction between humans and computers in multimedia systems can never be underestimated. We would like computers to be capable of understanding human intention and expression from audio, visual gestures, body movements, and so forth, as well as to imitate these actions. The multimodality research described in the previous sections is useful to tackle the understanding problem. For imitating human action, interactive human–computer vision (IHCV) may provide the solution.

Developing vision algorithms that can adapt processes designed to track, predict, and describe specific human–computer interactions in ways that are useful to the specific user in a given task is very important in multimedia systems. It enables augmented behaviors, such as augmented reality as an aid to human performance, by taking over tasks or making them easier.

In doing so, we do not necessarily require that these computational algorithms exactly correspond to how the brain enables perceptual and cognitive processes. Rather, these algorithms are designed to be useful to the user insofar as they reflect the actions or behavior, or provide important information to the user over the course of execution. These algorithms dynamically adapt to the behaviors of individual users to evolve into ever more useful and reliable systems.

IHCV is a learning problem. We develop learning algorithms that are expressive enough to track, predict, and describe how humans extract features and interpret images in different tasks. For example, we could have an iconic description of structures such as edges, textures, contours, etc. We can also have a symbolic description of structures such as mathematical formulas.

Two different examples of scene annotation that involve the IHCV approach are

- Tracking/prediction of human edge/feature labeling: Different tasks and image properties require the recognition of different types of edges/features.
- Learning to recognize human symbol drawing (e.g., equations): The recognition performance is invariant to size, orientation, position, and specific distortions.

To summarize, the objective of using MNNs in the task is to track what humans do and predict new cases.

In [132], a new approach to extract iconic structures was proposed. The iconic structures in images are those referred to as edges, textures, contours, etc. In IHCV, the issue that must be addressed properly is the adaptive extraction of those structures considered important for human perception. Typically, the factors to be considered are the varying illumination conditions of the background and the prototypes of the features representing the structures under a particular level of background illumination.

The DBNNs proposed by Kung and Taur [58] are particularly suited for such tasks. The motivation for using this architecture is that, in feature extraction, it would be more natural to adopt multiple sets of decision parameters and apply the appropriate set of parameters as a function of the local context, instead of adopting a single set of parameters across the whole image as in the traditional approaches.

The modular decision-based architecture thus constitutes a natural representation of the above adaptive decision process if we designate each subnetwork to represent a different background illumination level, and each unit in the subnetwork to represent different prototypes of features under the corresponding illumination level. When analyzing an input feature vector, a two-stage decision procedure is performed by a DBNN:

- Within a subnetwork, the units representing different prototypes under the corresponding illumination condition compete with one another. The unit giving the strongest output claims the identity of the input feature vector.

- The subnets then compete with one another, and the one with the largest output value will claim the identity of the input feature vector.

One very attractive feature of the DBNN is its robustness against noise and interference. Since the DBNN learns the average background information, the noise and interference are filtered out as random signals. Such robustness has been clearly demonstrated in edge detection [132].

6.4 Open Issues, Future Research Directions, and Conclusions

In this chapter, we have focused on the main attributes of neural networks relevant to their application to intelligent multimedia applications. Space limitations prohibit more exhaustive coverage of the subjects. More illustrative examples can be found in [92, 124] and numerous signal processing journals.

Although NNs have been quite successful in many applications of IMP, critical research topics remain to be solved. From the commercial system perspective, there are many promising application-driven research problems. These include analysis of multimodal scene change detection, facial expressions and gestures, fusion of gesture/emotion and speech/audio signals, automatic captioning for the hearing-impaired or second-language TV audiences, multimedia telephone, and interactive multimedia services for audio, speech, image, and video contents.

From a long-term research perspective, there is a need to establish a fundamental and coherent theoretical ground for intelligent multimedia technologies. A powerful preprocessing technique, capable of yielding salient object-based video representation, would provide a healthy footing for online object-oriented visual indexing. This suggests that a synergistic balance and interaction between representation and indexing must be carefully investigated. Another fundamental research subject requiring immediate attention is the modeling and evaluation of perceptual quality in multimodal human communication. For content-based visual query, incorporating user feedback in the interactive search process will also be a challenging but rewarding topic.

At the beginning of the chapter, we pointed out that integrating the three branches of computational intelligence may offer excellent design strategies for multimedia systems due to their synergistic power. The hierarchical FNN is one good example. However, such synergies have not been extensively explored in intelligent multimedia research. Investigation into this field will bring about new methodologies and techniques for future multimedia systems.

In conclusion, future telecommunication will place a major emphasis on media integration for human communication. Multimedia systems can achieve their potential only when they are truly integrated in three key ways: integration of content, integration with human users, and integration with other media systems [91]. Therefore, the following technologies will emerge to lead the future multimedia research [90]:

1. Technologies for generating any kind of cyberspace
2. Technologies for warping into cyberspace
3. Technologies for manipulating objects in cyberspace
4. Technologies for communicating with residents of cyberspace

To sum up, the research and application opportunities in intelligent multimedia processing are truly boundless. We must now explore further their vast benefits and enormous potential.

References

[1] J.A. Anderson, M.T. Gately, P.A. Penz, and D.R. Collins, "Radar signal categorization using a neural network," *Proceedings of IEEE,* vol. 78, pp. 1646–1657, 1990.

[2] J.A. Anderson and J.P. Sutton, "A network of networks: Computation and neurobiology," *World Congress of Neural Networks,* vol. 1, pp. 561–568, 1995.

[3] E. Andre, G. Herzog, and T. Rist, "From visual data to multimedia presentations," IEEE Colloquium Grounding Representations: Integration of Sensory Information in Natural Language Processing, Artificial Intelligence and Neural Networks, London, pp. 1–3, May 1995.

[4] M. Arbib, *The Handbook of Brain Theory and Neural Networks,* MIT Press, Cambridge, 1995.

[5] R. Bajaj and S. Chaudhury, "Signature verification using multiple neural networks," *Pattern Recognition,* vol. 30, no. 1, pp. 1–7, 1997.

[6] P.N. Belhumeur, J.P. Hespanha, and D.J. Kriegman, "Eigenfaces vs. Fisherfaces: Recognition using class specific linear projection," *IEEE Transactions on Pattern Analysis and Machine Intelligence,* vol. 19, no. 7, July 1997.

[7] H. Brandt, H.W. Lahmann, and R. Weber, "Quality control of saw blades based on neural networks and laser vibration measurements," Second International Conference on Vibration Measurements by Laser Techniques: Advances and Applications. *Proceedings of the SPIE — The International Society for Optical Engineering,* vol. 2868, pp. 119–124, Ancona, Italy, 1996.

[8] C. Bregler and Y. Konig, "Eigenlips for robust speech recognition," In *Proc. of the Int. Conf. on Acoustics, Speech and Signal Processing* (ICASSP'94), pp. 669–672, Adelaide, Australia, 1994.

[9] C. Bregler, S.M. Omohundro, and Y. Konig, "A hybrid approach to bimodal speech recognition," In *28th Asilomar Conf. on Signals, Systems, and Computers,* pp. 572–577, Pacific Grove, CA, 1994.

[10] J.S. Bridle, "Probabilistic interpretation of feedforward classification network outputs, with relationships to statistical pattern recognition," In *Neuro-computing: Algorithms, Architectures and Applications,* F. Fogelman-Soulie and J. Hérault, editors, pp. 227–236, Springer-Verlag, Berlin, 1991.

[11] R. Brunelli and T. Poggio, "Face recognition: Features versus templates," *IEEE Transactions on Pattern Analysis and Machine Intelligence,* vol. 15, pp. 1042–1052, 1993.

[12] J. Cao, M. Ahmad, and M. Shridhar, "A hierarchical neural network architecture for handwritten numeral recognition," *Pattern Recognition,* vol. 30, no. 2, pp. 289–294, 1997.

[13] Y. Chan, S.H. Lin, Y.P. Tan, and S.Y. Kung, "Video shot classification using human faces," in *IEEE International Conference on Image Processing 1996,* Lausanne, Switzerland.

[14] V. Chandrasekaran, M. Palaniswani, and T.M. Caelli, "Range image segmentation by dynamic neural network architecture," *Pattern Recognition,* vol. 29, no. 2, pp. 315–329, 1996.

[15] T. Chen, A. Katsaggelos, and S.Y. Kung, editors, "Content-based indexing and retrieval of visual information," in *IEEE Signal Processing Magazine,* pp. 45–48, July 1997.

[16] K. Chen, D. Xie, and H. Chi, "Text-dependent speaker identification using hierarchical mixture of experts," *Acta Scientiarum Naturalium Universitatis Pekinensis,* vol. 32, no. 3. pp. 396–404, May 1996.

[17] T. Chen, A. Katsaggelos, and S.Y. Kung, editors, "The past, present, and future of multimedia signal processing," *IEEE Signal Processing Magazine,* July 1997.

[18] Y.-K. Chen, Yunting Lin, and S.Y. Kung, "A feature tracking algorithm using neighborhood relaxation with multi-candidate pre-screening," in *Proceedings of IEEE International Conference on Image Processing,* vol. II, pp. 513–516, Lausanne, Switzerland, Sept. 1996.

[19] T. Chen and R. Rao, "Audio-visual interaction in multimedia communication," *Proceedings and ICASSP,* vol. 1, pp. 179–182, Munich, April 1997.

[20] G.I. Chiou and J.N. Hwang, "Image sequence classification using a neural network based active contour model and a hidden Markov model," *International Conference on Image Processing,* vol. III, pp. 926–930, Austin, Texas, November 1994.

[21] G.I. Chiou and J.N. Hwang, "A neural network based stochastic active contour model (NNS-SNAKE) for contour finding of distinct features," *IEEE Transactions on Image Processing,* vol. 4, no. 10, pp. 1407–1416, October 1995.

[22] L.D. Cohen and I. Cohen, "Finite-element methods for active contour models and balloons for 2-D and 3-D images," *IEEE Transactions on Pattern Analysis and Machine Intelligence,* vol. 15, no. 11, pp. 1131–1141, Nov. 1993.

[23] T.F. Cootes and C.J. Taylor, "Active shape models — smart snakes," In *Proceedings of British Machine Vision Conference,* pp. 266–275, Springer-Verlag, Berlin, 1992.

[24] J.M. Corridoni, A. del Bimbo, and L. Landi, "3D object classification using multi-object Kohonen networks," *Pattern Recognition,* vol. 29, no. 6, pp. 919–935, 1996.

[25] S.M. Courtney, L.H. Finkel, and G. Buchsbaum, "A multistage neural network for color constancy and color induction," *IEEE Transactions on Neural Networks,* vol. 6, no. 4, pp. 972–985, 1995.

[26] I.J. Cox, J. Ghosn, and P. Yianilos, "Feature-based face recognition using mixture distance," Tech. rep. 95-09, NEC Research Institute, 1995.

[27] P. Duchnowski, U. Meier, and A. Waibel, "See Me, Hear Me: Integrating automatic speech recognition and lipreading," *ICSLP'95,* Yokohoma, Japan, pp. 547–550, 1995.

[28] J.L. Elman, "Finding structure in time," *Cognitive Science,* vol. 14, pp. 179–211, 1990.

[29] D.B. Fogel, *Evolutionary Computation: Toward a New Philosophy of Machine Intelligence,* IEEE Press, Piscataway, NJ, 1995.

[30] K. Fukushima and N. Wake, "Handwritten alphanumerical character recognition by the neocognition," *IEEE Transactions on Neural Networks,* vol. 2, no. 3, pp. 355–365, 1991.

[31] A.J. Goldschen, O.N. Garcia, and E. Petajan, "Continuous optical automatic speech recognition by lipreading," in *28th Asilomar Conference on Signals, Systems, and Computers,* pp. 572–577, Pacific Grove, CA, 1994.

[32] L. Guan, "Image restoration by a neural network with hierarchical cluster architecture," *Journal of Electronic Imaging,* vol. 3, pp. 154–163, April 1994.

[33] L. Guan, S. Perry, R. Romagnoli, H.S. Wong, and H.S. Kong, "Neural vision system and applications in image processing and analysis," *Proceedings of IEEE International Conference on Acoustics, Speech and Signal Processing,* vol. II, pp. 1245–1248, Seattle, WA, 1998.

[34] D.H. Han, H.K. Sung, and H.M. Choi, "Nonlinear shape restoration based on selective learning SOFM approach," *Journal of the Korean Institute of Telematics and Electronics,* vol. 34C, no. 1, pp. 59–64, Jan. 1997.

[35] T. Harris, "Kohonen neural networks for machine and process condition monitoring," *Proceedings of the International Conference on Artificial Neural Nets and Genetic Algorithms,* Ales, France, pp. 3–4, April 1995.

[36] S. Haykin, *Neural Networks,* 2nd ed., Prentice-Hall, Englewood Cliffs, NJ, 1998.

[37] M.E. Hennecke, K.V. Prasad, and D.G. Stork, "Using deformable templates to infer visual speech dynamics," *Proceedings of 28th Annual Asilomar Conference,* vol. 1, pp. 578–582, Pacific Grove, CA, Nov. 1994.

[38] H. Hermansky, N. Morgan, A. Bayya, and P. Kohn, "RASTA-RLP speech analysis technique," *ICASSP'92,* pp. 121–124, San Francisco, CA, 1992.

[39] J. Hertz, A. Krogh, and R.G. Palmer, *Introduction to the Theory of Neural Computation,* ch. 7, pp. 163–196, 1991.

[40] T.S. Huang, C.P. Hess, H. Pan, and Z.-P. Liang, "A neuronet approach to information fusion," *Proceedings of IEEE First Workshop on Multimedia Signal Processing,* Princeton, NJ, June 1997.

[41] J.N. Hwang and H. Li, "A limited feedback time delay neural network," *International Joint Conference on Neural Networks,* Nagoya, Japan, pp. 271–274, October 1993.

[42] J.N. Hwang and E. Lin, "Mixture of discriminative learning experts of constant sensitivity for automated cytology screening," *1997 IEEE Workshop for Neural Networks for Signal Processing,* Amelia Island, FL, September 1997.

[43] J.N. Hwang, S.Y. Kung, M. Niranjan, and J.C. Principe, editors, "The past, present, and future of neural networks for signal processing," *IEEE Signal Processing Magazine,* November 1997.

[44] R.A. Jacobs, M.I. Jordan, S.J. Nowlan, and G.E. Hinton, "Adaptive mixtures of local experts," *Neural Computation,* vol. 3, pp. 79–87, 1991.

[45] D.L. James, "SARDNET: A self-organizing feature map for sequences," *Advances in Neural Information Processing Systems,* vol. 7, pp. 577–584, Nov. 1994.

[46] J. Jang, C.-T. Sun, and E. Mizutani, editors, *Neuro-Fuzzy and Soft Computing,* Prentice-Hall, NJ, 1997.

[47] M.I. Jordan and R.A. Jacobs, "Learning to control an unstable system with forward modeling," *Advances in NIPS '90,* pp. 325–331, 1990.

[48] M.I. Jordan and R.A. Jacobs, "Hierarchies of adaptive experts," *Witkin Neural Information Systems,* vol. 4, 1992.

[49] B.H. Juang and S. Katagiri, "Discriminative learning for minimum error classification," *IEEE Transactions on Signal Processing,* vol. 40, no. 12, pp. 3043–3054, 1992.

[50] M. Kass and D. Terzopoulos, "Snakes: Active contour models," *International Journal of Computer Vision,* pp. 321–331, 1988.

[51] J. Koh, M. Suk, and S.M. Bhandarkar, "A multilayer self-organizing feature map for range image segmentation," *Neural Networks,* vol. 8, no. 1, pp. 67–86, 1995.

[52] T. Kohonen, "Self-organized formation of topologically correct feature maps," *Biological Cybernetics,* vol. 43, pp. 59–69, 1982.

[53] T. Kohonen, *Self-Organization and Associative Memory,* 2nd edition, Springer-Verlag, Berlin, 1984.

[54] H. Kong and L. Guan, "A self-organizing tree map for eliminating impulse noise with random intensity distributions," *Journal of Electronic Imaging,* vol. 6, no. 1, pp. 36–44, 1998.

[55] H. Kong, "Self-organizing tree map and its applications in digital image processing," Ph.D. thesis, University of Sydney, Sept. 1998.

[56] B. Kosko, "Fuzzy systems are universal approximators," *Proceedings of IEEE International Conference on Fuzzy Systems,* pp. 1153–1162, San Diego, CA.

[57] S.Y. Kung, *Digital Neural Networks,* Prentice-Hall, Englewood Cliffs, NJ, 1993.

[58] S. Kung and J. Taur, "Decision-based neural networks with signal/image classification applications," *IEEE Transactions on Neural Networks,* vol. 6, no. 1, pp. 170–181, Jan. 1995.

[59] S.Y. Kung and J.-N. Hwang, "Neural networks for intelligent multimedia processing," *Proceedings of the IEEE,* vol. 86, no. 6, pp. 1244–1272, June 1998.

[60] S.Y. Kung, J.-S. Taur, and S.-H. Lin, "Synergistic modeling and applications of fuzzy neural networks," *Proceedings of the IEEE,* vol. 87, no. 8, August 1999.

[61] S.-H. Lai and M. Fang, "Robust and automatic adjustment of display window width and center for MR images," SCR Invention No. 97E7464, 1997.

[62] J. Lampinen and E. Oja, "Distortion tolerant pattern recognition based on self-organizing feature extraction," *IEEE Transactions on Neural Networks,* vol. 6, no. 3, pp. 539–547, 1995.

[63] R. Laganiere and P. Cohen, "Gradual perception of structure from motion: A neural approach," *IEEE Transactions on Neural Networks,* vol. 6, no. 3, pp. 736–748, 1995.

[64] K. Langer and F. Bodendorf, "Flexible user-guidance in multimedia CBT-applications using artificial neural networks and fuzzy logic," *International ICSC Symposia on Intelligent Industrial Automation and Soft Computing,* pp. B9–13, March 1996.

[65] F. Lavagetto, "Converting speech into lip movements: A multimedia telephone for hard of hearing people," *IEEE Transactions on Rehabilitation Engineering,* p. 114, March 1995.

[66] F. Lavagetto, S. Lepsoy, C. Braccini, and S. Curinga, "Lip motion modeling and speech driven estimation," *ICASSP'97,* pp. 183–186, Munich, Germany, April 1994.

[67] S.H. Lin, "Biometric identification for network security and access control," Ph.D. dissertation, Dept. of Electrical Engineering, Princeton University, Princeton, NJ, 1996.

[68] S.-H. Lin, S. Kung, and L.-J. Lin, "A probabilistic DBNN with applications to sensor fusion and object recognition," *Proceedings of 5th IEEE Workshop on Neural Networks for Signal Processing,* pp. 333–342, Aug. 1995.

[69] S.-H. Lin, Y. Chan, and S.Y. Kung, "A probabilistic decision-based neural network for location of deformable objects and its applications to surveillance system and video browsing," *IEEE International Conference on Acoustics, Speech and Signal Processing,* Atlanta, GA, 1996.

[70] S.-H. Lin, S.Y. Kung, and L.J. Lin, "Face recognition/detection by probabilistic decision-based neural networks," *IEEE Transactions on Neural Networks,* vol. 8, no. 1, pp. 114–132, Jan. 1997.

[71] S.-H. Lin, S.Y. Kung, and M. Fang, "A neural network approach for face/palm recognition," *Proceedings of 5th IEEE Workshop on Neural Networks for Signal Processing,* pp. 323–332, Aug. 1995.

[72] Z. Liu, J. Huang, Y. Wang, and T. Chen, "Extraction and analysis for scene classification," *Proceedings of IEEE First Workshop on Multimedia Signal Processing,* Y. Wang et al., editors, Princeton, NJ, June 1997.

[73] S.W. Lu and A. Szeto, "Hierarchical artificial neural networks for edge enhancement," *Pattern Recognition,* vol. 26, no. 8, pp. 1149–1163, 1993.

[74] H. McGurk and J. MacDonald, "Hearing lips and seeing voices," *Nature,* pp. 746–748, Dec. 1976.

[75] T. Mandl and H.C. Womser, "Soft computing — vague query handling in object oriented information systems," *Proceedings HIM'95,* pp. 277–291, Mase Konstanz, Germany, April 1995.

[76] M. Mangeas and A.S. Weigend, "First experiments using a mixture of nonlinear experts for time series prediction," *1995 World Congress on Neural Networks,* vol. 2, pp. 104–109, Washington, DC, July 1995.

[77] K. Mase and A. Pentland, "Automatic lipreading by optical-flow analysis," *Systems and Computers in Japan,* vol. 22, no. 6, pp. 67–76, 1991.

[78] Y. Matsuyama and M. Tan, "Multiply descent cost competitive learning as an aid for multimedia image processing," *Proceedings of 1993 International Joint Conference on Neural Networks,* Nagoya, Japan, pp. 2061–2064, Oct. 1993.

[79] M. McLuhan, *Understanding Media,* McGraw-Hill, New York, 1964.

[80] B. Moghaddam and A. Pentland, "Face recognition using view-based and modular eigenspaces," *SPIE,* vol. 2257, 1994.

[81] S.Y. Moon and J.N. Hwang, "Robust speech recognition based on joint model and feature space optimization of hidden Markov models," *IEEE Transactions on Neural Networks,* vol. 8, no. 2, pp. 194–204, March 1997.

[82] S. Morishima, K. Aizawa, and H. Harashima, "An intelligent facial image coding driven by speech and phoneme," *ICASSP,* pp. 1795–1978, 1989.

[83] Committee Draft of the Standard: ISO 11172-2, "Coding of moving pictures and associated audio for digital storage media at up to about 1.5 Mbits/s," Nov. 1991.

[84] Committee Draft of the Standard: ISO 13818, "MPEG-2 video coding standard," Nov. 1994.

[85] Special Issue of MPEG-4 Video Coding Standards, *IEEE Transactions on Circuits and Systems for Video Technology,* Feb. 1997.

[86] "Second draft of MPEG-7 applications document," ISO/IEC JTC1/SC29/WG11 Coding of Moving Pictures and Associated Audio MPEG97/N2666, Oct. 1997.

[87] "Third draft of MPEG-7 requirements," ISO/IEC JTC1/SC29/WG11 Coding of Moving Pictures and Associated Audio MPEG97/N2606, Oct. 1997.

[88] T. Murdoch and N. Ball, "Machine learning in configuration design," *(AI EDAM) Artificial Intelligence for Engineering Design, Analysis and Manufacturing,* vol. 10, no. 2, pp. 101–113, April 1996.

[89] Y. Nakagawa, E. Hirota, and W. Pedrycz, "The concept of fuzzy multimedia intelligent communication system (FuMICS)," *Proceedings of the Fifth IEEE International Conference on Fuzzy Systems,* pp. 1476–1480, New Orleans, LA, Sept. 1996.

[90] R. Nakatsu, "Media integration for human communication," *IEEE Signal Processing Magazine,* pp. 36–37, July 1997.

[91] C.L. Nikias, "Riding the new integrated media systems wave," *IEEE Signal Processing Magazine,* pp. 32–33, July 1997.

[92] *Neural Networks for Signal Processing, Proceedings of IEEE Workshops, 1991–1997,* vols. I–VII, IEEE Press.

[93] P.T.A. Nguyen, R. Romagnoli, P. Fekete, M.R. Arnison, L. Guan, and C. Cogswell, "A self-organizing map for extracting features of chromosomes in microscopy images," *Australian Journal of Intelligent Information Systems,* vol. 5, no. 1, pp. 34–38, 1998.

[94] E. Oja, "Principal component analysis, minor components, and linear neural networks," *Neural Networks,* vol. 5, no. 6, pp. 927–935, Nov.–Dec. 1992.

[95] A. Pedotti, G. Ferrigno, and M. Redolfi, "Neural network in multimedia speech recognition," *Proceedings of the International Conference on Neural Networks and Expert Systems in Medicine and Healthcare,* Plymouth, UK, pp. 167–173, Aug. 1994.

[96] A. Pentland, B. Moghaddam, and T. Starner, "View-based and modular eigenspaces for face recognition," *Proceedings of IEEE Conference on Computer Vision and Pattern Recognition,* pp. 84–91, June 1994.

[97] E. Petajan, B. Bischoff, D. Bodoff, and N. Brooke, "An improved automatic lipreading system to enhance speech recognition," *ACM SIGCHI,* pp. 19–25, 1988.

[98] T. Poggio and F. Girosi, "Networks for approximation and learning," *Proceedings of IEEE,* vol. 78, pp. 1481–1497, Sept. 1990.

[99] L.R. Rabiner and B.H. Juang, *Fundamentals of Speech Recognition,* Prentice-Hall, Englewood Cliffs, NJ, 1993.

[100] R.R. Rao and R.M. Mersereau, "Lip modeling for visual speech recognition," *Proceedings of 28th Annual Asilomar Conference,* vol. 1, pp. 587–590, Pacific Grove, CA, Nov. 1994.

[101] M.D. Richard and R.P. Lippmann, "Neural network classifiers estimate Bayesian a posteriori probabilities," *Neural Computation,* vol. 3, no. 4, pp. 461–483, 1991.

[102] J. Risch, R. May, J. Thomas, and S. Dowson, "Interactive information visualization for exploratory intelligence data analysis," *Proceedings of the IEEE 1996 Virtual Reality Annual International Symposium,* pp. 230–238, Santa Clara, CA, April 1996.

[103] R. Romagnoli, P.T.A. Nguyen, L. Guan, L. Cinque, and S. Levialdi, "Self-organizing map for segmenting 3D biological images," presented at the International Conference on Pattern Recognition, Brisbane, Australia, 1998.

[104] H.S. Ranganath, D.E. Kerstetter, and S.R.F. Sim, "Self partitioning neural networks for target recognition," *Neural Networks,* vol. 8, no. 9, pp. 1475–1486, 1995.

[105] L. Rothkrantz, V.R. Van, and E. Kerckhoffs, "Analysis of facial expressions with artificial neural networks," European Simulation Multiconference, Prague, Czech Republic, pp. 790–794, June 1995.

[106] D.E. Rumelhart, G.E. Hinton, and R.J. William, "Learning internal representation by error propagation," in *Parallel Distributed Processing: Explorations in the Micro-Structure of Cognition,* vol. 1, MIT Press, Cambridge, MA, 1986.

[107] P.L. Silsbee, "Sensory integration in audiovisual automatic speech recognition," *Proceedings of 28th Annual Asilomar Conference,* vol. 1, pp. 561–565, Pacific Grove, CA, Nov. 1994.

[108] P.L. Silsbee and A.C. Bovik, "Computer lipreading for improved accuracy in automatic speech recognition," *IEEE Transactions on Speech and Audio Processing,* vol. 4, no. 5, pp. 337–351, Sept. 1996.

[109] V. Shastri, L.C. Rabelo, and E. Onjeyekwe, "Device-independent color correction for multimedia applications using neural networks and abductive modeling approaches," 1996 IEEE International Conference on Neural Networks, Washington, DC, pp. 2176–2181, June 1996.

[110] D.F. Specht, "A general regression neural network," *IEEE Transactions on Neural Networks,* vol. 2, no. 6, pp. 569–576, 1991.

[111] N. Srinvasa and R. Sharma, "SOIM: A self-organizing invertable map with applications in active vision," *IEEE Transactions on Neural Networks,* vol. 8, no. 3, pp. 758–773, 1997.

[112] D.G. Stork, G. Wolff, and E. Levine, "Neural network lipreading system for improved speech recognition," *Proceedings of IJCNN,* pp. 285–295, 1992.

[113] Q. Summerfield, "Some preliminaries to a comprehensive account of audio-visual speech perception," in *Hearing by Eye: The Psychology of Lip-Reading,* B. Dodd and R. Campbell, editors, pp. 97–113, Lawrence Erlbaum, London, 1987.

[114] K. Sung and T. Poggio, "Learning human face detection in cluttered scenes," *Computer Analysis of Image and Patterns,* pp. 432–439, 1995.

[115] J.P. Sutton, J.S. Beis, and L.E.H. Trainor, "A hierarchical model of neuro-cortical synaptic organization," *Mathematical Computer Modeling,* vol. 11, pp. 346–350, 1988.

[116] J.P. Sutton, "Mean field theory of nested neural clusters," *Proceedings of the First AMSE International Conference on Neural Networks,* pp. 47–58, San Diego, CA, May 1991.

[117] F. Takeda and S. Omatu, "High speed paper currency recognition by neural networks," *IEEE Transactions on Neural Networks,* vol. 6, no. 1, pp. 73–77, 1995.

[118] Y.H. Tseng, J.N. Hwang, and F. Sheehan, "Three-dimensional object representation and invariant recognition using continuous distance transform neural networks," *IEEE Transactions on Neural Networks,* vol. 8, no. 1, pp. 141–147, Jan. 1997.

[119] L.H. Tung, I. King, and W.S. Lee, "Two-stage polygon representation for efficient shape retrieval in image databases," *Proceedings of the First International Workshop on Image Databases and Multi-Media Search,* pp. 146–153, Amsterdam, The Netherlands, 1996.

[120] M. Turk and A. Pentland, "Eigenfaces for recognition," *Journal of Cognitive Neuroscience,* vol. 3, pp. 71–86, 1991.

[121] J. Ukovich, "Image Retrieval in Multimedia Systems Using Neural Networks," B. Eng. thesis, University of Sydney, Dec. 1998.

[122] A. Waibel, T. Hanazawa, G. Hinton, K. Shikano, and K.J. Lang, "Phoneme recognition using time-delay neural networks," *IEEE Transactions on Acoustics, Speech, and Signal Processing,* vol. 37, no. 3, pp. 328–339, March 1989.

[123] L.X. Wang, "Fuzzy systems are universal approximators," *Proceedings of IEEE International Conference on Fuzzy Systems,* pp. 1163–1169, San Diego, CA.

[124] Y. Wang, A. Reibman, F. Juang, T. Chen, and S.Y. Kung, editors, *Proceedings of the IEEE Workshops on Multimedia Signal Processing,* IEEE Press, Princeton, NJ, 1997.

[125] S.R. Waterhouse and A.J. Robinson, "Constructive algorithms for hierarchical mixtures of experts," *Advances in Neural Information Processing,* vol. 8, pp. 584–590, Nov. 1995.

[126] S.R. Waterhouse, D. MacKay, and A.J. Robinson, "Bayesian methods for mixtures of experts," *Advances in Neural Information Processing,* vol. 8, pp. 351–357, Nov. 1995.

[127] S.R. Waterhouse and A.J. Robinson, "Non-linear prediction of acoustic vectors using hierarchical mixtures of experts," *Advances in Neural Information Processing Systems,* vol. 7, pp. 835–842, Nov. 1994.

[128] W.X. Wen, A. Jennings, and H. Liu, "Self-generating neural networks and their applications to telecommunications," *Proceedings of International Conference on Communication Technology,* pp. 222–228, Beijing, China, Sept. 1992.

[129] H. White, "Connectionist nonparametric regression: Multilayer feedforward networks can learn arbitrary mappings," *Neural Networks,* vol. 3, pp. 535–549, 1990.

[130] R.J. William and D. Zipser, "A learning algorithm for continually running fully recurrent neural networks," *Neural Computation,* vol. 1, no. 2, pp. 270–280, 1994.

[131] H.S. Wong and L. Guan, "Adaptive regularization in image restoration using evolutionary programming," *Proceedings of IEEE International Conference on Evolutionary Computation,* pp. 159–164, Anchorage, AK, 1998.

[132] H.S. Wong, T.M. Caelli, and L. Guan, "A model-based neural network for edge characterization," to appear in *Pattern Recognition,* 1999.

[133] W.A. Woods, "Language processing for speech understanding," in *Readings in Speech Recognition,* A. Waibel and K.F. Lee, editors, pp. 519–533, Morgan Kaufman, 1990.

[134] M.M. Yeung, B.L. Yeo, W. Wolf, and B. Liu, "Video browsing using clustering and scene transitions on compressed sequences," *Proceedings of SPIE, Multimedia Computing and Networking,* 1995.

[135] H.H. Yu and W. Wolf, "A hierarchical, multi-resolution method for dictionary-driven content-based image retrieval," *Proceedings of International Conference on Image Processing,* Santa Barbara, CA, Oct. 1997.

[136] B.P. Yuhas, M.H. Goldstein, T.J. Sejnowski, and R.E. Jenkins, "Neural networks models for sensory integration for improved vowel recognition," *Proceedings of IEEE,* vol. 78, no. 10, pp. 1658–1668, Oct. 1990.

[137] J. Zurada, R. Marks II, and C. Robinson, editors, *Computational Intelligence — Imitating Life,* IEEE Press, NJ, 1994.

Chapter 7

On Independent Component Analysis for Multimedia Signals

Lars Kai Hansen, Jan Larsen, and Thomas Kolenda

7.1 Background

Blind reconstruction of statistically independent source signals from linear mixtures is relevant to many signal processing contexts [1, 6, 8, 9, 22, 24, 36]. With reference to principal component analysis (PCA), the problem is often referred to as independent component analysis (ICA).[1]

The source separation problem can be formulated as a likelihood formulation (see, e.g., [7, 32, 35, 37]). The likelihood formulation is attractive for several reasons. First, it allows a principled discussion of the inevitable priors implicit in any separation scheme. The prior distribution of the source signals can take many forms and *factorizes* in the source index expressing the fact that we look for *independent* sources. Second, the likelihood approach allows for direct adaptation of the plethora of powerful schemes for parameter optimization, regularization, and evaluation of supervised learning algorithms. Finally, for the case of linear mixtures without noise, the likelihood approach is equivalent to another popular approach based on information maximization [1, 6, 27].

The source separation problem can be analyzed under the assumption that the sources either are time independent or possess a more general time-dependence structure. The separation problem for *autocorrelated* sequences was studied by Molgedey and Schuster [33]. They proposed a source separation scheme based on assumed nonvanishing temporal autocorrelation functions of the independent source sequences evaluated at a specific *time lag*. Their analysis was developed for sources mixed by square, nonsingular matrices. Attias and Schreiner derived a likelihood-based algorithm for separation of correlated sequences with a frequency domain implementation [2]–[4]. The approach of Molgedey and Schuster is particularly interesting as regards computational complexity because it forms a noniterative, constructive solution.

Belouchrani and Cardoso presented a general likelihood approach allowing for *additive noise* and nonsquare mixing matrices. They applied the method to separation of sources taking discrete values [7], estimating the mixing matrix using an estimate–maximize (EM) approach with both a deterministic and a stochastic formulation. Moulines et al. generalized the EM approach to separation of autocorrelated sequences in the presence of noise, and they explored a family of flexible source priors based on Gaussian mixtures [34]. The difficult problem

[1] There are a number of very useful ICA Web pages providing links to theoretical analysis, implementations, and applications. Follow links from the page `http://eivind.imm.dtu.dk/staff/lkhansen/ica.html`.

of noisy, overcomplete source models (i.e., more sources than acquired mixture signals) was recently analyzed by Lewicki and Sejnowski within the likelihood framework [28, 31].

In this chapter we study the likelihood approach and entertain two different approaches to the problem: a modified version of the Molgedey–Schuster scheme [15], based on time correlations, and a novel iterative scheme generalizing the mixing problem to separation of noisy mixtures of time-independent white sources [16]. The Molgedey–Schuster scheme is extended to the undercomplete case (i.e., more acquired mixture signals than sources), and further inherent erroneous complex number results are alleviated. In the noisy mixture problem we find a maximum posterior estimate for the sources that, interestingly, turns out to be nonlinear in the observed signal. The specific model investigated here is a special case of the general framework proposed by Belouchrani and Cardoso [7]; however, we formulate the parameter estimation problem in terms of the Boltzmann learning rule, which allows for a particular transparent derivation of the mixing matrix estimate.

The methods are applied within several multimedia applications: separation of sound, image sequences, and text.

7.2 Principal and Independent Component Analysis

PCA is a very popular tool for analysis of correlated data, such as temporal correlated image databases. With PCA the image database is decomposed in terms of "eigenimages" that often lend themselves to direct interpretation. A most striking example is face recognition, where so-called eigenfaces are used as orthogonal preprocessing projection directions for pattern recognition. The principal components (the sequence of projections of the image data onto the eigenimages) are also uncorrelated and, hence, perhaps the simplest example of independent components [9]. The basic tool for PCA is singular value decomposition (SVD).

Define the observed $M \times N$ signal matrix, representing a multichannel signal, by

$$X = \{X_{m,n}\} = \{x_m(n)\} = [\boldsymbol{x}(1), \boldsymbol{x}(2), \ldots, \boldsymbol{x}(N)] \tag{7.1}$$

where M is the number of measurements and N is the number of samples. $x_m(n)$, $n = 1, 2, \ldots, N$ is the mth signal and $\boldsymbol{x}(n) = [x_1(n), x_2(n), \cdots, x_M(n)]^\top$. In the case of image sequences, M is the number of pixels.

For the fixed choice of $P \leq M$, the SVD of $\boldsymbol{X}$ reads[2]

$$\boldsymbol{X} = \boldsymbol{U}\boldsymbol{D}\boldsymbol{V}^\top = \sum_{i=1}^{P} \boldsymbol{u}_i D_{i,i} \boldsymbol{v}_i^\top, \qquad X_{m,n} = \sum_{i=1}^{P} U_{m,i} D_{i,i} V_{n,i} \tag{7.2}$$

where $M \times P$ matrix $\boldsymbol{U} = \{U_{m,i}\} = [\boldsymbol{u}_1, \boldsymbol{u}_2, \ldots, \boldsymbol{u}_P]$ and $N \times P$ matrix $\boldsymbol{V} = \{V_{n,i}\} = [\boldsymbol{v}_1, \boldsymbol{v}_2, \ldots, \boldsymbol{v}_P]$ represent the orthonormal basis vectors (i.e., eigenvectors of the symmetric matrices $\boldsymbol{X}\boldsymbol{X}^\top$ and $\boldsymbol{X}^\top\boldsymbol{X}$, respectively). $\boldsymbol{D} = \{D_{i,i}\}$ is a $P \times P$ diagonal matrix of singular values. In terms of independent sources, SVD can identify a set of *uncorrelated* time sequences, the principal components: $D_{i,i}\boldsymbol{v}_i$, enumerated by the source index $i = 1, 2, \ldots, P$. That is, we can write the observed signal as a weighted sum of fixed eigenvectors (eigenimages) $\boldsymbol{u}_i$.

However, considering the likelihood for the time-correlated source density, we are often interested in a slightly more general separation of image sources that are *independent* in time

[2]Usually, SVD expresses $\boldsymbol{X} = \widetilde{\boldsymbol{U}}\widetilde{\boldsymbol{D}}\widetilde{\boldsymbol{V}}^\top$ where $\widetilde{\boldsymbol{U}}$ is $M \times M$, $\widetilde{\boldsymbol{D}}$ is $M \times N$, and $\widetilde{\boldsymbol{V}}$ is $N \times N$. $\boldsymbol{U}$ is the first P columns of $\widetilde{\boldsymbol{U}}$, $\boldsymbol{D}$ is the $P \times P$ upper-left submatrix of $\widetilde{\boldsymbol{D}}$, and $\boldsymbol{V}$ is the first P columns of $\widetilde{\boldsymbol{V}}$.

but not necessarily orthogonal in space (i.e., we would like to be able to perform a more general decomposition of the signal matrix),

$$\boldsymbol{X} = \boldsymbol{AS}, \qquad X_{m,n} = \sum_{i=1}^{P} A_{m,i} S_{i,n} \tag{7.3}$$

where $\boldsymbol{A}$ is a general mixing matrix of dimension $M \times P$ and $\boldsymbol{S}$ is a source data matrix with dimension $P \times N$ consisting of $P \leq M$ independent sources. Finding $\boldsymbol{A}$, $\boldsymbol{S}$ is often referred to as ICA (see, e.g., [6, 9]).

7.3 Likelihood Framework for Independent Component Analysis

Reconstruction of statistically independent components/sources from linear mixtures is relevant to many information processing contexts (see, e.g., [27] for an introduction and a recent review). We will derive a solution to the source separation based on the likelihood formulation (see, e.g., [7, 32, 37]). An additional benefit from working in the likelihood framework is that it is possible to discuss the *generalizability* of the ICA representation; in particular, we use the generalization error as a tool for optimizing the complexity of the representation (see also [14, 17]).

The noisy mixing model takes the form

$$\boldsymbol{X} = \boldsymbol{AS} + \boldsymbol{\mathcal{E}} \tag{7.4}$$

where $\boldsymbol{\mathcal{E}}$ is the $M \times N$ noise signal matrix. The noise is supposed to obey a specific zero mean, parameterized stationary probability distribution. The source signals are assumed to be stationary and mutually independent — that is, $p(s_i(k)s_j(n)) = p(s_i(k))p(s_j(n))$, $\forall i, j \in [1; M]$, $\forall n, k \in [1; N]$. The properties of the source signals are introduced by a parameterized prior probability density $p(\boldsymbol{S}|\boldsymbol{\psi})$, where $\boldsymbol{\psi}$ is the parameter vector. The likelihood of the parameters of the noise distribution, the parameters of the source distribution, and those of the mixing matrix is given by

$$L(\boldsymbol{A}, \boldsymbol{\theta}, \boldsymbol{\psi}) = p(\boldsymbol{X}|\boldsymbol{A}, \boldsymbol{\theta}, \boldsymbol{\psi}) = \int p(\boldsymbol{X} - \boldsymbol{AS}|\boldsymbol{\theta}) p(\boldsymbol{S}|\boldsymbol{\psi}) d\boldsymbol{S} \tag{7.5}$$

where $p(\boldsymbol{X} - \boldsymbol{AS}|\boldsymbol{\theta}) = p(\boldsymbol{\mathcal{E}}|\boldsymbol{\theta})$ is the noise distribution parameterized by the vector $\boldsymbol{\theta}$. We will assume that the noise can be modeled by i.i.d. Gaussian sequences with a common variance $\theta = \sigma^2$,

$$p(\boldsymbol{\mathcal{E}}|\sigma^2) = \frac{1}{(2\pi\sigma^2)^{MN/2}} \exp\left(-\frac{1}{2\sigma^2} \sum_{m=1}^{M} \sum_{n=1}^{N} \varepsilon_m^2(n)\right). \tag{7.6}$$

We will consider two different assumptions about the independent source distributions leading to different algorithms.

For the *time-independent white source* problem, the parameter-free source distribution of [32] is deployed:

$$p(\boldsymbol{S}) = \prod_{i=1}^{P} p(s_i) = \frac{1}{\pi^{NP}} \exp\left(-\sum_{n=1}^{N} \sum_{i=1}^{P} \log\cosh s_i(n)\right) \tag{7.7}$$

where $\boldsymbol{S}^\top = \{\boldsymbol{s}_1, \boldsymbol{s}_2, \cdots, \boldsymbol{s}_P\}$ and $\boldsymbol{s}_i = [s_i(1), s_i(2), \cdots s_i(N)]^\top$. In the *time-correlated* case, it is assumed that the sources are stationary, independent, possess time autocorrelation, have zero mean, and are Gaussian distributed:[3]

$$p(\boldsymbol{S}|\boldsymbol{\psi}) = \prod_{i=1}^{P} p(\boldsymbol{s}_i|\boldsymbol{\psi}_i) = \prod_{i=1}^{P} \frac{1}{(2\pi)^{N/2}\sqrt{\det(\boldsymbol{\Gamma}_{s_i})}} \exp\left(-\frac{1}{2}\boldsymbol{s}_i^\top \boldsymbol{\Gamma}_{s_i}^{-1} \boldsymbol{s}_i\right) \tag{7.8}$$

where $\boldsymbol{\psi} = [\boldsymbol{\psi}_1, \cdots, \boldsymbol{\psi}_P]$ and $\boldsymbol{\Gamma}_{s_i} = E[\boldsymbol{s}_i\boldsymbol{s}_i^\top] = \text{Toeplitz}([\gamma_{s_i}(0), \ldots, \gamma_{s_i}(N-1)])$[4] is the $N \times N$ Toeplitz autocorrelation matrix consisting of autocorrelation function values, $\gamma_{s_i}(m) = E[s_i(n)s_i(n+m)]$, $m = 0, 1, \ldots, N-1$. The autocorrelation matrix $\boldsymbol{\Gamma}_{s_i}$ is supposed to be parameterized by $\boldsymbol{\psi}_i$.

7.3.1 Generalization and the Bias-Variance Dilemma

The parameters of our blind separation model are estimated from a finite random sample, and therefore they also are random variables which inherit noise from the dataset on which they were trained. Within the likelihood formulation, the generalization error of a specific set of parameters is given by the average negative log-likelihood[5]

$$\begin{aligned} G(\boldsymbol{A}, \boldsymbol{\theta}, \boldsymbol{\psi}) &= \int -\log L(\boldsymbol{A}, \boldsymbol{\theta}, \boldsymbol{\psi}) \cdot p_*(\boldsymbol{X})\, d\boldsymbol{X} \\ &= \int [-\log \int p(\boldsymbol{X} - \boldsymbol{A}\boldsymbol{S}|\boldsymbol{\theta}) p(\boldsymbol{S}|\boldsymbol{\psi})\, d\boldsymbol{S}] \cdot p_*(\boldsymbol{X})\, d\boldsymbol{X} \end{aligned} \tag{7.9}$$

where $p_*(\boldsymbol{X})$ is the true distribution of data. The generalization error is a principled tool for model selection. In the context of blind separation, the optimal number of sources retained in the model is of crucial interest. We face a typical bias-variance dilemma [13]. If too few components are used, a structured part of the signal will be lumped with the noise, hence leading to a high generalization error because of "lack of fit." On the other hand, if too many sources are used, we expect "overfit" because the model will use the additional degrees of freedom to fit nongeneric details into the training data. The generalization error in (7.9) can be estimated using a test set of data *independent* of the training set.[6]

7.3.2 Noisy Mixing of White Sources

The specific model investigated here is a special case of the general framework proposed by Belouchrani and Cardoso [7]; however, we formulate the parameter estimation problem in terms of the Boltzmann learning rule, which allows for a particular transparent derivation of the mixing matrix estimate.

Let us first address the problem of estimating the sources if the mixing parameters are known (i.e., for given $\boldsymbol{A}$ and σ^2). Note that MacKay [32] showed that the gradient descent scheme

[3]By assuming stationarity, we implicitly neglect transient behavior due to initial conditions.

[4]Toeplitz($\cdot$) transforms a vector into a Toeplitz matrix.

[5]Note the close connection between generalization error and the Kullback–Leibler information (KL), as

$$\begin{aligned} \text{KL}(p_*(\boldsymbol{X}) : p(\boldsymbol{X}|\boldsymbol{A}, \boldsymbol{\theta}, \boldsymbol{\psi})) &= \int \log \frac{p_*(\boldsymbol{X})}{p(\boldsymbol{X}|\boldsymbol{A}, \boldsymbol{\theta}, \boldsymbol{\psi})} p_*(\boldsymbol{X})\, d\boldsymbol{X} \\ &= G(\boldsymbol{A}, \boldsymbol{\theta}, \boldsymbol{\psi}) + \int \log(p_*(\boldsymbol{X})) p_*(\boldsymbol{X})\, d\boldsymbol{X} \end{aligned}$$

[6]That is, we evaluate (7.9) on the test data by using $p_*(\boldsymbol{X}) = \delta(\boldsymbol{X} - \boldsymbol{X}_{\text{test}})$ where δ is the Dirac delta function and $\boldsymbol{X}_{\text{test}}$ are the test data.

for the likelihood problem, for vanishing noise variance, is equivalent to the Bell–Sejnowski rule [6]. Here we want to consider the more general noisy case. We use Bayes' formula $p(\boldsymbol{S}|\boldsymbol{X}) \propto p(\boldsymbol{X}|\boldsymbol{S})p(\boldsymbol{S})$ to obtain the posterior distribution of the sources

$$p(\boldsymbol{S}|\boldsymbol{X},\boldsymbol{A},\sigma^2) \propto \exp\left(-\frac{1}{2\sigma^2}\sum_{m=1}^{M}\sum_{n=1}^{N}\varepsilon_m^2(n) - \sum_{i=1}^{P}\sum_{n=1}^{N}\log\cosh s_i(n)\right)$$
$$= \exp\left(-\frac{1}{2\sigma^2}\sum_{m=1}^{M}\sum_{n=1}^{N}(\boldsymbol{X}-\boldsymbol{A}\boldsymbol{S})_{m,n}^2 - \sum_{i=1}^{P}\sum_{n=1}^{N}\log\cosh S_{i,n}\right). \quad (7.10)$$

The *maximum a posteriori* (MAP) source estimate is found by maximizing this expression w.r.t. $\boldsymbol{S}$[7], leading to the following nonlinear equation to solve iteratively for the MAP estimate $\widehat{\boldsymbol{S}}$,

$$-\boldsymbol{A}^\top\boldsymbol{A}\widehat{\boldsymbol{S}} + \boldsymbol{A}^\top\boldsymbol{X} - \sigma^2\tanh\widehat{\boldsymbol{S}} = \boldsymbol{0}\,. \quad (7.11)$$

There are two problems with equation (7.11). First, the equation is nonlinear — although only weakly nonlinear for low noise levels.[8] Second, $\boldsymbol{A}^\top\boldsymbol{A}$ may be ill conditioned or even singular. A useful rewriting that takes care of potential ill-conditioning of the system matrix leads to the iterative scheme,

$$\widehat{\boldsymbol{S}}^{(j+1)} = \left(\boldsymbol{A}^\top\boldsymbol{A}+\sigma^2\boldsymbol{I}\right)^{-1}\left(\boldsymbol{A}^\top\boldsymbol{X}+\sigma^2\left(\widehat{\boldsymbol{S}}^{(j)} - \tanh\left(\widehat{\boldsymbol{S}}^{(j+1)}\right)\right)\right) \quad (7.12)$$

where j denotes the iteration number and $\boldsymbol{I}$ is the identity matrix. This form suggests an approximate solution for low noise levels

$$\widehat{\boldsymbol{S}}^{(1)} = \boldsymbol{S}^{(0)} + \sigma^2\boldsymbol{H}^{-1}\left(\boldsymbol{S}^{(0)} - \tanh\boldsymbol{S}^{(0)}\right),$$
$$\boldsymbol{S}^{(0)} = \boldsymbol{H}^{-1}\boldsymbol{A}^\top\boldsymbol{X},\quad \boldsymbol{H} = \boldsymbol{A}^\top\boldsymbol{A}+\sigma^2\boldsymbol{I}\,, \quad (7.13)$$

exposing the fact that the presence of additive noise turns the otherwise linear separation problem into a nonlinear one. A nonlinear source estimate is also found in Lewicki and Sejnowski's analysis of the overcomplete problem [31].

Since the likelihood is of the hidden Gibbs form we can use a generalized Boltzmann learning rule to find the gradients of the likelihood of the parameters $\boldsymbol{A}$, σ^2. These averages can be estimated in a mean field approximation [16, 38] leading to recursive rules for $\boldsymbol{A}$ and σ^2,

$$\widehat{\boldsymbol{A}} = \boldsymbol{X}\widehat{\boldsymbol{S}}^\top\left(\widehat{\boldsymbol{S}}\widehat{\boldsymbol{S}}^\top+\beta\boldsymbol{I}\right)^{-1}, \quad (7.14)$$

$$\widehat{\sigma}^2 = \frac{1}{MN}\mathrm{Tr}\left(\boldsymbol{X}-\widehat{\boldsymbol{A}\boldsymbol{S}}\right)^\top\left(\boldsymbol{X}-\widehat{\boldsymbol{A}\boldsymbol{S}}\right) \quad (7.15)$$

where β is a regularization constant representing the lumped effect of neglected fluctuations in the mean field approach. β is estimated by

[7]Note in the case of zero noise, the posterior expression leads to the expression given in [32], and the solution is obtained by the Bell–Sejnowski algorithm [6].

[8]This expression is the gradient of the exponent of the posterior distribution. A globally convergent iterative solution can be assured if solving by gradient ascent $\nabla\boldsymbol{S} = \eta\cdot\partial\log p(\boldsymbol{S}|\boldsymbol{X},\boldsymbol{A},\sigma^2)/\partial\boldsymbol{S}$, with a sufficiently small step size, η. Here, however, we aim for a fast approximate solution for $\boldsymbol{S}$.

$$\beta = \widehat{\sigma}^2 \left(1 - \frac{1}{PN} \sum_{i=1}^{P} \sum_{n=1}^{N} \tanh^2 S_{i,n} \right) . \tag{7.16}$$

(See [16].)

Fluctuation corrections (hence the magnitude of β) can be derived in the low noise limit, based on a Gaussian approximation of the likelihood [16].

The overall algorithm then consists of iterating (7.13), (7.14)–(7.16), (7.12), (7.14)–(7.16), etc. Convergence of the algorithm is discussed in [16].

7.3.3 Separation Based on Time Correlation

Molgedey and Schuster [33] have proposed a simple noniterative source separation scheme based on assumed nonvanishing (time) autocorrelation functions of the independent sources that can be Gaussian distributed.[9] Their idea was developed for sources mixed by square, nonsingular $\boldsymbol{A}$ matrices. Here we generalize their approach in three ways:

- Handling the undercomplete case of more mixture signals than sources (i.e., $P \leq M$). In particular, the algorithm is well suited for cases where $P \ll M$.
- Alleviating inherent erroneous complex valued results.
- Allowing for simultaneous use of more cross-correlation matrix function values maintaining the simple noniterative solution.

Define the $M \times M$ cross-correlation function matrix for the mixture signals

$$\boldsymbol{C_x}(\tau) = E\left\{\boldsymbol{x}(n)\boldsymbol{x}^\top(n+\tau)\right\} = \left\{i, j \in [1; M] : x_i(n)x_j(n+\tau)\right\} \tag{7.17}$$

where $\tau = 0, \pm 1, \pm 2, \cdots$ is a time lag and $E\{\cdot\}$ is the expectation operator. Note for $\tau = 0$ we get the usual cross-correlation matrix, $\boldsymbol{C_x}(0) = E\{\boldsymbol{x}(n)\boldsymbol{x}^\top(n)\}$, which is positive semidefinite. Assume the noise-free model (7.3), $\boldsymbol{x}(n) = \boldsymbol{A}\boldsymbol{s}(n)$, where $\boldsymbol{s}(n) = [s_1(n), \ldots, s_P(n)]^\top, \boldsymbol{x}(n) = [x_1(n), \ldots, x_M(n)]^\top$ and further that the $M \times P$ mixing matrix has $\text{rank}(\boldsymbol{A}) = P \leq M$. Since $\boldsymbol{C_x}(0) = \boldsymbol{A}\boldsymbol{C_s}(0)\boldsymbol{A}^\top$ where $\boldsymbol{C_s}(0)$ is the $P \times P$ cross-correlation matrix for the source signals, and $\text{rank}(\boldsymbol{A}) = P$, then $\text{rank}(\boldsymbol{C_x}(0)) = P$. An eigenvalue decomposition of $\boldsymbol{C_x}(0)$ reads

$$\boldsymbol{C_x}(0) = \boldsymbol{Q}\boldsymbol{L}\boldsymbol{Q}^\top \tag{7.18}$$

where $\boldsymbol{Q} = [\boldsymbol{q}_1, \boldsymbol{q}_2, \ldots, \boldsymbol{q}_M]$ is the orthogonal matrix ($\boldsymbol{Q}^\top\boldsymbol{Q} = \boldsymbol{I}$) of eigenvectors $\boldsymbol{q}_i$ and $\boldsymbol{L} = \text{diag}(l_1, \ldots, l_M)$ is the diagonal matrix of eigenvalues $l_1 \leq l_2 \leq \cdots \leq l_P \leq 0$ and $l_{P+1} = l_{P+2} = \cdots = l_M = 0$. Consider projection onto the P-dimensional full rank subspace,

$$\widetilde{\boldsymbol{x}} = \widetilde{\boldsymbol{Q}}^\top \boldsymbol{x} \tag{7.19}$$

where $\widetilde{\boldsymbol{Q}} = [\boldsymbol{q}_1, \boldsymbol{q}_2, \ldots, \boldsymbol{q}_P]$ is the $M \times P$ projection matrix and $\widetilde{\boldsymbol{x}}$ is the $P \times 1$ projected mixture signal vector. Now define *quotient matrix*

$$\boldsymbol{K} = \boldsymbol{C}_{\widetilde{\boldsymbol{x}}}(\tau)\boldsymbol{C}_{\widetilde{\boldsymbol{x}}}^{-1}(0). \tag{7.20}$$

[9] At most, one source is allowed to be white.

Since $C_{\widetilde{x}}(\tau) = \widetilde{Q}^\top A C_s(\tau) A^\top \widetilde{Q}$, the quotient matrix can be expressed as[10]

$$K = (\widetilde{Q}^\top A) C_s(\tau) C_s^{-1}(0) (\widetilde{Q}^\top A)^{-1} \tag{7.21}$$

According to Appendix A, the quotient matrix has the eigenvalue decomposition $K = \Phi \Lambda \Phi^{-1}$ where Λ is a diagonal matrix of real eigenvalues and Φ are the associated real eigenvectors. Define a permutation matrix[11] $P = [e_{j_1}, \ldots, e_{j_P}]$ where $e_j = \{\delta_{ij}, i \in [1; P]\}$ are P-dimensional unit column vectors and $[j_1, j_2, \ldots, j_P]$ is a permutation of the numbers $[1; P]$. Note that $PP = I$. Further, define a diagonal scaling matrix $\Xi = \mathrm{diag}([\xi_1, \ldots, \xi_P])$ with $\xi_i \neq 0$. Comparing with (7.21) shows that eigenvalue decomposition of K can be used to identify the mixing matrix A, as shown by:

$$K = (\widetilde{Q}^\top A) C_s(\tau) C_s^{-1}(0) (\widetilde{Q}^\top A)^{-1} = \Phi \Xi P P \Xi^{-1} \Lambda \Xi^{-1} P P \Xi \Phi^{-1} \tag{7.22}$$

where P is a permutation matrix and Ξ a diagonal scaling matrix as defined in Section 7.2. Consequently,

$$\widetilde{Q}^\top A = \Phi \Xi P, \tag{7.23}$$

$$C_s(\tau) C_s^{-1}(0) = P \Xi^{-1} \Lambda \Xi^{-1} P. \tag{7.24}$$

Here we use the fact that $C_s(\tau)$ is diagonal due to independence of the source signals.

Consider measurements of the cross-correlation function matrix for T different τ's and define the extended quotient matrix:

$$K_{\mathrm{ext}} = \sum_{j=1}^{T} \alpha_j \cdot C_{\widetilde{x}}(\tau_j) C_{\widetilde{x}}^{-1}(0) \tag{7.25}$$

where α_j are scalar weights. Then eigenvalue decomposition of $K_{\mathrm{ext}} = \Phi \Lambda \Phi^{-1}$ leads to

$$\widetilde{Q}^\top A = \Phi \Xi P, \tag{7.26}$$

$$\sum_{j=1}^{T} \alpha_j \cdot C_s(\tau_j) C_s^{-1}(0) = P \Xi^{-1} \Lambda \Xi^{-1} P\,. \tag{7.27}$$

The generalized Molgedey–Schuster algorithm for identification of mixing and source signals up to scaling and permutations is thus summarized in the following steps:

1. Perform eigenvalue decomposition: $C_x(0) = Q L Q^\top$.
2. Compute projected mixing signals, $\widetilde{x} = \widetilde{Q}^\top x$.
3. Choose α_j and τ_j for $j = 1, 2, \ldots, T$ and compute the extended quotient matrix K_{ext}.
4. Perform eigenvalue decomposition: $K_{\mathrm{ext}} = \Phi \Lambda \Phi^{-1}$.
5. Up to scaling and permutations, the mixing matrix and sources are identified as:

$$A = \widetilde{Q} \Phi \tag{7.28}$$

$$S = \left(A^\top A\right)^{-1} A^\top X = \Phi^{-1} \widetilde{Q}^\top X\,. \tag{7.29}$$

[10]Note that $\widetilde{Q}^\top A$ has a full rank equal to P.

[11]WP gives a permutation of W's columns, whereas PW gives a permutation of the rows.

Estimation of Mixing Matrix and Source Signals

The procedure described above is based on true cross-correlation function matrices which in practice are estimated from available data. Consider the estimate

$$\widehat{C}_x(\tau) = \frac{1}{2N}\left(X_\tau X^\top + X X_\tau^\top\right) \tag{7.30}$$

where $X_\tau = \{x_m(n+\tau)\}$ is the time-shifted data matrix. Here we consider a cyclic permutation by τ time steps (i.e., $X_\tau = \{x_m((n+\tau)_N)\}$ where $(\cdot)_N$ denotes the argument modulo N). Equation (7.30) respects the fact that the true correlation matrix function $C_x(\tau)$ is symmetric.

Consider the SVD of $X = UDV^\top$ in (7.2) with P selected so that D consists of positive singular values only. When X_τ is formed by cyclic permutation, $XX^\top = X_\tau X_\tau^\top$; hence, $X_\tau = UDV_\tau^\top$ where V_τ is the cyclic permutation of V. The $P \times N$ projected mixture signal matrix is $\widetilde{X} = U^\top X = DV^\top$ and $\widetilde{X}_\tau = DV_\tau^\top$ as U is an estimate of $\widetilde{Q}$. The estimated quotient matrix is according to (7.20), given by

$$\begin{aligned}
\widehat{K} &= \widehat{C}_{\widetilde{x}}(\tau)\widehat{C}_{\widetilde{x}}^{-1}(0) \\
&= \frac{1}{2}\left(\widetilde{X}_\tau \widetilde{X}^\top + \widetilde{X}\widetilde{X}_\tau^\top\right)\left(\widetilde{X}\widetilde{X}^\top\right)^{-1} \\
&= \frac{1}{2}D\left(V_\tau^\top V + V^\top V_\tau\right)D\left(DV^\top V D\right)^{-1} \\
&= \frac{1}{2}D\left(V_\tau^\top V + V^\top V_\tau\right)D^{-1}\,.
\end{aligned} \tag{7.31}$$

The generalized Molgedey–Schuster ICA algorithm can be summarized in the following steps:

1. Perform SVD: $X = UDV^\top$ with P selected so that all singular values in D are positive. There is an option for *regularization* by discarding some of the smallest singular values, causing a reduction of P.

2. Perform eigenvalue decomposition of the estimated quotient matrix[12]

$$\begin{aligned}
\widehat{K} &= \frac{1}{2}D\left(V_\tau^\top V + V^\top V_\tau\right)D^{-1} \\
&= \widehat{\Phi}\widehat{\Lambda}\widehat{\Phi}^{-1}\,.
\end{aligned} \tag{7.32}$$

3. Estimate the mixing matrix and source signals:

$$\widehat{A} = U\widehat{\Phi}, \tag{7.33}$$

$$\widehat{S} = \widehat{\Phi}^{-1}DV^\top\,. \tag{7.34}$$

4. Cross-correlation matrix functions of the source signals are estimated as

$$\widehat{C}_s(0) = N^{-1}\widehat{S}\widehat{S}^\top = N^{-1}\cdot\widehat{\Phi}^{-1}D^2\widehat{\Phi}^{-\top}, \tag{7.35}$$

$$\widehat{C}_s(\tau) = \widehat{\Lambda}\widehat{C}_s(0)\,. \tag{7.36}$$

The fact that $\widehat{\Phi}$ is nonorthogonal in general implies that $\widehat{C}_s(0)$ and $\widehat{C}_s(\tau)$ are not diagonal. That is, finite sequence source signals cannot be expected to be *uncorrelated*. Unlike PCA, this scheme and other ICA schemes do not automatically produce a set of uncorrelated features.

[12] When $T > 1$ the term $(V_\tau^\top V + V^\top V_\tau)$ is replaced by $\sum_{j=1}^{T}\alpha_j(V_{\tau_j}^\top V + V^\top V_{\tau_j})$.

7.3.4 Likelihood

The major advantage of the Molgedey–Schuster algorithm is its noniterative nature; however, it is not directly guaranteed to minimize the likelihood. Still, the likelihood is a convenient tool for understanding the nature of the modeling. Deploying one τ ($T = 1$) is consistent with parameterizing the source distribution $p(\boldsymbol{S}|\boldsymbol{\psi})$ in (7.8) using one parameter per source. As more τ's are deployed, a more flexible parameterization of the likelihood applies.

The likelihood can be computed in a simple way using Fourier techniques. This also enables computation of validation/generalization error, and consequently a principled way to select optimal τ's aiming at achieving minimum generalization error. However, that discussion is beyond the scope of this chapter.

7.4 Separation of Sound Signals

In this example the aim is to demonstrate how ICA is applied to separation of sound signals. This could be thought of as a special case of blind signal separation in connection with the cocktail party problem illustrated in Figure 7.1.

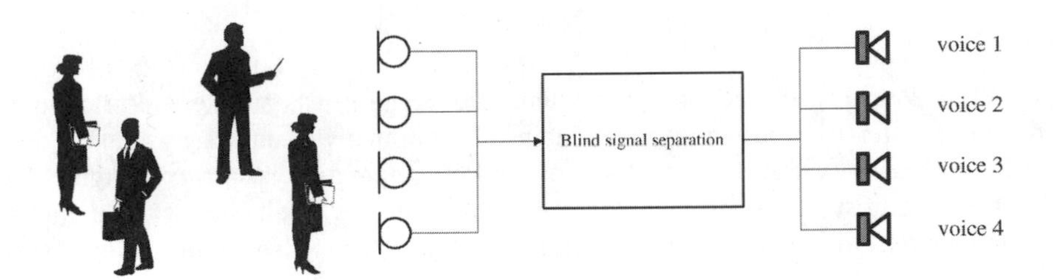

FIGURE 7.1
In the cocktail party problem, speech from a group of people is recorded by a number of microphones. Without prior knowledge of the dynamics in the voices, how they are mixed, or presence of additional noise sources, the goal is to separate the voices of the individual speakers into different output channels.

The present example deals with speech from three persons that are assumed statistically independent. The sampling frequency of the signals is 11,025 Hz and they consist of 50,000 samples each. A linear instantaneous mixing with a fixed known 3×3 mixing matrix is deployed and enables a quantitative evaluation of the ICA separation. The source and mixing signals are shown in Figure 7.2. In general these assumptions would not hold in real-world applications due to echo, noise, delay, and various nonlinear effects. In such cases more elaborate source separation is needed, as described, for example, in [2]–[4], [10]. In order to evaluate the results of the separation, we consider the so-called *system matrix* defined as

$$SM = \left(\widehat{\boldsymbol{A}}\widehat{\boldsymbol{C}}_s(0)^{1/2}\right)^{-1} \boldsymbol{P}\boldsymbol{A} \tag{7.37}$$

where $\widehat{\boldsymbol{A}}$ is the estimated mixing matrix, $\boldsymbol{P}$ is a permutation matrix, and $\widehat{\boldsymbol{C}}_s(0)$ is the cross-correlation matrix of the estimated source signals. If the separation is successful, the system matrix equals the identity matrix.

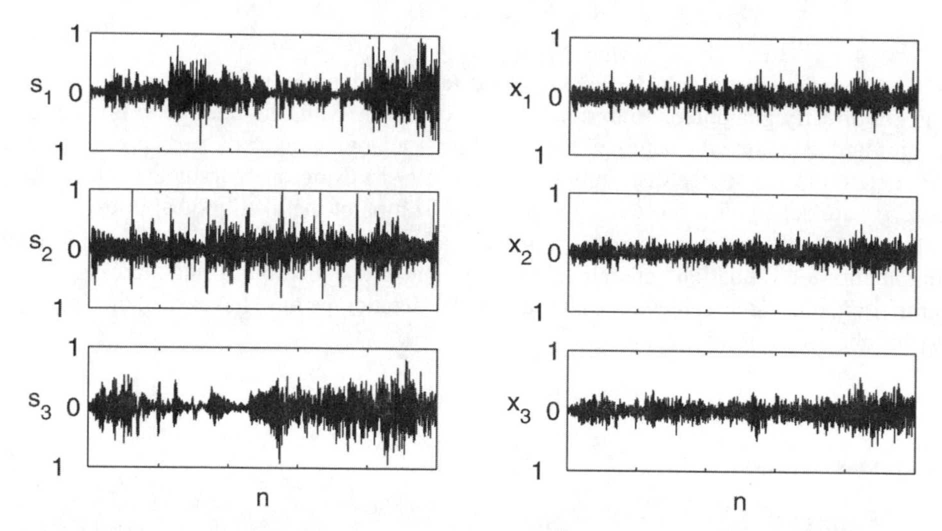

FIGURE 7.2
The original source sound signals $s_1(n)$, $s_2(n)$, and $s_3(n)$ consist of 50,000 samples and are assumed to be statistically independent. The mixture signals $x_1(n)$, $x_2(n)$, and $x_3(n)$ are linear instantaneous combinations of the source signals.

7.4.1 Sound Separation using PCA

The PCA described in Section 7.2 is often used because it is simple and relatively fast. Moreover, it offers the possibility of reducing the number of sources by ranking sources according to power (variance). The result of the PCA separation is shown in Figure 7.3 and the corresponding system matrix in Table 7.1. Obviously the result is poor when comparing estimated sources to the original sources in Figure 7.2. This is also confirmed by inspecting the system matrix in Table 7.1.

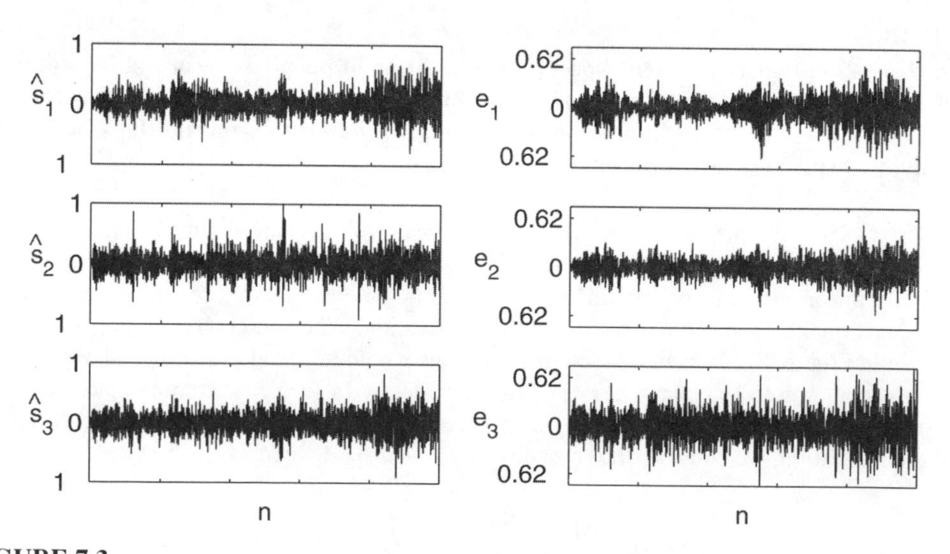

FIGURE 7.3
Separated sound source signals using PCA. Right panels show error signals, $e_i(n) = s_i(n) - \widehat{s_i}(n)$.

Table 7.1 System Matrix for the PCA Separation of Sound Signals

$$SM = \begin{bmatrix} 0.56 & 0.98 & 0.62 \\ 0.28 & 0.72 & 0.23 \\ 0.18 & 0.50 & 0.06 \end{bmatrix}$$

7.4.2 Sound Separation using Molgedey–Schuster ICA

The main advantage of the Molgedey–Schuster ICA algorithm is that it is noniterative and consequently very fast. A standard $T = 1$ ICA was employed, and the choice $\tau = 1$ gave the best performance. In Figure 7.4 the estimated sound signals from the separation are shown. Comparison with original source signals in Figure 7.2 indicates very good separation. The system matrix in Table 7.2 and an additional listening test also confirm this result.

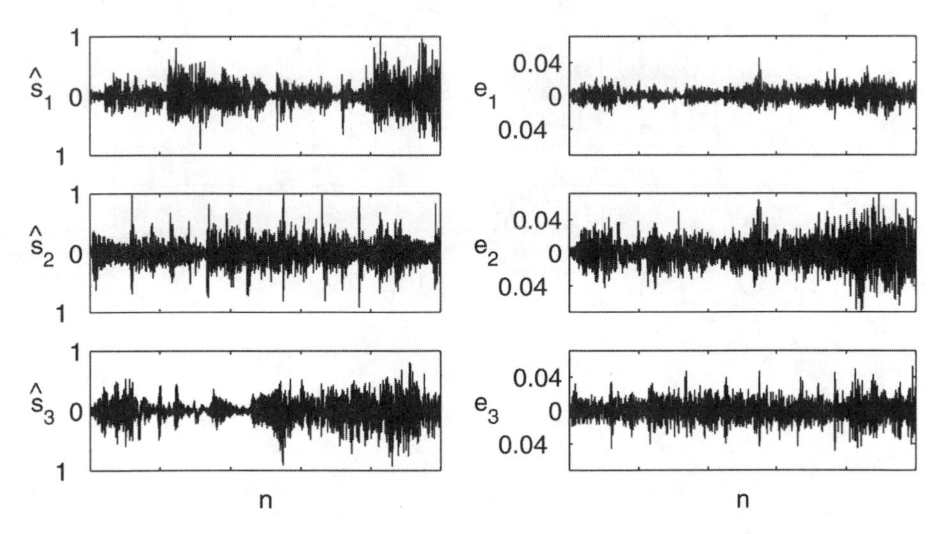

FIGURE 7.4
Separated sound source signals using Molgedey–Schuster ICA. Right panels show error signals, $e_i(n) = s_i(n) - \widehat{s_i}(n)$**.**

Table 7.2 System Matrix for the Molgedey–Schuster ICA Separation of Sound Signals

$$SM = \begin{bmatrix} 1.00 & 0.02 & 0.03 \\ 0.02 & 1.00 & -0.01 \\ -0.03 & -0.03 & -1.00 \end{bmatrix}$$

7.4.3 Sound Separation using Bell–Sejnowski ICA

The very commonly used Bell-Sejnowski ICA [6] is equivalent to maximum likelihood with assumptions like those presented in Section 7.3.2 in the case of zero noise. Bell–Sejnowski ICA iteratively computes an estimate of the mixing matrix by updating proportionally to the natural gradient of the likelihood. The step size (gradient parameter) was initially 10^{-4} and a line search was employed using bisection. The algorithm was terminated when the negative log-likelihood was below 10^{-12}. Due to the iterative nature, this algorithm is much more time consuming than the Molgedey–Schuster algorithm.

In Figure 7.5 and Table 7.3 the results of the separation are shown. Clearly, the system matrix is closer to the identity matrix than that of Molgedey–Shuster, at the expense of increased computational burden.

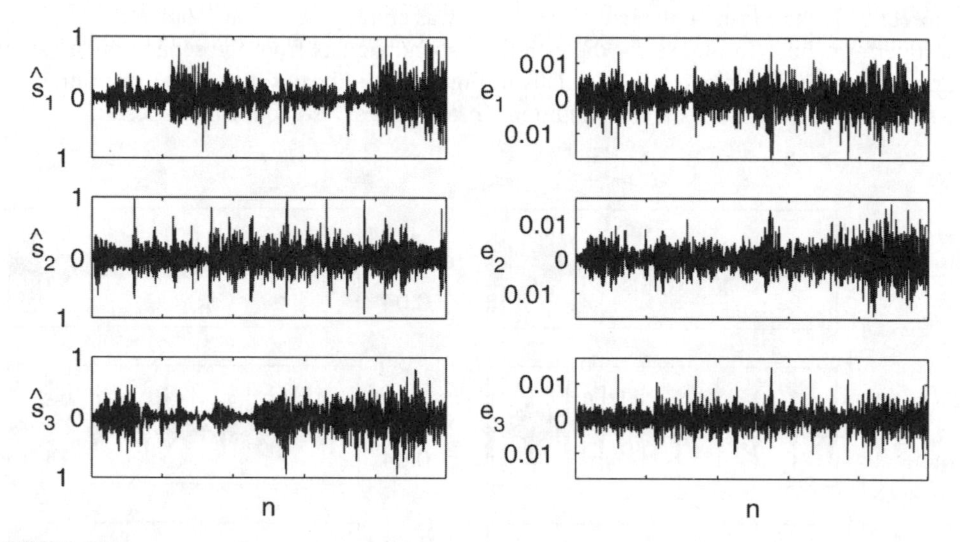

FIGURE 7.5
Separated sound source signals using Bell–Sejnowski ICA. Right panels show error signals, $e_i(n) = s_i(n) - \widehat{s_i}(n)$.

Table 7.3 System Matrix for the Bell–Sejnowski ICA Separation of Sound Signals

$$SM = \begin{bmatrix} 1.00 & -0.01 & 0.01 \\ 0.00 & 1.00 & -0.01 \\ 0.01 & 0.01 & 1.00 \end{bmatrix}$$

7.4.4 Comparison

Table 7.4 lists the norm of the system matrix deviation from the identity matrix as well as computation time.

Obviously, PCA was outperformed by both ICA algorithms due to very restricted separation capabilities. Both ICA algorithms performed very well. The major difference is computation time; MS-ICA was more than 200 times faster than BS-ICA. The advantage of the BS-ICA

Table 7.4 Norm of the System Matrix's Deviation from the Identity Matrix and Computation Time in Seconds

	$\|SM - I\|$	Computation Time (s)
PCA	1.21	0.25
MS-ICA[a]	0.05	0.25
BS-ICA[b], 22 iterations	0.05	56.10
BS-ICA[b], 56 iterations	0.01	152.18

[a] MS-ICA, Molgedey–Schuster ICA.
[b] BS-ICA, Bell–Sejnowksi ICA for 22 and 56 iterations, respectively.

algorithm is that the system matrix can be significantly closer to unity provided sufficient computation time. A hybrid of MS-ICA and BS-ICA in which MS-ICA is used to initialize BS-ICA seems obvious.

Listening to the separated signals, it was hardly impossible to tell the difference between the ICA results.

7.5 Separation of Image Mixtures

Applying ICA to images has been carried out in a number of applications ranging from face recognition to localizing activated areas in the brain (see, e.g., [5, 16, 19, 20, 29, 30]).

In this section we illustrate some of the basic features using ICA in contrast to or in combination with PCA for image segmentation. From a sequence of images, the objective is to extract sequence images where common features have been separated into different images. In the present case ICA is based on raw images; however, in principle, the segmentation can also be done from features extracted from the images. The simple dataset as shown in Figure 7.6 is used in this example. There are $P = 4$ original source images of $N = 9100$ (91 by 100) pixels rearranged into the $P \times N$ source matrix $\boldsymbol{S}$ so that each row represents an image. The $M \times N$ signal matrix $\boldsymbol{X}$ with $M = 6$ is generated by using the following $M \times P$ mixing matrix

$$\boldsymbol{A} = \begin{bmatrix} 1 & 1 & 0 & 1 \\ -1 & 1 & 0 & 1 \\ 1 & 1 & -2 & 1 \\ -1 & -1 & -2 & 1 \\ 1 & -1 & 0 & 1 \\ -1 & -1 & 0 & 1 \end{bmatrix}. \tag{7.38}$$

7.5.1 Image Segmentation using PCA

The result of applying PCA to the face dataset is shown in Figure 7.7. The number of nonzero eigenvalues is correctly determined to be 4. Notice that the eyebrow and mouth positions operate in pairs; when the mouth is "smiling" it cannot be "sad" and likewise for the eyebrows. PCA is able to detect this behavior but mixes both eyebrows and mouth pieces in sources 2 and 3. Further, only the nose is present in source 1. This is a typical effect in PCA because its decomposition is based on finding the directions with the most variance, which is not always well suited for the data.

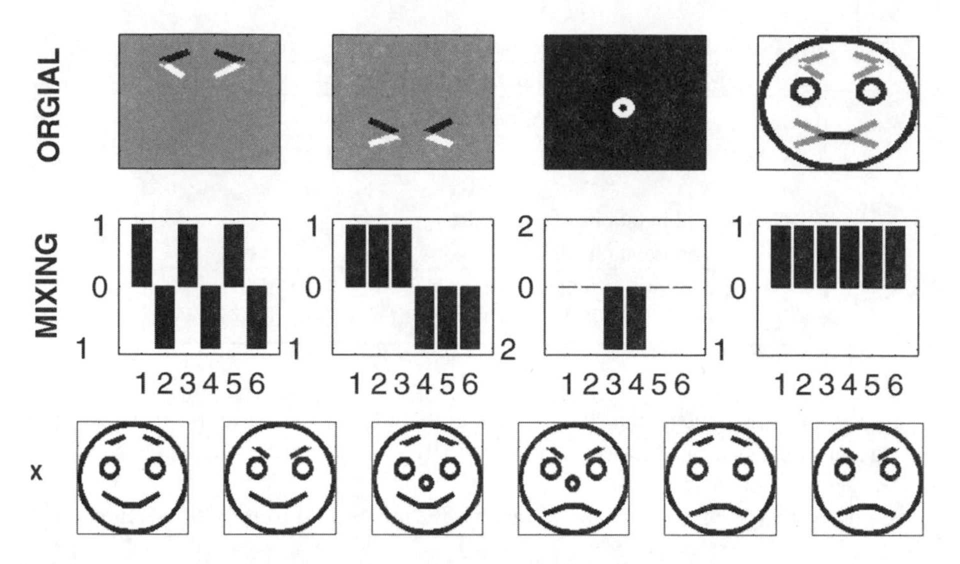

FIGURE 7.6
The artificial face dataset used for image segmentation. The top row shows the $P = 4$ sources of $N = 9100$ pixels, which is multiplied with the mixing A in the middle row to generate the signal matrix X with $M = 6$ components in the bottom row.

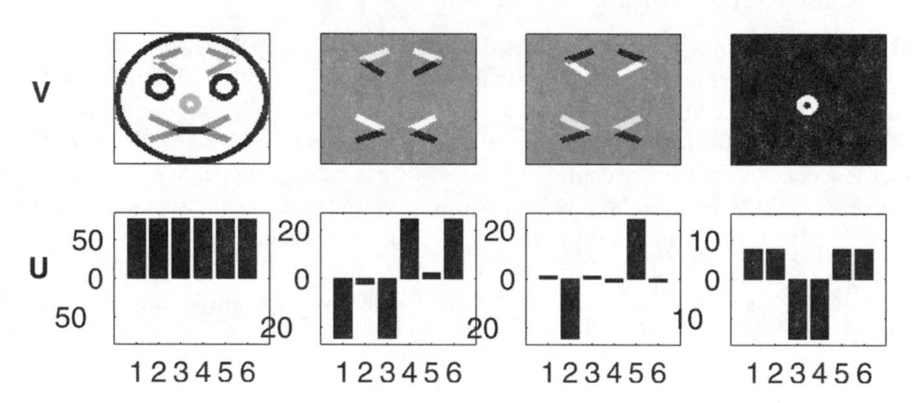

FIGURE 7.7
Applying PCA to the artificial face data. The sources V are shown in the top row and the corresponding mixing matrix estimate is shown in the bottom row. Unfortunately, PCA mixes the eyebrows and mouth pieces in sources 2 and 3. Further, only the nose is present in source 1.

7.5.2 Image Segmentation using Molgedey–Schuster ICA

ICA on images can be performed either to the signal matrix X or the transpose $X^\top$. In the first case N = number of pixels and M = number of images in sequence corresponding to assuming independence of pixels in the sources. In this case the sources are images and the mixing matrix is the time sequence. In the second case N = number of images in sequence and M = number of pixels corresponding to assuming independence driving time sequence sources. Thus, the mixing matrix corresponds to (eigen)images. This is summarized in Table 7.5.

Table 7.5 Two Ways of Performing ICA on Image Sequences

	Signal Matrix	
	X	$X^\top$
M	No. of images in sequence	No. of pixels
N	No. of pixels	No. of images in sequence
S	Images	Time sequence
A	Time sequence	Images
Assumption	Pixel independence	Time independence

The result when assuming pixel independence (i.e., using X as signal matrix) is shown in Figure 7.8. The result when assuming time independence (i.e., using $X^\top$ as signal matrix) is shown in Figure 7.9.

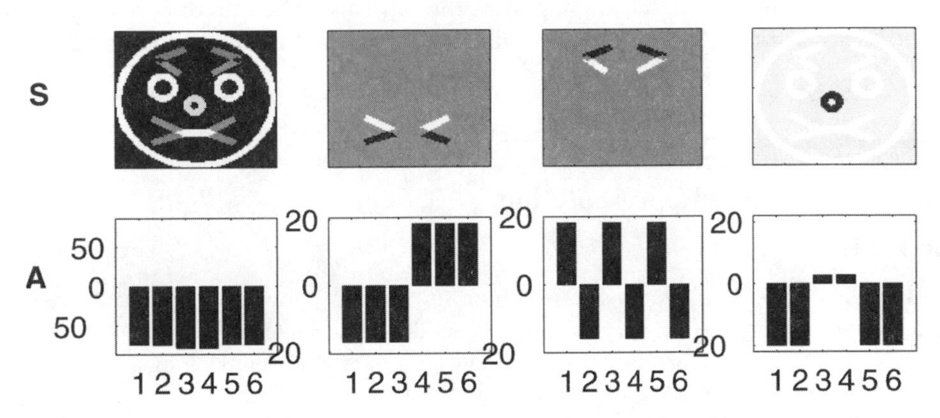

FIGURE 7.8
MS-ICA on the artificial face data with the *pixel-independence assumption* (i.e., X is the signal matrix). The estimated sources (eigenimages) are shown in the top row and the associated mixing matrix (time sequences) in the bottom row. Unlike PCA in Figure 7.7, MS-ICA does not mix eyebrows and mouths (i.e., the sources are almost perfect except for a small problem with the nose component in source 1). Also, the mixing matrix A is almost perfect in comparison with Figure 7.6.

7.5.3 Discussion

Real image applications often show preference toward ICA over PCA. This is mainly because ICA is able to produce a nonorthogonal basis and is not constrained by the variance ranking inherent in PCA. Using PCA as preprocessing to ICA in order to determine the number of sources has proven successful [6]. Also, the PCA estimate of the mixing matrix can be used as initialization for an iterative ICA scheme such as Bell–Sejnowski [6] and the algorithm of Section 7.3.2. Performing ICA using the Molgedey–Schuster algorithm gives better results than PCA, at comparable computational cost.

The choice of pixel independence vs. time independence is related to the problem at hand. In the image segmentation problem above, pixel independence gave the best result; however, other cases have shown preference to time independence (see, e.g., [15, 16]).

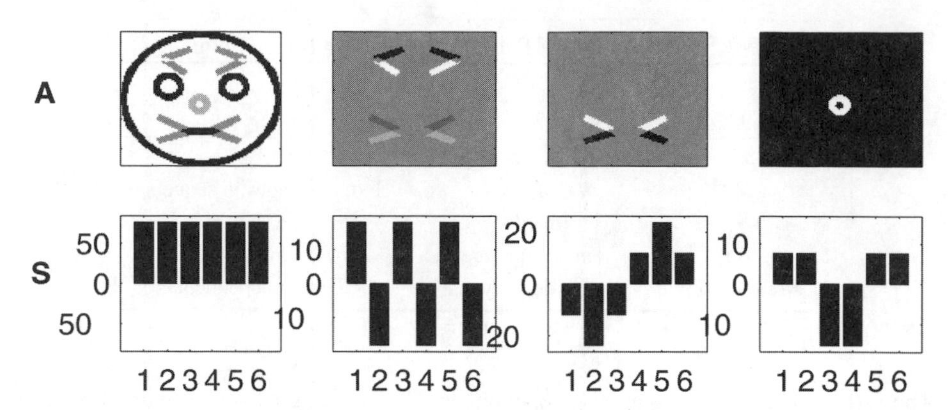

FIGURE 7.9
MS-ICA on the artificial face data with the *time-independence assumption* (i.e., $X^\top$ as signal matrix). The estimated sources (time sequences) are shown in the bottom row and associated mixing matrix (eigenimages) in the top row. The mouth is present in both eigenimages 2 and 3, thus producing a slightly worse result than that in Figure 7.8.

7.6 ICA for Text Representation

7.6.1 Text Analysis

The field of text analysis aims at searching for specific information and structure in text data, which has emerged rapidly in recent years due to the Internet and other massive text databases. The general ways of searching and grouping are usually Boolean[13] search and query[14] subset selection. These methods are straightforward but are not, however, based on statistical modeling. Due to the large amount of data, any statistical approach has been very difficult, and only in recent years has a serious effort been carried out.

The general idea behind many text analysis algorithms is the so-called N-gram histogram. The N-gram histogram is based on counting the simultaneous occurrence of N words or terms. We consider merely 1-gram histograms as higher order histograms that often have large areas of infinitesimal probability mass due to the infrequent occurrence of many word combinations. In Figure 7.10 a 1-gram histogram is shown and is referred to as the *term/document matrix*. The term/document matrix can contain features extracted from the documents and be used as a signal matrix X for PCA and ICA. Recently PCA and ICA have been applied to text analysis [21, 23, 25], and in the following we shall apply both PCA and ICA to the 1-gram histogram using the MED dataset [11]. The MED dataset is a commonly studied collection of medical abstracts. It consists of 1033 abstracts, of which 30 labels have been assigned to 696 of the documents. The goal is not to compare the performance of ICA to other unsupervised methods, but rather to demonstrate its capability in text analysis. Consequently, we restrict the study to 124 abstracts — that is, the first five groups/classes in the MED dataset that can be characterized by the following verbal descriptions:

1. The crystalline lens in vertebrates, including humans.

[13] A Boolean search operates from AND and OR operators.

[14] When a query is made, a subset of the data is selected. This can be done, for example, by a Boolean search — often found by SQL statements.

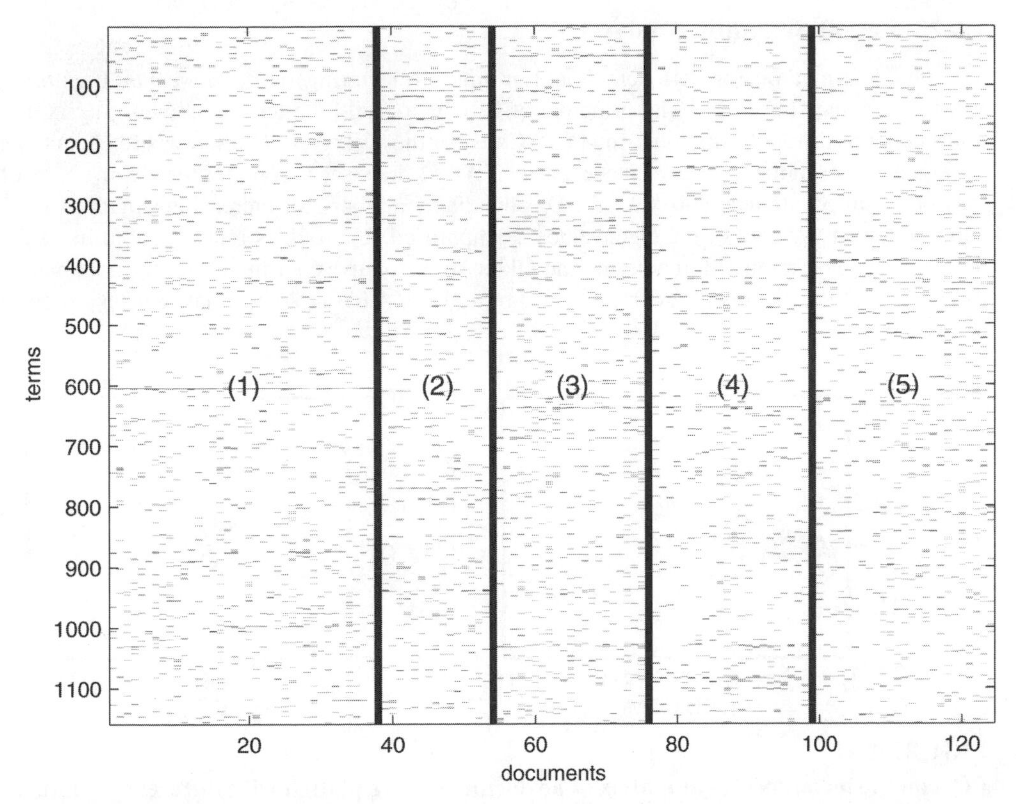

FIGURE 7.10
The term/document matrix X is a 1-gram histogram. The rows represent different words/terms appearing in a collection of text documents. In the present study we use $M = 1159$ terms. Each column represents the histogram for a specific document or text group. In the present example, $N = 124$ documents were used.

2. The relationship of blood and cerebrospinal fluid oxygen concentrations or partial pressures. A method of interest is polarography.

3. Electron microscopy of lung or bronchi.

4. Tissue culture of lung or bronchial neoplasms.

5. The crossing of fatty acids through the placental barrier. Normal fatty acid levels in placenta and fetus.

When constructing the histogram term/document matrix, words that occur in more than one abstract were chosen as term words. In order to facilitate the analysis, commonly used words[15] were removed; 1159 terms remained in the matrix. In summary, the term/document matrix $\boldsymbol{X}$ is $M = 1159$ by $N = 124$. The ICA algorithm used in this example is the noisy mixing algorithm described in Section 7.3.2.

[15] A stop word list was defined.

7.6.2 Latent Semantic Analysis — PCA

A classical method for both search and grouping (clustering) is *latent semantic analysis* (LSA), introduced by [11]. The principle of LSA is to build the term/document matrix and find a better basis representation using PCA. Consider the SVD $\boldsymbol{X} = \boldsymbol{U}\boldsymbol{D}\boldsymbol{V}^{\top}$ where $\boldsymbol{U}$ contains the eigenvectors of the term covariance matrix $\boldsymbol{X}\boldsymbol{X}^{\top}$. Likewise, $\boldsymbol{V}$ contains the eigenvectors of the document covariance matrix $\boldsymbol{X}^{\top}\boldsymbol{X}$. $\mathbf{D}$ is the diagonal matrix of increasing singular values equal to the square root of the eigenvalues. Paraphrased, $\boldsymbol{U}$ provides relative coordinates for the covariance between different terms and, likewise, $\boldsymbol{V}$ relative coordinates for the documents. In Figure 7.11 the documents are represented by a 3D PCA basis. A clear data cluster structure is noticed.

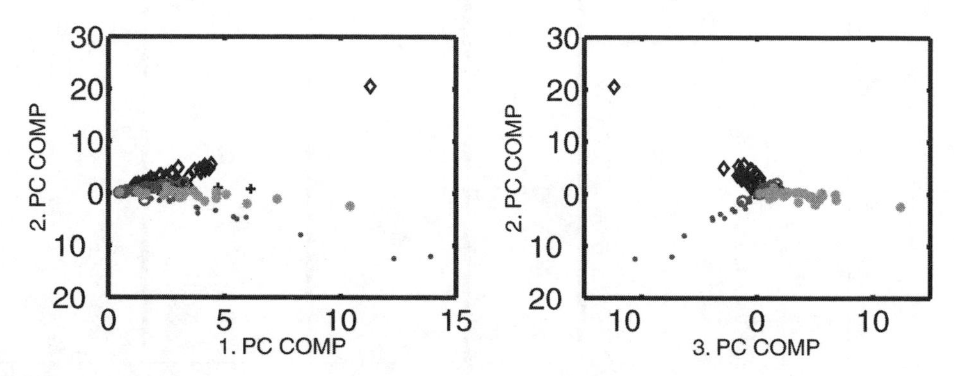

FIGURE 7.11
PCA on the term/document matrix. The documents are plotted with different signatures corresponding to the prelabeling into five classes. A clear cluster structure is noticed.

Using clustering techniques, the documents can now be clustered into groups of similar meaning. This also enables the characterization of a new document by projecting onto the identified PCA basis.

7.6.3 Latent Semantic Analysis — ICA

The objective of ICA in LSA is that it should serve as a clustering algorithm so that different semantic groups are represented by separate independent components. The ICA algorithm produces the mixing matrix $\boldsymbol{A}$ in which each column represents a histogram associated with a specific semantic cluster. The source matrix $\boldsymbol{S}$ expresses how the documents contribute to the semantic clusters.

Since we typically face problems with thousands of words in the terms list and possibly much fewer documents, this is a so-called extremely ill-posed learning problem, which can be remedied without loss of generality by PCA projection. The PCA decomposes the term/document matrix on eigen-histograms. These eigen-histograms are subject to an orthogonality constraint, being eigenvectors to a symmetric real matrix. We are interested in a slightly more general separation of sources that are independent as sequences, but not necessarily orthogonal in the word histogram; that is, we would like to be able to perform a more general decomposition of the data matrix, corresponding to the model in equation (7.4). Before performing the ICA we can make use of the PCA for simplification of the ICA problem. The approach here is similar to the so-called "cure for extremely ill-posed learning" [26] problem used to simplify supervised learning in short image sequences. We first note that the likelihood, considered as a function of the columns of $\boldsymbol{A}$ (histograms), can be split in two parts: part $\boldsymbol{A}_1$, orthogonal to the subspace spanned by the M rows of $\boldsymbol{X}$, and part $\boldsymbol{A}_2$, situated in the subspace spanned by the N columns of $\boldsymbol{X}$. The first part is trivially minimized for any nonzero configuration of sources by putting $\boldsymbol{A}_1 = \boldsymbol{0}$. It simply does not "couple" to data. The remaining part A_2 can be projected onto an N-dimensional hyperplane spanned by the documents. In this way we reduce the high-dimensional separation problem to the separation of a square (projected) data matrix of size $N \times N$. We note that it often may be possible to further limit the dimensionality of the PCA subspace, hence further reducing the histogram dimensionality M of the remaining problem. Using the "cure for extremely ill-posed learning" method, the problem is reduced to an $M = 124$ by $N = 124$ problem without loss of generality. However, we expect that even fewer components are needed for creating a generalizable model. In Figure 7.12 we show the test and training set errors evaluated on training sets of 104 patterns randomly chosen among the set of 124. The test set consists of the remaining 20 documents in each resample. The

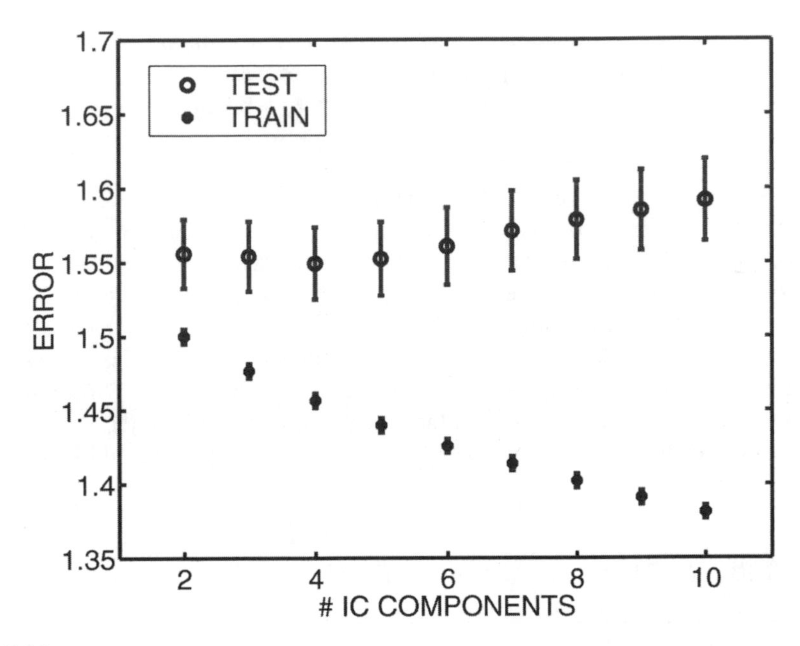

FIGURE 7.12
ICA analysis of the MED dataset. Training and test error as a function of the number of sources, or number of components P. The training set consists of 104 documents randomly chosen among the set of 124 possible, and the remaining 20 are used for test. The test curve shows a shallow minimum for $P = 4$ components, reflecting the bias-variance trade-off discussed in Section 7.3.1.

generalization error shows a shallow minimum for $P = 4$ independent components, reflecting the bias-variance trade-off (Section 7.3.1) as a function of the complexity of the estimated mixing matrix. In Figure 7.13 we show scatterplots in the most variant independent components. Although the distribution of documents forms a rather well-defined group structure in the PCA scatterplots, clearly the ICA scatterplots are much better axis aligned. We conclude that the nonorthogonal basis found by ICA better "explains" the group structure. To further illustrate this finding we have converted the ICA solution to a pattern recognition device by a simple heuristic. We assign a group label based on the magnitude of the recovered source signal. In Tables 7.6 and 7.7 we show that this device is quite successful in recognizing the group structure, although the ICA training procedure is completely unsupervised. For an ICA with three independent components, two are recognized perfectly and three classes are lumped together. The four-component ICA, which is the generalization optimal model, "recognizes" three of the

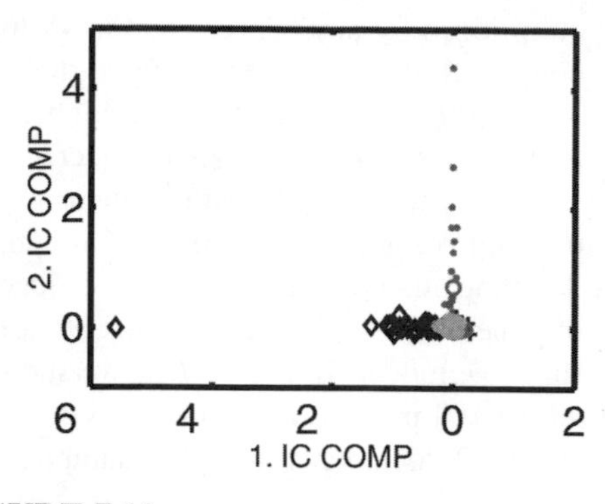

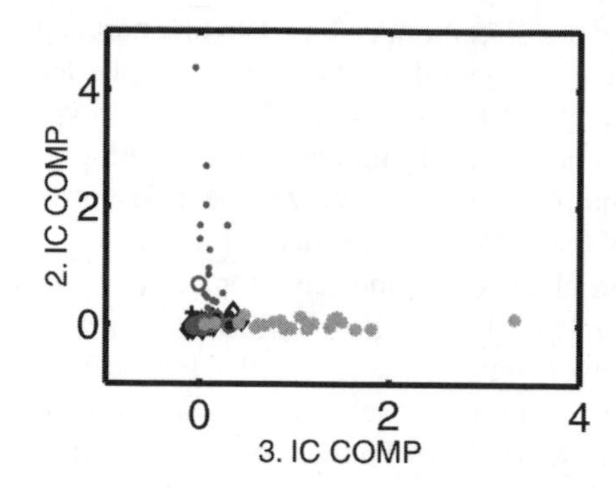

FIGURE 7.13
ICA applied on the term/document matrix. The documents are plotted with different signatures corresponding to the prelabeling into five classes. ICA projects the natural clusters along the basic vectors, making them easy to separate.

five classes almost perfectly and confuses the two classes 3 and 4. Inspecting the groups, we found that the two classes indeed are on very similar topics,[16] and investigating classifications for five or more ICA components did not resolve the ambiguity between them. The ability of the ICA classifier to identify the topic structure is further illustrated in Figure 7.14, where we show scatterplots coded according to ICA classifications. This shows that the ICA is better than PCA-based LSI in identifying relevant latent semantic structure. Finally, we inspect the histograms produced by ICA by back-projection using the PCA basis. Thresholding the ICA histograms, we find the salient terms for the given component. These terms are keywords for the given topic, as shown in Tables 7.6 and 7.7, and follow nicely the behavior of the confusion matrices.

Table 7.6 Confusion Matrix for a Simple Classifier Constructed from the Three-Component ICA

	Class					
	1	2	3	4	5	Keywords
IC_1	37	0	0	0	0	Lens protein
IC_2	0	16	1	1	0	Arterial blood cerebral oxygen rise
IC_3	0	0	21	22	26	Acid blood cell fatty free glucose insulin

Note: Two of the five MED classes are recovered, whereas the last independent component contains a mixture of the remaining three classes.

7.7 Conclusion

This chapter discussed the use of ICA for multimedia applications. In particular, we applied ICA to the separation of speech signals, segmentation of images, and text analysis/clustering.

[16]They both concern medical documents on diseases of the human lungs.

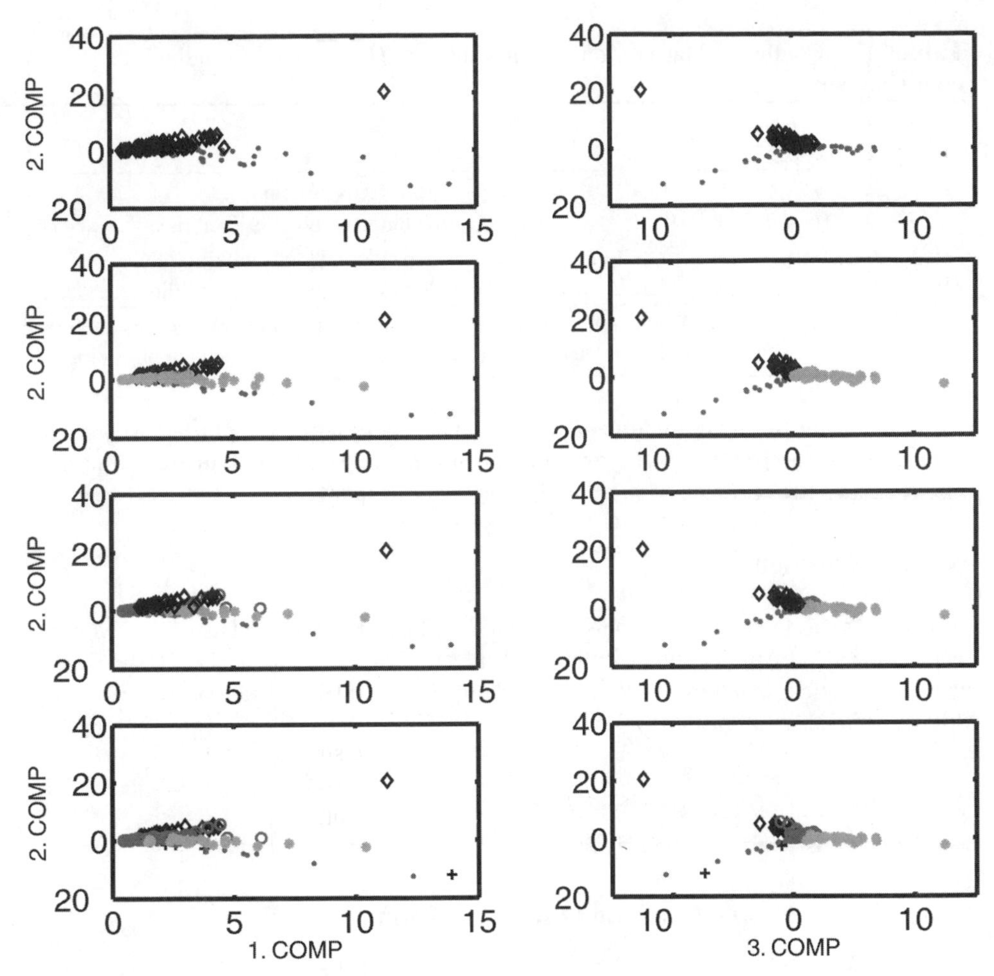

FIGURE 7.14
ICA analysis of the MED dataset. The dataset consists of 124 documents in five topics. The source signals recovered in the ICA have been converted to a simple classifier, and we have coded these classes by different shades. From top to bottom we show scatterplots in the principal component representation 1 vs. 2 and 3 vs. 2, with shading signifying the classification proposed by the ICA with 2, 3, 4, and 5 independent components, respectively.

A likelihood framework for ICA was presented and enables a unified view of different algorithms. Furthermore, this enables formulation of the generalization error, defined as the expected negative log-likelihood on independent examples. The generalization error is a principled tool for model optimization (e.g., number of sources retained in the model).

We focused on two ICA algorithms: separation based on time correlation and noisy mixing of white sources. In the first case we presented a generalized version of the Molgedey–Schuster algorithm, allowing for handling of undercomplete problems, alleviating inherent erroneous complex valued results, and allowing for simultaneous use of more cross-correlation measurements while maintaining the simple noniterative nature of the algorithm. In the noisy mixing case, a maximum a posteriori estimate for source estimation was employed, and the mixing matrix and noise variance were estimated via Boltzmann learning.

Table 7.7 Confusion Matrix for a Simple Classifier Constructed from the Four-Component ICA

	Class 1	2	3	4	5	Keywords
IC$_1$	31	0	0	0	0	Lens protein
IC$_2$	0	16	0	1	0	Arterial blood cerebral oxygen rise
IC$_3$	6	0	22	21	2	Alveolar cell lens lung
IC$_4$	0	0	0	1	24	Acid blood fatty free glucose insulin

Note: Three of the five MED classes are recovered, whereas the remaining two classes are mixed. The two unresolved classes are related because both make reference to the lung physiology.

Acknowledgment

This work was funded by the Danish Research Councils through the Distributed Multimedia Technologies and Applications within the Center for Multimedia and the THOR Center for Neuroinformatics. Andrew Back is acknowledged for valuable discussions concerning the Molgedey-Schuster algorithm.

Appendix A: Property of the Quotient Matrix

THEOREM 7.1

The quotient matrix $\boldsymbol{K} = \boldsymbol{C}_{\widetilde{x}}(\tau)\boldsymbol{C}_{\widetilde{x}}^{-1}(0)$ *has real eigenvalues and eigenvectors, and obtains the eigenvalue decomposition* $\boldsymbol{K} = \boldsymbol{\Phi}\boldsymbol{\Lambda}\boldsymbol{\Phi}^{-1}$.

PROOF $\boldsymbol{C}_{\widetilde{x}}(\tau)$ is symmetric since it can be expressed as $\boldsymbol{C}_{\widetilde{x}}(\tau) = \widetilde{\boldsymbol{Q}}^{\top}\boldsymbol{A}\boldsymbol{C}_s(\tau)\boldsymbol{A}^{\top}\widetilde{\boldsymbol{Q}}$. Further, $\boldsymbol{C}_{\widetilde{x}}(0)$ is positive definite, as $\boldsymbol{C}_s(0)$ is positive definite. A similarity transform of $\boldsymbol{K}$ is given by

$$\boldsymbol{K}_{\text{sim}} = \boldsymbol{C}_{\widetilde{x}}^{-1/2}(0)\boldsymbol{K}\boldsymbol{C}_{\widetilde{x}}^{1/2}(0) = \boldsymbol{C}_{\widetilde{x}}^{-1/2}(0)\boldsymbol{C}_{\widetilde{x}}(\tau)\boldsymbol{C}_{\widetilde{x}}^{-1/2}(0) \tag{7.39}$$

$\boldsymbol{K}_{\text{sim}}$ is thus symmetric with real eigenvalues and eigenvectors [18, Theorem 4.1.5], and obtains the eigenvalue decomposition $\boldsymbol{E}\boldsymbol{\Lambda}\boldsymbol{E}^{\top}$ where $\boldsymbol{E}$ is the orthogonal ($\boldsymbol{E}^{\top}\boldsymbol{E} = \boldsymbol{I}$) matrix of eigenvectors and $\boldsymbol{\Lambda}$ is a diagonal matrix of eigenvalues. Since $\boldsymbol{K}$ and $\boldsymbol{K}_{\text{sim}}$ are similar, they have the same eigenvalues, counting multiplicity [18, Corollary 1.3.4]. Finally, using the similarity transform $\boldsymbol{K} = \boldsymbol{C}_{\widetilde{x}}^{1/2}(0)\boldsymbol{K}_{\text{sim}}\boldsymbol{C}_{\widetilde{x}}^{-1/2}(0)$, $\boldsymbol{K}$ obtains the eigenvalue decomposition $\boldsymbol{K} = \boldsymbol{\Phi}\boldsymbol{\Lambda}\boldsymbol{\Phi}^{-1}$ where $\boldsymbol{\Phi} = \boldsymbol{C}_{\widetilde{x}}^{1/2}(0)\boldsymbol{E}$. ∎

References

[1] S. Amari, A. Cichocki, and H.H. Yang, "A New Learning Algorithm for Blind Signal Separation," in *Advances in Neural Information Processing Systems 8,* D. Touretzky, M. Mozer, and M. Hasselmo (eds.), MIT Press, Cambridge, MA, pp. 757–763, 1996.

[2] H. Attias and C.E. Schreiner, "Blind Source Separation and Deconvolution: The Dynamic Component Analysis Algorithm," *Neural Computation,* vol. 10, pp. 1373–1424, 1998.

[3] H. Attias and C.E. Schreiner, "Blind Source Separation and Deconvolution by Dynamic Component Analysis," in *Proceedings of the IEEE Workshop on Neural Networks for Signal Processing VII,* J. Principe et al. (eds.,) IEEE, Piscataway, NJ, pp. 456–465, 1997.

[4] H. Attias and C.E. Schreiner, "Blind Source Separation and Deconvolution: The Dynamic Component Analysis Algorithm," preprint to appear in *Neural Computation,* 1998. Available via `http://keck.ucsf.edu/~hagai/dca.ps`.

[5] M.S. Bartlett, H.M. Lades, and T.J. Sejnowski, "Independent Component Representations for Face Recognition," *Proceedings of the SPIE — The International Society for Optical Engineering,* vol. 3299, pp. 528–539, 1998.

[6] A. Bell and T.J. Sejnowski, "An Information-Maximization Approach to Blind Separation and Blind Deconvolution," *Neural Computation,* vol. 7, pp. 1129–1159, 1995.

[7] A. Belouchrani and J.-F. Cardoso, "Maximum Likelihood Source Separation by the Expectation-Maximization Technique: Deterministic and Stochastic Implementation," *Proceedings NOLTA,* pp. 49–53, 1995.

[8] J.-F. Cardoso and A. Soulourniac, "Blind Beamforming for Non Gaussian Signals," *IEEE Proceedings-F,* vol. 140, no. 6, pp. 362–370, 1993.

[9] P. Comon, "Independent Component Analysis: A New Concept," *Signal Processing,* vol. 36, pp. 287–314, 1994.

[10] G. Deco and D. Obradovic, *An Information-Theoretic Approach to Neural Computing,* Springer-Verlag, Berlin, 1996.

[11] S. Deerwester, S.T. Dumais, G.W. Furnas, T.K. Landauer, and R. Harshman, "Indexing by Latent Semantic Analysis," *Journal of the American Society for Information Science,* vol. 41, pp. 391–407, 1990.

[12] F. Ehlers and H.G. Schuster, "Blind Separation of Convolutive Mixtures and an Application in Automatic Speech Recognition in a Noisy Environment," *IEEE Transactions on Signal Processing,* vol. 45, no. 10, pp. 2608–2612, Oct. 1997.

[13] S. Geman, E. Bienenstock, and R. Doursat, "Neural Networks and the Bias/Variance Dilemma," *Neural Computation,* vol. 4, pp. 1–58, 1992.

[14] L.K. Hansen and J. Larsen, "Unsupervised Learning and Generalization," *Proceedings of the IEEE International Conference on Neural Networks 1996*, Washington, DC, vol. 1, pp. 25–30, 1996.

[15] L.K. Hansen and J. Larsen, "Source Separation in Short Image Sequences Using Delayed Correlation," *Proceedings of NORSIG'98,* Vigsø, Denmark, pp. 253–256, June 1998. `ftp://eivind.imm.dtu.dk/dist/1998/hansen.norsig98.ps.Z`.

[16] L.K. Hansen, "Blind Separation of Noisy Mixtures," Technical Report, Department of Mathematical Modelling, Techn. Univ. of Denmark, 1998. `http://eivind.imm.dtu.dk/staff/lkhansen/ica.html`.

[17] L.K. Hansen, J. Larsen, F.Å. Nielsen, S.C. Strother, E. Rostrup, R. Savoy, C. Svarer, and O.B. Paulson, "Generalizable Patterns in Neuroimaging: How Many Principal Components?" *NeuroImage*, vol. 9, pp. 534–544, 1999.

[18] R.A. Horn and C.R. Johnson, *Matrix Analysis,* Cambridge University Press: Cambridge, UK, 1994.

[19] J. Hurri, A. Hyvärinen, J. Kahunen, and E. Oja, "Image Feature Extraction Using Independent Component Analysis," *Proceedings of IEEE Nordic Conference on Signal Processing (NORSIG'96),* 1996.

[20] A. Hyvärinen, E. Oja, P. Hoyer, and J. Hurri, "Image Feature Extraction by Sparse Coding and Independent Component Analysis," *Proceedings of International Conference on Pattern Recognition 98,* Brisbane, Australia, pp. 1268–1273, 1998.

[21] C.L. Isbell, Jr. and P. Viola, "Restructuring Sparse High Dimensional Data for Effective Retrieval," *Advances in Neural Information Processing Systems 11,* MIT Press, Cambridge, MA, pp. 480–486, 1999.

[22] C. Jutten and J. Herault, "Blind Separation of Sources, An Adaptive Algorithm Based on Neuromimetic Architecture," *Signal Processing,* vol. 24, pp. 1–10, 1991.

[23] T. Kolenda and L.K. Hansen, "Independent Components in Text," Technical Report, Dept. of Mathematical Modelling, Tech. Univ. of Denmark, 1999. Available via `ftp,//eivind.imm.dtu.dk/dist/kolenda.nips99.ps.gz`.

[24] S.Y. Kung and C. Mejuto, "Extraction of Independent Components from Hybrid Mixture: KuicNet Learning Algorithm and Applications," *Proceedings of IEEE ICASSP98,* vol. 2, pp. 1209–1212, Seattle, WA, May 1998.

[25] T.K. Landauer, D. Laham, and P. Foltz, "Learning Human-Like Knowledge by Singular Value Decomposition, A Progress Report," *Advances in Neural Information Processing Systems 10,* MIT Press, Cambridge, MA, pp. 45–51, 1998.

[26] B. Lautrup, L.K. Hansen, I. Law, N. Mørch, C. Svarer, and S.C. Strother, "Massive Weight Sharing: A Cure for Extremely Ill-posed Problems," in *Supercomputing in Brain Research, From Tomography to Neural Networks,* World Scientific, H.J. Hermanet et al., (eds.), pp. 137–148, 1995.

[27] T.-W. Lee, M. Girolami, A.J. Bell, and T.J. Sejnowski, "A Unifying Information-Theoretic Framework for Independent Component Analysis," Salk Institute preprint, to appear in *International Journal on Computers and Mathematics with Applications,* in press, 1999. Available via `http,//www.cnl.salk.edu/~tewon/Public/ijmc99.ps.gz`.

[28] T.-W. Lee, M.S. Lewicki, M. Girolami, and T.J. Sejnowski, "Blind Source Separation of More Spruces Than Mixtures Using Overcomplete Representations," *Signal Processing Letters,* vol. 4, no. 4, April 1999. Available via `http://www.cnl.salk.edu/~tewon/Public/ocica99.ps.gz`.

[29] T.-W. Lee, M.S. Lewicki, and T.J. Sejnowski, "Unsupervised Classification, Segmentation, and De-Noising of Images using ICA Mixture Models," submitted for publication.

[30] T.-W. Lee, M.S. Lewicki, and T.J. Sejnowski, "ICA Mixture Models for Image Processing," *Proceedings of the 6th Joint Symposium on Neural Computation,* California Institute of Technology, Pasadena, CA, pp. 79–86, 1999.

[31] M.S. Lewicki and T.J. Sejnowski, "Learning Overcomplete Representations," Salk Institute preprint, *Neural Computation,* in press, 1999. Available via `http://www.cnl.salk.edu/~lewicki/papers/overcomplete.ps.gz`.

[32] D. MacKay, "Maximum Likelihood and Covariant Algorithms for Independent Components Analysis," Draft 3.7, 1996. Available via `ftp://mroa.cam.ac.uk/hello.ps.gz`.

[33] L. Molgedey and H. Schuster, "Separation of Independent Signals Using Time-Delayed Correlations," *Physical Review Letters,* vol. 72, no. 23, pp. 3634–3637, 1994.

[34] E. Moulines, J.-F. Cardoso, and E. Gassiat, "Maximum Likelihood for Blind Separation and Deconvolution of Noisy Signals Using Mixture Models," *Proceedings of ICASSP'97,* vol. 5, pp. 3617–3620, 1997.

[35] B.A. Olshausen, *Learning Linear, Sparse, Factorial Codes,* A.I. Memo 1580, MIT Press, Cambridge, MA, 1996.

[36] E. Oja, "PCA, ICA, and Nonlinear Hebbian Learning," *Proceedings of the International Conference on Artificial Neural Networks ICANN-95,* pp. 89–94, 1995.

[37] B.A. Pearlmutter and L.C. Parra, "Maximum Likelihood Blind Source Separation, A Context-Sensitive Generalization of ICA," in *Advances in Neural Information Processing Systems 9,* M.C. Mozer et al. (eds.), MIT Press, Cambridge, MA, pp. 613–619, 1997.

[38] C. Peterson and J.R. Anderson, "Mean Field Theory Learning Algorithm for Neural Networks," *Complex Systems,* vol. 1, pp. 995–1019, 1987.

Chapter 8

Image Analysis and Graphics for Multimedia Presentation

Tülay Adali and Yue Wang

8.1 Introduction

The success of multimedia applications is highly dependent on the effective representation of the information of interest from data that now come in a variety of forms. For the effective use of computer-reconstructed images, two steps are key: analysis of images through extraction of the key features of the image and the visualization of these features in a way that is suitable for the application at hand.

Model-based image analysis aims at capturing the intrinsic character of images with few parameters and is also instrumental in helping to understand the nature of the imaging process. Key issues in image analysis include model selection, parameter estimation, imaging physics, and the relationship of the image to the task (how the image is going to be utilized) [11, 28]. Stochastic model-based image analysis has been the most popular among the model-based image analysis methods because, most often, imaging physics can be modeled effectively with a stochastic model. For example, the suitability of standard finite normal mixture models has been verified for a number of medical imaging modalities [33, 73, 77]. In the first part of the chapter, we discuss a complete treatment of the stochastic model-based image analysis that includes model and model order selection, parameter estimation, and final segmentation. We focus on models that use finite normal mixtures and show examples in medical image segmentation and computer-aided diagnosis.

Computer graphics can play a central role in helping multimedia meet its challenges. Representing images in a form that matches our perceptual capabilities (mainly visual) and a problem's particular needs makes the process of getting information and digesting it easier and more effective. More specifically, good use of visualization and computer graphics in the multimedia environment can make a number of important tasks easier and more effective, such as,

1. Analyzing information on the images
2. Monitoring image content and changes
3. Interacting with image databases
4. Collaborating with other sites/groups
5. Handling video e-mail or browsing on the Web

In the second part of this chapter, we discuss how to use graphics modeling and visualization technologies to achieve this task. We address methods for graphical modeling and reconstruction and introduce deformable surface–spine models. We discuss applications in reconstruction of synthetic and range datasets and of 3D surgical prostate models.

8.2 Image Analysis

Stochastic model-based image analysis is a technique for partitioning an image into distinctive meaningful regions based on the statistical properties of both the gray-level and the context images. A good segmentation result depends on suitable model selection for the given image. For medical images, such as magnetic resonance (MR), positron emission tomography (PET), and radiographic images, model selection can be justified in terms of imaging physics, or alternatively, a better understanding of the imaging physics can be used to select a suitable model for a given imaging modality [33, 77]. Model selection refers to the determination of both the local statistical distributions of each region and the number of image regions.

In image analysis, we can treat pixel and context modeling separately, assuming that each pixel can be decomposed into a *pixel image* and a *context image*. Pixel image is defined as the observed gray level associated with the pixel, and finite mixture models have been the most popular pixel image models. In particular, standard finite normal mixtures (SFNMs) have been very widely used in statistical image analysis, and efficient algorithms are available for calculating the parameters of the model. Furthermore, by incorporating statistical properties of context images, where context image is defined as the membership of the pixel associated with different regions, a localized SFNM formulation can be used to impose local consistency constraints on context images in terms of a stochastic regularization scheme [74]. The next section describes the finite mixtures model and addresses identification of the model (i.e., estimation of the parameters of the model and the model order selection). In Section 8.2.2, we discuss approaches to modeling context. Also, it is important to note that, even though texture is an important property in the perception of images by humans, it is typically difficult to describe. It can be identified in terms of five perceptual dimensions: coarseness, contrast, directionality, line-likeness regularity, and roughness [60], and can be incorporated into the graphical representation discussed in Section 8.3.

8.2.1 Pixel Modeling

Given a digital image consisting of $N \equiv N_1 \times N_2$ pixels, assume that this image contains K regions and that each pixel is decomposed into a pixel image x and a context image l. By ignoring information regarding the spatial ordering of pixels, we can treat context images (i.e., pixel labels) as random variables and describe them using a multinomial distribution with unknown parameter π_k. Since this parameter reflects the distribution of the total number of pixels in each region, π_k can be interpreted as a prior probability of pixel labels determined by the global context information. Thus, the relevant (sufficient) statistics are the pixel image statistics for each component mixture and the number of pixels of each component. The marginal probability measure for any pixel image (i.e., the finite mixtures distribution) can be obtained by writing the joint probability density of x and l and then summing the joint density over all possible outcomes of l (i.e., by computing $p(x_i) \sum_l p(x_i, l)$), resulting in a sum of the following general form:

$$p(x_i) = \sum_{k=1}^{K} \pi_k p_k(x_i), \quad i = 1, \ldots, N \tag{8.1}$$

where x_i is the gray level of pixel i. $p_k(x_i)$'s are conditional region probability density functions (pdfs) with the weighting factor π_k, satisfying $\pi_k > 0$, and $\sum_{k=1}^{K} \pi_k = 1$. The generalized Gaussian pdf given region k is defined by [89]

$$p_k(x_i) = \frac{\alpha \beta_k}{2\Gamma(1/\alpha)} \exp\left[-\left|\beta_k(x_i - \mu_k)\right|^{\alpha}\right], \quad \alpha > 0, \quad \beta_k = \frac{1}{\sigma_k}\left[\frac{\Gamma(3/\alpha)}{\Gamma(1/\alpha)}\right]^{1/2} \tag{8.2}$$

where μ_k is the mean, $\Gamma(\cdot)$ is the gamma function, and β_k is a parameter related to the variance σ_k by

$$\beta_k = \frac{1}{\sigma_k}\left[\frac{\Gamma(3/\alpha)}{\Gamma(1/\alpha)}\right]^{1/2}. \tag{8.3}$$

When $\alpha \gg 1$, the distribution tends to be a uniform pdf; for $\alpha < 1$, the pdf becomes sharper; for $\alpha = 2.0$, one has the Gaussian pdf; and for $\alpha = 1.0$, the Laplacian pdf exists. Therefore, the generalized Gaussian model is a suitable model to fit the histogram distribution of those images whose statistical properties are unknown since the kernel shape can be controlled by selecting different α values. The finite Gaussian mixture model (FGGM) for $\alpha = 2$ is commonly referred to as the standard finite normal mixture model and has been the most frequently used form. It can be written as

$$p_k(x_i) = \sum_{k=1}^{K} \pi_k g\left(x_i \middle| \mu_k, \sigma_k^2\right) \quad i = 1, 2, \ldots, N \tag{8.4}$$

with

$$g\left(x_i \middle| \mu_k, \sigma_k^2\right) = \frac{1}{\sqrt{2\pi}\sigma_k} \exp\left(-\frac{(x_i - \mu_k)^2}{2\sigma_k^2}\right)$$

where μ_k and σ_k^2 are the mean and variance of the kth Gaussian kernel and K is the number of Gaussian components.

The whole image can be well approximated by an independent and identically distributed random field $\mathbf{X}$. The corresponding joint pdf is

$$P(\mathbf{x}) = \prod_{i=1}^{N} \sum_{k=1}^{K} \pi_k p_k(x_i) \tag{8.5}$$

where $\mathbf{x} = [x_1, x_2, \ldots, x_N]$ and $\mathbf{x} \in \mathbf{X}$. Based on the joint probability measure of pixel images, the likelihood function under finite mixture modeling can be expressed as $\mathcal{L}(\mathbf{r}) = \prod_{i=1}^{N} p_{\mathbf{r}}(x_i)$ where $\mathbf{r} : \{K, \alpha, \pi_k, \mu_k, \sigma_k, k = 1, \ldots, K\}$ denotes the model parameter set.

8.2.2 Model Identification

Once the model is chosen, identification addresses the estimation of the local region parameters $(\pi_k, \mu_k, \sigma_k, k = 1, \ldots, K)$ and the structural parameters (K, α). In particular the estimation of the order parameter, K, is referred to as model order selection.

Parameter Estimation

With an appropriate system likelihood function, the objective of model identification is to estimate the model parameters by maximizing the likelihood function, or equivalently minimizing the relative entropy between the image histogram $p_{\mathbf{x}}(u)$ and the estimated pdf $p_{\mathbf{r}}(u)$, where u is the gray level [2, 69]. There are a number of approaches to perform the maximum likelihood (ML) estimation of finite mixture distributions [66]. The most popular method is the expectation–maximization (EM) algorithm [18, 53]. The EM algorithm first calculates the posterior Bayesian probabilities of the data through the observations, obtains the current parameter estimates (E step), and then updates parameter estimates using generalized mean ergodic theorems (M step). The procedure cycles back and forth between these two steps. The successive iterations increase the likelihood of the model parameters. A neural network interpretation of this procedure is given in [49].

We can use relative entropy (the Kullback–Leibler distance) [31] for parameter estimation [i.e., we can measure the information theoretic distance between the histogram of the pixel images, denoted by $p_{\mathbf{x}}$, and the estimated distribution $p_{\mathbf{r}}(u)$, which we define as the global relative entropy (GRE)]:

$$D\left(p_{\mathbf{x}} \| p_{\mathbf{r}}\right) = \sum_{u} p_{\mathbf{x}}(u) \log \frac{p_{\mathbf{x}}(u)}{p_{\mathbf{r}}(u)} . \tag{8.6}$$

It can be shown that, when relative entropy is used as the distance measure, distance minimization is equivalent to the ML estimation of the model parameters [2, 69].

For the case of the FGGM model, the EM algorithm can be applied to the joint estimation of the parameter vector and the structural parameter α as follows [18]:

EM Algorithm

1. For $\alpha = \alpha_{\min}, \ldots, \alpha_{\max}$
   - $m = 0$, given initialized $\mathbf{r}^{(0)}$
   - E step: for $i = 1, \ldots, N,\ k = 1, \ldots, K$, compute the probabilistic membership

$$z_{ik}^{(m)} = \frac{\pi_k^{(m)} p_k(x_i)}{\sum_{k=1}^{K} \pi_k^{(m)} p_k(x_i)} \tag{8.7}$$

   - M step: for $k = 1, \ldots, K$, compute the updated parameter estimates

$$\begin{cases} \pi_k^{(m+1)} = \dfrac{1}{N} \displaystyle\sum_{i=1}^{N_1 N_2} z_{ik}^{(m)} \\ \mu_k^{(m+1)} = \dfrac{1}{N\pi_k^{(m+1)}} \displaystyle\sum_{i=1}^{N} z_{ik}^{(m)} x_i \\ \sigma_k^{2(m+1)} = \dfrac{1}{N\pi_k^{(m+1)}} \displaystyle\sum_{i=1}^{N} z_{ik}^{(m)} (x_i - \mu_k^{(m+1)})^2 \end{cases} \tag{8.8}$$

   - When $|\mathrm{GRE}^{(m)}(p_{\mathbf{x}} \| p_{\mathbf{r}}) - \mathrm{GRE}^{(m+1)}(p_{\mathbf{x}} \| p_{\mathbf{r}})| \le \epsilon$ is satisfied, go to step 2. Otherwise, $m = m + 1$ so go to E step.

2. Compute GRE, and go to step 1.

3. Choose the optimal $\hat{\mathbf{r}}$ that corresponds to the minimum GRE.

The EM algorithm, however, in general, has the reputation of being slow, because it has a first-order convergence in which new information acquired in the expectation step is not used immediately [84]. Recently, a number of online versions of the EM algorithm have been proposed for large-scale sequential learning (e.g., see [41, 47, 66, 69, 81]). Such a procedure obviates the need to store all the incoming observations, changing the parameters immediately after each data point, allowing for high data rates. Titterington [66] has developed a stochastic approximation procedure that is closely related to the probabilistic self-organizing mixture (PSOM) algorithm we are going to introduce here, and shows that the solution can be made consistent. Other similar formulations have been proposed by Marroquin et al. [41] and Weinstein et al. [81].

For the adaptive estimation of the SFNM model parameters, we can derive an incremental learning algorithm by the simple stochastic gradient descent minimization of $D(p_{\mathbf{x}} || p_{\mathbf{r}})$ [69, 73] given in (8.6) with the $p_{\mathbf{r}}$ given by (8.4):

$$\mu_k{}^{(t+1)} = \mu_k^{(t)} + a(t)\left(x_{t+1} - \mu_k^{(t)}\right) z_{(t+1)k}^{(t)}, \tag{8.9}$$

$$\sigma_k^{2(t+1)} = \sigma_k^{2(t)} + b(t)\left[\left(x_{t+1} - \mu_k^{(t)}\right)^2 - \sigma_k^{2(t)}\right] z_{(t+1)k}^{(t)},$$
$$k = 1, \ldots, K \tag{8.10}$$

where $a(t)$ and $b(t)$ are introduced as the learning rates, two sequences converging to zero, ensuring unbiased estimates after convergence. For details about derivation and the approximations, see [69, 70]. Based on generalized mean ergodic theorem [17], updates can also be obtained for the constrained regularization parameters, π_k, in the SFNM model. For simplicity, given an asymptotically convergent sequence, the corresponding mean ergodic theorem (i.e., the recursive version of the sample mean calculation) should hold asymptotically. Thus, we define the interim estimate of π_k by [71]:

$$\pi_k^{(t+1)} = \frac{t}{t+1}\pi_k^{(t)} + \frac{1}{t+1} z_{(t+1)k}^{(t)} . \tag{8.11}$$

Hence the updates given by (8.9), (8.10), and (8.11) together with evaluation of (8.7) using (8.4) provide the incremental procedure for computing the SFNM component parameters. Their practical use, however, requires strongly mixing conditions and a decaying annealing procedure (learning rate decay) [17, 25, 51]. In finite mixtures parameter estimation, algorithm initialization must be chosen carefully and appropriately. In [71], an adaptive Lloyd–Max histogram quantization (ALMHQ) algorithm is introduced for threshold selection which is also well suited to initialization in ML estimation. It can be used for initializing the network parameters, μ_k, σ_k^2, and $\pi_k, k, 1, 2, \ldots, K$.

Model Order Selection

Determination of the region parameter K directly affects the quality of the resulting model parameter estimation and, in turn, affects the results of segmentation. In a statistical problem formulation such as the one introduced in the previous section, the use of information theoretic criteria for the problem of model determination arises as a natural choice. Two popular approaches are Akaike's information criterion (AIC) [4] and Rissanen's minimum description length (MDL) [55]. Akaike proposed to select the model that gives the minimum AIC, which is defined by

$$\mathrm{AIC}\,(K_a) = -2\log\left(\mathcal{L}\left(\hat{\mathbf{r}}_{\mathrm{ML}}\right)\right) + 2K_a \tag{8.12}$$

where $\hat{\mathbf{r}}_{\text{ML}}$ is the maximum likelihood estimate of the model parameter set $\mathbf{r}$, and K' is the number of free adjustable parameters in the model [4, 33]. AIC selects the correct number of image regions K_0 when

$$K_0 = \arg\left\{\min_{1\le K\le K_{\max}} \text{AIC}(K)\right\} . \tag{8.13}$$

Rissanen addressed the problem from a quite different point of view. He reformulated the problem explicitly as an information coding problem in which the best model fitness was measured such that it assigned high probabilities to the observed data while at the same time the model itself was not too complex to describe [55]. The model is selected by minimizing the total description length defined by

$$\text{MDL}\left(K_a\right) = -\log\left(\mathcal{L}\left(\hat{\mathbf{r}}_{\text{ML}}\right)\right) + 0.5K_a\log(N) . \tag{8.14}$$

Similarly, the correct number of distinctive image regions K_0 can be estimated as

$$K_0 = \arg\left\{\min_{1\le K\le K_{\max}} \text{MDL}(K)\right\} . \tag{8.15}$$

A more recent formulation of information theoretic criterion, the minimum conditional bias and variance (MCBV) criterion [69, 75], selects a minimum conditional bias and variance model (i.e., if two models are about equally likely, MCBV selects the one whose parameters can be estimated with the smallest variance). The formulation is based on the fundamental argument that the value of the structural parameter cannot be arbitrary or infinite, because although such an estimate might be said to have low "bias," the price to be paid is high "variance" [23].

Since the joint maximum entropy is a function of K_a and $\hat{\mathbf{r}}$, by taking the advantage of the fact that model estimation is separable in components and structure, we define the MCBV criterion as

$$\text{MCBV}(K) = -\log\left(\mathcal{L}\left(\mathbf{x}|\hat{\mathbf{r}}_{\text{ML}}\right)\right) + \sum_{k=1}^{K_a} H\left(\hat{r}_{k\text{ML}}\right) \tag{8.16}$$

where $-\log(\mathcal{L}(\mathbf{x}|\hat{\mathbf{r}}_{\text{ML}}))$ is the conditional bias (a form of information theoretic distance) [17, 54] and $\sum_{k=1}^{K_a} H(\hat{r}_{k\text{ML}})$ is the conditional variance (a measure of model uncertainty) [51, 54] of the model. Because both of these terms represent natural estimation errors about their true models, they can be treated on an equal basis. A minimization of the expression in (8.16) leads to the following characterization of the optimum estimation:

$$K_0 = \arg\left\{\min_{1\le K\le K_{\max}} \text{MCBV}(K)\right\} . \tag{8.17}$$

That is, if the cost of model variance is defined as the entropy of parameter estimates, the cost of adding new parameters to the model must be balanced by the reduction they permit in the ideal code length for the reconstruction error. A practical MCBV formulation with code-length expression is further given by [17, 75]

$$\text{MCBV}(K) = -\log\left(\mathcal{L}\left(\mathbf{x}|\hat{\mathbf{r}}_{\text{ML}}\right)\right) + \sum_{k=1}^{K_a} \frac{1}{2}\log 2\pi e\text{Var}\left(\hat{\mathbf{r}}_{k\text{ML}}\right) \tag{8.18}$$

where the calculation of $H(\hat{\mathbf{r}}_{k\text{ML}})$ requires the estimation of the true ML model parameter values. It is shown that, for a sufficiently large number of observations, the accuracy of the

ML estimation tends quickly to be the best possible accuracy determined by the Cramer–Rao lower bounds (CRLBs) [51]. Thus, the CRLBs of the parameter estimates are used in the actual calculation to represent the "conditional" bias and variance [50]. We have found that, experimentally, the MCBV formulation for determining the value of K_0 exhibits very good performance consistent with both the AIC and the MDL criteria. It should be noted, however, that these are not the only plausible approaches to the problem of order selection; other approaches such as cross-validation techniques may also be quite useful [20, 36, 42, 48, 80].

8.2.3 Context Modeling

Once the pixel model is estimated, the segmentation problem is the assignment of labels to each pixel in the image. A straightforward way is to label pixels into different regions by maximizing the individual likelihood function $p_k(x)$ (i.e., to perform ML classification). Usually, this method may not achieve good performance because it does not use local neighborhood information in the decision. The CBRL algorithm [27] is one approach that can incorporate the local neighborhood information into the labeling procedure and thus improve the segmentation performance. The CBRL algorithm to perform/refine pixel labeling based on the localized FGGM model can be defined as follows [37]:

Let ∂i be the neighborhood of pixel i with an $m \times m$ template centered at pixel i. An indicator function is used to represent the local neighborhood constraints $R_{ij}(l_i, l_j) = I(l_i, l_j)$, where l_i and l_j are labels of pixels i and j, respectively. Note that pairs of labels are now either compatible or incompatible. Similar to the procedure in [27], one can compute the frequency of neighbors of pixel i that have the same label values k as at pixel i

$$\pi_k^{(i)} = p\left(l_i = k \middle| \mathbf{l}_{\partial i}\right) = \frac{1}{m^2 - 1} \sum_{j \in \partial i, j \neq i} I\left(k, l_j\right) \tag{8.19}$$

where $\mathbf{l}_{\partial i}$ denotes the labels of the neighbors of pixel i. Since $\pi_k^{(i)}$ is a conditional probability of a region, the localized FGGM pdf of gray-level x_i at pixel i is given by

$$p\left(x_i \middle| \mathbf{l}_{\partial i}\right) = \sum_{k=1}^{K} \pi_k^{(i)} p_k\left(x_i\right) \tag{8.20}$$

where $p_k(x_i)$ is given in (8.2). Assuming gray values of the image are conditional independent, the joint pdf of $\mathbf{x}$, given the context labels $\mathbf{l}$, is

$$P(\mathbf{x}|\mathbf{l}) = \prod_{i=1}^{N} \sum_{k=1}^{K} \pi_k^{(i)} p_k\left(x_i\right) \tag{8.21}$$

where $\mathbf{l} = (l_i : i = 1, \ldots, N)$.

It is important to note that the CBRL algorithm can obtain a consistent labeling solution based on the localized FGGM model (8.20). Since $\mathbf{l}$ represents the labeled image, it is consistent if $S_i(l_i) \geq S_i(k)$, for all $k = 1, \ldots, K$ and for $i = 1, \ldots, N$ [27], where

$$S_i(k) = \pi_k^{(i)} p_k\left(x_i\right) . \tag{8.22}$$

Now we can define

$$A(\mathbf{l}) = \sum_{i=1}^{N} \left(\sum_k I\left(l_i, k\right) S_i(k) \right) \tag{8.23}$$

as the average measure of local consistency and

$$LC_i = \sum_k I(l_i, k)\, S_i(k), \quad i = 1, \ldots, N \tag{8.24}$$

represents the local consistency based on $\mathbf{l}$. The goal is to find a consistent labeling $\mathbf{l}$ that can maximize (8.23). In the real application, each local consistency measure LC_i can be maximized independently. In [27], it has been shown that when $R_{ij}(l_i, l_j) = R_{ji}(l_j, l_i)$, if $A(\mathbf{l})$ attains a local maximum at $\mathbf{l}$, then $\mathbf{l}$ is a consistent labeling.

Based on the localized FGGM model, $l_i^{(0)}$ can be initialized by an ML classifier,

$$l_i^{(0)} = \arg\left\{\max_k \; p_k(x_i)\right\}, \quad k = 1, \ldots, K\,. \tag{8.25}$$

Then, the order of pixels is randomly permutated and each label l_i is updated to maximize LC_i — that is, classify pixel i into kth region if

$$l_i = \arg\left\{\max_k \; \pi_k^{(i)} p_k(x_i)\right\}, \quad k = 1, \ldots, K \tag{8.26}$$

where $p_k(x_i)$ is given in (8.2) and $\pi_k^{(i)}$ is given in (8.19). By considering (8.25) and (8.26), we can give a modified CBRL algorithm as follows [37]:

CBRL Algorithm

1. Given $\mathbf{l}^{(0)}$, m=0
2. Update pixel labels
   - Randomly visit each pixel for $i = 1, \ldots, N$
   - Update its label l_i according to

$$l_i^{(m)} = \arg\left\{\max_k \pi_k^{(i)(m)} p_k(x_i)\right\}$$

3. When $\frac{\sum(\mathbf{l}^{(m+1)} \oplus \mathbf{l}^{(m)})}{N_1 N_2} \leq 1\%$, stop; otherwise, $m = m + 1$, and repeat step 2.

8.2.4 Applications

Simulated Data

To verify the steps of the statistical image analysis framework we discussed, let us first consider a simulated image. Our example is the image shown in Figure 8.1a, which is made up of four overlapping normal components. Each component represents one local region. Noise levels are set to keep the same signal-to-noise ratio (SNR) between regions, where the SNR is defined by

$$\text{SNR} = 10 \log_{10} \frac{(\Delta\mu)^2}{\sigma^2} \tag{8.27}$$

where $\Delta\mu$ is the mean difference between regions and σ^2 is the noise power. The AIC, MDL, and MCBV curves as a function of the number of local clusters K are shown in Figure 8.1b. According to the information theoretic criteria, the minima of these curves indicate the correct number of local regions. From this experimental figure, it is clear that the number of

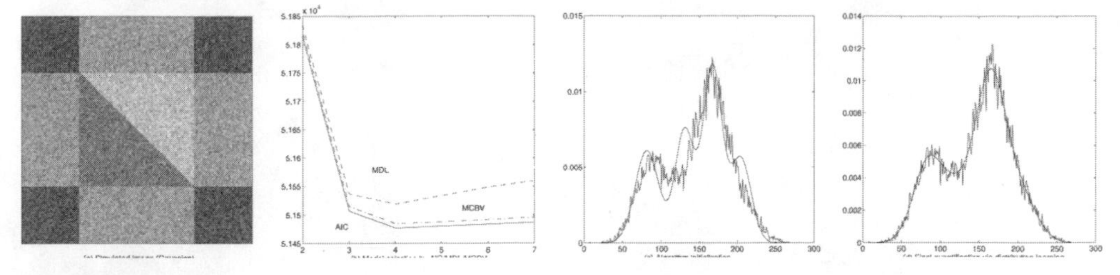

FIGURE 8.1
Experimental results of model selection, algorithm initialization, and final quantification on the simulated image: (a) original image with four components; (b) curves of the AIC/MDL/MCBV criteria where the minimum corresponds to $K_0 = 4$; (c) initial histogram learning by the ALMHQ algorithm; (d) final histogram learning by the PSOM algorithm.

Table 8.1 True Parameter Values and the Estimates for the Simulated Image of Figure 8.1

	True				Initial				Final			
k	1	2	3	4	1	2	3	4	1	2	3	4
π	0.25	0.125	0.5	0.125	0.234	0.234	0.364	0.185	0.23	0.135	0.48	0.157
μ	86	126	166	206	81	131	167	205	84	121	164	201
σ^2	400	400	400	400	235	158	157	177	354	365	373	463

local regions suggested by these criteria are all correct. After the algorithm initialization by ALMHQ [71], network parameters are finalized by the PSOM algorithm given in (8.9)–(8.11). The GRE value (8.6) is used as an objective measure to evaluate the accuracy of quantification. The results of the distribution learning by PSOM are shown in Figures 8.1c and d. The GRE in the initial stage achieves a value of 0.0399 nats, and after the final quantification by PSOM, is down to 0.008 nats. The numerical results are given in Table 8.1, where the units of μ and σ^2 simply represent the observed gray levels of the pixel images, whereas π is the probability measure. To simplify the representation, we omit their units as in [38, 56]. References [69, 70, 73] discuss these examples in more detail and present comparative results for parameter estimation using EM and PSOM, noting the advantages of PSOM due to its incremental nature.

Figure 8.2 shows the results of final image segmentation using the CBRL algorithm. We use the ML classifier to initialize the image segmentation (i.e., to initialize the quantified image by selecting the pixel label with the largest likelihood at each pixel) using equation (8.25). This gives a suitable starting point for relaxation labeling [74]. CBRL is then used to fine tune the image segmentation. Since the true scene is known in this experiment, the percentage of total classification error is used as the criterion for evaluating the performance of the segmentation technique. In Figure 8.2, the initial segmentation by the ML classification and the stepwise results of three iterations in PCRN are presented. In this experiment, algorithm initialization results in an average misclassification of 30%. It can be clearly seen that a dramatic improvement is obtained after several iterations of the CBRL by using local constraints determined by the context information. Also, note that the convergence is fast because after the first iteration most misclassifications are removed. The final percentage of classification errors for Figure 8.2 is about 0.7935%.

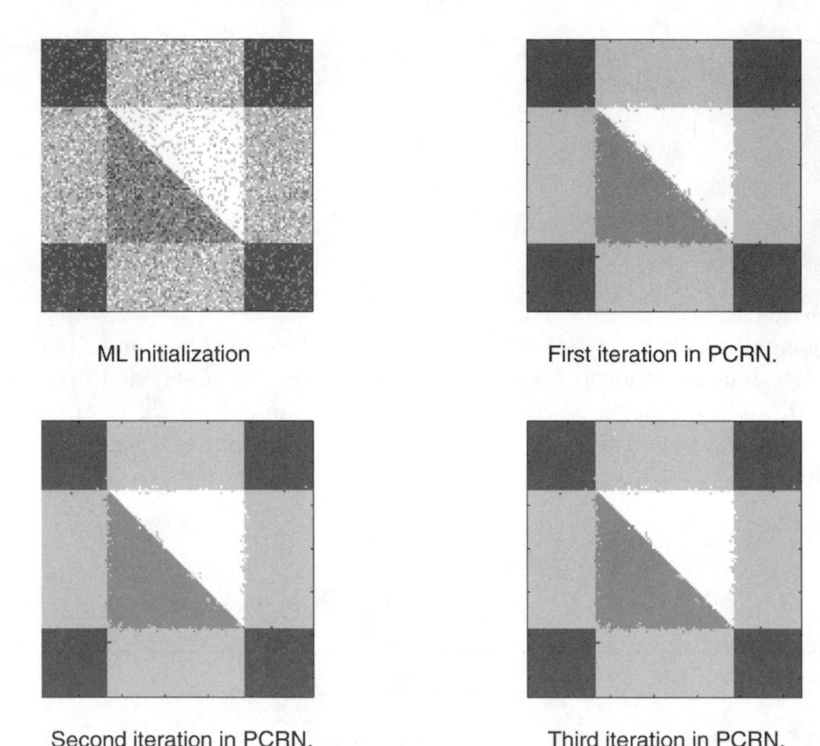

FIGURE 8.2
Image segmentation by PCRN on simulated image (with initialization by ML classification).

Brain MR Analysis

Quantitative analysis of brain tissues refers to the problem of estimating tissue quantities from a given image and segmentation of the image into contiguous regions of interest to describe the anatomical structures. The problem has recently received much attention largely due to the improved fidelity and resolution of medical imaging systems. Because of its ability to deliver high resolution and contrast, MR imaging (MRI) has been the dominant modality for research on this problem [14, 16, 38, 56, 83]. Based on the statistical properties of MR pixel images, use of an SFNM distribution is justified to model the image histogram, and it is shown that the SFNM model converges to the true distribution when the pixel images are asymptotically independent [73].

For this study, we use data consisting of three adjacent, T1-weighted MR images parallel to the AC–PC line. Since the skull, scalp, and fat in the original brain images do not contribute to the brain tissue, we edit the MR images to exclude nonbrain structures prior to tissue quantification and segmentation, as explained in [70, 74]. This also helps us to achieve better quantification and segmentation of brain tissues by delineation of other tissue types that are not clinically significant [38, 56, 83]. The extracted brain tissues are shown in Figure 8.3.

Evaluation of different image analysis techniques is a particularly difficult task, and dependability of evaluations by simple mathematical measures such as squared error performance is questionable. Therefore, most of the time, the quality of the quantified and segmented image usually depends heavily on subjective and qualitative judgments. Besides the evaluation performed by radiologists, we use the GRE value to reflect the quality of tissue quantification.

Based on the pre-edited MR brain image, the procedure for analysis of tissue types in a slice

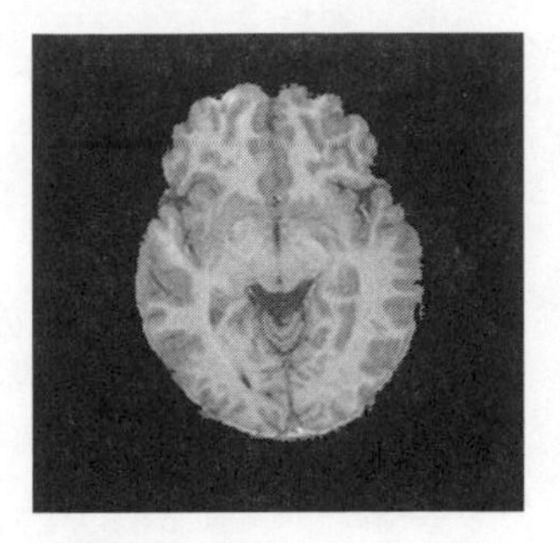
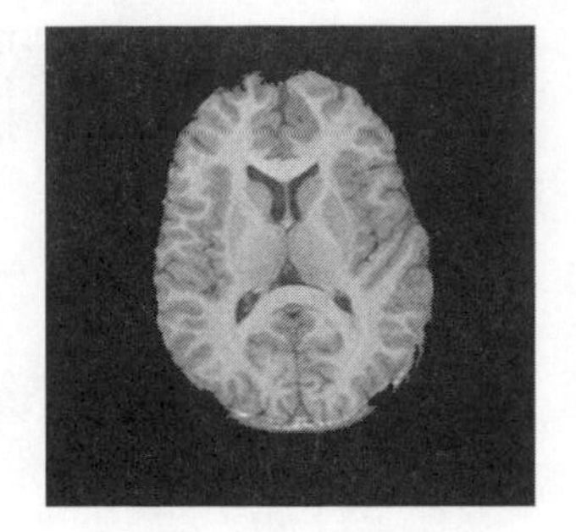
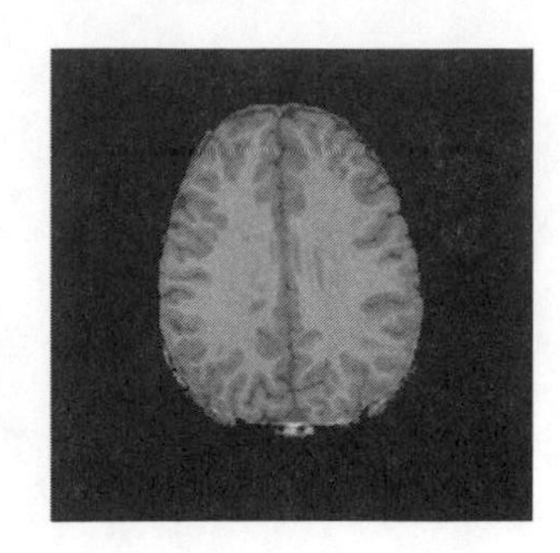

FIGURE 8.3
Three sample MR brain tissues.

is summarized as follows:

1. For each value of K (number of tissue types), $K = K_{\min}, \ldots, K_{\max}$, ML tissue quantification is performed by the PSOM algorithm [equations (8.9)–(8.11)].
2. Scan the values of $K = K_{\min}, \ldots, K_{\max}$, and use MCBV (8.16) to determine the suitable number of tissue types.
3. Select the result of tissue quantification corresponding to the value of K_0 determined in step 2.
4. Initialize tissue segmentation by ML classification (8.25).
5. Finalize tissue segmentation by CBRL [implementing (8.26)].

The performance of tissue quantification and segmentation is then evaluated in terms of the GRE value, convergence rate, computational complexity, and visual judgment.

The brain is generally composed of three principal tissue types: white matter (WM), gray matter (GM), and cerebrospinal fluid (CSF), plus their combinations, called the partial volume effect. We consider the pairwise combinations as well as the triple mixture tissue, defined as CSF–white–gray (CWG). More important, since the MRI scans clearly show the distinctive intensities at local brain areas, the functional areas within a tissue type need to be considered. In particular, the caudate nucleus and putamen are two important local brain functional areas because, in our complete image analysis framework, we allow the number of tissue types to vary from slice to slice (i.e., we do consider adaptability to different MR images). We let $K_{\min} = 2$ and $K_{\max} = 9$ and calculate AIC(K) [eq. (8.12)], MDL(K) [eq. (8.14)], and MCBV(K) [eq. (8.16)] for $K = K_{\min}, \ldots, K_{\max}$. The results with these three criteria are shown in Figure 8.4, which suggests that the three sample brain images chosen contain 6, 8, and 6 tissue types, respectively. According to the model fitting procedure using information theoretic criteria, the minima of these criteria indicate the most appropriate number of tissue types, which is also the number of hidden nodes in the corresponding PSOM (mixture components in SFNM). In the calculation of MCBV using (8.18), as discussed, one can use the CRLBs to represent the conditional variances of the parameter estimates, given by [50]:

$$\mathrm{Var}\left(\hat{\pi}_{k\mathrm{ML}}\right) = \frac{\pi_k(1-\pi_k)}{N}, \tag{8.28}$$

$$\mathrm{Var}\left(\hat{\mu}_{k\mathrm{ML}}\right) = \frac{\sigma_k^2}{N\pi_k}, \quad \text{and} \tag{8.29}$$

$$\mathrm{Var}\left(\hat{\sigma}_{k\mathrm{ML}}^2\right) = \frac{2\sigma_k^4(N\pi_k - 1)}{N^2\pi_k^2}. \tag{8.30}$$

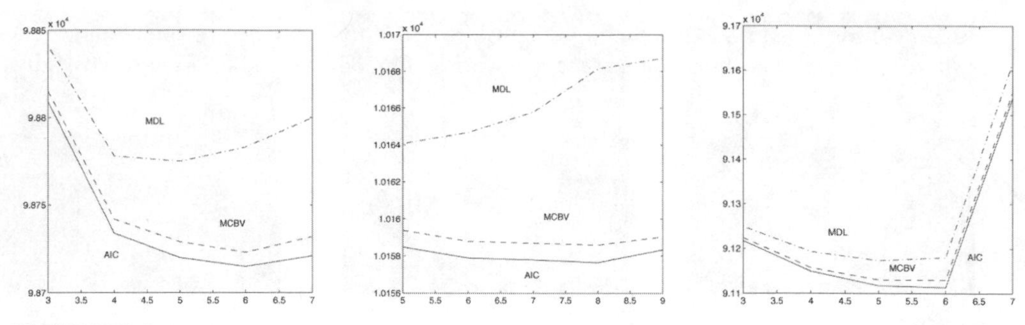

FIGURE 8.4
Results of model selection for slices 1–3 (K_0 = 6, 8, 6, left to right).

Note that since the true parameter values in the above equations are not available, their ML estimates are used to obtain the approximate CRLBs. From Figure 8.4, it is clear that, with real MR brain images, the overall performance of the three information theoretic criteria is fairly consistent. However, it is noted that AIC has a tendency to overestimate while MDL has a tendency to underestimate the number of tissue types [68], and MCBV provides a solution between those of AIC and MDL, which can be a desirable choice in terms of providing a balance between the bias and variance of the parameter estimates.

When performing the computation of the information theoretic criteria, we use PSOM to iteratively quantify different tissue types for each fixed K. The PSOM algorithm is initialized by the ALMHQ [71]. For slice 2, the results of final tissue quantification with $K_0 = 7, 8, 9$ are shown in Figure 8.5. Table 8.2 gives the numerical result of final tissue quantification for

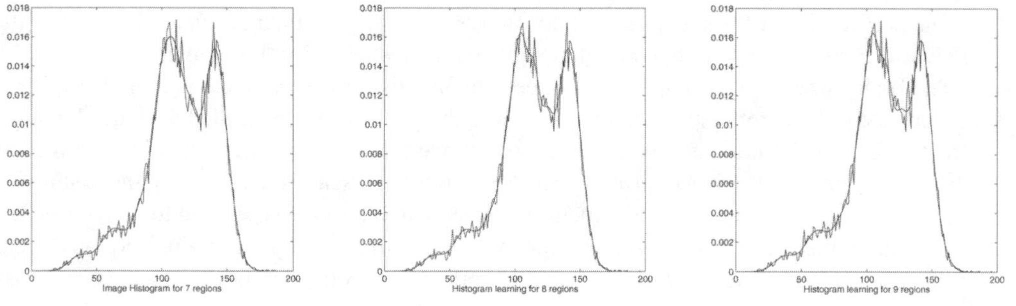

FIGURE 8.5
Histogram learning for slice 2 (K = 7, 8, 9 from left to right).

slice 2 corresponding to $K_0 = 8$, where a GRE value of 0.02 to 0.04 nats is achieved. These quantified tissue types agree with those of a physician's qualitative analysis results [69].

Table 8.2 Result of Parameter Estimation for Slice 2

Tissue Type	1	2	3	4	5	6	7	8
π	0.0251	0.0373	0.0512	0.071	0.1046	0.1257	0.2098	0.3752
μ	38.848	58.718	74.400	88.500	97.864	105.706	116.642	140.294
σ^2	78.5747	42.282	56.5608	34.362	24.1167	23.8848	49.7323	96.7227

The CBRL tissue segmentation for slice 2 is performed with $K_0 = 7, 8, 9$, and the algorithm is initialized by ML classification [eq. (8.25)] [66]. CBRL updates are terminated after 5 to 10 iterations since further iterations produced almost identical results. The segmentation results are shown in Figure 8.6. It is seen that the boundaries of WM, GM, and CSF are successfully

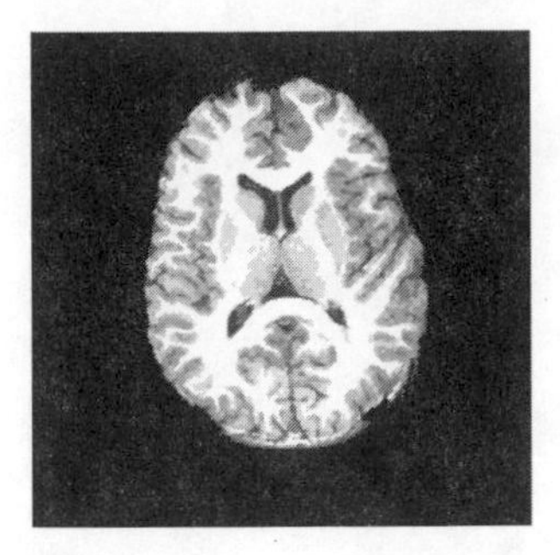
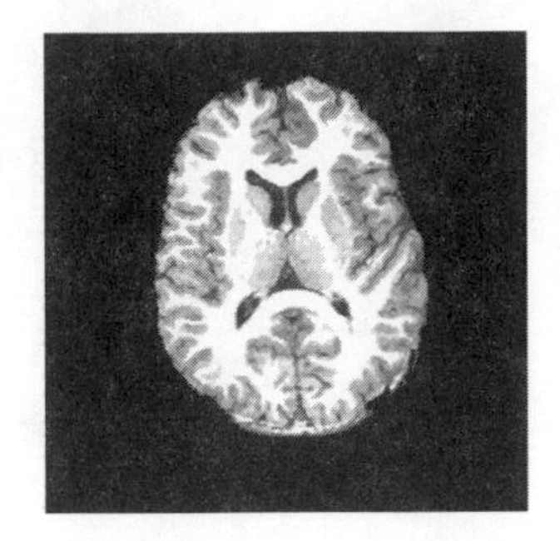
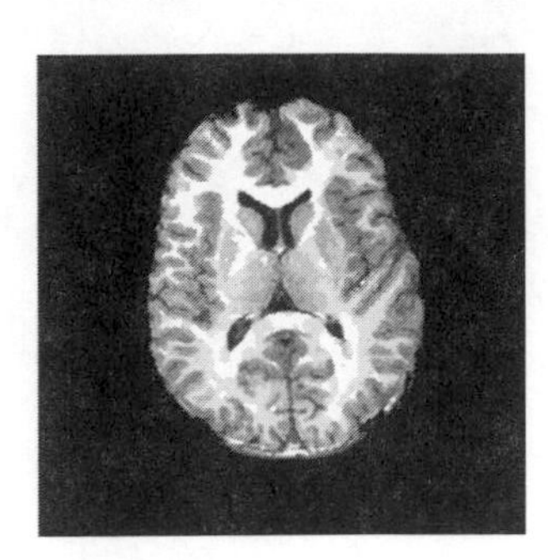

FIGURE 8.6
Results of tissue segmentation for slice 2 with $K_0 = 7, 8, 9$ (from left to right).

delineated. To see the benefit of using information theoretic criteria in determining the number of tissue types, the decomposed tissue type segments are given in Figure 8.7 with $K_0 = 8$. As can be observed in Figures 8.6 and 8.7, the segmentation with eight tissue types provides a very meaningful result. The regions with different gray levels are satisfactorily segmented, and the major brain tissues are clearly identified. If the number of tissue types were "underestimated" by one, tissue mixtures located within the putamen and caudate areas would be lumped into one component, but the results would still be meaningful. When the number of tissue types is "overestimated" by one, there is no significant difference in the quantification result, but the white matter would be divided into two components. For $K_0 = 8$, the segmented regions represent eight types of brain tissues: CSF, CG, CWG, GW, GM, putamen area, caudate area, and WM, as shown in Figure 8.7. These segmented tissue types again agree with the results of a radiologist's evaluation [69].

Mammogram Analysis

Another example application area for the image analysis framework we have introduced is in segmentation and extraction of suspicious mass areas from mammographic images. With an appropriate statistical description of various discriminate characteristics of both true and false candidates from the localized areas, an improved mass detection may be achieved in computer-aided diagnosis (Figure 8.8). Preprocessing can be an important tool for analysis depending on the application. In this example, one type of morphological operation is derived to enhance disease patterns of suspected masses by cleaning up unrelated background clutters, and then image segmentation is performed to localize the suspected mass areas using the stochastic relaxation labeling scheme [35, 37]. The mammograms for this study were selected from the Mammographic Image Analysis Society (MIAS) database and the Brook Army Medical Center (BAMC) database created by the Department of Radiology at Georgetown University Medical Center. The areas of suspicious masses were identified by an expert radiologist based on visual criteria and biopsy-proven results. The BAMC films were digitized with a laser film digitizer (Lumiscan 150) at a pixel size of $100 \times 100\mu$m and 4096 gray levels (12 bits). Before the method was applied, the digital mammograms were smoothed by averaging 4×4 pixels into 1 pixel. According to radiologists, the size of small masses is 3 to 15 mm in effective diameter. A 3-mm object in an original mammogram occupies 30 pixels in a digitized image with a 100-μm resolution. After the image size is reduced by four times, the object will occupy the range of about 7 to 8 pixels. An object the size of 7 pixels is expected to be detectable by

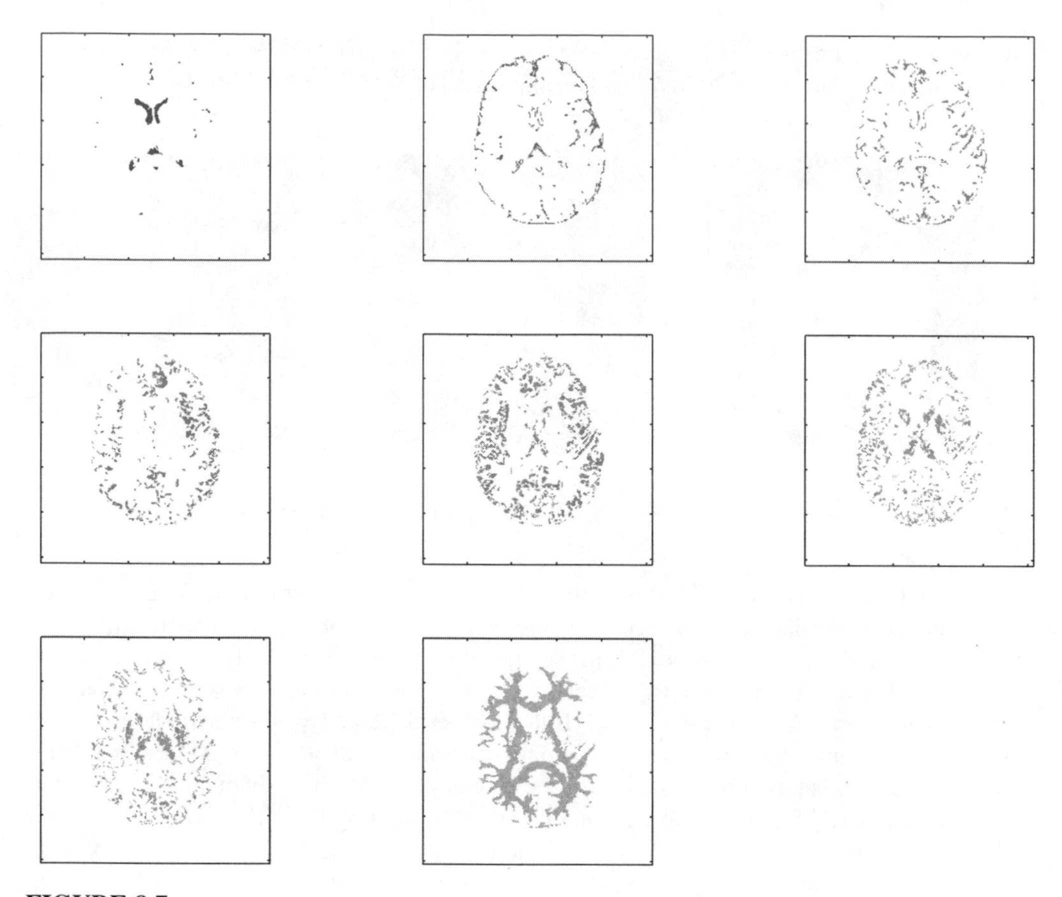

FIGURE 8.7
Result of tissue type decomposition for slice 2 that represents eight types of brain tissues: CSF, CG, CWG, GW, GM, putamen area, caudate area, and WM (left to right, top to bottom).

any computer algorithm. Therefore, the shrinking step is applicable for mass cases and can save computation time.

Consider the use of the FGGM model and the two information criteria — AIC and MDL — to determine the mixture number K. Tables 8.3 and 8.4 show the AIC and MDL values with different K and α of the FGGM model based on one original mammogram. As can be seen, although with different α, all AIC and MDL values achieve the minimum when $K = 8$. This indicates that AIC and MDL are relatively insensitive to the change of α. With this observation, we can decouple the relation between K and α and choose the appropriate value of one while fixing the value of the other. Figure 8.9a and b are two examples of AIC and MDL curves with different K and fixed $\alpha = 3.0$. Figure 8.9a is based on the original mammogram and Figure 8.9b is based on the enhanced mammogram. As we can see in Figure 8.9a, both criteria achieved the minimum when $K = 8$. It should be noted that although no ground truth is available in this case, our extensive numerical experiments have shown a very consistent performance of the model selection procedure and all the conclusions were strongly supported by the previous independent work reported by [5]. Figure 8.9b indicates that $K = 4$ is the appropriate choice for the mammogram enhanced by a dual morphological operation. This is believed to be reasonable since the number of regions decreases after background correction.

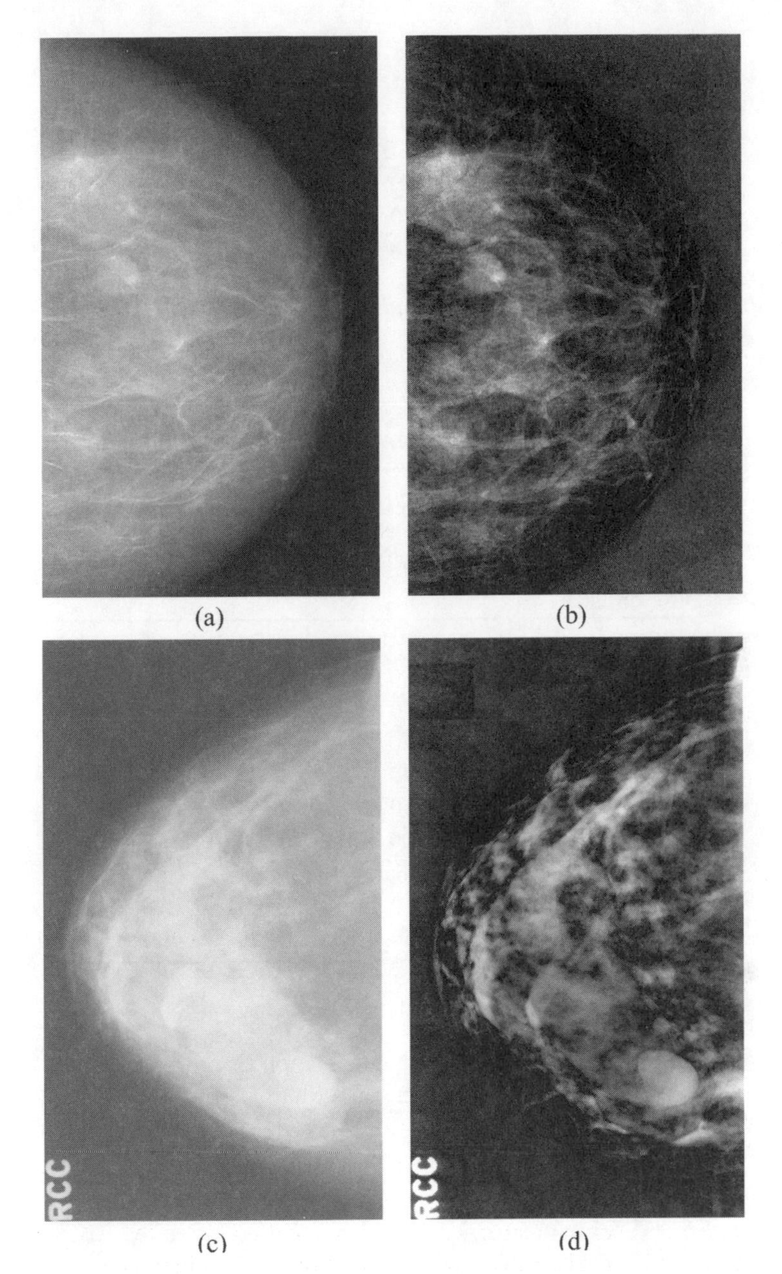

(a) (b) (c) (d)

FIGURE 8.8
Examples of mass enhancement: (a) original mammogram; (b) enhanced mammogram; (c) different original mammogram; (d) enhanced result of (c).

The order is then fixed at $K = 8$, and the value of α is changed for estimating the FGGM model parameters using the EM algorithm given in Section 8.2.2 with the original mammogram. The GRE value between the histogram and the estimated FGGM distribution is used as a measure of the estimation bias, and it is noted that the GRE achieved a minimum distance when the FGGM parameter $\alpha = 3.0$, as shown in Figure 8.10. A similar result was shown when the EM algorithm was applied to the enhanced mammogram with $K = 4$ (Figure 8.11). This indicated that the FGGM model might be better than the SFNM model ($\alpha = 2.0$) for

Table 8.3 Computed AICs for the FGGM Model with Different α

K	$\alpha = 1.0$	$\alpha = 2.0$	$\alpha = 3.0$	$\alpha = 4.0$
2	651250	650570	650600	650630
3	646220	644770	645280	646200
4	645760	644720	645260	646060
5	645760	644700	645120	646040
6	645740	644670	645110	645990
7	645640	644600	645090	645900
8	645550(min)	644570(min)	645030(min)	645850(min)
9	645580	644590	645080	645880
10	645620	644600	645100	645910

Table 8.4 Computed MDLs for the FGGM Model with Different α

K	$\alpha = 1.0$	$\alpha = 2.0$	$\alpha = 3.0$	$\alpha = 4.0$
2	651270	650590	650630	650660
3	646260	644810	645360	646350
4	645860	644770	645280	646150
5	645850	644770	645280	646100
6	645790	644750	645150	646090
7	645720	644700	645120	645930
8	645680(min)	644690(min)	645100(min)	645900(min)
9	645710	644710	645140	645930
10	645790	644750	645180	645960

Table 8.5 Comparison of Segmentation Error Resulting from Noncontextual and Contextual Methods

	Method		
	Soft classification	Bayesian classification	CBRL
GRE value	0.0067	0.4406	0.1578

mammographic images when the true statistical properties of the mammograms are generally unknown, although the SFNM has been successfully used in a large number of applications as well as in our previous example. Hence, the choice of the best model to describe the data depends on the nature of the data for the given problem.

After the determination of all model parameters, every pixel of the image is labeled to a different region (from 1 to K) based on the CBRL algorithm. Then, the brightest region, corresponding to label K, plus a criterion of closed isolated area, is chosen as the candidate region of suspicious masses. These results are noted to be highly satisfactory when compared to outlines of the lesions [37]. Also, similar to the previous example, GRE values can be used to assess the performance of the final segmentation. Table 8.5 shows our evaluation data from three different segmentation methods when applied to these real images.

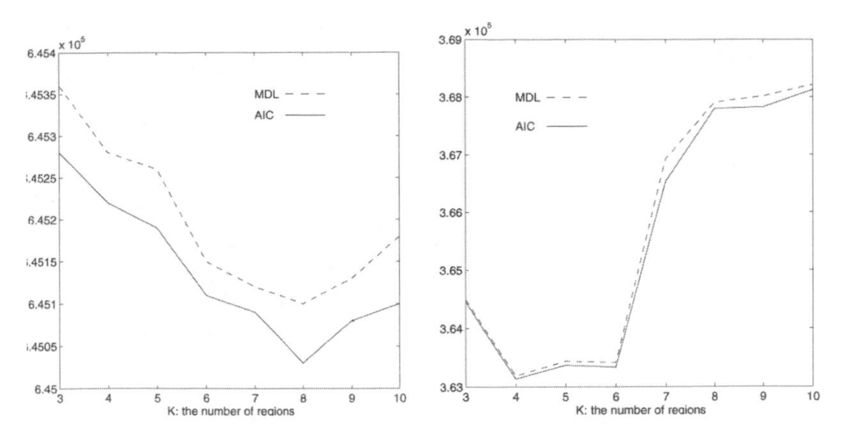

FIGURE 8.9
The AIC and MDL curves with different number of regions K. (a) The results based on the original mammogram, the optimal $K = 8$; (b) the results based on the enhanced mammogram, the optimal $K = 4$.

8.3 Graphics Modeling

Reconstruction of a 3D surface from a set of processed images is an important problem in the presentation and understanding of multimedia data. The data generated by imaging modalities such as 2D/3D camera, medical imaging, and other imaging devices provide a series of image slices of the object. The problem is then to infer a 3D representation of the object which will allow visualization as well as analysis of the geometry parameters of the object. In general, the 3D reconstruction process consists of three steps:

1. Extracting object contours from 2D cross-sectional images
2. Interpolating the intermediate contours between successive slices or among data points
3. Reconstructing surfaces or volumes from serial cross-sectional contours

Based on various image analysis algorithms, step 1 may be achieved through image segmentation or edge detection, which was discussed in the first part of this chapter and earlier in this book. In this section, we focus our discussions on steps 2 and 3.

Surface reconstruction is to form surfaces between contours of successive contours. If the interslice distances between the successive contours is small, the 3D structure of the object can be captured well by using surface reconstruction methods. However, if the contours are not closely spaced, the empty space between contours should be *filled* before surface reconstruction methods are applied. This procedure is usually referred to as *contour interpolation*. Many interpolation methods have been developed for various applications. For example, a linear interpolation algorithm is proposed in [82] to reconstruct prostatectomy specimens together with an enhanced extrapolation algorithm to overcome the difficulties in branching shapes and concave surfaces. This method is similar to the shape interpolation method described in [52] but has limitations in working with hemispherical shapes or round objects, primarily because of its linear characteristics. The elastic interpolation method given in [10] performs nonlinear contour interpolation by generating a series of intermediate contours filling the gap between the start and the goal contours. This method is based on Burr's dynamic elastic contour model [8] and can handle the branching situation very well by using union and/or intersection operators.

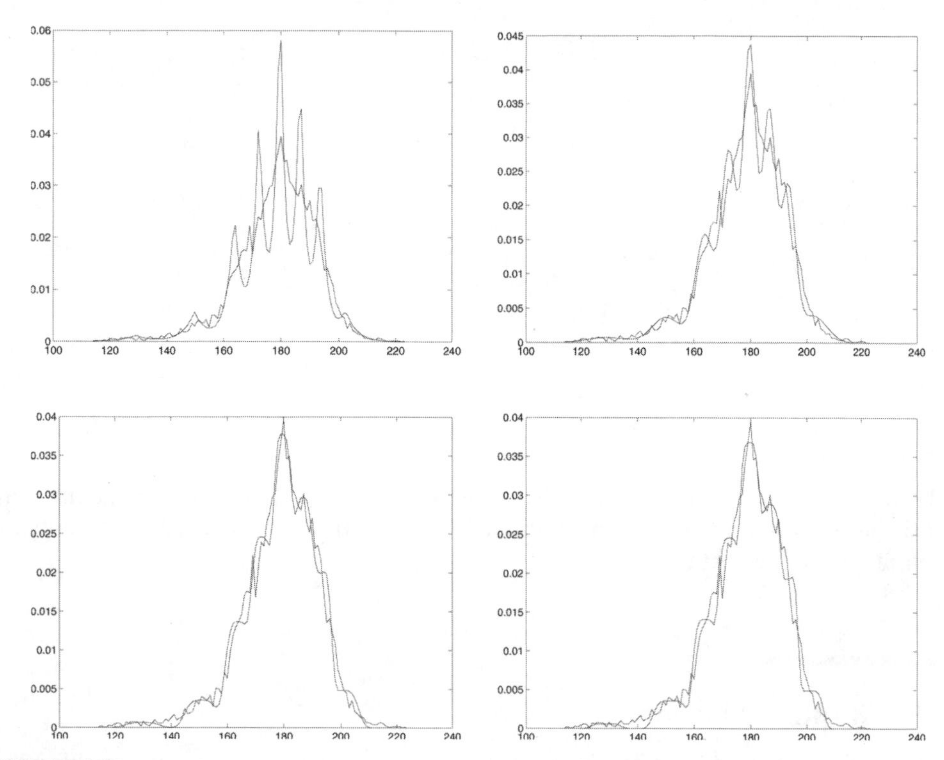

FIGURE 8.10
The comparison of learning curves and histogram of the original mammogram with different α; $K = 8$. The optimal $\alpha = 3.0$. (a) $\alpha = 1.0$, *GRE* $= 0.0783$; (b) $\alpha = 2.0$, *GRE* $= 0.0369$; (c) $\alpha = 3.0$, *GRE* $= 0.0251$; (d) $\alpha = 4.0$, *GRE* $= 0.0282$.

Many researchers have proposed using the elastic contour interpolation method for interpolating intermediate contours from the initial contours. The interpolation method assigns contours with the elastic property, and then, by applying forces onto them, deforms the start contour to conform to the goal contour. For example, a deformable surface–spine model has been proposed in [86, 87] to reconstruct the surface model from the interpolated contours. The deformable surface–spine model is a coupled dynamic system, where the surface and spine are confined in the following way: a deformable spine (axis) is determined from its contours, then all the surface patches are contracted to the spine through expansion/compression forces radiating from the spine while the spine itself is also confined to the surfaces. The surface refinement is governed by a second-order partial differential equation from Lagrangian mechanics, and the refining process is accomplished when the energy of this dynamic deformable surface–spine model reaches its minimum. A finite-element method is further used to solve the dynamic Lagrangian equation by constructing 9-degree-of-freedom (dof) triangular elements and 4-dof spine elements. In sum, both the elastic interpolation method and the deformable surface–spine model can be jointly used for building 3D graphics models for visualization and animation.

A contour on a plane $z = z_k$ can be defined as a linked list of vertices: $\vec{C} = \{(x_i, y_i), 1 \leq i \leq N\}$, or equivalently, defined as a concatenation of linked line segments where a line segment is represented by its two end vertices (x_i, y_i) and (x_{i+1}, y_{i+1}). Note that we drop z coordinates in the above expressions to simplify the notations. Given a start contour $\vec{C}_1 = \{(x_{1i}, y_{1i}), 1 \leq i \leq N_1\}$ and a goal contour $\vec{C}_2 = \{(x_{2i}, y_{2i}), 1 \leq i \leq N_2\}$, to interpolate between the start and goal contours, one must find a particular "force field" acting on the start contour and try to

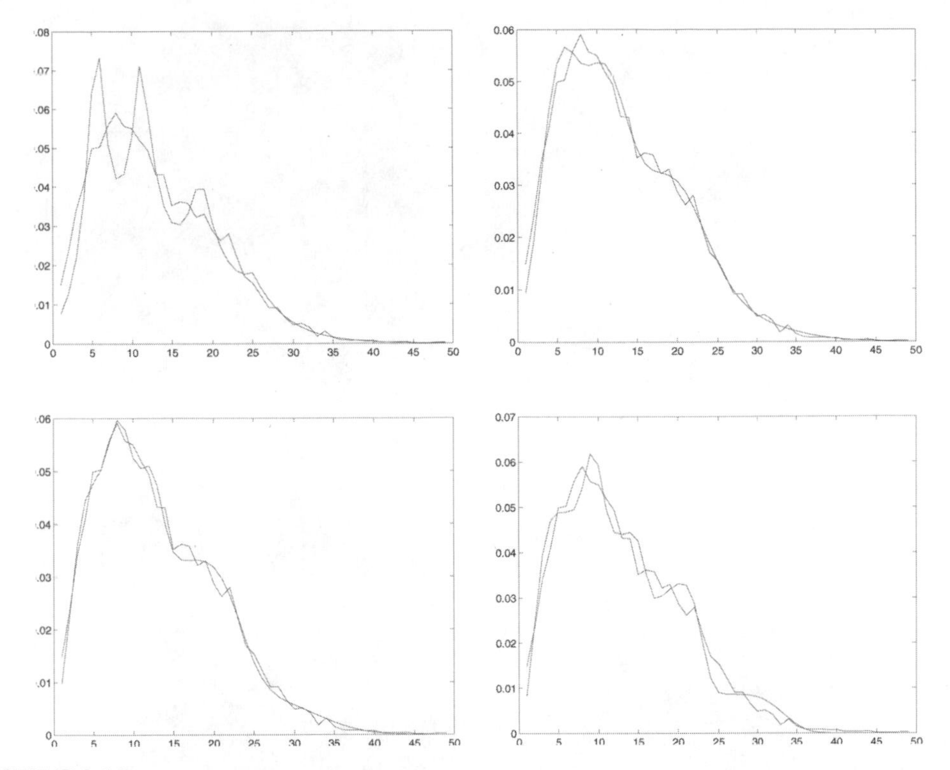

FIGURE 8.11
The comparison of learning curves and histogram of the enhanced mammogram with different α; $K = 4$. The optimal $\alpha = 3.0$. (a) $\alpha = 1.0$, *GRE* $= 0.0493$; (b) $\alpha = 2.0$, *GRE* $= 0.0126$; (c) $\alpha = 3.0$, *GRE* $= 0.0105$; (d) $\alpha = 4.0$, *GRE* $= 0.0676$.

deform it to conform to the goal contour. Thus, a three-step procedure is designed to achieve this task as described below [39].

1. Finding the Closest Line Segment

Let P_{1i} and $P_{1(i+1)}$ denote a line segment on the first contour C_1, and P_{2j} and $P_{2(j+1)}$ a line segment on the second contour C_2. In order to find the closest line segment of each vertex, a distance measure including the Euclidean distance and orientation property (the directional incompatibility) is used.

The directional incompatibility $\phi(i, j)$ between a vertex P_{1i} of C_1 and a line segment between the vertices P_{2j} and $P_{2(j+1)}$ of C_2 is defined as

$$\phi(i, j) = \frac{|(\vec{P}_{1(i+1)} - \vec{P}_{1i}) \times (\vec{P}_{2(j+1)} - \vec{P}_{2j})|}{|\vec{P}_{1(i+1)} - \vec{P}_{1i}|\,|\vec{P}_{2(j+1)} - \vec{P}_{2j}|}. \tag{8.31}$$

The above equation tells us that $\phi(i, j) = \sin\theta$, where θ is the angle between two vectors $\vec{P}_{1(i+1)} - \vec{P}_{1i}$ and $\vec{P}_{2(j+1)} - \vec{P}_{2j}$. The Euclidean distance from a vertex $\vec{P}_{1i}$ to a line segment of $\vec{P}_{2(j+1)} - \vec{P}_{2j}$ is $\eta(i, j) = |\vec{A} \times \vec{B}|/|\vec{B}|$, where $0 \leq \theta \leq \pi$, $\vec{A} = \vec{P}_{1i} - \vec{P}_{2j}$, and $\vec{B} = \vec{P}_{2(j+1)} - \vec{P}_{2j}$. Additionally, if $A\cos\theta < 0$ or $A\cos\theta > |\vec{B}|$, a term $\frac{|\vec{R}\cdot\vec{B}|}{|\vec{B}|}$ has to be included in $\eta(i, j)$, where $\vec{R} = \vec{P}_{1i} - \vec{P}'_{2j}$, and $\vec{P}'_{2j}$ is either the point $\vec{P}_{2j}$ or $\vec{P}_{2(j+1)}$ depending on which one is closer to $\vec{P}_{1i}$. Then, the total distance between a vertex $\vec{P}_{1i}$ and a line segment of $\vec{P}_{2(j+1)} - \vec{P}_{2j}$ is defined as the weighted sum, $d(i, j) = \phi(i, j) + \omega\,\eta(i, j)$, where ω is the

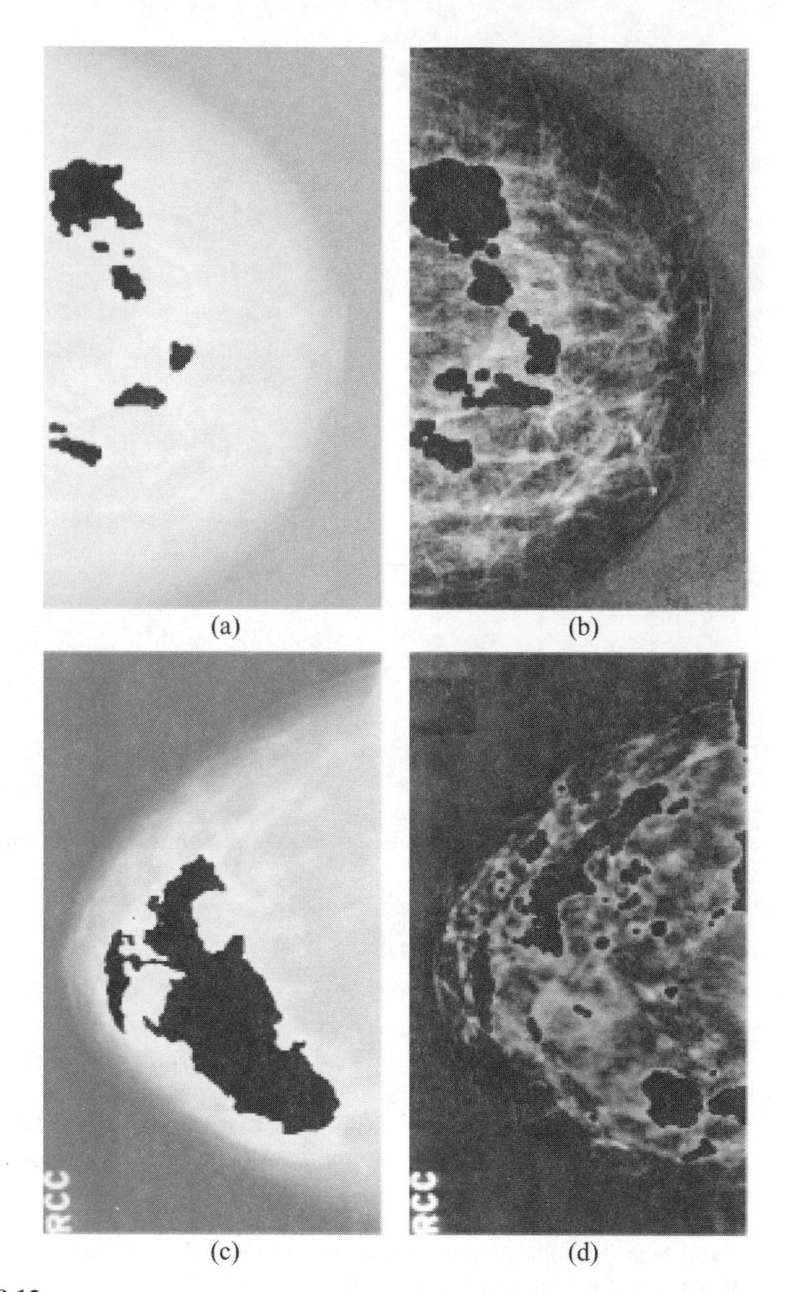

FIGURE 8.12

(a) The suspected mass segmentation results based on the original mammogram, (b) the results based on the enhanced mammogram, $K = 4$, $\alpha = 3.0$. (c) and (d) are the results based on another original mammogram and its enhanced image.

weight that in practice can be set to 1 in most cases. The closest line segment can be determined by finding the point index J_i giving minimum distance $d(i, j)$ — that is, $\min_j(i, j) = d(i, J_i)$.

2. Determining Displacement and Force Field

The displacement vector associated with a vertex $\vec{P}_{1i}$ and a line segment between the vertices $\vec{P}_{2J_i}$ and $\vec{P}_{2(J_i+1)}$ is defined as

$$\vec{D}_1(i, J_i) = \begin{cases} \vec{P}'_{2J_i} - \vec{P}_{1i}, & \text{if } A\cos\theta \geq |\vec{B}| \text{ or } A\cos\theta \leq 0 \\ A\sin\theta \, \frac{B_y\hat{i} - B_x\hat{j}}{|\vec{B}|}, & \text{if } 0 < A\cos\theta < |\vec{B}|, \end{cases} \tag{8.32}$$

where $\vec{P}'_{2J_i}$ is the point $\vec{P}_{2J_i}$ or $\vec{P}_{2(J_i+1)}$ depending on which is closer to $\vec{P}_{1i}$. Similarly, by reversing the roles of the start and the goal contours, the displacement vector $\vec{D}_2(j, I_j)$ can be determined at each vertex $\vec{P}_2$ of $\vec{C}_2$.

A force field is then defined as a function of the "pushing" and "pulling" forces:

$$\vec{F}(x, y) = \gamma^{-1} \left[\frac{\sum_{i=1}^{N_1} G_{1i} D_1(i, J_i)}{\sum_{i=1}^{N_1} G_{1i}} - \frac{\sum_{j=1}^{N_2} G_{2j} D_2(j, I_j)}{\sum_{j=1}^{N_2} G_{2j}} \right], \tag{8.33}$$

where G_{1i} and G_{2j} are designed to provide the effect that close neighbors have more influence than that of far neighbors, and they can be defined as Gaussian functions with covariance σ_k defined as $\sigma_k = \sigma_0 f^{-k}$, where f is a constant $1 \leq f \leq 2$, and γ can be regarded as a damping coefficient. For a discussion of how these parameters affect the dynamic behavior of the elastic contour model, see [39].

3. Generating Intermediate Contours

Consider a start contour $\vec{C}_1$ and a goal contour $\vec{C}_2$. One can compute the initial force field $\vec{F}^0$ according to the method described in step 2. Using $\vec{F}^0$, one defines the contour $\vec{I}^{k+1}$ from $\vec{I}^k$ interactively by providing $\vec{I}^0 = \vec{C}_1$; that is,

$$\vec{I}^{k+1} = \vec{I}^k + \vec{F}^k(x_{ki}, y_{ki}) . \tag{8.34}$$

8.3.1 Surface Reconstruction

Surface reconstruction is usually achieved by forming triangular patches between successive pairs of contours, which is often referred to as the tiling problem or triangulation problem in the literature. Solutions to the tiling problem can be categorized into two groups: (1) optimal approaches in some given criterion and (2) primarily heuristic approaches. Optimal methods provide the best triangulation in the sense of the given criterion and are often based on a graph description where a path in the graph defines a possible solution. A cost function (criterion) is assigned to each arc of the graph, and the optimal solution is obtained by finding the path with minimum or maximum cost function in the graph. For example, one can use *maximizing volume* as a cost function or use *minimizing area* instead. These two methods produce good results in practice, although the second method is preferred over the first because there is no need to deal separately with the convex and concave parts of objects. Heuristic approaches, on the other hand, are computationally less expensive and they usually define triangular patches one by one using only a local decision criterion. For instance, the triangular patches can be sequentially determined by choosing the shorter edge of two possible edges defining a patch. Most heuristic methods suffice when contours are similar in shape and orientation and are mutually centered. However, if contours are very different in shape, orientation, and position, heuristic methods can produce incorrect results.

In contrast to linear methods (tiling triangular patches), nonlinear surface reconstruction methods have been intensively proposed and studied. For example, a uniform B-spline approach has been developed to represent sectional contours and to further interpolate the surface between slices. A Hermite interpolation function with curvature sampling and a fast nearest mapping algorithm between two cross-sections is also proposed to perform nonlinear surface reconstruction using physically based deformable modeling. A more elegant approach to surface reconstruction using physical deformable models has been recently developed in the computer vision community and is now widely used in many areas such as computer graphics and animation, dynamics simulation, and modeling. We discuss this approach next.

8.3.2 Physical Deformable Models

Deformable models are based on variational principles of continuum mechanics. These dynamic principles are usually expressed in the form of dynamic differential equations. Elastic models [63] simulate nonlinear elastic materials. They incorporate deformation energies that are invariant with respect to rigid-body motions, impart no deformation, and grow monotonically with the magnitude of the deformation. The energy functionals are expressed as integral measures of the instantaneous deformation of a model away from its prescribed reference shape. The deformation is quantified in a convenient way using the fundamental forms of differential geometry (metrics, curvatures, etc.). Lagrange equations of motion balance the resulting elastic forces against inertial forces due to the mass distribution of the model, frictional damping forces, and externally applied forces. Elastically deformable models can efficiently model a variety of smooth objects with different shapes. They can also dynamically respond to external forces, which is very important in modeling human organs for the purpose of surgical planning and simulation in particular. Several deformable models (e.g., controlled-continuity splines under tension [61], symmetry-seeking models [64], and deformable superquadrics [61]), have been developed and applied to surface reconstruction [64], shape and motion recovery [44], and object recognition [57]. A dynamic finite-element surface model was proposed by Terzopoulos to track moving anatomical structures (e.g., the left ventricle) in 4D cardiac images for functional deformation analysis [59].

Inelastic models [91] are a powerful model-building medium. Unlike elastic models, which immediately regain their natural, undeformed shapes, inelastic models are commonly associated with high-polymer solids such as modeling clay or silicon putty. Consequently, inelastic models serve as a sort of freely sculptable *computational plasticine*. Free-form shapes may be created by interactively applying simulated forces on the inelastic model to stretch, squash, and mold it. Inelastic models tractably simulate three canonical inelastic behaviors — viscoelasticity, plasticity, and fracture. These behaviors may be incorporated into any of the elastic models described above by introducing internal processes that dynamically control resilience and fragility as a function of deformation.

Stochastic models combine deterministic deformable behaviors with random processes. This leads to the marriage of two well-known modeling techniques: splines and fractals. On one hand, spline shapes are easily constrained and are suitable for modeling smooth, man-made objects such as teapots, whereas fractals, although difficult to constrain, are suitable for synthesizing the various irregular shapes found in nature, such as a mountainous terrain. Constrained fractals are a class of deformable models that combine these seemingly opposed features by exploiting the remarkable relationship between fractals and generalized energy-minimizing splines, which may be derived through Fourier analysis. Constrained fractals are generated by a stochastic relaxation algorithm that *bombards* a spline subject to shape constraints with modulated white noise, letting the spline diffuse the noise into the desired fractal spectrum as it settles into equilibrium. In general, elastically deformable models are suitable to model relatively smooth objects, whereas inelastic models have the potential to model complex (moderately irregular) objects. On the other hand, stochastic deformable models are extremely important to model the various irregular shapes found in nature, such as mountainous terrain.

8.3.3 Deformable Surface–Spine Models

The surface and spine can be defined as geometric mappings from material (parametric) coordinate domains into 3D Euclidean space $\Re^3$. The surface can be defined by the following mapping M:

$$M: \quad (u, v) \mapsto \mathbf{x}(u, v, t) = (x(u, v, t), y(u, v, t), z(u, v, t)) , \tag{8.35}$$

where $(u, v) \in [0, 1]^2$ are the bivariate material coordinates; $x(u, v, t)$, $y(u, v, t)$, and $z(u, v, t)$ are the coordinates of a point on the surface in $\Re^3$; and t denotes the time-varying property of the deformable surface. Similarly, the spine can be defined by the mapping m:

$$m: \quad s \mapsto \mathbf{x}(s, t) = (x(s, t), y(s, t), z(s, t)) , \tag{8.36}$$

where $s \in [0, 1]$ is the univariate material coordinate and $x(s, t)$, $y(s, t)$, and $z(s, t)$ are the coordinates of a point on the spine in $\Re^3$.

The strain energy $\mathcal{E}$ can be found to characterize the deformable material of either the surface or the spine, which will be discussed in the next section as an instance of the spline function. Then the continuum mechanical equation

$$\mu \frac{\partial^2 \mathbf{x}}{\partial t^2} + \gamma \frac{\partial \mathbf{x}}{\partial t} + \frac{\delta \mathcal{E}(\mathbf{x})}{\delta \mathbf{x}} = \mathbf{f}(\mathbf{x}) \tag{8.37}$$

governs the nonrigid motion of the surface (spine) in response to an extrinsic force $\mathbf{f}(\mathbf{x})$, where μ is the mass density function of the deformable surface (spine) and γ is the viscosity function of the ambient medium. The third term on the left-hand side of the equation is the variational derivative of the strain energy functional $\mathcal{E}$, the internal elastic force of the surface (spine).

The deformable energy of surface $\mathbf{x}(u, v, t)$ can be defined by

$$\begin{aligned} \mathcal{E}_{\text{surface}}(u, v, t) = \int_0^1 \int_0^1 \Bigg(& w_{10} \left| \frac{\partial \mathbf{x}}{\partial u} \right|^2 + 2 w_{11} \left| \frac{\partial \mathbf{x}}{\partial u} \right| \times \left| \frac{\partial \mathbf{x}}{\partial v} \right| + w_{01} \left| \frac{\partial \mathbf{x}}{\partial v} \right|^2 \\ & + w_{20} \left| \frac{\partial^2 \mathbf{x}}{\partial u^2} \right|^2 + 2 w_{22} \left| \frac{\partial^2 \mathbf{x}}{\partial u \partial v} \right|^2 + w_{02} \left| \frac{\partial^2 \mathbf{x}}{\partial v^2} \right|^2 \Bigg) du\, dv , \end{aligned} \tag{8.38}$$

where the weights w_{10}, w_{11}, and w_{01} control the tensions of the surface and w_{20}, w_{22}, and w_{02} control its rigidities (bending energy). The deformable energy of spine $\mathbf{x}(u, t)$ is given by

$$\mathcal{E}_{\text{spine}}(s, t) = \int_0^1 \left(w_1 \left| \frac{d\mathbf{x}}{ds} \right|^2 + w_2 \left| \frac{d^2 \mathbf{x}}{ds^2} \right|^2 \right) ds . \tag{8.39}$$

The weight w_1 controls the tension along the spine (stretching energy), while w_2 controls its rigidity (bending energy).

To couple the surface with the spine, one should enforce $v \equiv s$, which maps the spine coordinate into the coordinate along the length of the surface. Then connect the spine with the surface by introducing the following forces on the surface and spine, respectively [64]:

$$\mathbf{f}^a_{\text{surface}}(u, s, t) = -(a/l) \left(\bar{\mathbf{x}}_{\text{surface}} - \mathbf{x}_{\text{spine}} \right) \tag{8.40}$$

$$\mathbf{f}^a_{\text{spine}}(s, t) = a \left(\bar{\mathbf{x}}_{\text{surface}} - \mathbf{x}_{\text{spine}} \right) \tag{8.41}$$

where a controls the strength of the forces; $\bar{\mathbf{x}}_{\text{surface}}$ is the centroid of the coordinate curve (s = constant) circling the surface and defined as $\bar{\mathbf{x}}_{\text{surface}} = \frac{1}{l} \int_0^1 \mathbf{x}_{\text{surface}} \left| \frac{\partial \mathbf{x}_{\text{surface}}}{\partial u} \right| du$, where l is the length given by $l = \int_0^1 \left| \frac{\partial \mathbf{x}_{\text{surface}}}{\partial u} \right| du$. In general, the above forces coerce the spine staying on an axial position of the surface. Further, if necessary, we can encourage the surface to be radially symmetric around the spine by introducing the following force:

$$\mathbf{f}^b_{\text{surface}} = b\,(\bar{\mathbf{r}} - |\mathbf{r}|)\,\hat{\mathbf{r}}\,, \tag{8.42}$$

where b controls the strength of the force; $\mathbf{r}$ is the radial vector of the surface with respect to the spine as $\mathbf{r}(u, s) = \mathbf{x}_{\text{surface}} - \mathbf{x}_{\text{spine}}$; the unit radial vector $\hat{\mathbf{r}}(u, s) = \mathbf{r}/|\mathbf{r}|$; and $\bar{\mathbf{r}}(s) = \frac{1}{l} \int_0^1 |\mathbf{r}| \frac{\partial \mathbf{x}_{\text{surface}}}{\partial u}\, du$, as the mean radius of the coordinate curve s = constant. Also, it is possible to provide control over expansion and contraction of the surface around the spine. This can be realized by introducing the following force:

$$\mathbf{f}^c_{\text{surface}} = c\hat{\mathbf{r}}\,, \tag{8.43}$$

where c controls the strength of the expansion or contraction force. The surface will inflate if $c > 0$ and deflate if $c < 0$.

Summing the above coupling forces in the motion equation associated with surface and spine, we obtain the following dynamic system describing the motion of the deformable surface–spine model:

$$\mu \frac{\partial^2 \mathbf{x}_{\text{surface}}}{\partial t^2} + \gamma \frac{\partial \mathbf{x}_{\text{surface}}}{\partial t} + \frac{\delta \mathcal{E}_{\text{surface}}}{\delta \mathbf{x}} = \mathbf{f}^{\text{ext}}_{\text{surface}} + \mathbf{f}^a_{\text{surface}} + \mathbf{f}^b_{\text{surface}} + \mathbf{f}^c_{\text{surface}}\,, \tag{8.44}$$

$$\mu \frac{\partial^2 \mathbf{x}_{\text{spine}}}{\partial t^2} + \gamma \frac{\partial \mathbf{x}_{\text{spine}}}{\partial t} + \frac{\delta \mathcal{E}_{\text{spine}}}{\delta \mathbf{x}} = \mathbf{f}^{\text{ext}}_{\text{spine}} + \mathbf{f}^a_{\text{spine}}\,, \tag{8.45}$$

where $\mathbf{f}^{\text{ext}}_{\text{surface}}$ is the external force applied on the surface and $\mathbf{f}^{\text{ext}}_{\text{spine}}$ is the external force applied on the spine.

Both the finite difference method and the finite element method can be used to compute the numerical solution to the surface $\mathbf{x}_{\text{surface}}$ and spine $\mathbf{x}_{\text{spine}}$. The finite difference method approximates the continuous function $\mathbf{x}$ as a set of discrete nodes in space. A disadvantage of the finite difference approach is that the continuity of the solution between nodes is not made explicitly. The finite-element method, on the other hand, provides continuous surface (or spine) approximation by approximating the unknown function $\mathbf{x}$ in terms of combinations of the basis functions. In the finite element method, we first tessellate the continuous material domain, (u, v) for the surface and s for the spine in our case, into a mesh of m element subdomains D_j, and then we approximate $\mathbf{x}$ as a weighted sum of continuous basis functions $\mathbf{N}_i$ (so-called shape functions): $\mathbf{x} \approx \mathbf{x}^h = \sum_i \mathbf{x}_i \mathbf{N}_i$, where $\mathbf{x}_i$ is a vector of nodal variables associated with mesh node i. The shape functions $\mathbf{N}_i$ are fixed in advance and the nodal variables $\mathbf{x}_i$ are the unknowns. The motion equation can then be discretized as

$$\mathbf{M}\frac{\partial^2 \mathbf{x}}{\partial t^2} + \mathbf{C}\frac{\partial \mathbf{x}}{\partial t} + \mathbf{K}\mathbf{x} = \mathbf{F}\,, \tag{8.46}$$

where $\mathbf{x} = [\mathbf{x}_1^T, \ldots, \mathbf{x}_i^T, \ldots, \mathbf{x}_n^T]$, $\mathbf{M}$ is the mass matrix, $\mathbf{C}$ the damping matrix, $\mathbf{K}$ the stiff matrix, and $\mathbf{F}$ the forcing matrix. $\mathbf{M}$, $\mathbf{C}$, and $\mathbf{F}$ can be obtained as follows:

$$\mathbf{M_j} = \int\int_{E_j} \mu \mathbf{N}_j^T \mathbf{N}_j du\, dv\,, \tag{8.47}$$

$$\mathbf{C_j} = \int\int_{E_j} \gamma \mathbf{N}_j^T \mathbf{N}_j du\, dv, \tag{8.48}$$

$$\mathbf{F_j} = \int\int_{E_j} \mathbf{N}_j^T \mathbf{f}_j du\, dv\,. \tag{8.49}$$

To compute **K**, use the following equation:

$$\mathbf{K}_j = \int\int_{E_j} \left(\mathbf{N}_b^T \boldsymbol{\beta} \mathbf{N}_b + \mathbf{N}_s^T \boldsymbol{\alpha} \mathbf{N}_s \right) du\, dv\ , \tag{8.50}$$

where

$$\mathbf{N}_b = \left[\frac{\partial^2 \mathbf{N}}{\partial u^2}, \frac{\partial^2 \mathbf{N}}{\partial u \partial v}, \frac{\partial^2 \mathbf{N}}{\partial v^2} \right]^T \tag{8.51}$$

$$\mathbf{N}_s = \left[\frac{\partial \mathbf{N}}{\partial u}, \frac{\partial \mathbf{N}}{\partial v} \right]^T \tag{8.52}$$

$$\boldsymbol{\alpha} = \begin{bmatrix} w_{02} & w_{22} \\ w_{22} & w_{20} \end{bmatrix} \tag{8.53}$$

$$\boldsymbol{\beta} = \begin{bmatrix} w_{01} & 0 & 0 \\ 0 & w_{11} & 0 \\ 0 & 0 & w_{10} \end{bmatrix} \tag{8.54}$$

The deformable surface consists of a set of connected triangular elements chosen for their ability to model a large range of topological shapes. Barycentric coordinates in two dimensions are the natural choice for defining shape functions over a triangular domain. Barycentric coordinates (L_1, L_2, L_3) are defined by the following mapping with material coordinates (u, v):

$$\begin{bmatrix} u \\ v \\ 1 \end{bmatrix} = \begin{bmatrix} u_1 & u_2 & u_3 \\ v_1 & v_2 & v_3 \\ 1 & 1 & 1 \end{bmatrix} \begin{bmatrix} L_1 \\ L_2 \\ L_3 \end{bmatrix}\ , \tag{8.55}$$

where (u_1, v_1), (u_2, v_2), and (u_3, v_3) are the coordinates of three vertex locations of the triangle.

We can use the 9-dof triangular element, which includes the position and its first parametric partial derivatives at each triangle vertex, as shown in Figure 8.14a. The shape functions of the first node in a 9-dof triangle are [91]:

$$\mathbf{N}_1^{9^T} = \begin{bmatrix} N_1 \\ N_2 \\ N_3 \end{bmatrix} = \begin{bmatrix} L_1 + L_1^2 L_2 + L_1^2 L_3 - L_1 L_2^2 - L_1 L_3^2 \\ c_3 \left(L_1^2 L_2 + 0.5 L_1 L_2 L_3 \right) - c_2 \left(L_1^2 L_3 + 0.5 L_1 L_2 L_3 \right) \\ -b_3 \left(L_1^2 L_2 + 0.5 L_1 L_2 L_3 \right) + b_2 \left(L_1^2 L_3 + 0.5 L_1 L_2 L_3 \right) \end{bmatrix} . \tag{8.56}$$

The triangle's symmetry in Barycentric coordinates can be used to generate the shape function for the second and third nodes in terms of the first. To generate $\mathbf{N}_2^9$, use the above equations but add a 1 to each index so that $1 \to 2$, $2 \to 3$, and $3 \to 1$. The $\mathbf{N}_3^9$ functions can be obtained by adding another 1 to each index. Note that the shape functions for a 9-dof triangle do not guarantee C^1 continuity between adjacent triangular elements. In [91], a 12-dof triangular element can be made C^1 continuous by adding 1 dof on each edge of the triangle (see [9] for details). An alternative to having a C^1 continuous triangular element is to use an 18-dof element which includes the nodal location, with its first and second partial derivatives evaluated at each node [43]. We use a 9-dof triangular element although the extension to 12- or 18-dof triangular elements is straightforward.

The finite element of the spine has 4 dof between two nodes located at the ends of the segment. The dof at each node correspond to its position and tangent. The spine segment can be approximated as the weighted sum of a set of Hermite polynomials: $\mathbf{x} \approx \mathbf{x}^h(s) = \sum_{i=0}^{3} \mathbf{x}_i N_i$, where $N_i, i = 0, \ldots, 3$ are given as follows:

$$\begin{aligned} N_0 &= 1 - 3(s/h)^2 + 2(s/h)^3, \\ N_1 &= h(s/h - 2(s/h)^2 + (s/h)^3), \\ N_2 &= 3(s/h)^2 - 2(s/h)^3, \\ N_3 &= h(-(s/h)^2 + (s/h)^3)\,, \end{aligned} \tag{8.57}$$

where h is the parametric element length.

8.3.4 Numerical Implementation

The deformable surface–spine model can be stabilized during the fitting process if its motion is critically damped to minimize vibrations. Critical damping can be achieved by appropriately balancing the mass and damping distributions. A simple way of eliminating vibration while preserving useful dynamics is to set the mass density in equation (8.46) to zero, thus reducing (8.46) to

$$\mathbf{C}\frac{\partial \mathbf{x}}{\partial t} + \mathbf{K}\mathbf{x} = \mathbf{F}\,. \tag{8.58}$$

This first-order dynamic system governs the model which has no inertia and comes to rest as soon as all the forces balance. We integrate equation (8.58) using an explicit first-order Euler method. The method begins with a simple forward difference approximation. Consider extrapolation from time level t to $t + \Delta t$ by forward differencing at t. The usual Taylor series expansion at time t has the form

$$\mathbf{x}(t + \Delta t) = \mathbf{x}(t) + \Delta t \frac{d\mathbf{x}}{\partial t}(t) + \frac{(\Delta t)^2}{2!}\frac{d^2\mathbf{x}}{\partial t^2}(\ell), \quad \ell \in (0, t)\,, \tag{8.59}$$

which yields the forward difference approximation

$$\frac{d\mathbf{x}}{\partial t} = \frac{\mathbf{x}(t + \Delta t) - \mathbf{x}(t)}{\Delta t} \tag{8.60}$$

and is only $O(\Delta t)$ accurate. Using this forward difference approximation and transposing terms involving $\mathbf{x}(t)$, we have

$$\mathbf{C}\mathbf{x}(t + \Delta t) = (\mathbf{C} - \Delta t\mathbf{K})\mathbf{x}(t) + \Delta t\mathbf{F}(t) \tag{8.61}$$

Thus, we obtain the updating formula for $\mathbf{x}$ from time t to $t + \Delta t$ as follows:

$$\mathbf{x}(t + \Delta t) = \left(\mathbf{I} - \Delta t\mathbf{C}^{-1}\mathbf{K}\right)\mathbf{x}(t) + \Delta t\mathbf{C}^{-1}\mathbf{F}(t) \tag{8.62}$$

It is well known that finite difference methods for initial-value systems yield expressions very similar to the above results obtained by finite element schemes. A noteworthy distinction is that the coefficient matrix $\mathbf{C}$ for finite differencing is diagonal in the usual difference approximation. This leads, in the forward difference approximation, to more efficient algorithms for solving the problem. In the finite element method, $\mathbf{C}$ is often sparse and ill conditioned, which causes difficulty in the computation of $\mathbf{C}^{-1}$. To obtain $\mathbf{C}^{-1}$, then, computationally complex singular decomposition methods have to be used. However, there is a physical solution in computational mechanics, called the *lumping* procedure, which overcomes the difficulty with the sparseness and ill-conditioning of matrix $\mathbf{C}$. The idea can be interpreted physically as replacing the continuous material with the distributed mass by the concentrated material with *lumped* mass (*beads*) at the nodes. In practice, there are several ways to perform such a lumping procedure, such as using modified shape functions or different numerical integral methods. Among those, the easiest way is keeping only the diagonal coefficients of **C** and discarding all the off-diagonal coefficients, which is the approach in solving the above-mentioned dynamic equation of the deformable surface–spine model.

8.3.5 Applications

In this section, we present applications of the algorithms we presented for graphical modeling, reconstruction, and representation with both discontinuity-embedded and smooth objects. The discontinuity-embedded deformable model defines a dynamic finite element representation with both continuous and discontinuous components as described in the previous section. First, we apply our new deformable model to reconstruct several synthetic and range datasets to illustrate its performance in recovering depth discontinuities in the final reconstructed surfaces (see Figure 8.13). In order to extract the contours of the object, we use Canny's edge operator to detect and locate depth discontinuities in the datasets (see Figure 8.14). It is not a trivial task in general to detect surface discontinuities; however, we assume such location information of depth discontinuities can be provided as a priori knowledge in our experiments (see Figure 8.15). We then initialize the discontinuity-embedded deformable model by a finite-element tessellation where the discontinuity path is identified within each element. The dataset acts as the external force to dynamically deform the model in order to fit the surface to the dataset. The final reconstructed surface is obtained when the dynamic motion equation reaches it equilibrium. Figure 8.16 shows the reconstructed object surface.

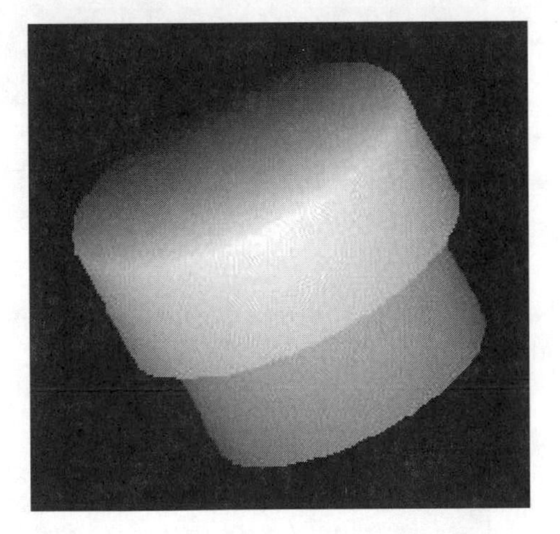

FIGURE 8.13
Range image of a simply synthetic object, where the dataset contains both smooth and discontinuous surfaces.

In Figures 8.17–8.21, we present the synthetic step data and the reconstruction results by the elastically deformable model and our discontinuity-embedded deformable model. As we can see, the elastically deformable model smooth over the depth discontinuity whereas the discontinuity-embedded deformable model recovers the depth discontinuity explicitly. With the discontinuity location information prescribed, the discontinuity-embedded deformable model incorporates a discontinuity component into its representation and dynamically conforms to the data in both continuous and discontinuous parts. The example is a tool synthetic dataset illustrated in Figure 8.17. With the location of the depth discontinuity, the

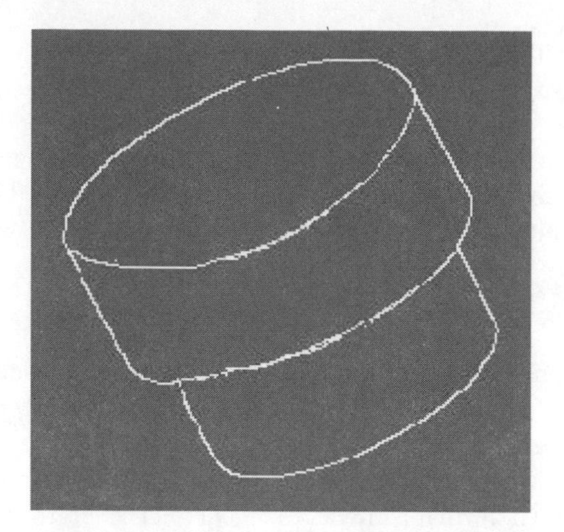

FIGURE 8.14
Frame representation of the synthetic object, where the boundaries can be extracted by various methods.

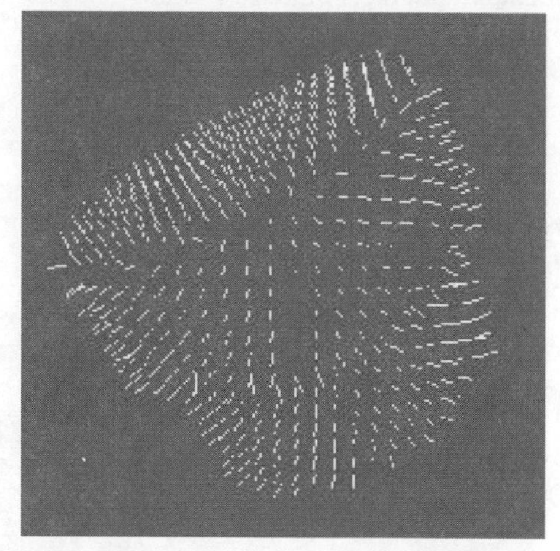

FIGURE 8.15
Gradient vector field of the synthetic object after an appropriate pre-processing step.

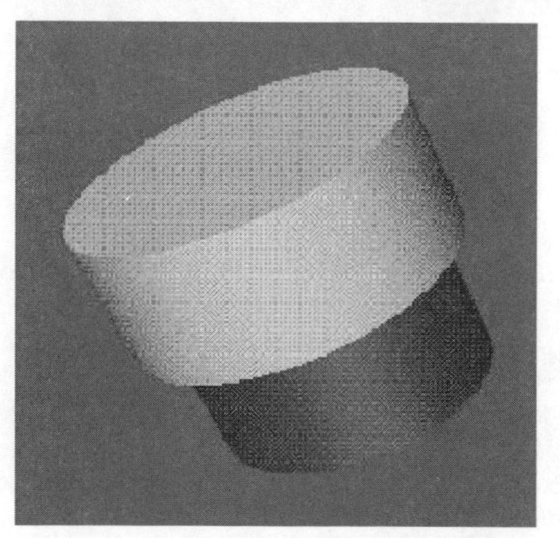

FIGURE 8.16
The reconstructed surface of the synthetic object, by incorporating the information regarding both continuous and discontinuous representations.

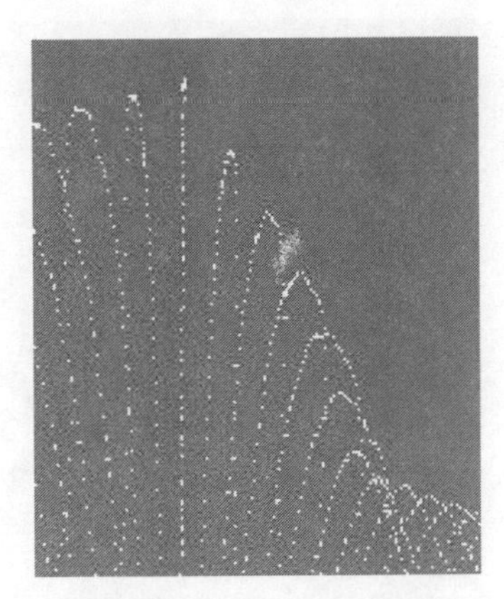

FIGURE 8.17
The contours of an object with deformable characteristics.

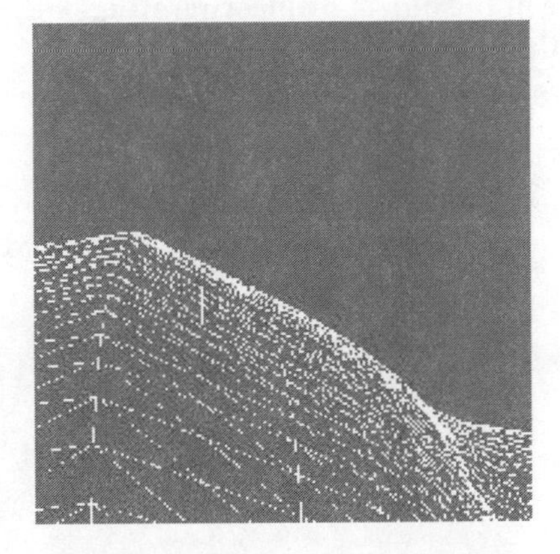

FIGURE 8.18
The reconstructed frame of the object by incorporating only the term representing the deformation property with smoothness constraint.

FIGURE 8.19
The reconstructed surface of the object showing a clear mismatch from the original contours.

FIGURE 8.20
The reconstructed frame of the object by incorporating both the terms representing the deformation and discontinuity properties.

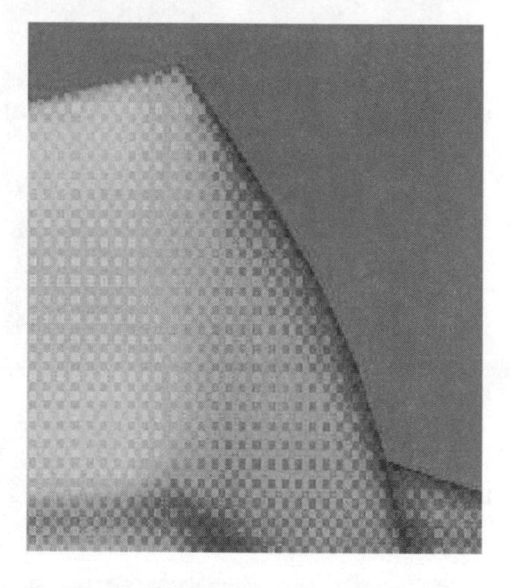

FIGURE 8.21
The reconstructed surface of the object showing a very satisfactory representation of the original object.

discontinuity-embedded deformable model is capable of recovering the jump height on the depth discontinuity (see Figure 8.20). As we see in the reconstructed surfaces by the elastically deformable model, the depth discontinuities are oversmoothed and hard to identify and localize (see Figures 8.18 and 8.19). The loss of discontinuities will obviously affect the outcome of those high-level processes such as object recognition. The depth discontinuities are well recovered by our discontinuity-embedded deformable model, as seen in Figure 8.21. With such a discontinuity-preserving surface reconstruction, high-level processes can easily extract the object boundary information to achieve the ultimate goal–object recognition. Since the discontinuity-embedded deformable model includes the conventional continuous component represented as in the elastically deformable model, all the advantages of the elastically deformable model are kept in the discontinuity-embedded deformable model for representing complex-structured smooth objects. Furthermore, the discontinuity-embedded deformable model can achieve more accurate representation of surfaces with discontinuities than that of the elastically deformable model.

We also applied our method to the reconstruction of the prostate model. A typical slice image of the surgical prostate is shown in Figure 8.22, with the contours of the prostate capsule as

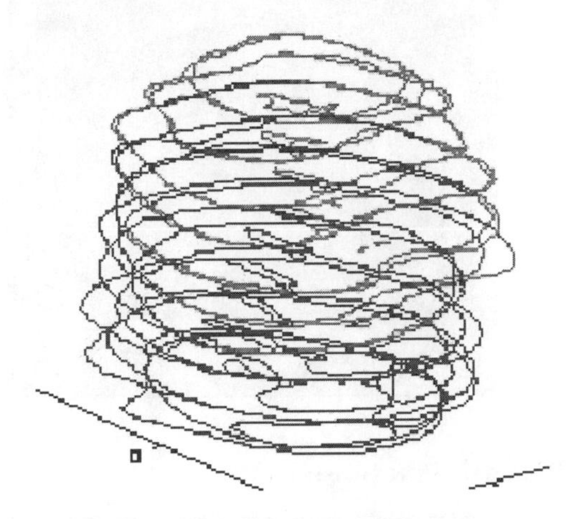

FIGURE 8.22
The contours of a three-dimensional prostate model consisting of multiple objects.

they are stacked in 3D. Next we can apply our elastic contour interpolation method to extracted contours and then use our deformable surface–spine model to reconstruct 3D surgical prostate models. We have developed reconstruction software with a graphical user interface (GUI) that can interact with users to specify a number of parameters such as number of slices to be inserted, the damping factor, and Gaussian smoothing variance in an elastic interpolation algorithm. Figure 8.23 shows the interpolated contours of the prostate capsule with six slices

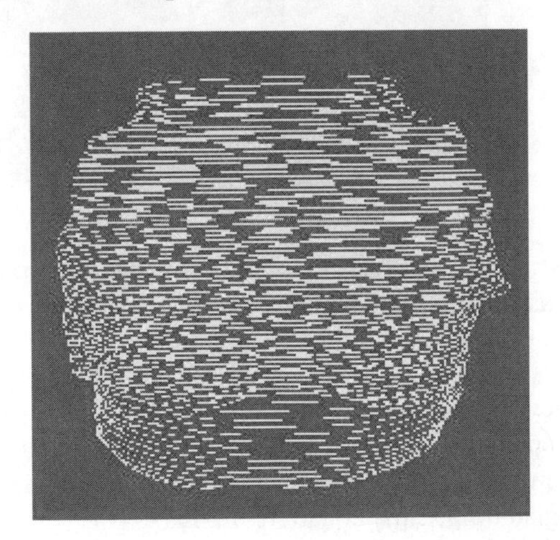

FIGURE 8.23
The 3D frame of the prostate model using an elastic interpolation method, where six virtual slices are generated between the two adjacent original slices.

being inserted using elastic contour models. As we can see, the nonlinear elastic interpolation method gives a very consistent result with original extracted contours. A triangular tiling algorithm has also been implemented in our reconstruction software using *minimizing area* as the cost function. The tiled triangular paths are then constructed, which result in a linear surface model of the prostate capsule.

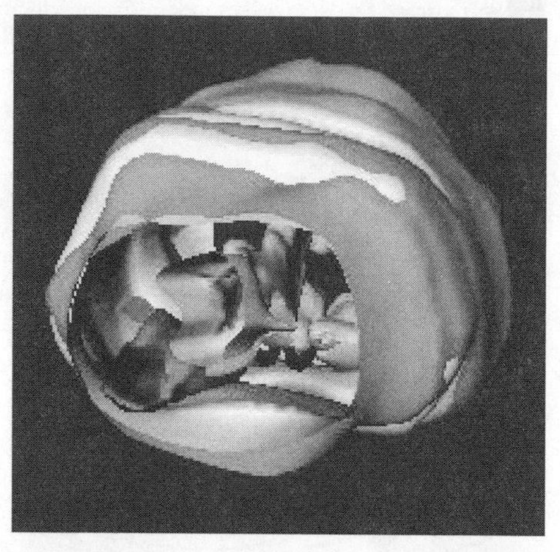

FIGURE 8.24
The reconstructed 3D model of the prostate.

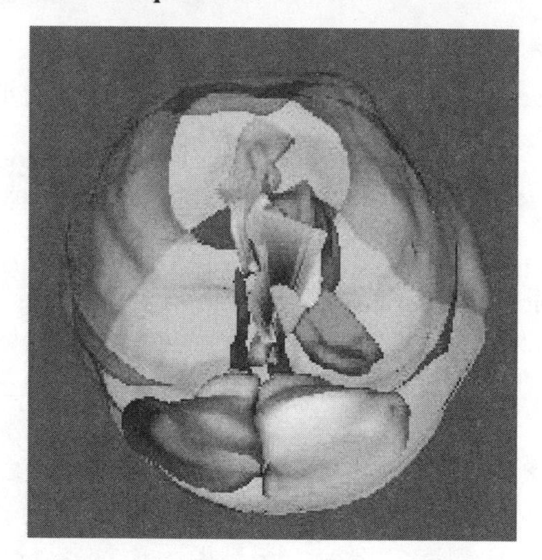

FIGURE 8.25
The complete reconstructed 3D prostate model rendered transparently with all anatomical structures such as the prostate capsule, urethra, seminal vesicles ejaculatory ducts, and carcinomas.

To improve accuracy in surface reconstruction, a sophisticated deformable surface–spine model has been developed using the finite element method to obtain a continuous surface representation. The dynamics of the deformable model are governed by continuum mechanics, known as the Lagrangian differential equation. In this experiment, external forces are determined by a sum of Gaussian weighted distances between goal contours and the surface of the model. The coupled forces between the surface and spine are also computed according to their relative positions to enforce the spine's staying on the axial position of the surface. The inflation or deflation forces are controlled in that we gradually reduce their strength. Initially the surface of the model is expanded or contracted largely to the goal contours, and when it is close to the goal contours we force the inflation or deflation forces to disappear. Since our finite element method uses the position and tangents of each node as nodal parameters, the reconstructed surface is of great smoothness, which outperforms the triangulation tiling approach. Figure 8.24 shows our final reconstructed surface model of the prostate capsule that possesses a very finely detailed surface description and a high fidelity to the original specimen. Figures 8.25 and 8.27 show a completely reconstructed 3D prostate model rendered with all anatomical structures such as the prostate capsule, urethra, seminal vesicles, ejaculatory ducts, and carcinomas.

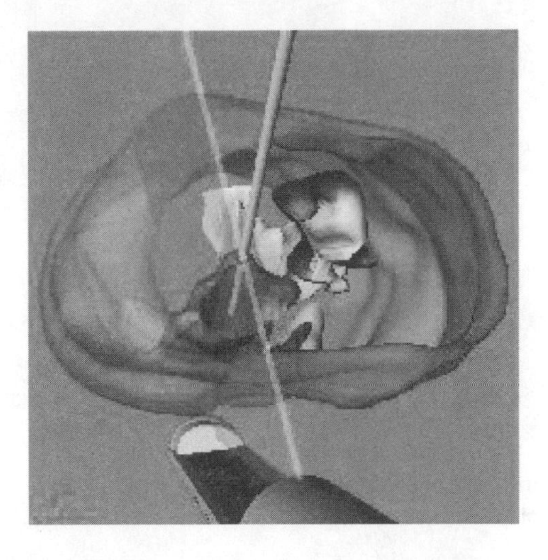

FIGURE 8.26
An transrectal ultrasound guided needle biopsy simulation interface, where the simulated ultrasound probe and image beam is integrated with the reconstructed 3D virtual reality scene.

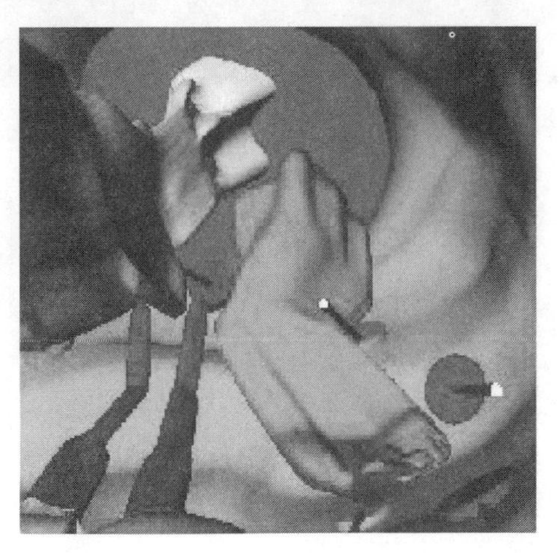

FIGURE 8.27
Virtual flying-through the complete reconstructed 3D prostate model, where a mis-targeted biopsy needle simulation is presented.

Interactive visualization of the 3D prostate model is achieved by using the state-of-the-art graphics toolkit, object-oriented *OpenInventor*. With a sophisticated set of various kinds of lights, 3D manipulators, and color and material editors, etc., we can examine the 3D prostate model in any viewpoint and interactively walk through it to better understand the relationships of the anatomical structures of the prostate and its tumors. The 3D stereo-glasses and a 3D mouse and trackball allow a full view of the 3D surgical prostate model, allowing the viewer (e.g., a surgeon) to examine the prostate. Furthermore, we have developed a system for image-guided needle biopsy simulation based on the reconstructed 3D surgical prostate model (see Figure 8.28). Different ultrasound-like imaging probes are simulated to provide axially and/or

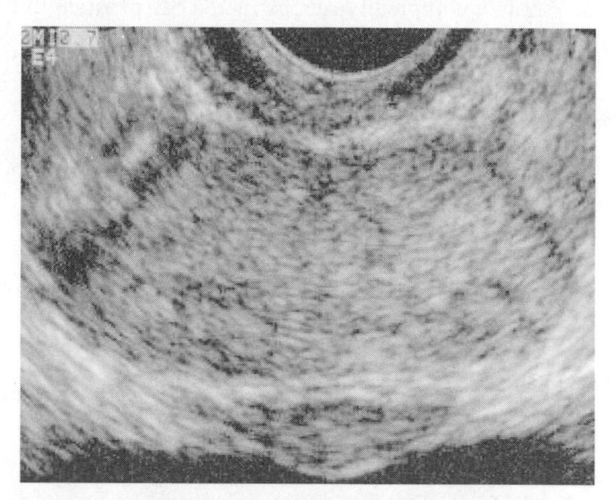

FIGURE 8.28
Display of an ultrasound image section.

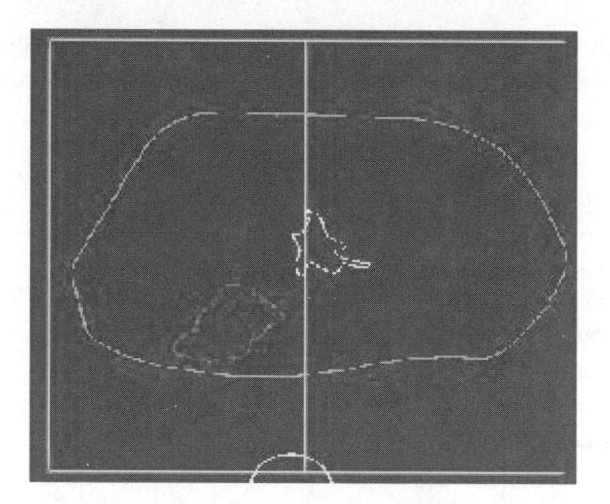

FIGURE 8.29
Graphical representation of the same scene as shown in Figure 8.28.

longitudinally oriented sectional images for efficiently planning needle pathways. Figures 8.28 and 8.29 show a 2D sectional image of contours and a 3D view of the inside prostate by removing the other half of the prostate. Needles with or without triggers are constructed and simulated to perform the actual biopsy on 3D computerized prostate models according to the planned needle pathways (see Figures 8.30 and 8.31). With an accurate 3D prostate model, realistic imaging probes, and needles provided by our virtual simulation system, a surgeon can sit before the computer to plan better needle paths and further to practice the actual biopsy procedure before he/she actually performs on a patient. More important, by analyzing outcomes of this simulation, we can validate the effectiveness of various biopsy techniques in prostate cancer detection and tumor volume estimation [87].

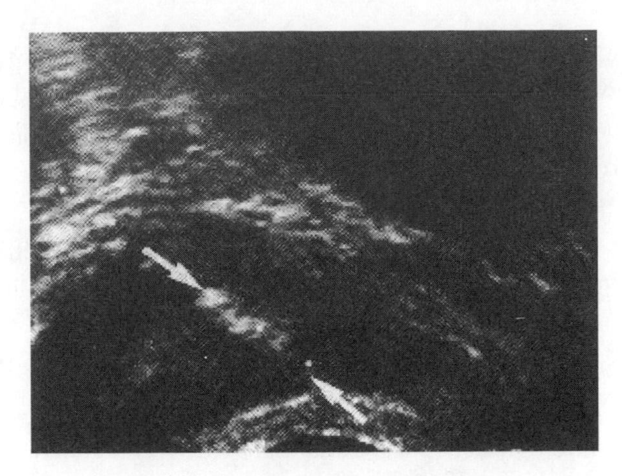

FIGURE 8.30
An ultrasound image showing the prostate tumor, pointed out by the two arrows.

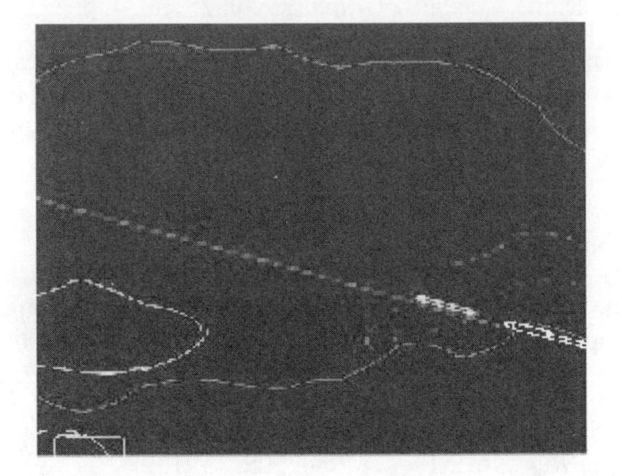

FIGURE 8.31
Demonstration of a planned needle biopsy procedure, including the target, needle tip, and biopsy pathway.

References

[1] T. Adali, Y. Wang, and N. Gupta, "A block-wise relaxation labeling scheme and its application to edge detection in cardiac MR image sequences," *Int. J. Imaging Systems,* Special Issue on Image Sequence Processing, August 1998.

[2] T. Adali, X. Liu, and M.K. Sönmez, "Conditional distribution learning with neural networks and its application to channel equalization," *IEEE Trans. Signal Processing,* vol. 45, no. 4, pp. 1051–1064, April 1997.

[3] T. Adali, M.K. Sönmez, and K. Patel, "On the dynamics of the LRE Algorithm: A distribution learning approach to adaptive equalization," in *Proc. IEEE Int. Conf. Acoust., Speech, Signal Processing,* Detroit, MI, 1995, pp. 929–932.

[4] H. Akaike, "A new look at the statistical model identification," *IEEE Trans. Automatic Control,* vol. 19, no. 6, December 1974.

[5] M.J. Bianchi, A. Rios, and M. Kabuka, "An algorithm for detection of masses, skin contours, and enhancement of microcalcifications in mammograms," *Proc. Symp. Computer Assisted Radiol.,* Winston-Salem, NC, June 1994, pp. 57–64.

[6] A.M. Bensaid, L.O. Hall, J.C. Bezdek, L.P. Clarke, M.L. Silbiger, J.A. Arrington, and R.F. Murtagh, "Validity-guided clustering with applications to image segmentation," *IEEE Trans. Fuzzy Systems,* vol. 4, no. 2, pp. 112–123, May 1996.

[7] C. Bouman and B. Liu, "Multiple resolution segmentation of texture images," *IEEE Trans. Pattern Anal. Machine Intell.,* vol. 13, no. 2, pp. 99–113, February 1991.

[8] D.J. Burr, "Elastic matching of line drawing," *IEEE Trans. Pattern Anal. Machine Intell.,* vol. 3, no. 6, pp. 102–112, Nov. 1981.

[9] G. Celniker and D. Gossard, "Deformable curve and surface finite–elements for free–form shape design," *Computer Graphics,* vol. 25, no. 4, pp. 257–266, July 1991.

[10] L. Chang, H. Chen, and J. Ho, "Reconstruction of 3D medical images: A nonlinear interpolation technique for reconstruction of 3D medical images," *CVGIP: Graphical Models Image Processing,* vol. 53, no. 4, pp. 382–391, July 1991.

[11] R. Chellappa, Q. Zheng, P. Burlina, C. Shekhar, and K. Eom, "On the positioning of multi–sensor imagery for exploitation and target recognition," *Proc. IEEE,* vol. 85, no. 1, pp. 120–138, January 1997.

[12] K.S. Cheng, J.S. Lin, and C.W. Mao, "The application of competitive Hopfield neural network to medical image segmentation," *IEEE Trans. Med. Imaging,* vol. 15, no. 4, pp. 560–567, August 1996.

[13] A. Chiarodo, "National Cancer Institute roundtable on prostate cancer: Future directions," *Cancer Res.,* vol. 51, pp. 2498–2505, 1991.

[14] H.S. Choi, D.R. Haynor, and Y. Kim, "Partial volume tissue classification of multichannel magnetic resonance images — A mixture model," *IEEE Trans. Med. Imaging,* vol. 10, pp. 395–407, September 1994.

[15] H.N. Cristiansen and T.W. Sederberg, "Conversion of complex contour line definition into polygonal element mosaics," *Computer Graphics*, vol. 12, no. 3, pp. 187–192, 1978.

[16] H.E. Cline, W.E. Lorensen, R. Kikinis, and R. Jolesz, "Three–dimensional segmentation of MR images of the head using probability and connectivity," *J. Computer Assisted Tomography,* vol. 14, pp. 1037–1045, 1990.

[17] T.M. Cover and J.A. Thomas, *Elements of Information Theory,* New York: John Wiley & Sons, 1991.

[18] A.P. Dempster, N.M. Laird, and D.B. Rubin, "Maximum likelihood from incomplete data via the EM algorithm," *J. R. Soc., Ser. B,* vol. 39, pp. 1–38, 1977.

[19] A.P. Dhawan and L. Arata, "Segmentation of medical images through competitive learning," *Computerized Methods Progress Biomed.,* vol. 40, pp. 203–215, 1993.

[20] J.H. Friedman, "On bias, variance, 0/1 — loss, and the curse-of-dimensionality," Technical report, Stanford University, 1996.

[21] H. Fuchs, Z.M. Kedem, and S.P. Uselton, "Optimal surface reconstruction from planar contours," *Commun. ACM,* vol. 20, no. 10, pp. 693–702, October 1977.

[22] M. Fuderer, "The information content of MR images," *IEEE Trans. Med. Imaging,* vol. 7, no. 4, pp. 368–380, 1988.

[23] S. Geman, E. Bienenstock, and R. Doursat, "Neural networks and the bias/variance dilemma," *Neural Computation,* vol. 4, pp. 1–52, 1992.

[24] L.O. Hall, A.M. Bensaid, L.P. Clarke, R.P. Velthuizen, M.S. Silbiger, and J.C. Bezdek, "A comparison of neural network and fuzzy clustering techniques in segmenting magnetic resonance images of the brain," *IEEE Trans. Neural Networks,* vol. 3, pp. 672–682, 1992.

[25] S. Haykin, *Neural Networks: A Comprehensive Foundation,* New York: Macmillan, 1994.

[26] A. Hoover, G.J. Baptoste, X. Jiang, P.J. Flynn, H. Bunke, D.B. Goldgof, K. Bowyer, D.W. Eggert, A. Fitzgibbon, and R.B. Fisher, "An experimental comparison of range image segmentation algorithms," *IEEE Trans. PAMI,* vol. 18, no. 7, pp. 673–688, July 1996.

[27] R.A. Hummel and S.W. Zucker, "On the foundations of relaxation labeling processes," *IEEE Trans. Pattern Anal. Machine Intell.,* vol. 5, no. 3, May 1983.

[28] A.K. Jain, "Markov random fields: Theory and application," Boston: Academic Press, 1993.

[29] E.T. Jaynes, "Information theory and statistical mechanics," *Physical Rev.,* vol. 108, no. 2, pp. 620–630/171–190, May 1957.

[30] E. Keppel, "Approximating complex surfaces by triangulation of contour lines," *IBM J. Res. Dev.,* vol. 19, pp. 2–11, January 1975.

[31] L. Kullback, and R.A. Leibler, "On information and sufficiency," *Ann. Math. Statistics,* vol. 22, pp. 79–86, 1951.

[32] M.A. Kupinski and M.L. Giger, "Automated seeded lesion segmentation on digital mammograms," *IEEE Trans. Med. Imaging,* vol. 17, no. 4, pp. 510–517, August 1998.

[33] T. Lei and W. Sewchand, "Statistical approach to X-ray CT imaging and its application in image analysis — Part II: A new stochastic model-based image segmentation technique for X-ray CT image," *IEEE Trans. Med. Imaging,* vol. 11, no. 1, pp. 62–69, March 1992.

[34] H. Li, Y. Wang, K.-J. Liu, and S.-H. Lo, "Morphological filtering and model–based segmentation of masses on mammographic images," *IEEE Trans. Med. Imaging,* 1998.

[35] H. Li, "Model-Based Image Processing Techniques for Breast Cancer Detection in Digital Mammography," Ph. D. dissertation, University of Maryland, May 1997.

[36] H. Li, S.C. Lo, Y. Wang, W. Hayes, M.T. Freedman, and S.K. Mun, "Detection of masses on mammograms using advanced segmentation techniques and an HMOE classifier," *Dig. Mammogr.,* New York: Elsevier, 1996.

[37] H. Li, Y. Wang, K.J.R. Liu, S.-C.B. Lo, "Statistical model supported approach to radiographic mass detection — Part I: Improving lesion characterization by morphological filtering and site segmentation," *IEEE Trans. Med. Imaging.*

[38] Z. Liang, J.R. MacFall, and D.P. Harrington, "Parameter estimation and tissue segmentation from multispectral MR images," *IEEE Trans. Med. Imaging,* vol. 13, no. 3, pp. 441–449, September 1994.

[39] W. Lin, C. Liang, and C. Chen, "Dynamic elastic interpolation for 3-D medical image reconstruction from serial cross section," *IEEE Trans. Med. Imaging,* vol. 7, no. 3, pp. 225–232, September 1988.

[40] W.C. Lin, E.C.K. Tsao, and C.T. Chen, "Constraint satisfaction neural networks for image segmentation," *Pattern Recognition,* vol. 25, pp. 679–693, 1992.

[41] J.L. Marroquin and F. Girosi, "Some extensions of the K-means algorithm for image segmentation and pattern classification," Technical report, MIT Artificial Intelligence Laboratory, January 1993.

[42] J.L. Marroquin, "Measure fields for function approximation," *IEEE Trans. Neural Networks,* vol. 6, no. 5, pp. 1081–1090, 1995.

[43] T. McInerney and D. Terzopoulos, "A dynamic finite element surface model for segmentation and tracking in multidimensional medical images with application to cardiac 4D image analysis," *Computerized Med. Imaging Graphics,* vol. 19, no. 1, pp. 69–83, 1995.

[44] D. Metaxas and D. Terzopoulos, "Shape and non–rigid motion estimation through physics–based synthesis," *IEEE Trans. Pattern Anal. Machine Intell.,* vol. 15. no. 6, pp. 580–591, June 1993.

[45] T.K. Moon, "The expectation-maximization algorithm," *IEEE Signal Processing Magazine,* pp. 47–60, November 1996.

[46] M. Morrison and Y. Attikiouzel, "A probabilistic neural network based image segmentation network for magnetic resonance images," *Proc. Conf. Neural Networks,* vol. 3, pp. 60–65, Baltimore, 1992.

[47] R.M. Neal and G.E. Hinton, "A view of the EM algorithm that justifies incremental, sparse, and other variants," in *Learning in Graphical Models,* M.I. Jordan (editor), pp. 355–368, Dordrecht: Kluwer Academic, 1998.

[48] L. Perlovsky, W. Schoendorf, B. Burdick, and D.M. Tye, "Model-based neural network for target detection in SAR images," *IEEE Trans. Image Processing,* vol. 6, no. 1, pp. 203–216, January 1997.

[49] L. Perlovsky and M. McManus, "Maximum likelihood neural networks for sensor fusion and adaptive classification," *Neural Networks,* vol. 4, pp. 89–102, 1991.

[50] L. I. Perlovsky, "Cramer-Rao bounds for the estimation of normal mixtures," *Pattern Recognition Lett.,* vol. 10, pp. 141–148, 1989.

[51] H.V. Poor, *An Introduction to Signal Detection and Estimation,* Berlin: Springer-Verlag, 1988.

[52] S.P. Raya and J.K. Udupa, "Shape-based interpolation of multidimensional object," *IEEE Trans. Med. Imaging,* vol. 9, pp. 32–42, 1990.

[53] R.A. Redner and N.M. Walker, "Mixture densities, maximum likelihood and the EM algorithm," *SIAM Rev.,* vol. 26, pp. 195–239, 1984.

[54] J. Rissanen, "Minimax entropy estimation of models for vector processes," *System Identification,* pp. 97–119, 1987.

[55] J. Rissanen, "A universal prior for integers and estimation by minimum description length," *Ann. Statistics,* vol. 11, no. 2, 1983.

[56] P. Santago and H.D. Gage, "Quantification of MR brain images by mixture density and partial volume modeling," *IEEE Trans. Med. Imaging,* vol. 12, no. 3, pp. 566–574, September 1993.

[57] S. Sclaroff and A.P. Pentland, "Modal matching for correspondence and recognition," *IEEE Trans. Pattern Anal. Machine Intell.,* vol. 17. no. 6, pp. 545–561, June 1995.

[58] I. Sesterhenn, F. Mostofi, R. Mattrey, J. Sands, C. Davis, and W. McCarthy, "Preliminary results of three-dimensional reconstruction of previously imaged prostate," *The Prostate Supplement,* vol. 4, pp. 33–41, 1992.

[59] A. Sunguroff and D. Greenberg, "Computer generated images for medical applications," *Comp. Graphics,* vol. 12, no. 3, pp. 196–202, 1978.

[60] H. Tamura, S. Mori, and T. Yamawaki, "Textural features corresponding to visual perception," *IEEE Trans. Systems, Man, Cybernetics,* vol. 8, pp. 460–473, 1978.

[61] D. Terzopoulos and D. Metaxas, "Dynamic 3D models with local and global deformations: Deformable superquadrics," *IEEE Trans. Pattern Anal. Machine Intell.,* vol. 13, no. 7, pp. 703–714, July 1991.

[62] D. Terzopoulos, "The computation of visible–surface representation," *IEEE Trans. Pattern Anal. Machine Intell.,* vol. 10, pp. 417–438, 1988.

[63] D. Terzopoulos and K. Fleischer, "Deformable models," *The Visual Computer,* vol. 4, pp. 306–331, 1988.

[64] D. Terzopoulos, A. Witkin, and M. Kass, "Symmetry–seeking models and 3D object reconstruction," *Int. J. Computer Vision,* vol. 1, pp. 211–221, 1987.

[65] D. Terzopoulos, "Regularization of inverse visual problems involving discontinuities," *IEEE Trans. Pattern Anal. Machine Intell.,* vol. 8, pp. 413–424, 1986.

[66] D.M. Titterington, A.F.M. Smith, and U.E. Markov, *Statistical Analysis of Finite Mixture Distributions,* New York: John Wiley, 1985.

[67] D.M. Titterington, "Comments on application of the conditional population-mixture model to image segmentation," *IEEE Trans. Pattern Anal. Machine Intell.,* vol. 6, no. 5, pp. 656–658, September 1984.

[68] Y. Wang, S.-H. Lin, H. Li, and S.-Y. Kung, "Data mapping by probabilistic modular network and information theoretic criteria," *IEEE Trans. Signal Processing,* pp. 3378–3397, December 1998.

[69] Y. Wang, T. Adali, S.-Y. Kung, and Z. Szabo, "Quantification and segmentation of brain tissue from MR images: A probabilistic neural network approach," *IEEE Trans. Image Processing,* Special Issue on Applications of Neural Networks to Image Processing, vol. 7, no. 8, pp. 1165–1181, August 1998.

[70] Y. Wang, T. Adali, C. Lau, and S.-Y. Kung, "Quantitative analysis of MR brain image sequences by adaptive self-organizing mixtures," *J. VLSI Signal Processing Systems Signal, Image, Video Technol.,* Special Issue on Neural Networks for Biomedical Image Processing, vol. 18, no. 3, pp. 219–239, April 1998.

[71] Y. Wang, T. Adali, and B. Lo, "Automatic threshold selection by histogram quantization," *SPIE J. Biomed. Optics,* vol. 2, no. 2, pp. 211–217, April 1997.

[72] Y. Wang, J. Xuan, I. Sesterhenn, W. Hayes, D. Ebert, J. Lynch, and S.K. Mun, "Statistical modeling and visualization of localized prostate cancer," *SPIE Med. Imaging,* Newport Beach, CA, February 1997.

[73] Y. Wang and T. Adali, "Efficient learning of finite normal mixtures for image quantification," in *Proc. IEEE Int. Conf. Acoust., Speech, Signal Processing,* Atlanta, GA, 1996, pp. 3422–3425.

[74] Y. Wang, T. Adali, M.T. Freedman, and S.K. Mun, "MR brain image analysis by distribution learning and relaxation labeling," *Proc. 15th South. Biomed. Eng. Conf.,* Dayton, OH, March 1996, pp. 133–136.

[75] Y. Wang, "Image quantification and the minimum conditional bias/variance criterion," *Proc. 30th Conf. Info. Sci. Systems,* Princeton, NJ, March 20–22, 1996, pp. 1061–1064.

[76] Y. Wang, T. Adali, C.M. Lau, and Z. Szabo, "Quantification of MR brain images by a probabilistic self–organizing map," *Radiology* (special issue), vol. 197, pp. 252–253, November 1995.

[77] Y. Wang, "MR imaging statistics and model-based MR image analysis," Ph. D. dissertation, University of Maryland, Baltimore, May 1995.

[78] Y. Wang and T. Adali, "Probabilistic neural networks for parameter quantification in medical image analysis," in *Biomedical Engineering Recent Development,* J. Vossoughi, Ed., 1994.

[79] Y. Wang and T. Lei, "A new stochastic model-based image segmentation technique for MR images," *Proc. 1st IEEE Int. Conf. Image Processing,* Austin, TX, 1994, pp. 182–185.

[80] M. Wax and T. Kailath, "Detection of signals by information theoretic criteria," *IEEE Trans. Acoust., Speech, Signal Processing,* vol. 33, no. 2, April 1985.

[81] E. Weinstein, M. Feder, and A.V. Oppenheim, "Sequential algorithms for parameter estimation based on the Kullback-Leibler information measure," *IEEE Trans. Acoust., Speech, Signal Processing,* vol. 38, no. 9, pp. 1652–1654, 1990.

[82] P.N. Werahera et al., "A 3-D reconstruction algorithm for interpolation and extrapolation of planar cross sectional data," *IEEE Trans. Med. Imaging,* vol. 14, no. 4, December 1995.

[83] A.J. Worth and D.N. Kennedy, "Segmentation of magnetic resonance brain images using analog constraint satisfaction neural networks," *Information Processing Med. Imaging,* pp. 225–243, 1993.

[84] L. Xu and M.I. Jordan, "On convergence properties of the EM algorithm for Gaussian mixture," Technical report, MIT Artificial Intelligence Laboratory, January 1995.

[85] J. Xuan and T. Adali, "Segmentation of magnetic resonance brain image: integrating region growing and edge detection," *Proc. IEEE Int. Conf. Image Processing (ICIP),* vol. 3, pp. 544–547, November 1995.

[86] J. Xuan, Y. Wang, T. Adali, Q. Zheng, W. Hayes, M.T. Freedman, and S.K. Mun, "A deformable surface–spine model for 3-D surface registration," *Proc. IEEE Int. Conf. Image Processing (ICIP),* Santa Barbara, CA, November 1997.

[87] J. Xuan, "Medical image understanding: Segmentation, modeling, and representation," Ph. D. dissertation, University of Maryland, Baltimore, May 1997.

[88] A.P. Zijdenbos, B.M. Dawant, R.A. Margolin, and A.C. Palmer, "Morphometric analysis of white matter lesions in MR images: method and validation," *IEEE Trans. Med. Imaging,* vol. 13, no. 4, pp. 716–724, December 1994.

[89] J. Zhang and J.W. Modestino, "A model-fitting approach to cluster validation with application to stochastic model-based image segmentation," *IEEE Trans. PAMI,* vol. 12, no. 10, pp. 1009–1017, October 1990.

[90] Y.J. Zhang, "A survey on evaluation methods for image segmentation," *Pattern Recognition,* vol. 29, no. 8, pp. 1335–1346, 1996.

[91] Zienkiewicz, *The Finite Element Method,* 3rd ed., New York: McGraw-Hill, 1967.

Chapter 9

Combined Motion Estimation and Transform Coding in Compressed Domain

Ut-Va Koc and K.J. Ray Liu

9.1 Introduction

The motion-compensated discrete cosine transform (DCT) video compression scheme (MC-DCT) is the basis of a number of international video coding standards, which are tabulated in Table 9.1, ranging from the low-bit-rate, and high-compression-rate videophone application to the high-end, high-bit-rate, and high-quality HDTV application requiring a modest compression rate. The MC-DCT scheme belongs to the class of hybrid spatial/temporal waveform-based video compression approaches [1, 2, 3]. As illustrated in Figure 9.1, the MC-DCT scheme employs motion estimation and compensation to reduce or remove temporal redundancy and then uses DCT to exploit spatial correlation among the pixels of the motion-compensated predicted frame errors (residuals). Efficient coding is accomplished by adding the quantization and variable-length coding steps after the DCT block. The coding model block rearranges the 2D DCT coefficients into a 1D order, usually in a zigzag manner. Basically all the standards in Table 9.1 follow this procedure with modifications of each step to reach different targeted bit rates and application goals. As these standards are becoming more and more prevalent in various forms of video products, efficient and cost-effective implementations of the standards become more important.

Table 9.1 DCT-Based Motion-Compensated Video Coding Standards

Standard	Application	Targeted Bit Rate	Remark
H.261	Teleconferencing over ISDN	$p \times 64$ kbps	
MPEG-1	Video on CD-ROM	$<$ 1.5 Mbps	
MPEG-2	Generic high-bit-rate applications	$>$ 1.5 Mbps	Also called H.262
HDTV	U.S. terrestrial broadcast	18 Mbps	Based on MPEG-2
DVB	European digital video broadcast	6–38 Mbps	Based on MPEG-2
H.263	Low-bit-rate communications over PSTN	$\sim$ 18 kbps	
MPEG-4	Object-based applications	$<$ 50 Mbps	

The implementation of a standard-compliant coder usually requires the conventional MC-DCT video coder structure, as shown in Figure 9.2a, where the DCT unit and the block-based motion estimation unit are two essential elements to achieve spatial and temporal compression,

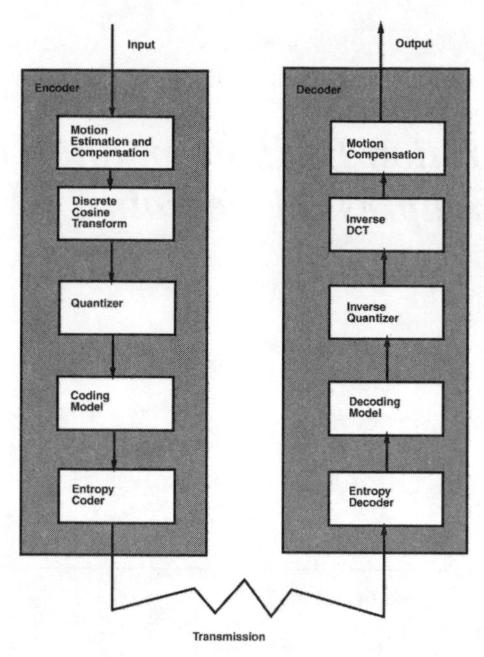

FIGURE 9.1
Motion-compensated DCT (MC-DCT) scheme.

respectively. The feedback loop for temporal prediction consists of a DCT, an inverse DCT

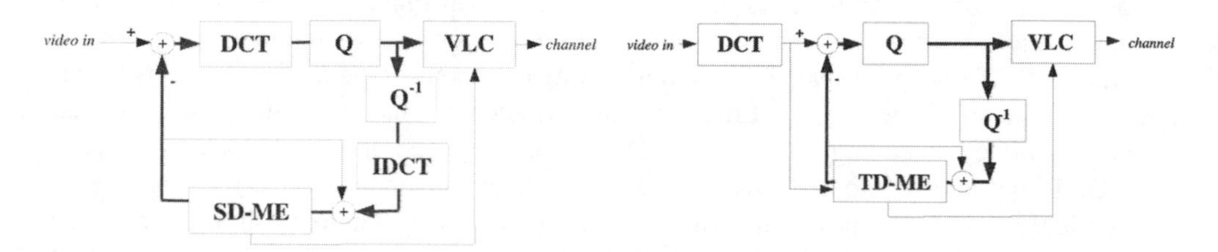

FIGURE 9.2
Different MC-DCT video coder structures: (a) motion estimation/compensation in the spatial domain; (b) motion estimation/compensation completely in the transform (DCT) domain.

(IDCT) and a spatial domain motion estimator (SD-ME), which is usually the full-search block-matching approach (BKM). This is undesirable. Besides adding complexity to the overall architecture, this feedback loop limits the throughput of the coder and becomes the bottleneck of a real-time high-end video codec. A compromise is to remove the loop and perform open-loop motion estimation based on original images instead of reconstructed images in sacrifice of the performance of the coder [2, 4]. The savings comes from the removal of the components IDCT and Q^{-1}, but the quality of the reconstructed images at the decode may gradually degrade because the decoder has no access to the original images, only the reconstructed images.

In this chapter, we propose a nonconventional coder structure, the fully DCT-based motion-compensated video coder structure, suitable for all the hybrid rectangularly shaped DCT motion-compensated video coding standards. In Section 9.2, we discuss the advantages of this proposed structure over the traditional architecture. The realization of this fully DCT-based architecture requires algorithms to estimate and compensate motions completely in the

DCT domain. In this way, we can combine motion estimation and transform coding completely in the compressed (transform) domain without the need to convert the DCT coefficients back to the spatial domain for motion estimation and compensation. In Section 9.4, we develop the DCT-based motion estimation algorithms based on the DCT pseudo-phase techniques covered in Section 9.3. Then we extend the DCT pseudo-phase techniques to the subpixel level in Section 9.5. From these subpixel techniques, we can estimate subpixel motion in the DCT domain without the need of image interpolation (Section 9.6). To complete the feedback loop of the structure, we explore and devise the DCT-based motion compensation algorithms (Section 9.7). Our conclusions are provided in Section 9.8.

9.2 Fully DCT-Based Motion-Compensated Video Coder Structure

In the conventional hybrid coder structure, as shown in Figure 9.2a, the presence of the IDCT block inside the feedback loop of the conventional video coder design comes from the fact that currently available motion estimation algorithms can estimate motion only in the spatial domain rather than directly in the DCT domain. Therefore, developing a transform domain motion estimation algorithm will eliminate this IDCT. Furthermore, the DCT block in the feedback loop is used to compute the DCT coefficients of motion-compensated residuals. However, for motion compensation in the DCT domain, this DCT block can be moved out of the feedback loop. From these two observations, an alternative solution without degradation of the performance is to develop motion estimation and compensation algorithms that can work in the DCT domain. In this way, the DCT can be moved out of the loop as depicted in Figure 9.2b and, thus, the operating speed of this DCT can be reduced to the data rate of the incoming stream. Moreover, the IDCT is removed from the feedback loop, which now has only two simple components Q and Q^{-1} (the quantizers) in addition to the transform domain motion estimator (TD-ME). This not only reduces the complexity of the coder but also resolves the bottleneck problem without any trade-off of performance. In this chapter, we demonstrate that all the essential components (DCT-based motion estimation and compensation) of this fully DCT-based coder provide comparable performance with less complexity than the pixel-based full-search approaches. Furthermore, different components can be jointly optimized if they operate in the same transform domain. It should be stressed that by using DCT-based estimation and compensation methods, standard-compliant bitstreams can be formed in accordance with the specification of any standard such as MPEG without the need to change the structure of any standard-compliant decoder. To realize this fully DCT-based video coder architecture to boost the system throughput and reduce the total number of components, we develop DCT-based algorithms that perform motion estimation and compensation directly on the DCT coefficients of video frames [1, 5, 6].

9.3 DCT Pseudo-Phase Techniques

As is well known, the Fourier transform (FT) of a signal $x(t)$ is related to the FT of its shifted (or delayed if t represents time) version, $x(t-\tau)$, by the equation

$$\mathcal{F}\{x(t-\tau)\} = e^{-j\omega\tau}\mathcal{F}\{x(t)\} , \tag{9.1}$$

where $\mathcal{F}\{\cdot\}$ denotes the Fourier transform. The phase of Fourier transform of the shifted signal contains the information about the amount of the shift, τ, which can easily be extracted. The phase correlation method was developed to estimate motion from the Fourier phase [3, 7]. However, the DCT or its counterpart, the discrete sine transform (DST), does not have any phase components as usually is found in the discrete Fourier transform (DFT); however, DCT (or DST) coefficients of a shifted signal do carry this shift information. To facilitate explanation of the DCT pseudo-phase techniques, let us first consider the case of one-dimensional discrete signals. Suppose that the signal $\{x_1(n);\ n \in \{0, \ldots, N-1\}\}$ is right shifted by an amount m (in our convention, a right shift means that $m > 0$) to generate another signal $\{x_2(n);\ n \in \{0, \ldots, N-1\}\}$. The values of $x_1(n)$ are all zero outside the support region $\mathcal{S}(x_1)$. Therefore,

$$x_2(n) = \begin{cases} x_1(n-m), & \text{for } n-m \in \mathcal{S}(x_1), \\ 0, & \text{elsewhere}. \end{cases}$$

The above equation implies that both signals have resemblance to each other except that the signal is shifted. It can be shown that, for $k = 1, \ldots, N-1$,

$$X_2^C(k) = Z_1^C(k)\cos\left[\frac{k\pi}{N}\left(m+\frac{1}{2}\right)\right] - Z_1^S(k)\sin\left[\frac{k\pi}{N}\left(m+\frac{1}{2}\right)\right], \tag{9.2}$$

$$X_2^S(k) = Z_1^S(k)\cos\left[\frac{k\pi}{N}\left(m+\frac{1}{2}\right)\right] + Z_1^C(k)\sin\left[\frac{k\pi}{N}\left(m+\frac{1}{2}\right)\right]. \tag{9.3}$$

Here X_2^S and X_2^C are DST (DST-II) and DCT (DCT-II) of the second kind of $x_2(n)$, respectively, whereas Z_1^S and Z_1^C are DST (DST-I) and DCT (DCT-I) of the first kind of $x_1(n)$, respectively, defined as follows [8]:

$$X_2^C(k) = \frac{2}{N}C(k)\sum_{n=0}^{N-1} x_2(n)\cos\left[\frac{k\pi}{N}(n+0.5)\right];\ k \in \{0, \ldots, N-1\}, \tag{9.4}$$

$$X_2^S(k) = \frac{2}{N}C(k)\sum_{n=0}^{N-1} x_2(n)\sin\left[\frac{k\pi}{N}(n+0.5)\right];\ k \in \{1, \ldots, N\}, \tag{9.5}$$

$$Z_1^C(k) = \frac{2}{N}C(k)\sum_{n=0}^{N-1} x_1(n)\cos\left[\frac{k\pi}{N}(n)\right];\ k \in \{0, \ldots, N\}, \tag{9.6}$$

$$Z_1^S(k) = \frac{2}{N}C(k)\sum_{n=0}^{N-1} x_1(n)\sin\left[\frac{k\pi}{N}(n)\right];\ k \in \{1, \ldots, N-1\}, \tag{9.7}$$

where

$$C(k) = \begin{cases} \frac{1}{\sqrt{2}}, & \text{for } k = 0 \text{ or } N, \\ 1, & \text{otherwise}. \end{cases}$$

The displacement, m, is embedded solely in the terms $g_m^s(k) = \sin[\frac{k\pi}{N}(m+\frac{1}{2})]$ and $g_m^c(k) = \cos[\frac{k\pi}{N}(m+\frac{1}{2})]$, which are called *pseudo-phases*, analogous to phases in the Fourier transform of shifted signals. To find m, we first solve (9.2) and (9.3) for the pseudo-phases and then use the sinusoidal orthogonal principles as follows:

$$\frac{2}{N}\sum_{k=1}^{N} C^2(k)\sin\left[\frac{k\pi}{N}\left(m+\frac{1}{2}\right)\right]\sin\left[\frac{k\pi}{N}\left(n+\frac{1}{2}\right)\right] = \delta(m-n) - \delta(m+n+1)\,, \tag{9.8}$$

$$\frac{2}{N}\sum_{k=0}^{N-1} C^2(k)\cos\left[\frac{k\pi}{N}\left(m+\frac{1}{2}\right)\right]\cos\left[\frac{k\pi}{N}\left(n+\frac{1}{2}\right)\right] = \delta(m-n) + \delta(m+n+1)\,. \tag{9.9}$$

Here $\delta(n)$ is the discrete impulse function, defined as

$$\delta(n) = \begin{cases} 1, & \text{for } n = 0, \\ 0, & \text{otherwise}. \end{cases} \tag{9.10}$$

Indeed, if we replace $\sin[\frac{k\pi}{N}(m+\frac{1}{2})]$ and $\cos[\frac{k\pi}{N}(m+\frac{1}{2})]$ by the computed sine and cosine pseudo-phase components, $\hat{g}_m^s(k)$ and $\hat{g}_m^c(k)$, respectively, in (9.8) and (9.9), both equations simply become IDST-II and IDCT-II operations on $\hat{g}_m^s(k)$ and $\hat{g}_m^c(k)$:

$$\text{IDST-II}\left(\hat{g}_m^s\right) = \frac{2}{N}\sum_{k=1}^{N} C^2(k)\hat{g}_m^s(k)\sin\left[\frac{k\pi}{N}\left(n+\frac{1}{2}\right)\right]\,, \tag{9.11}$$

$$\text{IDCT-II}\left(\hat{g}_m^c\right) = \frac{2}{N}\sum_{k=0}^{N-1} C^2(k)\hat{g}_m^c(k)\cos\left[\frac{k\pi}{N}\left(n+\frac{1}{2}\right)\right]\,. \tag{9.12}$$

The notation $\hat{g}$ is used to distinguish the computed pseudo-phase from the one in a noiseless situation (i.e., $\sin[\frac{k\pi}{N}(m+\frac{1}{2})]$ or $\cos[\frac{k\pi}{N}(m+\frac{1}{2})]$). A closer look at the right-hand side of (9.8) tells us that $\delta(m-n)$ and $\delta(m+n+1)$ have opposite signs. This property will help us detect the shift direction. If we perform an IDST-II operation on the pseudo-phases found, then the observable window of the index space in the inverse DST domain will be limited to $\{0, \ldots, N-1\}$. As illustrated in Figure 9.3, for a right shift, one spike (generated by the positive δ function) is pointing upward at the location $n = m$ in the gray region (i.e., the observable index space), whereas the other δ is pointing downward at $n = -(m+1)$ outside the gray region. In contrast, for a left shift, the negative spike at $n = -(m+1) > 0$ falls in the gray region but the positive δ function at $n = m$ stays out of the observable index space. It can easily be seen that a positive peak value in the gray region implies a right shift and a negative one means a left shift. This enables us to determine from the sign of the peak value the direction of the shift between signals.

The concept of pseudo-phases plus the application of sinusoidal orthogonal principles leads to the DCT pseudo-phase techniques, a new approach to estimate a shift or translational motion between signals in the DCT domain, as depicted in Figure 9.4a:

1. Compute the DCT-I and DST-I coefficients of $x_1(n)$ and the DCT-II and DST-II coefficients of $x_2(n)$.

2. Compute the pseudo-phase $\hat{g}_m^s(k)$ for $k = 1, \ldots, N$ by solving this equation:

$$\hat{g}_m^s(k) = \begin{cases} \frac{Z_1^C(k)\cdot X_2^S(k) - Z_1^S(k)\cdot X_2^C(k)}{[Z_1^C(k)]^2 + [Z_1^S(k)]^2}\,, & \text{for } k \neq N, \\ \frac{1}{\sqrt{2}}\,, & \text{for } k = N\,. \end{cases} \tag{9.13}$$

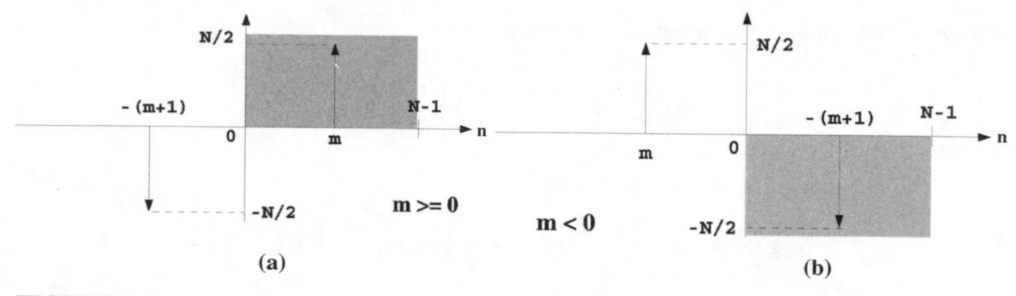

FIGURE 9.3
How the direction of shift is determined based on the sign of the peak value after application of the sinusoidal orthogonal principle for the DST-II kernel to pseudo-phases. (a) How to detect right shift. (b) How to detect left shift.

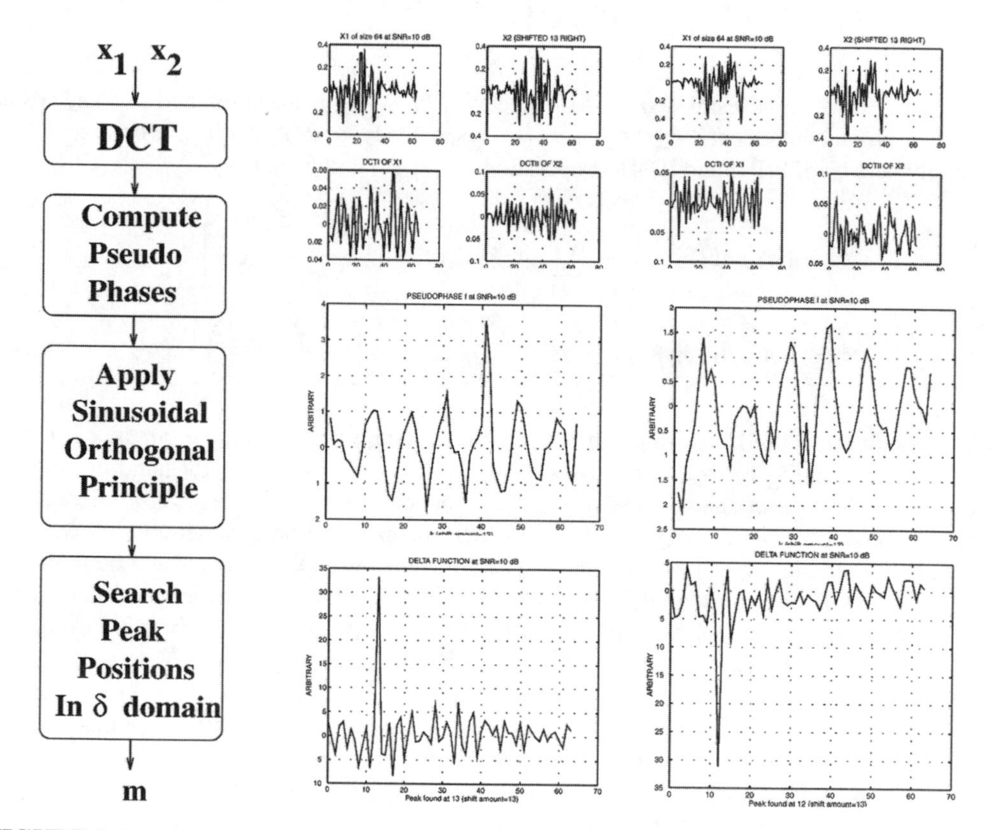

FIGURE 9.4
Illustration of one-dimensional DCT pseudo-phase techniques. (a) DCT Pseudo-Phase Techniques; (b) Right shift; (c) Left shift.

3. Feed the computed pseudo-phase, $\{\hat{g}^s_m(k);\ k = 1, \ldots, N\}$, into an IDST-II decoder to produce an output $\{d(n);\ n = 0, \ldots, N-1\}$, and search for the peak value. Then the estimated displacement $\hat{m}$ can be found by

$$\hat{m} = \begin{cases} i_p, & \text{if } d(i_p) > 0\,, \\ -(i_p + 1), & \text{if } d(i_p) < 0\,, \end{cases} \tag{9.14}$$

where $i_p = \arg\max_n |d(n)|$ is the index at which the peak value is located.

In step 1, the DCT and DST can be generated simultaneously with only $3N$ multipliers [9]–[11], and the computation of DCT-I can be easily obtained from DCT-II with minimal overhead, as will be shown later. In step 2, if noise is absent and there is only purely translational motion, $\hat{g}_m(k)$ will be equal to $\sin \frac{k\pi}{N}(m + 0.5)$. The output $d(n)$ will then be an impulse function in the observation window. This procedure is illustrated by two examples in Figure 9.4b and c with a randomly generated signal as input at a signal-to-noise ratio (SNR) of 10 dB. These two examples demonstrate that the DCT pseudo-phase techniques are robust even in an environment of strong noise.

9.4 DCT-Based Motion Estimation

The DCT pseudo-phase technique of extracting shift values from the pseudo-phases of the DCT of one-dimensional signals can be extended to the two-dimensional case. Let us confine the problem of motion estimation to this 2D translational motion model in which an object moves translationally by m_u in the x direction and m_v in the y direction as viewed on the camera plane and within the scope of a camera in a noiseless environment, as shown in Figure 9.5. Then by means of the DCT pseudo-phase technique, we can extract the displacement vector from the two consecutive frames of the images of that moving object by making use of the sinusoidal orthogonal principles (9.8) and (9.9). The resulting novel algorithm for this two-dimensional translational motion model is called the DXT-ME algorithm, which can estimate translational motion in the DCT domain.

9.4.1 The DXT-ME Algorithm

Based on the assumption of 2D translational displacements, we can extend the DCT pseudo-phase technique to the DXT-ME algorithm depicted in Figure 9.6. The previous frame x_{t-1} and the current frame x_t are fed into the 2D-DCT-II and 2D-DCT-I coders, respectively. A 2D-DCT-II coder computes four coefficients, DCCTII, DCSTII, DSCTII, and DSSTII, each of which is defined as a two-dimensional separable function formed by 1D-DCT/DST-II kernels:

$$X_t^{cc}(k,l) = \frac{4}{N^2} C(k)C(l) \sum_{m,n=0}^{N-1} x_t(m,n) \cos\left[\frac{k\pi}{N}(m+0.5)\right] \cos\left[\frac{l\pi}{N}(n+0.5)\right], \quad (9.15)$$

$$\text{for } k,l \in \{0,\ldots,N-1\},$$

$$X_t^{cs}(k,l) = \frac{4}{N^2} C(k)C(l) \sum_{m,n=0}^{N-1} x_t(m,n) \cos\left[\frac{k\pi}{N}(m+0.5)\right] \sin\left[\frac{l\pi}{N}(n+0.5)\right], \quad (9.16)$$

$$\text{for } k \in \{0,\ldots,N-1\}, l \in \{1,\ldots,N\},$$

$$X_t^{sc}(k,l) = \frac{4}{N^2} C(k)C(l) \sum_{m,n=0}^{N-1} x_t(m,n) \sin\left[\frac{k\pi}{N}(m+0.5)\right] \cos\left[\frac{l\pi}{N}(n+0.5)\right], \quad (9.17)$$

$$\text{for } k \in \{1,\ldots,N\}, l \in \{0,\ldots,N-1\},$$

$$X_t^{ss}(k,l) = \frac{4}{N^2} C(k)C(l) \sum_{m,n=0}^{N-1} x_t(m,n) \sin\left[\frac{k\pi}{N}(m+0.5)\right] \sin\left[\frac{l\pi}{N}(n+0.5)\right], \quad (9.18)$$

$$\text{for } k,l \in \{1,\ldots,N\},$$

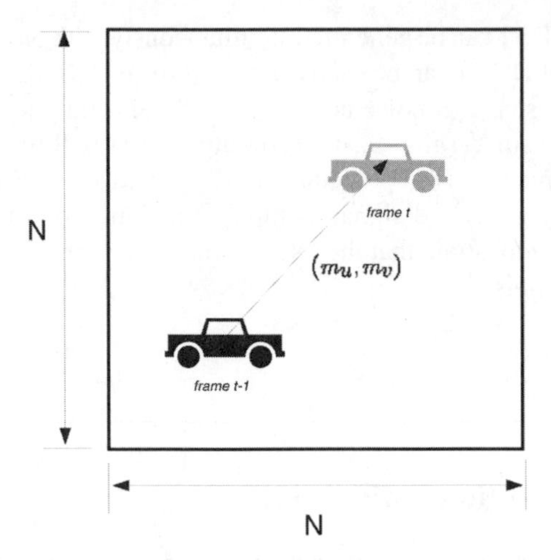

FIGURE 9.5
An object moves translationally by m_u in the x direction and m_v in the y direction, as viewed on the camera plane.

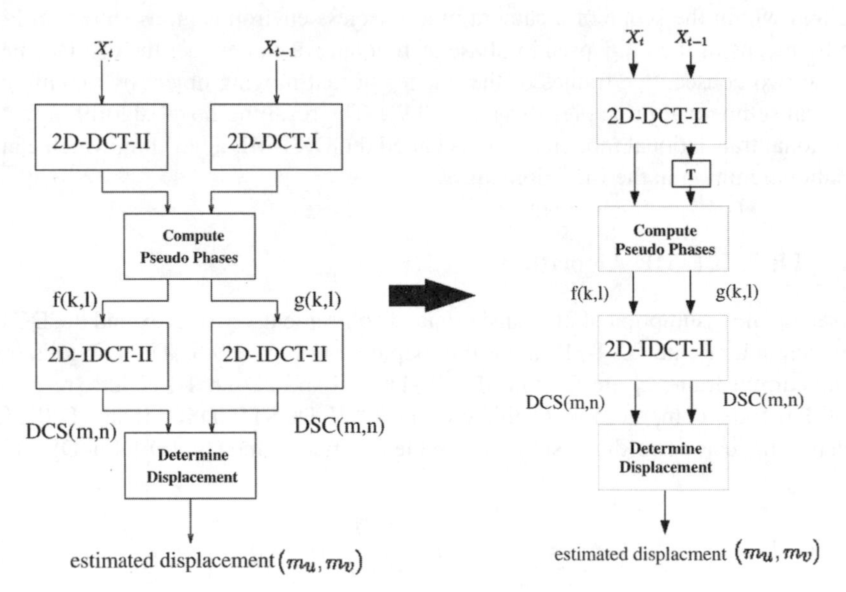

FIGURE 9.6
Block diagram of DXT-ME. (a) Flowchart; (b) structure.

or symbolically,

$$X_t^{cc} = \text{DCCTII}(x_t)\,, \quad X_t^{cs} = \text{DCSTII}(x_t)\,,$$
$$X_t^{sc} = \text{DSCTII}(x_t)\,, \quad X_t^{ss} = \text{DSSTII}(x_t)\ .$$

In the same fashion, the two-dimensional DCT coefficients of the first kind (2D-DCT-I) are calculated based on 1D-DCT/DST-I kernels:

$$Z_{t-1}^{cc}(k,l) = \frac{4}{N^2}C(k)C(l)\sum_{m,n=0}^{N-1} x_{t-1}(m,n)\cos\left[\frac{k\pi}{N}(m)\right]\cos\left[\frac{l\pi}{N}(n)\right], \quad \text{for } k,l \in \{0,\ldots,N\}, \tag{9.19}$$

$$Z_{t-1}^{cs}(k,l) = \frac{4}{N^2}C(k)C(l)\sum_{m,n=0}^{N-1} x_{t-1}(m,n)\cos\left[\frac{k\pi}{N}(m)\right]\sin\left[\frac{l\pi}{N}(n)\right], \quad \text{for } k \in \{0,\ldots,N\}, l \in \{1,\ldots,N-1\}, \tag{9.20}$$

$$Z_{t-1}^{sc}(k,l) = \frac{4}{N^2}C(k)C(l)\sum_{m,n=0}^{N-1} x_{t-1}(m,n)\sin\left[\frac{k\pi}{N}(m)\right]\cos\left[\frac{l\pi}{N}(n)\right], \quad \text{for } k \in \{1,\ldots,N-1\}, l \in \{0,\ldots,N\}, \tag{9.21}$$

$$Z_{t-1}^{ss}(k,l) = \frac{4}{N^2}C(k)C(l)\sum_{m,n=0}^{N-1} x_{t-1}(m,n)\sin\left[\frac{k\pi}{N}(m)\right]\sin\left[\frac{l\pi}{N}(n)\right], \quad \text{for } k,l \in \{1,\ldots,N-1\}, \tag{9.22}$$

or symbolically,

$$Z_{t-1}^{cc} = \text{DCCTI}(x_{t-1}),\ Z_{t-1}^{cs} = \text{DCSTI}(x_{t-1}),$$
$$Z_{t-1}^{sc} = \text{DSCTI}(x_{t-1}),\ Z_{t-1}^{ss} = \text{DSSTI}(x_{t-1}).$$

Similar to the one-dimensional case, assuming that only translational motion is allowed, one can derive a set of equations to relate the DCT coefficients of $x_{t-1}(m,n)$ with those of $x_t(m,n)$ in the same way as in (9.2) and (9.3).

$$\mathbf{Z}_{t-1}(k,l)\cdot\vec{\theta}(k,l) = \vec{\mathbf{x}}_t(k,l), \quad \text{for } k,l \in \mathcal{N}, \tag{9.23}$$

where $\mathcal{N} = \{1,\ldots,N-1\}$,

$$\mathbf{Z}_{t-1}(k,l) = \begin{bmatrix} Z_{t-1}^{cc}(k,l) & -Z_{t-1}^{cs}(k,l) & -Z_{t-1}^{sc}(k,l) & Z_{t-1}^{ss}(k,l) \\ Z_{t-1}^{cs}(k,l) & Z_{t-1}^{cc}(k,l) & -Z_{t-1}^{ss}(k,l) & -Z_{t-1}^{sc}(k,l) \\ Z_{t-1}^{sc}(k,l) & -Z_{t-1}^{ss}(k,l) & Z_{t-1}^{cc}(k,l) & -Z_{t-1}^{cs}(k,l) \\ Z_{t-1}^{ss}(k,l) & Z_{t-1}^{sc}(k,l) & Z_{t-1}^{cs}(k,l) & Z_{t-1}^{cc}(k,l) \end{bmatrix}, \tag{9.24}$$

$$\vec{\theta}(k,l) = \begin{bmatrix} g_{m_u m_v}^{CC}(k,l) \\ g_{m_u m_v}^{CS}(k,l) \\ g_{m_u m_v}^{SC}(k,l) \\ g_{m_u m_v}^{SS}(k,l) \end{bmatrix} = \begin{bmatrix} \cos\frac{k\pi}{N}(m_u+0.5)\cos\frac{l\pi}{N}(m_v+0.5) \\ \cos\frac{k\pi}{N}(m_u+0.5)\sin\frac{l\pi}{N}(m_v+0.5) \\ \sin\frac{k\pi}{N}(m_u+0.5)\cos\frac{l\pi}{N}(m_v+0.5) \\ \sin\frac{k\pi}{N}(m_u+0.5)\sin\frac{l\pi}{N}(m_v+0.5) \end{bmatrix}, \tag{9.25}$$

$$\vec{\mathbf{x}}_t(k,l) = \begin{bmatrix} X_t^{cc}(k,l) & X_t^{cs}(k,l) & X_t^{sc}(k,l) & X_t^{ss}(k,l) \end{bmatrix}^T. \tag{9.26}$$

Here $\mathbf{Z}_{t-1}(k,l) \in R^{4\times4}$ is the *system matrix* of the DXT-ME algorithm at (k,l). It can be easily shown that $\mathbf{Z}_{t-1}(k,l) \in R^{4\times4}$ is a unitary matrix [1]. At the boundaries of each block in the transform domain, the DCT coefficients of $x_{t-1}(m,n)$ and $x_t(m,n)$ have a much simpler one-dimensional relationship [5].

In a two-dimensional space, an object may move in four possible directions: northeast (NE: $m_u > 0, m_v > 0$), northwest (NW: $m_u < 0, m_v > 0$), southeast (SE: $m_u > 0, m_v < 0$), and southwest (SW: $m_u < 0, m_v < 0$). As explained in Section 9.3, the orthogonal equation for the DST-II kernel in (9.8) can be applied to the pseudo-phase $\hat{g}_m^s(k)$ to determine the sign of m

(i.e., the direction of the shift). In order to detect the signs of both m_u and m_v (or equivalently the direction of motion), it becomes obvious from the observation in the one-dimensional case that it is necessary to compute the pseudo-phases $\hat{g}^{SC}_{m_u m_v}(\cdot,\cdot)$ and $\hat{g}^{CS}_{m_u m_v}(\cdot,\cdot)$ so that the signs of m_u and m_v can be determined from $\hat{g}^{SC}_{m_u m_v}(\cdot,\cdot)$ and $\hat{g}^{CS}_{m_u m_v}(\cdot,\cdot)$, respectively. Taking the block boundary equations into consideration, we define two pseudo-phase functions as follows:

$$f_{m_u m_v}(k,l) = \begin{cases} \hat{g}^{CS}_{m_u m_v}(k,l), & \text{for } k,l \in \mathcal{N}, \\ \frac{1}{\sqrt{2}} \frac{Z^{cc}_{t-1}(k,l)X^{cs}_t(k,l) - Z^{cs}_{t-1}(k,l)X^{cc}_t(k,l)}{(Z^{cc}_{t-1}(k,l))^2 + (Z^{cs}_{t-1}(k,l))^2}, & \text{for } k=0, l \in \mathcal{N}, \\ \frac{1}{\sqrt{2}} \frac{Z^{cc}_{t-1}(k,l)X^{cs}_t(k,l) + Z^{sc}_{t-1}(k,l)X^{ss}_t(k,l)}{(Z^{cc}_{t-1}(k,l))^2 + (Z^{sc}_{t-1}(k,l))^2}, & \text{for } l=N, k \in \mathcal{N}, \\ \frac{1}{2} \frac{X^{cs}_t(k,l)}{Z^{cc}_{t-1}(k,l)}, & \text{for } k=0, l=N \end{cases} \tag{9.27}$$

$$g_{m_u m_v}(k,l) = \begin{cases} \hat{g}^{SC}_{m_u m_v}(k,l), & \text{for } k,l \in \mathcal{N}, \\ \frac{1}{\sqrt{2}} \frac{Z^{cc}_{t-1}(k,l)X^{sc}_t(k,l) - Z^{sc}_{t-1}(k,l)X^{cc}_t(k,l)}{(Z^{cc}_{t-1}(k,l))^2 + (Z^{sc}_{t-1}(k,l))^2}, & \text{for } l=0, k \in \mathcal{N}, \\ \frac{1}{\sqrt{2}} \frac{Z^{cc}_{t-1}(k,l)X^{sc}_t(k,l) + Z^{cs}_{t-1}(k,l)X^{ss}_t(k,l)}{(Z^{cc}_{t-1}(k,l))^2 + (Z^{cs}_{t-1}(k,l))^2}, & \text{for } k=N, l \in \mathcal{N}, \\ \frac{1}{2} \frac{X^{sc}_t(k,l)}{Z^{cc}_{t-1}(k,l)}, & \text{for } k=N, l=0 \end{cases} \tag{9.28}$$

In the computation of $f_{m_u m_v}(k,l)$ and $g_{m_u m_v}(k,l)$, if the absolute computed value is greater than 1, then this value is ill conditioned and should be discarded. This ill-conditioned situation occurs when the denominator in (9.27) and (9.28) is close to zero in comparison to the finite machine precision or set to zero after the quantization step in the feedback loop of the encoder, as shown in Figure 9.2. Due to the fact that neighboring image pixels are highly correlated, the high-frequency DCT coefficients of an image tend to be very small and can be regarded as zero after the quantization step, but the low-frequency DCT components usually have large values. Therefore, the ill-conditioned situation happens more likely when k and l are both large. It is desirable to set the value of $f_{m_u m_v}(k,l)$ (or $g_{m_u m_v}(k,l)$) as close as possible to the ideal value of $f_{m_u m_v}(k,l)$ (or $g_{m_u m_v}(k,l)$) with the infinite machine precision and no quantization. Since ideally $f_{m_u m_v}(k,l) = \cos\frac{k\pi}{N}(m_u+\frac{1}{2})\sin\frac{l\pi}{N}(m_v+\frac{1}{2})$,

$$\begin{aligned} f_{m_u m_v}(k,l) = \frac{1}{2}\bigg\{ &\sin\left[\frac{l\pi}{N}\left(m_v+\frac{1}{2}\right) + \frac{k\pi}{N}\left(m_u+\frac{1}{2}\right)\right] \\ &+ \sin\left[\frac{l\pi}{N}\left(m_v+\frac{1}{2}\right) - \frac{k\pi}{N}\left(m_u+\frac{1}{2}\right)\right]\bigg\}. \end{aligned}$$

For small values of m_u and m_v (slow motion) and large k and l (high-frequency DCT components), it is likely that

$$\frac{l\pi}{N}\left(m_v+\frac{1}{2}\right) - \frac{k\pi}{N}\left(m_u+\frac{1}{2}\right) \approx 0,$$

and the first term in $f_{m_u m_v}(k,l)$ is bounded by 1. Therefore, it is likely that $|f_{m_u m_v}(k,l)| \le 0.5$. Without any other knowledge, it is reasonable to guess that $f_{m_u m_v}(k,l)$ is closer to zero than to ± 1. A similar argument follows for the case of $g_{m_u m_v}(k,l)$. Thus in our implementation, we set the corresponding variable $f_{m_u m_v}(k,l)$ or $g_{m_u m_v}(k,l)$ to be zero when the magnitudes of the computed values exceed 1. This setting for ill-conditioned computed $f_{m_u m_v}(k,l)$ and $g_{m_u m_v}(k,l)$ values is found to improve the condition of $f_{m_u m_v}(k,l)$ and $g_{m_u m_v}(k,l)$ and also the overall performance of the DXT-ME algorithm.

These two pseudo-phase functions pass through 2D-IDCT-II coders (IDCSTII and IDSCTII) to generate two functions, DCS$(\cdot,\cdot)$ and DSC$(\cdot,\cdot)$, in view of the orthogonal property of DCT-II and DST-II in (9.8) and (9.9):

$$\begin{aligned}
\mathrm{DCS}(m,n) &= \mathrm{IDCSTII}\left(f_{m_u m_v}\right) \\
&= \frac{4}{N^2}\sum_{k=0}^{N-1}\sum_{l=1}^{N} C(k)C(l) f_{m_u m_v}(k,l)\cos\frac{k\pi}{N}\left(m+\frac{1}{2}\right)\sin\frac{l\pi}{N}\left(n+\frac{1}{2}\right) \\
&= \left[\delta\left(m-m_u\right)+\delta\left(m+m_u+1\right)\right]\cdot\left[\delta\left(n-m_v\right)-\delta\left(n+m_v+1\right)\right], \qquad (9.29)\\
\mathrm{DSC}(m,n) &= \mathrm{IDSCTII}\left(g_{m_u m_v}\right) \\
&= \frac{4}{N^2}\sum_{k=1}^{N}\sum_{l=0}^{N-1} C(k)C(l) g_{m_u m_v}(k,l)\sin\frac{k\pi}{N}\left(m+\frac{1}{2}\right)\cos\frac{l\pi}{N}\left(n+\frac{1}{2}\right) \\
&= \left[\delta\left(m-m_u\right)-\delta\left(m+m_u+1\right)\right]\cdot\left[\delta\left(n-m_v\right)+\delta\left(n+m_v+1\right)\right]. \qquad (9.30)
\end{aligned}$$

By the same argument as in the one-dimensional case, the 2D-IDCT-II coders limit the observable index space $\{(i,j): i,j=0,\ldots,N-1\}$ of DCS and DSC to the first quadrant of the entire index space, shown as gray regions in Figure 9.7, which depicts (9.29) and (9.30). Similar to the one-dimensional case, if m_u is positive, the observable peak value of $\mathrm{DSC}(m,n)$ will be positive regardless of the sign of m_v since $\mathrm{DSC}(m,n)=\delta(m-m_u)\cdot[\delta(n-m_v)+\delta(n+m_v+1)]$ in the observable index space. Likewise, if m_u is negative, the observable peak value of $\mathrm{DSC}(m,n)$ will be negative because $\mathrm{DSC}(m,n)=\delta(m+m_u+1)\cdot[\delta(n-m_v)+\delta(n+m_v+1)]$ in the gray region. As a result, the sign of the observable peak value of DSC determines the sign of m_u. The same reasoning may apply to DCS in the determination of the sign of m_v. The estimated displacement, $\hat{d}=(\hat{m}_u,\hat{m}_v)$, can thus be found by locating the peaks of DCS and DSC over $\{0,\ldots,N-1\}^2$ or over an index range of interest, usually $\Phi=\{0,\ldots,N/2\}^2$ for slow motion. How the peak signs determine the direction of movement is summarized in Table 9.2. Once the direction is found, $\hat{d}$ can be estimated accordingly:

Table 9.2 Determination of Direction of Movement (m_u, m_v) from the Signs of DSC and DCS

Sign of DSC Peak	sign of DCS Peak	Peak Index	Motion Direction
+	+	(m_u, m_v)	Northeast
+	−	$(m_u, -(m_v+1))$	Southeast
−	+	$(-(m_u+1), m_v)$	Northwest
−	−	$(-(m_u+1), -(m_v+1))$	Southwest

$$\hat{m}_u = \begin{cases} i_{\mathrm{DSC}} = i_{\mathrm{DCS}}, & \text{if } \mathrm{DSC}\left(i_{\mathrm{DSC}}, j_{\mathrm{DSC}}\right) > 0, \\ -\left(i_{\mathrm{DSC}}+1\right) = -\left(i_{\mathrm{DCS}}+1\right), & \text{if } \mathrm{DSC}\left(i_{\mathrm{DSC}}, j_{\mathrm{DSC}}\right) < 0, \end{cases} \qquad (9.31)$$

$$\hat{m}_v = \begin{cases} j_{\mathrm{DCS}} = j_{\mathrm{DSC}}, & \text{if } \mathrm{DCS}\left(i_{\mathrm{DCS}}, j_{\mathrm{DCS}}\right) > 0, \\ -\left(j_{\mathrm{DCS}}+1\right) = -\left(j_{\mathrm{DSC}}+1\right), & \text{if } \mathrm{DCS}\left(i_{\mathrm{DCS}}, j_{\mathrm{DCS}}\right) < 0, \end{cases} \qquad (9.32)$$

where

$$\left(i_{\mathrm{DCS}}, j_{\mathrm{DCS}}\right) = \arg\max_{m,n\in\Phi}\left|\mathrm{DCS}(m,n)\right|, \qquad (9.33)$$

$$\left(i_{\mathrm{DSC}}, j_{\mathrm{DSC}}\right) = \arg\max_{m,n\in\Phi}\left|\mathrm{DSC}(m,n)\right|. \qquad (9.34)$$

Normally, these two peak indices are consistent, but in noisy circumstances they may not agree. In this case, an arbitration rule must be made to pick the best index (i_D, j_D) in terms

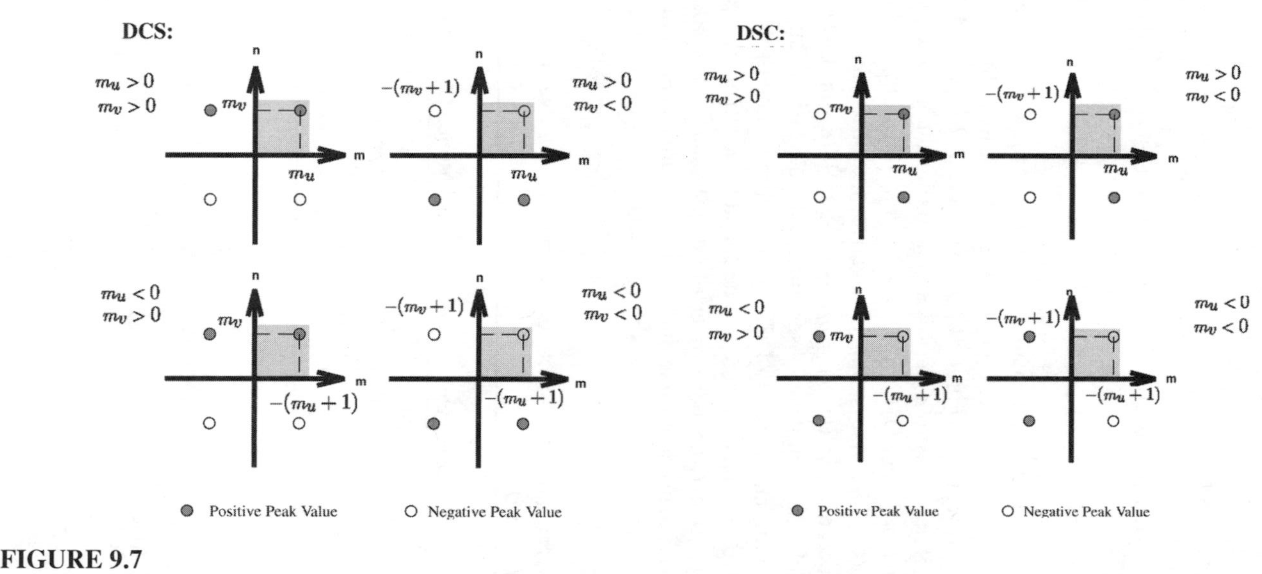

FIGURE 9.7
How the direction of motion is determined based on the sign of the peak value. (a) From DCS, (b) from DSC.

of minimum nonpeak-to-peak ratio (NPR):

$$(i_D, j_D) = \begin{cases} (i_{\mathrm{DSC}}, j_{\mathrm{DSC}}) & \text{if } \mathrm{NPR(DSC)} < \mathrm{NPR(DCS)}, \\ (i_{\mathrm{DCS}}, j_{\mathrm{DCS}}) & \text{if } \mathrm{NPR(DSC)} > \mathrm{NPR(DCS)}. \end{cases} \tag{9.35}$$

This index (i_D, j_D) will then be used to determine $\hat{d}$ by (9.31) and (9.32). Here NPR is defined as the ratio of the average of all absolute nonpeak values to the absolute peak value. Thus, $0 \leq \mathrm{NPR} \leq 1$, and for a pure impulse function, $\mathrm{NPR} = 0$. Such an approach to choose the best index among the two indices is found empirically to improve the noise immunity of this estimation algorithm.

In situations where slow motion is preferred, it is better to search the peak value in a zigzag way, as is widely done in DCT-based hybrid video coding [12, 13]. Starting from the index (0, 0), zigzagly scan all the DCS (or DSC) values and mark the point as the new peak index if the value at that point (i, j) is larger than the current peak value by more than a preset threshold θ:

$$(i_{\mathrm{DCS}}, j_{\mathrm{DCS}}) = (i, j) \text{ if } \mathrm{DCS}(i, j) > \mathrm{DCS}(i_{\mathrm{DCS}}, j_{\mathrm{DCS}}) + \theta, \tag{9.36}$$

$$(i_{\mathrm{DSC}}, j_{\mathrm{DSC}}) = (i, j) \text{ if } \mathrm{DSC}(i, j) > \mathrm{DSC}(i_{\mathrm{DSC}}, j_{\mathrm{DSC}}) + \theta. \tag{9.37}$$

In this way, large spurious spikes at the higher index points will not affect the performance and thus improve its noise immunity further. If there is no presence of slow motion in a fast-moving picture, then simply no slow motion is preferred and the estimator will be able to find a peak at the high-frequency region (i.e., large motion vector).

Figure 9.8 demonstrates the DXT-ME algorithm. Images of a rectangularly shaped moving object with arbitrary texture are generated (Figure 9.8a) and corrupted by additive white Gaussian noise at SNR = 10 dB (Figure 9.8b). The resulting pseudo-phase functions f and g, as well as DCS and DSC, are depicted in Figure 9.8c and d, correspondingly. Large peaks can be seen clearly in Figure 9.8d on rough surfaces caused by noise in spite of noisy input images. The positions of these peaks give us an accurate motion estimate (5, −3).

9.4.2 Computational Issues and Complexity

The block diagram in Figure 9.6a shows that a separate 2D-DCT-I is needed in addition to the standard DCT (2D-DCT-II). This is undesirable from the complexity viewpoint. However, this problem can be circumvented by considering the point-to-point relationship between the 2D-DCT-I and 2D-DCT-II coefficients in the frequency domain for $k, l \in \mathcal{N}$:

$$\begin{bmatrix} Z_{t-1}^{cc}(k,l) \\ Z_{t-1}^{cs}(k,l) \\ Z_{t-1}^{sc}(k,l) \\ Z_{t-1}^{ss}(k,l) \end{bmatrix} = \begin{bmatrix} +\cos\frac{k\pi}{2N}\cos\frac{l\pi}{2N} & +\cos\frac{k\pi}{2N}\sin\frac{l\pi}{2N} & +\sin\frac{k\pi}{2N}\cos\frac{l\pi}{2N} & +\sin\frac{k\pi}{2N}\sin\frac{l\pi}{2N} \\ -\cos\frac{k\pi}{2N}\sin\frac{l\pi}{2N} & +\cos\frac{k\pi}{2N}\cos\frac{l\pi}{2N} & -\sin\frac{k\pi}{2N}\sin\frac{l\pi}{2N} & +\sin\frac{k\pi}{2N}\cos\frac{l\pi}{2N} \\ -\sin\frac{k\pi}{2N}\cos\frac{l\pi}{2N} & -\sin\frac{k\pi}{2N}\sin\frac{l\pi}{2N} & +\cos\frac{k\pi}{2N}\cos\frac{l\pi}{2N} & +\cos\frac{k\pi}{2N}\sin\frac{l\pi}{2N} \\ +\sin\frac{k\pi}{2N}\sin\frac{l\pi}{2N} & -\sin\frac{k\pi}{2N}\cos\frac{l\pi}{2N} & -\cos\frac{k\pi}{2N}\sin\frac{l\pi}{2N} & +\cos\frac{k\pi}{2N}\cos\frac{l\pi}{2N} \end{bmatrix} \times \begin{bmatrix} X_{t-1}^{cc}(k,l) \\ X_{t-1}^{cs}(k,l) \\ X_{t-1}^{sc}(k,l) \\ X_{t-1}^{ss}(k,l) \end{bmatrix} \tag{9.38}$$

where X_{t-1}^{cc}, X_{t-1}^{cs}, X_{t-1}^{sc}, and X_{t-1}^{ss} are the 2D-DCT-II coefficients of the previous frame. A similar relation exists for the coefficients at the block boundaries. This observation results in the simple structure in Figure 9.6b, where block T is a coefficient transformation unit realizing (9.38).

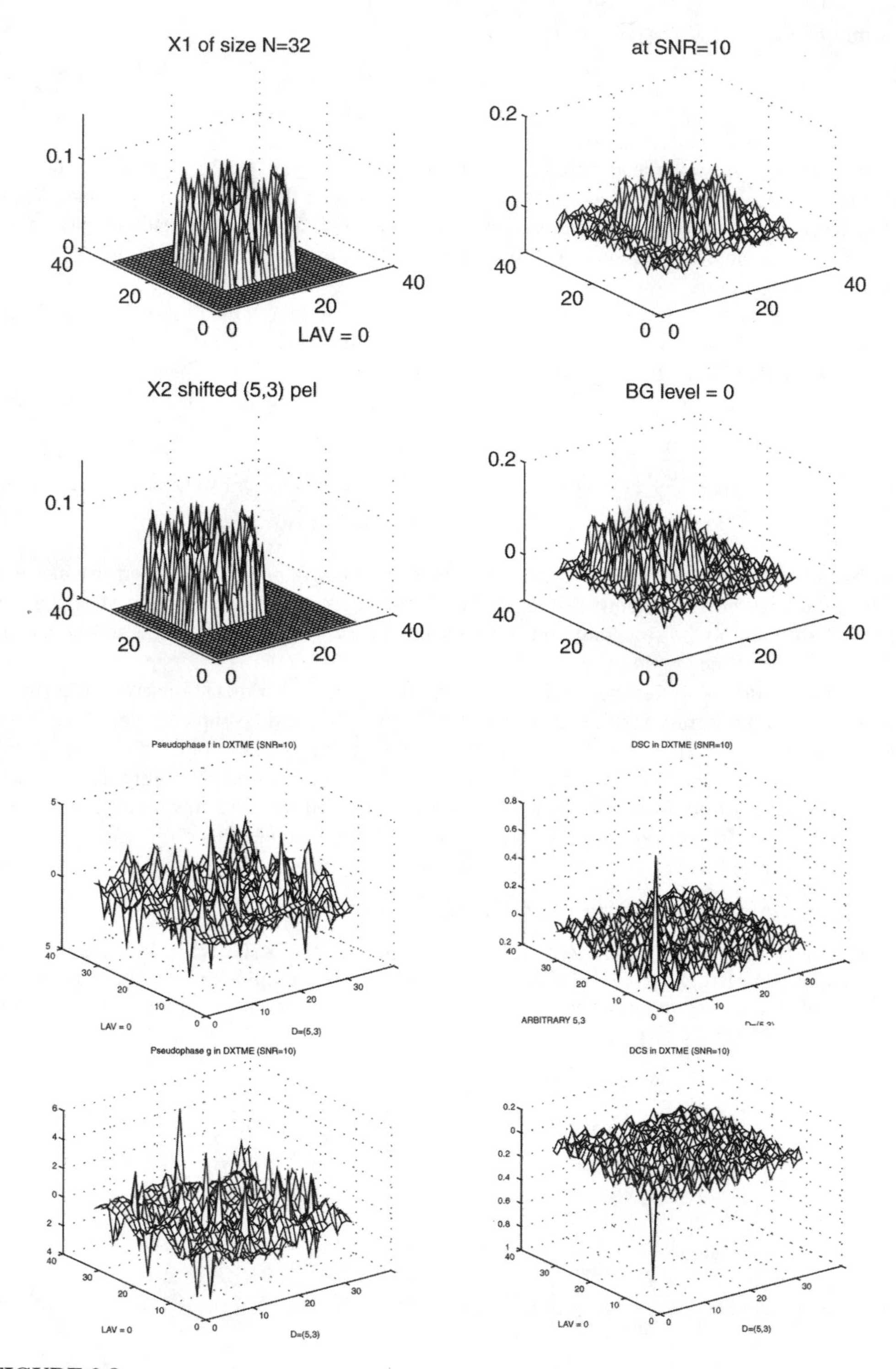

FIGURE 9.8

DXT-ME performed on the images of an object moving in the direction (5, −3) with additive white Gaussian noise at SNR = 10 dB. (a) Original inputs x_1 and x_2; (b) noise added; (c) f and g; (d) DSC and DCS.

Table 9.3 Computational Complexity of Each Stage in DXT-ME

Stage	Component	Computational Complexity
1	2D-DCT-II	$O_{dct} = O(N)$
	Coefficient transformation unit (T)	$O(N^2)$
2	Pseudo-phase computation	$O(N^2)$
3	2D-IDCT-II	$O_{dct} = O(N)$
4	Peak searching	$O(N^2)$
	Estimation	$O(1)$

If the DCT has computational complexity O_{dct}, the overall complexity of DXT-ME is $O(N^2) + O_{dct}$ with the complexity of each component summarized in Table 9.3. The computational complexity of the pseudo-phase computation component is only $O(N^2)$ for an $N \times N$ block and so is the unit to determine the displacement. For the computation of the pseudo-phase functions $f(\cdot, \cdot)$ in (9.27) and $g(\cdot, \cdot)$ in (9.28), the DSCT, DCST, and DSST coefficients (regarded as DST coefficients) must be calculated in addition to the DCCT coefficients (i.e., the usual 2D DCT). However, all these coefficients can be generated with little overhead in the course of computing 2D DCT coefficients. As a matter of fact, a parallel and fully pipelined 2D DCT lattice structure has been developed [9]–[11] to generate 2D DCT coefficients at a cost of $O(N)$ operations. This DCT coder computes DCT and DST coefficients dually due to its internal lattice architecture. These internally generated DST coefficients can be output to the DXT-ME module for pseudo-phase computation. This same lattice structure can also be modified as a 2D IDCT, which also has $O(N)$ complexity. To sum up, the computational complexity of this DXT-ME is only $O(N^2)$, much lower than the $O(N^4)$ complexity of BKM-ME.

Calculation of the actual number of computations, other than asymptotic complexity, requires the knowledge of specific implementations. In DCT-based motion-compensated video coding, DCT, IDCT, and peak searching are required; therefore, we will count only the number of operations required in the pseudo-phase computation. At each pixel position, we need to solve a 4×4 linear equation by means of the Gauss elimination method with 4 divisions, 40 multiplications, and 30 additions/subtractions. Therefore, the total number of operations is 18,944 for a 16×16 block and 75,776 for a corresponding overlapped block (32×32), while the BKM-ME approach requires 130,816 additions/subtractions for block size 16×16 and search area 32×32. Still, the number of operations required by the DXT-ME algorithm is smaller than BKM-ME. Further reduction of computations can be achieved by exploiting various properties in the algorithm.

A closer look at (9.27), (9.28), and (9.38), reveals that the operations of pseudo-phase computation and coefficient transformation are performed independently at each point (k, l) in the transform domain and therefore are inherently highly parallel operations. Since most of the operations in the DXT-ME algorithm involve mainly pseudo-phase computations and coefficient transformations in addition to DCT and IDCT operations, which have been studied extensively, the DXT-ME algorithm can easily be implemented on highly parallel array processors or dedicated circuits. This is very different from BKM-ME, which requires shifting of pixels and summation of differences of pixel values and hence discourages parallel implementation.

9.4.3 Preprocessing

For complicated video sequences in which objects may move across the border of blocks in a nonuniform background, preprocessing can be employed to enhance the features of moving objects and avoid violation of the assumption made for DXT-ME before feeding the images into the DXT-ME algorithm. Intuitively speaking, the DXT-ME algorithm tries to match the features of any object on two consecutive frames so that any translation motion can be estimated regardless of the shape and texture of the object as long as these two frames contain the significant energy level of the object features. Due to the matching property of the DXT-ME algorithm, effective preprocessing will improve the performance of motion estimation if preprocessing can enhance the object features in the original sequence. In order to keep the computational complexity of the overall motion estimator low, the chosen preprocessing function must be simple but effective in the sense that unwanted features will not affect the accuracy of estimation. Our study found that both edge extraction and frame differentiation are simple and effective schemes for extraction of motion information.

It is found that estimating the motion of an object from its edges is equivalent to estimating from its image projection [14]. Furthermore, since the DXT-ME algorithm assumes that an object moves within the block boundary in a completely dark environment, its edge information reduces the adverse effect of the object moving across the block boundary on the estimation accuracy. The other advantage of edge extraction is that any change in the illumination condition does not alter the edge information and in turn makes no false motion estimates by the DXT-ME algorithm. Since we only intend to extract the main features of moving objects while keeping the overall complexity low, we employ a very simple edge detection by convolving horizontal and vertical Sobel operators of size 3×3 with the image to obtain horizontal and vertical gradients, respectively, and then combine both gradients by taking the square root of the sum of the squares of both gradients [15]. Edge detection provides us the features of moving objects but also the features of the background (stationary objects), which is undesirable. However, if the features of the background have smaller energy than those of moving objects within every block containing moving objects, then the background features will not affect the performance of DXT-ME. The computational complexity of this preprocessing step is only $O(N^2)$ and thus the overall computational complexity is still $O(N^2)$.

Frame differentiation generates an image of the difference of two consecutive frames. This frame-differentiated image contains no background objects but the difference of moving objects between two frames. The DXT-ME estimator operates directly on this frame-differentiated sequence to predict motion in the original sequence. The estimate will be good if the moving objects are moving constantly in one direction in three consecutive frames. For 30 frames per second, the standard NTSC frame rate, objects can usually be viewed as moving at a constant speed in three consecutive frames. Obviously, this step also has only $O(N^2)$ computational complexity.

9.4.4 Adaptive Overlapping Approach

For fair comparison with BKM-ME, which has a larger search area than the block size, we adopt the adaptive overlapping approach to enlarge adaptively the block area. The enlargement of the block size diminishes the boundary effect that occurs when the displacement is very large compared to the block size. As a result, the moving objects may move partially or completely out of the block, making the contents in two temporally consecutive blocks very different. However, this problem also exists for other motion estimation algorithms. That is why we need to assume that objects in the scene are moving slowly. For rapid motion, it is difficult to track motion.

Earlier in this section we mentioned that we search for peaks of DSC and DCS over a fixed index range of interest $\Phi = \{0, \ldots, N/2\}^2$. However, if we follow the partitioning approach used in BKM-ME, then we may dynamically adjust Φ. At first, partition the whole current frame into $bs \times bs$ nonoverlapping reference blocks, shown as the shaded area in Figure 9.9a. Each reference block is associated with a larger search area (of size sa) in the previous frame

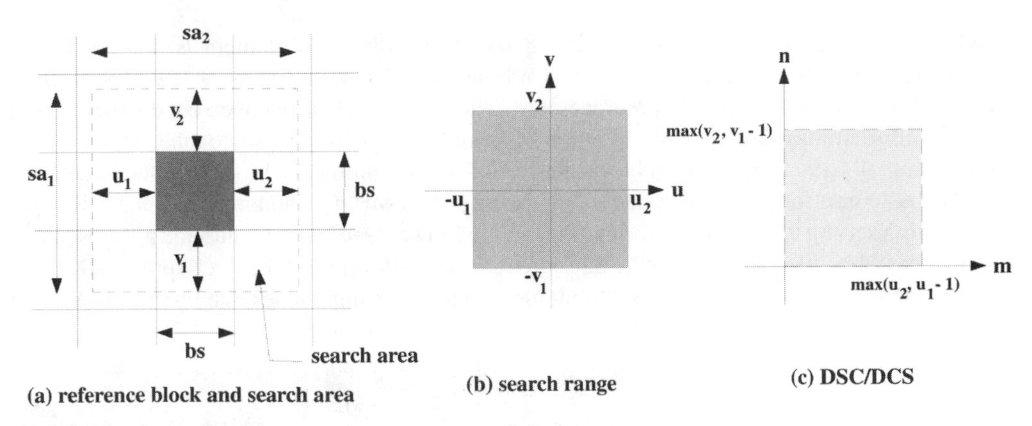

(a) reference block and search area (b) search range (c) DSC/DCS

FIGURE 9.9
Adaptive overlapping approach.

(the dotted region in the same figure) in the same way as for BKM-ME. From the position of a reference block and its associated search area, a search range $\mathcal{D} = \{(u, v) : -u_1 \leq u \leq u_2, \ -v_1 \leq v \leq v_2\}$ can then be determined as in Figure 9.9b. In contrast to BKM-ME, DXT-ME requires that the reference block size and the search area size be equal. Thus, instead of using the reference block, we use the block of the same size and position in the current frame as the search area of the previous frame. The peak values of DSC and DCS are searched in a zigzag way over this index range, $\Phi = \{0, \ldots, \max(u_2, u_1 - 1)\} \times \{0, \ldots, \max(v_2, v_1 - 1)\}$. In addition to the requirement that the new peak value be larger than the current peak value by a preset threshold, it is necessary to examine if the motion estimate determined by the new peak index lies in the search region $\mathcal{D}$. Since search areas overlap one another, the DXT-ME architecture utilizing this approach is called overlapping DXT-ME. Even though the block size required by the overlapping DXT-ME algorithm is larger than the block size for one DCT block, it is still possible to estimate motion completely in the DCT domain without going back to the spatial domain by concatenating neighboring DCT blocks directly in the DCT domain [16].

9.4.5 Simulation Results

A number of video sequences with different characteristics are used in our simulations to compare the performance of the DXT-ME algorithm [1, 5] with the full-search block-matching method (BKM-ME or BKM for the sake of brevity) as well as three commonly used fast-search block-matching approaches such as the logarithmic search method (LOG), the three-step search method (TSS), and the subsampled search approach (SUB) [17]. The performance of different schemes is evaluated and compared in terms of MSE (mean squared error per pel) and BPS (bits per sample) where MSE $= \frac{\sum_{m,n}[\hat{x}(m,n)-x(m,n)]^2}{N^2}$ and BPS is the ratio of the total number of bits required for each motion-compensated residual frame in JPEG format (BPS) converted by the image format conversion program ALCHEMY with quality = 32 to the number of pixels. As is widely used in the literature of video coding, all the block-matching methods adopt the conventional mean absolute difference (MAD) optimization criterion:

$$\hat{d} = (\hat{u}, \hat{v}) = arg \min_{(u,v) \in \mathsf{S}} \frac{\sum_{m,n} |x_2(m,n) - x_1(m-u, n-v)|}{N^2},$$

where S denotes the set of allowable displacements depending on which block-matching approach is in use.

The first sequence is the Flower Garden sequence, where the camera is moving before a big tree and a flower garden in front of a house, as shown in Figure 9.10a. Each frame has 352 × 224 pixels. Simple preprocessing is applied to this sequence: edge extraction or frame differentiation as depicted in Figure 9.10b and c, respectively. Since macroblocks, each consisting of 16 × 16 luminance blocks and two 8 × 8 chrominance blocks, are considered to be the basic unit for motion estimation/compensation in MPEG standards [13], the following simulation setting is adopted for simulations on the Flower Garden sequence and all subsequent sequences: 16 × 16 blocks on 32 × 32 search areas. Furthermore, the overlapping DXT-ME algorithm is used for fair comparison with block-matching approaches, which require a larger search area.

(a) Original (b) Edge extracted (c) Frame differentiated

FIGURE 9.10
Frame 57 in the Flower Garden sequence.

As can be seen in Figure 9.10b, the edge-extracted frames contain significant features of moving objects in the original frames so that DXT-ME can estimate the movement of the objects based on the information provided by the edge-extracted frames. Because the camera is moving at a constant speed in one direction, the moving objects occupy almost the whole scene. Therefore, the background features do not interfere with the operation of DXT-ME much but still affect the overall performance of DXT-ME as compared to the frame-differentiated preprocessing approach. The frame-differentiated images of the Flower Garden sequence, one of which is shown in Figure 9.10c, have residual energy strong enough for DXT-ME to estimate the motion directly on this frame-differentiated sequence due to the constant movement of the camera.

The performances for different motion estimation schemes are plotted in Figure 9.11 and summarized in Table 9.4 where the MSE and BPS values of different motion estimation approaches are averaged over the whole sequence from frame 3 to frame 99 for easy comparison. It should be noted that the MSE difference in Table 9.4 is the difference of the MSE value of the corresponding motion estimation scheme from the MSE value of the full-search block-matching approach (BKM), and the MSE ratio is the ratio of the MSE difference to the MSE of BKM. As indicated in the performance summary table (Table 9.5), the frame-differentiated DXT-ME algorithm is 28.9% worse in terms of MSE than the full-search block-matching approach, whereas the edge-extracted DXT-ME algorithm is 36.0% worse. Surprisingly, even though the fast-search block-matching algorithms (only 12.6% worse than BKM), TSS and LOG, have smaller MSE values than the DXT-ME algorithm, TSS and LOG have larger BPS values than the DXT-ME algorithm, as can clearly be seen in Table 9.4 and Figure 9.11. In other words, the motion-compensated residual frames generated by TSS and LOG require more bits than the DXT-ME algorithm to transmit/store after compression. This indicates that the DXT-ME algorithm is better than the logarithmic and three-step fast-search block-matching approaches for this Flower Garden sequence.

Another simulation is done on the Infrared Car sequence, which has the frame size 96 × 112 and one major moving object — the car moving along a curved road toward the camera fixed on the ground. In the performance summary table, Table 9.5, the frame-differentiated DXT-ME

Table 9.4 Performance Summary of the Overlapping DXT-ME Algorithm with Either Frame Differentiation or Edge Extraction as Preprocessing Against Full Search and Fast Search Block-Matching Approaches (BKM, TSS, LOG, SUB) over the Flower Garden Sequence

Approach	MSE	MSE Difference	MSE Ratio	BPF	BPS	BPS Ratio
BKM	127.021	0.000	0%	63726	0.808	0%
Frame-differentiated DXT-ME	163.712	36.691	28.9%	67557	0.857	6.0%
Edge-extracted DXT-ME	172.686	45.665	36.0%	68091	0.864	6.8%
TSS	143.046	16.025	12.6%	68740	0.872	7.9%
LOG	143.048	16.026	12.6%	68739	0.872	7.9%
SUB	127.913	0.892	0.7%	63767	0.809	1%

Note: MSE difference is the difference from the MSE value of full-search block-matching method (BKM), and MSE ratio is the ratio of MSE difference to the MSE of BKM.

Table 9.5 Performance Summary of the Overlapping DXT-ME Algorithm with Either Frame Differentiation or Edge Extraction as Preprocessing Against Full Search and Fast-Search Block-Matching Approaches (BKM, TSS, LOG, SUB) over the Infrared Car Sequence

Approach	MSE	MSE Difference	MSE Ratio	BPF	BPS	BPS Ratio
BKM	67.902	0.000	0%	10156	0.945	0%
Frame-differentiated DXT-ME	68.355	0.453	0.7%	10150	0.944	−0.1%
Edge-extracted DXT-ME	72.518	4.615	6.8%	10177	0.946	0.2%
TSS	68.108	0.206	0.3%	10159	0.945	0.0%
LOG	68.108	0.206	0.3%	10159	0.945	0.0%
SUB	68.493	0.591	0.9%	10159	0.945	0.0%

result is better than the SUB result in terms of MSE values and better than the full-search BKM result in terms of BPS values.

9.5 Subpixel DCT Pseudo-Phase Techniques

To further improve the compression rate, motion estimation with subpixel accuracy is essential because movements in a video sequence are not necessarily multiples of the sampling grid distance in the rectangular sampling grid of a camera. It is shown that significant improvement of coding gain can be obtained with motion estimation of half-pixel or finer accuracy [18]. Further investigation reveals that the temporal prediction error variance is generally decreased by subpixel motion compensation, but beyond a certain "critical accuracy" the possibility of further improving prediction by more accurate motion compensation is small [19]. As suggested in [18, 20], motion compensation with quarter-pel accuracy is sufficiently accurate for broadcast TV signals, but for videophone signals, half-pel accuracy is good enough. As a result, motion compensation with half-pel accuracy is recommended in MPEG standards [13, 21].

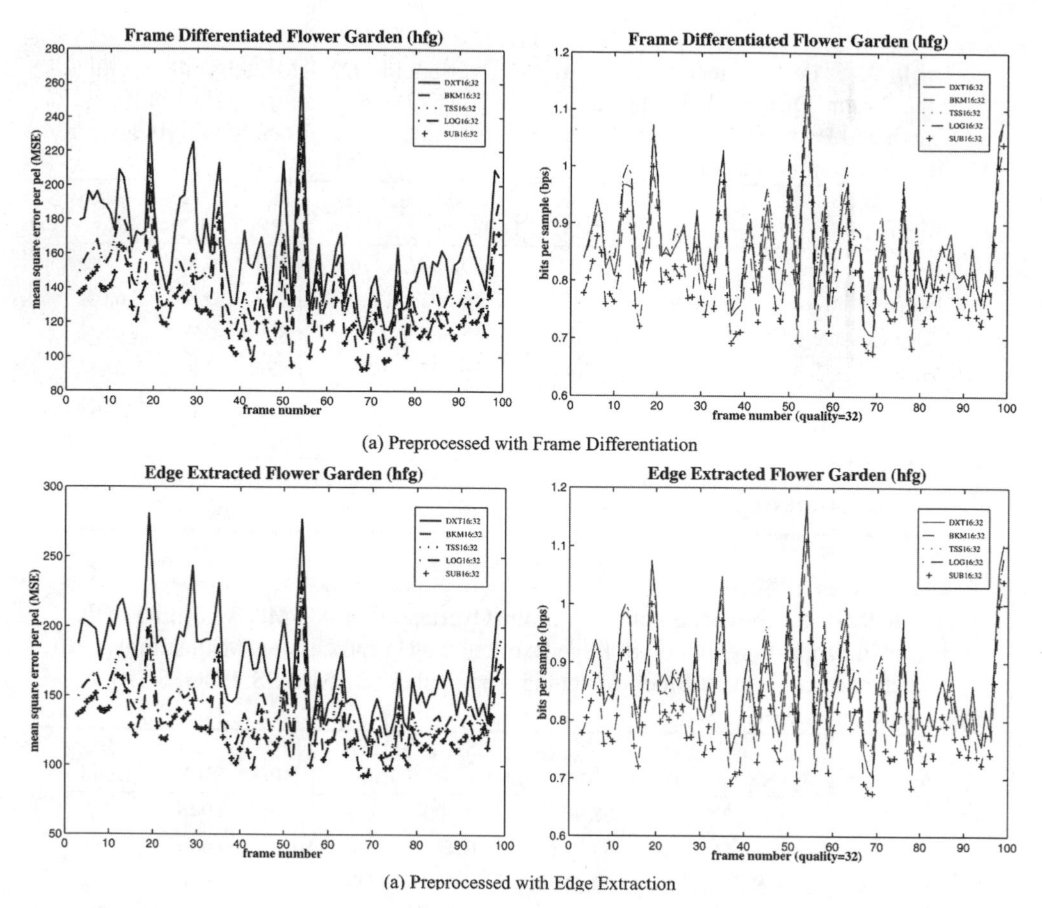

(a) Preprocessed with Frame Differentiation

(a) Preprocessed with Edge Extraction

FIGURE 9.11
Comparison of overlapping DXT-ME with block-matching approaches on the Flower Garden sequence.

Implementations of half-pel motion estimation now exist [22]–[24]. However, many of these implementations are based on the block-matching approach [20, 25, 26], which requires the interpolation of images through bilinear or other interpolation methods [27]. However, interpolation not only increases the complexity and data flow of a coder but also may adversely affect the accuracy of motion estimates from the interpolated images [20]. It is more desirable that subpixel accuracy of motion estimates be obtained without interpolating the images at a low computational cost in the DCT domain so that seamless integration of the motion compensation unit with the spatial compression unit is possible.

In this section, we extend the DCT pseudo-phase techniques discussed in Section 9.3 to the subpixel level and show that if the spatial sampling of images satisfies the Nyquist criterion, the subpixel motion information is preserved in the pseudo-phases of DCT coefficients of moving images. Furthermore, it can be shown that with appropriate modification, the sinusoidal orthogonal principles can still be applicable except that an impulse function is replaced by a sinc function whose peak position reveals subpixel displacement. Therefore, exact subpixel motion displacement can be obtained without the use of interpolation. From these observations, we can develop a set of subpixel DCT-based motion estimation algorithms that are fully compatible with the integer-pel motion estimator, for low-complexity and high-throughput video applications.

(a) Original

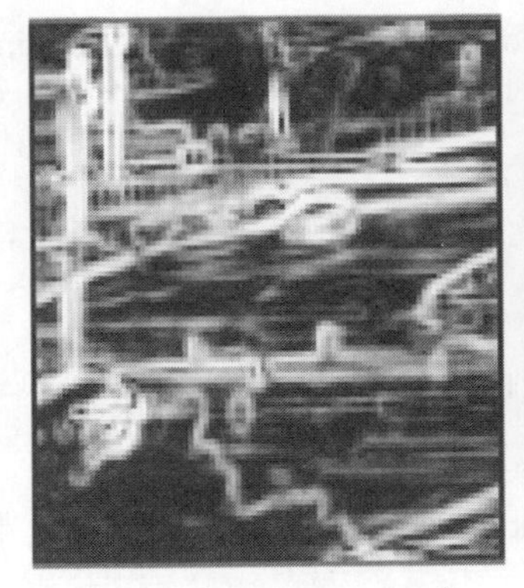

(b) Edge extracted

(c) Frame differentiated

FIGURE 9.12
Infrared Car sequence.

Without loss of generality, let us consider the one-dimensional model in which a continuous signal $x_c(t)$ and its shifted version $x_c(t-d)$ are sampled at a sampling frequency $1/T$ to generate two sample sequences $\{x_1(n) = x_c(nT)\}$ and $\{x_2(n) = x_c(nT-d)\}$, respectively. Let us define the DCT and DST coefficients as

$$X_i^C(k) \triangleq \text{DCT}\{x_i\} = \frac{2C(k)}{N}\sum_{n=0}^{N-1} x_i(n)\cos\frac{k\pi}{N}\left(n+\frac{1}{2}\right), \quad (9.39)$$

$$X_i^S(k) \triangleq \text{DST}\{x_i\} = \frac{2C(k)}{N}\sum_{n=0}^{N-1} x_i(n)\sin\frac{k\pi}{N}\left(n+\frac{1}{2}\right), \quad (9.40)$$

where $$C(k) = \begin{cases} \frac{1}{\sqrt{2}}, & \text{for } k = 0 \text{ or } N, \\ 1, & \text{otherwise}, \end{cases}$$

for $i = 1$ or 2. By using the sinusoidal relationship:

$$\cos\frac{k\pi}{N}\left(n+\frac{1}{2}\right) = \frac{1}{2}\left[e^{j\frac{k\pi}{N}\left(n+\frac{1}{2}\right)} + e^{-j\frac{k\pi}{N}\left(n+\frac{1}{2}\right)}\right], \quad (9.41)$$

$$\sin\frac{k\pi}{N}\left(n+\frac{1}{2}\right) = \frac{1}{2j}\left[e^{j\frac{k\pi}{N}\left(n+\frac{1}{2}\right)} - e^{-j\frac{k\pi}{N}\left(n+\frac{1}{2}\right)}\right], \quad (9.42)$$

we can show that the DCT/DST and DFT coefficients are related as follows:

$$X_i^C(k) = \frac{C(k)}{N}\left[\tilde{X}_i^Z(-k)e^{j\frac{k\pi}{2N}} + \tilde{X}_i^Z(k)e^{-j\frac{k\pi}{2N}}\right], \text{ for } k = 0, \ldots, N-1, \quad (9.43)$$

$$X_i^S(k) = \frac{C(k)}{jN}\left[\tilde{X}_i^Z(-k)e^{j\frac{k\pi}{2N}} - \tilde{X}_i^Z(k)e^{-j\frac{k\pi}{2N}}\right], \text{ for } k = 1, \ldots, N, \quad (9.44)$$

where $\{\tilde{X}_i^Z(k)\}$ is the DFT of the zero-padded sequence $\{x_i^Z(n)\}$ defined as

$$x_i^Z(n) = \begin{cases} x_i(n), & \text{for } n = 0, \ldots, N-1, \\ 0, & \text{for } n = N, \ldots, 2N-1, \end{cases} \quad (9.45)$$

so that

$$\tilde{X}_i^Z(k) \triangleq \text{DFT}\left\{x_i^Z\right\} = \sum_{n=0}^{N-1} x_i(n)e^{-j\frac{2k\pi n}{2N}}, \text{ for } k = 0, \ldots, 2N-1. \quad (9.46)$$

From the sampling theorem, we know that the discrete time Fourier transform (DTFT) of sequences $x_1(n)$ and $x_2(n)$ is related to the Fourier transform of $x_c(t)$, $X_c(\Omega)$, in the following way:

$$X_1(\omega) \triangleq \text{DTFT}\{x_1\} = \frac{1}{T}\sum_l X_c\left(\frac{\omega - 2\pi l}{T}\right), \tag{9.47}$$

$$X_2(\omega) \triangleq \text{DTFT}\{x_2\} = \frac{1}{T}\sum_l X_c\left(\frac{\omega - 2\pi l}{T}\right) e^{-j\left(\frac{\omega-2\pi l}{T}\right)d}. \tag{9.48}$$

Furthermore, if $X_c(\Omega)$ is bandlimited in the baseband $(-\frac{\pi}{T}, \frac{\pi}{T})$, then for $\Omega = \frac{\omega}{T} \in (-\frac{\pi}{T}, \frac{\pi}{T})$,

$$X_1(\Omega T) = \frac{1}{T}X_c(\Omega), \tag{9.49}$$

$$X_2(\Omega T) = \frac{1}{T}X_c(\Omega)\, e^{-j\Omega d}. \tag{9.50}$$

Thus, the DFT of $x_1(n)$ and $x_2(n)$ are

$$\begin{aligned}\tilde{X}_1(k) &\triangleq \text{DFT}\{x_1\} = \sum_{n=0}^{N-1} x_1(n)e^{-j\frac{2\pi kn}{N}} \\ &= X_1\left(\frac{2\pi k}{N}\right) = \frac{1}{T}X_c\left(\frac{2\pi k}{NT}\right),\end{aligned} \tag{9.51}$$

$$\begin{aligned}\tilde{X}_2(k) &\triangleq \text{DFT}\{x_2\} = \sum_{n=0}^{N-1} x_2(n)e^{-j\frac{2\pi kn}{N}} \\ &= X_2\left(\frac{2\pi k}{N}\right) = \frac{1}{T}X_c\left(\frac{2\pi k}{NT}\right) e^{-j\frac{2\pi kd}{NT}},\end{aligned} \tag{9.52}$$

whereas the DFT of $x_1^Z(n)$ and $x_2^Z(n)$ become

$$\tilde{X}_1^Z(k) = X_1\left(\frac{\pi k}{N}\right) = \frac{1}{T}X_c\left(\frac{\pi k}{NT}\right), \tag{9.53}$$

$$\tilde{X}_2^Z(k) = X_2\left(\frac{\pi k}{N}\right) = \frac{1}{T}X_c\left(\frac{\pi k}{NT}\right) e^{-j\frac{\pi kd}{NT}}. \tag{9.54}$$

Therefore,

$$X_2\left(\frac{\pi k}{N}\right) = X_1\left(\frac{\pi k}{N}\right) e^{-j\frac{\pi kd}{NT}}. \tag{9.55}$$

Substituting (9.55) back into (9.43) and (9.44), we get

$$X_2^C(k) = \frac{C(k)}{N}\left[\tilde{X}_1^Z(-k)e^{j\frac{k\pi d}{NT}}e^{j\frac{k\pi}{2N}} + \tilde{X}_1^Z(k)e^{-j\frac{k\pi d}{NT}}e^{-j\frac{k\pi}{2N}}\right], \quad \text{for } k = 0, \ldots, N-1, \tag{9.56}$$

$$X_2^S(k) = \frac{C(k)}{jN}\left[\tilde{X}_1^Z(-k)e^{j\frac{k\pi d}{NT}}e^{j\frac{k\pi}{2N}} - \tilde{X}_1^Z(k)e^{-j\frac{k\pi d}{NT}}e^{-j\frac{k\pi}{2N}}\right], \quad \text{for } k = 1, \ldots, N. \tag{9.57}$$

Using the sinusoidal relationship in (9.42) to change natural exponents back to cosine/sine, we finally obtain the relationship between $x_1(n)$ and $x_2(n)$ in the DCT/DST domain:

$$X_2^C(k) = \frac{2C(k)}{N} \sum_{n=0}^{N-1} x_1(n) \cos \frac{k\pi}{N}\left(n + \frac{d}{T} + \frac{1}{2}\right), \text{ for } k = 0, \ldots, N-1, \tag{9.58}$$

$$X_2^S(k) = \frac{2C(k)}{N} \sum_{n=0}^{N-1} x_1(n) \sin \frac{k\pi}{N}\left(n + \frac{d}{T} + \frac{1}{2}\right), \text{ for } k = 1, \ldots, N. \tag{9.59}$$

We conclude the result in the following theorem:

THEOREM 9.1

If a continuous signal $x_c(t)$ is $\frac{\pi}{T}$ bandlimited and the sampled sequences of $x_c(t)$ and $x_c(t-d)$ are $\{x_c(nT)\}$ and $\{x_c(nT-d)\}$, respectively, then their DCT and DST are related by

$$\mathrm{DCT}\{x_c(nT-d)\} = \mathrm{DCT}_{\frac{d}{T}}\{x_c(nT)\}, \tag{9.60}$$

$$\mathrm{DST}\{x_c(nT-d)\} = \mathrm{DST}_{\frac{d}{T}}\{x_c(nT)\}, \tag{9.61}$$

where

$$\mathrm{DCT}_\alpha\{x\} \triangleq \frac{2C(k)}{N} \sum_{n=0}^{N-1} x(n) \cos \frac{k\pi}{N}\left(n + \alpha + \frac{1}{2}\right), \tag{9.62}$$

$$\mathrm{DST}_\beta\{x\} \triangleq \frac{2C(k)}{N} \sum_{n=0}^{N-1} x(n) \sin \frac{k\pi}{N}\left(n + \beta + \frac{1}{2}\right), \tag{9.63}$$

are the DCT and DST with α and β shifts in their kernels, respectively. Here d is the shift amount and T is the sampling interval, but d/T is not necessarily an integer.

9.5.1 Subpel Sinusoidal Orthogonality Principles

The sinusoidal orthogonal principles in (9.8) and (9.9) are no longer valid at the subpixel level. However, we can extend these sinusoidal orthogonal equations suitable for subpixel motion estimation.

Let's replace, in (9.8) and (9.9), the integer variables m and n by the real variables u and v and define

$$\bar{L}_c(u, v) \triangleq \sum_{k=0}^{N-1} C^2(k) \cos \frac{k\pi}{N}\left(u + \frac{1}{2}\right) \cos \frac{k\pi}{N}\left(v + \frac{1}{2}\right), \tag{9.64}$$

$$\bar{L}_s(u, v) \triangleq \sum_{k=0}^{N-1} C^2(k) \sin \frac{k\pi}{N}\left(u + \frac{1}{2}\right) \sin \frac{k\pi}{N}\left(v + \frac{1}{2}\right). \tag{9.65}$$

Since

$$\begin{aligned} \bar{L}_c(u, v) &\triangleq \sum_{k=0}^{N-1} C^2(k) \cos \frac{k\pi}{N}\left(u + \frac{1}{2}\right) \cos \frac{k\pi}{N}\left(v + \frac{1}{2}\right) \\ &= -\frac{1}{2} + \frac{1}{2}[\xi(u-v) + \xi(u+v+1)], \end{aligned}$$

and

$$\bar{L}_s(u, v) \triangleq \sum_{k=1}^{N} C^2(k) \sin \frac{k\pi}{N}\left(u + \frac{1}{2}\right) \sin \frac{k\pi}{N}\left(v + \frac{1}{2}\right)$$
$$= \frac{1}{2} \sin\left[\pi\left(u + \frac{1}{2}\right)\right] \sin\left[\pi\left(v + \frac{1}{2}\right)\right] + \frac{1}{2}[\xi(u - v) - \xi(u + v + 1)] ,$$

we show that

$$\bar{L}_c(u, v) = -\frac{1}{2} + \frac{1}{2}[\xi(u - v) + \xi(u + v + 1)] , \quad (9.66)$$

$$\bar{L}_s(u, v) = \frac{1}{2} \sin\left[\pi\left(u + \frac{1}{2}\right)\right] \sin\left[\pi\left(v + \frac{1}{2}\right)\right] + \frac{1}{2}[\xi(u - v) - \xi(u + v + 1)] , \quad (9.67)$$

where

$$\xi(x) \triangleq \sum_{k=0}^{N-1} \cos\left(\frac{k\pi}{N}x\right) = \frac{1}{2}\left[1 - \cos\pi x + \sin\pi x \cdot \frac{\cos\frac{\pi x}{2N}}{\sin\frac{\pi x}{2N}}\right] . \quad (9.68)$$

If $\frac{(\pi x)}{(2N)}$ is so small that the second and higher order terms of $\frac{(\pi x)}{(2N)}$ can be ignored, then $\cos\frac{\pi x}{2N} \approx 1$, $\sin\frac{\pi x}{2N} \approx \frac{\pi x}{2N}$. Thus,

$$\xi(x) \approx \frac{1}{2}[1 - \cos\pi x] + N\text{sinc}(x) , \quad (9.69)$$

where $\text{sinc}(x) \triangleq \sin(\pi x)/(\pi x)$. For large N, $\xi(x)$ is approximately a sinc function whose largest peak can be identified easily at $x = 0$, as depicted in Figure 9.13a, where $\xi(x)$ closely resembles $N \cdot \text{sinc}(x)$, especially when x is small. The slope of $\xi(x)$ is also plotted in Figure 9.13b, which shows the sharpness of $\xi(x)$.

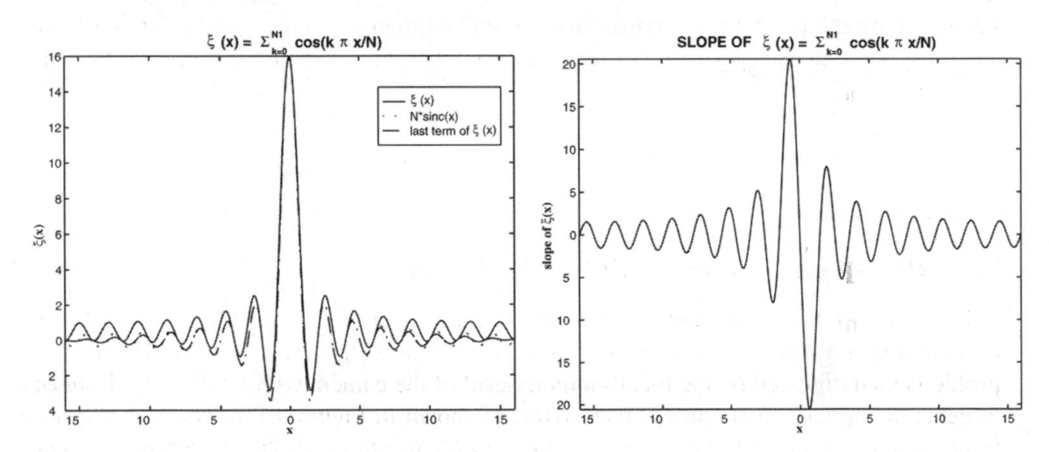

FIGURE 9.13
Plot of $\xi(x) = \sum_{k=0}^{N-1} \cos(\frac{k\pi}{N}x)$ and its slope for $N = 16$. Observe the similarity between the curves of N*sinc(x) and the last term of ξ. (a) $\xi(x)$; (b) slope of $\xi(x)$.

A closer look at (9.66) and (9.67) reveals that either $\bar{L}_c(u, v)$ or $\bar{L}_s(u, v)$ consists of ξ functions and one extra term, which is not desirable. In order to obtain a pure form of sinc functions similar to (9.8) and (9.9), we define two modified functions $L_c(u, v)$ and $L_s(u, v)$ as follows:

$$L_c(u, v) \triangleq \sum_{k=0}^{N-1} \cos\frac{k\pi}{N}\left(u+\frac{1}{2}\right)\cos\frac{k\pi}{N}\left(v+\frac{1}{2}\right), \tag{9.70}$$

$$L_s(u, v) \triangleq \sum_{k=1}^{N-1} \sin\frac{k\pi}{N}\left(u+\frac{1}{2}\right)\sin\frac{k\pi}{N}\left(v+\frac{1}{2}\right). \tag{9.71}$$

Then we can show that

$$L_c(u, v) = \frac{1}{2}[\xi(u-v) + \xi(u+v+1)], \tag{9.72}$$

$$L_s(u, v) = \frac{1}{2}[\xi(u-v) - \xi(u+v+1)]. \tag{9.73}$$

Equations (9.70)–(9.73) are the equivalent form of the sinusoidal orthogonal principles (9.8) and (9.9) at the subpixel level. The sinc functions at the right-hand side of the equations are the direct result of the rectangular window inherent in the DCT [28]. Figure 9.14a and b illustrate $L_s(x, -3.75)$ and $L_c(x, -3.75)$, respectively, where two ξ functions are interacting with each other but their peak positions clearly indicate the displacement. However, when the displacement v is small (in the neighborhood of -0.5), $\xi(u-v)$ and $\xi(u+v+1)$ move close together and addition/subtraction of $\xi(u-v)$ and $\xi(u+v+1)$ changes the shape of L_s and L_c. As a result, neither L_s nor L_c looks like two ξ functions and the peak positions of L_s and L_c are different from those of $\xi(u-v)$ and $\xi(u+v+1)$, as demonstrated in Figure 9.14c and d, respectively, where the peak positions of $L_s(x, -0.75)$ and $L_c(x, -0.75)$ are -1.25 and -0.5, differing from the true displacement -0.75. In the extreme case, $\xi(u-v)$ and $\xi(u+v+1)$ cancel out each other when the displacement is -0.5 such that $L_s(x, -0.5) \equiv 0$, as shown in Figure 9.14e.

Fortunately, we can eliminate the adverse interaction of the two ξ functions by simply adding L_c to L_s since $L_c(x, v) + L_s(x, v) = \xi(x-v)$, as depicted in Figure 9.14f, where the sum $L_c(x, -0.75) + L_s(x, -0.75)$ behaves like a sinc function and its peak position coincides with the displacement. Furthermore, due to the sharpness of this ξ function, we can accurately pinpoint the peak position under a noisy situation and in turn determine the motion estimate. This property enables us to devise flexible and scalable subpixel motion estimation algorithms in the subsequent sections.

9.6 DCT-Based Subpixel Motion Estimation

Consider a moving object casting a continuous intensity profile $I_t(u, v)$ on a camera plane of the continuous coordinate (u,v) where the subscript t denotes the frame number. This intensity profile is then digitized on the fixed sampling grid of the camera with a sampling distance d to generate the current frame of pixels $x_t(m, n)$ shown in Figure 9.15a where m and n are integers. Further assume that the displacement of the object between the frames $t-1$ and t is (d_u, d_v) such that $I_t(u, v) = I_{t-1}(u-d_u, v-d_v)$ where $d_u = (m_u + \nu_u)d = \lambda_u d$ and $d_v = (m_v + \nu_v)d = \lambda_v d$. Here m_u and m_v are the integer components of the displacement, and ν_u and $\nu_v \in [-\frac{1}{2}, \frac{1}{2}]$. Therefore,

$$x_t(m, n) = I_t(md, nd) = I_{t-1}(md - d_u, nd - d_v),$$

$$x_{t-1}(m, n) = I_{t-1}(md, nd),$$

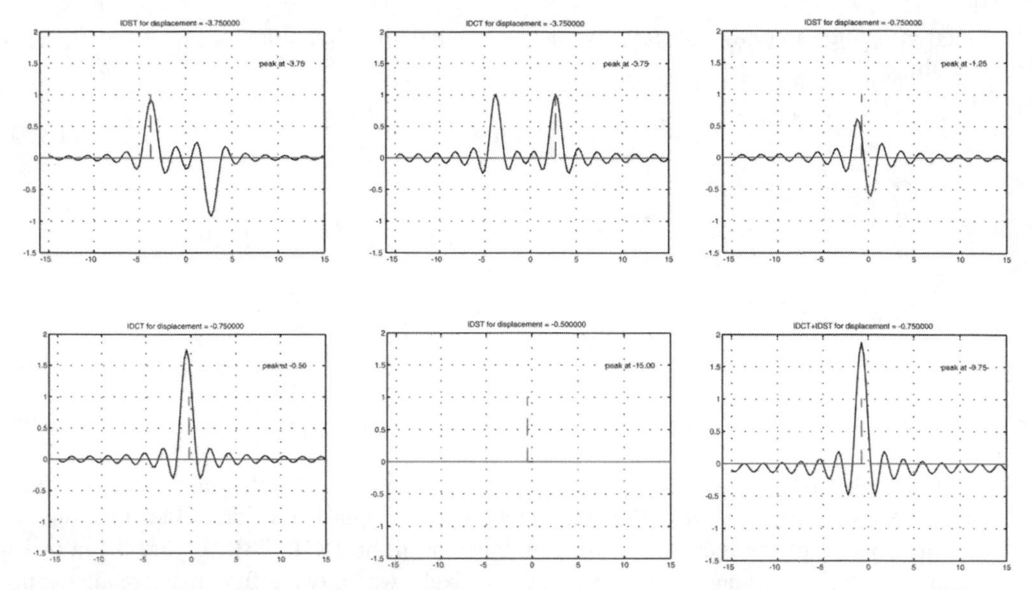

FIGURE 9.14
Illustration of sinusoidal orthogonal principles at the subpixel level for different displacements. (a) $L_s(x, -3.75)$; **(b)** $L_c(x, -3.75)$; **(c)** $L_s(x, -0.75)$; **(d)** $L_c(x, -0.75)$; **(e)** $L_s(x, -0.5)$; **(f)** $L_c(x, -0.75) + L_s(x, -0.75)$.

as in Figure 9.15b. Unlike the case of integer-pel movement, the displacement is not necessarily multiples of the sampling distance d. In other words, ν_u and ν_v do not necessarily equal zero.

For integer-pel displacements (i.e., $\lambda_u = m_u$ and $\lambda_v = m_v$), the pseudo-phases are computed by solving the *pseudo-phase motion equation* at (k, l):

$$\mathbf{Z}_{t-1}(k, l) \cdot \vec{\theta}_{m_u, m_v}(k, l) = \vec{\mathbf{x}}_t(k, l), \quad \text{for } k, l \in \mathcal{N} \tag{9.74}$$

where $\mathcal{N} = \{1, \ldots, N-1\}$, $\vec{\theta}_{m_u, m_v}$ is the pseudo-phase vector, and the 4×4 *system matrix* $\mathbf{Z}_{t-1}$ and the vector $\vec{\mathbf{x}}_t$ are composed from the 2D-DCT-II of $x_{t-1}(m, n)$ and the 2D-DCT-I of $x_t(m, n)$, respectively:

$$\mathbf{Z}_{t-1}(k, l) = \begin{bmatrix} Z_{t-1}^{cc}(k,l) & -Z_{t-1}^{cs}(k,l) & -Z_{t-1}^{sc}(k,l) & +Z_{t-1}^{ss}(k,l) \\ Z_{t-1}^{cs}(k,l) & +Z_{t-1}^{cc}(k,l) & -Z_{t-1}^{ss}(k,l) & -Z_{t-1}^{sc}(k,l) \\ Z_{t-1}^{sc}(k,l) & -Z_{t-1}^{ss}(k,l) & +Z_{t-1}^{cc}(k,l) & -Z_{t-1}^{cs}(k,l) \\ Z_{t-1}^{ss}(k,l) & +Z_{t-1}^{sc}(k,l) & +Z_{t-1}^{cs}(k,l) & +Z_{t-1}^{cc}(k,l) \end{bmatrix},$$

$$\vec{\mathbf{x}}_t(k, l) = \begin{bmatrix} X_t^{cc}(k,l) \\ X_t^{cs}(k,l) \\ X_t^{sc}(k,l) \\ X_t^{ss}(k,l) \end{bmatrix}, \quad \vec{\theta}_{m_u, m_v}(k, l) = \begin{bmatrix} g_{m_u m_v}^{CC}(k,l) \\ g_{m_u m_v}^{CS}(k,l) \\ g_{m_u m_v}^{SC}(k,l) \\ g_{m_u m_v}^{SS}(k,l) \end{bmatrix}.$$

Here the 2D-DCT-I of $x_{t-1}(m, n)$ and the 2D-DCT-II of $x_t(m, n)$ are defined in (9.15)–(9.18) and (9.19)–(9.23), respectively, where $\{Z_{t-1}^{xx}; xx = cc, cs, sc, ss\}$ can be obtained by a simple rotation (9.38) from $\{X_{t-1}^{xx}; xx = cc, cs, sc, ss\}$, which are computed and stored in memory in the previous encoding cycle.

However, for noninteger-pel movement, we need to use (9.60) and (9.61) in Theorem 9.1 to derive the system equation at the subpixel level. If the Fourier transform of the continuous

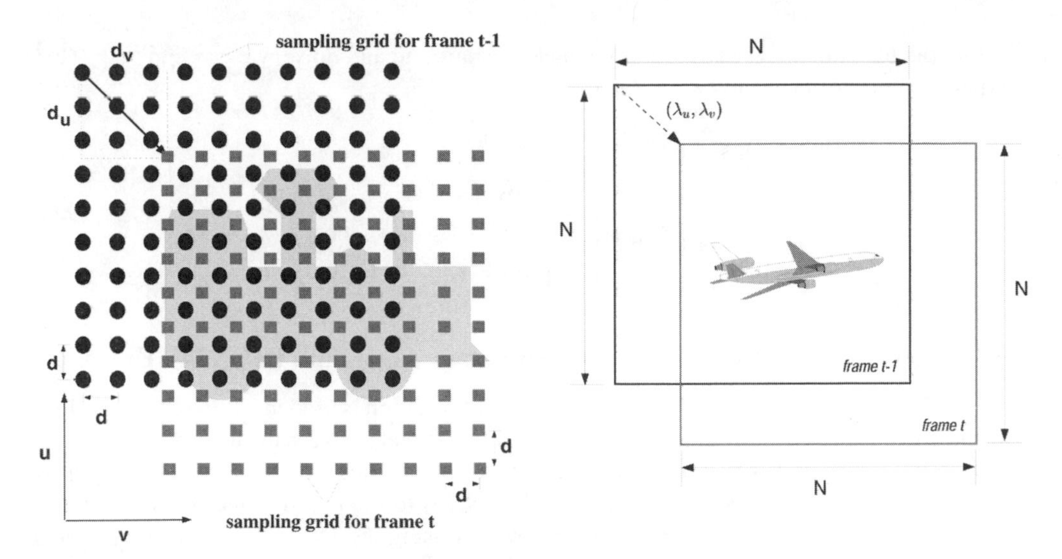

FIGURE 9.15
(a) The black dots and the gray squares symbolize the sampling grids for frames $I_{t-1}(u, v)$ and $I_t(u, v)$, respectively, at a sampling distance d. These two frames are aligned on the common object displaced by (d_u, d_v) in the continuous coordinate (u, v). (b) Two digitized images of consecutive frames, $x_{t-1}(m, n)$ and $x_t(m, n)$, are aligned on the common object moving $(\lambda_u, \lambda_v) = (d_u/d, d_v/d)$ pixels southeast.

intensity profile $I_t(u, v)$ is $\frac{\pi}{d}$ bandlimited and $I_t(u, v) = I_{t-1}(u - d_u, v - d_v)$, then according to Theorem 9.1, we can obtain the following 2D relations:

$$X_t^{cc}(k, l) = \frac{4}{N^2} C(k)C(l) \sum_{m,n=0}^{N-1} x_{t-1}(m, n) \cos\left[\frac{k\pi}{N}\left(m + \lambda_u + \frac{1}{2}\right)\right] \cos\left[\frac{l\pi}{N}\left(n + \lambda_v + \frac{1}{2}\right)\right]$$
$$\text{for } k, l \in \{0, \ldots, N-1\}, \tag{9.75}$$

$$X_t^{cs}(k, l) = \frac{4}{N^2} C(k)C(l) \sum_{m,n=0}^{N-1} x_{t-1}(m, n) \cos\left[\frac{k\pi}{N}\left(m + \lambda_u + \frac{1}{2}\right)\right] \sin\left[\frac{l\pi}{N}\left(n + \lambda_v + \frac{1}{2}\right)\right]$$
$$\text{for } k \in \{0, \ldots, N-1\}, l \in \{1, \ldots, N\}, \tag{9.76}$$

$$X_t^{sc}(k, l) = \frac{4}{N^2} C(k)C(l) \sum_{m,n=0}^{N-1} x_{t-1}(m, n) \sin\left[\frac{k\pi}{N}\left(m + \lambda_u + \frac{1}{2}\right)\right] \cos\left[\frac{l\pi}{N}\left(n + \lambda_v + \frac{1}{2}\right)\right]$$
$$\text{for } k \in \{1, \ldots, N\}, l \in \{0, \ldots, N-1\}, \tag{9.77}$$

$$X_t^{ss}(k, l) = \frac{4}{N^2} C(k)C(l) \sum_{m,n=0}^{N-1} x_{t-1}(m, n) \sin\left[\frac{k\pi}{N}\left(m + \lambda_u + \frac{1}{2}\right)\right] \sin\left[\frac{l\pi}{N}\left(n + \lambda_v + \frac{1}{2}\right)\right]$$
$$\text{for } k, l \in \{1, \ldots, N\}. \tag{9.78}$$

Thus, we can obtain the *pseudo-phase motion equation* at the subpixel level:

$$\mathbf{Z}_{t-1}(k, l) \cdot \vec{\theta}_{\lambda_u, \lambda_v}(k, l) = \vec{\mathbf{x}}_t(k, l), \text{ for } k, l \in \mathcal{N}, \tag{9.79}$$

where $\vec{\theta}_{\lambda_u, \lambda_v}(k, l) = [g^{\mathrm{CC}}_{\lambda_u, \lambda_v}(k, l), g^{\mathrm{CS}}_{\lambda_u, \lambda_v}(k, l), g^{\mathrm{SC}}_{\lambda_u, \lambda_v}(k, l), g^{\mathrm{SS}}_{\lambda_u, \lambda_v}(k, l)]^T$. A similar relationship between the DCT coefficients of $x_t(m, n)$ and $x_{t-1}(m, n)$ at the block boundary can be obtained in the same way.

In (9.79), the pseudo-phase vector $\vec{\theta}_{\lambda_u, \lambda_v}(k, l)$ contains the information of the subpixel movement (λ_u, λ_v). In an ideal situation where one rigid object is moving translationally within the block boundary without observable background and noise, we can find $\vec{\theta}_{\lambda_u,\lambda_v}(k,l)$ explicitly in terms of λ_u and λ_v as such:

$$\vec{\theta}_{\lambda_u,\lambda_v}(k,l) = \begin{bmatrix} g^{CC}_{\lambda_u,\lambda_v}(k,l) \\ g^{CS}_{\lambda_u,\lambda_v}(k,l) \\ g^{SC}_{\lambda_u,\lambda_v}(k,l) \\ g^{SS}_{\lambda_u,\lambda_v}(k,l) \end{bmatrix} = \begin{bmatrix} \cos\frac{k\pi}{N}(\lambda_u+\frac{1}{2}) & \cos\frac{l\pi}{N}(\lambda_v+\frac{1}{2}) \\ \cos\frac{k\pi}{N}(\lambda_u+\frac{1}{2}) & \sin\frac{l\pi}{N}(\lambda_v+\frac{1}{2}) \\ \sin\frac{k\pi}{N}(\lambda_u+\frac{1}{2}) & \cos\frac{l\pi}{N}(\lambda_v+\frac{1}{2}) \\ \sin\frac{k\pi}{N}(\lambda_u+\frac{1}{2}) & \sin\frac{l\pi}{N}(\lambda_v+\frac{1}{2}) \end{bmatrix} . \tag{9.80}$$

9.6.1 DCT-Based Half-Pel Motion Estimation Algorithm (HDXT-ME)

From (9.79), we know that the subpixel motion information is hidden, although not obvious, in the pseudo-phases. To obtain subpixel motion estimates, we can directly compute the pseudo-phases in (9.79) and then locate the peaks of the sinc functions after applying the subpixel sinusoidal orthogonal principles (9.70)–(9.73) to the pseudo-phases. Alternatively, we can have better flexibility and scalability by first using the DXT-ME algorithm to get an integer-pel motion estimate and then utilizing the pseudo-phase functions $f(k,l)$ and $g(k,l)$ computed in the DXT-ME algorithm to increase estimation accuracy to half-pel, due to the fact that (9.79) has exactly the same form as (9.74). Specifically, based on the subpixel sinusoidal orthogonal principles (9.70)–(9.73), the subpixel motion information can be extracted in the form of impulse functions with peak positions closely related to the displacement.

For the sake of flexibility and modularity in design and further reduction in complexity, we adopt the second approach to devise a motion estimation scheme with arbitrary fractional pel accuracy by applying the subpixel sinusoidal orthogonal principles to the pseudo-phase functions passed from the DXT-ME algorithm. The limitation of estimation accuracy will only be determined by the interaction effects of the ξ functions and the slope of the ξ function at and around zero and how well the subpixel motion information is preserved in the pseudo-phases after sampling.

We define $\overline{\text{DCS}}(u,v)$ and $\overline{\text{DSC}}(u,v)$ as follows:

$$\overline{\text{DCS}}(u,v) \triangleq \sum_{k=0}^{N-1}\sum_{l=1}^{N-1}\left[\frac{f(k,l)}{C(k)C(l)}\right]\cos\frac{k\pi}{N}\left(u+\frac{1}{2}\right)\sin\frac{l\pi}{N}\left(v+\frac{1}{2}\right), \tag{9.81}$$

$$\overline{\text{DSC}}(u,v) \triangleq \sum_{k=1}^{N-1}\sum_{l=0}^{N-1}\left[\frac{g(k,l)}{C(k)C(l)}\right]\sin\frac{k\pi}{N}\left(u+\frac{1}{2}\right)\cos\frac{l\pi}{N}\left(v+\frac{1}{2}\right). \tag{9.82}$$

Thus, from the subpixel sinusoidal orthogonal principles (9.70)–(9.73) and the definitions of $f(k,l)$ and $g(k,l)$, we can show that

$$\overline{\text{DCS}}(u,v) = \frac{1}{4}[\xi(u-\lambda_u)+\xi(u+\lambda_u+1)]\cdot[\xi(v-\lambda_v)-\xi(v+\lambda_v+1)], \tag{9.83}$$

$$\overline{\text{DSC}}(u,v) = \frac{1}{4}[\xi(u-\lambda_u)-\xi(u+\lambda_u+1)]\cdot[\xi(v-\lambda_v)+\xi(v+\lambda_v+1)]. \tag{9.84}$$

The rules to determine subpixel motion direction are summarized in Table 9.6 and are similar to the rules for determining integer-pel motion direction.

Figure 9.16 illustrates how to estimate subpixel displacements in the DCT domain. Figure 9.16c and d depict the input images $x_1(m,n)$ of size 16×16 (i.e., $N = 16$) and $x_2(m,n)$ displaced from $x_1(m,n)$ by $(2.5, -2.5)$ at SNR = 50 dB. These two images are sampled on a rectangular grid at a sampling distance $d = 0.625$ from the continuous intensity profile $x_c(u,v) = \exp(-(u^2+v^2))$ for $u, v \in [-5, 5]$ in Figure 9.16a, whose Fourier transform is

Table 9.6 Determination of Direction of Movement (λ_u, λ_v) from the Signs of $\overline{\text{DSC}}$ and $\overline{\text{DCS}}$

Sign of DSC Peak	Sign of DCS Peak	Peak Index	Motion Direction
+	+	(λ_u, λ_v)	Northeast
+	−	$(\lambda_u, -(\lambda_v + 1))$	Southeast
−	+	$(-(\lambda_u + 1), \lambda_v)$	Northwest
−	−	$(-(\lambda_u + 1), -(\lambda_v + 1))$	Southwest

bandlimited as in Figure 9.16b to satisfy the condition in Theorem 9.1. Figure 9.16e and f are the 3D plots of the pseudo-phases $f(k, l)$ and $g(k, l)$ provided by the DXT-ME algorithm, which also computes $\text{DSC}(m, n)$ and $\text{DCS}(m, n)$ as shown in Figure 9.16g and h, with peaks positioned at (3, 1) and (2, 2) corresponding to the integer-pel estimated displacement vectors (3, −2) and (2, −3), respectively, because only the first quadrant is viewed. As a matter of fact, $\text{DSC}(m, n)$ and $\text{DSC}(m, n)$ have large magnitudes at $\{(m, n); m = 2, 3,\ n = 1, 2\}$.

To obtain an estimate at half-pel accuracy, we calculate $\overline{\text{DSC}}(u, v)$ and $\overline{\text{DCS}}(u, v)$ in (9.81) and (9.82), respectively, for $u, v = 0 : 0.5 : N - 1$ as depicted in Figure 9.16i and j, where the peaks can clearly be identified at (2.5, 1.5) corresponding to the motion estimate (2.5, −2.5) exactly equal to the true displacement vector even though the two input images do not look alike. Note that the notation $a : r : b$ is an abbreviation of the range $\{a + i \cdot r \text{ for } i = 0, \ldots, \lfloor \frac{b-a}{r} \rfloor\} = \{a, a + r, a + 2r, \ldots, b - r, b\}$. For comparison, $\overline{\text{DSC}}(u, v)$ and $\overline{\text{DCS}}(u, v)$ are also plotted in Figure 9.16k and l, respectively, for $u, v = 0 : 0.25 : N - 1 = 0, 0.25, 0.5, \ldots, N - 1.25, N - 1$ where smooth ripples are obvious due to the ξ functions inherent in the $\overline{\text{DCS}}$ and $\overline{\text{DSC}}$ of (9.83) and (9.84) and have peaks also at (2.5, 1.5).

Therefore, the DCT-based half-pel motion estimation algorithm (HDXT-ME) comprises three steps:

1. The DXT-ME algorithm estimates the integer components of the displacement as $(\hat{m}_u, \hat{m}_v)$.

2. The pseudo-phase functions from the DXT-ME algorithm, $f(k, l)$ and $g(k, l)$, are used to compute $\overline{\text{DCS}}(u, v)$ and $\overline{\text{DSC}}(u, v)$ for $u \in \{\hat{m}_u - 0.5, \hat{m}_u, \hat{m}_u + 0.5\}$ and $v \in \{\hat{m}_v - 0.5, \hat{m}_v, \hat{m}_v + 0.5\}$ from (9.81) and (9.82), respectively.

3. Search the peak positions of $\overline{\text{DCS}}(u, v)$ and $\overline{\text{DSC}}(u, v)$ for the range of indices, $\overline{\Phi} = \{(u, v) : u \in \{\hat{m}_u - 0.5, \hat{m}_u, \hat{m}_u + 0.5\};\ v \in \{\hat{m}_v - 0.5, \hat{m}_v, \hat{m}_v + 0.5\}\}$, to find

$$\left(u_{\overline{\text{DCS}}}, v_{\overline{\text{DCS}}}\right) = arg \max_{u,v \in \overline{\Phi}} \left|\overline{\text{DCS}}(u, v)\right| , \tag{9.85}$$

$$\left(u_{\overline{\text{DSC}}}, v_{\overline{\text{DSC}}}\right) = arg \max_{u,v \in \overline{\Phi}} \left|\overline{\text{DSC}}(u, v)\right| . \tag{9.86}$$

These peak positions determine the estimated displacement vector $(\hat{\lambda}_u, \hat{\lambda}_v)$. However, if the absolute value of $\overline{\text{DSC}}(u, v)$ is less than a preset threshold $\epsilon_D > 0$, then $\hat{\lambda}_u = -0.5$; likewise, if $|\overline{\text{DCS}}(u, v)| < \epsilon_D$, $\hat{\lambda}_v = -0.5$. Therefore,

$$\hat{\lambda}_u = \begin{cases} u_{\overline{\text{DSC}}} = u_{\overline{\text{DCS}}}, & \text{if } |\overline{\text{DSC}}(u_{\overline{\text{DSC}}}, v_{\overline{\text{DSC}}})| > \epsilon_D , \\ -0.5, & \text{if } |\overline{\text{DSC}}(u_{\overline{\text{DSC}}}, v_{\overline{\text{DSC}}})| < \epsilon_D , \end{cases} \tag{9.87}$$

$$\hat{\lambda}_v = \begin{cases} v_{\overline{\text{DCS}}} = v_{\overline{\text{DSC}}}, & \text{if } |\overline{\text{DCS}}(u_{\overline{\text{DCS}}}, v_{\overline{\text{DCS}}})| > \epsilon_D , \\ -0.5, & \text{if } |\overline{\text{DCS}}(u_{\overline{\text{DCS}}}, v_{\overline{\text{DCS}}})| < \epsilon_D . \end{cases} \tag{9.88}$$

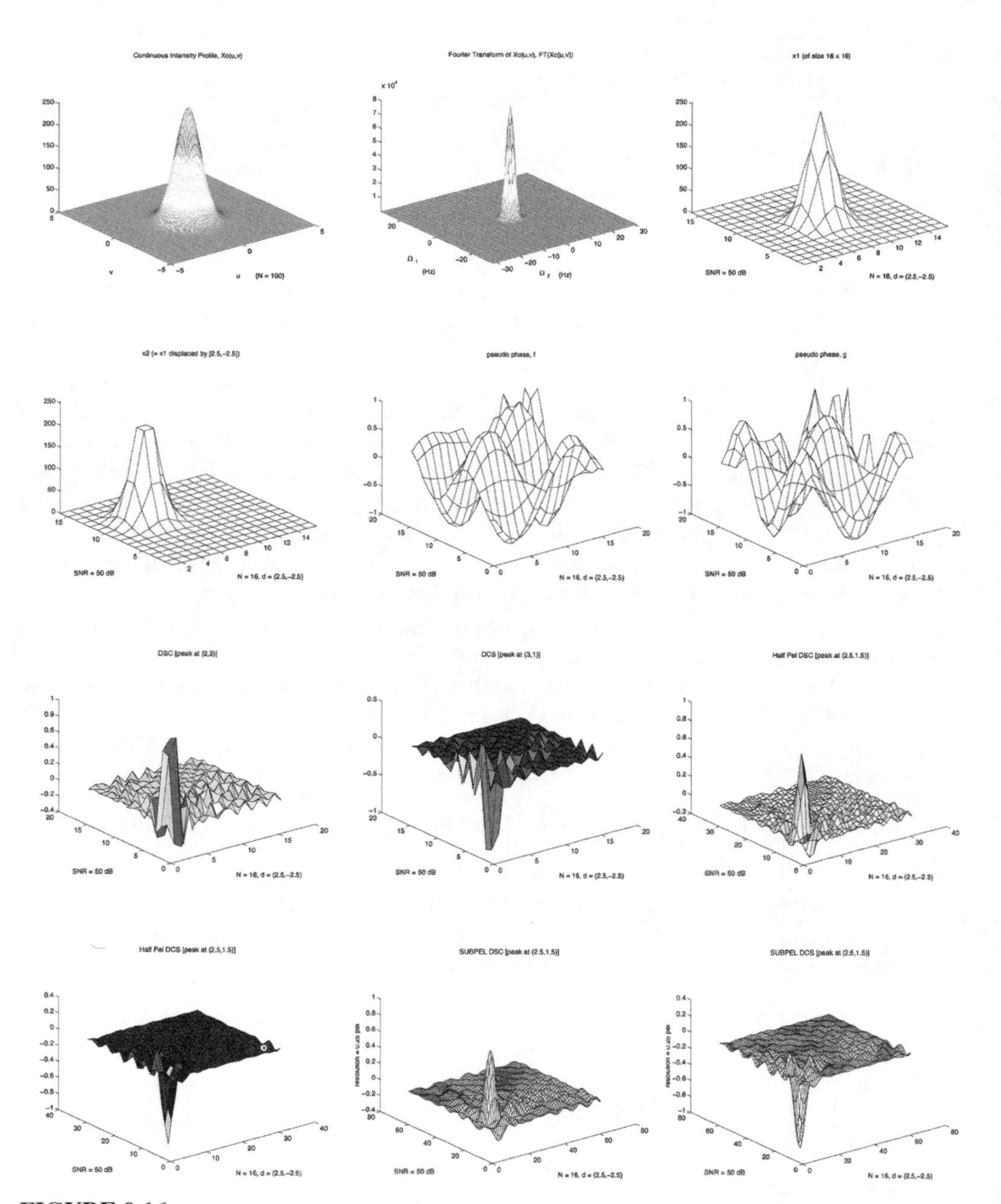

FIGURE 9.16

Illustration of DCT-based half-pel motion estimation algorithm (HDXT-ME). (a) Continuous intensity profile $x_c(u, v)$; (b) FT of $x_c(u, v)$, $X_c(\Omega_u, \Omega_v)$; (c) 16×16 $x_1(m, n)$; (d) 16×16 $x_2(m, n)$; (e) pseudo-phase $f(k, l)$; (f) pseudo-phase $g(k, l)$; (g) DSC(m, n); (h) DCS(m, n); (i) $\overline{\text{DSC}}(u, v)$ for $u, v = 0 : 0.5 : 15$; (j) $\overline{\text{DCS}}(u, v)$ for $u, v = 0 : 0.5 : 15$; (k) $\overline{\text{DSC}}(u, v)$ for $u, v = 0 : 0.25 : 15$; (l) $\overline{\text{DCS}}(u, v)$ for $u, v = 0 : 0.25 : 15$.

In step 2, only those half-pel estimates around the integer-pel estimate $(\hat{m}_u, \hat{m}_v)$ are considered due to the fact that the DXT-ME algorithm finds the nearest integer-pel motion estimate $(\hat{m}_u, \hat{m}_v)$ from the subpixel displacement. This will significantly reduce the number of computations without evaluating all possible half-pel displacements.

In step 3, the use of ϵ_D deals with the case of zero pseudo-phases when the displacement is -0.5. Specifically, if $\lambda_u = -0.5$, then $g^{SC}_{\lambda_u,\lambda_v}(k,l) = 0,\ \forall k,l$ which leads to $g(k,l) = 0$ and $\overline{\text{DSC}}(u,v) = 0$. However, in a noisy situation, it is very likely that $g(k,l)$ is not exactly zero and, thus, neither is $\overline{\text{DSC}}(u,v)$. Therefore, ϵ_D should be set very small but large enough to accommodate the noisy case. In our experiment, ϵ_D is empirically chosen to be 0.08. Similar consideration is made on $\overline{\text{DCS}}(u,v)$ for $\lambda_v = -0.5$. It is also possible that the peak positions of $\overline{\text{DCS}}(u,v)$ and $\overline{\text{DSC}}(u,v)$ differ in the noisy circumstances. In this case, the arbitration rule used in the DXT-ME algorithm may be applied.

To demonstrate the accuracy of this HDXT-ME algorithm, we use a 16×16 dot image x_1 in Figure 9.17a as input and displace x_1 to generate the second input image x_2 according to the true motion field $\{(\lambda_u, \lambda_v) : \lambda_u, \lambda_v = -5 : 0.5 : 4\}$ (shown in Figure 9.17b) through the bilinear interpolating function specified in the MPEG standard [13], which interpolates the value $x(m+u, n+v)$ from four neighboring pixel values for m, n being integers and $u, v \in [0, 1)$ in the following way:

$$x(m+u, n+v) = (1-u)\cdot(1-v)\cdot x(m,n) + (1-u)\cdot v\cdot x(m,n+1) + u\cdot(1-v)\cdot x(m+1,n) + u\cdot v\cdot x(m+1,n+1)\,. \quad (9.89)$$

Figure 9.17c shows the estimated motion field by the HDXT-ME algorithm, which is exactly the same as the true motion field.

Figure 9.18a–c further illustrate estimation accuracy for half-pel motion estimation schemes using peak information from $L_s(u,v)$, $L_c(u,v)$, and $L_c(u,v) + L_s(u,v)$, respectively. In Figure 9.18a, the "+" line indicates peak positions of $L_s(u,v)$ found in the index range $\{0 : 0.5 : 15\}$ for a block size $N = 16$ with respect to different true displacement values $\{-7 : 0.5 : 7\}$. The "o" line specifies the final estimates after determination of motion directions from the peak signs of $L_s(u,v)$ according to the rules in Table 9.6. These estimates are shown to align with the reference line $u = v$, implying their correctness. For the true displacement $= -0.5$, $L_s(-0.5, v) \equiv 0$ for all v and ϵ_D is used to decide whether the estimate should be set to -0.5. In Figure 9.18b, $L_c(u,v)$ is used instead of $L_s(u,v)$ but $L_c(u,v)$ is always positive, inferring that no peak sign can be exploited to determine motion direction. In Figure 9.18c, $L_c(u,v) + L_s(u,v)$ provides accurate estimates without adjustment for all true displacement values, but the index range must include negative indices (i.e., $[-15 : 0.5 : 15]$).

In the HDXT-ME algorithm, step 2 involves only nine $\overline{\text{DCS}}(u,v)$ and $\overline{\text{DSC}}(u,v)$ values at and around $(\hat{m}_u, \hat{m}_v)$. Since $\overline{\text{DCS}}(u,v)$ and $\overline{\text{DSC}}(u,v)$ are variants of inverse 2D-DCT-II, the parallel and fully pipelined 2D DCT lattice structure proposed in [9]–[11] can be used to compute $\overline{\text{DCS}}(u,v)$ and $\overline{\text{DSC}}(u,v)$ at a cost of $O(N)$ operations in N steps. Furthermore, the searching in step 3 requires $O(N^2)$ operations for one step. Thus, the computational complexity of the HDXT-ME algorithm is $O(N^2)$ in total.

9.6.2 DCT-Based Quarter-Pel Motion Estimation Algorithm (QDXT-ME and Q4DXT-ME)

The interaction of two ξ functions in $L_c(u,v)$ and $L_s(u,v)$ from (9.66) and (9.67) disassociates the peak locations with the displacement (λ_u, λ_v) for $\lambda_u,\ \lambda_v \in [-1.5, 0.5]$. In spite of this, in the HDXT-ME algorithm, we can still accurately estimate half-pel displacements by locating the peaks of $L_s(\lambda, v)$ for true displacements $\lambda = -N+1 : 0.5 : N-1$ and indices $v = 0 : 0.5 : N-1$ if ϵ_D is introduced to deal with the case for $\lambda = -0.5$. However, at

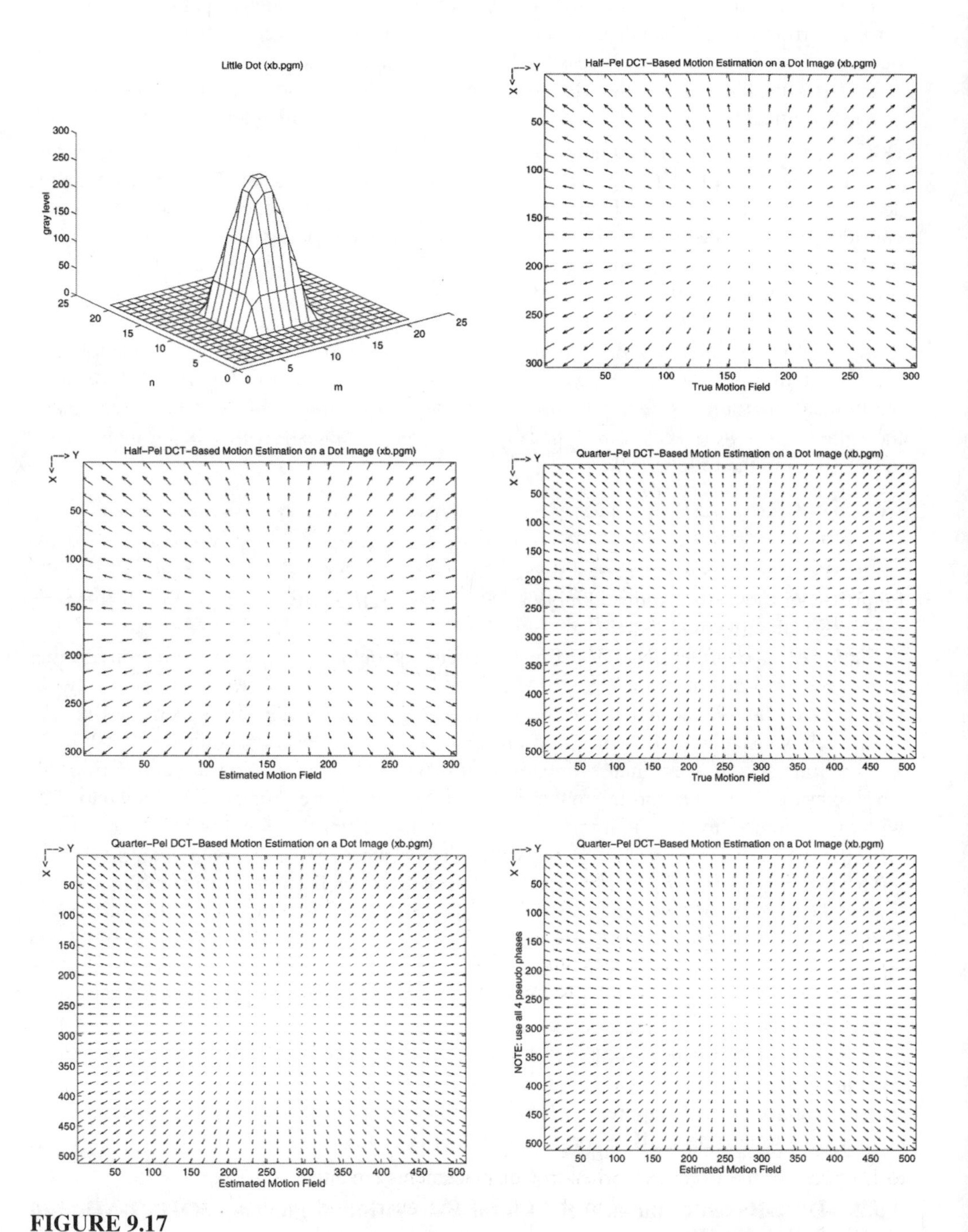

FIGURE 9.17
Estimated motion fields (c) and (e) of HDXT-ME and QDXT-ME by moving a dot image (a) according to the true motion fields (b) and (d).

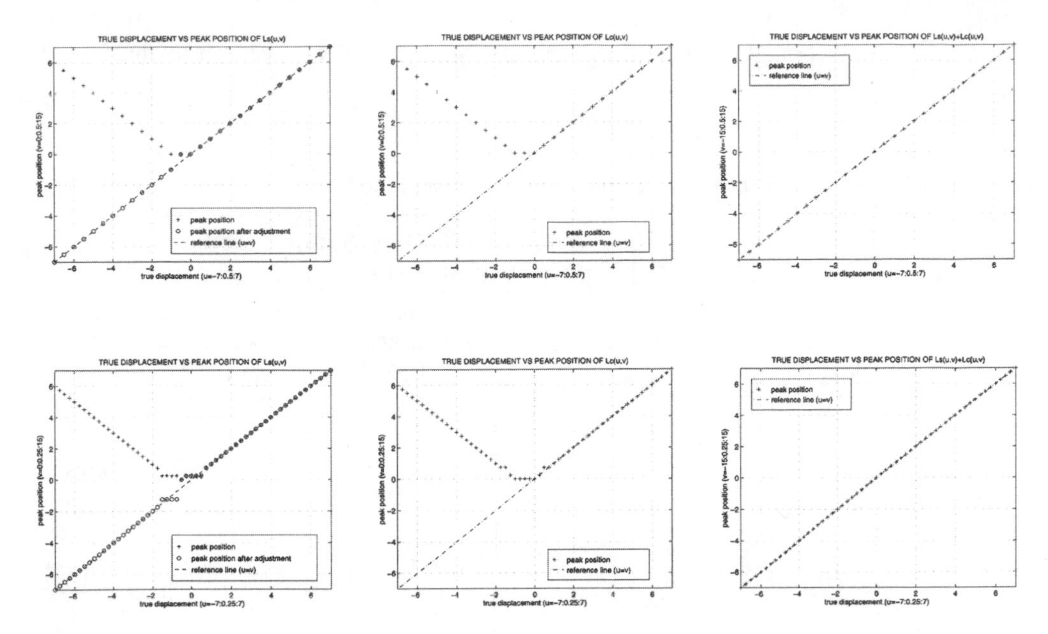

FIGURE 9.18
Relation between true displacements and peak positions for half-pel and quarter-pel estimation. The signs of peak values in $L_s(u, v)$ indicate the motion directions and are used to adjust the peak positions for motion estimates. (a) $L_s(u, v)$ for half-pel estimation; (b) $L_c(u, v)$ for half-pel estimation; (c) $L_c(u, v) + L_s(u, v)$ for half-pel estimation; (d) $L_s(u, v)$ for quarter-pel estimation; (e) $L_c(u, v)$ for quarter-pel estimation; (f) $L_c(u, v) + L_s(u, v)$ for quarter-pel estimation.

the quarter-pel level, it does cause estimation errors around $\lambda = -0.5$, as indicated in Figure 9.18d, where the peaks of $L_s(\lambda, v)$ stay at $v = 0$ for true displacements λ varying over $[-1, 0]$. The sum of $L_c(\lambda, v)$ and $L_s(\lambda, v)$ is a pure ξ function and thus the adverse interaction is eliminated. As a result, the peak position of this sum can be used to predict precisely the displacement at either the half-pel level or quarter-pel level, as demonstrated in Figure 9.18c and f, respectively. However, for two-dimensional images, $\overline{\text{DCS}}$ or $\overline{\text{DSC}}$ has four ξ functions as in (9.83) or (9.84). The DXT-ME algorithm provides two pseudo-phase functions $f(k, l)$ and $g(k, l)$, but only $\overline{\text{DCS}}$ and $\overline{\text{DSC}}$ are available for subpixel estimation. In this case, the sum of $\overline{\text{DCS}}$ and $\overline{\text{DSC}}$ can only annihilate two ξ functions, leaving two ξ functions as given by:

$$\overline{\text{DCS}}(u, v) + \overline{\text{DSC}}(u, v) = \frac{1}{2}\left[\xi\left(u - \lambda_u\right)\xi\left(v - \lambda_v\right) - \xi\left(u + \lambda_u + 1\right)\xi\left(v + \lambda_v + 1\right)\right] . \tag{9.90}$$

Even though this sum is not a single ξ function, the estimation error of using this sum is limited to 1/4 pixel for the worst case when true displacements are either -0.75 or -0.25.

The above discussion leads to the DCT-based quarter-pel motion estimation algorithm (QDXT-ME) as follows:

1. The DXT-ME algorithm computes the integer-pel estimate $(\hat{m}_u, \hat{m}_v)$.

2. $\overline{\text{DCS}}(u, v)$ and $\overline{\text{DSC}}(u, v)$ are calculated from $f(k, l)$ and $g(k, l)$ in (9.81) and (9.82), respectively, for the range of indices, $\overline{\Phi} = \{(u, v) : u = \hat{m}_u - 0.75 : 0.25 : \hat{m}_u + 0.75;\ v = \hat{m}_v - 0.75 : 0.25 : \hat{m}_v + 0.75\}$.

3. Search the peak position of $D_2(u,v) \triangleq \overline{\mathrm{DCS}}(u,v) + \overline{\mathrm{DSC}}(u,v)$ over $\overline{\Phi}$; that is,

$$(u_{D2}, v_{D2}) = arg \max_{u,v \in \overline{\Phi}} |D_2(u,v)| \ . \tag{9.91}$$

The estimated displacement vector is obtained as follows:

$$\left(\hat{\lambda}_u, \hat{\lambda}_v\right) = \begin{cases} (u_{D2}, v_{D2}), & \text{if } |D_2(u_{D2}, v_{D2})| > \epsilon_D \ , \\ (-0.5, -0.5), & \text{if } |D_2(u_{D2}, v_{D2})| < \epsilon_D \ . \end{cases} \tag{9.92}$$

Step 3 is based on the fact that $|D_2(\lambda_u, \lambda_v)| = 0$ if and only if $(\lambda_u, \lambda_v) = -0.5$. This QDXT-ME algorithm follows the same procedure as HDXT-ME except for the search region and using the sum of $\overline{\mathrm{DCS}}$ and $\overline{\mathrm{DSC}}$. Therefore, QDXT-ME has the same computational complexity, $O(N^2)$, as HDXT-ME.

If we modify the DXT-ME algorithm to provide the other two pseudo-phase functions g^{CC} and g^{SS} in addition to f and g, we can compute $\overline{\mathrm{DCC}}$ and $\overline{\mathrm{DSS}}$ in the following way:

$$\overline{\mathrm{DCC}}(u,v) \triangleq \sum_{k=0}^{N-1}\sum_{l=0}^{N-1} g^{CC}(k,l) \cos\frac{k\pi}{N}\left(u+\frac{1}{2}\right) \cos\frac{l\pi}{N}\left(v+\frac{1}{2}\right) , \tag{9.93}$$

$$\overline{\mathrm{DSS}}(u,v) \triangleq \sum_{k=1}^{N-1}\sum_{l=1}^{N-1} g^{SS}(k,l) \sin\frac{k\pi}{N}\left(u+\frac{1}{2}\right) \sin\frac{l\pi}{N}\left(v+\frac{1}{2}\right) . \tag{9.94}$$

Then we can show that

$$D_4(u,v) \triangleq \overline{\mathrm{DCC}}(u,v) + \overline{\mathrm{DCS}}(u,v) + \overline{\mathrm{DSC}}(u,v) + \overline{\mathrm{DSS}}(u,v) \tag{9.95}$$

$$= \xi(u-\lambda_u)\xi(v-\lambda_v) \ . \tag{9.96}$$

This sum contains only one ξ without any negative interaction effect whose peak is sharp at (λ_u, λ_v). This leads to another quarter-pel motion estimation algorithm (Q4DXT-ME), which can estimate accurately for all displacements at the quarter-pel or even finer level.

1. Find the integer-pel estimate $(\hat{m}_u, \hat{m}_v)$ by the DXT-ME algorithm.

2. Obtain four pseudo-phases g^{CC}, g^{CS}, g^{SC}, and g^{SS} from the modified DXT-ME algorithm. Compute $\overline{\mathrm{DCS}}(u,v)$, $\overline{\mathrm{DSC}}(u,v)$, $\overline{\mathrm{DCC}}(u,v)$, and $\overline{\mathrm{DSS}}(u,v)$ for the range of indices, $\overline{\Phi} = \{(u,v) : u = \hat{m}_u - 0.75 : 0.25 : \hat{m}_u + 0.75;\ v = \hat{m}_v - 0.75 : 0.25 : \hat{m}_v + 0.75\}$.

3. Search the peak position of $D_4(u,v)$ over $\overline{\Phi}$:

$$(u_{D4}, v_{D4}) = arg \max_{u,v \in \overline{\Phi}} |D_4(u,v)| \ .$$

The estimated displacement vector is then the peak position:

$$\left(\hat{\lambda}_u, \hat{\lambda}_v\right) = (u_{D4}, v_{D4}) \ .$$

9.6.3 Simulation Results

The Miss America image sequence, with slow head and shoulder movement accompanying occasional eye and mouth opening, is used for simulation [6]. The performance of the DCT-based algorithms is compared with BKM-ME and its subpixel counterparts in terms of MSE per pixel and BPS. For all the MSE values computed in the experiment, the bilinear interpolation in (9.89) is used for comparison to reconstruct images displaced by a fractional pixel because the bilinear interpolation is used in MPEG standards for motion compensation [13, 21].

As usual, the integer-pel BKM-ME algorithm minimizes the MAD function of a block over a larger search area. Its subpixel versions have two levels implemented: HBKM-ME and QBKM-ME. Both algorithms optimize the MAD value around the subpixel displacements around the integer-pel estimate. In addition, we also compare with three kinds of fast-search block-matching algorithms with integer-pel, half-pel, and quarter-pel accuracy: the three-step search algorithm (TSS, HTSS, QTSS), the logarithmic search algorithm (LOG, HLOG, QLOG), and the subsampled search algorithm (SUB, HSUB, QSUB) [17].

The simulation results are summarized by averaging over the sequence in terms of the MSE and BPS values in Table 9.7. The coding gain from subpixel motion estimation is obvious when we compare how much improvement we can have from integer-pel accuracy to half-pel and even quarter-pel accuracy:

- HBKM-ME has 47.03% less of MSE value or 12.24% less of BPS value than BKM-ME, whereas QBKM-ME has 60.76% less of MSE or 17.78% less of BPS than BKM-ME.
- Edge-extracted HDXT-ME has 45.36% less of MSE value or 12.95% less of BPS value than edge-extracted DXT-ME, whereas edge-extracted QDXT-ME has 59.79% less of MSE or 18.18% less of BPS.

Table 9.7 Performance Summary of DCT-Based Algorithms and Block-Matching Algorithms (BKM, TSS, LOG, SUB) at Different Accuracy Levels on Miss America Sequence

Approach	MSE	MSE Difference	MSE Ratio	BPF	BPS	BPS Ratio
Integer-Pel Accuracy						
BKM	7.187	0.000	0.0%	8686	0.343	0.0%
Frame-differentiated DXT	7.851	0.664	9.2%	8855	0.349	1.9%
Edge-extracted DXT	9.363	2.176	30.3%	9200	0.363	5.9%
TSS	7.862	0.675	9.4%	8910	0.352	2.6%
LOG	7.862	0.675	9.4%	8910	0.352	2.6%
SUB	7.202	0.015	0.2%	8684	0.343	0.0%
Half-Pel Accuracy						
HBKM	3.807	0.000	0.0%	7628	0.301	0.0%
Frame-differentiated HDXT	5.598	1.791	47.0%	8216	0.324	7.7%
Edge-extracted HDXT	5.116	1.308	34.4%	8000	0.316	4.9%
HTSS	3.877	0.070	1.8%	7676	0.303	0.6%
HLOG	3.877	0.070	1.8%	7676	0.303	0.6%
HSUB	3.810	0.002	0.1%	7628	0.301	0.0%
Quarter-Pel Accuracy						
QBKM	2.820	0.000	0.0%	7146	0.282	0.0%
Frame-differentiated QDXT	4.728	1.908	67.7%	7758	0.306	8.6%
Edge-extracted QDXT	3.899	1.079	38.3%	7578	0.299	6.0%
Frame-differentiated Q4DXT	4.874	2.054	72.8%	7785	0.307	8.9%
Edge-extracted Q4DXT	3.765	0.945	33.5%	7532	0.297	5.4%
QTSS	2.843	0.023	0.8%	7162	0.283	0.2%
QLOG	2.843	0.023	0.8%	7162	0.283	0.2%
QSUB	2.825	0.005	0.2%	7144	0.282	0.0%

9.7 DCT-Based Motion Compensation

Manipulation of compressed video data in the DCT domain has been recognized as an important component in many advanced video applications [29]–[33]. In a video bridge, where multiple sources of compressed video are combined and retransmitted in a network, techniques of manipulation and composition of compressed video streams entirely in the DCT domain eliminate the need to build a decoding/encoding pair. Furthermore, manipulation in the DCT domain provides flexibility to match heterogeneous quality of service requirements with different network or user resources, such as prioritization of signal components from low-order DCT coefficients to fit low-end communication resources. Finally, many manipulation functions can be performed in the DCT domain more efficiently than in the spatial domain [30] due to a much lower data rate and removal of the decoding/encoding pair. However, all earlier work has been focused mainly on manipulation at the decoder side.

To serve the purpose of building a fully DCT-based motion-compensated video coder, our aim is to develop the techniques of motion compensation in the DCT domain without converting back to the spatial domain before motion compensation. In [30], the method of pixelwise (integer-pel) translation in the DCT domain is proposed for extracting a DCT block out of four neighboring DCT blocks at an arbitrary position. Although addressing a different scenerio, this method can be applied after modification to integer-pel motion compensation in the DCT domain. For subpel motion compensation, we derive an equivalent form of bilinear interpolation in the DCT domain and then show that it is possible to perform other interpolation functions for achieving more accurate and visually better approximation in the DCT domain without increasing the complexity.

9.7.1 Integer-Pel DCT-Based Motion Compensation

As illustrated in Figure 9.19a, after motion estimation, the current block $\mathbf{C}$ of size $N \times N$ in the current frame I_t can be best predicted from the block displaced from the current block position by the estimated motion vector (d_u, d_v) in the spatial domain. This motion estimate determines which four contiguous predefined DCT blocks are chosen for the prediction of the current block out of eight surrounding DCT blocks and the block at the current block position. To extract the displaced DCT block in the DCT domain, a direct method is used to obtain separately from these four contiguous blocks four subblocks that can be combined to form the final displaced DCT block (as shown in Figure 9.19b, with the upper-left, lower-left, upper-right, and lower-right blocks from the previous frame I_{t-1} labeled as $\mathbf{B}_1$, $\mathbf{B}_2$, $\mathbf{B}_3$, and $\mathbf{B}_4$, respectively, in the spatial domain [30]). Subblocks $\mathbf{S}_i$ are extracted in the spatial domain from these four blocks by pre-multiplication and post-multiplication of the windowing/shifting matrices, $\mathbf{H}_i$ and $\mathbf{V}_i$:

$$\mathbf{S}_k = \mathbf{H}_k \mathbf{B}_k \mathbf{V}_k, \quad \text{for } k = 1, \ldots, 4, \tag{9.97}$$

where $\mathbf{H}_k$ and $\mathbf{V}_k$ are the $N \times N$ windowing/shifting matrices defined as

$$\mathbf{H}_1 = \begin{bmatrix} \mathbf{0} & \mathbf{I}_{h_1} \\ \mathbf{0} & \mathbf{0} \end{bmatrix}, \ \mathbf{V}_1 = \begin{bmatrix} \mathbf{0} & \mathbf{0} \\ \mathbf{I}_{v_1} & \mathbf{0} \end{bmatrix}, \ \mathbf{H}_2 = \begin{bmatrix} \mathbf{0} & \mathbf{0} \\ \mathbf{I}_{h_2} & \mathbf{0} \end{bmatrix}, \ \mathbf{V}_2 = \begin{bmatrix} \mathbf{0} & \mathbf{0} \\ \mathbf{I}_{v_2} & \mathbf{0} \end{bmatrix}, \tag{9.98}$$

$$\mathbf{H}_3 = \begin{bmatrix} \mathbf{0} & \mathbf{I}_{h_3} \\ \mathbf{0} & \mathbf{0} \end{bmatrix}, \ \mathbf{V}_3 = \begin{bmatrix} \mathbf{0} & \mathbf{I}_{v_3} \\ \mathbf{0} & \mathbf{0} \end{bmatrix}, \ \mathbf{H}_4 = \begin{bmatrix} \mathbf{0} & \mathbf{0} \\ \mathbf{I}_{h_4} & \mathbf{0} \end{bmatrix}, \ \mathbf{V}_4 = \begin{bmatrix} \mathbf{0} & \mathbf{I}_{v_4} \\ \mathbf{0} & \mathbf{0} \end{bmatrix}. \tag{9.99}$$

Here $\mathbf{I}_n$ is the $n \times n$ identity matrix (i.e., $\mathbf{I}_n = \text{diag}\{1, \ldots, 1\}$) and n is determined by the height/width of the corresponding subblock. These pre- and post-multiplication matrix operations can be visualized in Figure 9.19c, where the overlapped gray areas represent the extracted subblock. Then these four subblocks are summed to form the desired translated block $\hat{\mathbf{B}}_{ref}$.

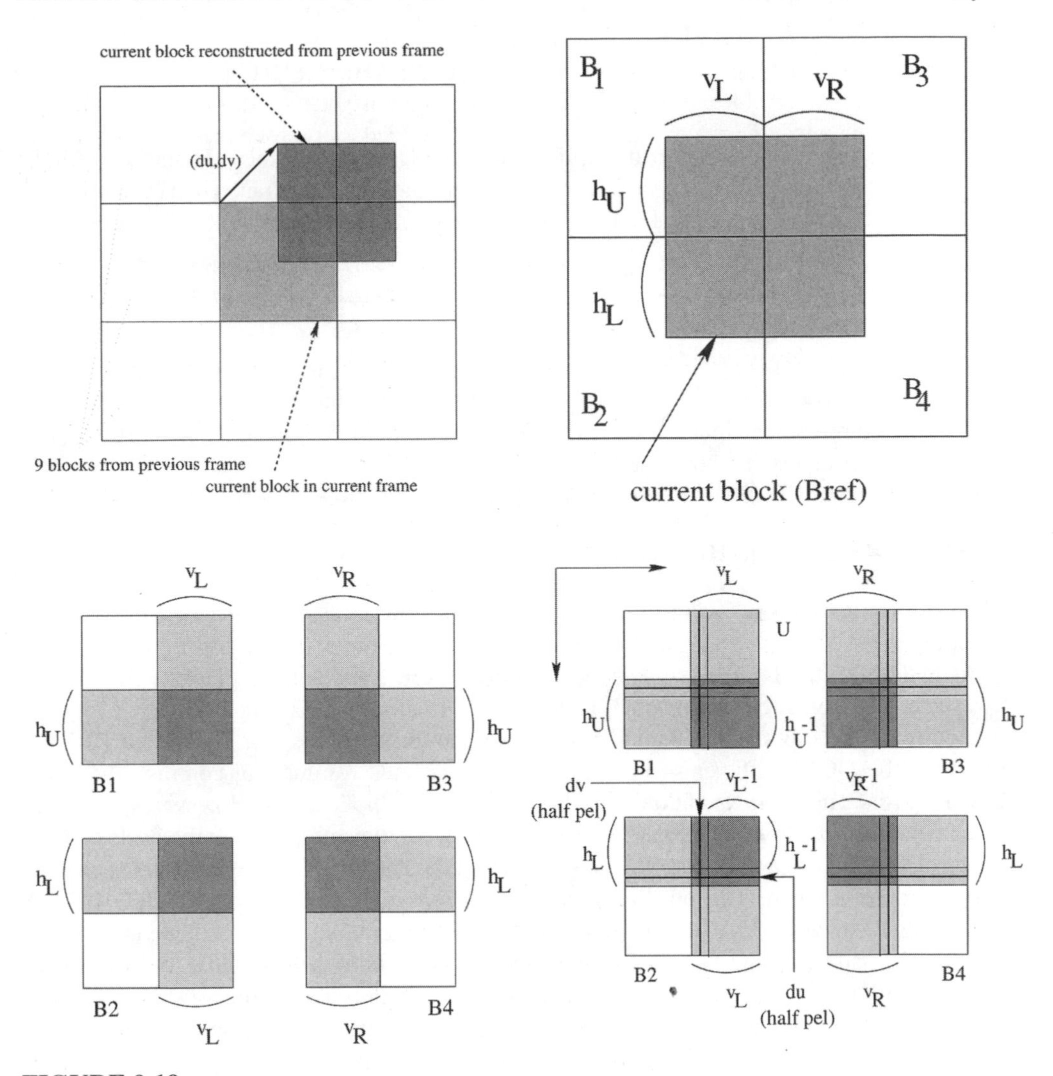

FIGURE 9.19
(a) Prediction of current block in current frame from four contiguous DCT blocks selected among nine neighboring blocks in previous frame based on the estimated displacement vector for current block. (b) Schematic diagram of how a pixelwise translated DCT block is extracted from four contiguous DCT blocks. (c) Decomposition of integer-pel DCT-based translation as four matrix multiplication operations. (d) Decomposition of half-pel DCT-based translation as four matrix multiplication operations.

If we define the DCT operation on an $N \times N$ matrix $\mathbf{B}$ as

$$\mathrm{DCT}\{\mathbf{B}\} = \mathbf{D}\mathbf{B}\mathbf{D}^T ,$$

where the (k, m) element of $\mathbf{D}$ is the DCT-II kernel:

$$\mathbf{D}(k, m) = \frac{2}{N} C(k) \cos \frac{k\pi}{N} \left(m + \frac{1}{2} \right), \quad \text{for } k, m = 0, \ldots, N-1 .$$

Therefore, $\mathbf{D}^T\mathbf{D} = \frac{2}{N}\mathbf{I}_N$. The formation of the DCT of $\hat{\mathbf{B}}_{ref}$ in the DCT domain can be described in this equation:

$$\mathrm{DCT}\{\hat{\mathbf{B}}_{ref}\} = \left(\frac{N}{2}\right)^2 \sum_{k=1}^{4} \mathrm{DCT}\{\mathbf{H}_k\}\mathrm{DCT}\{\mathbf{B}_k\}\mathrm{DCT}\{\mathbf{V}_k\} . \tag{9.100}$$

This corresponds to pre- and post-multiplication of the DCT transformed $\mathbf{H}_k$ and $\mathbf{V}_k$ with the DCT of $\mathbf{B}_k$ since DCT is a unitary orthogonal transformation and is guaranteed to be distributive to matrix multiplications. The DCT of the motion-compensated residual (displaced frame difference, DFD) for the current block $\mathbf{C}$ is, therefore,

$$\mathrm{DCT}\{DFD\} = \mathrm{DCT}\{\hat{\mathbf{B}}_{ref}\} - \mathrm{DCT}\{\mathbf{C}\} . \tag{9.101}$$

$\mathrm{DCT}\{\mathbf{H}_k\}$ and $\mathrm{DCT}\{\mathbf{V}_k\}$ can be precomputed and stored in the memory. Furthermore, many high-frequency coefficients of $\mathrm{DCT}\{\mathbf{B}_k\}$ or displacement estimates are zero (i.e., sparse and block-aligned reference blocks), making the actual number of computations in (9.100) small. In [30], simulation results show that the DCT domain approach is faster than the spatial domain approach by about 10 to 30%. Further simplication is also possible (as seen from Figure 9.19b):

$$\mathbf{H}_U = \mathbf{H}_1 = \mathbf{H}_3, \ \mathbf{H}_L = \mathbf{H}_2 = \mathbf{H}_4, \ \mathbf{V}_L = \mathbf{V}_1 = \mathbf{V}_2, \ \mathbf{V}_R = \mathbf{V}_3 = \mathbf{V}_4. \tag{9.102}$$

Therefore, only four windowing/shifting matrices need to be accessed from the memory instead of eight.

At first sight, the DCT-based approach requires more computations than the pixel-based approach. The pixel-based approach includes the IDCT and DCT steps. Thus, it requires n^2 additions (or subtractions) and 4 $n \times n$ matrix multiplications (for IDCT and DCT) to calculate the DCT coefficients of the motion-compensated residue. In contrast, the DCT-based approach needs $4n^2$ additions plus 8 $n \times n$ matrix multiplications. However, savings can be achieved by employing the sparseness of the DCT coefficients, which is the basis of DCT compression. More savings come at the subpixel level. The DCT-based approach requires no extra operations, whereas the pixel-based approach needs interpolation (i.e., to handle 4 (2×2) times bigger images). In [34], further savings in the computation of the windowing/shifting matrices is made by using fast DCT. A 47% reduction in computational complexity has been reported with fast DCT over the brute-force method without the assumption of sparseness, and a 68% reduction with only the top-left 4×4 subblocks being nonzero can be achieved with the use of fast DCT.

9.7.2 Subpixel DCT-Based Motion Compensation

For the case of subpixel motion, interpolation is used to predict interpixel values. According to the MPEG standards, bilinear interpolation is recommended for its simplicity in implementation and effectiveness in prediction [13, 21], although it is well known that a range of other interpolation functions, such as cubic, spline, Gaussian, and Lagrange interpolations, can provide better approximation accuracy and more pleasant visual quality [15, 27, 35, 36]. The complexity argument is true if the interpolation operation is performed in the spatial domain, but in the DCT domain, it is possible to employ better interpolation functions than the bilinear interpolation without any additional computational load increase.

Interpolation Filter

For simplicity of derivations, we start with the one-dimensional half-pel bilinear interpolation and then proceed to the two-dimensional case of quarter-pel accuracy with other

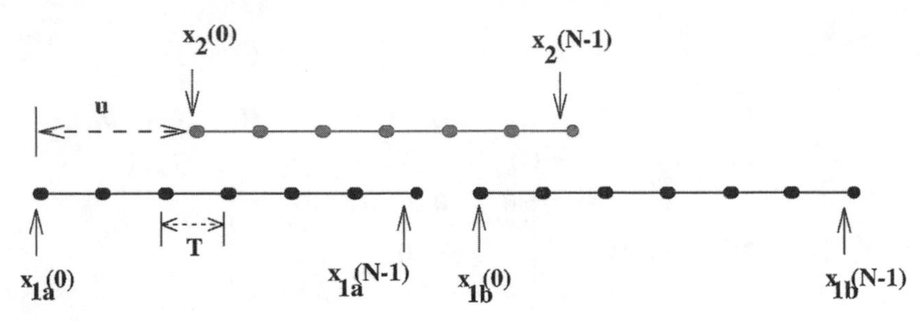

FIGURE 9.20
Illustration of extraction of the subpel displaced block, $x_2(n)$, from two adjacent 1D blocks, $x_{1a}(n)$ and $x_{1b}(n)$, with bilinear interpolation.

interpolation functions. Consider two one-dimensional adjacent blocks, $x_{1a}(n)$ and $x_{1b}(n)$ for $n = 0, \ldots, N-1$, as shown in Figure 9.20. We want to extract a block $\{x_2(n)\}_{n=0}^{N-1}$ of displaced u pixels to the right of $x_{1a}(0)$ where u is supposed to be an odd multiple of 0.5 (i.e., half-pel motion). Therefore, we can show that

$$x_2(n) = \begin{cases} \frac{1}{2}[x_{1a}(N+n-i) + x_{1a}(N+n-i+1)], & 0 \le n \le i-2\,, \\ \frac{1}{2}[x_{1a}(N-1) + x_{1b}(0)], & n = i-1\,, \\ \frac{1}{2}[x_{1b}(n-i) + x_{1b}(n-i+1)], & N-1 \ge n \ge i\,, \end{cases} \tag{9.103}$$

where $i = \lceil u \rceil$. In the matrix form,

$$\vec{\mathbf{x}}_2 = \mathbf{G}_{BL}(i)\vec{\mathbf{x}}_{1a} + \mathbf{G}_{BR}(i)\vec{\mathbf{x}}_{1b}\,, \tag{9.104}$$

where $\vec{\mathbf{x}}_2$, $\vec{\mathbf{x}}_{1a}$, and $\vec{\mathbf{x}}_{1b}$ are the column vectors of $x_2(n)$, $x_{1a}(n)$, and $x_{1b}(n)$, respectively, and $\mathbf{G}_{BL}(i)$ and $\mathbf{G}_{BR}(i)$ are defined as follows:

$$\begin{aligned} \mathbf{G}_{BL}(i) &= \frac{1}{2}\left\{\begin{bmatrix} \mathbf{0} & \mathbf{I}_i \\ \mathbf{0} & \mathbf{0} \end{bmatrix} + \begin{bmatrix} \mathbf{0} & \mathbf{I}_{i-1} \\ \mathbf{0} & \mathbf{0} \end{bmatrix}\right\}, \\ \mathbf{G}_{BR}(i) &= \frac{1}{2}\left\{\begin{bmatrix} \mathbf{0} & \mathbf{0} \\ \mathbf{I}_{N-i} & \mathbf{0} \end{bmatrix} + \begin{bmatrix} \mathbf{0} & \mathbf{0} \\ \mathbf{I}_{N-i+1} & \mathbf{0} \end{bmatrix}\right\}. \end{aligned} \tag{9.105}$$

In the DCT domain,

$$\text{DCT}\{\vec{\mathbf{x}}_2\} = \text{DCT}\{\mathbf{G}_{BL}(i)\}\text{DCT}\{\vec{\mathbf{x}}_{1a}\} + \text{DCT}\{\mathbf{G}_{BR}(i)\}\text{DCT}\{\vec{\mathbf{x}}_{1b}\}\,. \tag{9.106}$$

Here $\mathbf{G}_{BL}(i)$ and $\mathbf{G}_{BR}(i)$ can be regarded as bilinear interpolation filter matrices, which act as a linear filter or transform. Therefore, $\mathbf{G}_{BL}(i)$ and $\mathbf{G}_{BR}(i)$ can be replaced by any FIR filter or interpolation function of finite duration (preferably with the length much smaller than the block size N).

Bilinear Interpolated Subpixel Motion Compensation

For the 2D case, if (u, v) is the displacement of the reconstructed block $\hat{\mathbf{B}}_{ref}$ measured from the upper left corner of the block $\mathbf{B}_1$, then for $h_U = \lceil u \rceil$ and $v_L = \lceil v \rceil$,

$$\text{DCT}\{\hat{\mathbf{B}}_{ref}\} = \sum_{k=1}^{4} \text{DCT}\{\mathbf{H}_k\}\text{DCT}\{\mathbf{B}_k\}\text{DCT}\{\mathbf{V}_k\}\,, \tag{9.107}$$

where

$$\mathbf{H}_1 = \mathbf{H}_3 = \mathbf{H}_U = \mathbf{G}_{BL}(h_U),\ \mathbf{H}_2 = \mathbf{H}_4 = \mathbf{H}_L = \mathbf{G}_{BR}(h_U), \tag{9.108}$$

$$\mathbf{V}_1 = \mathbf{V}_2 = \mathbf{V}_L = \mathbf{G}^T_{BL}(v_L),\ \mathbf{V}_3 = \mathbf{V}_4 = \mathbf{V}_R = \mathbf{G}^T_{BR}(v_L)\,. \tag{9.109}$$

$$[\mathbf{G}_{BL}(h_U)\ \mathbf{G}_{BR}(h_U)] = \begin{bmatrix} 0 \cdots 0\ 0.5\ 0.5\ \ 0\ \ \cdots \\ \vdots\ \ddots\ \vdots\ \ 0\ \ \ddots\ \ \ddots\ \ 0 \\ 0 \cdots 0 \cdots\ \ 0\ \ 0.5\ 0.5 \end{bmatrix}. \tag{9.110}$$

Once again, $\mathbf{G}_{BL}(\cdot)$ and $\mathbf{G}_{BR}(\cdot)$ can be precomputed and stored in the memory as in the case of integer-pel motion compensation and thus the extra computational load for doing bilinear interpolation is eliminated.

Cubic Interpolated Subpixel Motion Compensation

Three different interpolation functions, namely cubic, cubic spline, and bilinear interpolations, are plotted in Figure 9.21a. As can be seen, the bilinear interpolation has the shortest filter length and the cubic spline has the longest ripple but the cubic spline also has the smallest

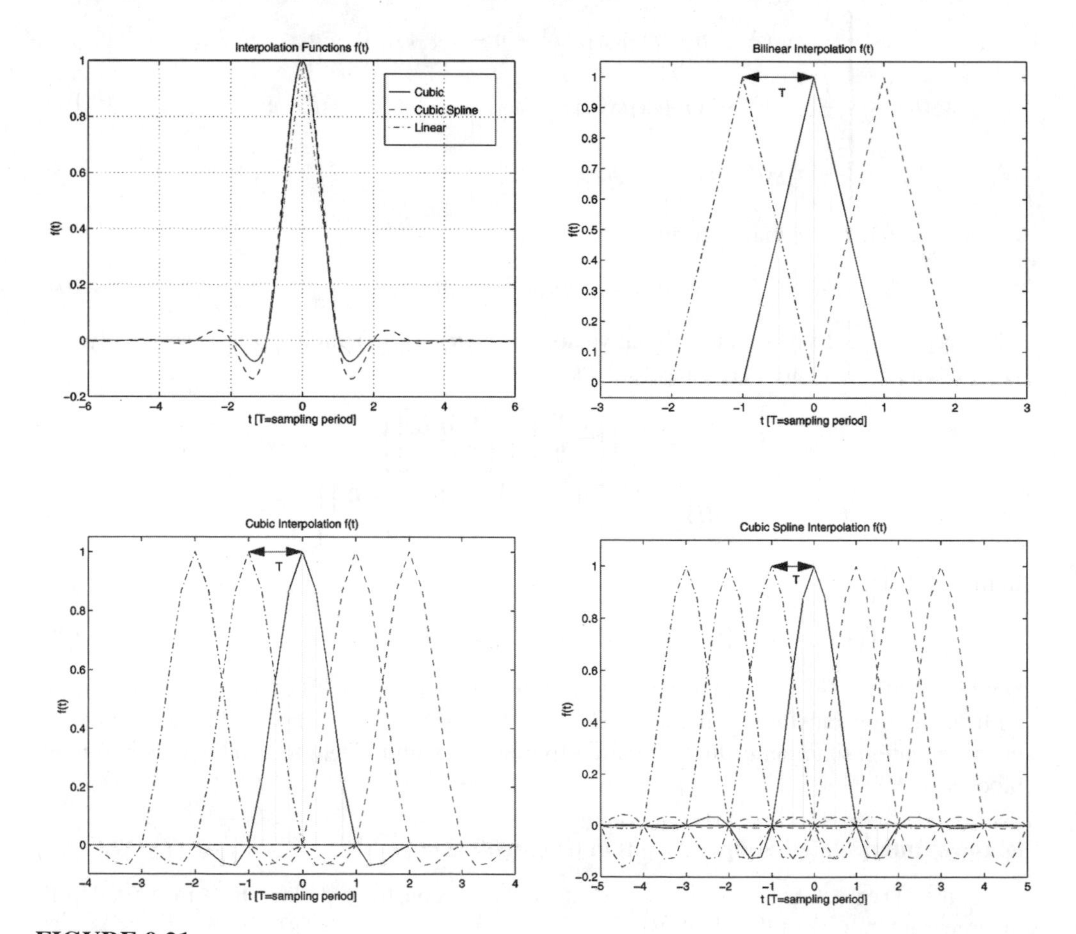

FIGURE 9.21
(a) Plots of different interpolation functions. (b), (c), and (d) depict how to form a pre- or post-multiplication matrix for half-pel or even quarter-pel DCT-based motion compensation.

approximation error of the three [36]. To compromise between filter length and approximation accuracy, we choose the cubic interpolation in the simulation. By choosing the resolution of the filter as half a pixel length, the bilinear interpolation $f_{hb}(n) = [0.5, 1, 0.5]$ and the cubic interpolation $f_{hc}(n) = [-0.0625, 0, 0.5625, 1.0000, 0.5625, 0, -0.0625]$. From Figure 9.21b, it is clear that the contributions at the half-pel position from all the pixel values are summed up and give rise to the bilinear filter matrices $\mathbf{G}_{BL}(\cdot)$ and $\mathbf{G}_{BR}(\cdot)$. In a similar way, as in Figure 9.21c, the cubic filter matrices $\mathbf{G}_{CL}(\cdot)$ and $\mathbf{G}_{CR}(\cdot)$ can be defined as

$$\begin{aligned}
\mathbf{G}_{CL}(i) &= \begin{bmatrix} \mathbf{0} & -0.0625\mathbf{I}_{i+1} \\ \mathbf{0} & \mathbf{0} \end{bmatrix} + \begin{bmatrix} \mathbf{0} & 0.5625\mathbf{I}_{i} \\ \mathbf{0} & \mathbf{0} \end{bmatrix} \\
&+ \begin{bmatrix} \mathbf{0} & 0.5625\mathbf{I}_{i-1} \\ \mathbf{0} & \mathbf{0} \end{bmatrix} + \begin{bmatrix} \mathbf{0} & -0.0625\mathbf{I}_{i-2} \\ \mathbf{0} & \mathbf{0} \end{bmatrix}, \\
\mathbf{G}_{CR}(i) &= \begin{bmatrix} \mathbf{0} & \mathbf{0} \\ -0.0625\mathbf{I}_{N-i-1} & \mathbf{0} \end{bmatrix} + \begin{bmatrix} \mathbf{0} & \mathbf{0} \\ 0.5625\mathbf{I}_{N-i} & \mathbf{0} \end{bmatrix} \\
&+ \begin{bmatrix} \mathbf{0} & \mathbf{0} \\ 0.5625\mathbf{I}_{N-i+1} & \mathbf{0} \end{bmatrix} + \begin{bmatrix} \mathbf{0} & \mathbf{0} \\ -0.0625\mathbf{I}_{N-i+2} & \mathbf{0} \end{bmatrix}.
\end{aligned}$$

Here $\mathbf{G}_{CL}(\cdot)$ and $\mathbf{G}_{CR}(\cdot)$ can be precomputed and stored. Therefore, its computational complexity remains the same as both integer-pel and half-pel bilinear interpolated DCT-based motion compensation methods. The reconstructed DCT block and the corresponding motion-compensated residual can be obtained in a similar fashion:

$$\mathrm{DCT}\{\hat{\mathbf{B}}_{ref}\} = \sum_{k=1}^{4} \mathrm{DCT}\{\mathbf{H}_k\}\mathrm{DCT}\{\mathbf{B}_k\}\mathrm{DCT}\{\mathbf{V}_k\}, \tag{9.111}$$

$$\mathrm{DCT}\{DFD\} = \mathrm{DCT}\{\hat{\mathbf{B}}_{ref}\} - \mathrm{DCT}\{\mathbf{C}\}, \tag{9.112}$$

where

$$\mathbf{H}_1 = \mathbf{H}_3 = \mathbf{H}_U = \mathbf{G}_{CL}(h_U), \ \mathbf{H}_2 = \mathbf{H}_4 = \mathbf{H}_L = \mathbf{G}_{CR}(h_U), \tag{9.113}$$

$$\mathbf{V}_1 = \mathbf{V}_2 = \mathbf{V}_L = \mathbf{G}_{CL}^T(v_L), \ \mathbf{V}_3 = \mathbf{V}_4 = \mathbf{V}_R = \mathbf{G}_{CR}^T(v_L). \tag{9.114}$$

This idea can be extended to other interpolation functions such as sharped Gaussian [35] and quarter-pel accuracy.

9.7.3 Simulation

Simulation is performed on the Infrared Car and Miss America sequences to demonstrate the effectiveness of our bilinear and cubic motion compensation methods.

The first set of simulations subsamples each picture $I_t(i, j)$ from the sequences (i.e., $y(i, j) = I_t(2*i, 2*j)$) and then this shrunken picture $y(i, j)$ is displaced by a half-pel motion vector (arbitrarily chosen as (2.5, 1.5)) with both bilinear and cubic interpolated motion compensation methods. The MSEs are computed as $MSE = \frac{\sum_{i,j}[\hat{x}(i,j)-x(i,j)]^2}{N^2}$ by treating the original unsampled pixels $I_t(2*i+1, 2*j+1)$ as the reference picture $x(i, j) = I_t(2*i+1, 2*j+1)$ where $\hat{x}(i, j)$ is the predicted pixel value from $y(i, j)$. As shown in Figure 9.22, the zero-order interpolation is also simulated for comparison. The zero-order interpolation, also called sample-and-hold interpolation, simply takes the original pixel value as the predicted half-pel pixel value [15]. As can be seen in Figure 9.22, both the bilinear and cubic methods have much lower MSE values than the zero-order method, and the cubic method performs much better than the bilinear counterpart without increased computational load.

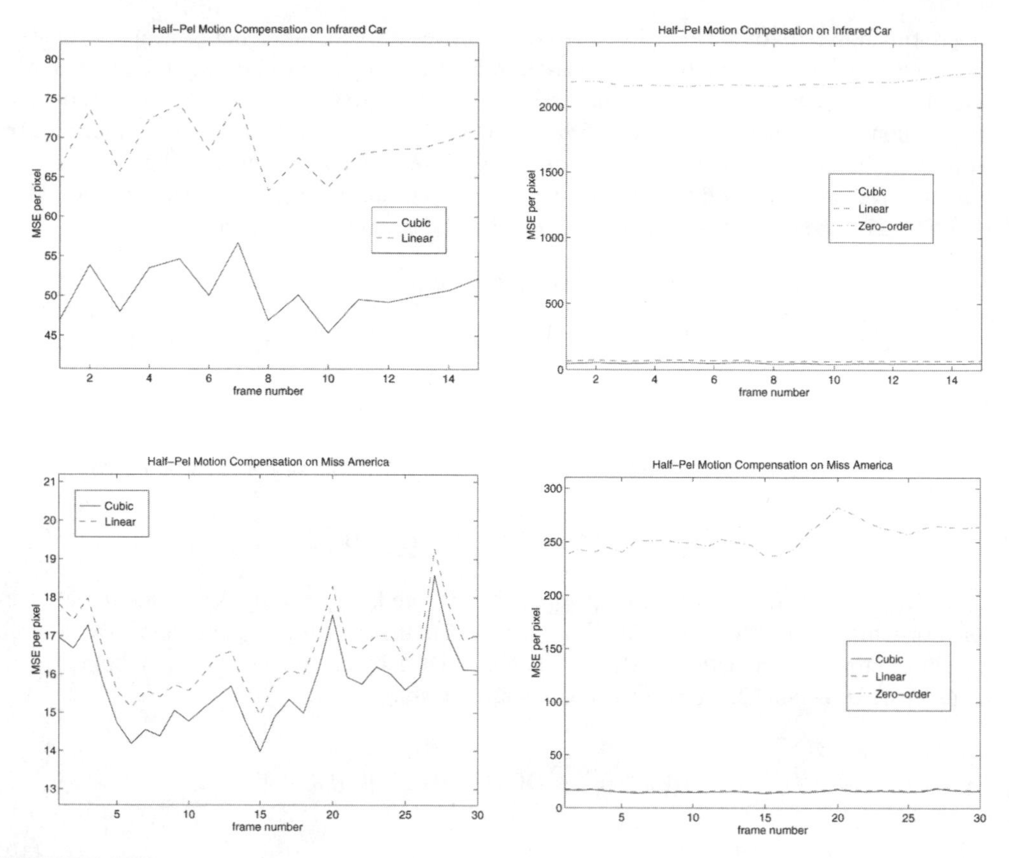

FIGURE 9.22
Pictures from the Infrared Car and Miss America sequences are subsampled and displaced by a half-pel motion vector with different motion compensation methods. The MSE-per-pixel values are obtained by comparing the original unsampled pixel values with the predicted pixel values of the motion-compensated residuals. Zero-order interpolation means replication of sampled pixels as the predicted pixel values.

Figure 9.23 shows the results of another set of simulations in which the subpixel DCT-based motion compensation algorithms generate motion-compensated residuals of the Infrared Car and Miss America sequences based on the displacement estimates of the full-search block-matching algorithm, where the residuals are used to compute the MSE and BPS values for comparison. It can be seen that the cubic interpolation approach achieves lower MSE and BPS values than the bilinear interpolation.

9.8 Conclusion

In this chapter, we propose a fully DCT-based motion-compensated video coder structure that not only is compliant with the hybrid motion-compensated DCT video coding standards but also has a higher system throughput and a lower overall complexity than the conventional video coder structure due to removal of the DCT and IDCT from the feedback loop. Therefore, it is

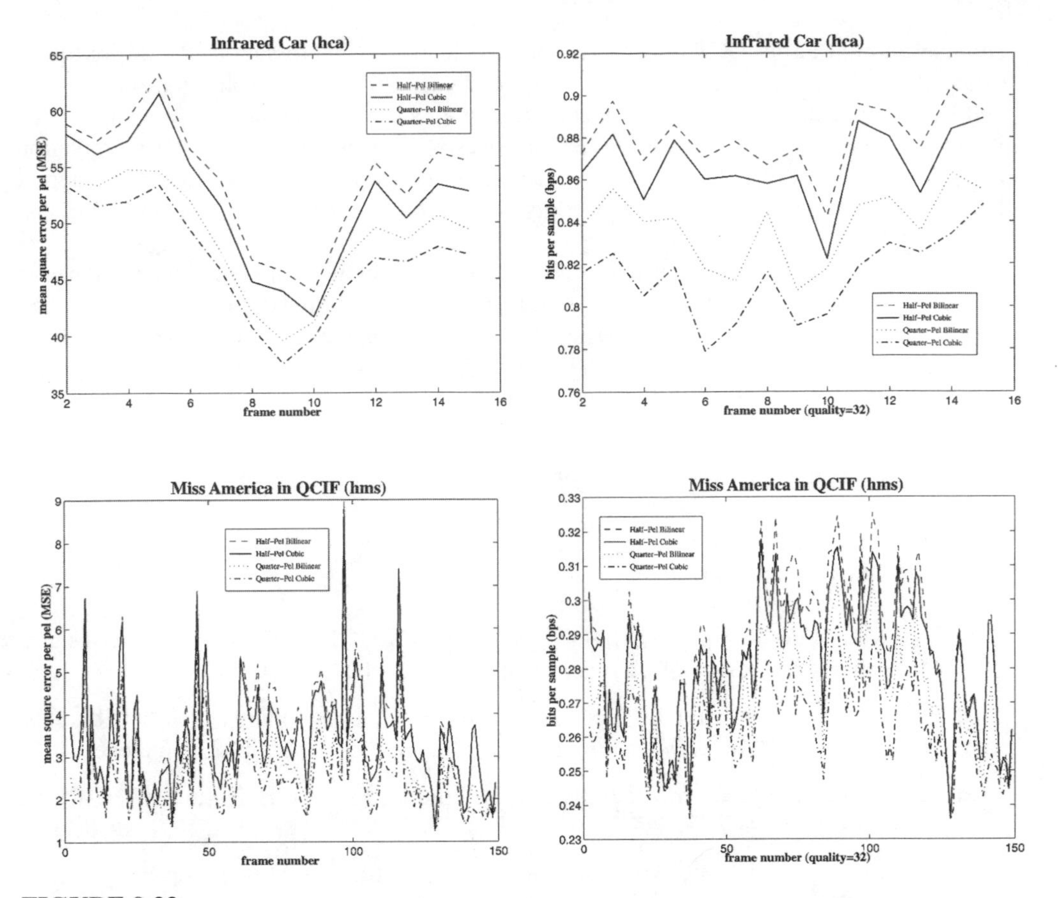

FIGURE 9.23
Pictures from the Infrared Car and Miss America sequences are subsampled and displaced by a half-pel motion vector with different motion compensation methods. The MSE-per-pixel values are obtained by comparing the original unsampled pixel values with the predicted pixel values of the motion-compensated residuals. Zero-order interpolation means replication of sampled pixels as the predicted pixel values.

more suitable for high-quality and high-bit-rate video applications such as HDTV or low-cost video coder implementation. To realize such a fully DCT-based coder, we develop DCT-based motion estimation and compensation algorithms.

The DCT pseudo-phase techniques that we develop provide us the means to estimate shifts in the signals in the DCT domain. The resulting DXT-ME algorithm has low computational complexity, $O(N^2)$, as compared to $O(N^4)$ for BKM-ME. Its performance over several image sequences is comparable with that achieved by BKM-ME and some fast-search approaches such as TSS, LOG, and SUB. Furthermore, its DCT-based nature enables us to incorporate its implementation with the DCT codec design to gain further savings in complexity and take advantage of advances in research on the DCT codec design. Finally, the DXT-ME algorithm has inherently highly parallel operations in computing the pseudo-phases and thus it is very suitable for VLSI implementation [37].

To deal with subpixel motion, we extend the DCT techniques to the subpixel level and derive the subpel sinusoidal orthogonal principles. We demonstrate that subpixel motion information is preserved in the DCT coefficients under the Nyquist condition. This fact enables us to develop the DCT-based half-pel and quarter-pel motion estimation algorithms to estimate subpixel motion in the DCT domain without any interpixel interpolation at a desired level of accuracy. This results in significant savings in computational complexity for interpolation and far less data flow compared to the conventional block-matching methods on interpolated images. Furthermore, it avoids the deterioration of estimation precision caused by interpolation and provides flexibility in the design in such a way that the same hardware can support different levels of required accuracy with a complexity $O(N^2)$, far less than $O(N^4)$ for BKM-ME and its subpixel versions.

Finally, we discuss the integer-pel DCT-based motion compensation method and develop the subpel DCT-based motion compensation schemes using the bilinear and cubic interpolation functions. We show that without increasing the number of computations, the cubic interpolated half-pel and quarter-pel schemes exhibit higher coding gain in terms of smaller MSE and BPS values than the bilinear interpolated counterparts.

References

[1] U.V. Koc, "Low-Complexity, High-Throughput Fully DCT-Based Video Codec Design," Ph.D. thesis, University of Maryland, College Park, MD, July 1996.

[2] H. Li, A. Lundmark, and R. Forchheimer, "Image sequence coding at very low bitrates: A review," *IEEE Trans. Image Processing,* vol. 3, no. 5, pp. 589–608, September 1994.

[3] M.A. Tekalp, *Digital Video Processing,* 1st edition, Prentice-Hall, Englewood Cliffs, NJ, 1995.

[4] J.S. McVeigh and S.-W. Wu, "Comparative study of partial closed-loop versus open-loop motion estimation for coding of HDTV," in *Proc. IEEE Workshop on Visual Signal Processing and Communications,* New Brunswick, September 1994, pp. 63–68.

[5] U.V. Koc and K.J.R. Liu, "DCT-based motion estimation," *IEEE Trans. Image Processing,* vol. 7, no. 7, pp. 948–965, July 1998.

[6] U.V. Koc and K.J.R. Liu, "Interpolation-free subpixel motion estimation techniques in DCT domain," *IEEE Trans. Circuits and Systems for Video Technology,* vol. 8, no. 4, pp. 460–487, August 1998.

[7] M. Ziegler, "Hierarchical motion estimation using the phase correlation method in 140 mbit/s HDTV-coding," in *Signal Processing of HDTV, II,* Turin, Italy, 1990, pp. 131–137.

[8] P. Yip and K.R. Rao, "On the shift property of DCT's and DST's," *IEEE Trans. Acoustics, Speech, and Signal Processing,* vol. ASSP-35, no. 3, pp. 404–406, March 1987.

[9] C.T. Chiu and K.J.R. Liu, "Real-time parallel and fully pipelined two-dimensional DCT lattice structures with applications to HDTV systems," *IEEE Trans. Circuits and Systems for Video Technology,* vol. 2, no. 1, pp. 25–37, March 1992.

[10] K.J.R. Liu and C.T. Chiu, "Unified parallel lattice structures for time-recursive discrete cosine/sine/hartley transforms," *IEEE Trans. Signal Processing,* vol. 41, no. 3, pp. 1357–1377, March 1993.

[11] K.J.R. Liu, C.T. Chiu, R.K. Kologotla, and J.F. JaJa, "Optimal unified architectures for the real-time computation of time-recursive discrete sinusoidal transforms," *IEEE Trans. Circuits and Systems for Video Technology,* vol. 4, no. 2, pp. 168–180, April 1994.

[12] CCITT Recommendation H.261, *Video Codec for Audiovisual Services at p × 64 kbit/s,* CCITT, August 1990.

[13] CCITT Recommendation MPEG-1, *Coding of Moving Pictures and Associated Audio for Digital Storage Media at up to about 1.5 Mbit/s,* ISO/IEC 11172, Geneva, Switzerland, 1993.

[14] A. Zakhor and F. Lari, "Edge-based 3-D camera motion estimation with application to video coding," *IEEE Trans. Image Processing,* vol. 2, no. 4, pp. 481–498, October 1993.

[15] A.K. Jain, *Fundamentals of Digital Image Processing,* Prentice-Hall, Englewood Cliffs, NJ, 1989.

[16] W.D. Kou and T. Fjallbrant, "A direct computation of DCT coefficients for a signal block taken from 2 adjacent blocks," *Signal Processing,* vol. 39, no. 7, pp. 1692–1695, 1991.

[17] B. Liu and A. Zaccarin, "New fast algorithms for the estimation of block motion vectors," *IEEE Trans. Circuits and Systems for Video Technology,* vol. 3, no. 2, pp. 148–157, April 1993.

[18] S.-L. Iu, "Comparison of motion compensation using different degrees of sub-pixel accuracy for interfield/interframe hybrid coding of HDTV image sequences," in *1992 IEEE Int. Conf. Acoustics, Speech, Signal Processing,* San Francisco, CA, 1992, vol. 3, pp. 465–468.

[19] B. Girod, "Motion compensation: Visual aspects, accuracy, and fundamental limits," in *Motion Analysis and Image Sequence Processing,* M.I. Sezan and R.L. Lagendijk, Eds., chapter 5, Kluwer Academic, 1993.

[20] B. Girod, "Motion-compensating prediction with fractional-pel accuracy," *IEEE Trans. Communications,* vol. 41, no. 4, pp. 604, April 1993.

[21] CCITT Recommendation MPEG-2, *Generic Coding of Moving Pictures and Associated Audio,* ISO/IEC 13818, Geneva, Switzerland, 1994, H.262.

[22] S.-I. Uramoto, A. Takabatake, and M. Yoshimoto, "A half-pel precision motion estimation processor for NTSC-resolution video," *IEICE Trans. Electronics,* vol. 77, no. 12, p. 1930, December 1994.

[23] T. Akiyama, H. Aono, K. Aoki, K.W. Ler, B. Wilson, T. Araki, T. Morishige, H. Takeno, A. Sato, S. Nakatani, and T. Senoh, "MPEG2 video codec using image compression DSP," *IEEE Trans. Consumer Electronics,* vol. 40, pp. 466–472, 1994.

[24] D. Brinthaupt, L. Letham, V. Maheshwari, J. Othmer, R. Spiwak, B. Edwards, C. Terman, and N. Weste, "A video decoder for H.261 video teleconferencing and MPEG stored interactive video applications," in *1993 IEEE International Solid-State Circuits Conference,* San Francisco, CA, 1993, pp. 34–35.

[25] G. Madec, "Half pixel accuracy in block matching," in *Picture Coding Symp.,* Cambridge, MA, March 1990.

[26] G. de Haan and W.A.C. Biezen, "Sub-pixel motion estimation with 3-D recursive search block-matching," *Signal Processing: Image Communication,* vol. 6, no. 3, pp. 229–239, June 1994.

[27] R.W. Schafer and L.R. Rabiner, "A digital signal processing approach to interpolation," *Proceedings of the IEEE,* pp. 692–702, June 1973.

[28] A. Papoulis, *Signal Analysis,* McGraw-Hill, New York, 1977.

[29] S.F. Chang and D.G. Messerschmitt, "A new approach to decoding and compositing motion-compensated DCT-based images," in *Proc. IEEE Int. Conf. Acoustics, Speech, Signal Processing,* 1993, vol. 5, pp. 421–424.

[30] S.-F. Chang and D.G. Messerschmitt, "Manipulation and compositing of MC-DCT compressed video," *IEEE Journal on Selected Areas in Communications,* vol. 13, no. 1, p. 1, January 1995.

[31] Y.Y. Lee and J.W. Woods, "Video post-production with compressed images," *SMPTE Journal,* vol. 103, pp. 76–84, February 1994.

[32] B.C. Smith and L. Rowe, "Algorithms for manipulating compressed images," *IEEE Comput. Graph. Appl.,* pp. 34–42, September 1993.

[33] J.B. Lee and B.G. Lee, "Transform domain filtering based on pipelining structure," *IEEE Trans. Signal Processing,* vol. 40, pp. 2061–2064, August 1992.

[34] N. Merhav and V. Bhaskaran, "A fast algorithm for DCT-domain inverse motion compensation," in *Proc. IEEE Int. Conf. Acoustics, Speech, Signal Processing,* 1996, vol. 4, pp. 2309–2312.

[35] W.F. Schreiber, *Fundamentals of Electronic Imaging Systems — Some Aspects of Image Processing,* 3rd edition, Springer-Verlag, Berlin, 1993.

[36] H. Hou and H.C. Andrews, "Cubic splines for image interpolation and digital filtering," *IEEE Trans. Acoustics, Speech, Signal Processing,* vol. 26, no. 6, pp. 508–517, 1978.

[37] J. Chen and K.J.R. Liu, "A complete pipelined parallel CORDIC architecture for motion estimation," *IEEE Trans. Circuits and Systems — II: Analog and Digital Signal Processing,* vol. 45, pp. 653–660, June 1998.

Chapter 10

Object-Based Analysis–Synthesis Coding Based on Moving 3D Objects

Jörn Ostermann

10.1 Introduction

For the coding of moving images with low data rates between 64 kbit/s and 2 Mbit/s, a block-based hybrid coder has been standardized by the ITU-T [8] where each image of a sequence is subdivided into independently moving blocks of size 16×16 picture elements (pels). Each block is coded by 2D motion-compensated prediction and transform coding [56]. This corresponds to a source model of "2D square blocks moving translationally in the image plane," which fails at boundaries of naturally moving objects and causes coding artifacts known as blocking and mosquito effects at low data rates.

In order to avoid these coding distortions, several different approaches to video coding have been proposed in the literature. They can be categorized into four methods: region-based coding, object-based coding, knowledge-based coding, and semantic coding.

Region-based coding segments an image into regions of homogeneous texture or color [36]. Usually, these regions are not related to physical objects. Regions are allowed to move and change their shape and texture over time. Some recent proposals merge regions with dissimilar texture but with similar motion into one entity in order to increase coding efficiency [12].

The concept of *object-based analysis–synthesis coding* (OBASC) aiming at a data rate of 64 kbit/s and below was proposed in [44]. A coder based on this concept divides an image sequence into moving objects. An object is defined by its uniform motion and described by motion, shape, and color parameters, where color parameters denote luminance and chrominance reflectance of the object surface. Those parts of an image that can be described with sufficient accuracy by moving objects require the transmission of motion and shape parameters only, since the texture of the previously coded objects can be used. The remaining image areas are called areas of model failure. They require the transmission of shape and texture parameters in order to generate a subjectively correct decoded image. This detection of model failures can be adapted to accommodate properties of the human visual system. OBASC in its basic form does not require any a priori knowledge of the moving objects. A first implementation of an OBASC presented in 1991 was based on the source model of "moving flexible 2D objects" (F2D) and was used for coding image sequences between 64 and 16 kbit/s [3, 18, 26]. Implementations of an OBASC based on the source models of "moving rigid 3D objects" (R3D) and "moving flexible 3D objects" (F3D) using 3D motion and 3D shape are presented in [48] and [49], respectively. In [13], an implicit 3D shape representation is proposed.

Knowledge-based coders [31] use a source model which is adapted to a special object. In contrast to OBASC, this allows the encoding of only a special object like a face. However, due to this adaptation of the source model to the scene contents, a more efficient encoding becomes possible. A recognition algorithm is required to detect the object in the video sequence. For encoding of faces, a predefined 3D face model gets adapted to the face in the sequence. Then, the motion parameters of the face are estimated and coded. Perhaps the most challenging task for a knowledge-based encoder is the reliable detection of the face position [20, 34, 60].

Semantic coders [15] are modeled after knowledge-based coders. Until now, semantic coding was mainly investigated for the encoding of faces using high-level parameters such as the facial action coding system [2, 11, 14, 21, 22, 68]. Alternatively, facial animation parameters as defined by MPEG-4 can be used [30, 53]. By using high-level parameters we limit the degrees of freedom of the object and achieve a higher data reduction.

Knowledge-based as well as semantic coders use three-dimensional source models. Since they can encode only a particular object, they require a different algorithm for encoding the image areas outside this object. An OBASC based on a 3D source model seems to be a natural choice.

The purpose of this chapter is twofold. First, the concept of object-based analysis–synthesis coding is reviewed. Second, different source models used for OBASC and their main properties are compared. In order to use 3D source models, a reliable motion estimation algorithm is required. Here, we develop a robust gradient-based estimator that is able to track objects. The coding efficiencies obtained with the source models F2D [27], R3D, and F3D are compared in terms of data rate required for the same picture quality. The coding schemes will be evaluated using videophone test sequences. As a well-known reference for picture quality, the block-based hybrid coder H.261 [8, 9] is used.

In Section 10.2, the principles of object-based analysis–synthesis coding are reviewed. In Section 10.3, different 2D and 3D source models for OBASC and their implementations in an OBASC are presented. Since the data rate of an OBASC depends mainly on how well the image analysis can track the moving objects, we present in Section 10.4 details of image analysis. A robust 3D motion estimator is developed. Model failure detection is discussed in more detail. In Section 10.5, we present an overview of parameter coding. The coding efficiency of the different source models is compared in Section 10.6. A final discussion concludes this book chapter.

10.2 Object-Based Analysis–Synthesis Coding

The goal of OBASC is the efficient encoding of image sequences. Each image of the sequence is called a *real image*. OBASC [44] subdivides each real image into moving objects called *real objects*. A real object is topologically connected and characterized by its uniform motion. A real object is modeled by a *model object* as defined by the source model of the encoder. Hence, one real object is described by one model object, whereas region-based coding describes regions with homogeneous textures as separate entities. Each model object m is described by three sets of parameters, $A^{(m)}$, $M^{(m)}$, and $S^{(m)}$, defining its motion, shape, and color, respectively. Motion parameters define the position and motion of the object; shape parameters define its shape. Color parameters denote the luminance as well as the chrominance reflectance on the surface of the object. In computer graphics, they are sometimes called texture. The precise meaning of the three parameter sets depends on the source model employed (see Section 10.3).

Figure 10.1 is used to explain the concept and structure of OBASC. An OBASC consists of five parts: image analysis, parameter coding, parameter decoding, parameter memory, and image synthesis. Instead of the frame memory used in block-based hybrid coding, OBASC requires a *memory for parameters* in order to store the coded, transmitted, and decoded parameters A', M', and S' for all objects. Whereas the double prime ($''$) symbol marks the transmitted parameters used to update the parameter memory, the prime ($'$) symbol marks the decoded parameters at the output of the parameter memory.

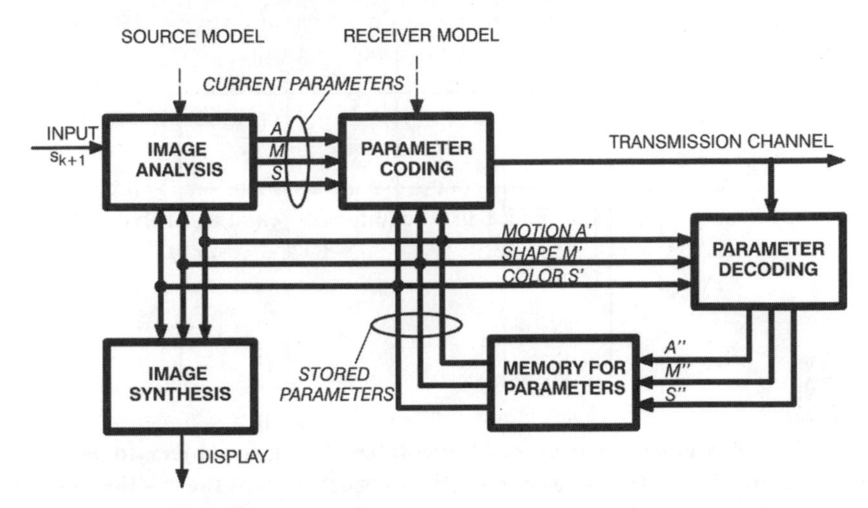

FIGURE 10.1
Block diagram of an object-based analysis–synthesis coder.

The parameter memories in the coder and decoder contain the same information. Evaluating these parameter sets, *image synthesis* produces a model image s'_k, which is displayed at the decoder. In order to avoid annoying artifacts at object boundaries, a shape-dependent antialiasing filter may be applied at object boundaries [58].

At time instant $k+1$, the image analysis has to evaluate the current image s_{k+1} considering the parameter sets A', M', and S' estimated for image s_k. The task of image analysis is to track each object known from previous frames and detect new moving objects. Each object m is described by three sets of parameters, $A_{m,k+1}$, $M_{m,k+1}$, and $S_{m,k+1}$. These parameter sets are available at the output of the image analysis in PCM format. Considering the previously estimated and coded parameter sets A', M', and S' creates a feedback loop in the encoder. This allows the image analysis to compensate for previous estimation errors as well as shape and motion quantization errors introduced by the lossy encoding of the parameters by parameter coding. Hence, an accumulation of estimation and quantization errors is avoided.

Figure 10.2 serves as an example to describe the parameter sets in the case of a source model of rigid 2D objects with 2D motion. The color parameters of an object can be covered and uncovered due to (1) camera motion, (2) a new object entering the scene, (3) motion of another object, or (4) egomotion. In this chapter, we focus on (2) to (4). The extension of OBASC to consider camera motion is straightforward on a conceptual level. Mech and Wollborn describe an implementation in [40]. In the example of Figure 10.2, areas of object 1 get uncovered due to the motion of object 2. Assuming that these parts of object 1 have not been visible before, the color parameters of these *uncovered areas* (UAs) have to be transmitted. Similarly, a rotating 3D object might uncover previously not visible areas for which the transmission of color parameters is required. The color parameters of an uncovered area can uniquely be associated with one object. Therefore, the color parameter $S_{m,k+1}$ of object m at time instant $k+1$ consists of its color parameter $S_{m,k}$ at time instant k and the color parameters $S^{UA}_{m,k+1}$ of

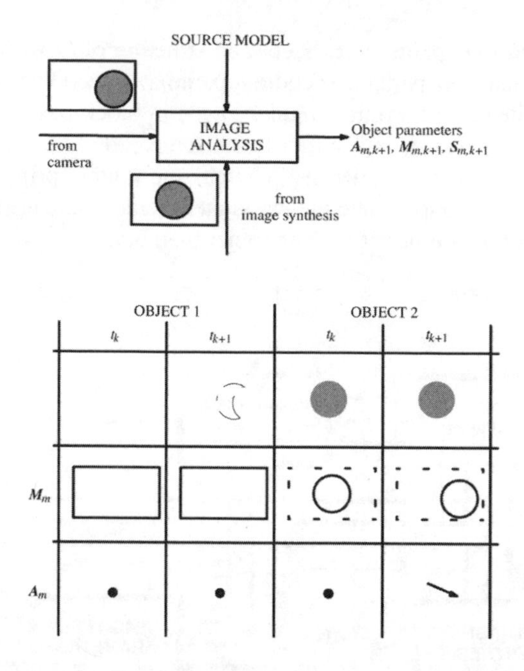

FIGURE 10.2
Image analysis demonstrated using rigid translaterally moving objects in the image plane. The dashed lines denote the image boundaries in order to show the positions of object 2. It is not necessary to know the motion parameters at time instant t_k. The texture parameters of object 1 change due to the uncovered areas.

its uncovered area:

$$S_{m,k+1} = S_{m,k} \cup S^{UA}_{m,k+1} \,. \tag{10.1}$$

If several objects are moving, an uncovered area can belong to a moving object. The shape of the uncovered area can be derived from object motion and object shape. In Figure 10.2, the shape of the uncovered area of object 1 is determined by the shape and motion of object 2 [64].

Most implementations of an image analysis for an OBASC assume moving objects in front of a static background. In the current image, moving and static objects are detected first by means of change detection [23, 45, 52, 64]. For moving objects, new motion and shape parameters are estimated in order to reuse most of the already transmitted color parameters $S_k^{'(m)}$. As pointed out in [25], the estimation of motion and shape parameters are mutually dependent problems. However, in the case of a static background, the correct estimation of motion is the more challenging task of the two [64]. Objects for which motion and shape parameters can be estimated successfully are referred to as *MC objects (model compliance)*. In the final step of image analysis, image areas that cannot be described by MC objects using the transmitted color parameters $S_k^{'(m)}$ and the new motion and shape parameters $A_{k+1}^{(m)} M_{k+1}^{(m)}$, respectively, are detected. Areas of *model failure* (MF) [46] are derived from these areas. They are defined by 2D shape and color parameters only and are referred to as *MF objects*. The detection of MF objects takes into account that small position and shape errors of the MC objects — referred to as *geometrical distortions* — do not disturb subjective image quality. Therefore, MF objects are limited to those image areas with significant differences between the motion- and shape-compensated prediction image and the current image s_{k+1}. They tend to be small in size. This allows coding of color parameters of MF objects with high quality, thus avoiding subjectively annoying quantization errors. Since the transmission of color parameters is expensive in terms of data rate, the total area of MF objects should not be larger than 4% of the image area, assuming 64 kbit/s, CIF (common intermediate format, 352×288 luminance and 176×144 chrominance pels/frame), and 10 Hz.

Depending on the *object class* MC/MF, the parameter sets of each object are coded by *parameter coding* using predictive coding techniques (Figure 10.3). Motion and shape parameters are encoded and transmitted for MC objects and shape and color parameters for MF objects. For MC objects, motion parameters are quantized and encoded. The motion information is used to predict the current shape of the MC object. After motion compensation of the shape, only the shape prediction error has to be encoded. The shape of uncovered areas is derived

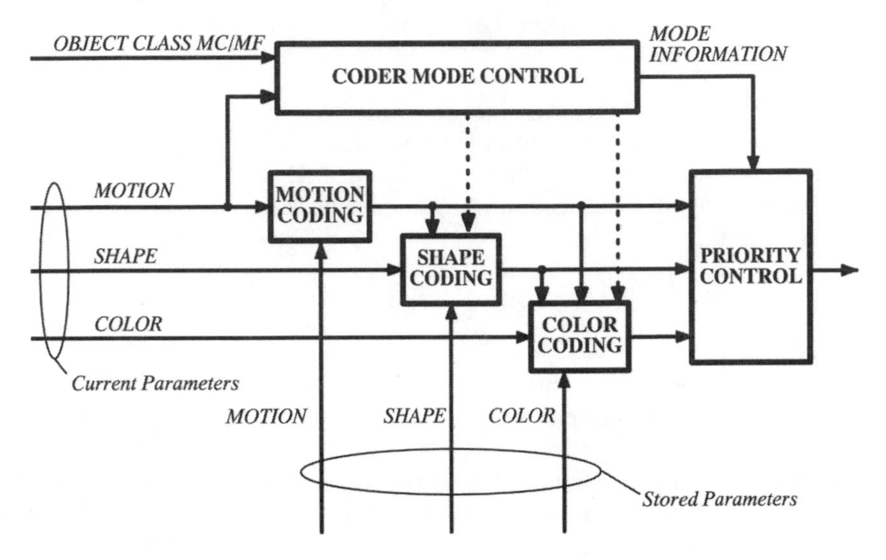

FIGURE 10.3
Block diagram of parameter coding.

from the shape and motion of the MC objects. In the case of uncovered areas being visible for the first time, color parameters have to be transmitted. For MF objects, a temporal prediction of the shape is not useful, since areas of model failure are not temporally correlated. Hence, the shape parameters of MF objects are encoded in intra mode. For the color parameters, the motion-compensated prediction error is computed using the motion parameters of the underlying MC object. Then the prediction error is quantized and encoded. Table 10.1 summarizes which parameter sets have to be transmitted for each object class. As in any block-based coder, color parameters have to be transmitted only in the case of a scene cut. Since the coding of color parameters generally requires a bit rate of more than 1 bit/pel in active areas, the size of the MF objects determines to a large extent the bit rate required for encoding an image sequence. Hence, image analysis should be optimized such that the size of the MF objects becomes small. Furthermore, parameter coding for MF and MC objects has to be optimized in order to minimize the overall data rate $R = R_A + R_M + R_S$ for coding all parameter sets [50].

Parameter decoding decodes the two parameter sets transmitted for each object class. In the *memory for parameters,* the position and shape of MC objects are updated. Furthermore, in areas of model failure, color parameters of MC objects are substituted by the color parameters of the transmitted MF objects. Therefore, only MC objects are available at the output of the parameter memory.

In OBASC, the suitability of source models can be judged by comparing the data rates required for coding the same image sequence with the same image quality. Image quality is influenced mainly by the algorithm for detecting model failures and by the bit rate available

Table 10.1 Parameter Sets That Must Be Coded for MC Objects, Uncovered Areas (UAs), and MF Objects and in the Case of a Scene Cut (SC)

Coder Mode	MC Object Mode		MF Object Mode	
Parameter Set	MC	UA	MF	SC
Motion parameters A	×			
Shape parameter M	×		×	
Color parameter S		×	×	×

for coding the color parameters of model failures. Assuming an image format of CIF with a reduced frame frequency of 10 Hz, an average area of MF objects of 4% of the image area should be sufficient in order to encode a videophone sequence with good subjective quality at a bit rate of 64 kbit/s.

10.3 Source Models for OBASC

In this section, the different source models applied to OBASC, their main properties, as well as some implementation details are presented. In order to highlight commonalities and differences between source models used for OBASC, it is useful to subdivide a source model into its main components, namely the camera model, the illumination model, the scene model, and the object model. The source model used here assumes a 3D *real world* that has to be modeled by a *model world.* Whereas the *real image* is taken by a *real camera* looking into the real world, a *model image* is synthesized using a *model camera* looking into the model world. A world is described by a *scene,* its *illumination,* and its camera. A scene consists of objects, their motion, and their relative position. Initially, the source models are distinguished from each other by the object model. For simplicity, we name the source models according to the name of their object model. Recent research also has focused on illumination models [62, 63].

The goal of the modeling is to generate a model world, W_k, with a model image identical to the real image, s_k, at a time instance k. This implies that the model objects may differ from the real objects. However, similarity between the real object and the model object generally helps in performing proper image analysis.

The following sections will describe the different parts of a source model. After the review of the camera, illumination, and scene model, different object models as applied to OBASC are explained. These object models are used to describe the real objects by means of MC objects. For each object model, parameter coding and some implementation details are highlighted.

10.3.1 Camera Model

The real camera is modeled by a static pinhole camera [65]. Whereas a real image is generated by reading the target of the real camera, a model image is read off the target of the model camera. Assuming a world coordinate system (x, y, z) and an image coordinate system (X, Y), this camera projects the point $\mathbf{P}^{(i)} = (P_x^{(i)}, P_y^{(i)}, P_z^{(i)})^T$ on the surface of an object in the scene onto the point $\mathbf{p}^{(i)} = (p_X^{(i)}, p_Y^{(i)})^T$ of the image plane according to

$$p_X^{(i)} = F \cdot \frac{P_x^{(i)}}{P_z^{(i)}}, \qquad p_Y^{(i)} = F \cdot \frac{P_y^{(i)}}{P_z^{(i)}} \tag{10.2}$$

where F is the focal length of the camera (Figure 10.4). This model assumes that the image plane is parallel to the (x, y) plane of the world coordinate system. For many applications, this camera model is of sufficient accuracy. However, in order to incorporate camera motion, a CAHV camera model [69] allowing for arbitrary camera motion and zoom should be used.

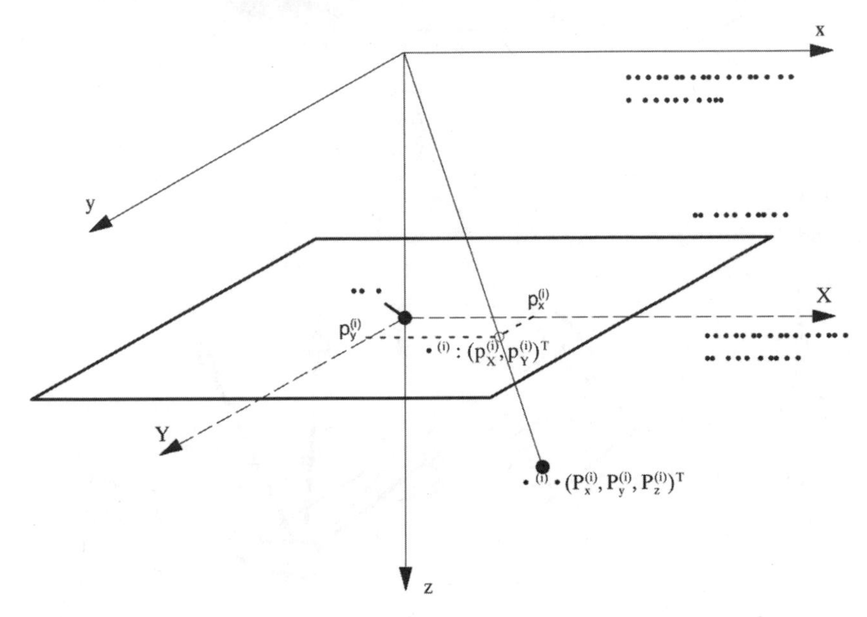

FIGURE 10.4
Camera model.

10.3.2 Scene Model

The scene model describes the objects of a world using an object model and the relationship between objects (Figures 10.2 and 10.5). It allows an explanation of the effects of covered and uncovered areas. In the case of uncovered areas, the relative position of the objects to each other and to the image plane allow the correct assignment of the area to one object.

10.3.3 Illumination Model

The illumination model describes the temporal changes in the video sequence caused by the changing illumination of the real world. The interaction of incident light from a light source with a point P of an object is described by the distribution $L_{r,\lambda}$ of reflected radiance from an object surface depending on the distribution $E_{i,\lambda}$ of incident irradiance and the object surface reflectance function R at this point according to

$$L_{r,\lambda}(L, V, N, P, \lambda) = R(L, V, N, P, \lambda) \cdot E_{i,\lambda}(L, N, \lambda) \tag{10.3}$$

Here N is the surface normal vector, L the illumination direction, V the viewing direction (to the focal point of the camera), and λ the wavelength of light (Figure 10.6). With simplifying

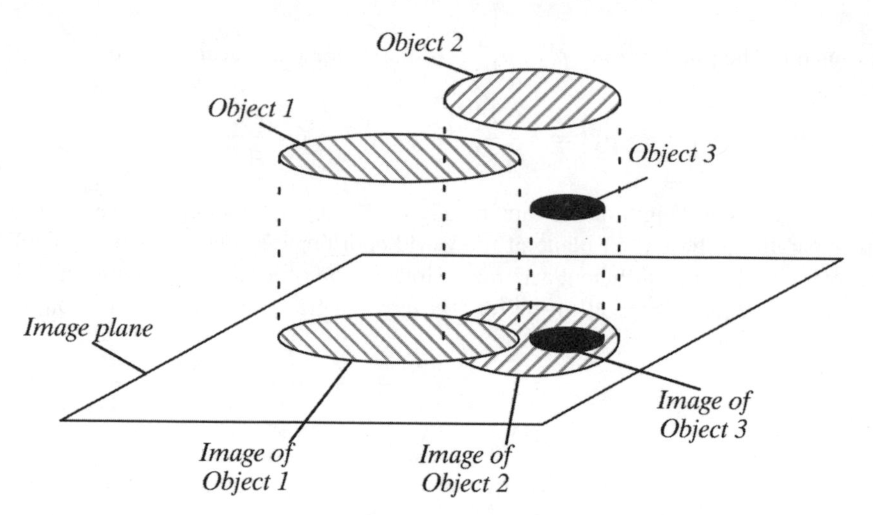

FIGURE 10.5
Scene model.

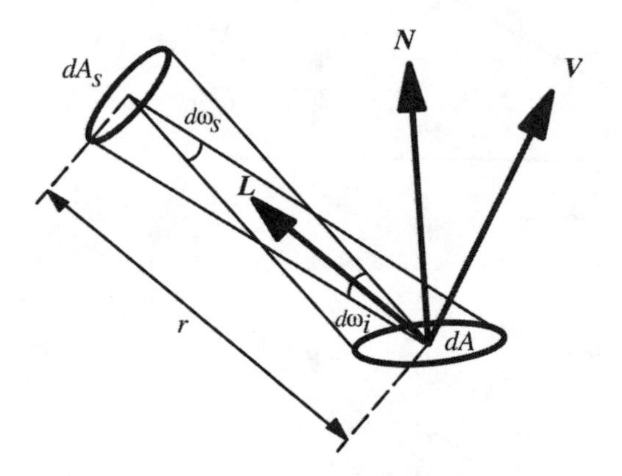

FIGURE 10.6
Surface patch dA with normal vector N illuminated from direction L by a point light source with the infinitesimal small area dA_s. The patch is viewed from direction V.

assumptions such as opaque object surfaces and temporally invariant illumination direction as well as viewing direction, (10.3) simplifies to

$$L_{r,\lambda}(N, P, \lambda) = R(N, P, \lambda) \cdot E_{i,\lambda}(N, \lambda) \tag{10.4}$$

Assuming the scene to be illuminated by a point light source and ambient diffuse light simplifies the description of the incident irradiance to the shading model of Phong used in early computer graphics [58]:

$$E_i(N) = c_{\text{ambient}} + c_{\text{lambert}} \cdot \max(0, LN) \tag{10.5}$$

with c_{ambient} the ambient irradiance and c_{lambert} the point light source irradiance. Assuming that N and therefore the object shape is known, this simple illumination model requires three parameters to be estimated: the ratio between ambient and direct irradiance, $c_{\text{ambient}}/c_{\text{lambert}}$, and the two angles describing the direction of the direct point light source irradiation. This model according to (10.5) has been implemented in an OBASC by Stauder [62]. In the image plane, Stander assumes that the luminance $l(p)$ of a point moving from p_k to p_{k+1} changes according to

$$l\left(p_{k+1}\right) = l\left(p_k\right) \cdot \frac{E_i(N_{k+1})}{E_i(N_k)} \ . \tag{10.6}$$

Pearson and others proposed to model the irradiance by a discrete irradiance map [55]. This map gives an irradiance value for each patch of the incident light on a Gaussian sphere. It is not restricted to any illumination situation. Several light sources can be handled. For a reasonable approximation of an illumination situation, the number of patches should be 9×9 or higher. This method is especially useful if the irradiance values can be measured.

Another simple illumination model assumes that the image signal $l(p)$ depends on the illumination $E(p)$ and the bidirectional reflection function $R(p)$. $R(p)$ accounts for the wavelength of the illumination, surface material, and the geometric arrangement of illumination, camera, and surface. The illumination $E_i(p)$ depends on ambient and direct light. Assuming diffuse illumination, diffuse reflecting surfaces, parallel projection, and a constant k_b, the image signal is given by the reflection model

$$l(p) = k_B \cdot E_i(p) \cdot R(p) \ . \tag{10.7}$$

In the image plane, the luminance $l(p)$ of a point moving from p_k to p_{k+1} changes according to

$$l\left(p_{k+1}\right) = k_b \cdot E\left(p_{k+1}\right) \cdot R\left(p_k\right) \ . \tag{10.8}$$

This reflection model indicates that illumination can be modeled by a multiplicative factor. This has proven to be useful in block matching, 3D motion estimation, and change detection [7, 19, 52, 63].

The simplest, yet most widely used illumination model simply assumes for the luminance of a moving point

$$l\left(p_{k+1}\right) = l\left(p_k\right) \ . \tag{10.9}$$

Sometimes, this model is referred to as the *constant intensity assumption*. The implicit assumptions are diffuse illumination, diffuse reflecting surfaces, and no temporal variation in the illumination. Here, we select the simple illumination model according to (10.9).

10.3.4 Object Model

The object model describes the assumptions of the source model about the real objects. In order to do so, shape, motion, and surface models are required. While all object models described here use the same surface model, they employ different motion and shape models as discussed below.

As far as the *surface model* is concerned, it is assumed that object surfaces are opaque and have a diffuse reflecting surface. The surface of an object m is described by the color parameters S_m. These color parameters contain the luminance as well as the chrominance reflectance.

Moving Rigid 2D Objects (R2D) with 3D Motion

Object Model

This object model assumes rigid 2D arbitrarily shaped objects. Hence, each object can be perceived as a part of a plane. Its projection into the image plane is the 2D silhouette of the object. Each object is allowed to move in 3D space. Allowing two parameters to describe the orientation of the plane in space and six parameters to describe object motion, the functional relationship between a point P on the object surface projected onto the image plane as $p = (X, Y)^T$ and $p' = (X', Y')^T$ before and after motion, respectively, is described by eight parameters $(a_1, \ldots, a_8)$ [23, 65]:

$$p' = (X', Y')^T = \left(\frac{a_1 X + a_2 Y + a_3}{a_7 X + a_8 + 1}, \frac{a_4 X + a_5 Y + a_6}{a_7 X + a_8 + 1} \right)^T . \tag{10.10}$$

Implementation

Image analysis for this source model estimates motion hierarchically for an object, which initially is the entire frame, s_{k+1}. Then, the motion-compensated prediction $\hat{s}$ of the object is computed using an image synthesis algorithm, which fetches the luminance for a point p' in frame s_{k+1} from point p of frame s_k according to the inverse of (10.10). Finally, the estimated motion parameters are verified by comparing the original image and the predicted image, and detecting those areas where the motion parameters do not allow for a sufficiently precise approximation of the original image. These areas are the objects where motion parameters are estimated in the next step of the hierarchy. This verification step allows the segmentation of moving objects using the motion as the segmentation criterion. Hötter [23] implemented three steps of the hierarchy. The image areas that could not be described by motion parameters at the end of this object segmentation and motion compensation process are the $\mathrm{MF}_{\mathrm{R2D}}$ objects.

In order to achieve robust estimates, Hötter allowed the motion model to reduce the number of parameters from eight to an affine transformation with six $(a_1, \ldots, a_6)$ parameters or to a displacement with two parameters (a_1, a_2). This adaptation is especially important as objects become smaller [23].

In order to increase the efficiency of parameter coding, especially the shape parameter coding, the segmentation of the current frame into moving objects takes the segmentation of the previous image as a starting point. In the image plane, this increases the temporal consistency of the 2D shape of the objects. Since eight parameters are estimated for each frame in which an object is visible, this model allows the object to change its orientation in space arbitrarily from one frame interval to the next. No temporal coherence for the orientation of the object is required or enforced.

Moving Flexible 2D Objects (F2D) with 2D Motion

Object Model

This source model assumes that the motion of a real object can be described by a homogeneous displacement vector field. This displacement vector field moves the projection of the real object into the image plane to its new position. Assuming a point P on the object surface moving from P to P', its projection into the image plane moves from p to p'. p and p' are related by the displacement vector $\vec{D}(p') = (D_X(p'), D_Y(p'))^T$:

$$p = p' - \vec{D}(p') . \tag{10.11}$$

The motion parameters of an MC object m are the displacement vectors of those points p' that belong to the projection of object m into the image plane. The shape of this object is defined by the 2D *silhouette* that outlines the projection of object m in the image plane.

Implementation

For estimating the changed area due to object motion, Hötter applies a change detector to the coded image s'_k and the real current image s_{k+1}. This change detector is initialized with the silhouette of the objects as estimated for image s_k. This allows object tracking and increases the coding efficiency for shape parameters. For displacement estimation, a hierarchical block-matching technique is used [4]. Experimental investigations show that an amplitude resolution of a half pel and a spatial resolution of one displacement vector for every 16×16 pel result in the lowest overall data rate for encoding. This implementation of image analysis is only able to segment moving objects in front of a static background. Since an OBASC relies on the precise segmentation of moving objects and their motion boundaries, this restriction on the image analysis limits the coding efficiency for scenes with complex motion.

To compute the motion-compensated prediction image, the texture of the previously decoded image can be used similar to the MPEG-1 and MPEG-2 standards. The displacement vector field is bilinearly interpolated inside an object. The vectors are quantized to half-pel accuracy. In order to accomplish prediction using these half-pel motion vectors, the image signal is bilinearly interpolated. In [27], the disadvantage of this *image synthesis by filter concatenation* was noted. Assuming that all temporal image changes are due to motion, the real image $s_k(p')$ is identical to $s_0(p)$. If displacement vectors with subpel amplitude resolution are applied, the displaced position does not necessarily coincide with the sampling grid of s_0. In that case, a spatial interpolation filter h is required to compute the missing sample. In Figure 10.7, the luminance value of $s_{k+1}(y_1)$ requires two temporal filter operations. Assuming bilinear

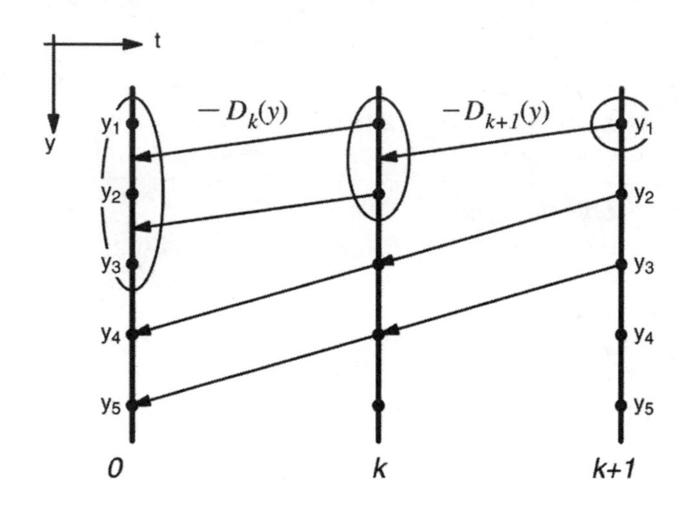

$$s'_{k+1}(y) = s_k(y) * h(y - D_{k+1}(y))$$

$$= s_0(y) * h(y - D_{k+1}(y)) * h(y - D_k(y))$$

FIGURE 10.7
Image synthesis by filter concatenation (one-dimensional case) [27].

interpolation and the displacement vector field as depicted in Figure 10.7, $s_{k+1}(y_1)$ is given by

$$s_{k+1}(y_1) = \frac{1}{4} s_0(y_1) + \frac{1}{2} s_0(y_2) + \frac{1}{4} s_0(y_3) \, . \tag{10.12}$$

The disadvantage of this method is that repeated interpolation results in severe low-pass filtering of the image. Hötter suggests *image synthesis by parameter concatenation* using an object memory for the texture parameters of each object (Figure 10.8). Thus, the interpolation filter h has to be applied only once. In order to synthesize the image $s_{k+1}(y_1)$, the total displacement

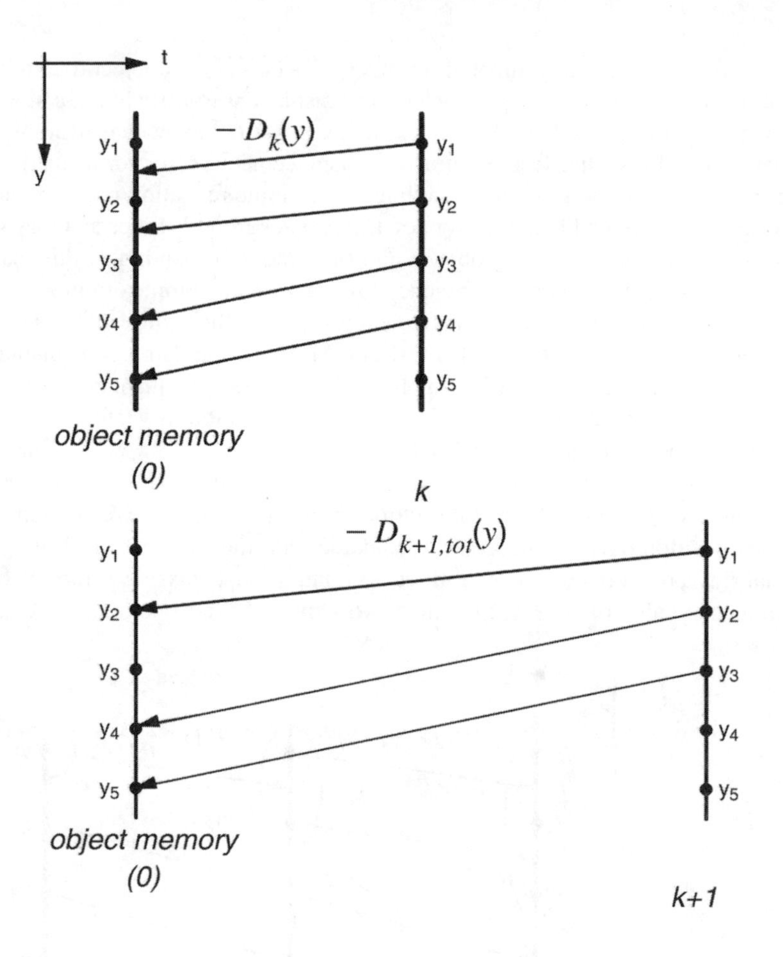

$$s'_{k+1}(y) = s_0(y) * h(y - D_{k+1,tot}(y))$$
$$\text{with } D_{k+1,tot}(y) = D_{k+1}(y) + D_k(y - D_{k+1}(y))$$

FIGURE 10.8
Image synthesis by parameter concatenation (one-dimensional case) using an object memory for color parameters [27].

between s_0 and s_{k+1} is computed by concatenating the displacement vectors. In this case, $s_{k+1}(y_1)$ is given by

$$s_{k+1}(y_1) = s_0(y_2) \ . \tag{10.13}$$

This is basically a method of texture mapping, as known from computer graphics, and OBASC based on 3D source models (see Section 10.3.4) [67]. Indeed, Hötter's implementation uses a mesh of triangles in order to realize this object memory, or texture memory as it is known in computer graphics (Figure 10.9). In [28], Hötter develops a stochastic model describing the synthesis errors due to spatial interpolation and displacement estimation errors. The model was verified by experiments. Use of this texture memory gives a gain of 1 dB in the signal-to-noise ratio for every 14 frames encoded. This gain is especially relevant since the improvement is only due to the less frequent use of the interpolation filter, thus resulting in significantly sharper images.

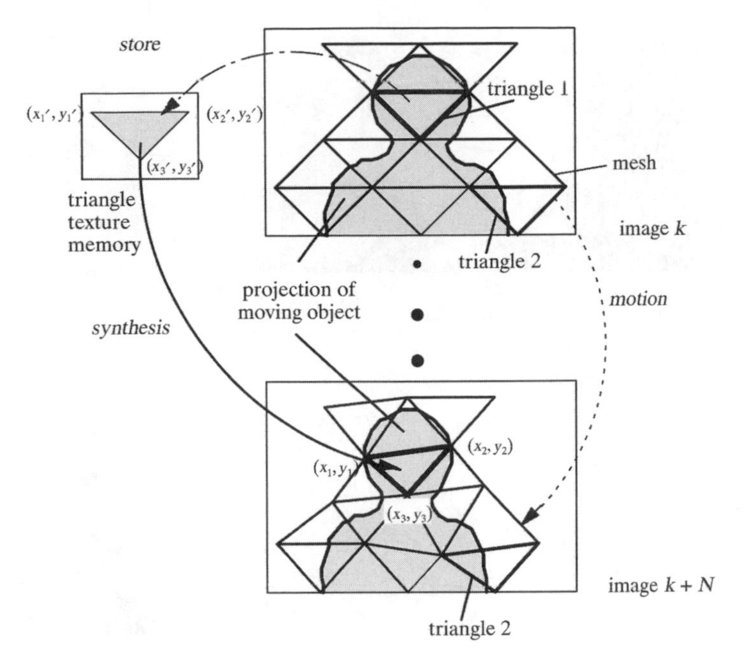

FIGURE 10.9
Image synthesis for MC objects using a triangle-based mesh as texture memory.

MF objects are detected as described in Section 10.4.3. The shape parameter coding is presented in Section 10.5.2. The 2D displacement vector fields are DPCM coded using spatial prediction.

Moving Rigid 3D Objects (R3D)

Object Model

In the model used here, the 3D shape is represented by a mesh of triangles, which is put up by vertices referred to as *control points*, $P_C^{(i)}$. The appearance of the model object surface is described by the color parameters $S^{(m)}$. In order to limit the bit rate for coding of shape parameters, the shape parameters $M^{(m)}$ of an object m represent a 2D binary mask, which defines the *silhouette* of the model object in the model image. During initialization, the 3D shape of an object is completely described by its 2D silhouette (i.e., there is an algorithm that computes a generalized 3D cylinder from a 2D silhouette (Figure 10.10) using a distance transform to determine the object depth (Figure 10.10b) [45]). The distance transform assigns two depth values $w_{F \pm Z}$ to each point h of the object silhouette. Each depth value depends on the Euclidian distance between point h and the silhouette boundary. Depending on the application, an appropriate mapping $d \rightarrow w_{F \pm Z}$ can be selected. Using a mapping derived from an ellipse is suitable for the modeling of head and shoulder scenes. The object width b and the object depth h are related according to (Figure 10.11):

$$w_{F \pm Z}(d) = F \pm \begin{cases} \frac{h}{b}\sqrt{d(b-d)} & \text{for } d < \frac{b}{2} \\ \frac{h}{2} & \text{otherwise .} \end{cases} \tag{10.14}$$

In order to determine the object width, we use the four measurements b_1, b_2, b_3, and b_4 according to Figure 10.12 in order to determine the maximum elongation of the object in the direction of the x and y axes as well as the two diagonals. We determine the object width b as

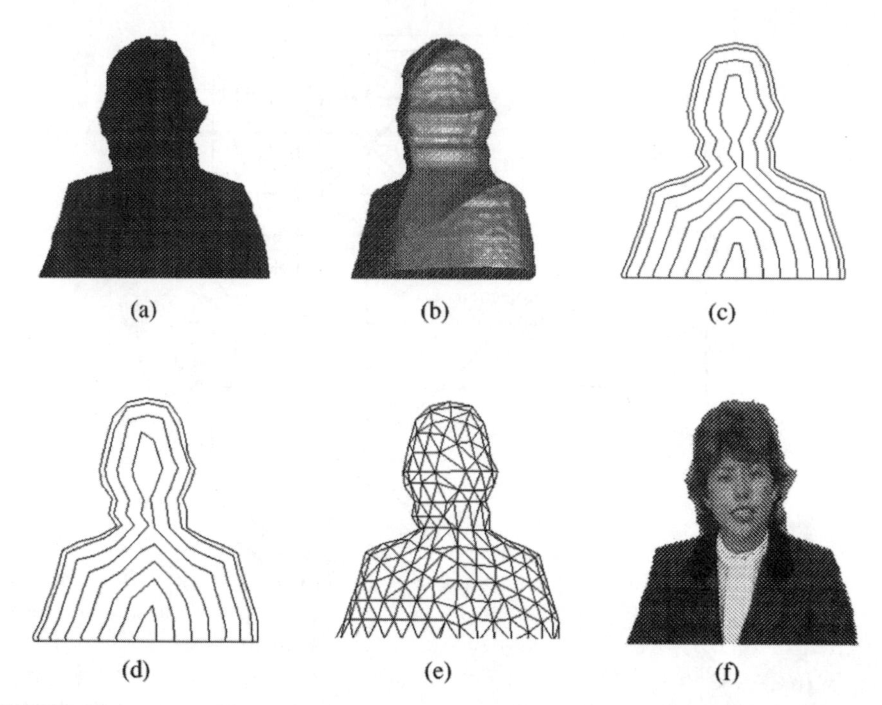

FIGURE 10.10
Processing steps from object silhouette to model object: (a) object silhouette; (b) 3D object shape with required silhouette rotated by 30° and illuminated; (c) contour lines approximating the object shape; (d) polygons approximating the contour lines; (e) mesh of triangles using polygon points as vertices; (f) model object with color parameters projected onto it.

the minimum of the four values according to

$$b = \min(b_1, b_2, b_3, b_4) \; . \tag{10.15}$$

$w_{F\pm Z}$ is constant for $d > b/2$. Hence, the surface of the model object is parallel to the image plane where the object is wide. In order to automatically adapt the object depth to the width of different objects, we set the ratio $\beta = b/h$ instead of h to a fixed value:

$$w_{F\pm Z}(d) = F \pm \begin{cases} \frac{1}{\beta}\sqrt{d(b-d)} & \text{for } d < \frac{b}{2} \\ \frac{b}{2\beta} & \text{otherwise} \; . \end{cases} \tag{10.16}$$

In Figure 10.10, the ratio β between object width and object depth is set to 1.5. The maximum distance between the estimated silhouette and the silhouette of the model object does not exceed $d_{\max} \leq 1.4$ pel (Figure 10.10d). After initialization, the shape parameters $M^{(m)}$ are used as update parameters to the model object shape.

An object may consist of one, two, or more rigid *components* [6]. The subdivision of an object into components is estimated by image analysis. Each component has its own set of motion parameters. Since each component is defined by its control points, the components are linked by those triangles of the object having control points belonging to different components. Due to these triangles, components are flexibly connected. Figure 10.13 shows a scene with the objects Background and Claire. The model object Claire consists of the two components Head and Shoulder.

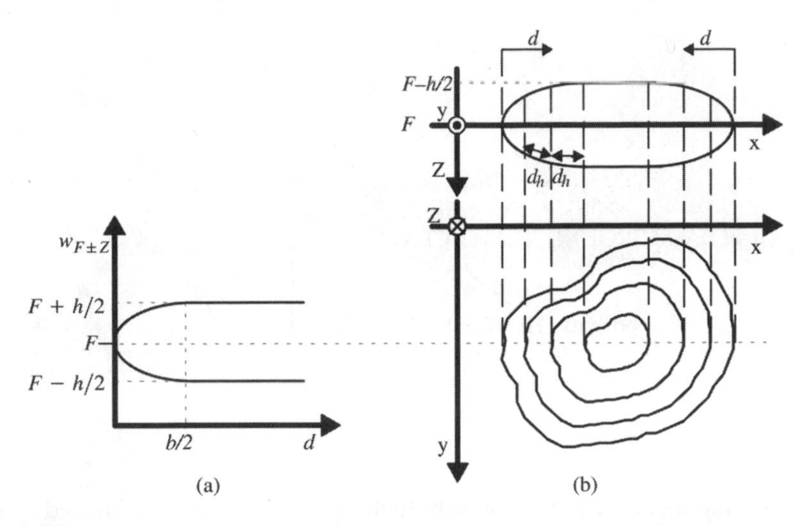

FIGURE 10.11
A 3D shape symmetric to the image plane $Z = F$ is created. (a) Distance transform according to (10.14); d is the smallest distance to the border of the object silhouette, b is set according to (10.15), and $\beta = b/h = 1.5$. In the example, the object width is larger than b according to 10.15. (b) Cut through a model object (top); view from the focal point of the camera onto the contour lines (bottom). For computing object depth, we always measure the distance d to the closest boundary point [51].

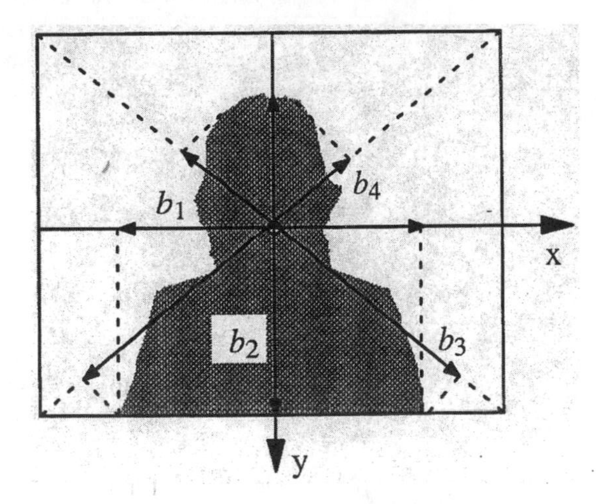

FIGURE 10.12
Determining the object width. The distances b_1, b_2, b_3, and b_4 are measured by determining the parallel projection of the silhouette onto the x and y axes and onto the image diagonals. We use the minimum as object width, here b_1.

3D motion is described by the parameters $A^{(m)} = (T_x^{(m)}, T_y^{(m)}, T_z^{(m)}, R_x^{(m)}, R_y^{(m)}, R_z^{(m)})$ defining translation and rotation. A point $P^{(i)}$ on the surface of object m with N control points $P_C^{(i)}$ is moved to its new position $P'^{(i)}$ according to

$$P'^{(i)} = \left[R_C^{(m)}\right] \cdot \left(P^{(i)} - C^{(m)}\right) + C^{(m)} + T^{(m)} \tag{10.17}$$

with the translation vector $T^{(m)} = (T_x^{(m)}, T_y^{(m)}, T_z^{(m)})^T$, the object center $C = (C_x, C_y, C_z) =$

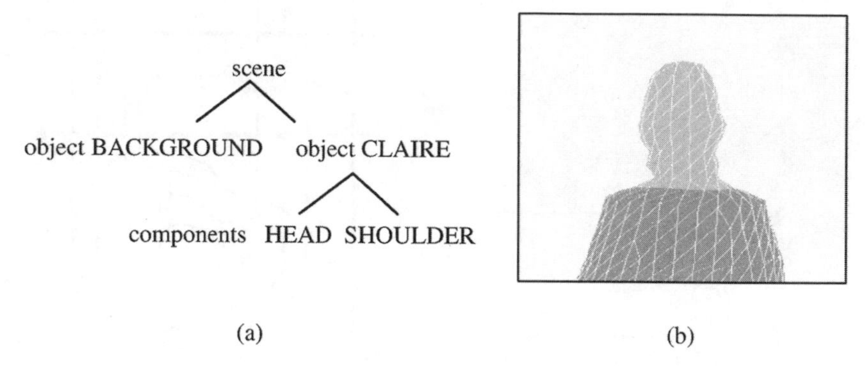

FIGURE 10.13
Model scene and model object Claire subdivided into two flexibly connected components: (a) scene consisting of two objects; (b) components of model object Claire.

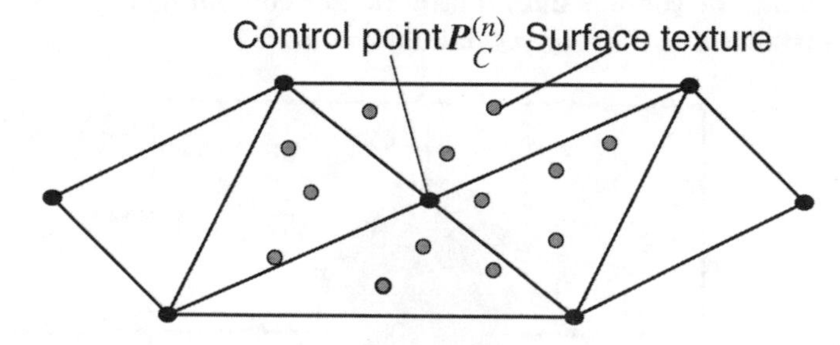

FIGURE 10.14
Triangular mesh with color parameter on the skin of the model object.

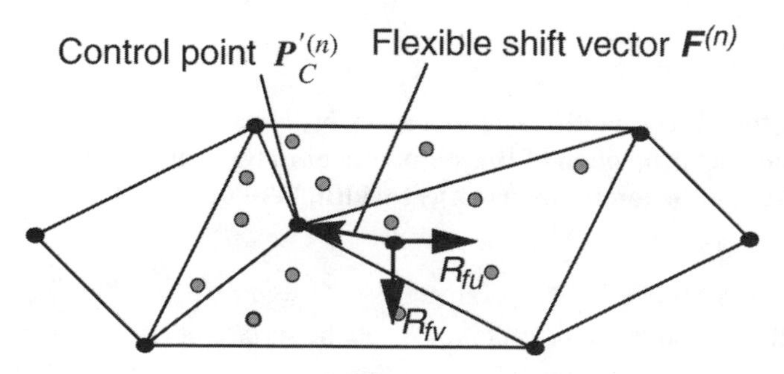

FIGURE 10.15
Triangular mesh after flexible shape compensation by flexible shift vector $F^{(n)}$.

$\frac{1}{N}\sum_{i=1}^{N} P_C^{(i)}$, the rotation angles $R_C = (R_x^{(C)}, R_y^{(C)}, R_z^{(C)})^T$, and the rotation matrix $[R_C]$ defining the rotation in the mathematically positive direction around the x, y, and z axes with the rotation center C:

$$[R_C] = \begin{bmatrix} \cos R_y \cos R_z, & \sin R_x \sin R_y \cos R_z - \cos R_x \sin R_z, & \cos R_x \sin R_y \cos R_z + \sin R_x \sin R_z \\ \cos R_y \sin R_z, & \sin R_x \sin R_y \sin R_z + \cos R_x \cos R_z, & \cos R_x \sin R_y \sin R_z - \sin R_x \cos R_z \\ -\sin R_y, & \sin R_x \cos R_y, & \cos R_x \cos R_y \end{bmatrix} \tag{10.18}$$

Implementation

An overview of the image analysis and a detailed description of motion estimation is given in Section 10.4. Parameter coding is presented in Section 10.5.

Moving Flexible 3D Objects (F3D)

Object Model

In addition to the properties of the source model R3D, the source model F3D allows for local flexible shifts on the surface of the model object shape. This is modeled by a flexible skin (Figure 10.14). This flexible skin can be moved tangentially to the surface of the object (Figure 10.15). It allows modeling of local deformations. In the model world, the flexible surface is modeled by a shift of control points $P_C^{(n)}$ in the tangential surface plane. The normal vector to this tangential surface plane is computed by averaging the normal vectors $n_{Dj}^{(n)}$ to the J triangles to which the control point $P_C^{(n)}$ belongs:

$$n_t^{(n)} = \sum_{j=1}^{J} n_{Dj}^{(n)} \tag{10.19}$$

This tangential surface plane with the normal vector $n_t^{(n)}$ is spanned by $R_{fu}^{(n)}$ and $R_{fv}^{(n)}$. These vectors are of unit length and are orthogonal to each other. For each control point $P_C^{(n)}$, two *flexible shape parameters* $S_f^{(n)} = (S_{fu}^{(n)}, S_{fv}^{(n)})^T$ have to be estimated.

$$\begin{aligned} P_C^{'(n)} &= P_C^{(n)} + F^{(n)} \\ P_C^{'(n)} &= P_C^{(n)} + S_{fu}^{(n)} R_{fu}^{(n)} + S_{fv}^{(n)} R_{fv}^{(n)} \end{aligned} \tag{10.20}$$

with F the *flexible shift vector* and $P_C^{(n)} = (P_x, P_y, P_z)^T$ and $P_C^{'(n)} = (P'_X, P'_Y, P'_Z)^T$ being a control point before and after shift, respectively. The flexible shift vectors $F^{(n)}$ can be interpreted as local motion parameters, in contrast to the global motion parameters, $R_C^{(m)}$ and $T^{(m)}$.

Implementation

The flexible shift vectors require additional data rates for encoding. Hence they are estimated and transmitted only for those image areas that cannot be described with sufficient accuracy using the source model R3D (see Section 10.4.1) [49]. These areas are the MF_{R3D} objects. The shift parameters are estimated for one MF_{R3D} object at a time. For all control points of an MC object that are projected onto one MF_{R3D} object, the shift parameters are estimated jointly because one control point affects the image synthesis of all the triangles to which it belongs. For estimation, the image signal is approximated by a Taylor series expansion and the parameters are estimated using a gradient method similar to the 3D motion estimation algorithm described in Section 10.4.2. Since the robust 3D motion estimation as described in Section 10.4.2 is not influenced by model failures, it is not necessary to estimate the 3D motion parameters again or to estimate them jointly with the flexible shift parameters.

10.4 Image Analysis for 3D Object Models

The goal of image analysis is to gain a compact description of the current real image s_{k+1}, taking the transmitted parameter sets $A_k^{'(m)}, M_k^{'(m)}, and S_k^{'(m)}$ and subjective image quality requirements into account. The image analysis consists of the following parts: image synthesis, change detection, 3D motion estimation, detection of object silhouettes, shape adaptation, and model failure detection. Whereas Section 10.4.1 gives a short overview of image analysis, the sections that follow describe 3D motion estimation and detection of model failures in more detail.

10.4.1 Overview

Figure 10.16 shows the structure of image analysis. The inputs to image analysis are the current real image, s_{k+1}, and the model world, W_k', described by its parameters $A_k^{'(m)}$, $M_k^{'(m)}$, and $S_k^{'(m)}$ for each object m. First, a model image, s_k', of the current model world is computed by means of image synthesis.

In order to compute the change detection mask, B_{k+1}, the change detection evaluates the images s_k' and s_{k+1} on the hypothesis that moving real objects generate significant temporal changes in the images [23, 64], that they have occluding contours [45], and that they are opaque [45, 63]. This mask B_{k+1} marks the projections of moving objects and the background uncovered due to object motion as changed. Areas of moving shadows or illumination changes are not marked as changed because illumination changes can be modeled by semitransparent objects [62].

Since change detection accounts for the silhouettes of the model objects, the changed areas in mask B_{k+1} will be at least as large as these silhouettes. Figure 10.17 gives the change detection mask B_{k+1} for frames 2 and 18 of the test sequence Claire [10]. Due to little motion of Claire at the beginning of the sequence, only parts of the projection of the real object are detected during the first frames. The other parts of the person are still considered static background.

In order to compensate for real object motion and to separate the moving objects from the uncovered background, 3D motion parameters are estimated for the model objects [1], [32]–[35], [37, 38, 48]. The applied motion estimation algorithm requires motion, shape, and color parameters of the model objects and the current real image s_{k+1} as input. Motion parameters A_{k+1} are estimated using a Taylor series expansion of the image signal, linearizing the rotation matrix (10.18) assuming small rotation angles and maximum likelihood estimation (see Section 10.4.2) [29].

The resulting motion parameters are used to detect the uncovered background, which is included in mask B_{k+1}. The basic idea for the detection of uncovered background is that the projection of the moving object before and after motion has to lie completely in the changed area [23]. Subtracting the uncovered background from mask B_{k+1} gives the new silhouette C_{k+1} for all model objects (Figure 10.17).

The silhouette of each model object m is then compared and adapted to the real silhouette $C_{k+1}^{(m)}$ [45]. Differences occur either when parts of the real object start moving for the first time or when differences between the shape of the real and the model object become visible

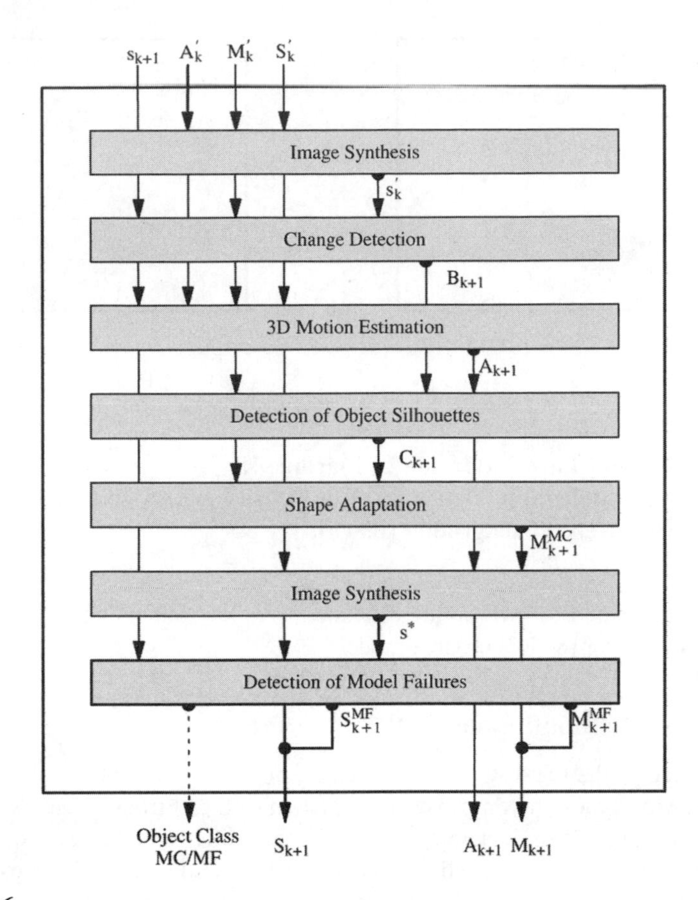

FIGURE 10.16

Block diagram of image analysis: A'_k, M'_k, and S'_k stored motion, shape, and color parameters; s_{k+1} real image to be analyzed; s'_k, s^* model images; B_{k+1} change detection mask; C_{k+1} object silhouettes; M_{k+1} shape parameters for MC and MF objects; S_{k+1} color parameters for MC and MF objects. Arrows indicate the information used in various parts of image analysis. Semicircles indicate the output of processing steps.

during rotation. In order to compensate for the differences between the silhouettes of the model objects and C_{k+1}, the control points close to the silhouette boundary are shifted perpendicular to the model object surface such that the model object gets the required silhouette. This gives the new shape parameters M^{MC}_{k+1}, where MC denotes model compliance.

For the detection of model failures, a model image s^* is synthesized using the previous color parameters S'_k and the current motion and shape parameters A_{k+1} and M^{MC}_{k+1}, respectively. The differences between the images s^* and s_{k+1} are evaluated for determining the areas of model failure. The areas of model failure cannot be compensated for using the source model of "moving rigid 3D objects." Therefore, they are named *rigid model failures* (MF_{R3D}) and are represented by MF_{R3D} objects. These MF objects are described by 2D shape parameters M^{MF}_{k+1} and color parameters S^{MF}_{k+1} only.

In case the source model F3D is used, three more steps have to be added to the image analysis (Figure 10.18). As explained in Section 10.3.4, flexible shift parameters are estimated only for those parts of the model object that are projected onto areas of MF_{R3D}. Following the example of Figure 10.23, the estimation is limited to control points that are projected onto the eye, mouth, and right ear area. After estimation of the shift parameters, a model image is

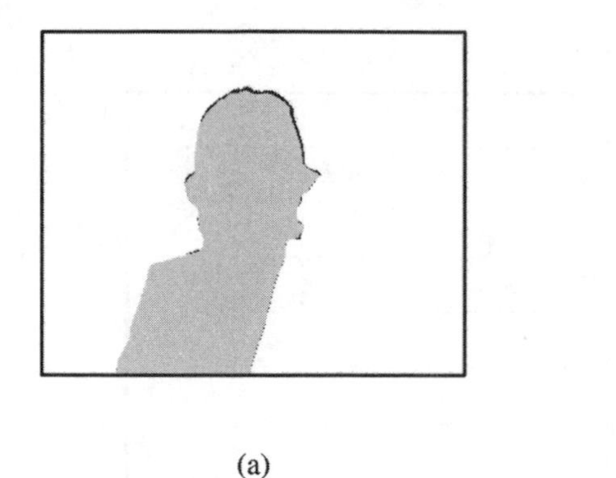

(a)

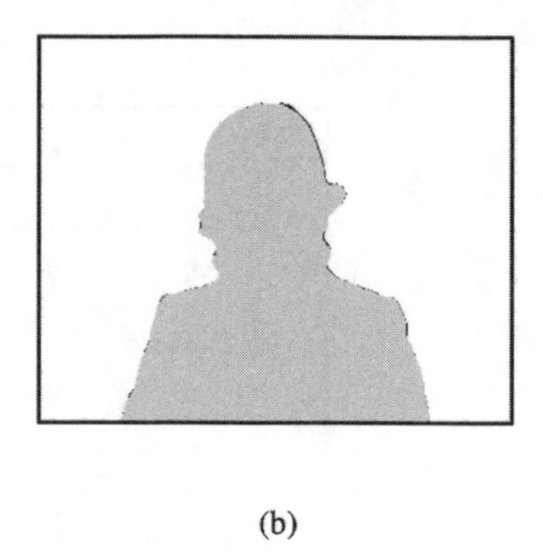

(b)

FIGURE 10.17
Object silhouettes of Claire: (a) frame 2; (b) frame 18. Gray and black areas are marked changed by change detection. Detection of object silhouettes gives the object silhouette (gray) and the uncovered background (black).

synthesized and the areas of MF_{F3D} are estimated using the same algorithm for the detection of model failures as for the R3D source model.

10.4.2 Motion Estimation for R3D

Image analysis of the OBASC must compensate for the motion of the objects in the scene in order to provide a motion-compensated prediction of the current image s_{k+1}. The 3D real objects are modeled by 3D model objects. Usually, the motion and the shape of the real objects are unknown. In order to ensure a reliable and robust estimation, methods for robust estimation are used. In Section 10.4.2, the basic motion estimation algorithm, which enables tracking of real objects with model objects, is reviewed [48]. In Section 10.4.2, methods for robust estimation are developed and compared. Since the shape of the real objects is not known and these objects tend not to be completely rigid due to facial expressions and hair motion, we developed a gradient-based motion estimator instead of a feature-based estimator.

Basic Motion Estimation

In order to derive the motion estimation algorithm, it is assumed that differences between two consecutive images s_k and s_{k+1} are due to object motion only. In order to estimate these motion parameters, a gradient method is applied here.

During motion estimation, each object is represented by a set of observation points. Each observation point $O^{(j)} = (Q^{(j)}, g^{(j)}, I^{(j)})$ is located on the model object surface at position $Q^{(j)}$ and holds its luminance value $I^{(j)}$ and its linear gradients $g_Q^{(j)} = (g_x^{(j)}, g_y^{(j)})^T$. g_Q are the horizontal and vertical luminance gradients from the image which provided the color parameters for the object. For simplicity, we use s_k here. The gradients are computed by convoluting the image signal with the Sobel operator

$$E = \frac{1}{8} \cdot \begin{bmatrix} 1 & 0 & -1 \\ 2 & 0 & -2 \\ 1 & 0 & -1 \end{bmatrix} \tag{10.21}$$

giving the gradients

$$\begin{aligned} g_x(x, y) &= l(x, y)^* E \\ g_y(x, y) &= l(x, y)^* E^T . \end{aligned} \tag{10.22}$$

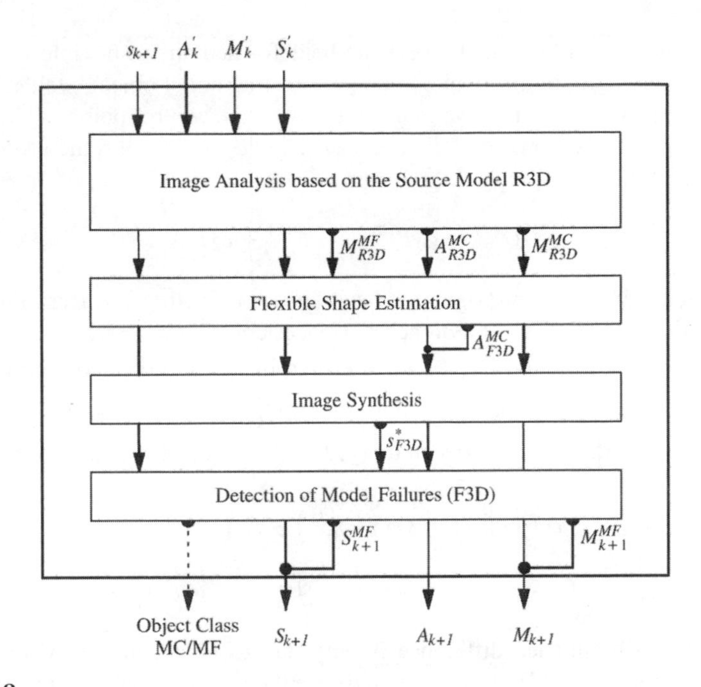

FIGURE 10.18
Block diagram of image analysis: A'_k, M'_k, and S'_k stored motion, shape, and color parameters; s_{k+1} real image to be analyzed; s^* model image; $A_{k+1} = A^{MC}_{R3D} + A^{MC}_{F3D}$ global (R, T) and local (S_f) motion parameters of MC objects; $M_{k+1} = M^{MC}_{R3D} + M^{MF}_{k+1}$ shape parameters of MC and MF objects; $S_{k+1} = S'_k + S^{MF}_{k+1}$ color parameters of MC and MF objects. Arrows indicate the information used in various parts of image analysis. Semicircles indicate the output of processing steps.

The measure for selecting observation points is a high spatial gradient. This adds robustness against noise to the estimation algorithm (see Section 10.4.2). Figure 10.19 shows the location of all observation points belonging to the model object Claire. If parts of the object texture are exchanged due to MF objects, the observation points for the corresponding surface of the object are updated. The observation points are also used for the estimation of flexible shift parameters.

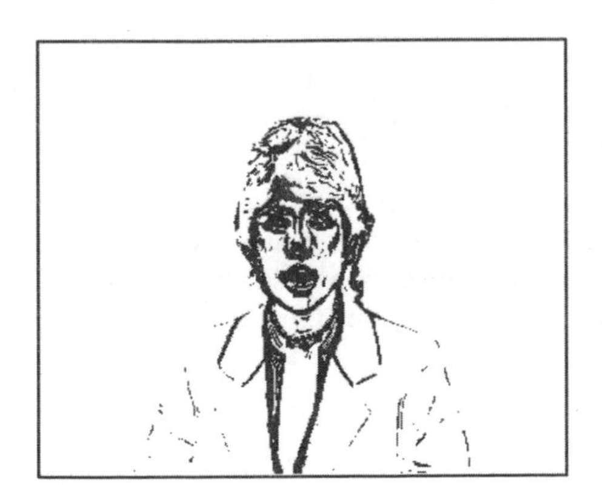

FIGURE 10.19
The position of all observation points of model object Claire.

Since some triangles of a model object can be deformed due to flexible shifts or due to its control points belonging to different components of the model object, we define the position of an observation point $Q^{(j)}$ relative to the position of the control points $P^{(0)}$, $P^{(1)}$, and $P^{(2)}$ of its triangle using barycentric coordinates c_0, c_1, and c_2 of the coordinate system defined by $P^{(0)}$, $P^{(1)}$, and $P^{(2)}$:

$$Q^{(j)} = c_0 P^{(0)} + c_1 P^{(1)} + c_2 P^{(2)} . \tag{10.23}$$

It is assumed that objects are rigid and have diffuse reflecting surfaces. Furthermore, diffuse illumination of the scene is assumed.[1] Hence, color parameters are constant. With an observation point $O_k^{(j)} = (Q_k^{(j)}, g_Q^{(j)}, I^{(j)})$ at time instant k projected onto the image plane at $q_k^{(j)}$ and the same observation point after motion $O_{k+1}^{(j)} = (Q_{k+1}^{(j)}, g^{(j)}, I^{(j)})$ projected onto $q_{k+1}^{(j)}$, the luminance difference between image k and image $k+1$ at position $q_k^{(j)}$ is

$$\begin{aligned} \Delta I\left(q_k^{(j)}\right) &= s_{k+1}\left(q_k^{(j)}\right) - s_k\left(q_k^{(j)}\right) \\ &= s_{k+1}\left(q_k^{(j)}\right) - s_{k+1}\left(q_{k+1}^{(j)}\right) , \end{aligned} \tag{10.24}$$

assuming that the luminance difference is only due to object motion with $s_{k+1}(p_{k+1}) = s_k(p_k) = I$. According to [4], we can approximate the image signal using a Taylor expansion of second order without explicitly computing the second derivative. We compute the second-order gradient $\bar{g}$ by averaging the linear gradients of the observation point and the image signal

$$\bar{g} = \frac{1}{2}\left(g_Q + g_{k+1}(q_k)\right) . \tag{10.25}$$

Approximating the image signal with a Taylor series and stopping after the linear term gives

$$s_{k+1}\left(q_{k+1}\right) \approx s_{k+1}\left(q_k\right) + \bar{g} \cdot \left(q_{k+1} - q_k\right) . \tag{10.26}$$

Now, we can express the luminance difference according to (10.24) as

$$\Delta I\left(q_k\right) = -\bar{g} \cdot \left(q_{k+1} - q_k\right) = \left(g_x^{(j)}, g_y^{(j)}\right)^T \cdot \left(q_{k+1}^{(j)} - q_k^{(j)}\right) . \tag{10.27}$$

Substituting image coordinates by model world coordinates with equation (10.2) yields

$$\Delta I^{(j)} = F \cdot g_x^{(j)} \left(\frac{Q_{x,k+1}^{(j)}}{Q_{z,k+1}^{(j)}} - \frac{Q_{x,k}^{(j)}}{Q_{z,k}^{(j)}} \right) + F \cdot g_y^{(j)} \left(\frac{Q_{y,k+1}^{(j)}}{Q_{z,k+1}^{(j)}} - \frac{Q_{y,k}^{(j)}}{Q_{z,k}^{(j)}} \right) \tag{10.28}$$

The position $Q_k^{(j)}$ of the observation point $O^{(j)}$ is known. By relating Q_k to Q_{k+1} by means of the motion equation (10.17), a nonlinear equation with the known parameters ΔI, g, and F and the six unknown motion parameters results. This equation is linearized by linearizing the rotation matrix R_C (10.18), assuming small rotation angles

$$\left[R'_C\right] = \begin{bmatrix} 1 & -R_z & R_y \\ R_z & 1 & -R_x, \\ -R_y & R_x & 1 \end{bmatrix} \tag{10.29}$$

[1] See [7] and [62] on how to consider illumination effects.

giving

$$Q_{k+1} = \left[R'_C\right] \cdot (Q_k - C) + C + T \tag{10.30}$$

Substituting (10.30) into (10.28), the linearized equation for one observation point is

$$\begin{aligned}
\Delta I = & F \cdot g_x / Q_z \cdot T_x \\
& + F \cdot g_y / Q_z \cdot T_y \\
& - \left[\left(Q_x g_x + Q_y g_y \right) F / Q_z^2 + \Delta I / Q_z \right] \cdot T_z \\
& - \Big[\left[Q_x g_x \left(Q_y - C_y \right) + Q_y g_y \left(Q_y - C_y \right) + Q_z g_y \left(Q_z - C_z \right) \right] F / Q_z^2 \\
& + \Delta I / Q_z \left(Q_y - C_y \right) \Big] \cdot R_x \\
& + \Big[\left[Q_y g_y \left(Q_x - C_x \right) + Q_x g_x \left(Q_x - C_x \right) + Q_z g_x \left(Q_z - C_z \right) \right] F / Q_z^2 \\
& + \Delta I / Q_z \left(Q_x - C_x \right) \Big] \cdot R_y \\
& - \left[g_x \left(Q_y - C_y \right) - g_y \left(Q_x - C_x \right) \right] F / Q_z \cdot R_z
\end{aligned} \tag{10.31}$$

with the unknown motion parameters $T = (T_x, T_y, T_z)^T$ and $R_C = (R_x, R_y, R_z)^T$ and the observation point $O_k = (Q_k, g, I)$ at position $Q_k = (Q_x, Q_y, Q_z)^T$. In order to get reliable estimates for the six motion parameters, equation (10.31) has to be established for many observation points, resulting in an overdetermined system of linear equations

$$A \cdot x - b = r \ , \tag{10.32}$$

with the residual $r = (r_1, \ldots, r_J)^T$, $x = (T_X, T_Y, T_Z, R_X, R_Y, R_Z)^T$, $b = (\Delta I(q^{(1)}), \ldots, \Delta I(q^{(J)})^T$, and $A = (a_1, \ldots, a_J)^T$, and a_j according to (10.31). The equations are solved by minimization of r:

$$|r|^2 = r^T \cdot r \underset{x}{\rightarrow} \min \ , \tag{10.33}$$

which corresponds to a minimization of the prediction error of the observation points

$$\sum_{0^{(j)}} \left(\Delta I^{(j)} \right)^2 \rightarrow \min \ . \tag{10.34}$$

The motion parameters are given by

$$\hat{x} = \left(A^T \cdot A \right)^{-1} \cdot A^T \cdot b \tag{10.35}$$

In order to avoid the inversion of large matrices, we do not compute A but immediately compute the 6×6 matrix $A^T \cdot A$.

Due to the linearizations in (10.26) and (10.29), motion parameters have to be estimated iteratively for each model object. After every iteration, the model object is moved according to (10.17) using the estimated motion parameters $\hat{x}$. Then, a new set of motion equations is established, giving new motion parameter updates. Since the motion parameter updates approach zero during the iterations, the introduced linearizations do not harm motion estimation. The iteration process terminates if the decrease of the residual error $|r|^2$ becomes negligible.

Robust Motion Estimation

Equation (10.32) is solved such that the variance of the residual errors ΔI is minimized. However, this approach is sensitive to measurement errors [41]. Measurement errors occur because (10.32) is based on several model assumptions and approximations that tend to be valid for the majority of observation points but not all. Observation points that violate these assumptions are named *outliers* [59]. When using (10.34) for solving (10.32), outliers have a significant influence on the solution. Therefore, we have to take measures that limit the influence of these outliers on the estimation process [51]. Sometimes, the following assumptions are not valid:

1. Rigid real object
2. Quadratic image signal model
3. Small deviations of model object shape from real object shape

Each of these cases is discussed below.

If parts of the real object are nonrigid (i.e., the object is flexible), we have image areas that cannot be described by the current motion and shape parameters A_k and M_k, respectively, and the already transmitted color parameters S'_k. These image areas can be detected due to their potentially high prediction error ΔI. Observation points in these areas can be classified as outliers. For iteration i of (10.34), we will consider only observation points for which the following holds true:

$$\Delta I_i^{(j)} < \sigma_{\Delta I} \cdot T_{ST} \tag{10.36}$$

with

$$\sigma_{\Delta I} = \sqrt{\frac{1}{J} \sum_{j=0}^{J} \left(\Delta I_i^{(j)}\right)^2} \,. \tag{10.37}$$

The threshold T_{ST} is used to remove the outliers from consideration.

According to (10.26), motion estimation is based on the gradient method, which allows for estimating only small local displacements $(q_{k+1}^{(j)} - q_i^{(j)})$ in one iteration step [43]. Given an image gradient $g_i^{(j)}$ and a maximum allowable displacement $V_{\max} = |v_{\max}| = |(v_{x,\max}, v_{y,\max})^T|$, we can compute a maximum allowable frame difference $\Delta I_{\text{limit}}(q_i^{(j)})$ at an image location $q_i^{(j)}$

$$\Delta I_{\text{limit}}\left(q_i^{(j)}\right) = \left|v_{\max} \cdot g_i^{(j)}\right| \,. \tag{10.38}$$

Observation points with $|\Delta I(q_i^{(j)})| > |\Delta I_{\text{limit}}(q_i^{(j)})|$ are excluded from consideration for the ith iteration step. We assume that they do not conform to the image signal model assumption.

Considering image noise, we can derive an additional criterion for selecting observation points. Assuming white additive camera noise n, we measure the noise of the image difference signal as

$$\sigma_{\Delta I}^2 = 2 \cdot \sigma_n^2 \,. \tag{10.39}$$

According to (10.27), we represent the local displacement $(q_{k+1}^{(j)} - q_i^{(j)})$ as a function of the noiseless luminance signal. Therefore, the luminance difference $\Delta I(q^{(j)})$ and the gradient

$g^{(j)}$ should have large absolute values in order to limit the influence of camera noise. Hence, we select as observation points only points with a gradient larger than a threshold T_G:

$$\left|g^{(j)}\right| > T_G \,. \tag{10.40}$$

Relatively large gradients allow also for a precise estimation of the motion parameters. Summarizing these observations, we conclude that we should select observation points with large absolute image gradients according to (10.40). Equations (10.36) and (10.38) are the selection criteria for the observation points we will use for any given iteration step.

Instead of using the binary selection criteria for observation points according to (10.36) and (10.38), we can use continuous cost functions to control the influence of an observation point on the parameter estimation. We use the residuum r according to (10.32) as measure for the influence of an observation point [57, 70]. Assuming that the probability density function $f(r)$ of the residuals r_j according to (10.32) is Gaussian, (10.34) is a maximum-likelihood estimator or M estimator [29].

Now, we will investigate how different assumptions about $f(r)$ influence the M estimator. Let us assume that one $f(r)$ is valid for all observation points. A critical point for selecting an appropriate probability density function is the treatment of outliers. Ideally, we want outliers to have no influence on the estimated motion parameters.

The M estimator minimizes the residuum r_j according to (10.32) using a cost function $\varrho(r_j)$:

$$\sum_{j=1}^{J} \varrho\left(r_j\right) \underset{x}{\rightarrow} \min \,. \tag{10.41}$$

With

$$\Psi\left(r_j\right) = \frac{\delta(\varrho(r_j))}{\delta x} \tag{10.42}$$

the solution of (10.41) becomes

$$\sum_{j=1}^{J} \Psi\left(r_j\right) = 0 \,. \tag{10.43}$$

Equation (10.43) becomes an M estimator for the probability density function $f(r)$ if we set

$$\varrho(r) = -\log f(r) \,. \tag{10.44}$$

This M estimator is able to compute the correct solution for six motion parameters with up to 14% of the observation points being outliers [70]. Some authors report success with up to 50% outliers [38].

Let us assume that ε% of our measurement data represents outliers that do not depend on the observable motion. Now, we can choose a separate probability density function for the residuals of the outliers. Let us assume that the residuals of the *inliers* (non-outliers) are Gaussian distributed and that the outliers have an arbitrary Laplace distribution. We can approximate the probability density function of the residuals with [29, 70]

$$f(r) = \begin{cases} \frac{1-\varepsilon}{\sqrt{2\pi}} e^{-\frac{r^2}{2}} & \text{if } |r| < a \\ \frac{1-\varepsilon}{\sqrt{2\pi}} e^{-(a|r|-\frac{a^2}{2})} & \text{otherwise} \end{cases} \,. \tag{10.45}$$

The cost function is

$$\varrho(r) = \begin{cases} \frac{r^2}{2} & \text{for } |r| < a \\ a \cdot |r| - \frac{a^2}{2} & \text{otherwise} \end{cases} \tag{10.46}$$

with the associated M estimator

$$\Psi(r_j) = \max\left[-a, \min(r_j, a)\right] , \tag{10.47}$$

where a is the threshold for detecting outliers. In order to adapt the outlier detection to the image difference signal ΔI, we select a proportional to $\sigma_{\Delta I}$ (10.37).

Often, the probability density function $f(r)$ is unknown. Therefore, heuristic solutions for $\Psi(r)$ such as the cost function $(1 - r^2/b^2)^2$ according to Turkey were found [38, 70]:

$$\Psi(r_j) = \begin{cases} r_j \cdot \left(1 - \frac{r_j^2}{b^2}\right)^2 & \text{if } |r_j| < b \\ 0 & \text{otherwise} . \end{cases} \tag{10.48}$$

The cost $(1 - r^2/b^2)^2$ increases to 1 when $|r|$ decreases. Observation points with $|r| \geq b$ are excluded from the current iteration; b is the threshold for detecting outliers. In order to adapt the outlier detection to the image difference signal ΔI, we select b proportional to $\sigma_{\Delta I}$ (10.37).

The shape difference between the model object and the real object can be modeled by means of a spatial uncertainty of an observation point along the line of sight. This can be considered using a Kalman filter during motion estimation [39].

Experimental Results

Ideally, each iteration i of our motion estimation would consist of four steps. First, we solve (10.35) with all observation points. Then we select the observation points that fulfill the criteria for robust motion estimation. In the third step, we estimate the motion parameters using (10.35) again with these selected observation points. Finally, we use these parameters for motion compensation according to (10.17). In order to avoid solving (10.35) twice, we use the observation points that fulfilled the criteria for robust estimation in iteration $i - 1$. Hence, each iteration i consists of three steps: (1) solve (10.35) using the observation points selected in iteration $i - 1$, (2) motion compensate the model object with the estimated motion parameters, and (3) select the new observation points to be used in iteration $i + 1$.

In order to evaluate the motion estimation algorithm and the robust estimation methods, we test the algorithm on synthetic image pairs that were generated with known parameters [51]. First, we create a model object using the test sequence Claire (Figure 10.10f). We create the first test image by projecting the model image into the image plane. Before synthesizing the second test image, we move the model object and change its facial expression, simulating motion and model failures, respectively. Finally, we add white Gaussian noise with variance σ_n^2 to the test images. In the following tests, we use the model object that corresponds to the first test image. We estimate the motion parameters that resulted in the second test image.

As a first quality measure, we compute the average prediction error variance σ_{diff}^2 inside the model object silhouette between the motion-compensated model object and the second test image. As a second quality measure, we compute the error of the model object position, measured as the average position error d_{mot} of all vertices between the estimated position $\hat{P}^{(n)}$ and the correct position $P^{(n)}$:

$$d_{mot} = \frac{1}{N} \sum_{n=1}^{N} \left| P^{(n)} - \hat{P}^{(n)} \right| . \tag{10.49}$$

With (10.49), we can capture all motion estimation errors. The smaller d_{mot} gets, the better the estimator works in estimating the true motion between two images. Estimation of true motion is required in order to be able to track an object in a scene. We do not compare directly the estimated motion parameters because they consist of translation and rotation parameters that are not independent of each other, thus making it difficult to compare two motion parameter sets.

Figure 10.20 shows the prediction error σ_{diff}^2 in relation to the image noise for three thresholds. Using the threshold $T_{ST} = \infty$ according to (10.36) allows us to consider every observation point. We use only observation points with a small luminance difference when setting $T_{ST} = 1$. Alternatively, we can use (10.38). Hence, we select only observation points that

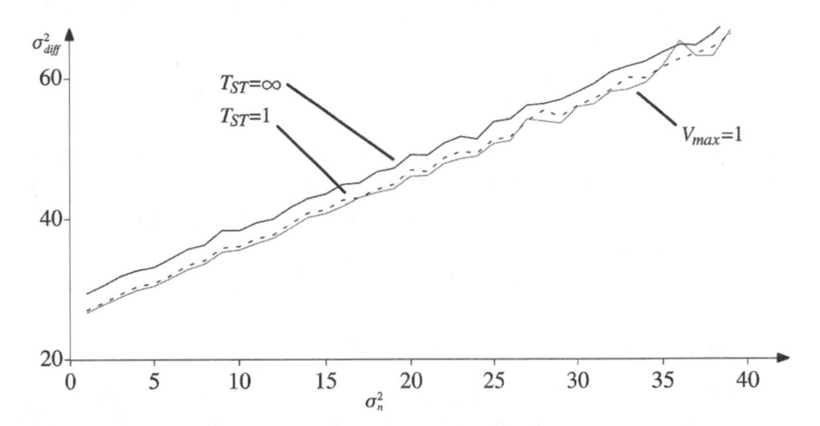

FIGURE 10.20
Prediction error variance as a function of image noise. The different curves were measured using different criteria for selecting observation points: $V_{\max}$ according to (10.38), **T_{ST} according to** (10.36).

indicate a small local motion. As can be seen, removing the outliers from the measurement data reduces the prediction error. Due to the model failure, the prediction error variance σ_{diff}^2 does not decrease to 0. As can be seen in Figure 10.20, the prediction error increases with the image noise.

If we compare the average position error d_{mot}, we see it decreases significantly when we use the criteria according to (10.36) and (10.38) (Figure 10.21). d_{mot} is an important criterion for estimating the true motion of an object. According to the experiments shown in Figure 10.21, using $\Delta I(q^{(j)})$ with (10.36) and the threshold $T_{ST} = 1$ as the control criterion results in the smallest position error d_{mot}.

Using an M estimator further decreases the position error d_{mot}. Using the probability density function (10.45) with

$$a = \sigma_{\Delta I} \cdot T_{ST} , \tag{10.50}$$

and the prediction error variance $\sigma_{\Delta I}^2$ of all observation points during iteration i, gives the best results for $T_{ST} = 0.2$. The most precise estimation was measured using the cost function according to Turkey, (10.48), with

$$b = \sigma_{\Delta I} \cdot T_{ST} , \tag{10.51}$$

the prediction error variance $\sigma_{\Delta I}^2$ of all observation points during iteration i, and $T_{ST} = 1$. The position error d_{mot} is not influenced much by image noise. This is because the noise is Gaussian and the model object covers a relatively large area of the image (30% for Claire).

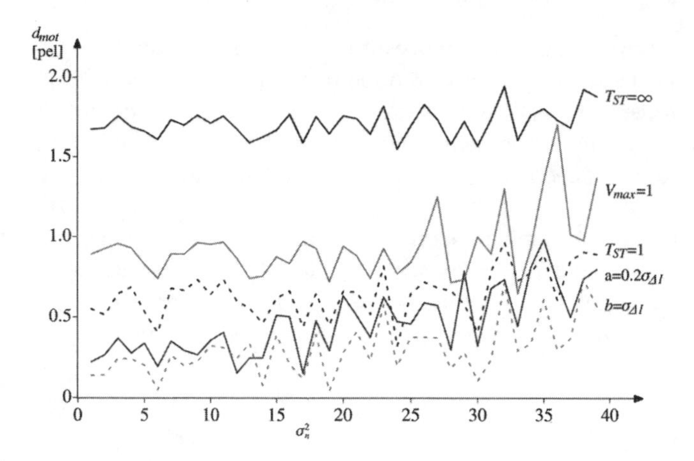

FIGURE 10.21
Average deviation of control point position d_{mot} according to (10.49) as a function of image noise σ_n. The curves were created using different criteria: $V_{\max}$ according to (10.19), T_{ST} according to (10.17), a according to (10.45) and (10.50), b according to (10.48) and (10.51), and $\sigma_{\Delta I}$ according to (10.37).

Segmentation into Components

The initial segmentation of moving objects into components was developed by Busch [6]. After motion compensating the entire rigid model object, the segmentation algorithm clusters neighboring triangles with similar 2D motion parameters. If these clusters allow for an improved motion compensation, a model object is subdivided into flexibly connected components. This decision is based on the evaluation of two consecutive images only.

In [39], Martínez proposes a different approach to segmenting an object into components. The 3D motion is measured for each triangle. In order to achieve a reliable estimate for the triangles, a Kalman filter modeling model object shape errors and the camera noise is adopted. Triangles with similar 3D motion are stored in a cluster memory. As soon as a cluster in the memory is stable for several images, this cluster is used to define a component of the object. This algorithm yields segmentation of persons into head, shoulders, and arms. Further improvements in motion estimation are achieved by enforcing spatial constraints due to spherical joints between components.

After segmenting an object into rigid components, 3D motion is estimated iteratively for each individual component as well as for the entire object.

10.4.3 MF Objects

MF objects are not related to real objects. They are just used to cover the deficiencies of the object and illumination models as well as estimation errors. Therefore, MF objects are always detected at the end of image analysis. This procedure can be seen as the final verification step of image analysis. According to the scene model, MF objects exist only in the model image plane.

Detection of MF Objects

MF objects are estimated by comparing the current real image with the model image s^*_{k+1}, which is synthesized using previously transmitted color parameters S'_k and the current motion and shape parameters A^{MC}_{k+1} and M^{MC}_{k+1}, respectively (Figures 10.16 and 10.18). As a result of this comparison, we will segment those image areas that cannot be described with sufficient subjective quality using MC objects as defined by the source model. Each MF object is described by its 2D silhouette and the color parameters inside the silhouette.

The detection of MF objects implies a receiver model. The following list gives some qualitative properties of the receiver model. It is assumed that the subjective image quality is not disturbed by:

1. Camera noise
2. Small position errors of the moving objects
3. Small shape errors of the moving objects
4. Small areas with erroneous color parameters inside a moving object

The errors listed as items (2) to (4) are referred to as geometrical distortions. Properties of the human visual system such as the modulation transfer function and spatiotemporal masking are not considered.

The following algorithm implicitly incorporates the above-mentioned assumptions. The difference image between the prediction image s^*_{k+1} and the current image s_{k+1} is evaluated by binarizing it using an adaptive threshold T_e such that the error variance of the areas that are not declared as synthesis errors is below a given allowed noise level N_e. $N_e = 6/255$ is a commonly used threshold. The resulting mask is called the synthesis error mask. Figure 10.22a and b show a scaled difference image and the resulting synthesis error mask, respectively.

The synthesis error mask marks those pels of image s^*_{k+1} which differ significantly from the corresponding pels of s_{k+1}. Since the areas of synthesis errors are frequently larger than 4% of the image area, it is not possible to transmit color parameters for these areas with a sufficiently high image quality (i.e., visible quantization errors would occur). However, from a subjective point of view it is not necessary to transmit color parameters for all areas of synthesis errors. Due to the object-based image description, the prediction image s^*_{k+1} is subjectively pleasant. There are no block artifacts, and object boundaries are synthesized properly.

There are two major reasons for synthesis errors. First of all, synthesis errors are due to position and shape differences between a moving real object and its corresponding model object. These errors are caused by motion and shape estimation errors. They displace contours in the image signal and will produce line structures in the synthesis error mask. Due to the feedback of the estimated and coded motion and shape parameters into image analysis (Figure 10.1), these estimation errors tend to be small and unbiased and they do not accumulate. Therefore, it is reasonable to assume that these errors do not disturb subjective image quality. They are classified as geometrical distortions. As a simple detector of geometrical distortions, a median filter of size 5×5 pel is applied to the mask of synthesis errors (Figure 10.22c).

Second, events in the real world that cannot be modeled by the source model will contribute to synthesis errors. Using the source model R3D, it is not possible to model changing human facial expressions or specular highlights. Facial expressions in particular are subjectively important. In order to be of subjective importance, it is assumed that an erroneous image region has to be larger than 0.5% of the image area (Figure 10.22c). Model failures are those image areas where the model image s^*_{k+1} is subjectively wrong (Figure 10.22d). Each area of model failure is modeled by an $\mathrm{MF}_{\mathrm{R3D}}$ object, defined by color and 2D shape parameters (Figure 10.23a and b). Applying the F3D object model to the $\mathrm{MF}_{\mathrm{R3D}}$ objects (Figure 10.18) can compensate for some of the synthesis errors such that the $\mathrm{MF}_{\mathrm{F3D}}$ objects tend to be smaller (Figure 10.23).

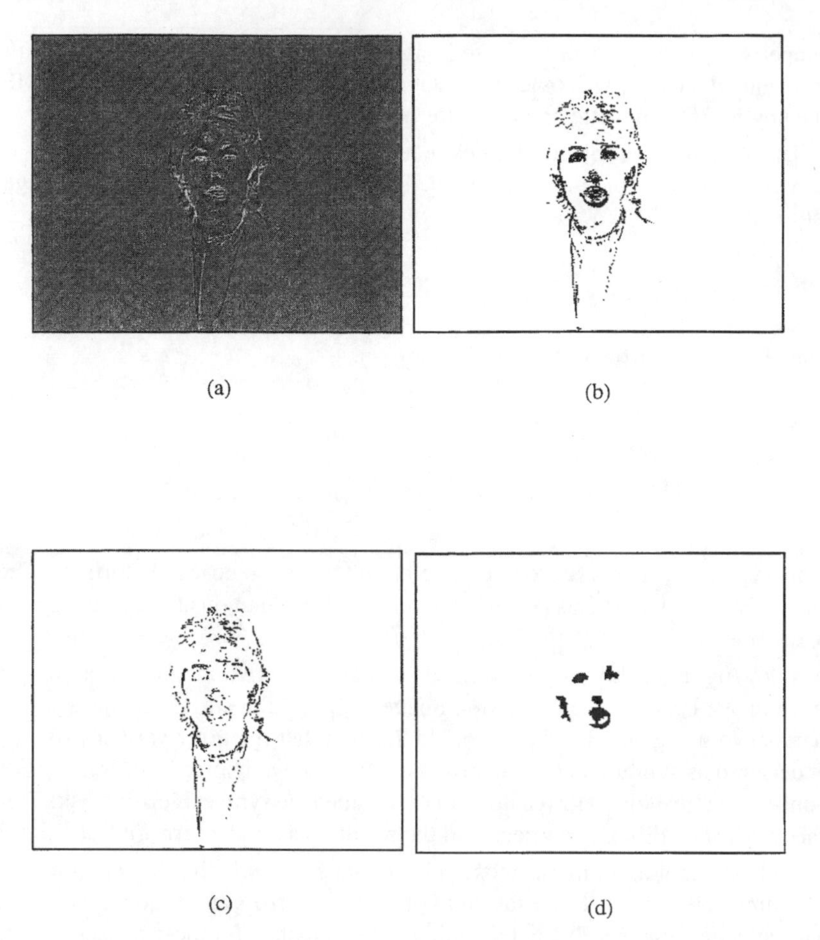

FIGURE 10.22
Detection of model failures: (a) scaled difference image between real image s_{k+1} and model image after motion and shape compensation s^*_{k+1}; (b) synthesis error mask; (c) geometric distortions and perceptually irrelevant regions; (d) mask $\mathrm{MF_{R3D}}$ with model failures of the source model rigid 3D object.

10.5 Optimization of Parameter Coding for R3D and F3D

The task of parameter coding is the efficient coding of the parameter sets motion, shape, and color provided by image analysis. Parameter coding uses a coder mode control to select the appropriate parameter sets to be transmitted for each object class. The priority of the parameter sets is arranged by a priority control.

10.5.1 Motion Parameter Coding

The unit of the estimated object translation $T = (t_x, t_y, t_z)^T$ is pel. The unit of the estimated object rotation $R_C = (R_x^{(C)}, R_y^{(C)}, R_z^{(C)})^T$ is degree. These motion parameters are PCM coded by quantizing each component with 8 bit within an interval of ± 10 pel and degrees, respectively. This ensures a subjectively lossless coding of motion parameters.

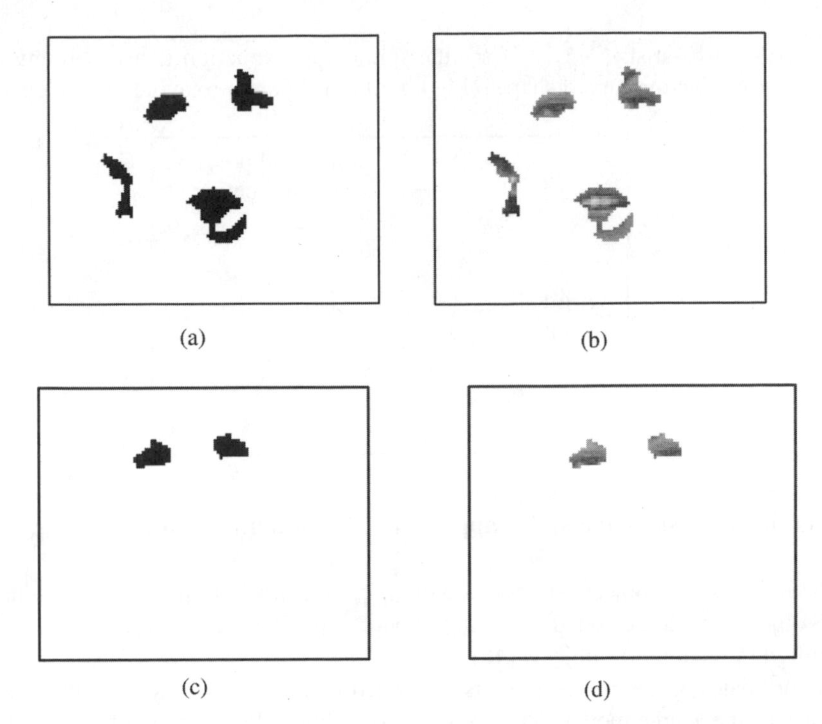

FIGURE 10.23
Detection of model failures: MF_{R3D} objects with (a) shape and (b) color parameters. After the source model F3D is applied to the MF_{R3D} objects, the MF_{F3D} objects are detected with the (c) shape and (d) texture parameters. (a) is an enlargement of Figure 10.22d.

10.5.2 2D Shape Parameter Coding

Since the model object shape is computed and updated from its silhouette, shape parameters are essentially 2D. The principles for coding the shape parameters of MF and MC objects are identical. Shape parameters are coded using a polygon/spline approximation developed by Hötter [24]. A measure d_{max} describes the maximum distance between the original and approximated shape. First, an initial polygon approximation of the shape is generated using four points (Figure 10.24a). Where the quality measure d^*_{max} is not satisfied, the approximation

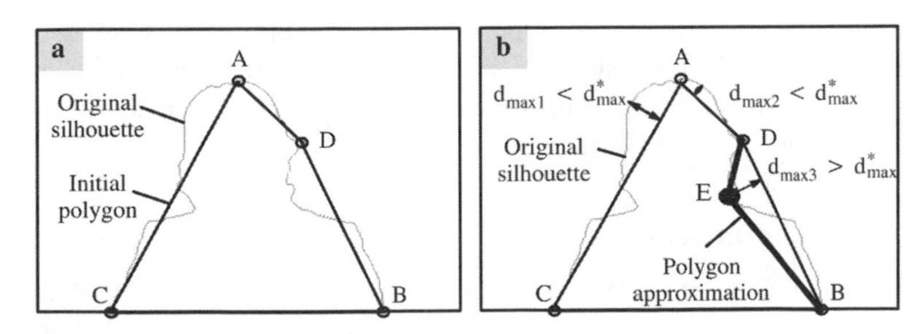

FIGURE 10.24
Polygon approximation: (a) initial polygon; (b) insertion of a new polygon point.

is iteratively refined through insertion of additional polygon points until the measure fulfills $d_{max} \leq d^*_{max}$ (Figure 10.24b). In case the source model F2D is used, we check for each line of the polygon, whether an approximation of the corresponding contour piece by a spline approximation also satisfies d^*_{max}. If so, the spline approximation is used, giving a natural shape approximation for curved shapes (Figure 10.25). In order to avoid visible distortions at

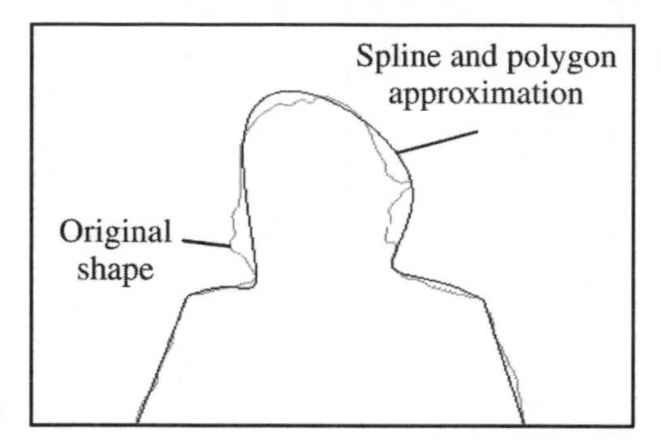

FIGURE 10.25
Combination of polygon and spline approximation for a quality measure $d^*_{max} = 15$.

object boundaries, MC objects are coded with $d^*_{max} = 1.4$ pel. Experimental results showed that MF objects should be coded with d^*_{max} about 2.1 pel in order to minimize the overall bit rate required for coding MF objects [46].

The coordinates of the polygon points are coded relative to their perspective predecessor. In the case of the source model F2D, the curve type line/spline is coded for each line of the polygon.

The data rate for coding shape parameters of MC objects is cut to half by using the motion-compensated coded silhouette of the last image as a prediction of the current silhouette. Starting with this approximation, only shape update parameters have to be transmitted.

10.5.3 Coding of Component Separation

The split of an object into components is defined on a triangle basis. Whenever a new component is defined by the encoder, its shape is encoded losslessly with a flag for each visible triangle. The encoder and decoder then define the shape of the component by connecting the visible triangles with the ones that they occlude at the back of the object.

10.5.4 Flexible Shape Parameter Coding

A list of all currently visible control points having flexible shape parameters $S^n_f \neq 0$ is transmitted using a run-length code. In a second step, the components of the corresponding vectors S^n_f are linearly quantized using 16 representation levels within an interval of ± 5 pel. The quantized vector components are entropy coded.

10.5.5 Color Parameters

Conventional DCT is not suitable for the coding of color parameters of arbitrarily shaped regions. New algorithms have been developed for this application [17, 61]. Here the special type of DCT for arbitrarily shaped regions developed by Gilge [17] is improved by applying a segmentation of the color parameters into homogeneous regions prior to transform coding [47]. The segmentation is based on the minimum spanning tree [42] using the signal variance as criterion. The boundaries of the regions are coded using a chain code [16]. The DCT coefficients are quantized with a linear quantizer of signal-dependent step size. The advantage of this scheme using segmentation prior to transform coding is that errors due to coarse quantization are mainly concentrated at the boundaries of the segmented regions, where they are less visible due to masking of the human visual system in areas of high local activity.

10.5.6 Control of Parameter Coding

Due to limited data rate, a transmission of all parameter sets cannot be guaranteed. Coder control is used to overcome this difficulty. It consists of coder mode control and priority control. Coder mode control selects the relevant parameter sets and coder adjustments for each object, and priority control arranges these parameter sets for transmission (see Table 10.1).

Depending on the model object class MF or MC, the coder mode control selects two parameter sets for transmission. For MC objects, only motion $A_{k+1}^{(m)}$ and shape update parameters $M_{k+1}^{(m)}$ are coded. Coding of color parameters is not necessary because the existing color parameters $S_k^{(m)}$ of the model objects are sufficient to synthesize the image properly. 2D shape parameters, defining the location of the model failures in the image plane, and color parameters are coded for MF objects.

Priority control guarantees that the motion parameters of all MC objects are transmitted first. In a second step, the shape parameters of the MC objects are transmitted. Finally, the shape and color parameters of the MF objects are transmitted until the available data rate is exhausted.

10.6 Experimental Results

The object-based analysis–synthesis coder based on the source models R3D and F3D is applied to the test sequences Claire [10] and Miss America [5], with a spatial resolution corresponding to CIF and a frame rate of 10 Hz. The results are compared to those of an H.261 coder [8, 9] and OBASC based on the source model F2D as presented by Hötter [26]–[28]. As far as detection of model failures and coding of shape parameters are concerned, the same algorithms and coder adjustments are applied. Parameter coding aims at a data rate of approximately 64 kbit/s. However, the bit rate of the coder is not controlled and no buffer is implemented. In the experiments, the allowed noise level N_e for detection of model failures is set to 6/255. Color parameters of model failures are coded according to Section 10.5.5 with a peak signal-to-noise ratio (PSNR) of 36 dB. In all experiments the coders are initialized with the first original image of the sequence (i.e., the frame memory is initialized with the first original image for the block-based coder H.261). For the two object-based analysis–synthesis coders, the model object Background in the memory for parameters is initialized with the first original image.

For head and shoulder scenes the 3D model object is usually divided into two to three components. Applying the estimated motion parameter sets to the model object gives a natural impression of object motion. This indicates that the estimated motion parameters are close to the real motion parameters and that the distance transform applied to the object silhouette for generating the 3D model object shape is suitable for the analysis of head and shoulder scenes.

The area of rigid model failures MF_{R3D} is on average less than 4% of the image area (Figure 10.26). Generalizing this, for head and shoulder scenes the rigid model failures can perhaps be expected to be less than 15% of the moving area. The exact figures of model failure area as related to moving area are 12% for Claire and 7% for Miss America. The test sequence Claire seems to be more demanding, due to the fast rotation of the subject's head, whereas Miss America's motion is almost 2D.

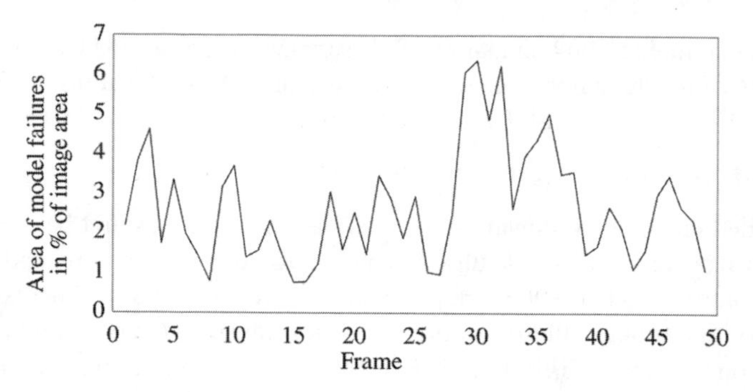

FIGURE 10.26

Area of rigid model failures MF_{R3D} in pel for the test sequence Claire. The total area is 101,376 pel. The average area of model failures is 3.5% of the image area.

Table 10.2 compares the average bit rate for the different parameter sets motion, shape, and color and the source models F2D, R3D, and F3D. Coding of the head and shoulder test sequences Claire and Miss America and the MPEG-4 test sequence Akiyo will not exceed the data rates given in Table 10.2. The source models F2D and R3D need approximately the same data rate. Due to the displacement vector field, OBASC based on the source model F2D requires a relatively high amount of motion information. Shape parameters include the shape of MC and MF objects. Shape parameters of MC objects require similar data rates for both source models. However, the source model F2D causes only a few large MF objects, whereas the source model R3D causes smaller but more MF objects. This larger number of MF_{R3D} objects is due to the applied source model assuming rigid shapes. Shape differences between real and model objects as well as small flexible motion on the surface of real objects cannot be compensated for. These effects cause small local position errors of the model objects. If texture with high local activity is displaced for more than 0.5 pel, model failures are detected due to the simple filter for the detection of geometric distortions. Since these small position errors can be compensated for when using the source model F2D, the overall data rate for shape parameters is 750 bit higher for the source model R3D. Since this is due to the more local motion model of F2D, we have to also look at comparing the sum of motion and shape data rates. Here, the source model R3D requires 7.5% fewer bits than the source model F2D.

Table 10.2 Average Bit Rate of Parameter Sets for Different Source Models

Source Model	Motion: R_A (bit/frame)	Shape of MC Objects: $R_{M,MC}$ (bit/frame)	Shape of MF Objects: $R_{M,MF}$ (bit/frame)	Area of MF Objects and Uncovered Background (% of image area)
F2D	1100	900	900	4%
R3D	200	500	1150	4%
F3D	200	950	1000	3%

Note: The coders use the same algorithm for detection of model failures.

When comparing F3D with R3D, we notice that the average area of MF objects decreases from 4 to 3%. For CIF sequences, this results in bit savings of at least 1000 bit/frame for the texture parameters of MF objects. At the same time, the data rate for MF object shape decreases from 1150 to 1000 bit/frame. This is mainly due to the smaller number of MF objects in the case of F3D. Since the use of F3D requires the transmission of the flexible shift vectors as an additional dataset, the rate for MC object shape parameters increases by 450 bit/frame.

This indicates that by spending an additional 450 bit/frame on the shape parameters of an MC object, we save 150 bit/frame on the MF object shape and reduce the area of MF objects by 1%. Therefore, the bit savings is significantly higher than the costs for this additional parameter set of F3D.

Figure 10.27 shows part of the 33rd decoded frame of the test sequence Claire using the source models F2D, R3D, F3D, and H.261. Subjectively, there is no difference between the source models F2D, R3D, and F3D. However, the source model F3D requires only 56 kbit/s instead of 64 kbit/s. When compared to decoded images of an H.261 coder [9], picture quality is improved twofold (Figure 10.28). At the boundaries of moving objects, no block or mosquito artifacts are visible, due to the introduction of shape parameters. Image quality in the face is improved, because coding of color parameters is limited to model failures, which are mainly located in the face. Since the average area of model failures (i.e., the area where color parameters have to be coded) covers 4% of the image area, color parameters can be coded at a data rate higher than 1.0 bit/pel. This compares to 0.1 to 0.4 bit/pel available for coding of color parameters with an H.261/RM8 encoder.

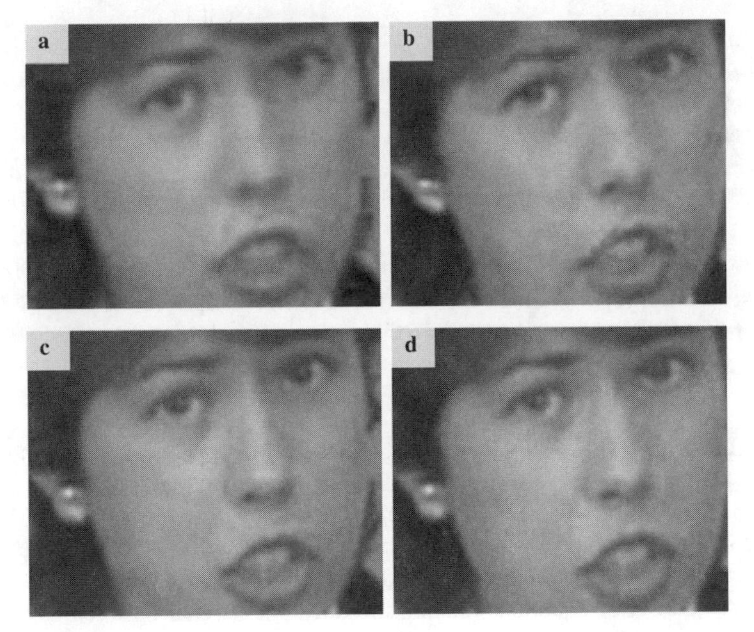

FIGURE 10.27
Part of the 33rd decoded frame of test sequence Claire at a data rate of 64 kbit/s: (a) block-based hybrid coder H.261 (RM8), (b) F2D, (c) R3D, (d) F3D at 56 kbit/s.

10.7 Conclusions

In this chapter the concept and implementation of an object-based analysis–synthesis coder based on the source models of moving rigid 3D objects (R3D) and moving flexible 3D objects (F3D) aiming at a data rate of 64 kbit/s have been presented. An OBASC consists of five parts: image analysis, parameter coding and decoding, image synthesis, and memory for parameters. Each object is defined by its uniform 3D motion and is described by motion, shape, and color parameters. Moving objects are modeled by 3D model objects.

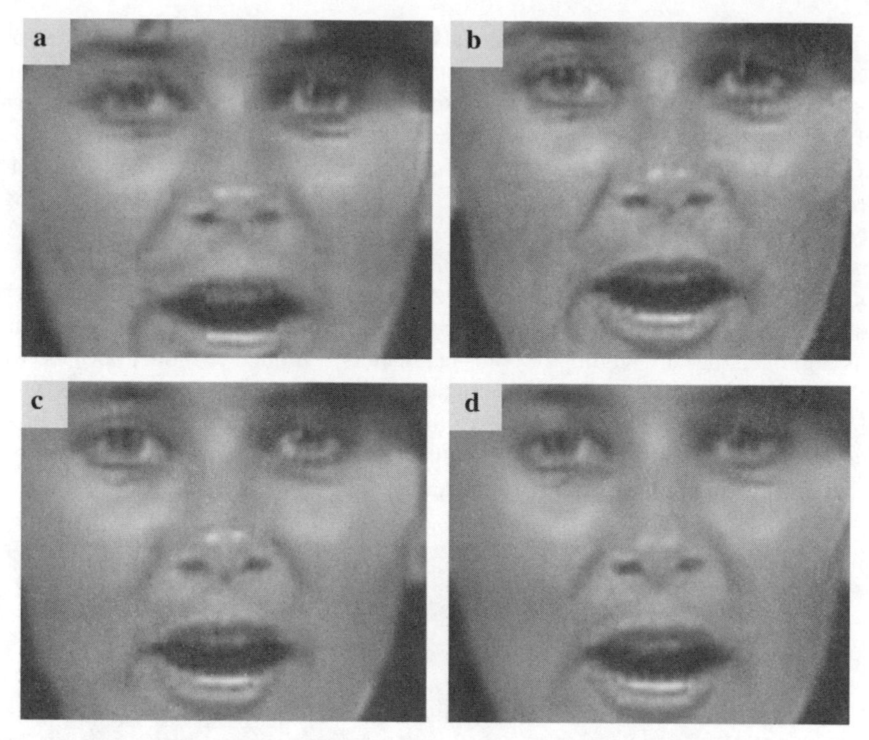

FIGURE 10.28
Part of the decoded frame 8 of the test sequence Miss America, to demonstrate blocking artifacts at 64 kbit/s: (a) block-based hybrid coder H.261 (RM8), (b) F2D, (c) R3D, (d) F3D at 56 kbit/s.

The goal of image analysis is to arrive at a compact parametric description of the current image of a sequence, taking already transmitted parameter sets into account. Image analysis includes shape and motion estimation as well as detection of model failures. Moving objects are segmented using temporal change detection and motion parameters. The algorithm for estimating these motion parameters is based on previous work on gradient-based motion estimation. A new set of equations relating the difference signal between two images to 3D motion parameters has been established. Using robust motion estimation algorithms enables us to track rigid as well as flexible objects such as head and shoulders throughout a video scene, thus enabling an efficient object-based video coder. The 3D shape of a model object is computed by applying a distance transformation, giving object depth, to the object silhouette. Since the estimated motion parameters applied to these 3D model objects give a natural impression of motion, this distance transform is very suitable for analysis of head and shoulder scenes.

Those areas of an image that cannot be modeled by the applied source model are referred to as model failures and are modeled by MF objects. They are described by color and 2D shape parameters only. Model failures are detected taking subjective criteria into account. It is assumed that geometrical distortions such as small position and shape errors of the moving objects do not disturb subjective image quality. Due to these subjective criteria, the average area of model failures is less than 4% of the image area for typical videophone test sequences.

Flexible shift parameters of the source model F3D are estimated only for those parts of an object that cannot be described using the source model R3D. This limits the additional data rate required for the flexible shift parameters and increases the overall efficiency of this source model F3D over R3D.

With respect to coding, shape and color parameters are coded for MF objects, whereas motion and shape update parameters have to be coded for MC objects. Motion parameters are PCM coded and shape parameters are coded using a polygon approximation. Prior to coding, color parameters are segmented into homogeneous regions. Then a DCT for arbitrarily shaped regions is applied.

The presented coder has been compared to OBASC based on the source model of moving flexible 2D objects (F2D). With regard to typical head and shoulder videophone test sequences, it is shown that the picture quality at the average bit rate of 64 kbit/s is the same regardless of whether the source model R3D or F2D is applied. When using F3D, the data rate shrinks from 64 to 56 kbit/s for the same picture quality. When compared to images coded according to H.261, there are no mosquito or block artifacts because the average area for which color parameters are transmitted is 10% of the image area for H.261 and 4% for OBASC. Therefore, OBASC allows coding of color parameters for MF objects with a data rate higher than 1.0 bit/pel. At the same time, MC objects are displayed without subjectively annoying artifacts.

This chapter has demonstrated the feasibility of OBASC using 3D models. As shown in [31], this coder can be used as a basis for knowledge-based and semantic coders. However, image analysis as presented here is not yet able to describe with sufficient accuracy scenes with a lot of motion, such as gesticulating hands. In [39], the 3D motion estimation algorithm is extended to enable the segmentation of flexibly connected rigid objects and components. In [40], an OBASC with an image analysis is presented that enables camera motion compensation. The concept of MF detection is viable for any coder that uses smooth motion vector fields and knows the approximate location of object boundaries [54, 66]. It enables reduction of the data rate of a block-based coder without decreasing the subjective image quality.

References

[1] J.K. Aggarwal, N. Nandhakumar, "On the computation of motion from sequences of images — a review," *Proc. IEEE,* vol. 76, no. 8, pp. 917–935, August 1988.

[2] K. Aizawa, H. Harashima, T. Saito, "Model-based analysis–synthesis image coding (MBASIC) system for a person's face," *Signal Processing: Image Communication,* vol. 1, no. 2, pp. 139–152, October 1989.

[3] P. van Beek, A.M. Tekalp, "Object-based video coding using forward tracking 2D mesh layers," *Proc. SPIE,* vol. 3024, pt. 2, pp. 699–710, 1997.

[4] M. Bierling, "Displacement estimation by hierarchical blockmatching," *3rd SPIE Symposium on Visual Communications and Image Processing,* Cambridge, MA, S. 942–951, November 1988.

[5] British Telecom Research Lab (BTRL), "Test sequence *Miss America,* CIF, 10 Hz, 50 frames," Martlesham, Great Britain.

[6] H. Busch, "Subdividing non rigid 3D objects into quasi rigid parts," *IEEE 3rd Int. Conf. on Image Processing and Applications,* IEE Publ. 307, pp. 1–4, Warwick, UK, July 1989.

[7] G. Bozdagi, A.M. Tekalp, L. Onural, "3-D motion estimation and wireframe adaptation including photometric effects for model-based coding of facial image sequences," *IEEE Trans. Circuits and Systems for Video Technology,* vol. 4, no. 3, pp. 246–256, June 1994.

[8] ITU-T Recommendation H.261. Video codec for audio visual services at p $\times$ 64 kbit/s, March 1993.

[9] CCITT SG XV, Doc. #525, Description of Ref. Model 8 (RM8), 1989.

[10] Centre National d'Etudes des Telecommunication (CNET), "Test sequence *Claire,* CIF, 10 Hz, 156 frames," Paris, France.

[11] C.S. Choi, T. Takebe, H. Harashima, "Three-dimensional (3-D) facial model-based description and synthesis of facial expressions," *IEICE Trans. Electronics and Communications in Japan,* Part 3, vol. 74, no. 7, pp. 12–23, 1991.

[12] G. Chuang, M. Kunt, "Contour simplification and motion compensated coding," *Signal Processing Image Communication,* vol. 7, no. 4–6, pp. 279–296, November 1995.

[13] N. Diehl, "Object-oriented motion estimation and segmentation in image sequences," *Signal Processing: Image Communication,* vol. 3, no. 1, pp. 23–56, February 1991.

[14] P. Eckman, V.W. Friesen, *Facial Coding System,* Consulting Psychologists Press, Palo Alto, CA, 1977.

[15] R. Forchheimer, O. Fahlander, T. Kronander, "A semantic approach to the transmission of face images," *Picture Coding Symposium (PCS'84),* Cesson-Sevigne, France, no. 10.5, July 1984.

[16] M. Eden, M. Kocher, "On the performance of a contour coding algorithm in the context of image coding part 1: Contour segment coding," *Signal Processing,* vol. 8, no. 4, pp. 381–386, June 1985.

[17] M. Gilge, T. Engelhardt, R. Mehlan," Coding of arbitrarily shaped image segments based on a generalized orthogonal transform," *Signal Processing: Image Communication,* vol. 1, no. 2, pp. 153–180, October 1989.

[18] P. Gerken, "Object-based analysis–synthesis coding of image sequences at very low bit rates," *IEEE Trans. Circuits and Systems for Video Technology,* vol. 4, no. 3, pp. 228–235, June 1994.

[19] M. Gilge, "Motion estimation by scene adaptive block matching (SABM) and illumination correction," *Proc. SPIE,* vol. 1244, pp. 355–366, 1990.

[20] H.P. Graf, E. Cosatto, M. Potamianos, "Robust recognition of faces and facial features with a multi-modal system," *Computational Cybernetics and Simulation,* vol. 3, pp. 2034–2039, 1997.

[21] H. Harashima, K. Aizawa, T. Saito, "Model-based analysis synthesis coding of videotelephone images — conception and basic study of intelligent image coding," *Trans. IEICE,* vol. E 72, no. 5, pp. 452–459, May 1989.

[22] H. Harashima, F. Kishino, "Intelligent image coding and communications with realistic sensations — recent trends," *Trans. IEICE,* vol. E 74, no. 6, pp. 1582–1592, June 1991.

[23] M. Hötter, R. Thoma, "Image segmentation based on object oriented mapping parameter estimation," *Signal Processing,* vol. 15, no. 3, pp. 315–334, October 1988.

[24] M. Hötter, "Predictive contour coding for an object-oriented analysis–synthesis coder," *IEEE Int. Symp. on Information Theory,* San Diego, CA, p. 75, January 1990.

[25] M. Hötter, "Object-oriented analysis–synthesis coding based on moving two-dimensional objects," *Signal Processing: Image Communication,* vol. 2, no. 4, pp. 409–428, December 1990.

[26] M. Hötter, "Optimization of an object-oriented analysis–synthesis coder based on the model of flexible 2D-objects," *Picture Coding Symposium (PCS '91)*, Tokyo, Japan, no. 10.4, September 1991.

[27] M. Hötter, "Optimization and efficiency of an object-oriented analysis–synthesis coder," *IEEE Trans. Circuits and Systems for Video Technology,* vol. 4, no. 2, April 1994.

[28] M. Hötter, "Objektorientierte Analyse-Synthese-Codierung basierend auf dem Modell bewegter, zweidimensionaler Objekte," *Fortschr.-Ber. VDI Reihe,* vol. 10, no. 217, 1992.

[29] P. Huber, *Robust Statistics,* Wiley, New York, 1981.

[30] ISO/IEC IS 14496-2 Visual, 1999 (MPEG-4 Visual).

[31] M. Kampmann, J. Ostermann, "Automatic adaptation of a face model in a layered coder with an object-based analysis–synthesis layer and a knowledge-based layer," *Signal Processing: Image Communication,* vol. 9, no. 3, pp. 201–220, March 1997.

[32] F. Kappei, C.-E. Liedtke, "Modelling of a natural 3-D scene consisting of moving objects from a sequence of monocular TV images," *SPIE,* vol. 860, 1987.

[33] F. Kappei, G. Heipel, "3D model based image coding," *Picture Coding Symposium (PCS '88),* Torino, Italy, p. 4.2, September 1988.

[34] R. Koch, "Adaptation of a 3D facial mask to human faces in videophone sequences using model based image analysis," *Picture Coding Symposium (PCS '91),* Tokyo, Japan, pp. 285–288, September 1991.

[35] R. Koch, "Dynamic 3-D scene analysis through synthesis feedback control," *IEEE T-PAMI,* vol. 15, no. 6, pp. 556–568, June 1993.

[36] M. Kunt, "Second-generation image-coding techniques," *Proc. IEEE,* vol. 73, no. 4, pp. 549–574, April 1985.

[37] H. Li, P. Roivainen, R. Forchheimer, "3-D motion estimation in model-based facial image coding," *IEEE T-PAMI,* vol. 15, no. 6, pp. 545–555, June 1993.

[38] H. Li, R. Forchheimer, "Two-view facial movement estimation," *IEEE Trans. Circuits and Systems for Video Technology,* vol. 4, no. 3, pp. 276–287, June 1994.

[39] G. Martínez, "Shape estimation of articulated 3D objects for object-based analysis–synthesis coding (OBASC)," *Signal Processing: Image Communication,* vol. 9, no. 3, pp. 175–199, March 1997.

[40] R. Mech, M. Wollborn, "A noise-robust method for 2D shape estimation of moving objects in video sequences considering a moving camera," *Signal Processing,* vol. 66, no. 2, pp. 203–217, April 1998.

[41] P. Meer, D. Mintz, D.Y. Kim, "Robust regression methods in computer vision: a review," *International Journal of Computer Vision,* no. 6, pp. 59–70, 1991.

[42] O.J. Morris, M. de J. Lee, A.G. Constantinides, "Graph theory for image analysis: an approach based on the shortest spanning tree," *IEEE Proc.,* no. 2, April 1986.

[43] H.G. Musmann, P. Pirsch, H.J. Grallert, "Advances in picture coding," *Proc. IEEE,* vol. 73, no. 4, pp. 523–548, April 1985.

[44] H.G. Musmann, M. Hötter, J. Ostermann, "Object-oriented analysis–synthesis coding of moving images," *Signal Processing: Image Communication,* vol. 3, no. 2, pp. 117–138, November 1989.

[45] J. Ostermann, "Modelling of 3D-moving objects for an analysis–synthesis coder," *Proc. SPIE,* vol. 1260, pp. 240–250, 1990.

[46] J. Ostermann, H. Li, "Detection and coding of model failures in an analysis–synthesis coder based on moving 3D-objects," *3rd International Workshop on 64 kbit/s Coding of Moving Video,* Rotterdam, The Netherlands, no. 4.4, September 1990.

[47] J. Ostermann, "Coding of color parameters in an analysis–synthesis coder based on moving 3D-objects," *Picture Coding Symposium (PCS '91),* Tokyo, Japan, no. 10.5, September 1991.

[48] J. Ostermann, "Object-based analysis–synthesis coding based on the source model of moving rigid 3D objects," *Signal Processing: Image Communication,* no. 6, pp. 143–161, 1994.

[49] J. Ostermann, "Object-oriented analysis–synthesis coding (OOASC) based on the source model of moving flexible 3D objects," *IEEE Trans. Image Processing,* vol. 3, no. 5, pp. 705–711, September 1994.

[50] J. Ostermann, "The block-based coder mode in an object-based analysis–synthesis coder," *Proceedings of 1994 28th Asilomar Conference on Signals, Systems and Computers,* vol. 2, pp. 960–964, Pacific Grove, CA, 31 Oct.–2 Nov. 1994.

[51] J. Ostermann, "Analyse-Synthese-Codierung basierend auf dem Modell bewegter, dreidimensionaler Objeckte," *Fortschr.-Ber. VDI Reihe,* vol. 10, no. 391, 1995.

[52] J. Ostermann, "Segmentation of image areas changed due to object motion considering shadows," *Multimedia Communications and Video Coding,* Y. Wang (ed.), Plenum Press, New York, 1996.

[53] J. Ostermann, "Animation of synthetic faces in MPEG-4," *Proc. Computer Animation '98,* pp. 49–55, Philadelphia, PA, 8–10 June 1998.

[54] J. Ostermann, "Feedback loop for coder control in a block-based hybrid coder with mesh-based motion compensation," *1997 IEEE Int. Conf. on Acoustics, Speech, and Signal Processing,* vol. 4, pp. 2673–2676, Munich, Germany, 21–24 April 1997.

[55] D.E. Pearson, "Texture mapping in model-based image coding," *Image Communication,* vol. 2, no. 4, pp. 377–395, December 1990.

[56] R. Plomben, Y. Hatori, W. Geuen, J. Guichard, M. Guglielmo, H. Brusewitz, "Motion video coding in CCITT SG XV — The video source coding," *Proc. IEEE GLOBECOM,* vol. II, pp. 31.2.1–31.2.8, December 1988.

[57] W.H. Press et al., *Numerical Recipes in C,* Cambridge University Press, London, 1992.

[58] D.F. Rogers, *Procedural Elements for Computer Graphics,* McGraw-Hill, New York, 1985.

[59] P.J. Rousseeuw, A.M. Leroy, *Robust Regression and Outlier Detection,* John Wiley and Sons, New York, 1987.

[60] T. Sasaki, S. Akamatsu, H. Fukamachi, Y. Suenaga, "A color segmentation method for image registration in automatic face recognition," *Advances in Color Vision, Optical Society of America,* Proc. of 1992 Technical Digest Series, vol. 4, SaB17, pp. 179–181, January 1992.

[61] H. Schiller, M. Hötter, "Investigations on colour coding in an object-oriented analysis–synthesis coder," *Signal Processing: Image Communication,* vol. 5, no. 4, pp. 319–326, October 1993.

[62] J. Stauder, "Estimation of point light source parameters for object-based coding," *Signal Processing: Image Communication,* vol. 7, no. 4–6, pp. 355–379, November 1995.

[63] J. Stauder, R. Mech, J. Ostermann, "Detection of moving cast shadows for object segmentation," *IEEE Trans. Multimedia,* vol. 1, no. 1, pp. 65–76, March 1999.

[64] R. Thoma, M. Bierling, "Motion compensating interpolation considering covered and uncovered background," *Signal Processing: Image Communication,* vol. 1, no. 2, pp. 191–212, October 1989.

[65] R.Y. Tsai, T.S. Huang, "Estimating three–dimensional motion parameters of a rigid planar patch," *IEEE Trans. Acoustics, Speech, and Signal Processing,* vol. ASSP-29, no. 6, pp. 1147–1152, December 1981.

[66] Y. Wang, J. Ostermann, "Evaluation of mesh-based motion estimation in H.263-like coders," *IEEE Trans. Circuits and Systems for Video Technology,* vol. 8, no. 3, pp. 243–252, June 1998.

[67] Y. Wang, O. Lee, "Active mesh — a feature seeking and tracking image sequence representation scheme," *IEEE Trans. Image Processing,* vol. 3, no. 5, pp. 610–624, September 1994.

[68] B. Welsh, "Model-based coding of images," Ph.D. thesis, Essex University, Great Britain, January 1991.

[69] Y. Yakimovsky, R. Cunningham, "A system for extracting three-dimensional measurements from a stereo pair of TV cameras," *Computer Graphics and Image Processing,* vol. 7, no. 2, pp. 195–210, April 1978.

[70] X. Zhuang, T. Wang, P. Zhang, "A highly robust estimator through partially likelihood function modelling and its application in computer vision," *IEEE T-PAMI,* vol. 14, no. 1, pp. 19–35, January 1992.

Chapter 11

Rate-Distortion Techniques in Image and Video Coding

Aggelos K. Katsaggelos and Gerry Melnikov

11.1 The Multimedia Transmission Problem

One of the central issues in multimedia communications is the efficient transmission of information from source to destination. This broad definition includes a large number of target applications. They may differ in the type (i.e., text, voice, image, video, etc.) and form (analog or digital) of the data being transmitted. Applications also differ by whether lossless or lossy data transmission is employed and by the kind of channel the transmitted data passes through on its way from an encoder to a decoder or a storage device. Additionally, there are wide variations in what represents acceptable loss of data. For example, whereas certain loss of information is acceptable in compressing video signals, it is not acceptable in compressing text signals.

In general, the goal of a communications system is the transfer of data in a way that is resource efficient, minimally prone to errors, and without significant delay. In a particular application we may know the target operating environment (bit rate, channel characteristics, acceptable delay, etc.). The challenge is to arrive at the data representation satisfying these constraints. For example, if the application at hand is transmission of digital video over a noisy wireless channel, the task is to allocate the available bits within and across frames optimally based on the partially known source statistics, the assumed channel model, and the desired signal fidelity at the receiver.

In problems where some information loss is inevitable, whether caused by quantizing a deterministic signal or through channel errors, rate distortion theory provides some theoretical foundations for deriving performance bounds. The fundamental issue in rate distortion theory is finding the bound for the fidelity of representation of a source of known statistical properties under a given description length constraint. The dual problem is that in which a given fidelity requirement is prescribed and the source description length is to be minimized.

Although there is no argument that the count of bits should be the metric for rate or the description length, the choice of the corresponding metric for distortion is not straightforward and may depend on the particular application. The set of all possible distortion metrics can be broadly categorized into unquantifiable perceptual metrics and metrics for which a closed-form mathematical representation exists. The latter class, particularly the mean squared error (MSE) metric, and, to a lesser extent, the metric based on the *maximum* operator [34], has been primarily used in the image and video coding community partly due to the ease of computation and partly for historical reasons.

The well-known Shannon's source and channel coding theorems [35] paved the way for separating, without loss of optimality, the processes of removing redundancy from data (source coding) and the transmission of the resulting bitstream over a noisy channel (channel coding). Shannon also established a lower (entropy) bound on the performance of lossless source compression algorithms and subsequent error-free transmission over a channel of limited capacity. Hence, most practical communication systems have the structure shown in Figure 11.1.

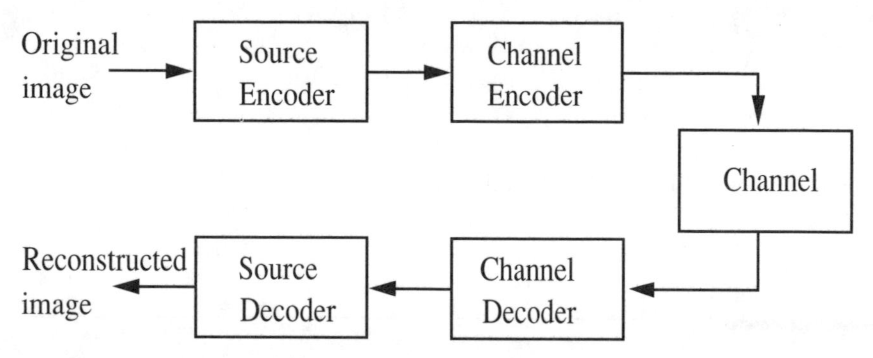

FIGURE 11.1
High-level view of a video communications system.

Many applications demand data compression beyond entropy (i.e., with some loss of information). Wireless transmission of video is one example when the original data stream must be compressed by factors of hundreds or thousands, due to limited channel capacity. Clearly, with lossy compression, there is a trade-off between the transmission of fewer bits through a channel and the quality of the reconstructed signal. The central issue in the rate-distortion theory is how much redundancy can be removed from a given data source while satisfying a distortion constraint. Conversely, the constraint can be imposed on the rate, in which case a lower bound on distortion is sought. The theory is based on knowledge of the source statistics and not on any specific source coding scheme. Hence it provides a lower bound on the rate, which clearly may not be achieved by a particular encoder implementation. In addition to not being constructive, this bound assumes perfect knowledge of the source statistics which, in most cases, is not available. Source model complexity is yet another dimension in rate-distortion theory. The trade-off here is between a very simple source model (e.g., an i.i.d. Gaussian model), which is relatively easy to treat analytically, on one hand, and a model that is better adapted to a given source but is more complex. Clearly, any performance bounds obtained with the help of the rate-distortion theory are only as good as the source models and may be of little use if they are inaccurate.

Even though, traditionally, rate-distortion theory was applied only to situations where distortion was caused deterministically through quantization at the encoder, it is applied equally well to the case when the distortion is introduced stochastically in a noisy channel. In this case, however, the tightness of the bound also depends on the accuracy of the channel model.

The applicability of Shannon's separation principle between source and channel coding for video transmission, under some real-world constraints, was studied in [20] and the references therein. The conclusion is that for applications such as the Internet, with its packetization and delay constraints, for multicast and layered applications, where the decoding must be possible for any subset of the received bitstream, as well as for applications requiring unequal error protection, utilizing the separation principle yields results inferior to those obtained by joint source–channel coding. Practical application of Shannon's bound is also hampered by such factors as the infinite block size (symbol length) assumption, and the corresponding time delay.

As discussed in Section 11.3, some of these real-world constraints can be easily incorporated into a rate-distortion optimized coding algorithm.

Lossless data compression techniques capitalize on redundancies present in the source and do not involve any trade-offs between rate and distortion. When it comes to designing lossy data compression algorithms, however, the challenge is to represent the original data in such a way that the application of coarse quantizers, which is what makes compression beyond entropy possible, costs the least in terms of the resulting degradation in quality.

The objective of rate-distortion theory is to find a lower bound on the rate such that a known source is represented with a given fidelity, or conversely. Thus, if the rate (R) and distortion (D) are quantified, varying the maximum distortion or fidelity continuously, and finding the corresponding bound R, is equivalent to tracing a curve in the R–D space. This curve is called the rate-distortion function (RDF). Even assuming that the source model is correct, points on this curve, however, may be difficult to achieve not only because of the infinite block length and computational resources assumption, but also because they represent a lower bound among *all* possible encoders under the sun.

11.2 The Operational Rate-Distortion Function

In the previous section we discussed the classical rate-distortion theory, the primary purpose of which is to establish performance bounds for a given data source across all possible encoders, operating at their "best." In practice, however, we often deal with one fixed encoder structure and try to optimize its free parameters. Finding performance bounds in such a setup is the subject matter of the operational rate-distortion (ORD) theory.

Operating within the framework of a fixed encoder simplifies the problem a lot. Solving the problem of optimal bit allocation in this restricted case, however, does not guarantee operation at or near the bound established by the RDF. As an example, let us consider the simplest image encoder, which approximates the image by the DC values of blocks derived from a fixed partition. Clearly, finding the best possible quantization scheme for these coefficients, while making the encoder optimal, will not do a good job of compressing the image or come close to the RDF, since the encoder structure itself is very simple.

ORD theory is based on the fact that every encoder maps the input data into independent or dependent sources of information which need to be represented efficiently. The finite number of modes the encoder can select for each of these data subsources can be thought of as the admissible set of quantizers. That is, a particular quantization choice for each of the subsources in the image or video signal, for example, constitutes one admissible quantizer. A quantizer, in this case, is defined in the most general terms. It may operate on image pixels, blocks, or coefficients in a transformed domain. The encoder, in addition to being responsible for selecting quantizers (reconstruction levels) for each of the subsources, must arrive at a good partition of the original signal into subsources. Let us call a particular assignment of quantizers to all subsources a *mode of operation* at the encoder. Each mode can be categorized by a rate R and a distortion D. The set of all (R, D) pairs, whose cardinality is equal to the number of admissible modes, constitutes the quantizer function (QF). Figure 11.2, in which the QF is shown with the $\times$ symbol, illustrates this concept.

The ORD curve, denoted by the dotted line in Figure 11.2, is a subset of the QF and represents modes of desirable operation of the encoder. Mathematically, the set of points on the ORD curve is defined as follows:

$$\text{ORD} = \left\{ i \in \text{QF} : R_i \leq R_j \text{ or } D_i \leq D_j, \forall j \in \text{QF}, i \neq j \right\}, \tag{11.1}$$

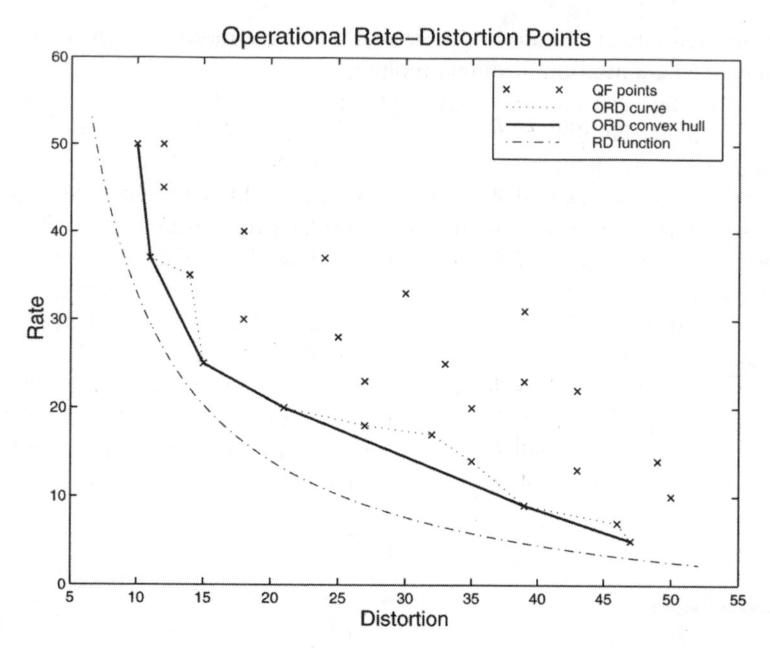

FIGURE 11.2
The operational rate-distortion (ORD) curve.

where R_i, D_i and R_j, D_j are the rate-distortion pairs associated with modes i and j, respectively. For convenience, it is customary to connect consecutive points of the ORD set, thus forming the ORD curve. Clearly, it is desirable to operate on the ORD curve, and modes not on the curve are not optimal in the sense that within the same encoder, a smaller distortion is achievable for the same or smaller rate, or vice versa.

As stated, the concept of operating on the ORD curve is quite generic and can be applied to a variety of applications. Practically, the idea of rate-distortion optimization is equivalent to bit budget allocation among different subsources of information in a given compression or channel transmission framework. When an allocation scheme results in a mode belonging to the ORD set, an algorithm is operating at its optimum.

The framework of bit allocation with the goal of reaching a point on the ORD curve applies equally well in cases when the source is stochastic in nature and is known only through the model of its probability density function (*pdf*), or in cases when transmission through a noisy channel, rather than compression, is the problem. In these cases, $E(D)$, the expected distortion, is used instead of distortion, and the tightness of the bound established by the ORD curve is sensitive to source or channel model accuracy.

11.3 Problem Formulation

The central problem in ORD optimization is to select appropriately the modes of the given algorithm such that a point on the ORD curve is reached. This is equivalent to saying that no other selection of parameters would lead to a better distortion performance for the same bit rate. Following the notation of [28], if B is a code belonging to the set of all possible modes S_B generated by the given algorithm as a code for the specified data source, and $R(\cdot)$ and $D(\cdot)$

are the associated rate and distortion functions, we seek a mode B^*, which is the solution to the following constrained optimization problem,

$$\min_{B \in S_B} D(B), \quad \text{subject to:} \quad R(B) \leq R_{\max} . \tag{11.2}$$

It turns out that the problem dual to that of (11.2), in which the source fidelity or distortion is constrained, can be solved using the same tools (described in Section 11.4) as the rate-constrained problem. The dual problem can be expressed as follows:

$$\min_{B \in S_B} R(B), \quad \text{subject to:} \quad D(B) \leq D_{\max} . \tag{11.3}$$

As stated, these problems are general enough to include many possible distortion metrics and many parameter encoding schemes, including differential ones. Since the concept of rate and distortion is typically associated with quantizers, it often helps to think of solutions to (11.2) and (11.3) as optimal bit allocations among (possibly dependent) quantizers.

11.4 Mathematical Tools in RD Optimization

In this section we discuss two powerful optimization methods: the Lagrangian multiplier method and dynamic programming (DP). These techniques are very suitable for the kind of problems discussed here, that is, the allocation of resources among a finite number of dependent quantizers. For an overview of optimization theory the reader is referred to [24]. In all examples presented in this chapter, these two tools are used in conjunction with each other. First, the Lagrangian multiplier method is used to convert a constrained optimization problem into an unconstrained one. Then, the optimal solution is found by subdividing the whole problem into parts with the help of DP.

11.4.1 Lagrangian Optimization

The Lagrangian multiplier method described here is the tool used to solve constrained optimization problems. The idea behind the approach is to transfer one or more constraints into the objective function to be minimized. In the context of image and video coding, the most commonly used objective function is the distortion, with the bit rate being the constraint. As stated, this problem is difficult because it provides no quantifiable measure by which an encoder can make a local decision of selecting the best quantizer among several available quantizers. The Lagrangian multiplier method solves this problem by adding the rate constraint to the objective function, thereby redefining it.

Mathematically, this idea can be stated as follows. Finding the optimal solution $B^*(\lambda)$ of

$$\min_{B \in S_B} \left(D(B) + \lambda \cdot R(B) \right) , \tag{11.4}$$

where λ is a positive real number, is equivalent to solving the following constrained optimization problem:

$$\min_{B \in S_B} D(B), \quad \text{subject to:} \quad R(B) \leq R_{\max} . \tag{11.5}$$

Clearly, the optimal solution B^* is a function of λ, the Lagrangian multiplier. It is worth noting that the converse is not always true. Not every solution to the constrained problem can be found with the unconstrained formulation. That is, there may be values of $R_{\max}$, achievable optimally by exhaustive search or some other method, for which the corresponding λ does not exist and, therefore, is not achievable by minimizing (11.4).

Since, in practice, the constrained problem needs to be solved for a given $R_{\max}$, a critical step in this method is to select λ appropriately, so that $R(B^*(\lambda)) \approx R_{\max}$. Choosing such a λ can be thought of as determining the appropriate trade-off between the rate and the distortion, which is application specific.

A graphical relationship between points on the ORD curve and line segments in the first quadrant with slope $-\frac{1}{\lambda}$ can be established, based on the fact that the rate and distortion components of points on the ORD curve are a nonincreasing and a nondecreasing function of λ, respectively [26, 28]. That is, as Figure 11.3 shows, if we start with a line of that slope passing through the origin and keep moving in the northeast direction, the sweeping line will first intersect the ORD curve at the point(s) corresponding to the rate and the distortion that are the optimal solutions to (11.4) when the trade-off of λ is used.

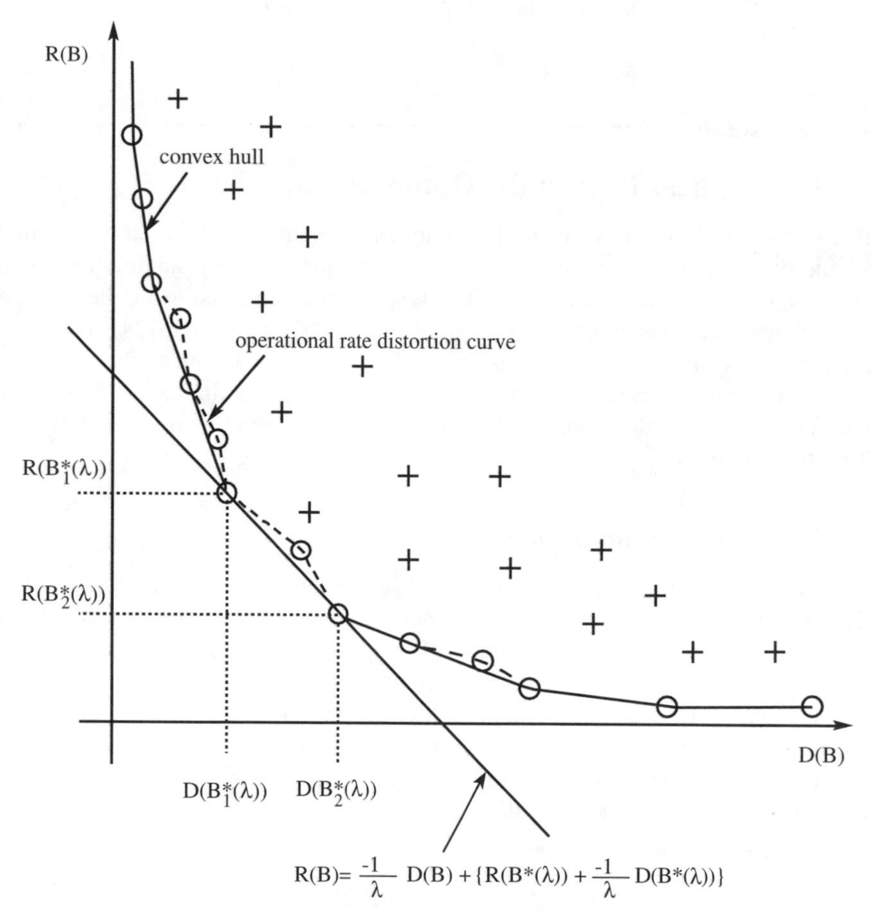

FIGURE 11.3
The line of slope $-\frac{1}{\lambda}$ intersects the ORD curve at points having the rate-distortion trade-off of λ.

There exist two fundamentally different approaches to finding λ_{optimal} meeting the $R_{\max}$ budget. The first approach assumes a continuous model $D(R)$ for the ORD function, for example, a decaying exponential. Then, λ_{optimal} is approximated through $\lambda = -\frac{dD}{dR}$, evaluated at $R = R_{\max}$. A model-based technique, however, is only as good as the model itself.

The second approach, which is also based on the monotonicity of $R(\lambda)$ and $D(\lambda)$, uses an iterative search to find λ_{optimal}. The process consists of running the encoder for two different values of λ, λ_l and λ_u, corresponding to the beginning and the end of the interval to which λ_{optimal} belongs. This interval is iteratively redefined either with the bisection method or based on a fast Bezier curve search technique [28].

11.4.2 Dynamic Programming

Dynamic programming (DP) is a tool that is typically applied to optimization problems in which the optimal solution involves a finite number of decisions, and after one decision is made, problems of the same form, but of a smaller size, arise [2]. It is based on the principle that the optimal solution to the overall problem consists of optimal solutions to its subparts or subproblems. In problems of this type exhaustive search solves the same subproblems over and over as it tries to find the global solution at once. By contrast, DP solves each subproblem just once and their solutions are stored in memory.

In the image and video coding applications considered in this chapter, optimality is achieved by finding the ordered sequence of quantizers $[q_0^*, \ldots, q_N^*]$, minimizing the overall cost function,

$$J^*\left(q_0^*, \ldots, q_N^*\right) = \min_{q_0, \ldots, q_N} J\left(q_0, \ldots, q_N\right) , \tag{11.6}$$

where both the quantizers and their number, N, have to be determined by the encoder. In the context of rate- or distortion-constrained optimization, DP is combined with the Lagrangian multiplier method, in which case the cost function is often written as J_λ to emphasize its dependence on λ. The DP method is applicable because the total cost function minimization can be broken down into subproblems in a recursive manner as follows:

$$J^*\left(q_0^*, \ldots, q_N^*\right) = \min_i \left[J^*\left(q_0^*, \ldots, q_i^*\right) + J^*\left(q_{i+1}^*, \ldots, q_N^*\right)\right], \quad 0 \leq i \leq N . \tag{11.7}$$

It should be noted, however, that straightforward application of DP to problems of very large dimensions may be impractical due to coding delay considerations. Even though DP results in solutions of significantly lower complexity than exhaustive search, it nevertheless performs the equivalent of exhaustive search on the local level. In such cases, a greedy suboptimal matching pursuit approach, based on incremental return, may be called for. This was done in [18] and [6] in the context of low-bit-rate video and fractal compression, respectively.

11.5 Applications of RD Methods

In this section we describe the application of RD-based methods to several different areas of image and video processing. These include motion estimation, motion-compensated interpolation, object shape coding, fractal image compression, and quad-tree (QT)-based video coding. The common theme shared by these applications is that they all can be stated as resource allocation problems among dependent quantizers (i.e., in the forms shown in Section 11.3). Operationally optimal solutions are obtained in each case and are shown to significantly outperform traditional heuristic approaches. In all these applications the mathematical tools described in Section 11.4 are used in the optimization process.

11.5.1 QT-Based Motion Estimation and Motion-Compensated Interpolation

Most of the resources in present-day multimedia communication systems are devoted to digital video, because its three-dimensional nature is inherently more complex than that of speech signals and text. The function of a video codec is unique for several reasons. First, the large amount of raw data necessitates high compression efficiency. Second, a typical video waveform exhibits correlation in the spatial and, to a greater extent, in the temporal direction. Furthermore, the human visual system is more sensitive to errors in the temporal direction.

Motion compensation (MC) is the technique most commonly used in video coding to capitalize on temporal correlation. In this context, some frames in a video sequence are encoded with still image coding techniques. These frames are called I frames. The second type of frames, P frames, are predicted from their reference frames using motion information, which has to be estimated, usually on a block-by-block basis. A set of motion vectors, defining for each pixel in the current frame its location in the reference frame, is called the displacement vector field (DVF).

Due to its simplicity and easy hardware implementation, block matching is the most popular method of motion estimation. Although the use of irregularly shaped regions potentially allows for better local adaptivity of the DVF to a frame's spatial segmentation, its application in video coding is hampered by the need to transmit shape information. By allowing segmentation into blocks of variable sizes, a compromise between a compact representation (blocks of a fixed size) of the DVF and its local adaptivity (complex object-oriented approach) is achieved. That is, large blocks can be used for the background and smaller blocks can be used for areas in motion. The QT structure is an efficient way of segmenting frames into blocks of different sizes, and it was used to represent the inhomogeneous DVF in [11, 28]. In addition, the QT approach enables a tractable search for the joint and optimal segmentation and motion estimation, which is not possible with the complex object-based segmentation [28].

Optimal Motion Estimation

In this section an operationally optimal motion estimator is derived [28, 32]. The overall problem solved here can be stated as that of minimizing the displaced frame difference (DFD) for a given maximum bit rate, and with respect to a given intra-coded reference frame. It should be noted, however, that some efforts have been made to jointly optimize the anchor and the motion-compensated frame encoding [10]. The QT-based frame decomposition used here is achieved by recursively subdividing a $2^N \times 2^N$ image into subimages of halved size until the block size $2^{n_0} \times 2^{n_0}$ is reached. This decomposition results in an $(N - n_0 + 1)$-level hierarchy, where blocks at the nth level are of size $2^n \times 2^n$.

Within this decomposition, each square block $b_{l,i}$ (lth level in the QT, ith block in that level) is associated with $M_{l,i}$ — the set of all admissible motion vectors, of which $m_{l,i}$ is a member. Then a local state $s_{l,i} = [l, i, m_{l,i}] \in S_{l,i} = \{l\} \times \{i\} \times M_{l,i}$ can be defined for each block $b_{l,i}$. Consequently, a global state x, representing the currently chosen local state, is defined as $x \in X = \bigcup_{l=N}^{n_0} \bigcup_{i=0}^{4^{N-l}-1} S_{l,i}$, where X is the set of all admissible global state values.

A complete description of the DVF in the rate-distortion framework requires that the individual block states $s_{l,i}$ be enumerated sequentially because the rate function may involve arbitrary-order dependencies resulting from differential encoding of parameters. Hence, the code for a predicted frame consists of a global state sequence $x_0, \ldots, x_{N_\Theta - 1}$, which represents the left-to-right ordered leaves of a valid QT Θ.

In this context, the frame distortion is an algebraic sum of block distortions $d(x_j)$, where the blocks correspond to the leaves of the chosen QT decomposition Θ. That is,

$$D\left(x_0, \ldots, x_{N_\Theta - 1}\right) = \sum_{j=0}^{N_\Theta - 1} d\left(x_j\right) . \tag{11.8}$$

The individual block distortion metric chosen here is the MSE of the DFD projected on the block.

Encoding the motion vectors of the DVF is a challenging task in itself. On the one hand, high coding efficiency can be achieved by considering long codewords composed of many motion vectors $m_{l,i}$ along the scanning path. On the other hand, as explained in Section 11.4, the complexity of the optimal solution search is directly related to the order of dependency in the differential encoding of parameters. As a compromise between the two, a first-order DPCM scheme is used, allowing a first-order dependency between the leaves along the scanning path.

Another challenge is posed by the fact that a typical image exhibits intensity and motion vector correlation in two dimensions, whereas a scanning path is inherently one dimensional. A scan according to the Hilbert curve was shown in [28, 32] to possess certain space-filling properties and create a representation of the 2D data, which is more correlated than that resulting from a raster scan. It can be generated in a recursive fashion and is natural for QT-decomposed images.

Based on the chosen first-order DPCM along a Hilbert scanning path, the overall frame rate can be expressed as follows:

$$R\left(x_0, \ldots, x_{N_\Theta - 1}\right) = \sum_{j=0}^{N_\Theta - 1} r\left(x_{j-1}, x_j\right) , \tag{11.9}$$

where $r(x_{j-1}, x_j)$ is the block bit rate, which is a function of the quantizers used for encoding the current and the previous blocks.

Having defined the distortion and the rate, the problem of motion estimation is posed as a constrained optimization problem as follows:

$$\min_{x_0, \ldots, x_{N_\Theta - 1}} \quad D\left(x_0, \ldots, x_{N_\Theta - 1}\right) , \quad \text{subject to:} \quad R\left(x_0, \ldots, x_{N_\Theta - 1}\right) \leq R_{\max} . \tag{11.10}$$

Since the process of approximating a block $b_{l,i}$ in the current frame by another block of equal size in the reference frame, using motion vector $m_{l,i}$, can be viewed as quantization of the block $b_{l,i}$, with which a certain rate and a certain distortion are associated, the problem of rate-constrained motion estimation can be viewed as an optimal bit allocation problem among the blocks of a QT with leaf dependencies or dependent quantizers. Hence, the methods of Section 11.4 (Lagrangian multiplier-based unconstrained optimization using DP within a trellis) are applicable.

The search for the best block match under a motion vector is by far the most computationally expensive part of the optimization. To make the search faster, a slightly suboptimal clustering scheme is used, in which only a subset of possible motion vectors is considered [28].

Motion Estimation Results

The QT-based optimal motion estimation scheme compares favorably with TMN4 (an implementation of the H.263 standard), using the same quantizers to encode the DVF. For the QCIF video sequence, the smallest block size of 8×8 was chosen and the Hilbert scan was modified [28] to cover nonsquare frames. The predicted frame 180 of the Mother and Daughter sequence and its corresponding scanning path are shown in Figures 11.4 and 11.5, respectively.

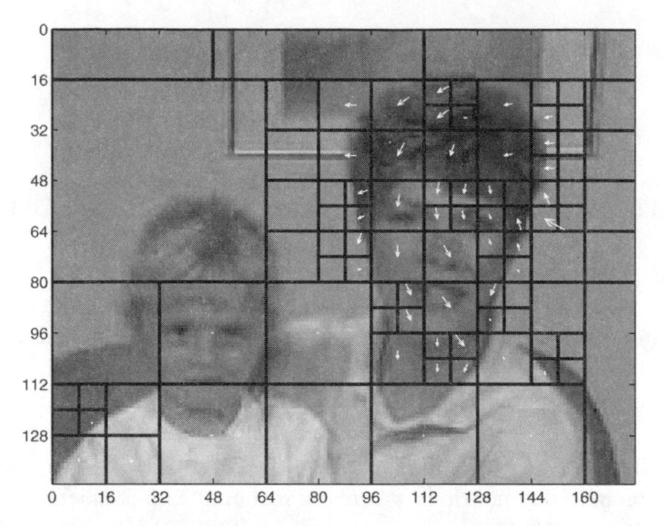

FIGURE 11.4
Segmentation and DVF of the predicted frame.

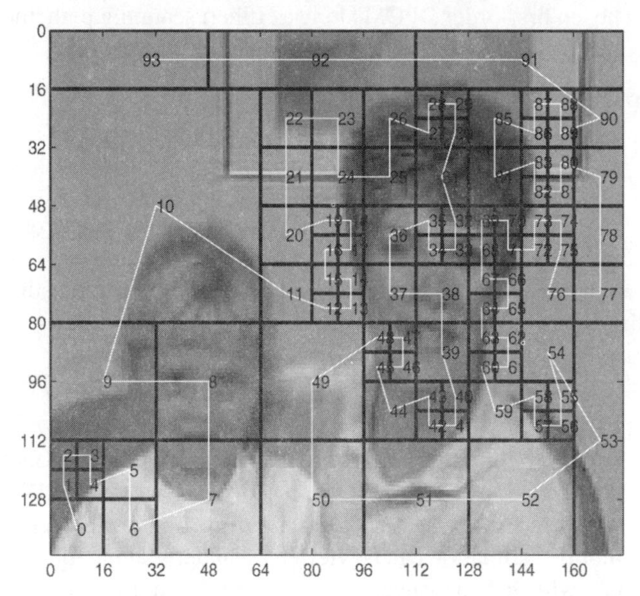

FIGURE 11.5
The overall scanning path.

When compared to the original frame 180, the DFD peak signal-to-noise ratio (PSNR), in the case of optimal QT motion estimation, is 31.31 dB. The corresponding bit rate is 472 bits. Compared to the TMN4 encoder, operating at the same rate and resulting in the PSNR of 30.65 dB, the optimal scheme represents a 0.67-dB improvement in PSNR. The results are more dramatic if, rather than the rate, the PSNR is matched between the two algorithms (at 30.65 dB) (i.e., the dual problem is solved). In this case, the optimal QT-base motion estimator encodes frame 180 with 344 bits, and that is a 26.8% improvement over the 470 bits required by TMN4.

Motion-Compensated Interpolation

The problem of frame interpolation is very important in very-low-bit-rate video coding. The transmission channel capacity, by imposing an upper limit on the bit rate, necessitates a two-pronged approach to video compression: some frames are encoded and transmitted; other frames are not encoded at all, or dropped, and must be interpolated at the decoder. It is common for video codecs to operate at the rate of 7.5 or 10 frames per second (fps) by dropping every third or fourth frame. Without frame interpolation, or with zero-hold frame interpolation, the reconstructed video sequence will appear jerky to a human observer.

The problem of interpolation is ill posed because not enough information is given to establish a metric by which to judge the goodness of a solution. In our context, some frame in the future, $\hat{f}_N$, and some frame in the past, $\hat{f}_0$, of the interpolated frame is all that is available at the decoder. Therefore, since the original skipped frame is not available at the decoder, the MSE minimization-based interpolation is not possible.

What makes this problem solvable is the underlying assumption that changes in the video scene are due to object motion, in particular linear motion. In contrast to the traditional approach, where the motion is estimated for frame $\hat{f}_N$ with respect to frame $\hat{f}_0$ and then projected onto the interpolated frame, here the motion is estimated directly for the interpolated frame. Thus, the problems related to the fact that not all pels of the interpolated frame are associated with a motion vector are avoided. Figure 11.6 demonstrates this idea.

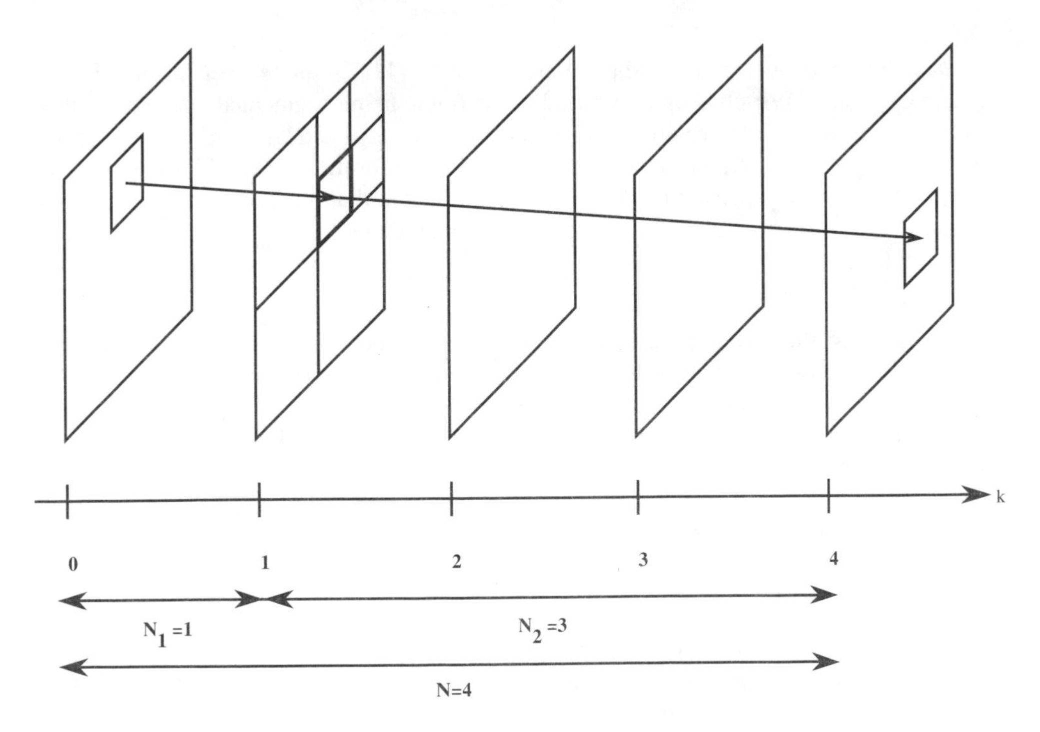

FIGURE 11.6
Motion-compensated interpolation (of frame 1 from frame 0 and frame 4).

The problem at hand is to find the segmentation of the interpolated frame into blocks $b_{l,i}$ (QT decomposition) and the associated motion vectors $m_{l,i}$. The motion, however, is with respect to both reference frames ($\hat{f}_0$ and $\hat{f}_N$). A block $b_{l,i}$ in the QT decomposition is interpolated as follows:

$$\hat{f}_n(x, y) = \frac{N_2 \hat{f}_0(x - N_1 m_{j,x}, y - N_1 m_{j,y}) + N_1 \hat{f}_N(x + N_2 m_{j,x}, y + N_2 m_{j,y})}{N_1 + N_2}$$
$$\forall (x, y) \in b_{l,i} , \tag{11.11}$$

where $0 \leq n \leq N$, $m_{j,x}$ and $m_{j,y}$ are the x and y coordinates of the motion vector m_j of the block $b_{l,i}$, and N_1, N_2 are the temporal distances of the interpolated frame to frames f_0 and f_N, respectively. This weighted definition leads to a smooth transition of the interpolated frame toward the closer reference frame as the distance between them decreases, causing less jerkiness.

Let x_j denote the global system state, corresponding to the interpolated block $b_{l,i}$ undergoing motion m_j. Consistent with the block interpolation formula defined in (11.11), the associated block distortion is defined as follows:

$$d\left(x_j\right) = \sum_{(x,y) \in b_{l,i}} \left(\hat{f}_0\left(x - N_1 m_{j,x}, y - N_1 m_{j,y}\right) - \hat{f}_N\left(x + N_2 m_{j,x}, y + N_2 m_{j,y}\right)\right)^2 , \tag{11.12}$$

and the overall distortion is the algebraic sum of block distortions corresponding to the leaves in the chosen QT decomposition. That is,

$$D\left(x_0, \ldots, x_{N_\Theta - 1}\right) = \sum_{j=0}^{N_\Theta - 1} d\left(x_k\right) . \tag{11.13}$$

Clearly, minimizing the frame distortion defined in (11.13) alone over all possible QT segmentations and DVF choices would lead to the frame being segmented into blocks of the smallest size possible. The resulting DVF would be very noisy and have little resemblance to the underlying object motion in a video scene. It is desired for the estimated DVF to possess a measure of smoothness present in the real DVF. It turns out that this goal can be achieved by regularizing the objective function with the total bit rate, that is,

$$\min_{x_0, \ldots, x_{N_\Theta - 1}} \left(D\left(x_0, \ldots, x_{N_\Theta - 1}\right) + \lambda \cdot R\left(x_0, \ldots, x_{N_\Theta - 1}\right)\right) , \tag{11.14}$$

where λ is the regularization parameter. Minimizing the above objective function leads to a smooth DVF because, with a differential encoding scheme, there is a strong correlation between the smoothness of the DVF and the bit rate necessary for its encoding. Hence, smoothness is achieved for motion vectors along a scanning path. With the Hilbert scanning path employed, this translates into DVF smoothness in all directions.

The optimal solution to the regularized optimization problem of (11.14) is then found with the help of DP applied to a trellis structure, as explained in Section 11.4.

Interpolation Results

The described algorithm has been applied to the compressed Mother and Daughter sequence at the frame rate of 15 fps and a constant frame PSNR of 34.0 dB. Every second frame in this sequence is dropped, thus resulting in a 7.5 fps sequence. Then these dropped frames are reconstructed from the two neighboring frames using the operationally optimal motion-compensated interpolation scheme described in the preceding section.

The issue of selecting the proper regularization parameter λ is addressed in [5]. Here, a λ of 0.01 is used. Figure 11.7 shows the reconstructed frame 86, which was interpolated from frame 84 and frame 88. The resulting DVF and QT segmentation are also overlaid on this figure. The interpolated frame is very similar to the original frame 86 and is only 1 dB lower in PSNR than the reconstructed version, had it been transmitted.

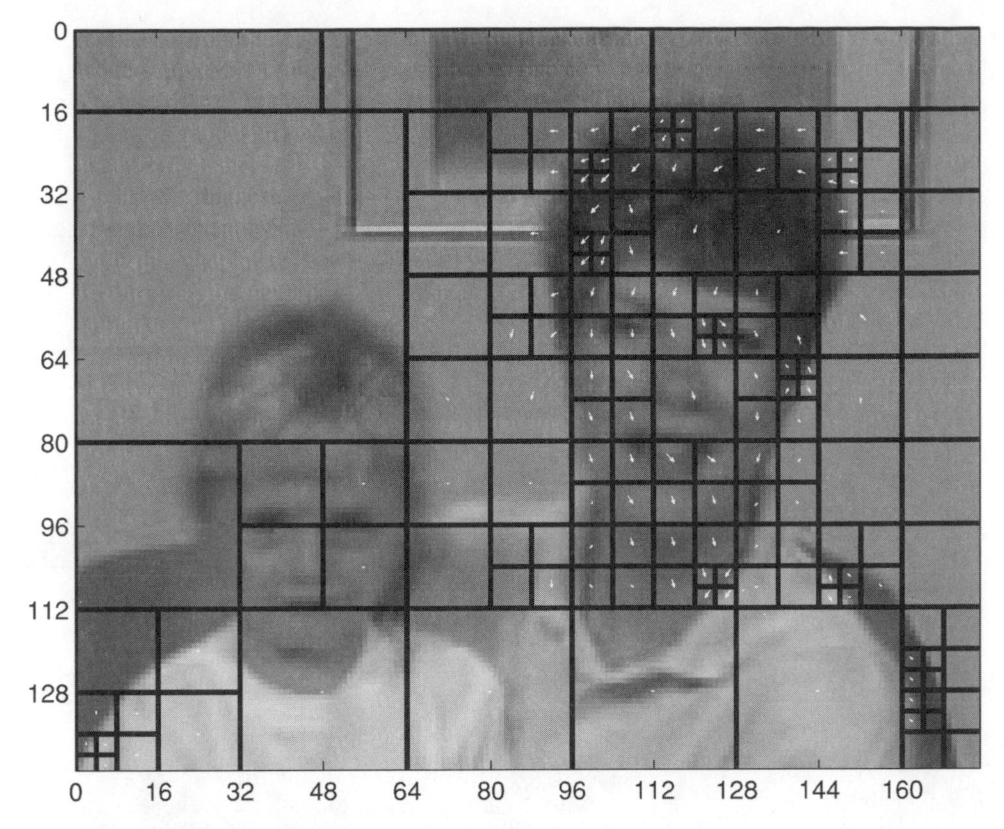

FIGURE 11.7
Segmentation and DVF of the interpolated frame.

11.5.2 QT-Based Video Encoding

In this subsection the operationally optimal bit allocation scheme among QT segmentation, DVF, and DFD is presented. The overall problem solved here can be stated as that of minimizing the distortion between the original and the reconstructed frames for a given bit budget $R_{\max}$, where motion is estimated with respect to a given reference frame. Hence, the job of the encoder is to optimally allocate the available bit budget to segmentation, motion, and error quantization components.

Code Structure

As in the case with optimal motion estimation and interpolation, discussed in the previous subsection, here the QT structure is used for frame segmentation due to its being a compromise between a fixed-block-size approach and a scene-adaptive object-based segmentation. Thus, the original $2^N \times 2^N$ image is decomposed into a hierarchy of square blocks, with the smallest block size $2^{n_0} \times 2^{n_0}$.

Using the same notation as in Section 11.5.1, each square block $b_{l,i}$ is associated with $M_{l,i}$ — the set of all admissible motion vectors, of which $m_{l,i}$ is a member — and $Q_{l,i}$ — the set of all admissible residual error quantizers, of which $q_{l,i}$ is a member. Then a local state $s_{l,i} = [l, i, q_{l,i}, m_{l,i}] \in S_{l,i} = \{l\} \times \{i\} \times M_{l,i} \times Q_{l,i}$ can be defined for block $b_{l,i}$. Consequently, a global state x, representing the currently chosen local state, is defined as $x \in X = \cup_{l=N}^{n_0} \cup_{i=0}^{4^{N-1}-1} S_{l,i}$, where X is the set of all admissible global state values.

For completeness of description, individual block states $s_{l,i}$ must be enumerated sequentially because the rate function may involve arbitrary-order dependencies resulting from differential encoding of parameters. Hence the code for a predicted and motion-compensated frame consists of a global state sequence $x_0, \ldots, x_{N_\Theta-1}$, which represents the left-to-right ordered leaves of a valid QT Θ.

In this context, the frame distortion is an algebraic sum of block distortions $d(x_j)$, implemented with the MSE metric, where the blocks correspond to the leaves of the chosen QT decomposition Θ. That is,

$$D\left(x_0, \ldots, x_{N_\Theta-1}\right) = \sum_{j=0}^{N_\Theta-1} d\left(x_j\right) . \tag{11.15}$$

Again, as a compromise between complexity and efficiency, a first-order DPCM scheme is used for encoding motion vectors, allowing a first-order dependency between the leaves along the scanning path. The scanning path itself is, for reasons described in Section 11.5.1, the Hilbert curve, recursively generated to fill the frame space [28].

Based on the chosen first-order DPCM along the scanning path, the overall frame rate can be expressed as follows:

$$R\left(x_0, \ldots, x_{N_\Theta-1}\right) = \sum_{j=0}^{N_\Theta-1} r\left(x_{j-1}, x_j\right) , \tag{11.16}$$

where $r(x_{j-1}, x_j)$ is the block bit rate, which depends on the encoding of the current and the previous blocks.

In (11.16) we assume that the total frame rate can be distributed among its constituent blocks. This assumption is intuitive in the case of the rate associated with the motion vector component $r^{DVF}(x_{j-1}, x_j)$ and the residual error quantization component $r^{DFD}(x_j)$. It turns out that the QT segmentation rate can also be distributed on the block basis. With QT decomposition, only 1 bit is required to signal a splitting decision at each level. Hence, smaller blocks carry the segmentation costs of all of their predecessors. With the sequential scanning order of blocks belonging to the same parent in the QT, the first scanned block is arbitrarily assigned the cost $r^{SEG}(x_j)$ associated with decomposition up to that level. Thus, the segmentation component of the rate can also be defined on the block basis.

To complete the discussion of the rate, we must also take into account the fact that some blocks in the QT decomposition are not predicted, but rather intra-coded. This may be applicable to newly appearing or uncovered objects that are not found in the reference frame or when the motion model fails to find a good match for a block. When that happens, the intra-coded block's DC coefficient can be predicted from its predecessor's along the scanning path. Therefore, the $r^{DC}(x_{j-1}, x_j)$ component must be added to the total rate. Clearly, it is equal to 0 for predicted blocks, and the task of deciding which blocks are coded intra and which blocks are coded inter is a part of the optimization process at the encoder.

In summary, the block rate, corresponding to the transition from state x_{j-1} to state x_j is expressed as follows:

$$r\left(x_{j-1}, x_j\right) = r^{SEG}\left(x_j\right) + r^{DFD}\left(x_j\right) + r^{DVF}\left(x_{j-1}, x_j\right) + r^{DC}\left(x_{j-1}, x_j\right) . \tag{11.17}$$

Having defined the distortion and the rate, the problem of joint segmentation, motion estimation, and residual error encoding is posed as a constrained optimization problem as follows:

$$\min_{x_0, \ldots, x_{N_\Theta-1}} D\left(x_0, \ldots, x_{N_\Theta-1}\right), \quad \text{subject to:} \quad R\left(x_0, \ldots, x_{N_\Theta-1}\right) \leq R_{\max} . \tag{11.18}$$

This problem can be viewed as the optimal bit allocation problem among the blocks of a QT with leaf dependencies and, hence, the methods of Section 11.4 apply. In particular, it is converted into the unconstrained minimization problem using the Lagrangian multiplier method,

$$J_\lambda\left(x_0, \ldots, x_{N_\Theta-1}\right) = D\left(x_0, \ldots, x_{N_\Theta-1}\right) + \lambda \cdot R\left(x_0, \ldots, x_{N_\Theta-1}\right) , \tag{11.19}$$

and dynamic programming, also discussed in Section 11.4, is used to find the optimal solution. The resulting optimal solution is a sequence of states $[x_0^*, \ldots, x_{N_\Theta-1}^*]$.

Graphically, the DP algorithm is illustrated by Figure 11.8. In it, each node (black circle) represents a particular state x_j the corresponding block $b_{l,i}$ is in. The lines connecting these states correspond to the possible scanning orders of a Hilbert curve, and weights $j_\lambda(x_{j-1}, x_j) = d(x_j)+\lambda \cdot r(x_{j-1}, x_j)$ are associated with each transition. Then the problem of optimal resource allocation among segmentation, motion, and DFD quantization can be stated as that of finding the shortest path in the trellis from S to T, the two auxiliary states. In the implementation, the

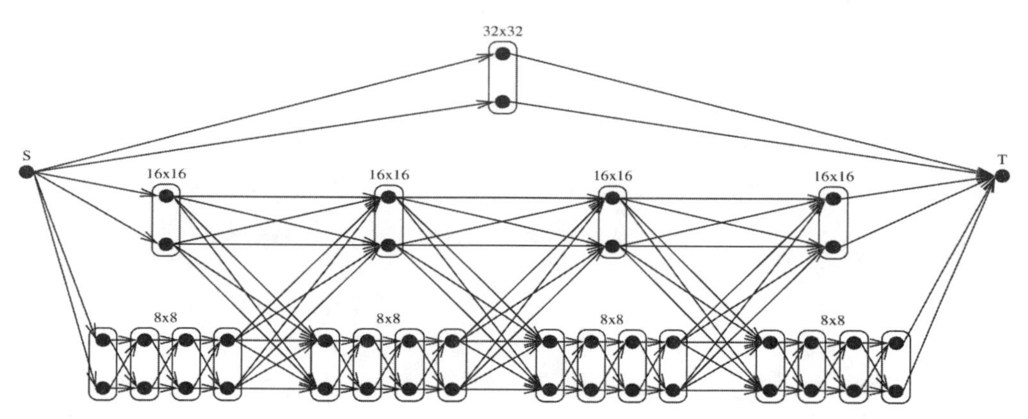

FIGURE 11.8
The trellis structure.

maximum block size was 32×32 and the minimum block size was 8×8, with a consequence that no segmentation information for blocks larger than 32 and smaller than 8 needed to be sent to the decoder. Other implementation details can be found in [28]–[30].

Results

The encoder presented here is compared with TMN4, which is an H.263 standard implementation. Both algorithms were tested on 200 frames of the Mother and Daughter sequence, down-sampled by a factor of 4 in the time axis, with the quantizer step size QP set to 10 in TMN4. In the first test, the λ parameter of the optimal algorithm was adjusted for each frame so that the resulting distortion matched that produced by TMN4 at every encoded frame. The resulting curves are shown in Figure 11.9. In the second test, the λ parameter of the optimal algorithm was adjusted for each frame so that the resulting rate matched that produced by TMN4 at every encoded frame. The resulting curves are shown in Figure 11.10. Clearly, the optimal approach significantly outperforms H.263 in both experiments. In the matched distortion case, the average frame bit rate was reduced by about 25%, and, in the matched rate case, the average PSNR distortion was increased by about 0.72 dB.

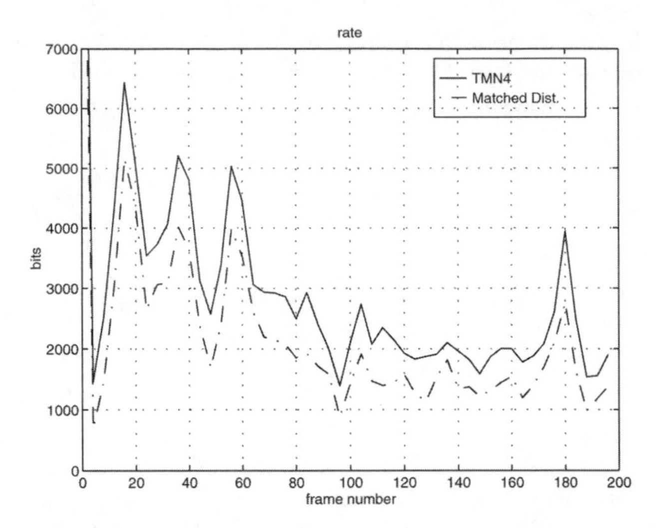

FIGURE 11.9
The matched distortion result.

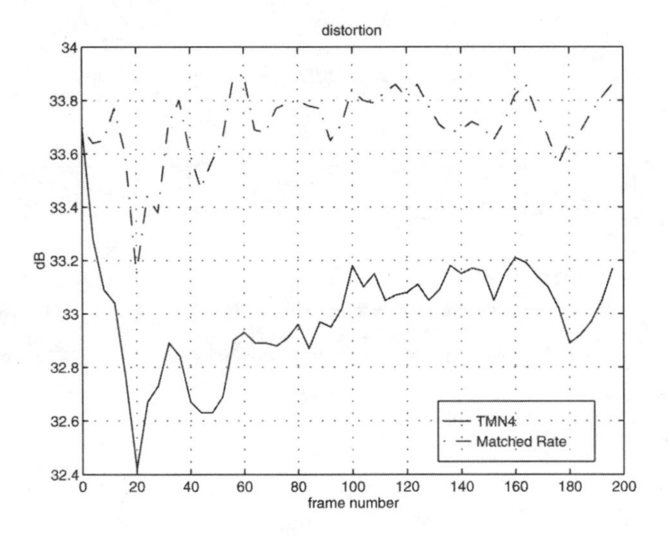

FIGURE 11.10
The matched rate result.

11.5.3 Hybrid Fractal/DCT Image Compression

This section describes the application of the rate-distortion techniques to the hybrid fractal/DCT image compression. Drawing on the ability of DCT to remove interpixel redundancies and on the ability of fractal transforms to capitalize on long-range correlations in the image, the hybrid coder performs an optimal, in the rate-distortion sense, bit allocation among coding parameters. An orthogonal basis framework is used within which an image segmentation and a hybrid block-based transform are selected jointly.

Problem Formulation

Within the chosen fractal/DCT framework, the problem to be solved is that of simultaneous segmentation of an input image into blocks of variable sizes, and, for each, to find a code in such a way that any other choice of segmentation and coding parameters would result in a greater distortion for the same rate, or vice versa. Problems of this type are discussed in Section 11.2 and the corresponding solution tools in Section 11.4. In this context, for a given image x, we want to solve the following optimization problem:

$$\min_{s \in S, c \in C_s} D\left(x_{s,c}, x\right) \quad \text{subject to:} \quad R\left(x_{s,c}\right) \leq R_{\max} , \tag{11.20}$$

where $x_{s,c}$ is the encoded image; D the distortion metric; s a member of the set of all possible image segmentations S; c a member of C_s, the set of all possible codes given segmentation s; R the bit rate associated with segmentation s and code c; and $R_{\max}$ the target bit budget. The distortion metric chosen here is the MSE.

Fractal Basics

Fractal image coding takes advantage of image self-similarities on different scales. That is, instead of sending quantization indices of transformed image subblocks, a fractal coder describes the image as a collection of nonexpansive transformations onto itself. Most fractal algorithms, beginning with Jacquin's implementation [8], break an image into nonoverlapping square regions, called ranges. Each range block r_i is encoded by a nonexpansive transformation T_i^* that operates on the entire original image x and maps a domain block d_i, twice the size of the range block and located elsewhere in the image, onto r_i. The job of the encoder is to find a transformation T_i that minimizes the collage error. That is,

$$T_i^* = \arg \min_{T_i \in \Theta} \|r_i - T_i(x)\| , \tag{11.21}$$

where Θ is the pool of available transforms. The whole transformation T is a sum of partitioned transformations,

$$T(x) = \sum_{i=1}^{N} T_i^*(x), \quad x = \sum_{i=1}^{N} r_i , \tag{11.22}$$

where N is the number of partitions or ranges.

The Collage Theorem establishes an upper bound on the reconstruction error of the decoded image as a function of the collage error and s, the contractivity of T. Specifically,

$$d\left(x, x_T\right) \leq \frac{1}{1-s} \cdot d(x, Tx) , \tag{11.23}$$

where x_T is the decoded image under transformation T.

Transformations T_i are restricted to a set of discrete contractive affine transformations operating on x. Following the notation used in [21], each T_i has the following structure:

$$T_i(x) = \beta_i \; P_i \; D_i \; I_i \; Fet_i \; x + t_i , \tag{11.24}$$

where Fet_i is a transformation matrix that fetches the correct domain block, I_i applies one of the standard eight isometries, D_i is the decimation operator that shrinks the domain block to the range block size, P_i is the place operator that places the result in the correct region occupied by the range block, β_i is a scalar, and t_i is a constant intensity block. Hence, in analogy with vector quantization (VQ), various permutations of Fet_i, I_i, and β_i represent the codebook. Implementation details on the specific choice of the variable-length coding (VLC) for the various parameters of (11.24) can be found in [13].

Blocks for which a good approximation, under a contractive transformation, can be found elsewhere in the image, can be efficiently encoded using the fractal transform. The self-similarity assumption, which is central to fractals, however, may not be justified for all blocks.

In this case, spending more bits on the fractal transform by employing more isometries or finer quantizers is not efficient [4, 25].

The discrete cosine transform (DCT) has been the transform of choice for most codecs due to its decorrelation and energy compaction properties. Complicated image features, however, require a significant number of DCT coefficients to achieve good fidelity. The coarse quantization of these coefficients results in blocking artifacts and unsharp edges.

The coder described here is a hybrid in that it not only adaptively selects which transform (fractal or DCT) to use on any given block, but also can use them jointly and in any proportion. Thus it capitalizes on the ability of the fractal transform to decorrelate images on the block level and on the ability of the DCT to decorrelate pixels within each block. The optimization techniques of Section 11.4 are applied within the chosen framework, resulting in a code that is optimal in the operational sense.

The decoder reconstructs an approximation to the original image by iterative application of the transformation defined by (11.22) to any arbitrary image x_0. Although the Collage Theorem, expressed by (11.23), guarantees the eventual convergence of this process to an approximation with a bounded error, the number of these iterations may be quite large.

Segmentation

The goal of image segmentation in the context of compression is to adapt to local characteristics of the image. Segmenting the image into very small square blocks or into objects of nonsquare shapes, while accomplishing this goal, is also associated with a high cost of description. Here, the set of all possible segmentations S is restricted to be on the QT lattice as a compromise between local adaptivity and simplicity of description. For a 256×256 input image it is a three-level QT with a maximum block size of 16×16 pixels and a minimum size of 4×4 pixels. At each level of the QT only 1 bit is required to signal a splitting decision, with no such bit required at the lowest level.

Code Structure

As mentioned in Section 11.5.3, the transform employed for encoding range blocks consists of the fractal and DCT components. The hybrid approach allows for more flexibility at the encoder, and the various components of the overall transform are designed to complement each other. This idea leads naturally to the concept of orthogonality.

The presence of the DCT component in the overall transform, as well as the fact that the frequency domain interpretation lends itself naturally to the concept of orthogonality and energy compactness, makes it more convenient to perform the collage error minimization of (11.21) in the DCT domain. This is done by applying the DCT to both the range block r_i under consideration and all candidate-decimated domain blocks.

It is convenient to cast the problem of finding the overall transform as that of vector space representation. Vectors are formed by zigzag-scanning square blocks, as shown in Figure 11.11. Hence, minimizing the collage error with respect to a range vector $\bar{r_i}$ is equivalent to finding its best approximation in a subspace spanned by a combination of transformed domain vectors and some fixed (image-independent) vectors $\bar{f_k}$.

In agreement with the terminology used in [23], let the vector $\bar{r}_i$, of size $M^2 \times 1$, represent the DCT coefficients of range block i, of size $M \times M$, scanned in zigzag order. Similarly, the vector $\bar{d_i}$ comes from the chosen domain block, after decimation, DCT, and the application of isometry operators. The fractal component of the overall transform can then be expressed as follows:

$$\bar{d_i} = P_i \; Z_i \; DCT_i \; D_i \; I_i \; Fet_i \; x \; , \tag{11.25}$$

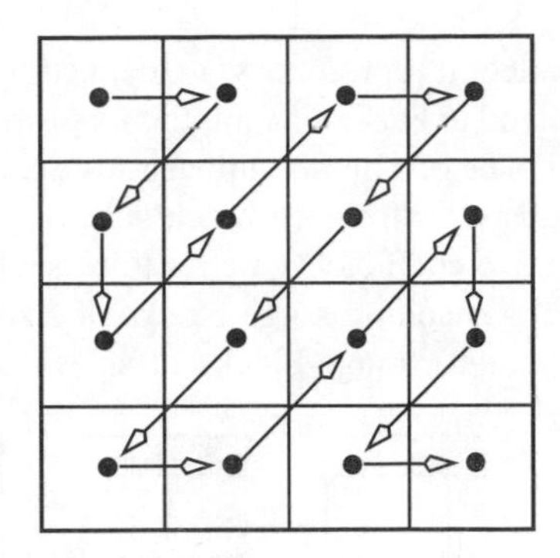

FIGURE 11.11
Zigzag scan of a square block.

where P_i, D_i, I_i, and Fet_i are defined as in (11.24) and DCT_i and Z_i are the block DCT and zigzag-scan operators, respectively. Let the intensity translation term be represented by a linear combination of $N_i - 1$ fixed vectors $\bar{f}_{ik}$, of size $M^2 \times 1$, where the subscript i indicates that the pool of available fixed vectors and the total number of them may be locally adaptive to the range vector. Mathematically, the range vector $\bar{r}_i$ is then approximated by N_i vectors as follows:

$$\bar{r}_i \approx \beta_i \cdot \bar{d}_{io} + \sum_{k=0}^{N_i-2} c_{ik} \cdot \bar{f}_{ik} , \tag{11.26}$$

where $\bar{d}_{io}$ is the projection of $\bar{d}_i$ onto the orthogonal complement of the subspace spanned by vectors $\bar{f}_{ik}$ for $k = 0, \ldots, N_i - 2$, which are themselves orthogonal to each other. Making components of the overall transform orthogonal to each other carries many benefits [16, 22]. These include fast convergence at the decoder, noniterative determination of scaling and intensity translation parameters, no restriction on the magnitude of the scaling coefficient, and continuity of the magnitude of the translation term between neighboring blocks.

In this implementation a bank of fixed subspaces is used to model a range vector of a given size. To illustrate how a fixed subspace is formed from the coefficients of a block DCT, let us, for simplicity, consider a 2×2 block of DCT coefficients. A subspace of dimension 3 (with the full space of dimension 4) is then formed from the low-frequency coefficients as shown in Figure 11.12. Each $\bar{f}_{ik}$ corresponds to one coefficient in the two-dimensional DCT of size $M \times M$. Each vector $\bar{f}_{ik}$ has zeros in all positions, except the one corresponding to the

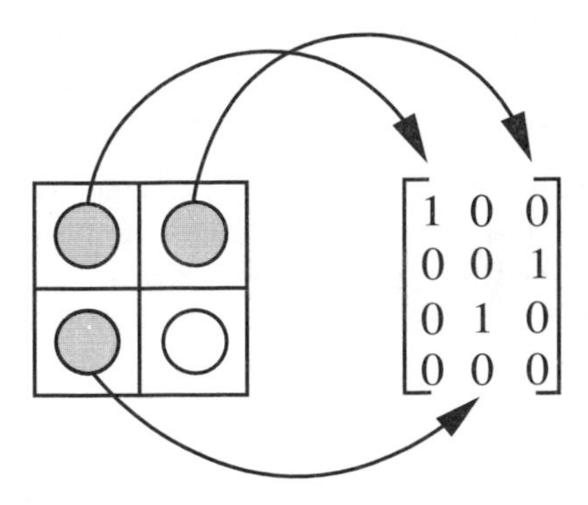

FIGURE 11.12
Mapping of selected DCT coefficients into basis vectors.

order in which the DCT coefficient it represents was scanned, where it has 1. Since in larger blocks more DCT coefficients tend to be significant, the fixed space used for the encoding of a 16×16 range block is allowed to be of a higher dimension than that of a 4×4 block. With the limited number of available subspaces for each block size, only the subspace index, and not the positions of individual nonzero coefficients, needs to be sent to the decoder. Figure 11.13 shows the subspaces allowed for encoding range blocks of size 4×4 (range vectors of size 16×1). The banks of subspaces for range blocks of sizes 8×8 and 16×16 are defined

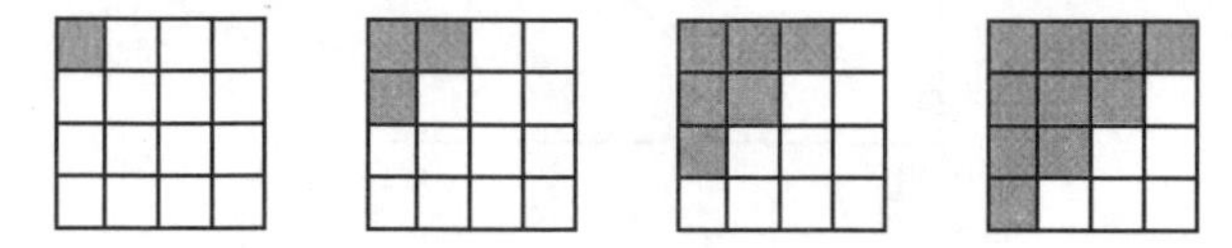

FIGURE 11.13
The 4 subspaces for block size of 4.

similarly and the details on this, as well as on the VLCs used, can be found in [16].

The job of the encoder is then, for each range vector $\bar{r}_i$, to select the fixed space dimension $N_i - 1$, the set of coefficients c_{ik}, the domain block to be fetched by Fet_i, the isometry I_i, and the scaling coefficient β_i. As a result of orthogonalization, the fixed subspace coefficients c_{ik} will carry low-frequency information and fractal component parameters will carry high-frequency information.

Directed Acyclic Graph (DAG) Solution

The DC coefficients of neighboring blocks in an image decomposition exhibit high correlation. In this formulation, coefficient c_{i0} corresponds to the quantized DC value of the range block in question. Hence, differential encoding must be introduced between adjacent blocks to take advantage of this redundancy. The Hilbert curve is known to satisfy certain adjacency requirements [28] and is efficient for predictive coding. For a 256×256 input image, a sixth-order Hilbert curve is used.

If we let each node represent a block in the QT decomposition of the image, and define a transition cost $g_{i,j}$ as the cost of encoding range block r_i with range block r_j as its predecessor, then the overall problem of finding the optimal segmentation and the hybrid fractal/DCT code can be posed as that of finding the shortest path through the leaves of the QT decomposition or trellis, with each leaf having one to three possible codes, corresponding to one to three possible predecessors of a block in our Hilbert curve.

Clearly, the optimal scanning path possesses the optimal substructure property (i.e., it consists of optimal segments). This motivates the use of dynamic programming in the solution. Refer to [28] and Section 11.4 for more details on the use of DP in problems of this type.

Fractal Compression Results

Performance of the operationally optimal hybrid fractal/DCT algorithm is compared to JPEG, which is one of the most popular DCT-based compression schemes. Figure 11.14 shows the ORD curve obtained with this approach when compressing a 256×256 Lena image. JPEG's ORD curve is also shown on this plot. An improvement of 1.5 to 3.0 dB is achieved across the range of bit rates.

Figures 11.15 and 11.16 demonstrate the improvement in quality, over JPEG, for the same bit rate (0.20 bpp). Figure 11.17 shows the optimal segmentation as determined by this encoder. Efficiency is achieved by using larger block sizes in relatively uniform areas and smaller block sizes in edgy areas. Overall, the fractal component of the transform, representing high-frequency information, used about 30% of the available bit budget. This, coupled with

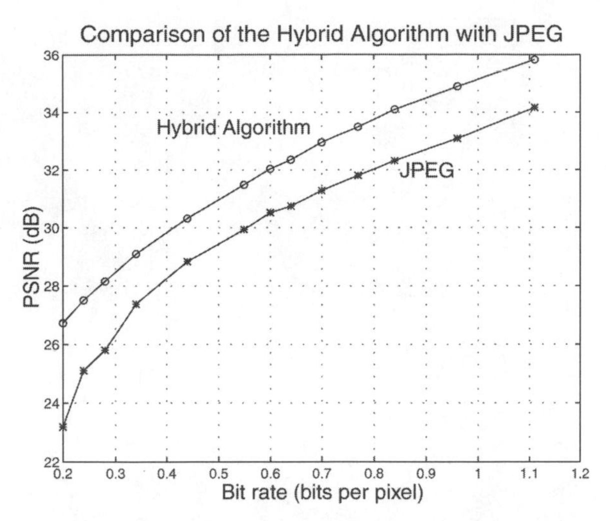

FIGURE 11.14
Hybrid fractal/DCT algorithm vs. JPEG.

FIGURE 11.15
Hybrid algorithm ($R = 0.20$ bpp, PSNR $= 26.71$ dB).

the rate-distortion optimized scanning path, segmentation, and code selection, resulted in significant gains in the quality of the reconstructed image.

11.5.4 Shape Coding

In this section we show how rate-distortion operationally optimal techniques can be applied to the problem of object shape coding. Interest in this problem is motivated by a growing explosion in new multimedia applications, including, but not limited to, video conferencing,

FIGURE 11.16
JPEG (R = 0.20 bpp, PSNR = 23.19 dB).

interactive multimedia databases, film authoring, etc. They all have in common the requirement that video information be accessible on an object-by-object basis.

Commercial video compression standards, such as MPEG-1, MPEG-2, H.261, and H.263, are block-based codecs. They segment a video scene into fixed blocks of predetermined sizes and achieve compression by quantizing texture and motion vectors. This format of representation is not natural for the above-mentioned applications, since there is no clear separation of one object from another and from the background in a generated bitstream. It leads to an unnecessary waste of bits describing a background that carries no information. But it is also inefficient in terms of accessing encoded information. For example, in pattern recognition applications objects are detected through their boundaries, which are not explicitly available in block-based bitstreams.

The emerging MPEG-4 and MPEG-7 multimedia coding standards are designed to address these new requirements. Although it is not clear whether object-based treatment of a video scene is justified in terms of coding efficiency, it is in some cases a requirement from the application point of view. In an object-oriented coder, bits must be efficiently allocated among bitstream components (segmentation, motion, shape, texture) and then within each component. Although ORD optimal joint resource allocation among and within these components in an object-oriented coder remains an ellusive goal, here we address the shape coding aspect of this problem.

Shape information plays the central role in object description. Efforts at its efficient representation, which intensified as a result of the MPEG-4 standardization, can be classified into two categories [9]. The first consists of bitmap-based coders, which can be further broken down into context-based [1] and modified read fax-like [36] coders. The baseline-based shape coder [12] and vertex-based polynomial coder [7, 19] belong to the second category. However, arguably, the bitmap-based coders defeat the goal of object orientation, since in them the shape information is not explicit.

In what follows we describe an intra-mode vertex-based boundary encoding scheme and

FIGURE 11.17
Optimal segmentation ($R = 0.44$ bpp, PSNR = 30.33 dB).

how it is optimized using the techniques of Section 11.4. Implementation details can be found in [33]. An inter-mode vertex-based boundary encoding scheme is derived in [15].

Algorithm

The original boundary is approximated by second-order connected spline segments. A spline segment is completely defined by three consecutive control points (p_{u-1}, p_u, p_{u+1}). It is a parametric curve (parameterized by t), which starts at the midpoint between p_{u-1} and p_u and ends at the midpoint between p_u and p_{u+1}, as t sweeps from 0 to 1. Mathematically, the second-order spline segment used is defined as follows:

$$\vec{Q}_u\left(p_{u-1}, p_u, p_{u+1}, t\right) = \begin{bmatrix} t^2 & t & 1 \end{bmatrix} \cdot \begin{bmatrix} 0.5 & -1.0 & 0.5 \\ -1.0 & 1.0 & 0.0 \\ 0.5 & 0.5 & 0.0 \end{bmatrix} \cdot \begin{bmatrix} p_{u-1,x} & p_{u-1,y} \\ p_{u,x} & p_{u,y} \\ p_{u+1,x} & p_{u+1,y} \end{bmatrix}, \quad (11.27)$$

where $p_{i,x}$ and $p_{i,y}$ are, respectively, the vertical and horizontal components of point p_i.

Besides solving the interpolation problem at the midpoints, the definition of the spline used makes it continuously differentiable everywhere, including the junction points. Segment continuity is ensured because the next spline segment, (p_u, p_{u+1}, p_{u+2}), will originate at the end of the current segment. Placing the control points appropriately, a great variety of shapes can be approximated, including straight lines and curves, a property that makes splines a very attractive building block for the contour approximation problem. The operationally optimal straight line shape approximation was derived in [31].

In order to fit a continuous spline segment to the support grid of the original boundary, spline points are quantized toward the nearest integer value. Thus the solution to our shape approximation problem is an ordered set of control points, which, based on a particular definition of the distortion and rate (discussed below), and for a given rate-distortion trade-off λ, results in a rate-distortion pair (R, D).

Theoretically, the ordered set of control points can be composed of points located anywhere in the image. Most of them, however, are highly unlikely to belong to the solution, as they are too far from the original boundary, and distortions of more than several pixels are not tolerable in most applications. For this reason, and also to decrease computational complexity, we exclude those points from consideration. What remains is the region of space, shown in gray in Figure 11.18, centered around the original boundary and termed the admissible control point band, to which candidate control points must belong. Pixels in this band are labeled by

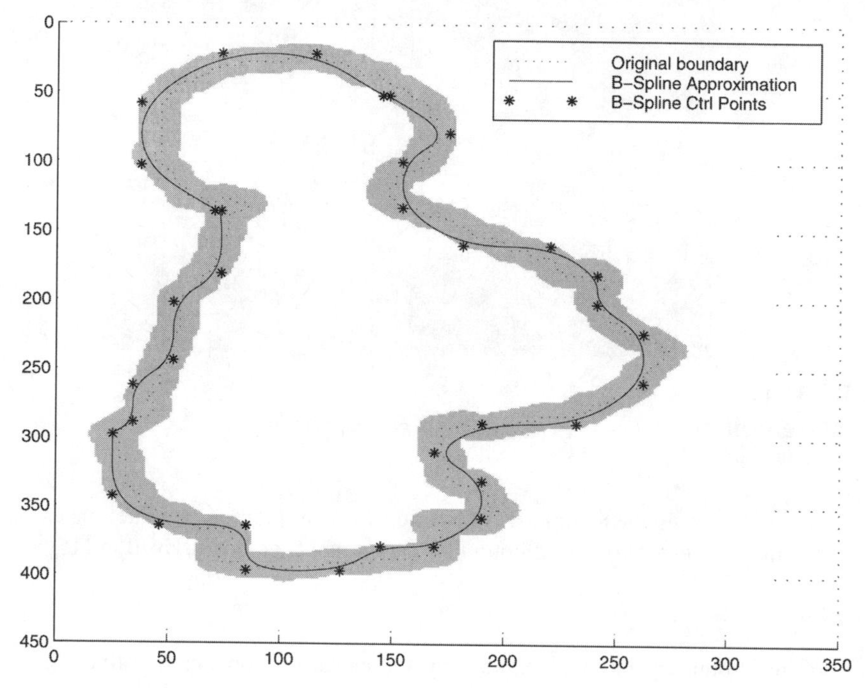

FIGURE 11.18
Admissible control point band.

the index of the closest original boundary pixel (boundary pixels themselves are ordered and labeled). Consecutive control points must be of the increasing index, thus ensuring that the approximating curve can only go forward along the original boundary.

Distortion

Measuring the distortion between an original and approximating boundary is a nontrivial problem. In the minimum–maximum distortion problem [31], stated in Section 11.3, the metric is a binary function, evaluating to zero in case the approximation is inside the distortion band of width $D_{\max}$, and evaluating to infinity in case some portion of the approximating curve lies outside this band. This metric may be useful when fidelity of approximation must be guaranteed for all pixels.

The minimum total distortion problem, solved optimally in [14, 17], is based on a global distortion measure, where local errors are not explicitly constrained. This measure was used in the MPEG-4 standardization process to compare performances of competing algorithms and is also used here explicitly in the optimization process. It is defined as follows:

$$D = \frac{\text{number of pixels in error}}{\text{number of interior pixels}} \tag{11.28}$$

A pixel is judged to be in error if it is in the interior of the original boundary but not in the interior of the approximating boundary, or vice versa. It should be noted that the application of this distortion metric in the optimization process is a stark departure from techniques proposed previously in the literature, in which ad hoc algorithms were simply evaluated after their execution, with equation (11.28). A variation of this definition was used in [14], where in the numerator the area between the original boundary and its continuous approximation was used.

Regardless of the distortion criterion, in order to define the total boundary distortion, a segment distortion needs to be defined first. For that, the correspondence between a segment of the approximating curve and a segment of the original boundary must be established. Figure 11.19 illustrates this concept. In it, the midpoints of the line segments (p_{u-1}, p_u) and (p_{u+1}, p_u),

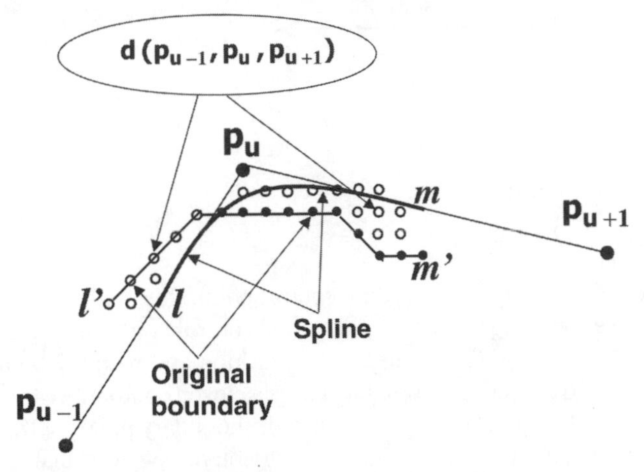

FIGURE 11.19
Area between the original boundary segment and its spline approximation (circles).

l and m, respectively, are associated with the points of the boundary closest to them, l' and m'. When more than one boundary pixel is a candidate, we select the one with the larger index. This ensures that the starting boundary pixel of the next segment coincides with the last boundary pixel of the current segment. That is, the segment of the original boundary (l', m') is approximated by the spline segment (l, m). In order to ensure that some pixels on the edge between two adjacent distortion areas are not counted twice, we exclude points on the line (m, m') from being counted among the distortion pixels (shown in circles).

Let us now define by $d(p_{u-1}, p_u, p_{u+1})$ the segment distortion, as shown in Figure 11.19. Based on the segment distortions, the total boundary distortion is therefore defined by

$$D\left(p_0, \ldots, p_{N_P-1}\right) = \sum_{u=0}^{N_p} d\left(p_{u-1}, p_u, p_{u+1}\right) , \tag{11.29}$$

where $p_{-1} = p_{N_p+1} = p_{N_p} = p_0$ and N_p is the number of control points.

Rate

Consecutive control point locations along an approximating curve are decorrelated using a second-order prediction model [9]. Each control point is described in terms of the relative angle α it forms with respect to the line connecting two previously encoded control points, and by the run length β (in pixels), as shown in Figure 11.20A. The range of values of the angle α is taken from the set $\{-90°, -45°, 45°, 90°\}$, thus requiring only 2 bits (Figure 11.20B). The rationale for excluding the angle of 0° is that that orientation is unlikely, since it can be

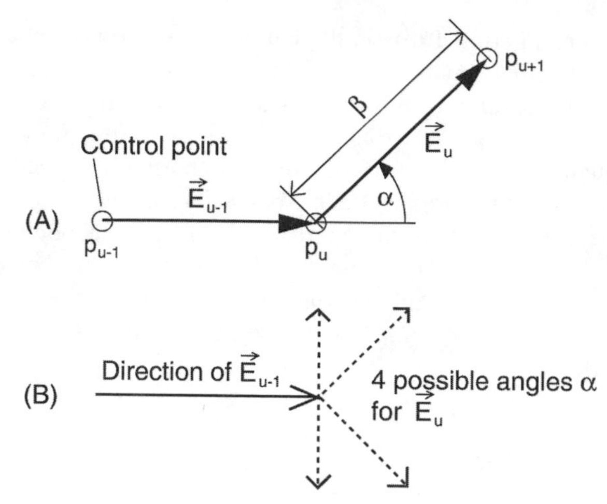

FIGURE 11.20
Encoding of a spline control point.

achieved by properly placing the preceding control point, thus with fewer bits. The exception to this scheme is the encoding of the first and second control points, for which predictive coding does not exist. They are encoded in an absolute fashion and with a 3-bit angle, respectively. To ensure that a closed contour is approximated by a closed contour, we force the control point band of the last boundary pixel to collapse to just the boundary pixel itself, thereby making the approximating curve pass through it. It should be noted, however, that, in general, this way of encoding consecutive control points (*run, angle*) is somewhat arbitrary and other predictive schemes or a different range of α could have been used without loss of generality.

If $r(p_{u-1}, p_u, p_{u+1})$ denotes the segment rate for representing p_{u+1} given control points p_{u-1}, p_u, then the total rate is given by

$$R\left(p_0, \ldots, p_{N_P-1}\right) = \sum_{u=0}^{N_p-1} r\left(p_{u-1}, p_u, p_{u+1}\right) . \tag{11.30}$$

Here we do not restrict the possible locations of p_{u+1}, given p_{u-1} and p_u, and let the encoder determine the locally most efficient VLC for the vector $\overline{p_{u+1} - p_u}$.

DAG Solution

Having defined the distortion and the rate, we now solve the following rate-constrained optimization problem:

$$\min_{p_0,\ldots,p_{N_P-1}} D\left(p_0, \ldots, p_{N_P-1}\right), \quad \text{subject to:} \quad R\left(p_0, \ldots, p_{N_P-1}\right) \leq R_{\max} , \tag{11.31}$$

where both the location of the control points p_i and their overall number N_P have to be determined.

Let us define an incremental cost of encoding one spline segment as

$$w\left(p_{u-1}, p_u, p_{u+1}\right) = d\left(p_{u-1}, p_u, p_{u+1}\right) + \lambda \cdot r\left(p_{u-1}, p_u, p_{u+1}\right) . \tag{11.32}$$

The overall Lagrangian cost function can then be written as

$$J_\lambda\left(p_0, \ldots, p_{N_P-1}\right) = \sum_{u=1}^{N_P-1} w\left(p_{u-1}, p_u, p_{u+1}\right) + w\left(p_{N_P-1}, p_{N_P}, p_{N_P+1}\right)$$
$$= J_\lambda\left(p_0, \ldots, p_{N_P-2}\right) + w\left(p_{N_P-1}, p_{N_P}, p_{N_P+1}\right) . \quad (11.33)$$

This problem now can be cast as the shortest path problem in a graph with each consecutive pair of control points playing the role of a vertex and incremental costs $w()$ serving as the corresponding weights [9]. The problem is then efficiently solved using the techniques of Section 11.4.

VLC Optimization

The operationally optimal shape-coding algorithm described here can claim optimality only with respect to the chosen representation of control points of the curve. That is, our solution is operationally optimal when the encoding structure (*run, angle*) and its associated VLC are fixed, which is equivalent to solving the following optimization problem:

$$\left\{p_0^*, \ldots, p_{N_P-1}^*\right\} = \arg\left[\min_{p_0, \ldots, p_{N_P-1}} J_\lambda^*\left(p_0, \ldots, p_{N_P-1}\right) | VLC\right] . \quad (11.34)$$

Here, we take the operationally optimal approach one step further, and remove the dependency of the ORD on an ad hoc VLC used. Our goal is to compress the source whose alphabet consists of tuples (*run, angle*), exhibiting first-order dependency, close to its entropy. Therefore, the problem can be stated as follows:

$$\left\{p_0^*, \ldots, p_{N_P-1}^*\right\} = \arg \min_{p_0, \ldots, p_{N_P-1}; f \in F} J_\lambda^*\left(p_0, \ldots, p_{N_P-1}\right) , \quad (11.35)$$

where the code is operationally optimal over all f belonging to the family of the probability mass functions, F, associated with the code symbols. Hence, two problems need to be solved jointly: the distribution model, f, and the boundary approximation based on that model. Clearly, as f changes in the process of finding the underlying probability model, symbols generated by the operationally optimal coder are also changed. As is typically done with such codependent problems, the two solutions are arrived at in an iterative fashion [27]. The overall iterative procedure is shown in Figure 11.21. Iterations begin with the optimal boundary encoding algorithm, described in the preceding sections, compressing the input boundaries based on some initial conditional distribution of the (run_i, angle_i) symbol, conditioned on the previously encoded run_{i-1}.

Having encoded the input sequence at iteration k, based on the probability mass function $f^k(\cdot)$, we use the frequency of the output symbols to compute $f^{k+1}(\cdot)$, and so on. It is straightforward to show that the total Lagrangian cost is a nonincreasing function of the iteration k. Thus each iteration brings f closer to the local minimum of $J_\lambda(\cdot)$, and f^k converges to f^M, where M is the number of the last iteration. The iterations stop when $|J_\lambda^k(\cdot) - J_\lambda^{k-1}(\cdot)| \leq \epsilon$. The local minimum in this context should be understood in the sense that a small perturbation of the probability mass function f will result in increases in the cost function $J_\lambda(\cdot)$.

Shape-Encoding Results

Figure 11.22 shows the ORD curve resulting from the application of the described iterative algorithm to the SIF sequence, Kids. As shown in Figure 11.21, after convergence the symbols were arithmetically encoded. For comparison purposes, ORD curves with no VLC optimization and the rate-distortion performance of the baseline method, which is the most

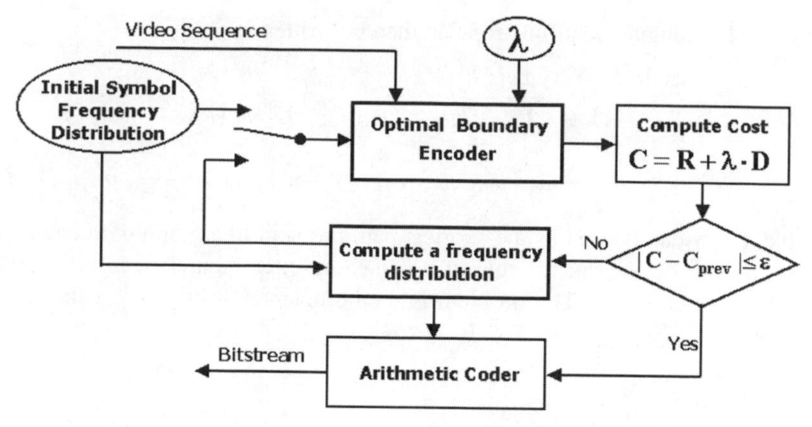

FIGURE 11.21
The entropy encoder structure.

efficient method among competing algorithms in MPEG-4, are also shown. The distortion axis represents the average of the D's defined in (11.28) for one frame, over 100 frames. The approach described here is by far superior to both the contour-based and pixel-based algorithms in [9]. It also outperforms the fixed-VLC area-based approach [14] (shown with squares) and the fixed-VLC pixel-based encoding (shown as the VLC1 and VLC3 curves).

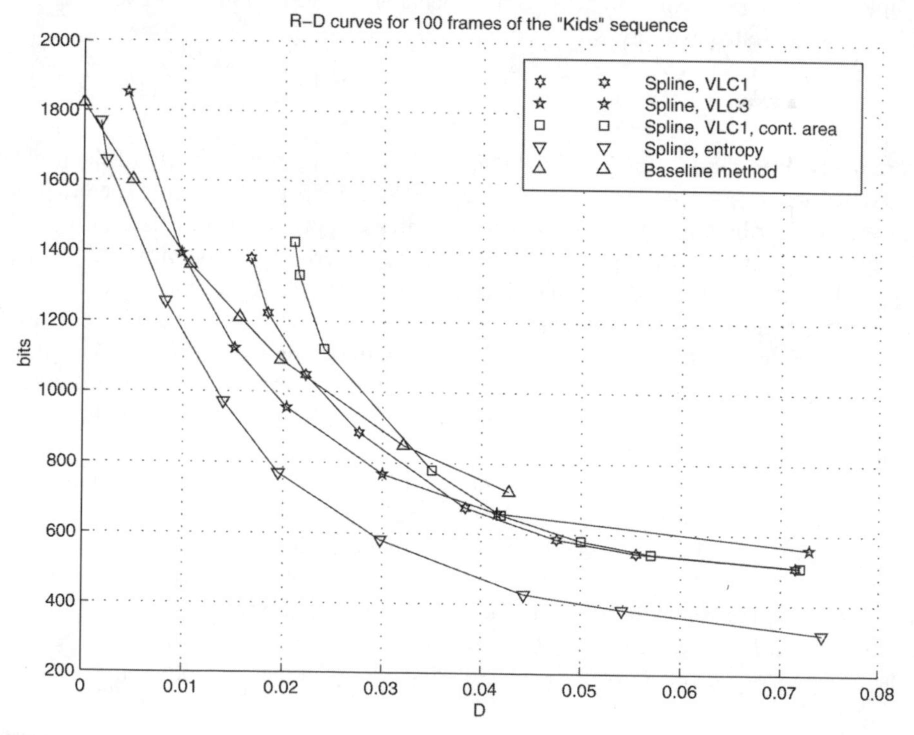

FIGURE 11.22
Rate-distortion curves.

Figure 11.23 shows a sample frame in the sequence and the error associated with the optimal solution. If the uncompressed boundary requires 3 bits per boundary pixel, the approximation

FIGURE 11.23
Shape approximation errors (white pixels).

shown is a 7.1:1 compression. It can be seen that the optimal way to encode some small objects is not to encode them at all, as demonstrated by the space between the legs of the kid on the left, shown in white as an erroneous area. Algorithms not optimized in the rate-distortion sense lack the ability to discard small or noise-level objects, which is required when operating at very low bit rates.

11.6 Conclusions

In this chapter we investigated the application of rate-distortion techniques to image and video processing problems. The ORD theory was presented along with several useful mathematical tools commonly used to solve discrete optimization problems. We concluded by providing several examples from different areas of multimedia research in which successful application of RD techniques led to significant gains in performance.

References

[1] N. Brady, F. Bossen, N. Murphy, "Context-based arithmetic encoding of 2D shape sequences," *Proc. ICIP-97,* pp. I-29–32, 1997.

[2] T.H. Cormen, C.E. Leiserson, R.L. Rivest, *Introduction to Algorithms*. MIT Press, Cambridge, pp. 301–328, 1990.

[3] T.M. Cover, J.A. Thomas, *Elements of Information Theory.* John Wiley & Sons, New York, 1991.

[4] Y. Fisher, *Fractal Image Compression — Theory and Applications.* Springer-Verlag, New York, 1994.

[5] N.P. Galatsanos, A.K. Katsaggelos, "Methods for choosing the regularization parameter and estimating the noise variance in image restoration and their relation," *IEEE Trans. on Image Processing,* vol. 1, pp. 322–336, July 1992.

[6] M. Gharavi-Alkhansari, T.S. Huang, "Fractal image coding using rate-distortion optimized matching pursuit," *Proc. SPIE: Visual Communications and Image Processing,* vol. 2727, pp. 1386–1393, 1996.

[7] M. Hötter, "Object-oriented analysis–synthesis coding based on moving two-dimensional objects," *Signal Processing: Image Communications,* vol. 2, pp. 409–428, Dec. 1990.

[8] A.E. Jacquin, "A fractal theory of iterated Markov operators with applications to digital image coding," Ph.D. thesis, Georgia Institute of Technology, 1989.

[9] A.K. Katsaggelos, L. Kondi, F.W. Meier, J. Ostermann, G.M. Schuster, "MPEG-4 and rate distortion based shape coding techniques," *Proc. IEEE,* pp. 1126–1154, June 1998.

[10] L.P. Kondi, A.K. Katsaggelos, "On the encoding of the anchor frame in video coding," *Proc. IEEE Intl. Conf. on Consumer Electronics,* pp. 18–19, Chicago, IL, June 11–13, 1997.

[11] J. Lee, "Optimal quad-tree for variable block size motion estimation," *Proc. ICIP-95,* vol. 3, pp. 480–483, Oct. 1995.

[12] S. Lee et al., "Binary shape coding using 1-D distance values from baseline," *Proc. ICIP-97,* pp. I-508–511, 1997.

[13] G. Melnikov, "Hybrid fractal/DCT image compression algorithms using an orthogonal basis and non-square partitions," M.S. thesis, Northwestern University, 1997.

[14] G. Melnikov, P.V. Karunaratne, G.M. Schuster, A.K. Katsaggelos, "Rate-distortion optimal boundary encoding using an area distortion measure," *Proc. ISCAS-98,* pp. 289–292, 1998.

[15] G. Melnikov, G.M. Schuster, A.K. Katsaggelos, "Jointly optimal inter-mode shape coding and VLC selection," *Proc. ICIP-99,* Oct. 1999.

[16] G. Melnikov, A.K. Katsaggelos, "A non uniform segmentation optimal hybrid fractal/DCT image compression algorithm," *Proc. ICASSP-98,* vol. 5, pp. 2573–2576, 1998.

[17] G. Melnikov, G.M. Schuster, A.K. Katsaggelos, "Simultaneous optimal boundary encoding and variable-length code selection," *Proc. ICIP-98,* pp. I-256–260, Oct. 1998.

[18] R. Neff, A. Zakhor, "Matching pursuit video coding at very low bit rates," *DCC'95: Data Compression Conference,* March 1995.

[19] K.J. O'Connell, "Object-adaptive vertex-based shape coding method," *IEEE Trans. on Circuits and Systems for Video Technology,* vol. 7, pp. 251–255, Feb. 1997.

[20] A. Ortega, K. Ramchandran, "Rate-distortion methods for image and video compression," *IEEE Signal Processing Magazine,* pp. 23–50, Nov. 1998.

[21] G.E. Øien, S. Lepsoy, T. Ramstad, "An inner product space approach to image coding by contractive transformations," *Proc. ICASSP-91,* pp. 2773–2776, 1991.

[22] G.E. Øien, S. Lepsoy, "Fractal-based image coding with fast decoder convergence," *Signal Processing,* vol. 40, pp. 105–117, 1994.

[23] B.-B. Paul, M.H. Hayes III, "Video coding based on iterated function systems," *Proc. ICASSP-95,* vol. 4, pp. 2269–2272, 1995.

[24] D.A. Pierre, *Optimization Theory and Applications.* Dover Publications, 1986.

[25] D.C. Popescu, A. Dimca, H. Yan, "A nonlinear model for fractal image coding," *IEEE Trans. on Image Processing,* vol. 6, no. 3, March 1997.

[26] K. Ramchandran, M. Vetterli, "Best wavelet packet bases in a rate-distortion sense," *IEEE Trans. on Image Processing,* vol. 2, pp. 160–175, April 1993.

[27] D. Saupe, "Optimal piecewise linear image coding," *Proc. SPIE Conf. on Visual Communications and Image Processing,* vol. 3309, pp. 747–760, 1997.

[28] G.M. Schuster, A.K. Katsaggelos, *Rate-Distortion Based Video Compression, Optimal Video Frame Compression and Object Boundary Encoding.* Kluwer Academic Press, 1997.

[29] G.M. Schuster, A.K. Katsaggelos, "Video compression scheme with optimal bit allocation among segmentation, displacement vector field, and displaced frame difference," *IEEE Trans. on Image Processing,* vol. 6, no. 11, pp. 1487–1502, Nov. 1997.

[30] G.M. Schuster, A.K. Katsaggelos, "A video compression scheme with optimal bit allocation between displacement vector field and displaced frame difference," *IEEE Journal on Selected Areas in Communications,* vol. 15, no. 9, pp. 1739–1751, Dec. 1997.

[31] G.M. Schuster, A.K. Katsaggelos, "An optimal polygonal boundary encoding scheme in the rate-distortion sense," *IEEE Trans. on Image Processing,* vol. 7, no. 1, pp. 13–26, Jan. 1998.

[32] G.M. Schuster, A.K. Katsaggelos, "An optimal quadtree-based motion estimation and motion-compensated interpolation scheme for video compression," *IEEE Trans. on Image Processing,* vol. 7, no. 11, pp. 1505–1523, Nov. 1998.

[33] G.M. Schuster, G. Melnikov, A.K. Katsaggelos, "Optimal shape coding techniques," *IEEE Signal Processing Magazine,* pp. 91–108, Nov. 1998.

[34] G.M. Schuster, G. Melnikov, A.K. Katsaggelos, "A review of the minimum maximum criterion for optimal bit allocation among dependent quantizers," *IEEE Trans. on Multimedia,* vol. 1, no. 1, pp. 3–17, March 1999.

[35] C.E. Shannon, "Coding theorems for a discrete source with a fidelity criterion," *IRE National Convention Record, Part 4,* pp. 142–163, 1959.

[36] N. Yamaguchi, T. Ida, T. Watanabe, "A binary shape coding method using modified MMR," *Proc. ICIP-97,* pp. I-504–508, 1997.

Chapter 12

Transform Domain Techniques for Multimedia Image and Video Coding

S. Suthaharan, S.W. Kim, H.R. Wu, and K.R. Rao

12.1 Coding Artifacts Reduction

12.1.1 Introduction

Block-based transform coding is used for compressing digital video that normally requires a large bandwidth during transmission. Compression of digital video is vital for the reduction of bandwidth for effective storage and transmission. However, this results in coding artifacts in the decoded video, especially at low bit rates. Techniques, in either the spatial domain or the transform domain, can be developed to reduce these artifacts. Several methods have been proposed in the spatial domain to reduce the so-called *blocking artifact*. However, none of these methods can reduce all the coding artifacts [1] at the same time. Some methods are image enhancement techniques and others are intrinsically iterative, which makes them impossible for real-time applications [2]. Also, they do not completely eliminate the artifact. Hence, researchers are investigating new approaches.

The objective of the proposed approach is to present a new transform domain filtering technique to reduce a number of coding artifacts [1] including the well-known blocking artifact. In digital video coding, compression is achieved by first transforming the digital video from the spatial and temporal domains into the frequency domain using the block-based discrete cosine transform (DCT) and then applying quantization to the transform coefficients followed by variable-length coding [3].

The DCT is a block-based transform, and at low bit rates the noise caused by the coarse quantization of transform coefficients is visible in the form of a blocking artifact. In order to reduce this artifact while maintaining compatibility with current video coding standards, various spatial domain postfiltering techniques such as low-pass filtering (LPF) [4], projection onto convex sets (POCS) [5], maximum a posteriori (MAP) [6] filters, and adaptive low-pass filters (ALPFs) [7, 8] have been introduced.

Because the quantization of transform coefficients is the main error source for the coding artifact at low bit rates, it would be much more effective in tackling the coding artifacts in the transform domain than in the spatial domain. Recently, a weighted least square (WLS) [2] method has been introduced in the transform domain to estimate the transform coefficients from their quantized versions. It assumes the uniform probability density function for the quantization errors and estimates their variances from the step size of the corresponding quantizer. It also estimates the variances of signal and noise separately and thus increases the computational complexity as well as computational errors. Therefore, it is more sensible to estimate the signal-to-noise ratio (SNR) as a single entity because this ratio plays a major role in a number of image quality restoration techniques.

We have recently proposed an improved Wiener filter (IWF) [9] by estimating the SNR in an image restoration problem. This IWF is also suitable to reduce the coding artifact in compressed images. In this section, the IWF and the WLS is investigated and modified and a new approach is developed in the transform domain to reduce the coding artifacts. First, the methods WLS and approximated WLS* are investigated and implemented to reduce the blocking artifacts. Second, the method IWF is further investigated and the SNR of the quantized transform coefficients is estimated. This estimated SNR is used with the IWF to reduce the noise in the transform coefficients. This noise is the source of coding artifacts. Reducing such noise results in a corresponding reduction of the coding artifacts.

12.1.2 Methodology

Let us first discuss the mathematical model used in image and video transform coding schemes. It is clear from the digital image and video coding literature that block-based transform coding can be modeled as follows:

$$Y = Q(T(x)) \,, \tag{12.1}$$

where x, T, Q, and Y represent the input image, the discrete cosine transform, the quantization process, and the quantized DCT coefficients, respectively.

Using this transform coding mechanism, a linear model for the quantization error can be developed and written as follows:

$$n = Y - T(x) \,, \tag{12.2}$$

where n is the quantization error, which introduces the signal-independent noise on the transform coefficients $T(x)$ and results in Y. Therefore, n is often called the quantization noise. To simplify the above linear noise model, we write it as follows:

$$Y = X + n \,, \tag{12.3}$$

where $X = T(x)$ and n is a zero-mean additive noise introduced by the quantization process of the DCT coefficients. Without loss of generality it can be assumed that the noise n is also uncorrelated with transform coefficients X.

With the a priori information about n and X, the latter can be estimated from Y using the Wiener filter technique [10, 11]. That is,

$$\hat{X} = IFFT\left\{\frac{1}{1+|n|^2/2} \cdot FFT(Y)\right\} \tag{12.4}$$

where FFT and $IFFT$ represent the fast Fourier transform and inverse fast Fourier transform, respectively. The symbol $|\cdot|^2$ represents the power spectrum, and the ratio $|n|^2/|X|^2$ is called the noise-to-signal power ratio (which is the a priori representation of the SNR).

As we can see from (12.4), computation of X-*hat* requires the a priori knowledge of power spectra of the nonquantized transform coefficient (X) and the quantization noise (n). In reality such information is rarely available. We propose two approaches to estimate the noise-to-signal power ratio as a whole and use it in (12.4) to restore the transform coefficients from the corrupted version.

APPROACH I (IWF): Assuming Noise Power Spectrum is Known

In this approach, we use the assumption suggested by Choy et al. [2] that the quantization noise has a uniform probability density function. Thus quantization noise variance is given by $\sigma_n^2 = q^2/12$, where q is the known step size of the quantizer applied to the transform coefficients X to get Y.

Since the quantization noise n is assumed zero-mean uncorrelated with X, using (12.3) we can derive [9, 12]:

$$|Y|^2 = |X|^2 + |n|^2 , \tag{12.5}$$

where $|n|^2 = \sigma_n^2$, which can be calculated from the quantization step size.

It gives

$$|n|^2/|X|^2 = |n|^2/\left(|Y|^2 - |n|^2\right) . \tag{12.6}$$

The noise-to-signal power ratio can be calculated from the power spectra of the quantized transform coefficients Y and the quantization noise n.

Using this ratio and (12.4), we can reduce the quantization error in the transform coefficients and in turn can reduce a number of coding artifacts including the blocking artifact in the decompressed images.

The advantage of this proposed method over WLS is that it uses only one assumption (noise variance), whereas WLS uses one approximation (mean of X) and one assumption (noise variance). Because no approximation is used in our proposed approach, it gives better results in terms of peak signal-to-noise ratio (PSNR) as well as visual quality.

APPROACH II (IWF*): Estimate the Noise-to-Signal Power Ratio

In this approach, we propose a technique to estimate the noise-to-signal power ratio as a whole and then use it in the Wiener filter equation (12.4). We recently proposed the IWF technique to handle the image restoration problem [9]. In this technique, we do not need the a priori information about the noise-to-signal power ratio in order to apply the Wiener filter; instead, it is estimated from the given degraded image. It has been successfully used in the image restoration problem. We use this approach to remove the coding artifacts introduced by the coarse quantization of DCT coefficients.

The IWF method needs two versions of an image so that the noise-to-signal power ratio can be estimated. In digital video, we have a sequence of images (frames) and the consecutive frames have very few differences, except when the scene changes. Therefore, the decoded frames can be different because of the different quantization scalers used and thus the quantization error (noise) can be different. Let us assume that the DCT coefficients of the ith frame are to be restored from its quantization noise; then we can model the quantized DCT coefficients of the ith and $(i+1)$th frames as follows [refer to (12.3)]:

$$\begin{aligned} Y_i &= X + n_i \\ Y_{i+1} &\approx X + n_{i+1} \end{aligned} \tag{12.7}$$

The restriction of the method is that the adjacent frames cannot be quantized with the same quantization scaler. In the above equations, it has been assumed that there is no scene change and thus the approximation of the second equation is valid. Therefore, Y_{i+1} cannot be used when the scene changes. To overcome this problem, we suggest using the previous frame as the second one when the scene changes; thus, we can write:

$$\begin{aligned} Y_i &= X + n_i \\ Y_{i-1} &\approx X + n_{i-1} \end{aligned} \tag{12.8}$$

To implement IWF we need only two versions of an image, and in digital video coding we can have two images using either the previous frame or the next frame for the second image as shown in (12.7) and (12.8). In case of scene changes on the previous and the next frames, we can construct the second frame ($i+1$ or $i-1$) from the ith frame using the methods discussed in [10]. If we represent the quantized DCT coefficients of the frame (that is to be restored) by Y_1 and the second one by Y_2, then we have

$$\begin{aligned} Y_1 &= X + n_1 \\ Y_2 &\approx X + n_2 \end{aligned} \tag{12.9}$$

In video compression, the quantization scaler is used as a quantization parameter, and it is valid to assume a linear relationship between the quantization noises n_1 and n_2. Thus, the linear relationship $n_2 = an_1$ is acceptable (where a is a constant).

Therefore, equation (12.9) can be written as follows:

$$\begin{aligned} Y_1 &= X + n_1 \\ Y_2 &\approx X + a \cdot n_1 \end{aligned} \tag{12.10}$$

From these two equations and using the definition of a power spectrum [12], we can easily derive

$$\begin{aligned} |Y_1 - Y_2|^2 &\approx (1-a)^2\, |n_1|^2 \\ |Y_2 - a \cdot Y_1|^2 &\approx (1-a)^2 |X|^2 \end{aligned} \tag{12.11}$$

Dividing the first equation by the second, the noise-to-signal power ratio of Y_1 can be approximated as follows:

$$\frac{|n_1|^2}{|X|^2} \approx \frac{|Y_1 - Y_2|^2}{|Y_2 - a \cdot Y_1|^2}, \tag{12.12}$$

where the constant a can be calculated during the encoding process as follows:

$$a = \frac{Var(Y_2 - X)}{Var(Y_1 - X)}, \tag{12.13}$$

and a single value for each frame can be transmitted to the decoder. Although an extra overhead on the bit rate is created, compared to the PSNR improvement and visual quality of the images, it is marginal. This overhead can be reduced by transmitting the differences in a.

On the other hand, to avoid this overhead, the constant a can be approximated to the corresponding ratio calculated from the decoded versions of Y_1 and Y_2 as follows:

$$a = \frac{Var(\text{decoded}(Y_2))}{Var(\text{decoded}(Y_1))}. \tag{12.14}$$

The Wiener filter in (12.4), replacing Y by Y_1 and n by n_1 along with the above expressions, has been used to restore the transform coefficients X of Y_1.

The important point to note here is that the proposed methods have been used in the transform domain to reduce the noise presented in the quantized DCT coefficients. This in turn reduces the blocking artifacts in the decompressed images. Also, note that if we cannot assume that the consecutive frames have very little differences, we can still use equation (12.7) by constructing the second image from the first according to the method discussed in [9, 10] by Suthaharan, and thus equation (12.9) is valid.

12.1.3 Experimental Results

In the simulation, the proposed methods (IWF and IWF*) have been implemented using a number of test video sequences, which include Flower Garden, Trevor, Footy, Calendar, and Cameraman. We show here the effectiveness of the proposed methods with the images of Trevor, Footy, and Cameraman. Figure 12.1 displays the original Trevor image, its MPEG coded image, and the processed images by the WLS method and IWF* method, respectively. In Figure 12.1b the coding artifacts can be clearly seen. Although the images in Figure 12.1c and 12.1d look similar, a closer look on a region-by-region (and along the edges) basis shows that many of the coding artifacts have been removed in Figure 12.1d, and it is supported by the higher PSNR value. Figures 12.2 and 12.3 show similar results for the Footy and Cameraman images. All these figures show that the proposed methods significantly reduce the coding artifacts by removing some noise introduced by the quantization process. This improves the PSNR by more than 1 dB and in turn improves the visual quality of the images. Note that in our simulation we have used the MPEG quantizer with an 8×8 block-based DCT [13].

It is clear from [2] that Choy et al.'s proposed filter WLS gives better PSNR values than the LPF [4] and POCS filter [5]. In our experiment, we have compared the results of IWF and IWF* with those of WLS and WLS*, respectively (Table 12.1).

Table 12.1 PSNR Improvements of the Images Reconstructed Using WLS and IWF

MPEG Encoded Images	bpp	PSNR (dB)	PSNR Improvements over POCS [5]			
			WLS	WLS*	IWF	IWF*
Flower Garden	0.3585	23.1401	0.9668	0.5317	1.1439	1.2537
Trevor	0.1537	31.8739	0.8074	0.6009	1.0338	1.0239
Footy	0.2325	26.4812	0.9769	0.5958	1.0324	1.0253
Calendar	0.3398	22.4223	0.8777	0.6232	1.2363	1.1867
Cameraman	0.2191	27.1588	0.5485	0.3568	1.1002	0.9100

From the PSNR improvements shown in Table 12.1, we can conclude that the proposed methods give better restoration of the transform coefficients than those of the WLS and WLS* methods and yield a better visual quality of images.

It is evident from our simulation that our proposed methods restore the transform coefficients from the quantized versions and, thus, they can reduce a number of coding artifacts, such as

1. Blocking artifact: This is due to coarse quantization of the transform coefficients. The blocking artifacts can be clearly seen in all the images.

2. DCT basis images effect: This is due to the appearance of DCT basis images. For example, it can be seen in certain blocks on the background of the Footy image and ground of the Cameraman image. In the filtered images of Footy and Cameraman, this effect has been reduced.

3. Ringing effect: This is due to quantization of the AC DCT coefficients along the high-contrast edges. It is prominent in the images of Trevor and Cameraman along the arm and shoulder, respectively, and in the filtered images it has been reduced.

4. Staircase effect: This is due to the appearance of DCT basis images along the diagonal edges. It appears due to the quantization of higher order DCT basis images and fails to be muted with lower order basis images.

FIGURE 12.1
Images of Trevor. (a) Original image; (b) MPEG coded image (0.1537 bpp, 31.8739 dB); (c) processed by WLS (32.6813 dB); *(Continued)*.

FIGURE 12.1
***(Cont.)* Images of Trevor. (d) Processed by the proposed IWF* method (32.9178 dB).**

12.1.4 More Comparison

In this section we have compared our proposed techniques with a recently published blocking artifact reduction method proposed by Kim et al. [14]. In their method the artifact reduction operation has been applied to only the neighborhood of each block boundary in the wavelet transform at the first and second scales. The technique removes the blocking component that reveals stepwise discontinuity at block boundaries. It is a blocking artifact reduction technique and does not necessarily reduce the other coding artifacts mentioned above. It is evident from the images in Figure 12.4 that this method still blurs the image (Figure 12.4c) significantly and thus some edge details that are important for visual perception have been lost.

We have used the JPEG-coded Lena image provided by Kim et al. [14] to compare our results. This Lena image is JPEG coded with a 40:1 compression ratio. The enlarged portion of the original and JPEG-coded Lena images are given in Figures 12.4a and b. In Figure 12.4c, the processed image of Figure 12.4b by the method proposed by Kim et al. is presented. As can be seen in Figure 12.4c, their proposed method still blurs the image significantly and thus the sharpness of the image is lost. In addition, there are a number of other obvious problems: (1) the ringing effect along the right cheek edge, (2) the blurred stripes on the hat, and (3) the blurred edge between the hat and the forehead, to name just a few. The image processed by our proposed method (IWF*) is presented in Figure 12.4d. In this image we can clearly see the sharpness of the edges, while reducing a number of coding artifacts, and an overall improvement in the visual quality of the image.

12.2 Image and Edge Detail Detection

12.2.1 Introduction

The recent interest of the Moving Pictures Expert Group (MPEG) is object-based image representation and coding. Compared to the conventional frame-based compression techniques, the object-based coding enables MPEG-4 to cover a wide range of emerging applications including multimedia. The MPEG-4 supports new tools functionality not available in existing standards.

FIGURE 12.2
Images of Footy. (a) Original image; (b) MPEG coded image (0.2325 bpp, 26.4812 dB); (c) processed by WLS (27.4581 dB); *(Continued)*.

FIGURE 12.2
***(Cont.)* Images of Footy. (d) Processed by the proposed IWF* method (27.6095 dB).**

One of the important tools needed to enhance and broaden its applications is to introduce effective methods for image segmentation. Image segmentation techniques not only enhance the MPEG standards but also are needed for many computer vision and image processing applications. The goal of image segmentation is to find regions that represent objects or meaningful parts of objects. Therefore, image segmentation methods will look for objects that either have some measure of homogeneity within them or have some measure of contrast with the objects or their border.

In order to carry out image segmentation we need effective image and edge detail detection and enhancement algorithms. Since edges often occur at image locations representing object boundaries, edge detection is extensively used in image segmentation when we want to divide the image into areas corresponding to different objects. The first stage in many edge detection algorithms is a process of enhancement that generates an image in which ridges correspond to statistical evidence for an edge [15]. This process is achieved using linear operators, including Roberts, Prewitt, and Canny. These are called edge convolution enhancement techniques. They are based on convolution and are suitable for detecting the edges for still images. These techniques do not adapt any visual perception properties, but use only the statistical behavior of the edges. Thus, they cannot detect the edges that might contribute to the edge fluctuations and coding artifacts that could occur in the temporal domain [16, 17].

In this section a transform domain technique has been introduced to detect image and edge details that are suitable for image segmentation and reduction of coding artifacts in digital video coding. This method uses the perceptual properties and edge contrast information of the transform coefficients and, thus, it gives meaningful edges that are correlated with human visual perception. Also, the method allows users to select suitable edge details from different levels of edge details detected by the method for different applications in which human visual quality is an important factor.

12.2.2 Methodology

Let us consider an image I of size $N_1 n \times N_2 n$, where I is divided into $N_1 \times N_2$ blocks with each block having $n \times n$ pixels. In transform coding, a block is transformed into the transform domain using a two-dimensional separable unitary transform such as the DCT, and this process can be expressed by

$$Y = T * X * T' \tag{12.15}$$

FIGURE 12.3
Images of Cameraman. (a) Original image; (b) MPEG coded image (0.2191 bpp, 27.1588 dB); (c) processed by WLS (27.8365 dB); *(Continued)*.

FIGURE 12.3

(Cont.) **Images of Cameraman. (d) Processed by the proposed IWF* method (28.0718 dB).**

where X and Y represent a block of I and its transform coefficients, respectively; T is the $n \times n$ unitary transform matrix; and T' represents the *transpose* of the matrix T. The operator $*$ represents the matrix multiplication. The 8×8 block-based DCT is used in the experiment, and it is also the one used in the many international image and video coding standards.

Let a be the DC coefficient of X and U be the image that is obtained using the AC coefficients of X and zero DC coefficient. Then we have the following:

$$a = \frac{1}{n} \sum_{i=1}^{n} \sum_{j=1}^{n} x(i, j) \tag{12.16}$$

$$X = T' * U * T + \frac{a}{n} \begin{bmatrix} 1 \ldots 1 \\ \ldots\ldots \\ 1 \ldots 1 \end{bmatrix}_{n \times n} \tag{12.17}$$

where $x(i, j)$ is the (i, j)th intensity value of image block X. It is known that two-dimensional transform coefficients, in general, have different visual sensitivity and edge information. Thus, transform coefficients in U can be decomposed into a number of regions based on frequency level and edge structures, and they are called *frequency distribution decomposition* and *structural decomposition*, respectively [3] (see Figure 12.5). Using these regions of transform coefficients, we treat the edge details corresponding to the low- (and medium) frequency transform coefficients separately from the edge details corresponding to the high-frequency coefficients.

It is clear from a visual perception viewpoint that the low- (and medium) frequency coefficients are much more sensitive than the high-frequency coefficients. In our proposed algorithm, we use this convention and separate the edge details falling in the low- (and medium) frequency coefficients from the edge details falling in the high-frequency coefficients. To carry out this task, let us first define a new image block X_1 from the transform coefficients of X using the following equation, similar to equation (12.17):

$$X_1 = T' * (L \cdot *U) * T + \frac{\alpha \cdot a}{n} \begin{bmatrix} 1 \ldots 1 \\ \ldots\ldots \\ 1 \ldots 1 \end{bmatrix}_{n \times n} \tag{12.18}$$

where L is called an *edge-enhancement* matrix and α can be chosen to adjust the DC level of X_1 to obtain different levels of edge details with respect to the average intensity of the block

FIGURE 12.4
Lena image coded by JPEG with a 40:1 compression ratio and the processed images. (a) Original image; (b) coded image; (c) processed by Kim et al. method; *(Continued)*.

FIGURE 12.4
***(Cont.)* Lena image coded by JPEG with a 40:1 compression ratio and the processed images. (d) Processed by our proposed method.**

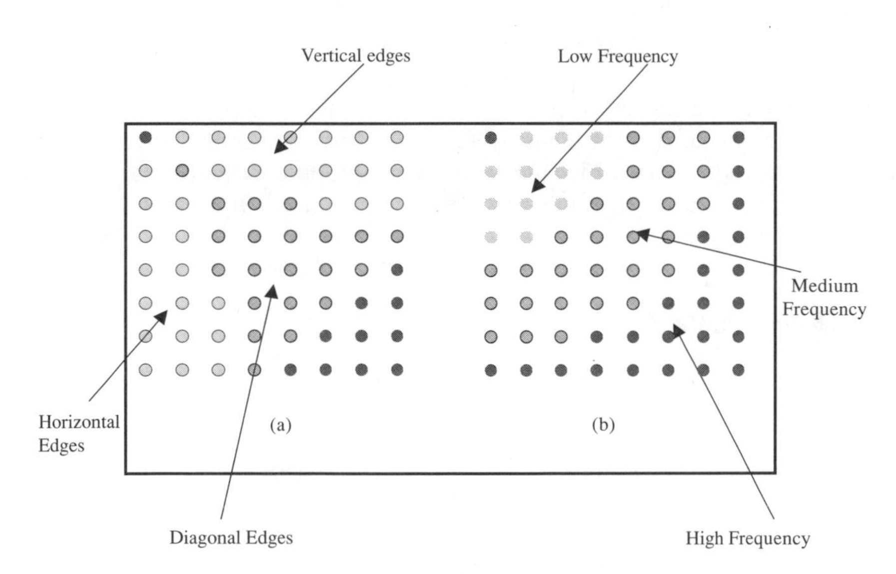

FIGURE 12.5
Decomposition of transform coefficients. (a) Structural decomposition based on edge details; (b) frequency distribution decomposition.

X_1. The operator $\cdot *$ represents the element-by-element matrix multiplication as defined in the MATLAB package [18].

Selection of L and α is up to the user's application and thus it forms a flexible algorithm. Suitable selection of L and α for different applications is leading to new research. In this approach we used the JPEG-based quantization table as the edge-enhancement matrix, and it is given in the following equation:

$$L = \begin{bmatrix} \alpha & 60 & 70 & 70 & 90 & 120 & 255 & 255 \\ 60 & 60 & 70 & 96 & 130 & 255 & 255 & 255 \\ 70 & 70 & 80 & 120 & 200 & 255 & 255 & 255 \\ 70 & 96 & 120 & 145 & 255 & 255 & 255 & 255 \\ 90 & 130 & 200 & 255 & 255 & 255 & 255 & 255 \\ 120 & 255 & 255 & 255 & 255 & 255 & 255 & 255 \\ 255 & 255 & 255 & 255 & 255 & 255 & 255 & 255 \\ 255 & 255 & 255 & 255 & 255 & 255 & 255 & 255 \end{bmatrix} \tag{12.19}$$

One can select different weighting for the edge-enhancement matrix L. During the selection of weights $L(i, j)$ we treat DC and AC coefficients differently, because the DC coefficient contains information about an average (spatial) of the intensity values of an image block whereas the AC coefficients contain information about the edges in an image. Using the above L and $\alpha = 1$, the proposed method can easily identify vertical, horizontal, and diagonal edges that are vital to improving image quality with respect to visual perception. Significant separation of these two regions using appropriate L can cause intensity of certain edge details (based on their contrast) in the low- (and medium) frequency coefficients to be pushed down below zero while keeping the intensity of the edge details in the high-frequency coefficients above zero.

In order to get different levels of edge details based on their contrast, the α can be adjusted above 1. By increasing α from 1, the average edge intensity can be lifted and low- (and medium) contrast edges can be lifted above zero while keeping the high-contrast edge details below zero. Thus, we can obtain different levels of edge detail with respect to the average edge details in a block.

12.2.3 Experimental Results

Experiments were carried out using a number of images to evaluate the performance of the proposed method; however, only the results of the Cameraman and Flower Garden images are given in this chapter. In Figures 12.6a and 12.7a the original images of the Cameraman and Flower Garden are given, respectively.

In Figures 12.6b and 12.7b, the edge details are obtained from the images in Figures 12.6a and 12.7a using the proposed method with the edge-enhancement matrix L in equation (12.19) and $\alpha = 1$. We see from these images that the proposed method detects the edges and also enhances much of the image details. For example, in the Cameraman image we can see gloves details, pockets in the jacket, windows in the tower, and so forth. Similarly, in the Flower Garden image we can see roof textures, branches in the trees, and many other sensitive details.

The images in Figures 12.6c and d and 12.7c and d show the results with the same L in equation (12.19) but now with different $\alpha = 25$ and 50, respectively. This proves that the increase in α leaves only high-contrast edge details. We can see from the Cameraman image that the proposed method can even highlight the mouth edge detail.

In general, edge detection algorithms give edge details that are suitable for determining the image boundary for segmentation. The proposed method not only gives edge details for segmentation but also provides a flexible (adaptive to the DC level) edge detail identification algorithm that is suitable for coding artifact reduction in the transform-coding scheme and allows user control on the DC level to obtain different levels of edge details for different applications.

FIGURE 12.6

Images of Cameraman. (a) Original image; (b) edge details with L and $\alpha = 1$; (c) edge details with L and $\alpha = 25$; *(Continued)*.

FIGURE 12.6
***(Cont.)* Images of Cameraman. (d) Edge details with L and $\alpha = 50$.**

FIGURE 12.7
Images of Flower Garden. (a) Original image; (b) edge details with L and $\alpha = 1$; *(Continued)*.

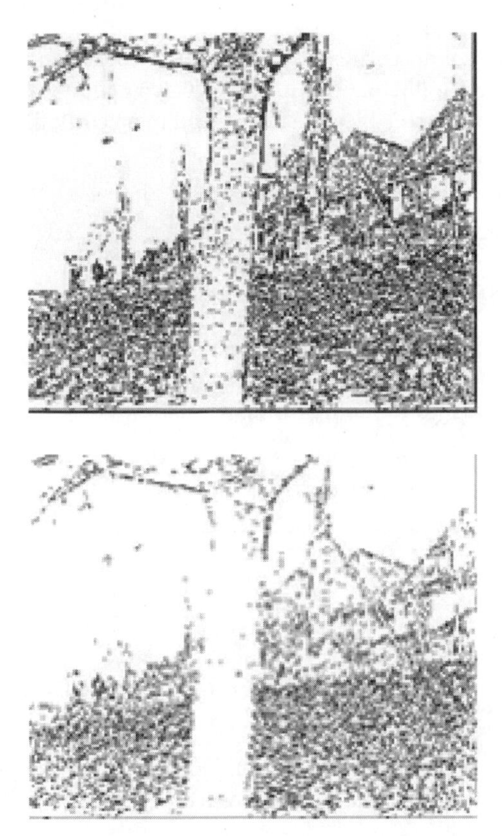

FIGURE 12.7
***(Cont.)* Images of Flower Garden. (c) Edge details with L and $\alpha = 25$; (d) edge details with L and $\alpha = 50$.**

12.3 Summary

In Section 12.1, we introduced two approaches in the transform domain to estimate the quantization error (noise) in DCT coefficients. The IWF method uses the Wiener filter assuming uniform quantization noise, whereas the IWF* uses the improved Wiener filter in the transform domain in order to correct the quantization error that causes the coding artifacts in the decoded images. The advantage of these proposed methods over WLS and WLS* is that they use a lesser number of approximations or assumptions compared to WLS and WLS* and give better results even for low-bit-rate coded images. The IWF* method does not directly depend on the noise characteristics and thus can be used for any type of noise. The proposed methods give better results than the other two methods in terms of PSNR, bit rate, and visual image quality. In addition, they also give better results than the method recently introduced by Kim et al. [14].

In Section 12.2, a new approach was proposed to identify image and edge details that are suitable for image segmentation and coding artifacts reduction in digital image and video coding. In contrast to existing methods, which are mainly edge convolution enhancement techniques and use statistical properties of the edges, the proposed method uses human perceptual information and edge details in the transform coefficients. The proposed method is suitable for many applications, including low-level image processing, medical imaging, image/video coding and transmission, multimedia image retrieval, and computer vision. It identifies the edge details based on the human visual properties of the transform coefficients, and these edges can contribute to the temporal edge fluctuations in digital video coding, which can lead to unpleasant artifacts. Identification of such edge details can lead to an early fix during encoding.

Acknowledgement

We thank Nam Chul Kim (Visual Communication Laboratory, Dept. of Electrical Engineering, Kyungpook National University, Korea, who has kindly given us the simulation results of the JPEG-coded Lena images for comparison.

References

[1] M. Yuen and H.R. Wu, "A survey of hybrid MC/DPCM/DCT video coding distortions," *Signal Processing EURASIP J.*, vol. 70, pp. 247–278, Oct. 1998.

[2] S.S.O. Choy, Y.-H. Chan, and W.-C. Siu, "Reduction of block-transform image coding artifacts by using local statistics of transform coefficients," *IEEE Signal Processing Lett.*, vol. 4, no. 1, Jan. 1997.

[3] K.R. Rao and J.J. Hwang, *Techniques and Standards for Image, Video, and Audio Coding,* Prentice-Hall, Englewood Cliffs, NJ, 1996.

[4] H.C. Reeves and J.S. Lim, Reduction of blocking effects in image coding, *Opt. Eng.*, vol. 23, pp. 34–37, Jan./Feb. 1984.

[5] Y. Yang, N.P. Galatsanos, and A.K. Katsaggelos, "Regularized reconstruction to reduce blocking artifacts of block discrete cosine transform compressed images," *IEEE Trans. CSVT,* vol. 3, pp. 421–432, Dec. 1993.

[6] R.L. Stevenson, "Reduction of coding artifacts in transform image coding," *Proc. ICASSP,* vol. 5, pp. 401–404, 1993.

[7] S. Suthaharan and H.R. Wu, "Adaptive-neighbourhood image filtering for MPEG-1 coded images," *Proc. ICARCV'96: Fourth Int. Conf. on Control, Automation, Robotics and Vision,* pp. 1676–1680, Singapore, Dec. 1996.

[8] S. Suthaharan, "Block-edge reduction in MPEG-1 coded images using statistical inference," *1997 IEEE ISCAS,* pp. 1329–1332, Hong Kong, June 1997.

[9] S. Suthaharan, "New SNR estimate for the Wiener filter to image restoration," *J. Electronic Imaging,* vol. 3, no. 4, pp. 379–389, Oct. 1994.

[10] S. Suthaharan, "A modified Lagrange's interpolation for image restoration," *Austr. J. Intelligent Information Processing Systems,* vol. 1, no. 2, pp. 43–52, June 1994.

[11] J.M. Blackledge, *Quantitative Coherent Imaging,* Academic Press, New York, 1989.

[12] J.S. Lim, *Two-Dimensional Signal and Image Processing,* Prentice-Hall, Englewood Cliffs, NJ, 1990.

[13] Secretariat ISO/IEC JTC1/SC29, ISO CD11172-2, "Coding of moving pictures and associated audio for digital storage media at up to about 1–5 Mbit/s," MPEG-1, ISO, Nov. 1991.

[14] N.C. Kim, I.H. Jang, D.H. Kim, and W.H. Hong, "Reduction of blocking artifact in block-coded images using wavelet transform," *IEEE Trans. CSVT,* vol. 8, no. 3, pp. 253–257, June 1998.

[15] S.E. Umbaugh, *Computer Vision and Image Processing: A Practical Approach Using CVIP Tools,* Prentice-Hall, Englewood Cliffs, NJ, 1998.

[16] S.M. Smith and J.M. Brady, "SUSAN — a new approach to low level image processing," *Int. J. Computer Vision,* vol. 23, no. 1, pp. 45–78, May 1997.

[17] I. Sobel, "An isotropic 3 × 3 image gradient operator." In *Machine Vison for Three-Dimensional Scenes,* H. Freeman, ed., pp. 376–379, Academic Press, New York, 1990.

[18] MATLAB (version 5), "Getting started with MATLAB," The Math Works Inc., 1996.

Chapter 13

Video Modeling and Retrieval

Yi Zhang and Tat-Seng Chua

13.1 Introduction

Video is the most effective media for capturing the world around us. By combining audio and visual effects, it achieves a very high degree of reality. With the widely accepted MPEG (Moving Pictures Expert Group) [12] digital video standard and low-cost hardware support, digital video has gained popularity in all aspects of life.

Video plays an important role in entertainment, education, and training. The term "video" is used extensively in the industry to represent all audiovisual recording and playback technologies. Video has been the primary concern of the movie and television industry. Over the years, that industry has developed detailed and complete procedures and techniques to index, store, edit, retrieve, sequence, and present video materials. The techniques, however, are mostly manual in nature and are designed mainly to support human experts in creative moviemaking. They are not set up to deal with the large quantity of video materials available. To manage these video materials effectively, it is necessary to develop automated techniques to model and manage large quantities of videos.

Conceptually, the video retrieval system should act like a library system for the users. In the library, books are cataloged and placed on bookshelves according to well-defined classification structures and procedures. All the particular information about a book, such as the subject area, keywords, authors, ISBN number, etc., are stored to facilitate subsequent retrievals. At the higher level, the classification structure acts like a conceptual map in helping users to browse and locate related books. Video materials should be modeled and stored in a similar way for effective retrieval. A combination of techniques developed in the library and movie industries should be used to manage and present large quantities of videos.

This chapter discusses the modeling of video, which is the representation of video contents and the corresponding contextual information in the form of a conceptual map. We also cover the indexing and organization of video databases. In particular, we describe the development of a system that can support the whole process of logging, indexing, retrieval, and virtual editing of video materials.

13.2 Modeling and Representation of Video: Segmentation vs. Stratification

There are many interesting characters, events, objects, and actions contained in a typical video. The main purpose of video modeling is to capture such information for effective representation and retrieval of video clips. There are two major approaches to modeling video, namely, the segmentation [5, 15] and stratification [1] approaches.

In the *segmentation* approach, a video sequence is segmented into physical chunks called *shots*. By definition, a video shot is the smallest sequence of frames that possesses a simple and self-contained concept. In practice, video shots are identified and segmented using scene change detection techniques [21]. A change in scene will mark the end of the previous shot and the start of the next one. Figure 13.1 shows the way in which a video sequence is segmented into shots.

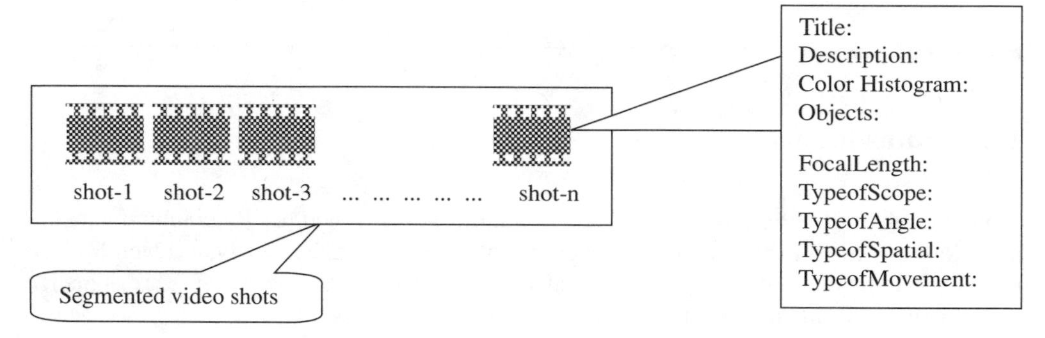

FIGURE 13.1
Segmentation: modeling and representation of video shots.

In the segmentation approach, a video shot is an atomic unit that can be manipulated for representation and retrieval. Figure 13.1 also shows how a video shot can be represented. Important attributes such as title, description, audio dialogues, color histogram, and visual objects should be extracted and logged. In addition, we can capture cinematic information [8], such as the focal length, scope, and angle of shots, for the purpose of sequencing video shots for presentation. By combining all the attributes mentioned above, a good representation of video shots can be achieved.

Although the segmentation approach is easy to understand and implement, it has several drawbacks. In such a model, the granularity of the information is a shot which typically contains a self-contained but high-level concept. Once the video is segmented, the shot boundaries are fixed and it is not possible for users to access or present video contents within the shot boundaries. This inflexibility limits the users' ability to model more complicated events in a way not originally intended by the authors. Because of the diverse contents and meaning inherited in the video, certain events are partially lost. For example, a shot may contain multiple characters and actions; it is thus not possible for authors to anticipate all the users' needs and log all interesting events. Moreover, the segment boundaries may be event dependent. This may result in a common story being fragmented across multiple shots, which will cause discontinuity to users looking for the story. For these reasons, automated description of video contents at the shot level is generally not possible.

To overcome this problem, we investigate an object-based video indexing scheme to model the content of video based on the occurrences of simple objects and events (known as entities).

This is similar to the stratification approach proposed in [1]. In this scheme, an entity can be a concept, object, event, category, or dialogue that is of interest to the users. The entities may occur over many, possibly overlapping, time periods. A strand of an entity and its occurrences over the video stream are known as a *stratum*. Figure 13.2 shows the modeling of a news video with strata such as the object "Anchor-Person-A"; categories like "home-news," "international news," "live reporting," "finance," "weather," and "sports"; and audio dialogues. The meaning of the video at any time instance is simply the union of the entities occurring during that time, together with the available dialogue.

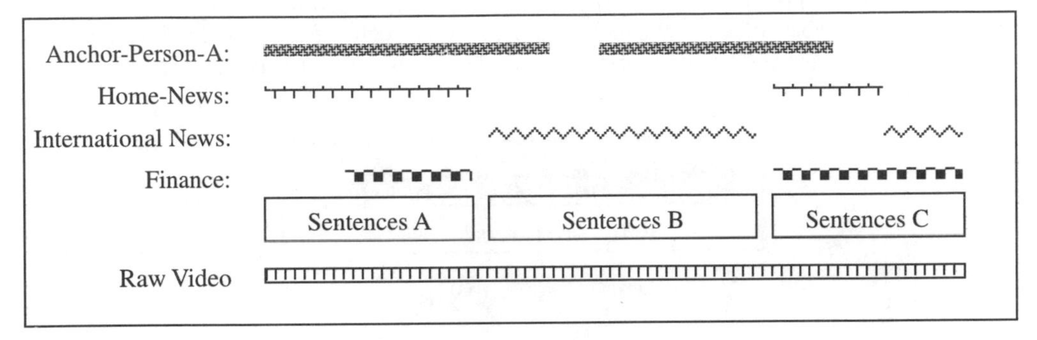

FIGURE 13.2
Object-based modeling of a news video.

Object-based modeling of video has several advantages. First, because strata tend to contain only simple entities, the pseudo-object models together with relevance feedback (RF) [7] and other learning methods may be adopted to identify and track entities automatically. Second, the meaning of video at any instance is simply the union of strata occurring at that time. Thus the system can flexibly compose portions of video sequences whose meanings closely match that of the query. Finally, the strata information provides the meta-information for the video. Such information can be used to support innovative functions such as the content-based fast-forwarding and summarization [7] of video.

13.2.1 Practical Considerations

The main advantage of the stratification approach over segmentation is that it is possible to automate the process of indexing. However, when the project was started several years ago, this advantage was not realizable because of the lack of content-based analysis tools for automated indexing. Without such tools, it is extremely tedious to index video content using the stratification approach. Because of this, coupled with the simplicity of the segmentation approach in supporting our initial research goals in video retrieval, virtual editing, and summarization, we adopted the segmentation approach. This project is an extension of the work done in Chua and Ruan [5] using digital video with more advanced functionalities for video manipulation, browsing, and retrieval. We are currently working on the development of a system based on the stratification approach to model video in the news domain.

This chapter describes the design and implementation of our segmentation model for video retrieval. The domain of application is documentary videos.

13.3 Design of a Video Retrieval System

This section considers the modeling of video using the segmentation approach. There are several processes involved in modeling video sequences to support retrieval and browsing. The two main steps are video parsing and indexing. The term *parsing* refers to the process of segmenting and logging video content. It consists of three tasks: (1) *temporal segmentation* of video material into elementary units called segments or shots; (2) *extraction of content* from these segments; and (3) *modeling of context* in the form of a concept hierarchy. The indexing process supports the storage of extracted segments, together with their contents and context, in the database. Retrieval and browsing rely on effective parsing and indexing of raw video materials. Figure 13.3 summarizes the whole process of video indexing, retrieval, and browsing.

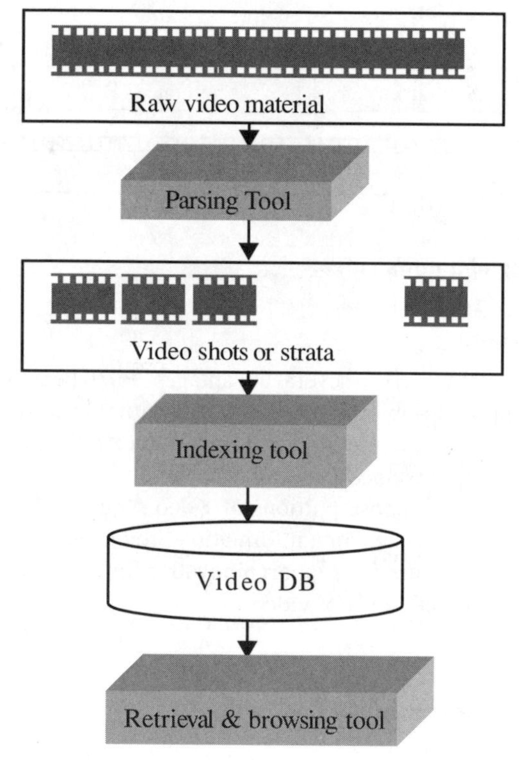

FIGURE 13.3
Parsing and indexing video for retrieval and browsing.

13.3.1 Video Segmentation

The success of the segmentation approach depends largely on how well the video materials are divided into segments or shots. This involves identifying suitable criteria to decide which points in a video sequence constitute segment boundaries. Manual segmentation has been done effectively in the movie industry. However, manual segmentation is time consuming and prone to error. Moreover, its results are biased toward the authors' intents.

With advances in image processing techniques, computer-aided segmentation is now possible [21]. The main objective is to detect the joining of two shots in the video sequence and locate the exact position of these joins. These joins are made by the video editing process, and they can be of the following types based on the techniques involved in the editing process:

- Abrupt cut
- Dissolve
- Fade in/fade out
- Curtain and circle

In an abrupt cut, the video editor does nothing but simply concatenate the two shots together. The other three joins are generally known as gradual transitions, in which the video editor uses a special technique to make the join appear visually smooth.

Thus, segmentation becomes a matter of finding those features that constitute transitions. A lot of work has been done on detecting abrupt cuts, and many techniques have been developed to handle gradual transitions in both the compressed and the decompressed domain [11, 13, 19]. We detect the cut transition by sequentially measuring successive inter-frame differences based on features such as the color histogram. When the difference is above the global threshold, a cut transition is declared.

When dealing with gradual transitions, Zhang et al. [21] used two thresholds. They computed accumulated differences of successive frames when the inter-frame difference was above a lower threshold. When this accumulated difference exceeded the high threshold, a gradual transition was declared. Other approaches employ template matching [20], model-based [10], statistical [18], and feature-based [19] methods. Most of the methods employed for detecting gradual transitions require careful selection of the threshold for the method to work effectively [11]. This is a difficult problem. Lin et al. [14] proposed a multi-resolution temporal analysis approach based on wavelet theory to detect all transitions in a consistent manner.

In this chapter, we employ the Zhang et al. [21] approach of computing accumulated differences of color histogram in successive frames to detect both abrupt cut and gradual transitions. We employ this approach because it is simple to implement and has been found to work well. Our tests show that it could achieve a segmentation accuracy of greater than 80%. Based on the set of segments created, a visual interface is designed to permit the authors to review and fine-tune the segment boundaries. The interface of the Shot Editor is given in Figure 13.9. The resulting computer-aided approach has been found to be satisfactory.

13.3.2 Logging of Shots

After the video is segmented, each shot is logged by analyzing its contents. Logging is the process of assigning meanings to the shots. Typically, each shot is manually assigned a title and text description, which are used in most text-based video retrieval systems. Increasingly, speech recognition tools are being used to extract text from audio dialogues [16], and content-based analysis tools are being used to extract pseudo-objects from visual contents [6]. These data are logged as shown in Figure 13.1. The combination of textual and content information enables different facets of content to be modeled. This permits more accurate retrieval of the shots.

Because video is a temporal medium, in addition to retrieving the correct shots based on the query, the shots retrieved must be properly sequenced for presentation. Thus, the result of a query should be a dynamically composed video sequence, rather than a list of shots. To do this, we need to capture cinematic information [8]. Typical information that is useful for sequencing purposes includes the focal length and angle of shots. Such information permits the sequencing of video by gradually showing the details (from far shots to close-up shots), a typical presentation technique used in documentary video. The complete information logged for the shot thus includes both content and cinematic information, as shown in Figure 13.1.

Video is a temporal medium that contains a large amount of other time-related information. This includes relationships between objects and motion, the order in which main characters appear, and camera motions/operations. However, there is still a lack of understanding of how object motions are to be queried by users. Also, automated detection of motion and object relationships is not feasible with the current technology. These topics are open for research and exploration. Thus, temporal information is not included as part of the shot representation.

13.3.3 Modeling the Context between Video Shots

In the segmentation approach, although the information logged provides a good representation for individual video shots, the context information between video shots is not properly represented. Without contextual information, it is hard to deduce the relationships between shots. For example, in the domain of an animal, the contextual information permits different aspects of the life of an animal, say a lynx, to be linked together (see Figure 13.10). It also provides high-level information such as the knowledge that lion and lynx belong to the same cat family. The context information can be captured in the structured modeling approach using a two-layered model as shown in Figure 13.4.

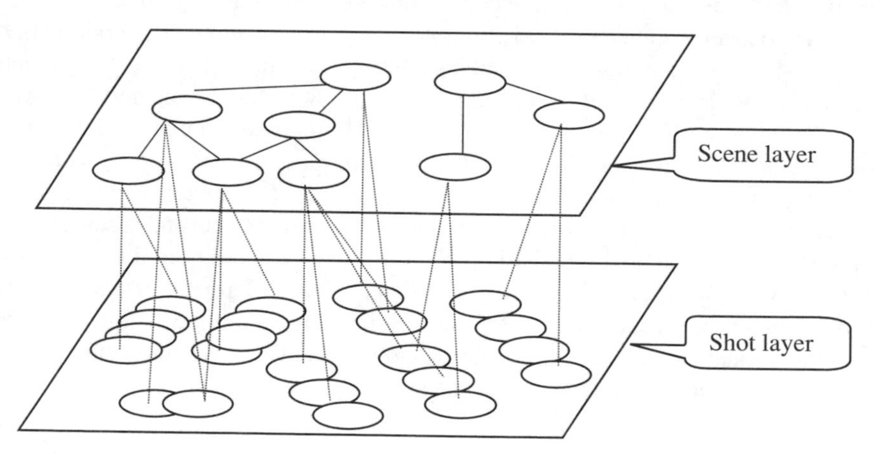

FIGURE 13.4
A two-layered model for representing video context.

The layered model consists of a lower shot layer and the scene layer. The shot layer contains all the shots indexed, whereas the scene layer models the video context information. The context information can be modeled as a scene hierarchy as shown in Figure 13.4. Maintaining the context information only at the scene layer permits more than one scene hierarchy to be modeled above the same set of shots. This facilitates multiple interpretation of the shot layer and leads to a more flexible creation of scenes.

Typical information encoded at the scene includes the title and textual descriptions, together with parent–child relations. Because the scene hierarchy captures the knowledge of the set of video shots, it can be used to support concept-based retrieval of the video shots. A retrieval of a set of video shots can be viewed as the retrieval of an appropriate scene (or concept) in the scene hierarchy. This will be illustrated in a later section.

13.4 Retrieval and Virtual Editing of Video

13.4.1 Video Shot Retrieval

The shots and scenes permit the video to be represented in a form that can be stored and manipulated by the computer. Because the main semantic information captured for the shot is the title, text descriptions, and dialogues, we store these as free text so that a free-text retrieval technique can be employed to retrieve video shots using free-text queries. Conceptually, each shot is stored as a free-text document in an inverted file. We employ the vector space information retrieval (IR) model [17] to retrieve the shots. In the vector space IR model, each shot is represented as a vector of dimension t of the form:

$$\underline{S_i} = (d_{i1}\, d_{i2}\, d_{i3} \ldots d_{it}) \tag{13.1}$$

where d_{ij} is the weight of term j in the ith shot S_i, and t is the number of text terms used. By representing the query $\underline{Q}$ in the same way, that is,

$$\underline{Q} = (q_1\, q_2\, q_3 \ldots q_t) \tag{13.2}$$

the similarity between query $\underline{Q}$ and shot $\underline{S_i}$ can be computed using the cosine similarity formula [17], given by:

$$Sim\left(\underline{S_i}, \underline{Q}\right) = \frac{\sum_{k=1}^{t}(d_{ik} * q_k)}{\sqrt{\sum_{k=1}^{t}(d_{ik})^2 * \sum_{k=1}^{t}(q_k)^2}} \tag{13.3}$$

Equation (13.3) is used as the basis to rank all shots with respect to the query.

13.4.2 Scene Association Retrieval

To permit more effective retrieval of shots, the knowledge built into the scene structure should be used. The scene is used to link shots semantically related to each other under the same scene hierarchy. Many of these shots may not have common text descriptions and thus may not be retrievable together at the same time. To improve higher retrieval accuracy, the idea here is to retrieve higher level scenes as much as possible so that these related shots may be retrieved by the same query. A scene association algorithm is developed for this purpose. The algorithm is best illustrated using Figure 13.5.

After the user issues a query, the scene association algorithm proceeds as follows:

1. *Shot retrieval:* First we compute the similarities between the query and all the shots in the shot layer using equation (13.3). The shots are ranked based on the similarity values assigned to them. The list of shots whose similarity values are above a predefined threshold is considered to be relevant to the query. This is known as the *initial relevant shot list*. Shots that appear on the initial relevant shot list are assigned a value of 1; all other shots are assigned a value of 0. In the example given in Figure 13.5, shots S_1, S_3, S_4, S_7, and S_8 are given a value of 1 because they are on the initial relevant shot list. Shots S_2, S_5, S_6, and S_9 are assigned a value of 0 because they are not.

2. *Association:* We then compute a similarity value for each leaf node in the scene structure. This is done by computing the percentage of its children that are assigned the value of 1. For example, scene leaf node C_4 is given a value of 1/3.

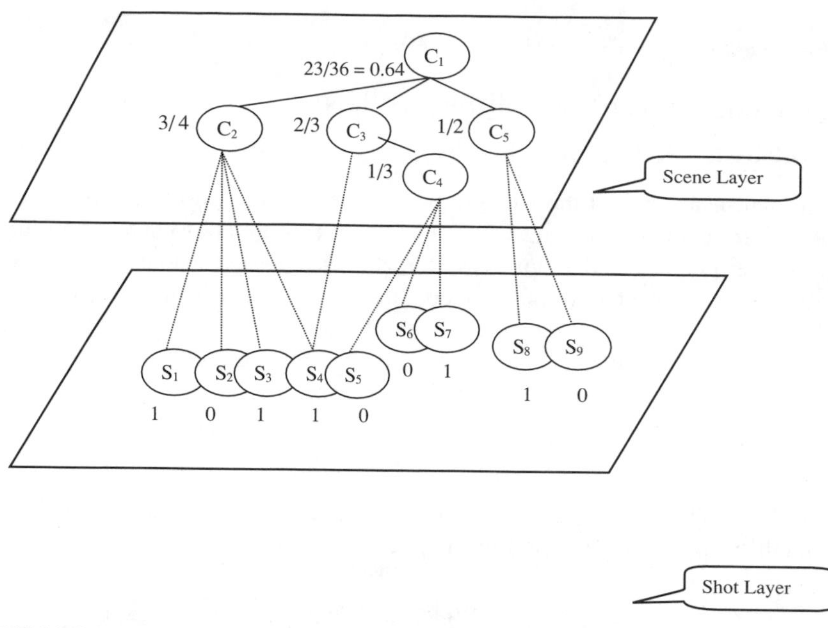

FIGURE 13.5
Scene association to retrieve video shots.

3. *Propagation:* The similarity values of the scene leaf nodes are then propagated up the scene hierarchy. Each parent node is given a value equal to the average of the similarity values of its child nodes. For example, C_1 is assigned a value that is the average of the similarity values of all its child nodes C_2, C_3, and C_5.

4. *Selection:* At the end of propagation, we select the scene node with the highest value as the most relevant scene. If the similarity value of this scene node exceeds a predefined threshold $\mathbf{T_s}$, then all shots under this scene together with the initial relevant shot list are returned. Otherwise, only the initial relevant shot list will be returned.

In Figure 13.5, video shots S_1, S_3, S_4, S_7, and S_8 are selected in the initial relevant shot list. After the scene association step, scene C_2 is selected as the most relevant scene. All the shots under scene C_2 and those in the initial relevant shot list are retrieved and returned to the users. Notice that by using scene association, shot S_2, which is not in the initial relevant shot list, can also be retrieved. Thus, by giving higher priority to retrieving higher level scenes, we are able to retrieve scenes that represent more general concepts and cover a greater number of scenes and shots. This will result in better recall in video shot retrieval.

13.4.3 Virtual Editing

Because video is a temporal medium, it has the additional problem of sequencing the retrieved video shots before presentation. In the computer industry, the automatic creation of a video sequence from video clips is called *virtual editing* [3]. Cinematic rules are used as the basis to sequence the retrieved video shots in order to form a meaningful sequence. This process is more art than science.

Parameters for Virtual Editing

Before virtual editing can be carried out, a set of cinematic parameters must be recorded at indexing time. Normally, the parameters logged include:

- *Focal length*

  ELS — extreme long shot
  LS — long shot
  DFS — deep focus shot
  MLS — medium long shot
  MS — medium shot
  CU — close-up shot
  ECU — extreme close-up shot

- *Type of scope*

  1-shot — only one main character in the shot
  2-shot
  3-shot
  group-shot — many characters in the shot

- *Type of angle*

  Bird view
  Eye-level angle
  High angle
  Low angle
  Oblique angle

- *Type of spatial*

  Inclusion
  Exclusion

- *Type of movement*

  Dolly
  Hand-held
  Pan
  Tilt
  Zoom
    Zoom in
    Zoom out

These parameters are sufficient to generate satisfactory documentary-style movies. In fact, the "focal length," "type of angle," and "type of movement" should be enough to generate most typical presentation sequences that start from a general view before zooming into the details or vice versa. Such styles are effective in presenting information.

Cinematic Rules for Virtual Editing

Cinematic rules have existed since the birth of film-making [2, 9]. These rules are widely accepted by film editors and have been applied to generate a tremendous number of films. Generally speaking, there are two classes of rules:

1. Basic rules that are common sense based
2. Complex rules that are normally subject matter based and rely on factors such as psychology, aesthetics, and so forth.

The rules provide only general guidelines for how a movie can be made. The making of a memorable film relies largely on the editor's creativity in applying the rules. Because there is a general lack of understanding of how films are edited and sequenced, it is hard to automate the creative aspects of film-making. Thus, for practical reasons, only the simple and easy-to-understand rules will be automated. The rules that are commonly studied and implemented are:

- *Concentration rule:* An actor, object, or topic is introduced in the sequence from long shots, to medium shots, to close-up shots. The relationships between the objects and the environment are also evolved through this sequence.
- *Enlarge rule:* This is the inverse of the concentration rule.
- *General rule:* This is a combination of the concentration and enlarge rules. It intends to present an intact action in a sequence that supports better understanding.
- *Parallel rule:* This aims to present two different themes alternately. Normally, the two themes are similar to or contrast with each other. Examples of such themes include the alternating shots of two actors walking toward each other, or the typical chase scenes of police and villains.
- *Rhythm rule:* The shots are chosen based on the rhythm requested. Longer duration shots are used for slower rhythms, whereas shorter duration shots are used for faster rhythms. The rhythm of the intra-shot action should be matched with that of the presentation.
- *Sequential rule:* Shots are ordered chronologically.
- *Content rule:* Shots in a sequence share common content attributes.

From Concept to Video Sequence

To generate a video sequence from a concept, the user needs to specify not just the query but also presentation information such as the time constraint, the cinematic rule to apply, and the necessary parameters for the cinematic rule. This can be achieved by using a script, as shown in Figure 13.6. The script, as a template, is used to record necessary information for sequencing purposes.

Script title: lynx family
Cinematic rule: sequential
Time duration: 20
Query: lynx family together
Scene (optional): lynx

FIGURE 13.6
An example of a script for virtual editing.

A general script class is defined. A specific script for each cinematic rule is implemented. Each script has its own method to sequence the retrieved video shots. The most important cinematic parameter used is the focal length, which is recorded for the main character. It is used by the script to generate effects of the concentration rule, enlarge rule, and general rule. Video shots are sorted based on this parameter to form the appropriate presentation sequence. Other parameters such as the number of frames help in restricting the duration of the presentation.

Given a script, the following steps are followed to generate a meaningful video sequence:

1. User defines the script, which contains all necessary information for sequencing, including script title, cinematic rule, query terms, time duration, etc.

2. The query is first used to retrieve the relevant video shots using, if appropriate, the scene association method.

3. The shots retrieved are ordered using the cinematic parameter(s) determined by the cinematic rule chosen. For example, if the enlarge rule is chosen, the shots are ordered from long shots to close-up shots.

4. The sequence is formed by selecting a suitable number of shots in each cinematic category, subject to the time limit.

5. Before presentation, user may edit the sequence by adding new shots and/or cutting shots from the sequence.

6. When the result is satisfactory, the final sequence is presented to the user.

13.5 Implementation

This section describes the implementation of the digital video retrieval system. The overall design of the system is given in Figure 13.7. The three major components of the system are:

- *Shot Editor:* Caters to the indexing and logging of video shots.

- *Scene Composer:* Supports the creation and editing of scenes. It uses a retrieval tool to help users semiautomate the creation of scenes by issuing queries. A subset of this tool supports the browsing of scenes by end users.

- *Video Presentor:* Supports the retrieval and virtual editing of video materials.

Figure 13.7 also shows the interactions between the users and the system, and among different components of the system. The system is designed to support two types of users: (1) high-level users or indexers who will use the sophisticated tools to index and edit video shots and scenes, and (2) normal end users who will interact only with the Scene Browser to browse through the scenes, and with the Video Presentor to query and sequence video shots. The functionalities provided to these two types of users are quite different. Figure 13.8 gives the main user interface and the subset of functionality provided to the indexers and end users. The specific interfaces and functions of the components are described separately below.

Shot Editor

The Shot Editor is a tool that helps users index and log the raw video shots. Figure 13.9 shows the interface of the Shot Editor. The video is viewed on the top-left window, which provides a VCR-like interface for users to browse the contents of the shots and/or to fine-tune

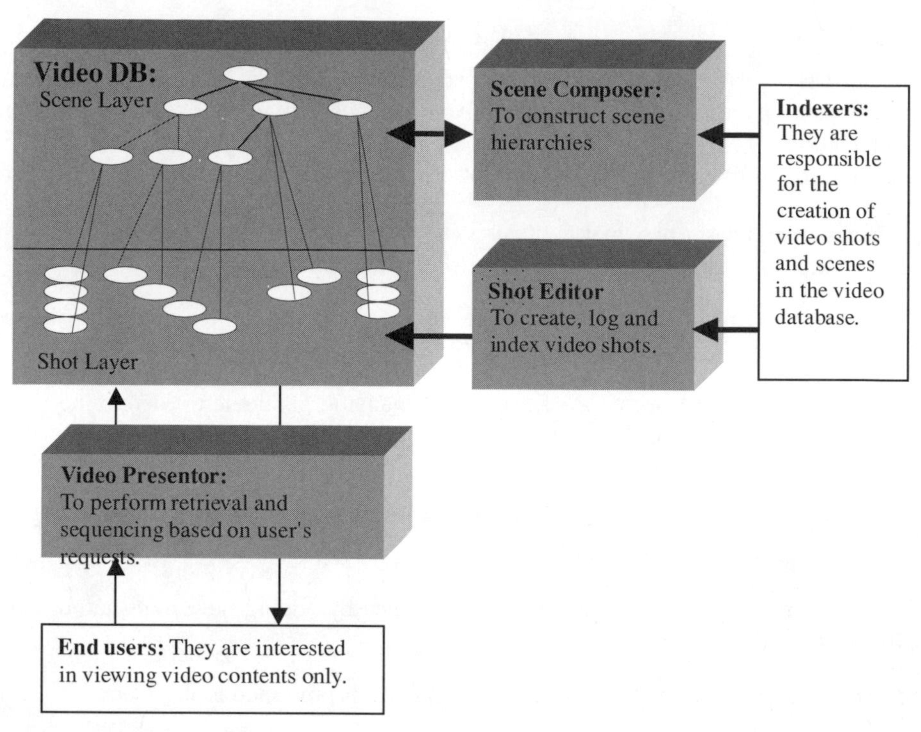

FIGURE 13.7
System components of the digital video retrieval system.

the shot boundaries. Users may log or edit the semantic information for the shots, such as the title, text descriptions, and dialogues, and cinematic parameters. The list of shots created is shown in the top-right list browser.

Scene Composer and Browser

The Scene Composer is used to compose scenes that capture the context information and domain knowledge of the video shots. Through the interface given in Figure 13.10, users may create a new scene by clicking on the "New" button at the top menu bar. The edit panel at the bottom permits users to view the list of existing shots and scenes that can be used as members of the new scene. Because the possible set of shots and scenes may be large, we may limit the set by using the shot retrieval and scene association modules (see Section 13.3) to filter out only those that match the content of the new scene being created. This facilitates the task of composing complex scenes and opens up the possibility of creating some of the scenes automatically. Users may browse and edit existing scenes at any time. Users may also view the video sequence generated from the scene by clicking on the "Play" button.

A simpler interface, without all the editing functions, is provided to end users to browse the content of the scenes. Because the scene structures provide rich context information and domain knowledge of the underlying interface, it is extremely beneficial for new users to have a good understanding of the overall structure before accessing the video database.

Video Presentor

The Video Presentor is the main interface users interact with to perform video retrieval and virtual editing. Figure 13.11 shows the interface together with its child windows. To retrieve

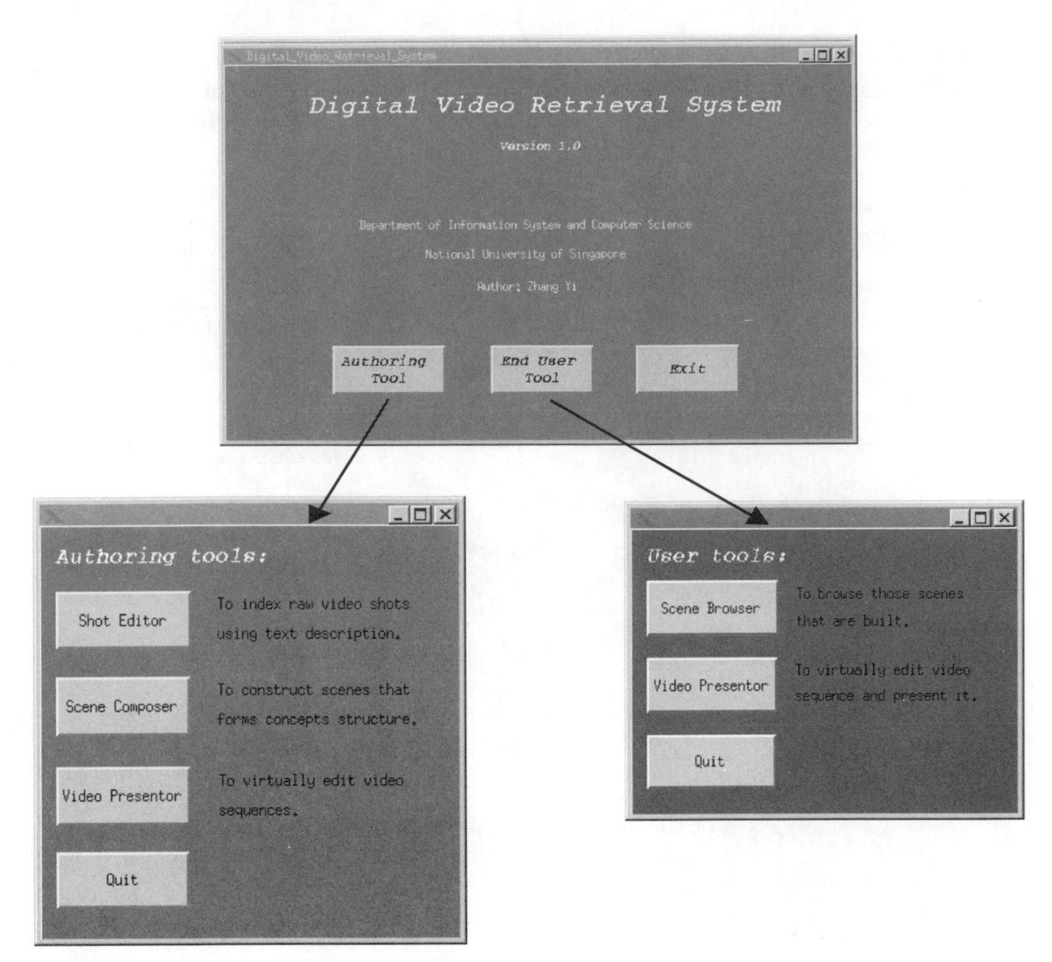

FIGURE 13.8
Main user interface of the digital video retrieval system.

the video sequence, users need to pose a free-text query. Users may optionally enter a time limit as well as cinematic rules and parameters. Based on the query script, the system first retrieves a relevant set of shots (given in the Retrieved List Browser) by using only the scene association retrieval function. It then performs the virtual editing function to arrive at a final sequence (listed in the Sequenced List Browser) that meets the users' presentation specifications. Users may view the generated sequence automatically by manipulating the VCR-like interface at the top-left window. Users may also choose to view each sequenced shot individually or edit the sequence manually. The final sequence may be saved for future viewing.

Instead of issuing a query, users may choose to view an existing sequence generated previously.

13.6 Testing and Results

The system was developed at the Multimedia Information Laboratory at the National University of Singapore. It was implemented on the Sun Solaris using C++ and employed the MPEG-TV tool to manipulate and display digital video in MPEG-1. We chose a documentary video in the domain of animal for testing. We selected more than 24 minutes of video materials in 7 different animal categories. The video was analyzed and logged using the procedures as outlined in Section 13.3. Altogether, 164 video shots and 45 scenes were created. A summary of the video shots created in different categories is given in Table 13.1.

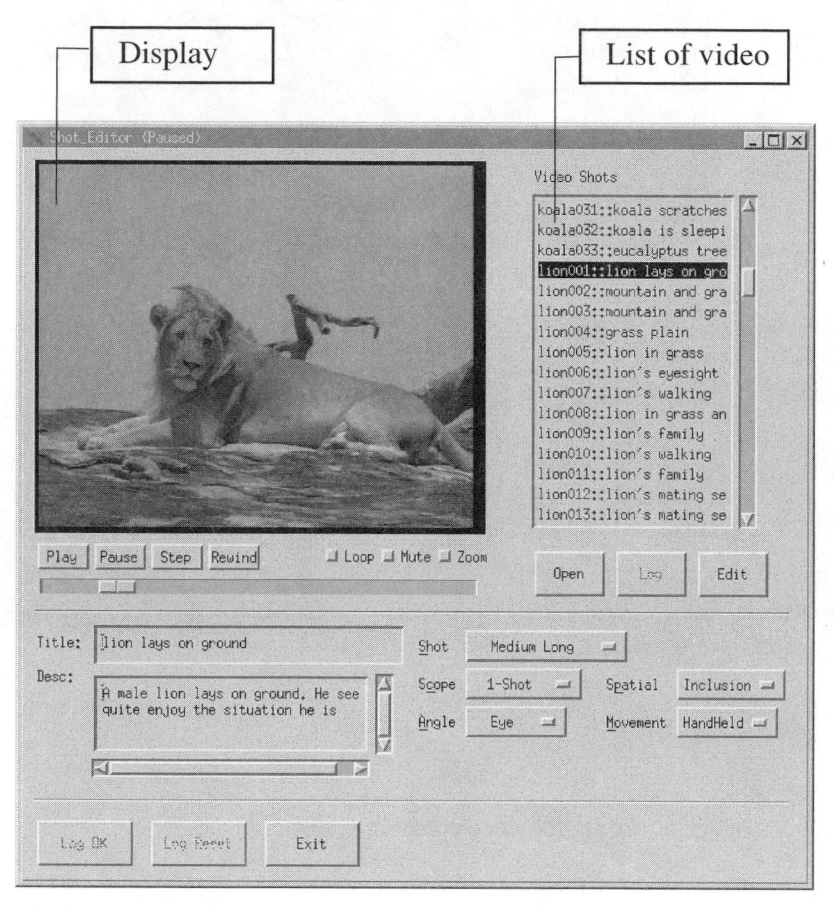

FIGURE 13.9
User interface of the Shot Editor.

Table 13.1 Characteristics of Video Materials Used

Category	Total Duration (s)	Number of Shots Created	Number of Scenes Created
Koala	219.9	33	8
Lion	222.9	27	7
Lynx	213.2	22	7
Manatee	184.4	16	4
Manta ray	242.2	16	5
Marine turtle	225.7	23	6
Ostrich	205.7	27	8
TOTAL	1447.2	164	45

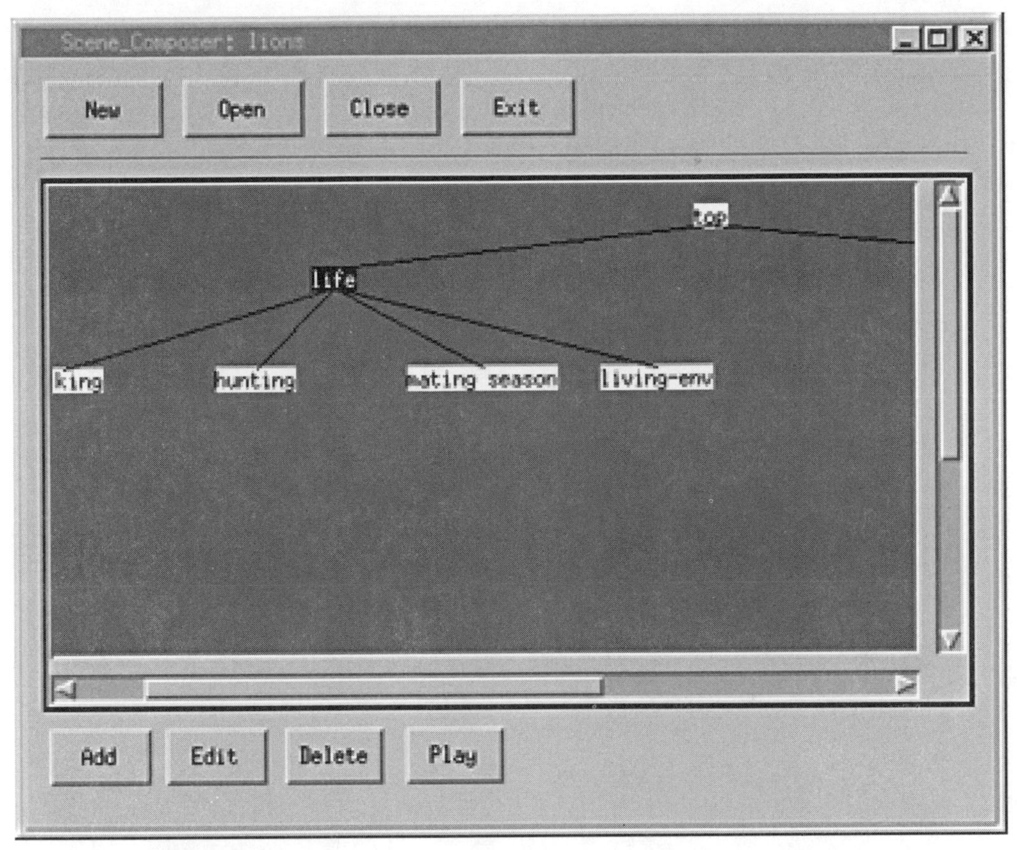

FIGURE 13.10
User interface of the Scene Composer.

A set of seven queries was chosen to test the retrieval effectiveness. The queries, together with the list of relevant shots, are summarized in Table 13.2. The results of retrieval, using the shot retrieval technique, are presented in Table 13.3 in terms of normalized recall and precision. From Table 13.3, it can be observed that the shot retrieval technique is quite effective. Further tests using the scene association technique demonstrated that we could achieve a further 10 to 12% improvement in retrieval performance.

To evaluate the effectiveness of virtual editing, we conducted a series of retrievals using similar queries but with different cinematic rules. The results are shown in Figure 13.12. Figures 13.12a–c show the results of issuing the query "koala sleeps" to retrieve the set of relevant video shots using the scene association technique but applying different cinematic rules to sequence the shots. The time limit chosen is 25 seconds in all three cases. The sequences generated using concentration, enlarge, and general rules are shown in Figures 13.12a–c, respectively. The final sequence shown in Figure 13.12d is generated by using the query "koala scratches" with the parallel cinematic rule and longer duration.

From the results (Figures 13.12a–c), it can be seen that our system is able to generate sequences that conform to the desired cinematic rules within the time limit. Figure 13.12(d) also shows that the system is able to generate complicated sequences based on parallel themes. These results clearly meet the expectations of users and demonstrate that both our retrieval and virtual editing functions are effective.

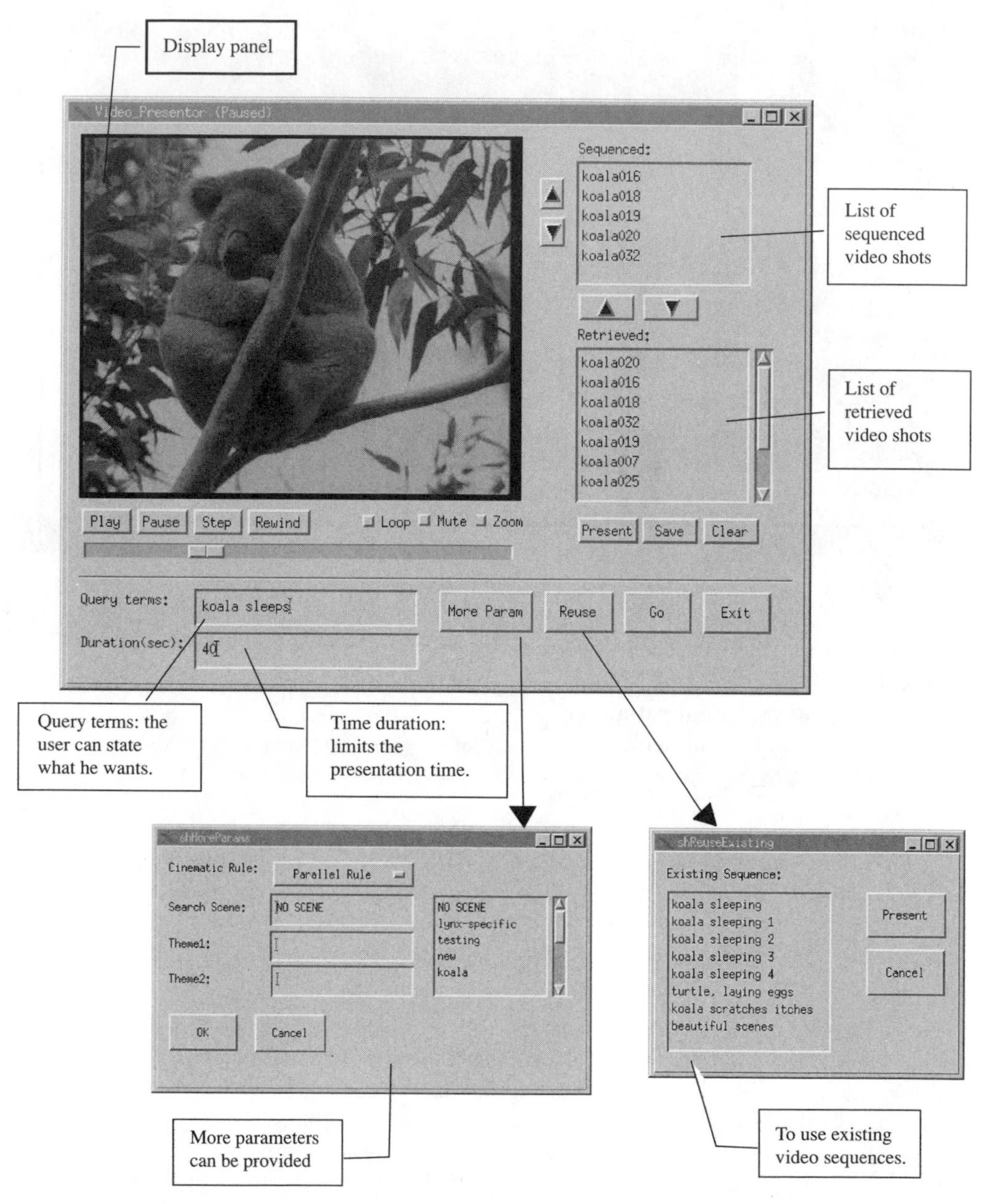

FIGURE 13.11
User interface of the Video Presentor.

FIGURE 13.12
Results of virtual editing using the same query but different cinematic rules. (a) query: koala sleeps; rule: Concentration; duration: 25 sec (b) query: koala sleeps; rule: Enlarge; duration: 25 sec (c) query: koala sleeps; rule: General; duration: 25 sec (d) query: koala scratches itches; rule: Parallel; Theme1: body; Theme2: head; duration: 30 sec.

13.7 Conclusion

This chapter discussed the modeling, representation, indexing, retrieval, and presentation of digital video. It also reviewed the design and implementation of a digital video retrieval system to accomplish the above task. The effectiveness of the system was demonstrated using a documentary video in the domain of animals. The results showed that virtual editing can be carried out effectively using simple cinematic rules.

With advances in automated (pseudo) object identification and transcription of audio, we are beginning to research ways to automate the process of logging video shots. We are also developing a new system based on the stratification approach, centered on advanced functionalities for retrieval, interaction, semantic-based fast-forwarding, and video summarization.

Table 13.2 Queries and Set of Relevant Video Shots

Query	Relevant Video Shots
Koala on a tree	koala001.mpg.txt, koala003.mpg.txt, koala007.mpg.txt, koala008.mpg.txt, koala009.mpg.txt, koala010.mpg.txt, koala015.mpg.txt, koala021.mpg.txt, koala022.mpg.txt, koala024.mpg.txt
Lion walking	lion007.mpg.txt, lion009.mpg.txt, lion010.mpg.txt
Lynx hunting	lynx013.mpg.txt, lynx014.mpg.txt, lynx015.mpg.txt, lynx016.mpg.txt
Manatee family	manatee008.mpg.txt, manatee009.mpg.txt, manatee010.mpg.txt, manatee011.mpg.txt, manatee013.mpg.txt
Manta ray flying	manta_ray001.mpg.txt, manta_ray010.mpg.txt, manta_ray012.mpg.txt
Marine turtle giving birth	marine_t012.mpg.txt, marine_t013.mpg.txt, marine_t014.mpg.txt, marine_t015.mpg.txt, marine_t016.mpg.txt, marine_t018.mpg.txt, marine_t019.mpg.txt
Ostrich running	ostrich003.mpg.txt, ostrich014.mpg.txt, ostrich015.mpg.txt

Table 13.3 Normalized Recall and Precision of Retrieval

Query	Recall (normalized)	Precision (normalized)
Koala on a tree	0.816388	0.662276
Lion walking	0.798734	0.616213
Lynx hunting	0.882279	0.655542
Manatee family	0.847985	0.668782
Manta ray flying	0.871069	0.743911
Marine turtle giving birth	0.692742	0.514673
Ostrich running	0.753145	0.592238
Average	0.808906	0.636234

References

[1] Aguierre Smith, T.G. and Pincever, N.C. (1991): Parsing Movies in Context. *USENIX*, pp. 157–168.

[2] Balazs, B. (1952): *Theory of the Film,* Dennis Dobson Ltd., London.

[3] Bloch, G.R. (1988): From Concepts to Film Sequences. *RIAO'88,* pp. 761–767.

[4] Chua, T.S. and Kankanhalli, M.S. (1998): Towards Pseudo Object Models for Content-Based Visual Information Retrieval. *Proceedings of International Symposium on Multimedia Information Processing,* ChengLi, Taiwan, pp. 182–192.

[5] Chua, T.S. and Ruan, L.Q. (1995): A Video Retrieval and Sequencing System. *ACM Transactions of Information System,* vol. 13, no. 4, pp. 373–407, October.

[6] Chua, T.S. and Chu, C.X. (1998): Color-Based Pseudo Object Model for Image Retrieval with Relevance Feedback. *Proceedings of International Conference on Advanced Multimedia Content Processing,* Osaka, Japan, pp. 148–162, November.

[7] Chua, T.S., Low, W.C., and Chu, C.X. (1998): Relevance Feedback Techniques for Color-Based Image Retrieval. *Proceedings of Multimedia Modeling'98,* Lausanne, Switzerland, pp. 24–31, October.

[8] Davenport, G, Aguierre-Smith, T., and Pincever, N. (1991): Cinematic Primitives for Multimedia. *IEEE Computer Graphics and Applications,* pp. 67–74, July.

[9] Dyer, M.R. (1966): *Film: A Montage of Theories,* Dutton, New York.

[10] Hampapur, A., Jain, R., and Weymouth, T.E. (1995): Production Model Based Digital Video Segmentation. *Multimedia Tools and Applications,* vol. 1, no. 1, pp. 9–46.

[11] Idris, F. and Panchanathan, S. (1997): Review of Image and Video Indexing Techniques, *Journal of Visual Communication and Image Representation,* vol. 8, no. 2, pp. 146–166.

[12] ISO/IEC 11172, Moving Pictures Expert Group Committee (1993): Information Technology — Coding of Moving Pictures and Associated Audio for Digital Storage Media at up to about 1.5 Mbit/s. ISO/IEC 11172 — 1,2,3,4.

[13] Jiang, H., Helal, A., Elmagarimid, A.K., and Joshi, A. (1998): Scene Change Detection Techniques for Video Database System. *ACM Multimedia Systems,* vol. 6, pp. 186–195.

[14] Lin Yi, M., Kankanhalli. S., and Chua, T.-S. (1999): Temporal Multi-Resolution Analysis for Video Segmentation. Technical report, School of Computing, National University of Singapore.

[15] Rubin, B. and Davenport, G. (1989): Structured Content Modeling for Cinematic Information. *SIGCHI Bulletin,* vol. 21, no. 2, pp. 78–79.

[16] Rudnicky, A.I., Lee, K.-F., and Hauptmann, A.G. (1994): Survey of Current Speech Technology. *Communications of the ACM,* vol. 37, no. 3, pp. 52–57.

[17] Salton, G. and McGill, M. (1983): *Introduction to Modern Information Retrieval.* McGraw-Hill, New York.

[18] Sethi, I.K. and Patel, N.A. (1995): Statistical Approach to Scene Change Detection. *SPIE Conference on Storage and Retrieval for Image and Video Database III,* vol. 2420, pp. 329–338.

[19] Zabih, R., Miller, J., and Mai, K. (1995): A Feature-Based Algorithm for Detecting and Classifying Scene Breaks. *Fourth ACM Multimedia Conference,* pp. 189–200.

[20] Zhang, H.J., Kankanhalli, A., and Smoliar, S.W. (1993): Automatic Partitioning of Full-Motion Video. *ACM Multimedia Systems,* vol. 1, no. 1, pp. 10–28, July.

[21] Zhang, H.J., Low, C.Y., Smoliar, S.W., and Wu, J.H. (1995): Video Parsing, Retrieval and Browsing: An Integrated and Content-Based Solution. *ACM Multimedia '95,* pp. 15–23.

Chapter 14

Image Retrieval in Frequency Domain Using DCT Coefficient Histograms

Jose A. Lay and Ling Guan

14.1 Introduction

As an increasing amount of multimedia data is distributed, used, and stored in the compressed format, an intuitive approach for lowering the computational complexity toward the implementation of an efficient content-based retrieval application is to propose a scheme that is able to perform retrieval directly in the compressed domain. In this chapter, we show how energy histograms of the lower frequency discrete cosine transform coefficients (LF-DCT) can be used as features for the retrieval of JPEG images and the parsing of MPEG streams. We also demonstrate how the feature set can be designed to cope with changes in image representation due to several common transforms by harvesting manipulation techniques known to the DCT domain.

14.1.1 Multimedia Data Compression

Loosely speaking, multimedia data compression includes every aspect involved in using a more economical representation to denote multimedia data. Hence, to appreciate the significance of multimedia data compression, the types and characteristics of a multimedia datum itself need to be examined.

Multimedia is a generic term used in computing environments. In the digital information processing domain, it refers to a data class assembled by numerous independent data types, such as text, graphics, audio, still images, moving pictures, and other composite data types. A composite data type is a derivative form created when instances of two or more independent data types are combined to form a new medium. Multimedia data are used in a broad range of applications. Sample applications include telemedicine, video telephony, defense object tracking, and Web TV.

Multimedia data can be categorized in numerous ways. With respect to the authoring process, multimedia data can be classified as synthesized and captured media. The classification can also be based on the requirement of presentation where multimedia data are grouped into discrete and continuous media.

Multimedia data have several characteristics. They are generally massive in size, which requires considerable processing power and large storage supports. Storage requirements for several uncompressed multimedia data elements are listed in Table 14.1. Where continuous media are used, multimedia data will also be subjected to presentation constraints such as media synchronization and the requirement for continuity in operation [1]. Consequently, the bandwidth requirement for a synchronized media unit will be the aggregate bandwidth along with the synchronization overhead. For instance, a musical video clip in common intermediate format (CIF) may consist of a series of synchronized video, text, and audio data. This media may need to be continuously played for 3 min, and the aggregate bandwidth requirement excluding the overhead will be a steady throughput of 74.4 Mbps over the 3-min time span. Thus, a system with a storage space of not less than 1.67 GB and a data bus of sustainable 9.3 MB/s will be required to present this media. Although the storage requirement may not seem to be too ambitious, the data rate is certainly beyond the capability of the fastest CD-ROM drive embedded in the current PCs.

Table 14.1 Storage Requirements for Several Uncompressed Multimedia Data Elements

Multimedia Data Type	Sampling Details	Storage
Stereo audio (20–20 KHz)	44,000 samples/s × 2 channels × 16 bit/sample	1.41 Mbps
VGA image (640 × 480)	640 × 480 pixels × 24 bit/pixel	7.37 Mbit/image
Digitized video (NTSC)	720 × 576 pixels/frame × 24 bit/pixel × 30 frames/s	298.59 Mbps
Web TV video (CIF)	352 × 288 pixels/frame × 24 bit/pixel × 30 frames/s	72.99 Mbps

Furthermore, as the Internet facilitates the multimedia data from a workstation onto the widely networked environment, the network bandwidth will also need to match the throughput requirement. The transmission rates for several network standards are listed in Table 14.2. It is clear that the uncompressed multimedia data are scarcely supported by local area networks (LANs), let alone a connection established through the public switched telephone network (PSTN).

Table 14.2 Sample Data Rates of Several Network Standards

Network Technology	Data Rate
Public switched telephone network (PSTN)	0.3–56 Kbps
Integrated services digital network (ISDN)	64–144 Kbps
Telecommunication T-1	1.5 Mbps
Telecommunication T-3	10 Mbps
Local area network (Ethernet)	10 Mbps/100 Mbps

To overcome the bulky storage and transmission bandwidth problems, a compressed form of multimedia data was introduced. Extensive compression may significantly reduce the bandwidth and storage need; however, compression may also degrade the quality of multimedia data, and often the loss is irreversible. Therefore, multimedia data compression may be viewed as a trade-off of the efficiency and quality problem [2, 3]. The data rates of several compressed multimedia data are listed in Table 14.3.

For the last decade, three very successful compression standards on multimedia data elements have been JPEG, MPEG-1, and MPEG-2. JPEG has facilitated the vast distribution of images

Table 14.3 Sample Data Rates of Compressed Data

Compressed Data	Sampling Details	Bandwidth
Channel audio for MPEG	64/128/192 Kbps/channel × N channel	N × 64/128/192 Kbps
Color JPEG image (640 × 480)	640 × 480 pixels × 3 color components × (0.25–2) bit/component sample	0.23–1.84 Mbit/image
MPEG-1 video	360 × 345 pixels/frame × 30 frames/s	1.5 Mbps or higher
H.261 video (CIF)	352 × 288 pixels/frame × (15–30) frames/s	56 Kbps–2 Mbps

on the Internet. MPEG-1 has made possible the storage of a movie title onto a couple of video compact disc (VCD) media. MPEG-2 then extended this achievement to the forms of DVD and HDTV. Their success stories have been prominent ones [4]. Not only have they created many new opportunities in business, but they have also changed the computing experience by shortening the path in multimedia content authoring and usage. Digital images are now directly attainable from digital cameras in the JPEG format. Likewise, MPEG-1 and -2 streams can be straightforwardly obtained from digital videocameras. In short, many more people have been enabled to create and use multimedia contents.

14.1.2 Multimedia Data Retrieval

The various aspects concerned with the enabling of multimedia data accessing are generally termed multimedia data retrieval (MDR). As increasing amounts of multimedia data or their elements become available in the digital format, information technology is expected to provide maximum usability of these data. However, the established text-based indexing schemes have not been feasible to capture the rich content of multimedia data, as subjective annotation may lead to undetectable similarities in the retrieval process. Consequently, content-based retrieval (CBR) was proposed. In addition to textual descriptors, multimedia data are described using their content information; color, texture, shape, motion vector, pitch, tone, etc., are used as features to allow searching to be based on rich content queries. The use of textual descriptors will still be desirable, because they are needed in identifying information that cannot be automatically extracted from multimedia contents, such as name of the author, date of production, etc.

Three basic modules of a CBR system are feature extraction, feature description, and proximity evaluation. Feature extraction deals with how the specific traits of content information can be identified and extracted from the content-rich data. Feature description specifies how those features can be described and organized for efficient retrieval processing. Lastly, proximity evaluation provides the specification in which similarities among contents can be measured based on their features.

The advancement of CBR studies has been remarkable. The current challenge has been the network-wide implementation of these techniques. In past years, many studies on CBR have been conducted, notably in the image and video retrieval domain. Numerous features and their associated description and proximity evaluation schemes have been introduced. A handful of proprietary CBR systems have also been developed. Content-based image and video retrieval is now considered a developed field. However, searching content information across the Internet has not been viable, because no unified feature description scheme has been commonly adopted. For this reason, an effort to standardize the description of content information was initiated. The work was named Multimedia Content Description Interface but is better known as MPEG-7. Surveys on CBR systems and their research issues are given in [5]–[7].

MPEG-7 aims to extend the capabilities of the current CBR systems by normalizing a standard set of descriptors that can be used to describe multimedia contents. MPEG-7 also intends to standardize ways to define other descriptors and their description schemes. MPEG-7 will also standardize a language to specify description schemes [8]. Information on MPEG-7 is available through the MPEG Web site [9].

Figure 14.1 depicts the scope of MPEG-7 [8]. Because the focal interest of MPEG-7 is on the interfacing of descriptors, the feature extraction process and how the features are used in searching on a database will remain an open area for industry competition, since their normalization is not required to allow interoperability.

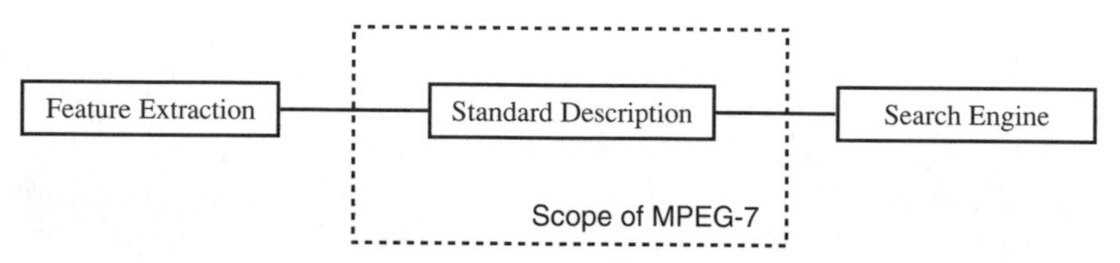

FIGURE 14.1
Diagram block of the proposed MPEG-7 scope.

14.1.3 About This Chapter

The chapter has been written to be self-contained. Background information is included to provide newcomers with the underlying concepts. Readers familiar with those concepts may skim through or skip them. Subjects are discussed within a general perspective. While the primary aim of this chapter concentrates on compressed domain image and video retrieval using the energy histogram features, efforts have been made to present the discussion under the broader theme of MDR, which allows relation with MPEG-7 to be easily established.

More studies on retrieval and processing models are needed to optimize the applicability of MPEG-7. Because only the interfacing of features is covered in MPEG-7, many more research opportunities and needs are left open for feature extraction and search engine studies. Furthermore, while feature extraction has been an active research subject in CBR studies, works on supporting retrieval models have been scarce. Therefore, it is sensible that many more studies on retrieval and processing models are needed before an MPEG-7 optimum search engine can be materialized. A brief discussion on retrieval and processing models as well as the perceived MPEG-7 search engine is consecutively presented in Sections 14.3.1, 14.3.2, and 14.3.3.

Meanwhile, as multimedia data are distributed, used, and stored in the compressed format, the compressed domain technique is highly relevant. The compressed domain technique deals with data directly (or through partial decompression) in their compressed domain, hence avoiding (or reducing) the decompression overhead found in many uncompressed domain schemes.

The DCT domain features are at the center of compressed domain techniques. A noteworthy observation on the many successful multimedia standards is that most of them are based on transform coding using the DCT. Logically, exploitations of DCT domain features are essential in supporting the realization of an efficient current CBR application. The underlying concepts of DCT, the DCT coefficients in JPEG and MPEG data, and the energy histogram are presented in Section 14.2, while an overview of the DCT domain features reported in recent studies is given in Section 14.3.5.

The rest of Section 14.3 will be used to present various aspects related to the proposed retrieval scheme. An MPEG-7 optimum search engine construct is presented in Section 14.3.3. The DCT domain manipulation techniques are covered in Section 14.3.4, while the energy histograms of the LF-DCT coefficient features are presented in Section 14.3.5. Several proximity evaluation schemes are discussed in Section 14.3.6, and the experimental results for the retrieval of JPEG images and the parsing of MPEG streams are provided in Section 14.3.7. Conclusions are given in Section 14.4.

14.2 The DCT Coefficient Domain

Before we describe how the DCT coefficients can be used as potent features for retrieving a JPEG image and/or parsing an MPEG stream, we shall first enunciate the basic concept underlying the DCT domain. We will embark by giving an alternative explanation of the DCT using the matrix notation, then go on to show how the DCT coefficients reside in the JPEG and MPEG data, and finally articulate the notion of the energy histograms of the DCT coefficients.

14.2.1 A Matrix Description of the DCT

DCT was first introduced in 1974 [10]. It is now the major building block of many very popular image and video compression standards. Together with the vast development of semiconductors, DCT-powered compression standards have delivered a magnificent computing environment, an internetworked world rich with multimedia contents.

The 8×8 block forward and inverse 2D **DCT**s used in JPEG and MPEG are given by

Forward 2D DCT:

$$f(u, v) = \frac{1}{4} CuCv \sum_{i=0}^{7} \sum_{j=0}^{7} s(i, j) \cos \frac{(2i+1)u\pi}{16} \cos \frac{(2j+1)v\pi}{16}$$

Inverse 2D DCT:

$$s(i, j) = \frac{1}{4} \sum_{u=0}^{7} \sum_{v=0}^{7} CuCvf(u, v) \cos \frac{(2i+1)u\pi}{16} \cos \frac{(2j+1)v\pi}{16}$$

where $C\tau = 1/\sqrt{2}$ for $\tau = 0$ and $C\tau = 1$ for $\tau \neq 0$. The $f(u, v)$ are the so-called DCT coefficients and $s(i, j)$ are the values of the i, j input samples.

Since the 2D DCT is attainable by concatenating two 1D DCTs, we will use the latter to convey the purpose of this section. Thus, we can reveal the concept without dealing with too many indices in the equation. The formal definition for an 8-element 1D DCT is given by

Forward 1D DCT:

$$f(u) = \frac{1}{2} Cu \sum_{i=0}^{7} s(i) \cos \frac{(2i+1)u\pi}{16}$$

where $Cu = 1/\sqrt{2}$ for $u = 0$ and 1 otherwise, $f(u)$ are the 8-element 1D DCT coefficients, and $s(i)$ are the 8 input elements.

Thinking in vector terms, we can rewrite the transform using a matrix notation by arranging $f(u)$ and $s(i)$ and substituting values for $(2i + u)$:

$$\begin{bmatrix} f0 \\ f1 \\ f2 \\ f3 \\ f4 \\ f5 \\ f6 \\ f7 \end{bmatrix} = \frac{1}{2} Cu \cos \begin{bmatrix} 0 & 0 & 0 & 0 & 0 & 0 & 0 & 0 \\ \frac{\pi}{16} & \frac{3\pi}{16} & \frac{5\pi}{16} & \frac{7\pi}{16} & \frac{9\pi}{16} & \frac{11\pi}{16} & \frac{13\pi}{16} & \frac{15\pi}{16} \\ \frac{2\pi}{16} & \frac{6\pi}{16} & \frac{10\pi}{16} & \frac{14\pi}{16} & \frac{18\pi}{16} & \frac{22\pi}{16} & \frac{26\pi}{16} & \frac{30\pi}{16} \\ \frac{3\pi}{16} & \frac{9\pi}{16} & \frac{15\pi}{16} & \frac{21\pi}{16} & \frac{27\pi}{16} & \frac{33\pi}{16} & \frac{39\pi}{16} & \frac{45\pi}{16} \\ \frac{4\pi}{16} & \frac{12\pi}{16} & \frac{20\pi}{16} & \frac{28\pi}{16} & \frac{36\pi}{16} & \frac{44\pi}{16} & \frac{52\pi}{16} & \frac{60\pi}{16} \\ \frac{5\pi}{16} & \frac{15\pi}{16} & \frac{25\pi}{16} & \frac{35\pi}{16} & \frac{45\pi}{16} & \frac{55\pi}{16} & \frac{65\pi}{16} & \frac{75\pi}{16} \\ \frac{6\pi}{16} & \frac{18\pi}{16} & \frac{30\pi}{16} & \frac{42\pi}{16} & \frac{54\pi}{16} & \frac{66\pi}{16} & \frac{78\pi}{16} & \frac{90\pi}{16} \\ \frac{7\pi}{16} & \frac{21\pi}{16} & \frac{35\pi}{16} & \frac{49\pi}{16} & \frac{63\pi}{16} & \frac{77\pi}{16} & \frac{91\pi}{16} & \frac{105\pi}{16} \end{bmatrix} \begin{bmatrix} s0 \\ s1 \\ s2 \\ s3 \\ s4 \\ s5 \\ s6 \\ s7 \end{bmatrix}$$

Let us denote the $f(u)$ vector with **f**, the cosine function matrix as **K**, and the $s(i)$ vector with **s**. We have:

$$\mathbf{f} = \frac{1}{2} Cu \mathbf{K} \mathbf{s} .$$

Note that we have chosen to write **f** and **s** as column vectors in the equation. Intuitively, the matrix notation shows that a DCT coefficient $f(u)$ is simply a magnitude obtained by multiplying a signal vector (**s**) with several scaled discrete cosine values distanced at certain multiples of $\pi/16$ frequency. Therefore, calculating the DCT coefficients of a particular signal is essentially carrying out the frequency decomposition [12] or, in a broader sense, the content decomposition of that signal.

Each row in the cosine function matrix represents the basis function of a specific decomposition frequency set. To help visualize this concept, we will reconstruct the cosine matrix by explicitly calculating their cosine values. Furthermore, since $\pi/16$ is a factor common to all elements, the trigonometric rules allow the new matrix to be rewritten using only references to the values of the first quadrant components. By doing so, we hope to communicate the idea without getting involved with long decimal elements in the matrix. We denote the first quadrant components of the **K** matrix as:

$\cos 0$	$\cos \pi/16$	$\cos 2\pi/16$	$\cos 3\pi/16$	$\cos 4\pi/16$	$\cos 5\pi/16$	$\cos 6\pi/16$	$\cos 7\pi/16$
$a0$	$a1$	$a2$	$a3$	$a4$	$a5$	$a6$	$a7$

The new matrix may be rewritten as:

$$\begin{bmatrix} f0 \\ f1 \\ f2 \\ f3 \\ f4 \\ f5 \\ f6 \\ f7 \end{bmatrix} = \frac{1}{2} Cu \begin{bmatrix} a0 & a0 & a0 & a0 & a0 & a0 & a0 & a0 \\ a1 & a3 & a5 & a7 & -a7 & -a5 & -a3 & -a1 \\ a2 & a6 & -a6 & -a2 & -a2 & -a6 & a6 & a2 \\ a3 & -a7 & -a1 & -a5 & a5 & a1 & a7 & -a3 \\ a4 & -a4 & -a4 & a4 & a4 & -a4 & -a4 & a4 \\ a5 & -a1 & a7 & a3 & -a3 & -a7 & a1 & -a5 \\ a6 & -a2 & a2 & -a6 & -a6 & a2 & -a2 & a6 \\ a7 & -a5 & a3 & -a1 & a1 & -a3 & a5 & -a7 \end{bmatrix} \begin{bmatrix} s0 \\ s1 \\ s2 \\ s3 \\ s4 \\ s5 \\ s6 \\ s7 \end{bmatrix}$$

Note that the occurrence of sign changes increases as we move downward along the matrix rows. Row 0 has no sign changes, since $a0 = \cos 0 = 1$ for every element in that row. However, row 1 has one sign change, row 2 has two sign changes, and so on. The sign changes within a basis function basically indicate the zero-crossings of the cosine waveform. Thus, as

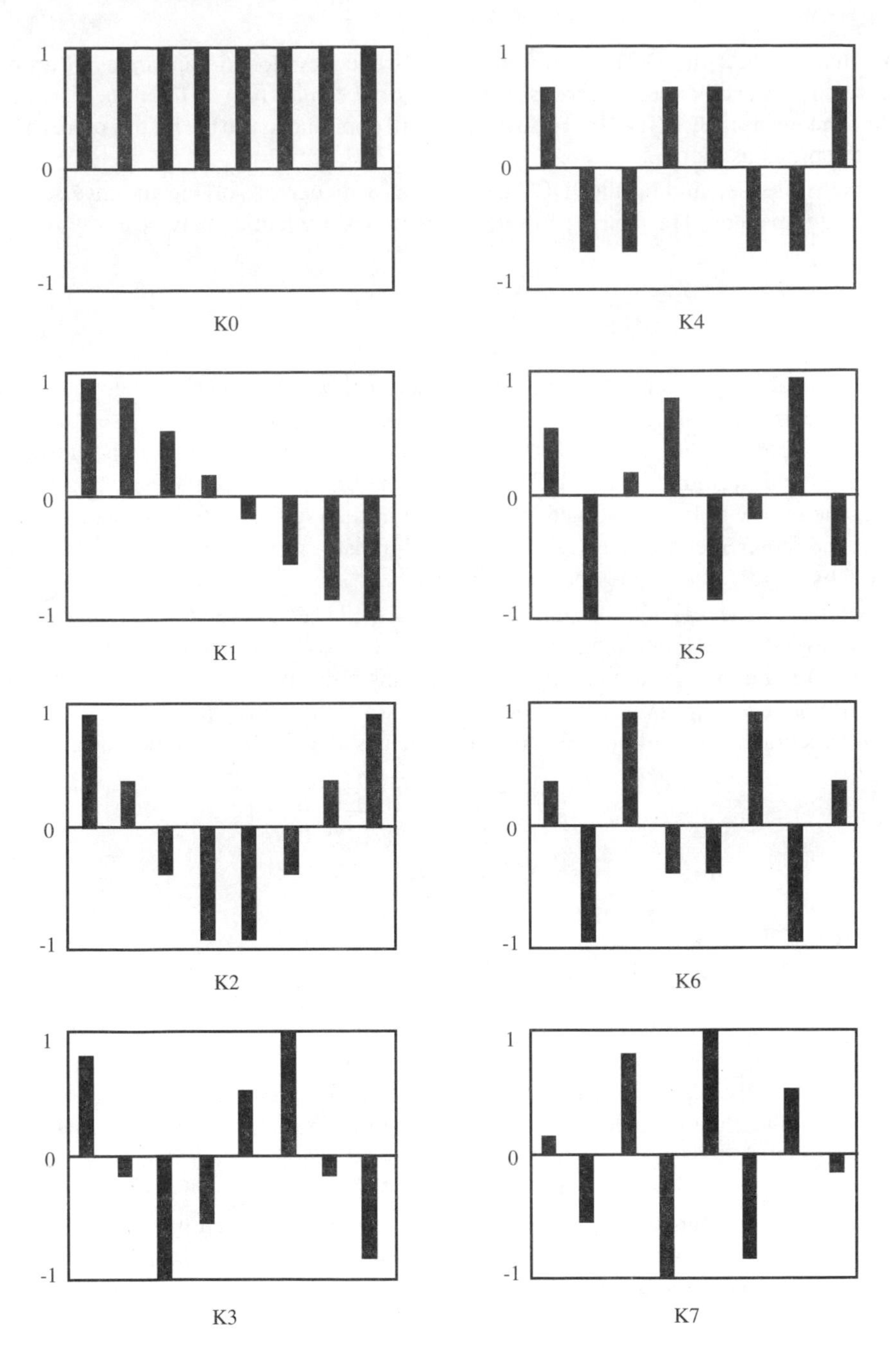

FIGURE 14.2
Eight DCT cosine basis function waveforms.

the occurrence of the sign change intensifies, the frequency of the waveform increases. The eight basis function waveforms associated with matrix **K** are shown in Figure 14.2.

Since the first cosine basis function (K0) has no alternating behavior, the DCT coefficient associated with this basis function is usually dubbed as the DC coefficient, referring to the abbreviation used in electrical engineering for the direct current. Consequently, the other DCT coefficients are called AC coefficients.

Now that we view the DCT coefficients as the frequency domain apparatus of a time or spatial signal, we shall expect to reconstruct the original signal from its DCT coefficients, that is, to find the inverse DCT (IDCT). To do this, we will continue to use the matrix representation built in the previous section.

In a matrix sense, finding the IDCT can be just a matter of solving the inverse for the transforming operator. The matrix equivalent for the IDCT equation may be easily written as

$$\mathbf{s} = \left(\frac{1}{2}Cu\mathbf{K}\right)^{-1}\mathbf{f}$$

which is basically a problem of finding an inverse for the scaled matrix. Note that we have neglected the $1/2Cu$ scaling constants in the discussion so far. We do so because it really does not inhibit us from passing on the ideas of frequency or content decomposition. However, as we move on with formulating the IDCT equation, the importance of these scaling constants will become relevant. In fact, a major reason for introducing the scaling constants is largely based on the requirement for having an orthogonal transforming matrix. So one can achieve the IDCT by merely transposing the matrix $1/2Cu\mathbf{K}$.

To demonstrate the role of the scaling constants, we will need to explore some characteristics of the basis functions. Examining the matrix **K**, we can see that each of the basis functions (matrix-row) is orthogonal to the others. This means a dot product of any two rows in the matrix will yield zero. However, none of the basis functions is orthonormal. In other words, the basis functions are not of unit vectors. Nevertheless, the results are almost as good.

$$\mathbf{V} = \mathbf{K}\mathbf{K}^T = \begin{bmatrix} 8 & 0 & 0 & 0 & 0 & 0 & 0 & 0 \\ 0 & 4 & 0 & 0 & 0 & 0 & 0 & 0 \\ 0 & 0 & 4 & 0 & 0 & 0 & 0 & 0 \\ 0 & 0 & 0 & 4 & 0 & 0 & 0 & 0 \\ 0 & 0 & 0 & 0 & 4 & 0 & 0 & 0 \\ 0 & 0 & 0 & 0 & 0 & 4 & 0 & 0 \\ 0 & 0 & 0 & 0 & 0 & 0 & 4 & 0 \\ 0 & 0 & 0 & 0 & 0 & 0 & 0 & 4 \end{bmatrix}$$

Note that each of the diagonal elements in **V** equals the dot product of a basis function with itself. $V_{0,0}$ is directly attainable in the form of the sum of squares of the first basis function ($K0$), whereas the others are computed by application of trigonometric identities.

Equally important are the zero-value elements in **V**, which indicate that orthogonality does exist among the basis function vectors. However, having orthogonal row vectors alone is not sufficient for **K** to become an orthogonal matrix. An orthogonal matrix also requires all of its row vectors (column vectors) to be of unit length, so the product of the matrix with its transpose ($\mathbf{K}\mathbf{K}^T$) can produce an identity matrix (**I**).

Since **K** is a square matrix and the orthogonal property exists among its basis function vectors, the only task left is to turn them into unit vectors, which can be realized simply by scaling each of the basis functions with its length:

$$\mathbf{K}_{iu} = \frac{\mathbf{K}i}{\|\mathbf{K}i\|} \qquad (i = 0, 1, \ldots, 7) \quad \text{and} \quad \|\mathbf{K}_{iu}\| = 1 .$$

Simple arithmetic yields $\|\mathbf{k}0\| = \sqrt{8} = 2\sqrt{2}$ and $\|\mathbf{k}1\| \ldots \|\mathbf{k}7\| = \sqrt{4} = 2$. Therefore, to make **K** orthogonal, we shall scale the first basis function ($K0$) by a factor of $1/\left(2\sqrt{2}\right)$, while the others need only to be divided by 2. Separating the $1/2$ factor from the elements, $\mathbf{K}_c$ can be devised as:

$$\mathbf{K}_c = \frac{1}{2}\begin{bmatrix} c0 & c0 & c0 & c0 & c0 & c0 & c0 & c0 \\ a1 & a3 & a5 & a7 & -a7 & -a5 & -a3 & -a1 \\ a2 & a6 & -a6 & -a2 & -a2 & -a6 & a6 & a2 \\ a3 & -a7 & -a1 & -a5 & a5 & a1 & a7 & -a3 \\ a4 & -a4 & -a4 & a4 & a4 & -a4 & -a4 & a4 \\ a5 & -a1 & a7 & a3 & -a3 & -a7 & a1 & -a5 \\ a6 & -a2 & a2 & -a6 & -a6 & a2 & -a2 & a6 \\ a7 & -a5 & a3 & -a1 & a1 & -a3 & a5 & -a7 \end{bmatrix}$$

with $c0 = a0/\sqrt{2} = 1/\sqrt{2} = a4$. $\mathbf{K}_c$ is an orthogonal matrix, where $\mathbf{K}_c^{-1} = \mathbf{K}_c^T$, so $\mathbf{K}_c\mathbf{K}_c^T = \mathbf{K}_c^T\mathbf{K}_c = \mathbf{I}$. It is quite interesting to see that we have just turned the highly sophisticated DCT into the down-to-earth $\mathbf{A}\boldsymbol{x} = \boldsymbol{b}$ linear equation:

$$\begin{aligned} &\text{Forward DCT:} && \mathbf{f} = \mathbf{K}_c\mathbf{s} \\ &\text{Inverse DCT:} && \mathbf{s} = \mathbf{K}_c^{-1}\mathbf{f} \Rightarrow \mathbf{s} = \mathbf{K}_c^T\mathbf{f} \end{aligned}$$

Treating the DCT as linear equations not only allows us to eagerly derive the IDCT term, but also clearly presents some of its most prominent properties by using solely the linear algebra concept.

From the linear algebra point of view, the DCT is an orthogonal linear transformation. A linear transformation means the DCT can preserve the vector addition and scalar multiplication of a vector space. Thus, given any two vectors ($\mathbf{p}$ and $\mathbf{q}$) and a scalar (α), the following relations are true:

$$f(\mathbf{p}+\mathbf{q}) = f(\mathbf{p}) + f(\mathbf{q}),$$
$$f(\alpha\mathbf{p}) = \alpha f(\mathbf{p}).$$

Linearity is useful when dealing with frequency domain image manipulation. Several simple techniques of the DCT frequency domain image manipulation are discussed in Section 14.3.3. Meanwhile, the orthogonal term implies that the lengths of the vectors will be preserved subsequent to a DCT transformation. For an input vector $\mathbf{s} = [a, b, c, d, e, f, g]$, and the transformed vector $\mathbf{f}$, we have

$$\|\mathbf{s}\| = \|\mathbf{f}\| = \sqrt{a^2+b^2+c^2+d^2+e^2+f^2+g^2+h^2}$$

$$\text{as} \quad \|\mathbf{f}\|^2 = \|\mathbf{K}_c\mathbf{s}\|^2 = (\mathbf{K}_c\mathbf{s})^T(\mathbf{K}_c\mathbf{s}) = \mathbf{s}^T\left(\mathbf{K}_c^T\mathbf{K}_c\right)\mathbf{s} = \mathbf{s}^T\mathbf{s} = \|\mathbf{s}\|^2.$$

This characteristic is often referred to as the energy conservation property of the DCT.

Another important property of being an orthogonal transform is that the product of two or more orthogonal matrices will also be orthogonal. This property then enables a higher dimensional DCT and its inverse to be performed using less complicated lower dimensional DCT operations. An example for performing the 8×8 block 2D DCT and its inverse through the 8-element DCT is given below:

Forward 2D DCT:

$$\mathbf{f}_{2D} = \mathbf{K}_c\mathbf{s}\,\mathbf{K}_c^T$$

Inverse 2D DCT:

$$\mathbf{s}_{2D} = \mathbf{K}_c^T\mathbf{f}\,\mathbf{K}_c$$

Now that we have described how the 2D DCT can be accomplished by successive operations of 1D DCT, we shall turn our attention to the most interesting behavior of the DCT known as the energy packing property, which indeed has brought the DCT coefficients into the heart of several widely adapted compression techniques of the decade.

Even though the total energy of the samples remains unaffected subsequent to a DCT transform, the distribution of the energy will be immensely altered. A typical 8×8 block transform will have most of the energy relocated to its upper-left region, with the DC coefficient ($\mathbf{f}_{00}$) representing the scaled average of the block and the other AC coefficients denoting the intensity of edges corresponding to the frequency of the coefficients. Figure 14.3 depicts the energy relocation that occurred in the transform of a typical 8×8 image data block.

$$\underbrace{\begin{bmatrix} 10 & 10 & 10 & 10 & 10 & 10 & 10 & 10 \\ 1 & 1 & 1 & 1 & 1 & 1 & 1 & 1 \\ 1 & 1 & 1 & 1 & 1 & 1 & 1 & 1 \\ 1 & 1 & 1 & 1 & 1 & 1 & 1 & 1 \\ 1 & 1 & 1 & 1 & 1 & 1 & 1 & 1 \\ 1 & 1 & 1 & 1 & 1 & 1 & 1 & 1 \\ 1 & 1 & 1 & 1 & 1 & 1 & 1 & 1 \\ 10 & 10 & 10 & 10 & 10 & 10 & 10 & 10 \end{bmatrix}}_{\text{An Image Data Block}} \underset{\text{2D DCT}}{\Rightarrow} \underbrace{\begin{bmatrix} 26 & 0 & 0 & 0 & 0 & 0 & 0 & 0 \\ 0 & 0 & 0 & 0 & 0 & 0 & 0 & 0 \\ 24 & 0 & 0 & 0 & 0 & 0 & 0 & 0 \\ 0 & 0 & 0 & 0 & 0 & 0 & 0 & 0 \\ 18 & 0 & 0 & 0 & 0 & 0 & 0 & 0 \\ 0 & 0 & 0 & 0 & 0 & 0 & 0 & 0 \\ 10 & 0 & 0 & 0 & 0 & 0 & 0 & 0 \\ 0 & 0 & 0 & 0 & 0 & 0 & 0 & 0 \end{bmatrix}}_{\text{DCT Coefficients}}$$

FIGURE 14.3
Energy packing property.

Now that we have presented the fundamental idea of the DCT in the matrix dialect, we would like to end this section by taking just another step to rewrite the IDCT in its formal vein:

Inverse 1D DCT:

$$\mathbf{s} = \mathbf{K}_c^{-1}\mathbf{f} = \mathbf{K}_c^T\mathbf{f}$$

$$s(i) = \sum_{i=0}^{7} \frac{1}{2} Cuf(u) \cos \frac{(2i+1)u\pi}{16} \qquad \text{where} \qquad \begin{cases} Cu = \frac{1}{\sqrt{2}} & \text{for} \quad u = 0 \\ Cu = 1 & \text{for} \quad u > 0 \end{cases}$$

14.2.2 The DCT Coefficients in JPEG and MPEG Media

Video can be viewed as a sequence of images updated at a certain rate. This notion is also valid in the compressed data domain. For instance, a sequence of JPEG images can be used to constitute a motion JPEG (M-JPEG) video stream. In fact, many popular video compression standards including MPEG-1, MPEG-2, H.261, and H.263 are built upon the DCT transform coding techniques developed and used in JPEG. Therefore, the topic of DCT coefficients in MPEG media will be best described after that of JPEG is presented.

The JPEG standard acknowledges two classes of encoding and decoding processes known as lossy and lossless JPEG. Lossy JPEG is based on the energy packing characteristic of DCT and includes mechanisms where a certain amount of information may be irreversibly lost subsequent to its coding processes. Lossy JPEG is able to achieve substantial compression rates. Its modes of operation are further divided into baseline, extended progressive, and extended hierarchical. Conversely, lossless JPEG is based on predictive algorithms where content information can be fully recovered in a reconstructed image. However, lossless JPEG can attain only a moderate compression rate. Because lossless JPEG is based on non-DCT-based algorithms, only lossy JPEG is of interest in this chapter.

Figure 14.4 shows a block diagram of the lossy JPEG codec structure. In the encoding process the spatial image data are grouped into a series of 8 (pixel) $\times$ 8 (pixel) blocks. Each of these blocks is then fed into a forward 2D DCT to produce the 64 DCT coefficients. The blocks are processed in a sequence from left to right and from top to bottom. The DCT coefficients are then scalarly quantized using a quantization factor set in a quantization table [12]:

$$f_q(u, v) = \text{round}\left[\frac{f(u, v)}{Q(u, v)}\right]$$

where $f(u, v)$, $f_q(u, v)$, and $Q(u, v)$ are the DCT coefficients being quantized, their quantized values, and the quantization factors provided in the quantization table, respectively.

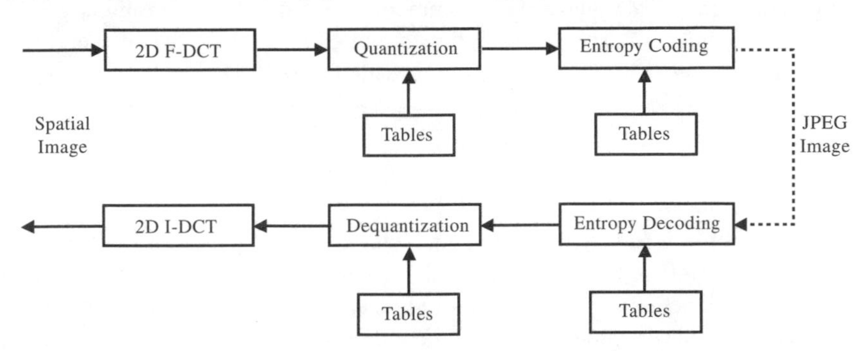

FIGURE 14.4
A JPEG codec structure.

The quantization step is the lossy part of lossy JPEG. Quantization is primarily employed to prune the higher frequency coefficients by dividing them with larger factors. Thus, variations of quantization tables can be used to tune the desirable compression ratio. However, because a rounding-off operation is involved in every quantization process, quantized coefficients may be subjected to irreversible loss of information. Therefore, quantization tables need to be specifically designed so that the quality degradation is still in the tolerable range. Separate quantization tables are used for the luminance and chrominance components of JPEG data. Two quantization tables provided in the JPEG standard are tabulated in Figure 14.5.

$$\begin{bmatrix} 16 & 11 & 10 & 16 & 24 & 40 & 51 & 61 \\ 12 & 12 & 14 & 19 & 26 & 58 & 60 & 55 \\ 14 & 13 & 16 & 24 & 40 & 57 & 69 & 56 \\ 14 & 17 & 22 & 29 & 51 & 87 & 80 & 62 \\ 18 & 22 & 37 & 56 & 68 & 109 & 103 & 77 \\ 24 & 35 & 55 & 64 & 81 & 104 & 113 & 92 \\ 49 & 64 & 78 & 87 & 103 & 121 & 120 & 101 \\ 72 & 92 & 95 & 98 & 112 & 100 & 103 & 99 \end{bmatrix}$$

Luminance Quantization Table

$$\begin{bmatrix} 17 & 18 & 24 & 47 & 99 & 99 & 99 & 99 \\ 18 & 21 & 26 & 66 & 99 & 99 & 99 & 99 \\ 24 & 26 & 56 & 99 & 99 & 99 & 99 & 99 \\ 47 & 66 & 99 & 99 & 99 & 99 & 99 & 99 \\ 99 & 99 & 99 & 99 & 99 & 99 & 99 & 99 \\ 99 & 99 & 99 & 99 & 99 & 99 & 99 & 99 \\ 99 & 99 & 99 & 99 & 99 & 99 & 99 & 99 \\ 99 & 99 & 99 & 99 & 99 & 99 & 99 & 99 \end{bmatrix}$$

Chrominance Quantization Table

FIGURE 14.5
Quantization tables.

Upon quantization, the 8×8 DCT coefficients within a block are arranged in a zigzag order. Since the DC coefficients tend to be highly correlated among the adjacent data blocks, the difference of two consecutive DC coefficients (rather than an actual DC value) is coded to enhance the compression ratio, whereas the other AC coefficients are run-length coded to remove the "zeros" redundancy. These semicoded coefficients are then further entropy coded using Huffman or arithmetic coding techniques to produce the final JPEG bits. Conversely, on the decoding side, inverse operations of the encoding processes are performed.

In addition to the DCT-based transform coding mechanisms, color subsampling is used in JPEG to further enhance the compression rate. It is understood that human eyes are more sensitive to brightness (luminance) than to color (chrominance). Therefore, certain color information may be arbitrarily reduced from a color image without generating significant quality losses to human perceptions. Consequently, the YUV or YC_bC_r (rather than RGB) color representation system is adopted in JPEG and MPEG. The luminance (Y) component represents a gray-scale version of the image. The chrominance components (UV) are used to add color to the gray-scale image. One commonly used subsampling ratio is 4:2:2, which means that the luminance of each pixel is sampled while the chrominance of every two pixels is sampled. Several useful JPEG resources are provided in [13]–[15].

Because the updating rate of a video sequence is normally not less than tens of images per second, adjacent images in a video stream may be expected to be in high correlation. Therefore, temporal coding techniques can be used on top of the spatial coding to further enhance the compression performance.

Relying on both, MPEG adopted the intra- and inter-coding schemes for its data. An MPEG stream consists of I (intra), P (predictive), and B (bidirectional) coded frames. An I frame is an intra-coded independent frame. Spatial redundancy on independent frames is removed by the DCT coding where a coded image can be independently decoded. P and B frames are inter-coded reference frames. Temporal redundancy on reference frames is detached by the means of motion estimation. A P frame is coded based on its preceding I or P frame, while a B frame is coded using both of the preceding and the following I and/or P frames. Therefore, decoding a reference frame may depend on one or more related frames. A fine survey of the current and emerging image and video coding standards is presented in [16].

M-JPEG is an extension of JPEG to cope with moving pictures where each frame of a video stream is compressed individually using the JPEG compression technique. The independent compression allows easy random access to be performed on an M-JPEG stream, thus enabling M-JPEG to enjoy much popularity in the nonlinear video editing application.

14.2.3 Energy Histograms of the DCT Coefficients

Histogram techniques were originally introduced into the field of image retrieval in the form of color histograms [17]. A color (gray-level) histogram of a digital image is formed by counting the number of times a particular color (intensity) occurs in that image.

$$h[i] = n_i \qquad \begin{cases} h[i] = \text{color histogram of color } i \\ n_i = \text{number of times color } i \text{ occurs in the image} \end{cases}$$

Since color images are normally presented in a multidimensional color space (e.g., RGB or YUV), color histograms can be defined using either a multidimensional vector or several one-dimensional vectors.

The color histogram of an image is a compelling feature. As a global property of color distribution, color histograms are generally invariant to translation and perpendicular rotations. They can also sustain modest alterations of viewing angle, changes in scale, and occlusion [17]. Their versatility may also be extended to include scaling invariance through the means of normalization [18]. However, color histograms are intolerant to the changes of illumination. A small perturbation in the illumination may contribute to considerable differences in histogram data.

Similar to color histograms, an energy histogram of the DCT coefficients is obtained by counting the number of times an energy level appears in the DCT coefficient blocks of a DCT compressed image. Thus, the energy histograms for a particular color component (h_c) in an 8

$\times$ 8 DCT data block can be written as

$$h_c[t] = \sum_{u=0}^{7} \sum_{v=0}^{7} \begin{cases} 1 & \text{if } E(f[u,v]) = t \\ 0 & \text{otherwise} \end{cases}$$

where $E(f[u, v])$ denotes the energy level of the coefficient at the (u, v) location and t is the particular energy bin.

As with color histograms, energy histograms are generally tolerant to rotations and modest object translations.

14.3 Frequency Domain Image/Video Retrieval Using DCT Coefficients

Frequency domain CBR offers twofold advantages. Computational complexity issues introduced by the discrepancy of spatial domain (uncompressed) feature schemes and frequency domain (compressed) data can be a hindrance in implementing an efficient CBR application, especially on the real-time platform. The frequency domain CBR approach is able to reduce the complexity by processing the compressed data directly or through partial decompression in their frequency domain. Furthermore, direct processing of compressed data allows the retrieval system to operate on rather compact data, which are beneficial in terms of computation resources and network-wide processing.

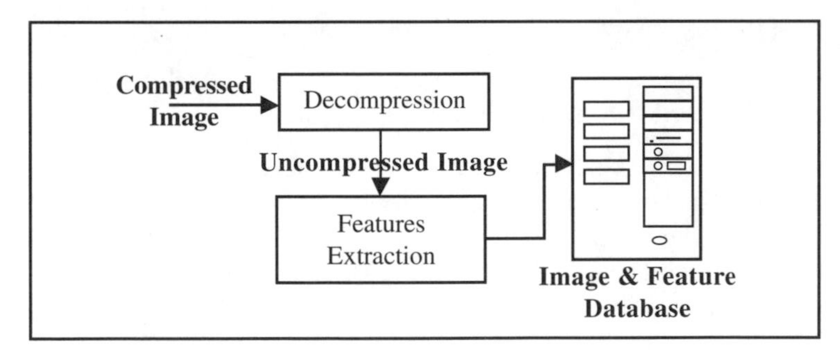

FIGURE 14.6
Extracting uncompressed domain features from a compressed image.

Nevertheless, many more studies are needed before a full-fledged frequency domain CBR system can be materialized. Because conventional features and processing techniques may not be directly accessible in the compressed domain, exploration of new frequency domain features and processing techniques is becoming mandatory.

Repetition and modified variations of data are common in network environments. Since data sharing is likely in network computing, modified copies of a content are expected across network-based databases. For instance, image databases may contain many images that may differ only in their visual representation. These images are often of basic transformed operations (e.g., mirroring, transposing, or rotating). Therefore, detecting similarities on those transformed images is pertinent to a network-based CBR system.

Later in this section, the energy histograms of LF-DCT coefficients are used as features for retrieval of JPEG images (based on the query by model method) as well as for parsing of MPEG videos. The targeted retrieval scheme is desired to be able to support network- and MPEG-7-based implementation. Therefore, current content-based retrieval and processing models and their requirements are studied. An MPEG-7 optimum search engine construct is also presented. Image manipulation techniques in the DCT domain are examined with regard to the building of limited transformed variant proof features.

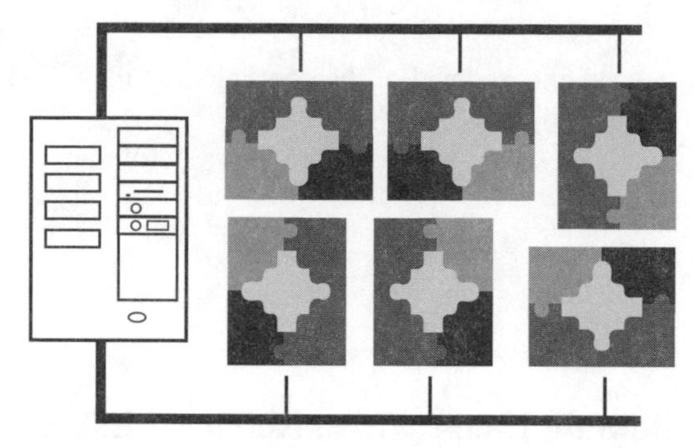

FIGURE 14.7
Transformed variants are common in network databases.

In video applications, studies have shown that the DC coefficients can be used to detect abrupt scene changes. However, the use of DC coefficients alone does not provide a robust method for parsing of more complex video sequences such as ones with luminance changes and/or dissolving transitions. The energy histogram features are used to enhance the segmentation of DCT-based video. Experimental results for video parsing of MPEG streams along with the retrieval of JPEG images are presented in Section 14.3.7, while a CBR model for content-based video retrieval is briefly described in Section 14.3.1.

14.3.1 Content-Based Retrieval Model

The current CBR model is characterized by a separate feature database. To avoid the high computational cost posted by the uncompressed feature techniques, many of the current CBR systems are built on a dual-database model where a pair of independent databases are used to catalogue features and data [18, 19]. Figure 14.8 shows the dual-database CBR model used in image retrieval applications. The independent feature database is built in addition to the image database itself during the setup phase. Proximity evaluation can be performed by contrasting the extracted features of a query with the records maintained in the feature database. When matches are obtained, the associated image data are returned from the image database. Therefore, the dual-database model is also known as the off-line or indexing model, because off-line feature extraction and pre-indexing processing are required during a database formation.

The dual-database model is advantageous from several perspectives. Since the structure of the dual-database model is comparable to the general indexing system used in text-based databases, this model may enjoy the support of many established techniques and developed tools. Furthermore, because features are pre-extracted during database creation, conventional spatial domain techniques may be used without causing high computational complexities at run time. The dual-database model also fits well with the needs of the video retrieval application, where features from key frames representing the segmented video shots are extracted for indexing use. The content-based video retrieval model is discussed later in this section.

Nevertheless, there are also drawbacks attached to the dual-database model. Because searching in a dual-database CBR system is performed on the pre-extracted feature sets, the query's features have to conform with the feature scheme used by the feature database. Consequently, choices of features are determined by the in-search feature database. Moreover, universal

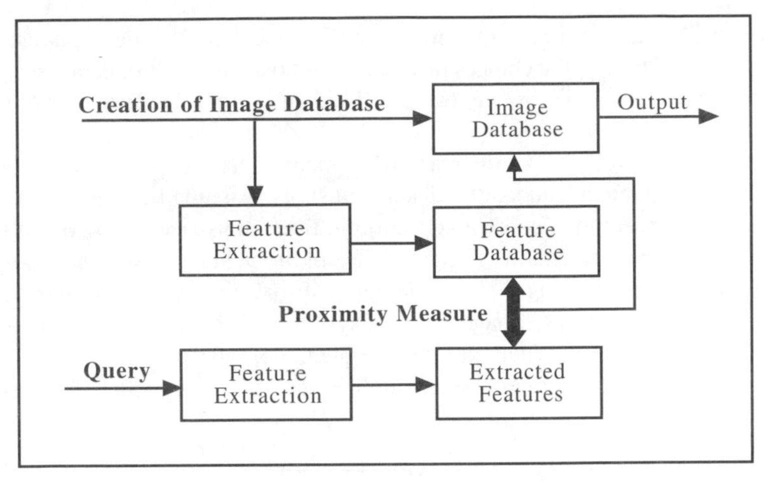

FIGURE 14.8
The dual-database content-based image retrieval model.

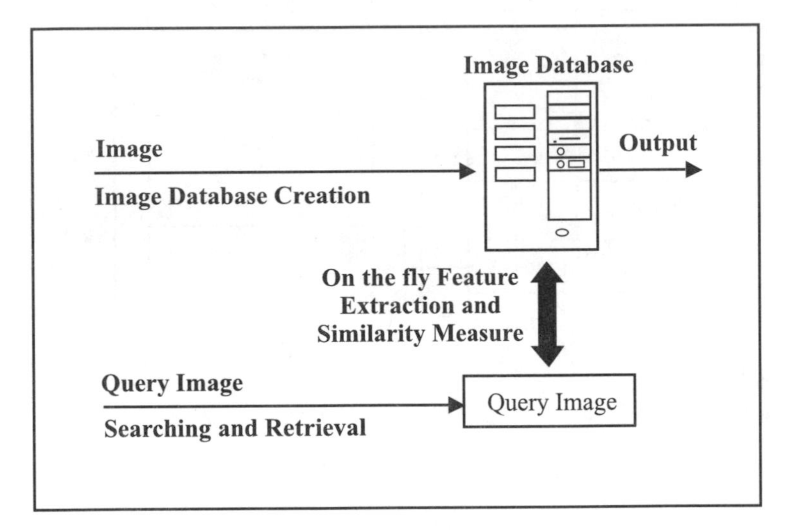

FIGURE 14.9
The single-database content-based image retrieval model.

searching across the Internet would also be impracticable until the unified description sought by MPEG-7 is widely implemented.

Alternatively, the single-database CBR model [20] can be employed. The single-database CBR model used in image retrieval applications is illustrated in Figure 14.9. In such a model, no preprocessing is required during database construction. Features are extracted on the fly within a retrieval cycle directly from data. Therefore, rapid feature extraction and proximity evaluation are obligatory to the single-database systems. Because feature extraction and proximity evaluation are executed on the fly within a retrieval cycle, the single-database CBR model is also known as the online CBR model.

As with the dual-database model, the single-database model also has upsides and downsides. It is practical for compressed domain-based retrieval (pull application) and filtering (push application), especially when content-based coding such as that of MPEG-4 is used. It also supports ad hoc Internet-wide retrieval implementations because raw compressed data can be read and processed locally at the searcher machine. This local processing of feature extraction will unlock the restriction of the choices of features imposed by feature databases. However, sending raw compressed data across a network is disadvantageous, because it tends to generate high traffic loads.

Video retrieval is generally more efficient to implement with the dual-database model. Video streams are segmented into a number of independent shots. An independent shot is a sequence of image frames representing a continuous action in time and space. Subsequent to the segmentation, one or more representative frames of each of the segmented sequences are extracted for use as key frames in indexing the video streams. Proximity evaluations can then be performed as that of a dual-database image retrieval system (i.e., by contrasting the query frame with each of the key frames). When matches are obtained, relevant video shots are returned from the video database. The structure of a simplified content-based video database is shown in Figure 14.10.

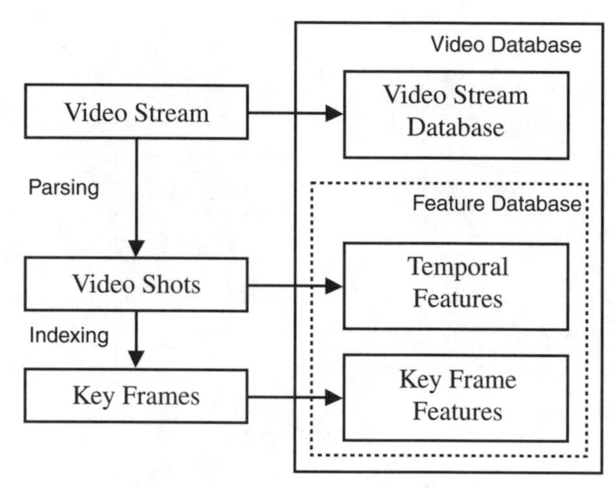

FIGURE 14.10
Structure of a simplified content-based video database.

14.3.2 Content-Based Search Processing Model

From the search processing perspective, two fundamental models can be associated with the dual-database and single-database CBR applications. For the dual-database systems, search processing is normally performed on and controlled by the database-in-search. Thus, the associated search processing model is termed the archivist processing model, since proximity evaluation is executed on the archivist environments. The model is also known as the client–server model, because all the processing and know-how is owned and offered by the archivist (server) to the searcher (client). Conversely, on a single-database system, search processing is normally performed on and controlled by the search initiator. Therefore, the associated search processing model is termed the searcher processing model. The archivist processing model and the searcher processing model are illustrated in Figures 14.11a and b, respectively.

The current search processing models are unsatisfactory. The client–server model is undesirable because all the knowledge on how a search is performed is owned and controlled by the archivist server. The searcher processing model is impractical because its operation may involve high network traffic.

Alternatively, a paradigm called the search agent processing model (SAPM) [22] can be employed. The SAPM is a hybrid model built upon the mobile agent technology. Under

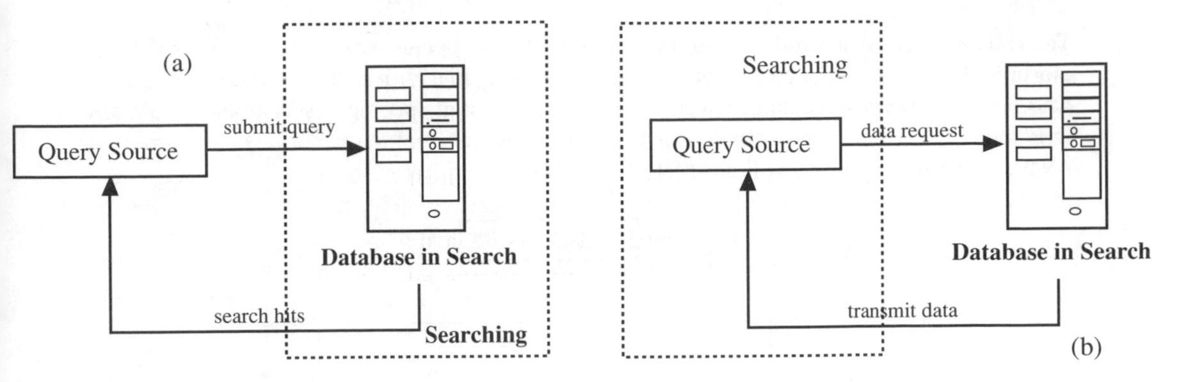

FIGURE 14.11
(a) Archivist processing model; (b) searcher processing model.

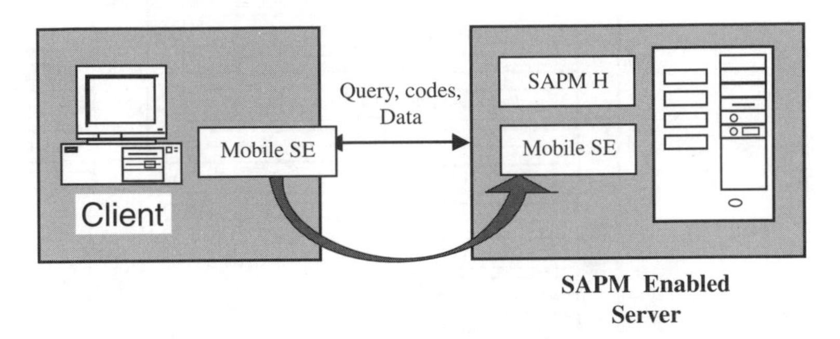

FIGURE 14.12
The search agent processing model (SAPM).

the SAPM, an agent (a traveling program) can be sent to perform feature extraction and/or proximity evaluation on remote databases. Figure 14.12 illustrates the sending of a mobile search engine to an SAPM-enabled database host.

14.3.3 Perceiving the MPEG-7 Search Engine

One way to perceive the characteristics of an MPEG-7 optimum search engine is to build on the objectives, scope, requirements, and experimental model of MPEG-7 itself. Since the aims and scope have already been presented in Section 14.1.2, only the remaining issues are covered in this section.

MPEG-7 is intended to be generic. It will support pull and push applications in both real-time and non-real-time platforms. In push applications, the MPEG-7 description can be used to filter information contents such as in automatic selection of programs based on a user profile. Likewise, in pull applications, the description can be used to locate multimedia data stored on distributed databases based on rich queries.

Although MPEG-7 aims to extend the proprietary solution of content-based applications, the descriptions used by MPEG-7 will not be of the content-based features alone. Because MPEG-7 is going to address as broad a range of applications as possible, a large number of description schemes will be specified and further amendment will be accommodated. In general, description of an individual content can be classified into content-based and content identification categories [39]. The content-based description includes descriptors (Ds) and description schemes (DSs) that represent features that are extracted from the content itself.

The content identification description covers the Ds and DSs that represent features that are closely related but cannot be extracted from the content. In addition, there will also be Ds and DSs for collection of contents as well as for the application-specific descriptions. The many DSs included in the current version of the generic audiovisual description scheme (generic AVDS) [38] are depicted in Figure 14.13.

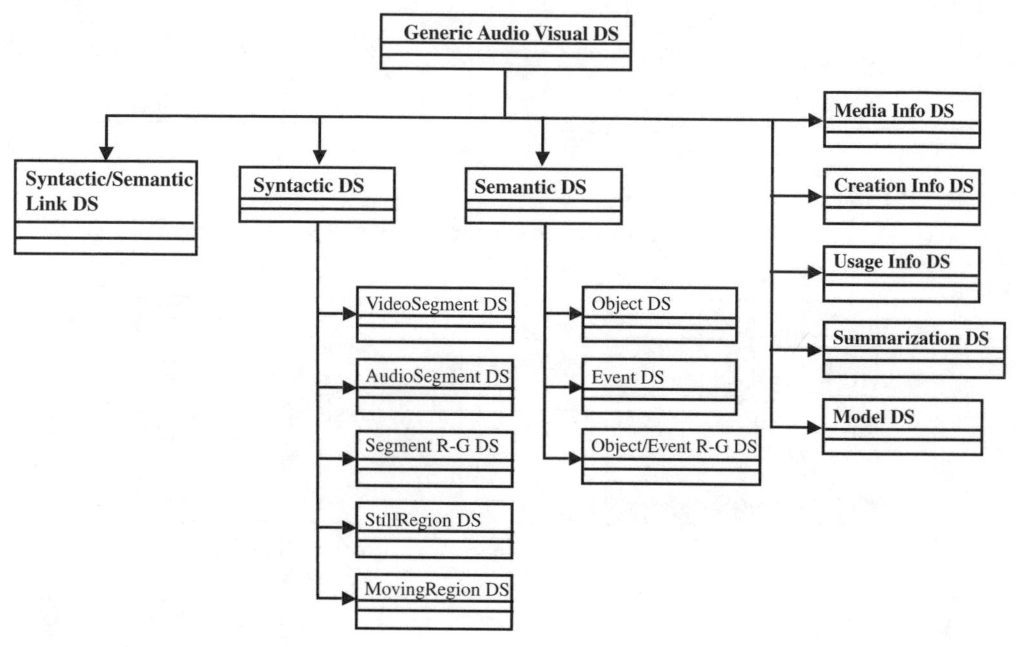

FIGURE 14.13
MPEG-7's generic audiovisual description scheme (AVDS).

The syntactic DS is used to specify the physical structures and signal properties of an image or a multimedia stream. Features such as shots, regions, color, texture, and motion are described under this DS category. The semantic DS is used to specify semantic features that appear in an image or a multimedia stream. Semantic notions such as object and event are described in this DS group. The relations between the syntactic and semantic descriptions are established using the syntactic/semantic link DS. Meta-information relating to media (storage, format, coding, etc.), creation (title, authors, etc.), and usage information (rights, publication, cost of usage, etc.) are, respectively, described in the media info DS, creation info DS, and usage info DS. The summarization DS is used to specify a set of summaries to allow fast browsing of a content. And the model DS is used to provide a way to denote the relation of syntactic and semantic information in which the contents are closely related to interpretation through models.

A wide-ranging choice of description schemes will allow a content to be described in numerous fashions. The same content is likely to be differently described according to the application and/or the user background. A content may also be described in the multiple-level description approach. Therefore, it will remain a challenging task for the MPEG-7 search engine to infer similarity among versions of the description flavors.

The descriptions of a content will be coded and provided as an MPEG-7 file or stream [40]. This file may be co-located or separately maintained with respect to the content. Likewise, the MPEG-7 stream may be transmitted as an integrated stream, in the same medium, or through a different mechanism with regard to the associated content. Access to partial descriptions is intended to take place without full decoding of the stream. The MPEG-7 file component is clearly represented in the experimental model (XM) architecture shown in Figure 14.14. Note that the shaded blocks are the normative components.

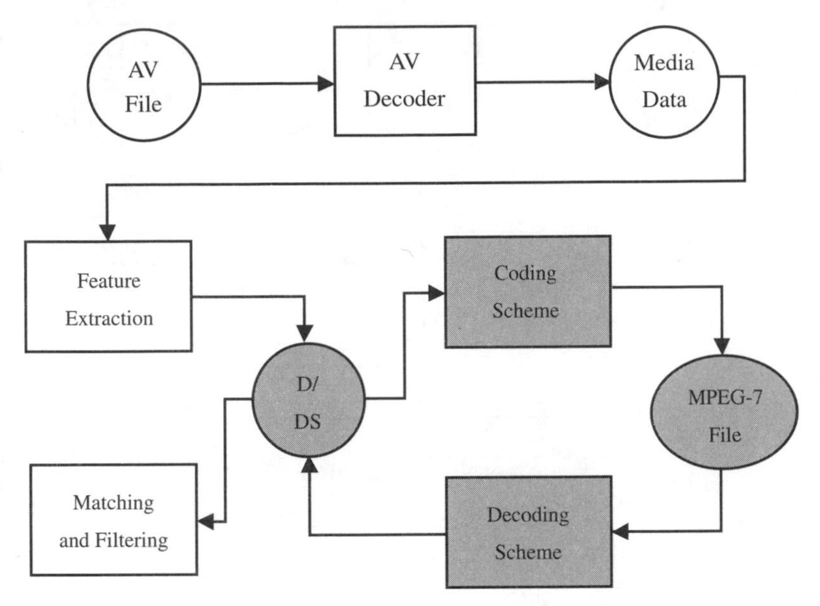

FIGURE 14.14
The MPEG-7 experimental model architecture.

The XM is the basis for core experiments in MPEG-7. It is also the reference model for the MPEG-7 standard [41]. Therefore, it is apparent that the indexing or dual-database model will be a better fit, since descriptions may be independently maintained in MPEG-7.

MPEG-7 will also specify mechanisms for the management and protection on intellectual property of its descriptions. It is possible that only requests equipped with proper rights will be allowed access to certain descriptions in an MPEG-7 stream.

A candidate model for the MPEG-7 proper search engine can be based on the meta-search engine [42]–[44]. However, the meta-search engine model is lacking many functions needed in coping with the complexity of the MPEG-7 description. It also lacks efficient mechanisms for controlling the search on a remote search engine. The integration of computational intelligence tools is by no means easy. Therefore, a new search engine type will be needed. Given the distributed characteristic of the MPEG-7 databases, there has been consideration to base the MPEG-7 XM on the COM/DCOM and CORBA technologies. The meta-search engine and the MPEG-7 optimum search tool (MOST) [23] models are shown in Figures 14.15 and 14.16, respectively.

14.3.4 Image Manipulation in the DCT Domain

In Section 14.2.1, it was presented that the DCT is just a linear transform. The linear property allows many manipulations on the DCT compressed data to be performed directly in the DCT domain. In this section, we show how certain transformations of JPEG images can be accomplished by manipulating the DCT coefficients directly in the frequency domain. In addition, a brief description of the algebraic operations will first be presented. Several works on the DCT compressed domain-based manipulation techniques are given in [31]–[33].

Compressed domain image manipulation is relevant, because it avoids the computationally expensive decompression (and recompression) steps required in uncompressed domain processing techniques. Furthermore, in applications where lossy operators are involved, as in

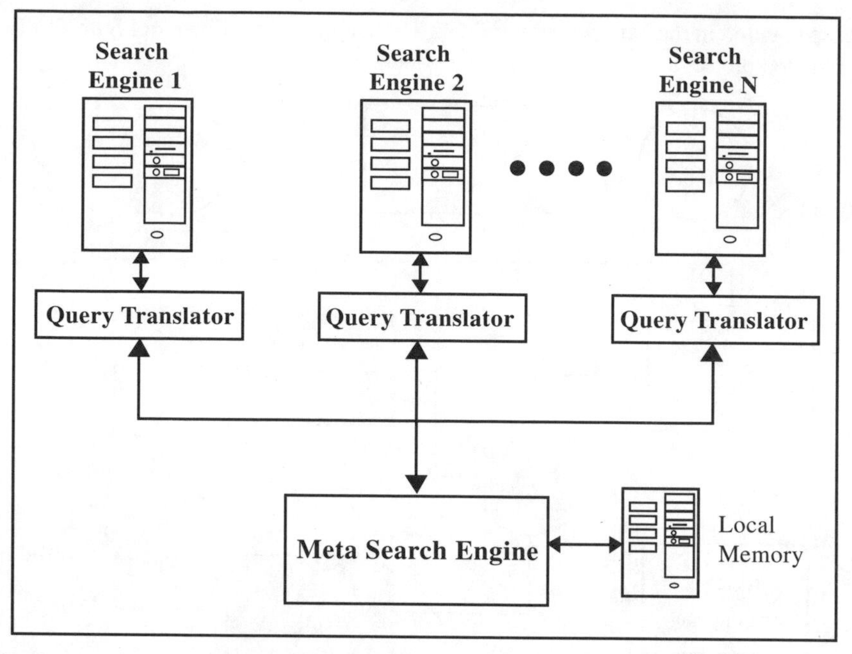

FIGURE 14.15
The meta-search engine model.

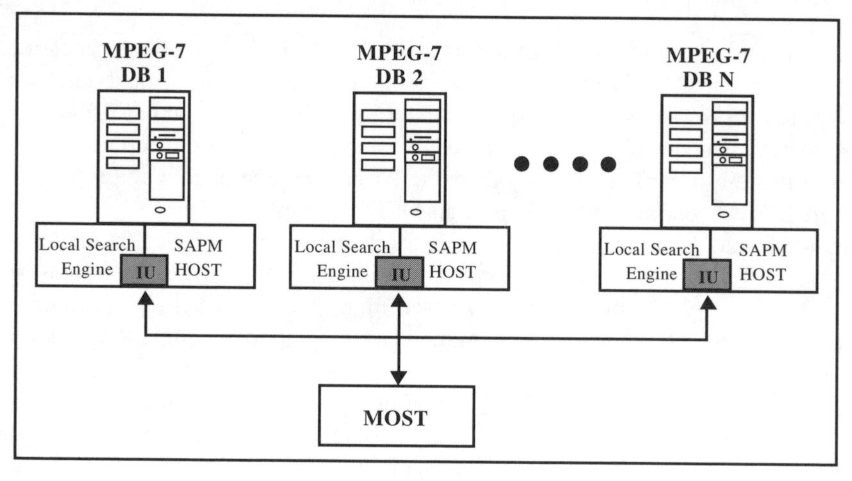

FIGURE 14.16
The MPEG-7 optimum search tool (MOST) model.

the baseline JPEG, avoiding the recompression step is crucial to spare the image from further degradation due to lossy operations in the recompression process. Thus, direct compressed domain manipulation is lossless in nature. Lossless manipulation is highly appreciated, as it is in accordance with the preservation characteristic of the digital data.

Retracting the linear property in Section 14.2.1, we now show how the algebraic operations of images can be attained directly in the DCT coefficient domain. Let **p** and **q** be the uncompressed images with (i, j) as the spatial domain indices, while **P** and **Q** represent the corresponding DCT compressed images with (u, v) as the DCT frequency domain indices, and α and β as scalars. Several algebraic operations can be written as follows:

Pixel addition:

$$
\begin{aligned}
f(\mathbf{p}+\mathbf{q}) &= f(\mathbf{p}) + f(\mathbf{q}) \\
\mathbf{p}[i,j] + \mathbf{q}[i,j] &\Rightarrow \mathbf{P}[u,v] + \mathbf{Q}[u,v]
\end{aligned}
$$

Scalar addition:

$$
\mathbf{p}[i,j] + \beta \Rightarrow \begin{cases} \mathbf{P}[u,v] + 8\beta & \text{for } [u,v] = (0,0) \\ \mathbf{P}[u,v] & \text{for } [u,v] \neq (0,0) \end{cases}
$$

Scalar multiplication:

$$
\begin{aligned}
f(\alpha\mathbf{p}) &= \alpha f(\mathbf{p}) \\
\alpha\mathbf{p}[i,j] &\Rightarrow \alpha\mathbf{P}[u,v]
\end{aligned}
$$

Note that f stands for the forward DCT operator. The addition of a constant to the uncompressed image data ($\mathbf{p}$) will only affect the DC coefficient in $\mathbf{P}$, since no intensity change is introduced to the image data by the scalar addition. The algebraic operations of addition and multiplication for each of the scalar and pixel functions for the JPEG data are provided in [31].

Additionally, using the DCT coefficients, several transformations such as mirroring or flipping, rotating, transposing, and transversing of a DCT compressed image can be realized directly in the DCT frequency domain. The transformation is realized by rearranging and adjusting the DCT coefficients with several simple linear operators such as permutation, reflection, and transpose matrices.

Let $\mathbf{Q}$ be the JPEG compressed DCT coefficient block, $\mathbf{D}$ be a diagonal matrix of $\{1, -1, 1, -1, 1, -1, 1, -1\}$, and $\mathbf{Q}_\theta$, $\mathbf{Q}_{\text{HM}}$, and $\mathbf{Q}_{\text{VM}}$, respectively, stand for the θ angle rotated, horizontal mirror, and vertical mirror of $\mathbf{Q}$. The various rotational and mirroring operations are given in [33] as

Mirroring:

$$
\begin{aligned}
\mathbf{Q}_{HM} &= \mathbf{QD} \\
\mathbf{Q}_{VM} &= \mathbf{DQ}
\end{aligned}
$$

Rotation:

$$
\begin{aligned}
\mathbf{Q}_{90} &= \mathbf{DQ}^T \\
\mathbf{Q}_{-90} &= \mathbf{Q}^T\mathbf{D} \\
\mathbf{Q}_{180} &= \mathbf{DQD}
\end{aligned}
$$

Horizontal and vertical mirroring of a JPEG image can be obtained by swapping the mirror pairs of the DCT coefficient blocks and accordingly changing the sign of the odd-number columns or rows within each of the DCT blocks. Likewise, transposition of an image can be accomplished by transposing the DCT blocks followed by numerous internal coefficient transpositions. Furthermore, transverse and various rotations of a DCT compressed image can be achieved through the combination of appropriate mirroring and transpose operations. For instance, a 90° rotation of an image can be performed by transposing and horizontally mirroring the image. The JPEG lossless image rotation and mirroring processing is also described in [34]. A utility for performing several lossless DCT transformations is provided in [13].

14.3.5 The Energy Histogram Features

Before we proceed with the energy histogram features, several DCT domain features common to CBR applications such as color histograms, DCT coefficient differences, and texture features will be presented in this section. An overview of the compressed domain technique is given in [27].

Color histograms are the most commonly used visual feature in CBR applications. Since the DC coefficient of a DCT block is the scale average of the DCT coefficients in that DCT block, counting the histogram of the DC coefficient is a direct approximation of the color histogram technique in the DCT coefficient domain. The DC coefficient histograms are widely used in video parsing for the indexing and retrieval of M-JPEG and MPEG video [28, 29].

Alternatively, the differences of certain DCT coefficients can also be employed. In [30], 15 DCT coefficients from each of the DCT blocks in a video frame are selected to form a feature vector. The differences of the inner product of consecutive DCT coefficient vectors are used to detect the shot boundary.

Texture-based image retrieval based on the DCT coefficients has also been reported. In [24], groups of DCT coefficients are employed to form several texture-oriented feature vectors; then a distance-based similarity evaluation measure is applied to assess the proximity of the DCT compressed images. Several recent works involving the use of DCT coefficients are also reported in [20], [24]–[26].

In this work the energy histogram features are used. Because one of the purposes in this work has been to support real-time capable processing, computational inefficiency should be avoided. Therefore, instead of using the full-block DCT coefficients, we propose to use only a few LF-DCT coefficients in constructing the energy histogram features. Figure 14.17 shows the LF-DCT coefficients used in the proposed feature set.

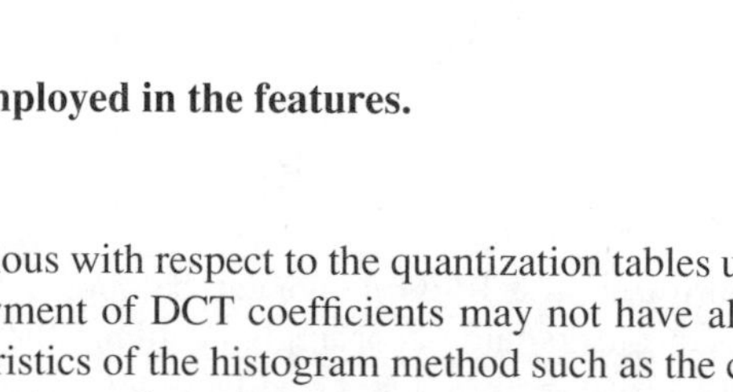

F1F	DC	AC_{01}	AC_{02}	AC_{03}
F2F	AC_{10}	AC_{11}	AC_{12}	AC_{13}
F3F	AC_{20}	AC_{21}	AC_{22}	AC_{23}
F4F	AC_{30}	AC_{31}	AC_{32}	AC_{33}

FIGURE 14.17
LF-DCT coefficients employed in the features.

The reduction is judicious with respect to the quantization tables used in JPEG and MPEG. However, partial employment of DCT coefficients may not have allowed inheritance of the many favorable characteristics of the histogram method such as the consistency of coefficient inclusion in an overall histogram feature. Since it is also our aim to have the proposed retrieval system capable of identifying similarities in changes due to common transformations, the invariant property has to be acquired independently. To achieve this aim, we utilized the lossless transformation properties discussed in Section 14.3.4.

Bringing together all previous discussions, six square-like energy histograms of the LF-DCT coefficient features were selected for the experiment:

Feature	Construction Components
F1	F1F
F2A	F2F
F2B	F1F+F2F
F3A	F2F+F3F
F3B	F1F+F2F+F3F
F4B	F1F+F2F+F3F+F4F

The square-like features have been deliberately chosen for their symmetry to the transpose operation, which is essential to the lossless DCT operations discussed in Section 14.3.4. Low-frequency coefficients are intended as they convey a higher energy level in a typical DCT coefficient block. F1 contains a bare DC component, whereas F2B, F3B, and F4B resemble the 2×2, 3×3, and 4×4 upper-left region of a DCT coefficient block. F2A and F3A are obtained by removing the DC coefficient from the F2B and F3B blocks. F2B, F3A, and F3B are illustrated in Figures 14.18a, b, and c, respectively. Note that counting the F1 energy histograms alone resembles the color histogram technique [17] in the DCT coefficient domain. The introduction of features F2A and F3A is meant to explore the contribution made by numerous low-frequency AC components, while the use of F2B, F3B, and F4B is intended for evaluating the block size impact of the combined DC and AC coefficients.

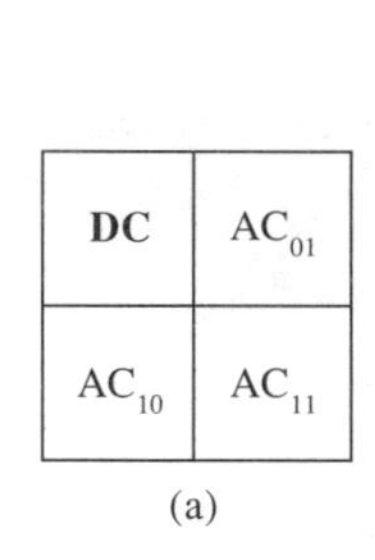

FIGURE 14.18
Samples of the square-like features.

14.3.6 Proximity Evaluation

The use of energy histograms as retrieval features is also an advantageous approach from the perspective of proximity evaluation. In many cases, a computationally inexpensive distance-based similarity measure can be employed. Figure 14.19 illustrates the distance-based similarity measure among pairs of the histogram bins.

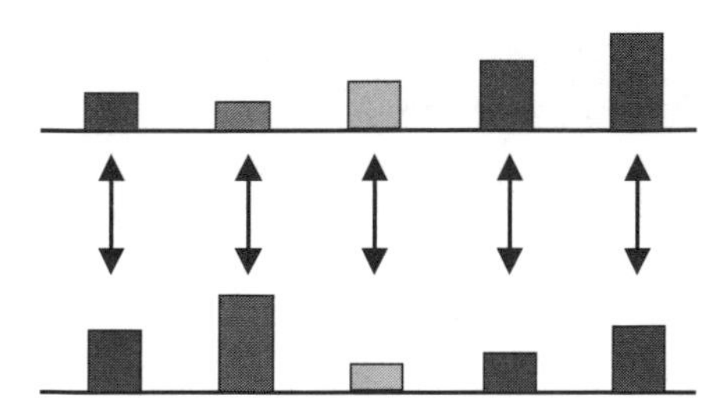

FIGURE 14.19
Bin-wise similarity measure.

Several widely used proximity evaluation schemes such as the Euclidean distance, the city block distance, and the histogram intersection method will be described below. However, the underlying notion of the histogram space will be characterized first.

Because histograms are discretely distributed, each bin of the histogram can be thought of as a one-dimensional feature component (or coordinate) of the n-dimensional feature space ($\Re^n$), where n is the number of bins in the histogram. Furthermore, if we define the n-dimensional feature space ($\Re^n$) as the histogram space (H^n), then every n-bin histogram feature can be represented as a point in that histogram space [17]. Consequently, for any pair of the histogram features, h_j and h_k, the distance between the two histograms can be perceived as the distance of the two representative points in the histogram space. Thus, the distance between h_j and h_k, $D(h_j, h_k)$, can be defined to satisfy the following criteria [35]:

1	$D(h_j, h_j) = 0$	The distance of a histogram from itself is zero.
2	$D(h_j, h_k) \geq 0$	The distance of two histograms is never a negative value.
3	$D(h_j, h_k) = D(h_k, h_j)$	The distance of two histograms is independent of the order of the measurement (symmetry).
4	$D(h_j, h_l) \leq D(h_j, h_k) + D(h_k, h_l)$	The distance of two histograms is the shortest path between the two points.

Figure 14.20 illustrates two 2-bin histogram features and their distance, respectively, represented as two points and a straight line on the two-dimensional histogram space.

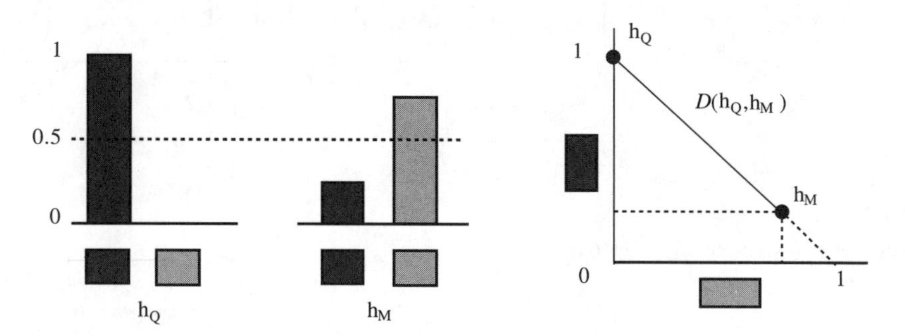

FIGURE 14.20
Histogram features and their distance on the histogram space.

In linear algebra, if **Q** and **M** are the feature vectors in the n-dimensional Euclidean space ($\Re^n$): $\mathbf{Q} = (q_1, q_2, q_3, \ldots, q_n)$ and $\mathbf{M} = (m_1, m_2, m_3, \ldots, m_n)$, the Euclidean distance (d_E) between the Q and M, written $d_E(Q, M)$, is defined by:

$$d_E(Q, M) = \sqrt{(q_1 - m_1)^2 + (q_2 - m_2)^2 + (q_3 - m_3)^2 + \cdots + (q_n - m_n)^2} \, .$$

Correspondingly, the Euclidean distance of any pair of the histogram features can be defined in the histogram space (H^n):

$$d_E(Q, M) = \left[\left(h_Q - h_M\right)^T \left(h_Q - h_M\right) \right]^{1/2} \quad \text{or}$$

$$d_E^2(Q, M) = \sum_{t=1}^{n} \left(h_Q[t] - h_M[t]\right)^2$$

where $d_E(Q, M)$ is the Euclidean distance between the two images, h_Q and h_M are the histograms of the two images, and $h_Q[t]$ and $h_M[t]$ represent the pairs of the histogram bins.

In general, Euclidean distance provides a very effective proximity measurement in image retrieval application. However, it is also a computationally expensive technique, especially when the floating point data type is involved, as with the DCT coefficient. The $d_E(Q, M)$ requires n floating point multiplications, where n is the number of bins in the histogram feature.

To overcome the high computational problem, the city block distance is usually employed. The city block distance (d_{CB}) is defined by

$$d_{CB}(Q, M) = \sum_{t=1}^{n} \left| h_Q[t] - h_M[t] \right|$$

where $d_{CB}(Q, M)$ denotes the city block distance between the two histogram features, $h_Q[t]$ and $h_M[t]$ are the pairs of the histogram bins, and n is the number of bins in the histogram.

Figure 14.21 depicts an extreme condition where the two 2D histogram features under measurement are maximally apart. The Euclidean distance and the city block distance for the two feature points on the histogram space are computed below.

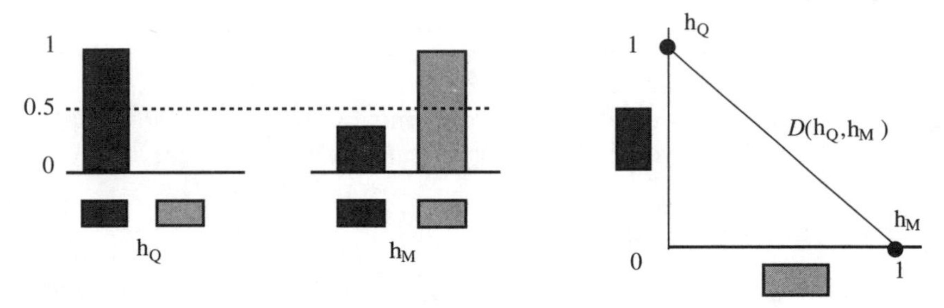

FIGURE 14.21
Maximum distances over two 2D histogram features.

The Euclidean distance is given by

$$d_E^2(Q, M) = \sum_{t=1}^{2} \left(h_Q[t] - h_M[t] \right)^2 = 1 + 1 = 2$$
$$d_E(Q, M) = \sqrt{2}$$

and the city block distance is

$$d_{CB}(Q, M) = \sum_{t=1}^{2} \left| h_Q[t] - h_M[t] \right| = 1 + 1 = 2$$
$$d_{CB}(Q, M) = 2 \, .$$

Alternatively, the histogram proximity evaluation scheme used in [17] can be employed. The histogram intersection method is used to locate the common parts of objects by intersecting the two images under similarity evaluation. For two images Q (query) and M (model) composed of n-bin histograms, the histogram intersection of the two images is defined by

$$\sum_{t=1}^{n} \min \left(h_Q[t], h_M[t] \right)$$

where $h_Q[t]$ and $h_M[t]$ denote the particular pair of the histogram bins. Accordingly, the normalized distance of the histogram intersection can be written as

$$d_h(Q, M) = 1 - \frac{\sum_{t=1}^{n} \min\left(h_Q[t], h_M[t]\right)}{\min\left(\Gamma_Q, \Gamma_M\right)}$$

where $d_h(Q, M)$ denotes the distance metric of the histogram intersection, $h_Q[t]$ and $h_M[t]$ denote a particular pair of the histogram bins, and

$$\Gamma_x = \sum_{t=1}^{n} h_x[t] \qquad \text{with} \quad x = (Q, M)\ .$$

Normalization can be used to add the scale invariance property to the histogram features [18]. Let $h_Q[t]$ be an n-bin histogram feature of an image. Then the normalized n-bin histogram feature, written as $h^n_Q[t_n]$, is defined as

$$h^n_Q\,[t_n] = \frac{h_Q[t]}{\sum_{t=1}^{n} h_Q[t]} \qquad \text{for} \quad t = 1, 2, 3, \ldots, n\ .$$

The city block distance is used for the experiments in this work because it provides the lowest computational complexity and easy implementation.

14.3.7 Experimental Results

Experimental results for the use of energy histogram features on the retrieval of JPEG images and the parsing of MPEG video are presented in this section.

Image Retrieval

The experiment on image retrieval [20] was based on a single database containing nearly 4700 uncategorized, uniformly sized JPEG photographs of a broad range of real-life subjects. The collection is produced and maintained by Media Graphics International [36]. The DCT coefficients were extracted by utilizing the library provided in [13]. Dequantization steps were performed because quantization tables used in JPEG could vary among images. The retrieval system was built on the single-database model.

Some 40 query images were preselected by hand to ensure images appearing similar to the human visual system were properly identified in the database. Thresholding was not considered, because the purpose was to reveal the prerecognized images' position on the retrieved image list. A sample of two query images and their similar associates are shown in Figure 14.22. Two query images, 36_238.JPG and 49_238.JPG, are used here to illustrate the features contribution on the retrieval performance. The first image group consists of three very similar images taken with slight camera movement, and the second group contains two similar images with slightly different background. The results for the experiment are tabulated in Table 14.4.

We observed that energy histograms based exclusively on the DC coefficient (F1) might only perform well on the retrieval of images with high similarity in colors. The results of F2A and F3A suggested that histograms of low-frequency AC coefficients, which carry the texture and edge information, are contributory to the similarity measure. Thus, the combination of DC and numerous low-level AC coefficients (F2B, F3B, F4B) yielded better results on both

FIGURE 14.22
Sample of query images and their similarity sets.

Table 14.4 Retrieval Hit for Queries

	F1	F2A	F2B	F3A	F3B	F4
Q: 36_238						
Rank 1 (best)	37_238	37_238	37_238	37_238	37_238	37_238
Rank 2	35_238	X	35_238	35_238	35_238	35_238
Rank 3	X	X	X	X	X	X
Rank 4	X	35_238	X	X	X	X
Q: 49_238						
Rank 1 (best)	X	X	48_238	X	48_238	48_238
Rank 2	X	X	X	48_238	X	X
Rank 3	X	48_238	X	X	X	X
Rank 4	X	X	X	X	X	X
Rank 5	48_238	X	X	X	X	X

of the lists. On comparison of block size effect, further examination using other queries in our experiment shows that in general F2B and F3B are much preferable to F4B. This may be due to the fact that as the feature block grows larger, heavily quantized coefficients are also taken into consideration; hence erroneous results may be generated.

We also noticed that the retrieval performance for features F2B and F3B is relatively unaffected by translation of small objects in globally uniform images. A retrieval sample is shown in Figure 14.23.

As for retrieval of lossless DCT transformed images, a query was chosen and transformed using the JPEGtran utility provided in [13]. The transformed images were then added into the image database prior to the retrieval test. We observed that all features are able to recognize the transformed images. An image group used in the experiment is shown in Figure 14.24.

On the issue of computational cost, we noticed that the complexity could be significantly reduced through attentive selection of features in the compressed domain. A 2×2 block feature (F2B) may reduce the feature complexity by a factor of 1/16 with respect to the overall color histogram method [17].

FIGURE 14.23
Retrieval of images with translated object contents.

  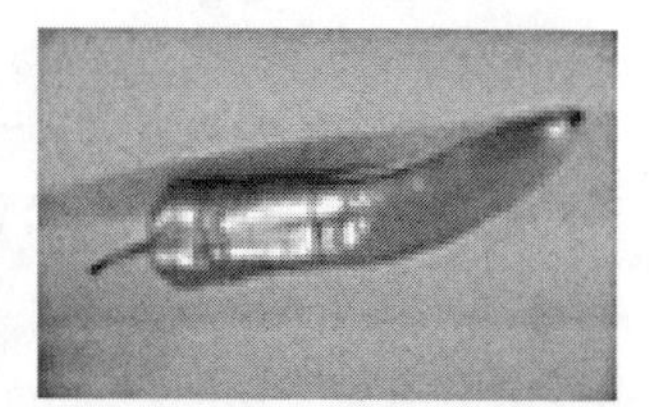

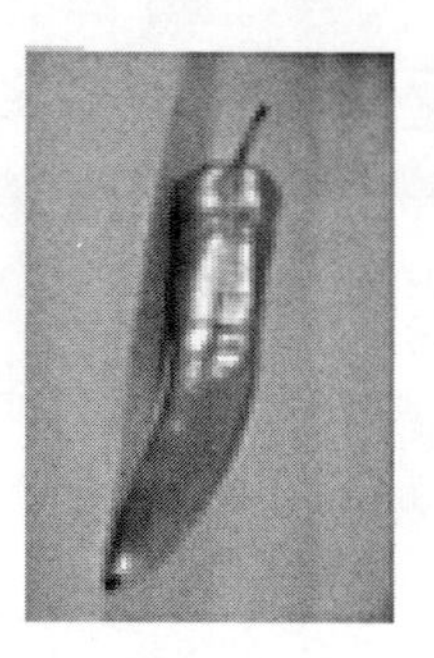 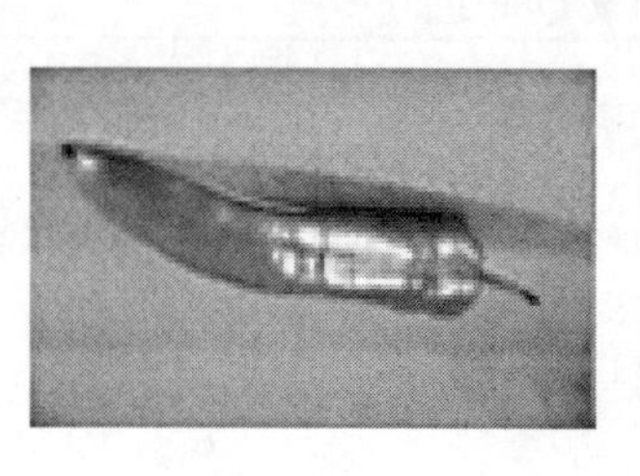 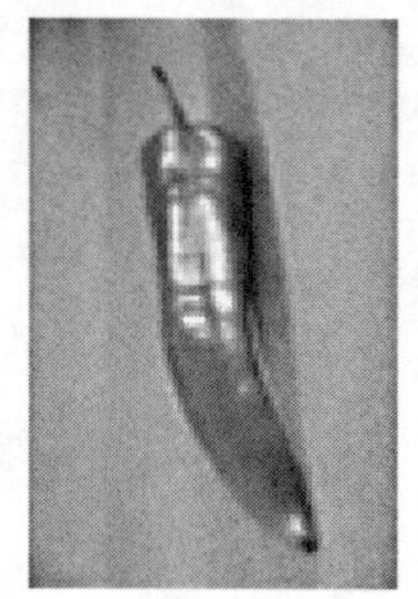 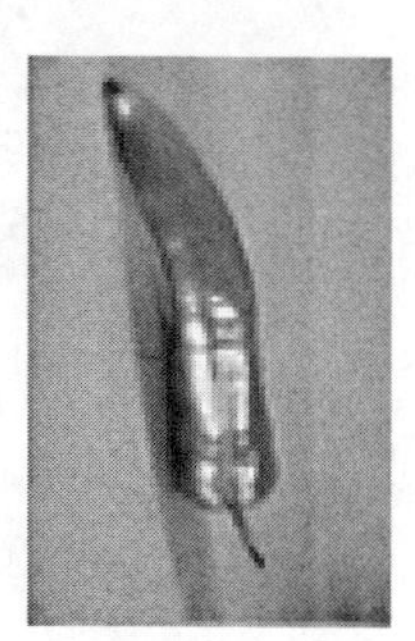

FIGURE 14.24
Retrieval of transformed images.

Video Parsing

The experiment on video parsing [21] was based on several MPEG streams selected for several testing purposes. Opera House is a simple stream showing scenic views around the Sydney Opera House. It is used to examine the feature performance on parsing of video containing only sharp transitions. Aquarium and Downtown were obtained from the Internet [37]. Aquarium is a relatively simple video sequence, whereas Downtown is a fast-changing video sequence with rapid camera movements and dissolve transitions. Blink was produced in a controlled environment in order to investigate the effects of a flashing and blinking light source. A short description of the video streams is provided in Table 14.5, and several sample frames from Opera House are shown in Figure 14.25.

For relatively simple video sequences such as the Opera House, the use of DC coefficients alone is sufficient to detect abrupt scenery changes. However, for more complex dissolve transitions and other effects, the use of the DC coefficient alone can lead to false detection. In general, features based on F1F and F2F yield better results than the F1F alone. False detection

FIGURE 14.25
Sample frames from Opera House.

Table 14.5 Details of the Test Streams

Name	No. of Frames	Transitions
Opera House	2200	5 sharp
Aquarium	750	1 dissolve, 2 sharp
Downtown	240	5 dissolve
Blink	225	No transition

also occurs on features based on F3F and F4F.

The block diagrams for detecting sharp and dissolve transitions are, respectively, shown in Figures 14.22 and 14.26. Since a sharp transition occurs between two frames, the distance of the energy histogram features is usually very large for all YUV components. Therefore, a sharp transition can be detected by using any one of the YUV components. Otherwise, the combined distance of the three component distances can be used.

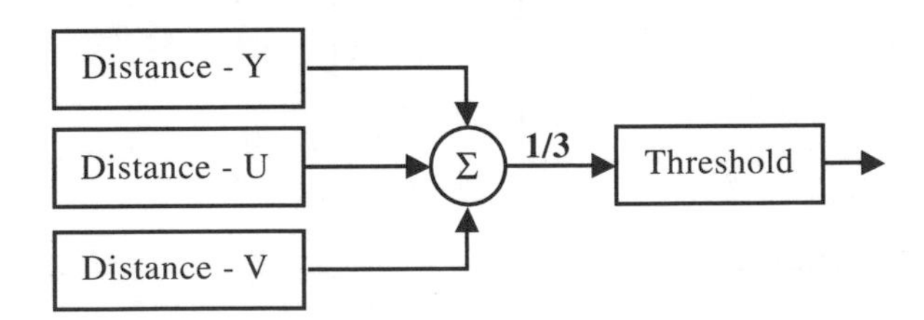

FIGURE 14.26
Sharp transition detection with combined component distances.

For a dissolve transition, the frames of the leading shot will fade out and gradually lose their color information, whereas the frames of the second shot will gradually increase in color. This characteristic can be used to assist video parsing. Thus, the features based on U and V components alone can be used to detect dissolve transitions. However, this results in a rather noisy distance graph.

Because the frames are changing gradually, the distances between energy histograms produced from DCT coefficients are small. A median filter is applied to enhance the characteristic. The filtered distances may show large pulses at nontransitional frames; however, this rarely happens at the same time for all three YUV components. Various methods such as averaging were used to combine the filtered distance values of the three components. An effective method found was to take the product of the three component distances. In fact, using the distance graph of the Y component to attenuate the unwanted noise yielded promising results.

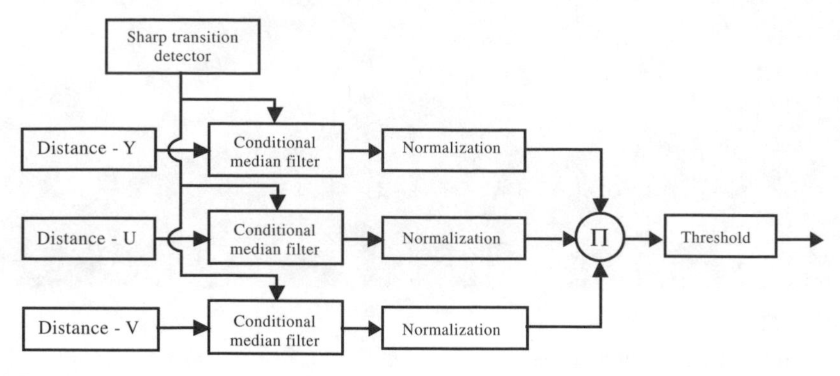

FIGURE 14.27
Diagram block for dissolve transition detection.

The results obtained for Aquarium are shown in Figure 14.28. The graphs indicate a sharp increase at points where sharp transitions occur. The lower-right graph shows the results of combining all the distances after applying a median filter; the crest near frame 600 is caused by ripples on the water surface in the video. The ripples introduce perturbation into UV components. Note that in the normalized total distance plots, the effects of sharp transitions have been removed by ignoring the distance at sharp transitions in the process of combining the three components.

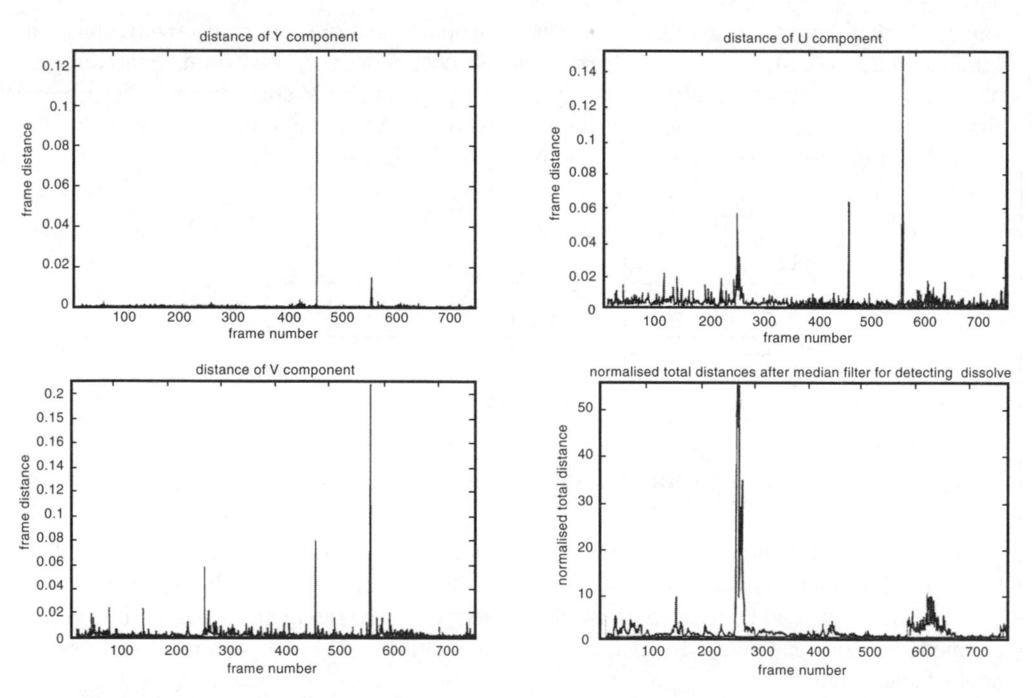

FIGURE 14.28
Component and normalized distances for Aquarium using F2B.

The normalized total distance for Downtown is given in Figure 14.29. Downtown was parsed using the same configuration as in the experiment for Figure 14.23. An interesting fact is that for fast camera movement and the zooming in and zooming out effects, the features display characteristics similar to dissolve transitions. Zooming out at frame 105 and rapid camera movement at frame 165 in Downtown resulted in large peaks in the graph. This can be seen in Figure 14.7 near frames 110 and 170.

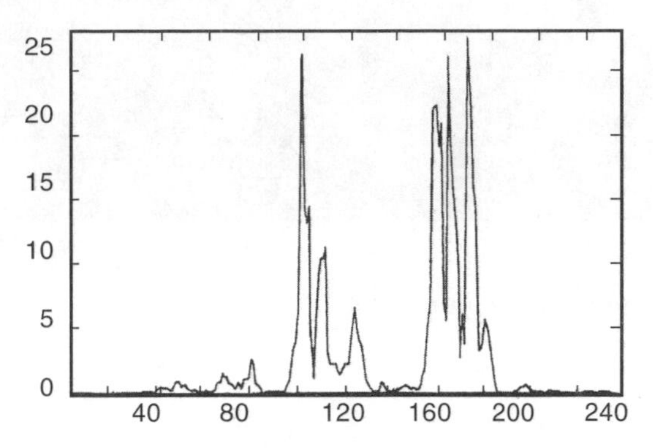

FIGURE 14.29
Normalized distances for Downtown using F2B.

The effects of the change of illumination were investigated using the Blink stream. The results obtained are shown in Figure 14.30. The Blink stream contains many blinks from a light bulb followed by a flash. It is observed that the changes of illumination caused by the light result in great perturbation on the energy histogram of the UV components. Frames containing the blinking effect were correctly parsed using the features, whereas the flash effect was parsed as a false transition. An illumination change of a video shot is illustrated in Figure 14.31.

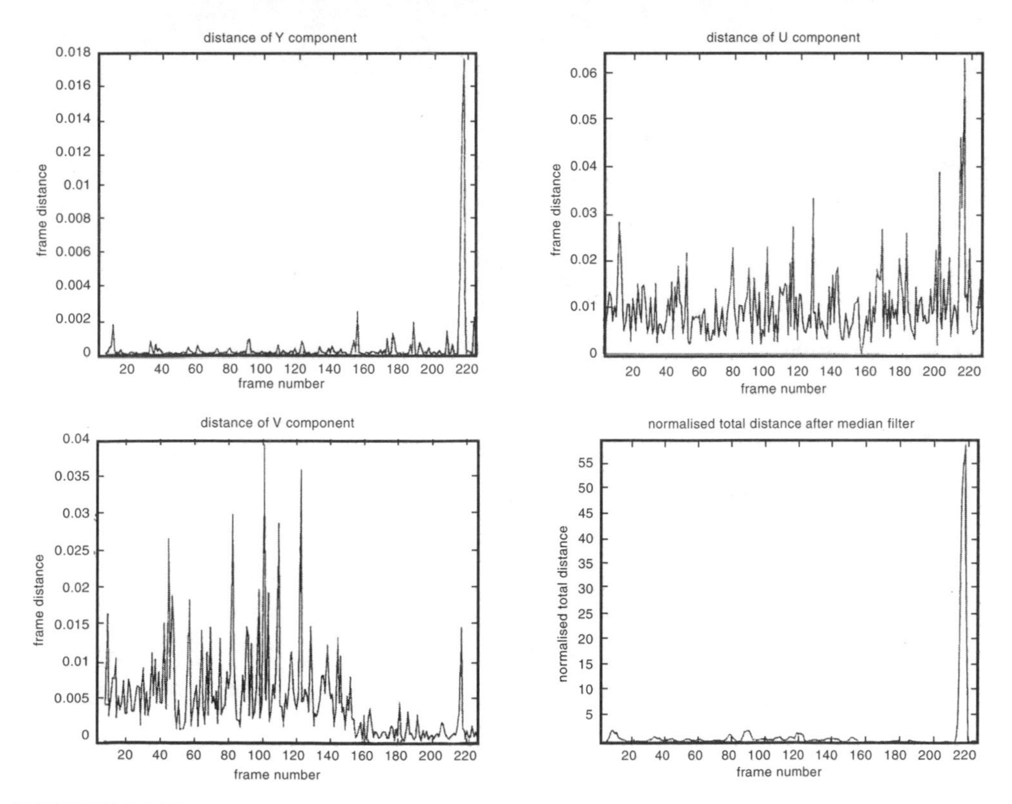

FIGURE 14.30
Component and normalized distances for Blink using F2B.

FIGURE 14.31
Change of illumination within a video shot.

14.4 Conclusions

We have presented the use of energy histograms of the LF-DCT coefficient features for the retrieval of JPEG images and the parsing of MPEG videos. We have shown how the features can be used to perform retrieval on medium-size databases as well as to parse relatively complex shot transitions. We have also shown that by introducing the transpose symmetry, a vigorous feature set can be built to accommodate the DCT domain-based lossless transforms.

References

[1] D.A. Adjeroh and K.C. Nwosu, Multimedia Database Management-Requirements and Issues, *IEEE Multimedia,* July–Sep. 1997, pp. 24–33.

[2] L.D. Davidson and R.M. Gray, *Data Compression,* Halsted Press, PA, 1976.

[3] J.D. Gibson et al., *Digital Compression for Multimedia: Principles and Standards,* Morgan Kaufmann, CA, 1998.

[4] C. Chiariglione, Impact of MPEG Standards on Multimedia Industry, *Proc. of the IEEE,* vol. 86, no. 6, June 1998, pp. 1222–1227.

[5] T.S. Huang and Y. Rui, Image Retrieval: Past, Present, and Future, *Proc. of Int. Symposium on Multimedia Information Processing,* Taiwan, Dec. 1997.

[6] D. Doermann, The Indexing and Retrieval of Document Images: A Survey, *Computer Vision and Image Understanding,* vol. 70, no. 3, June 1998, pp. 287–298.

[7] R.W. Picard, Light-Years from Lena: Video and Image Libraries of the Future, *Proceedings of ICIP,* 1995, pp. 310–313.

[8] MPEG-7: Context, Objectives and Technical Roadmap, vol. 11, ISO/IEC JTC1/SC29/WG11 N2729, Seoul, Korea, March 1999.

[9] Multimedia Content Description Interface (MPEG-7), `http://drogo.cselt.stet.it/mpeg/`.

[10] N. Ahmed, T. Natarajan, and K.R. Rao, Discrete Cosine Transform, *IEEE Transactions on Computers,* Jan. 1974, pp. 90–93.

[11] ISO/IEC 10918-1: Information Technology — Digital Compression and Coding of Continuous-Tone Still Images, ISO-IEC, 1994.

[12] W.B. Pennebaker and J.L. Mitchell, *JPEG Still Image Data Compression Standard,* Van Nostrand-Reinhold, 1993, ch. 4.

[13] T. Lane, Independent JPEG Group (IJG) JPEG Library, `ftp://ftp.uu.net/graphics/jpeg/jpegsrc.v6b.tar.gz`.

[14] A.C. Hung, Stanford University Portable Video Research Group's (PRVG) PRVG-JPEG Software Codec, `ftp://havefun.stanford.edu:/pub/jpeg/JPEGv1.2.1.tar.Z`.

[15] Cornell lossless JPEG implementation, `ftp://ftp.cs.cornell.edu:/pub/multimed/ljpg.tar.Z`.

[16] B.G. Haskell et al., Image and Video Coding — Emerging Standards and Beyond, *IEEE Transactions on Circuits and Systems for Video Technology,* vol. 8, no. 7, Nov. 1998.

[17] M.J. Swain and D.H. Ballard, Indexing Via Color Histograms, *Proc. of 3rd Int. Conf. on Computer Vision,* Osaka, Japan, 1990, pp. 390–393.

[18] A.K. Jain and A. Vailaya, Image Retrieval Using Color and Shape, *Pattern Recognition,* vol. 29, no. 8, 1996, pp. 1233–1244.

[19] V.N. Gudivada, and V.V. Raghavan, Modeling and Retrieving Images by Content, *Information Processing & Management,* vol. 33, no. 4, 1997, pp. 427–452.

[20] J.A. Lay and L. Guan, Image Retrieval Based on Energy Histograms of the Low Frequency DCT Coefficients, *Proc. of the 1999 ICASSP,* pp. 3009–3012.

[21] O.K.-C. Bao, J.A. Lay, and L. Guan, Compressed Domain Video Parsing Using Energy Histogram of the Low Frequency DCT Coefficients, *Proceedings of SPIE on Storage and Retrieval for Image and Video Databases VIII,* vol. 3972, San Jose, Jan. 2000.

[22] J.A. Lay and L. Guan, Processing Models for Distributed Content Based Multimedia Retrieval Systems, *Proceedings of the Fifth International/National Conference on Digital Image Computing Techniques and Applications (DICTA),* Perth, 1999.

[23] J.A. Lay and L. Guan, Searching for an MPEG-7 Optimum Search Engine, *Proceedings of SPIE on Storage and Retrieval for Image and Video Databases VIII,* vol. 3972, San Jose, Jan. 2000.

[24] H.J. Bae and S.H. Jung, Image Retrieval Using Texture Based on DCT, *Proc. of Int. Conf. on Information, Comm., and Signal Processing,* Singapore, 1997, pp. 1065–1068.

[25] A. Vellaikal and C.-C.J. Kuo, Joint Spatial-Spectral Indexing for Image Retrieval, *Proceedings of the 1996 IEEE Int. Conf. on Image Processing,* Lausanne, Switzerland, part 3, pp. 867–870.

[26] S.-S. Yu, J.-R. Liou, and W.-C. Chen, Computational Similarity Based on Chromatic Barycenter Algorithm, *IEEE Trans. on Consumer Electronics,* vol. 42, no. 2, 1995, pp. 216–220.

[27] S.F. Chang, Compressed-Domain Techniques for Image/Video Indexing and Manipulation, *Proceedings of the 1995 IEEE Int. Conf. on Image Processing,* part 1, pp. 314–317.

[28] M.M. Yeung and B. Liu, Efficient Matching and Clustering of Video Shots, *Proceedings of the 1995 IEEE Int. Conf. on Image Processing,* part 1, pp. 338–341.

[29] K. Shen and E.J. Delp, A Fast Algorithm for Video Parsing Using MPEG Compressed Sequences, *Proceedings of IEEE Int. Conf. on Image Processing,* 1995, part 2, pp. 252–255.

[30] F. Arman, A. Hsu, and M.Y. Chiu, Image Processing on Compressed Data for Large Video Databases, *Proceedings of ACM Multimedia 1993,* Anaheim, CA, pp. 267–272.

[31] B.C. Smith and L.A. Rowe, Algorithms for Manipulating Compressed Images, *IEEE Computer Graphics and Applications,* Sept. 1993, pp. 34–42.

[32] S.F. Chang and D.G. Messerschmidt, Manipulating and Compositing of MC-DCT Compressed Video, *IEEE Journal on Selected Areas in Communication,* vol. 13, Jan. 1995, pp. 1–11.

[33] R. de Queiroz, Processing JPEG-Compressed Images, *Proceedings of the 1997 Int. Conf. on Image Processing,* vol. 2, pp. 334–337.

[34] B. Shen and I.K. Sethi, Scanline Algorithms in the JPEG DCT Compressed Domain, *Journal of Electronic Imaging,* vol. 5, no. 2, April 1996, pp. 182–190.

[35] J.R. Smith, Integrated Spatial and Feature Image Systems: Retrieval, Analysis and Compression, Ph.D. thesis, Columbia University, 1997.

[36] Media Graphics International, Photo Gallery 5,000 vol. 1 CD-ROM, `http://www.media-graphics.net.`

[37] Microsoft Corp., MPEG Sample Videos, `http://www.microsoft.com/Theater/Download/mpeg_download.asp.`

[38] MPEG, ISO/IEC JTC1/SC29/WG11 N2966, MPEG-7 Generic AV Description Scheme (V0.7), Melbourne, Oct. 1999.

[39] MPEG, ISO/IEC JTC1/SC29/WG11 N2998, Overview of MPEG-7 Description Tools: Descriptors and Description Schemes (V0.2), Melbourne, Oct. 1999.

[40] MPEG, ISO/IEC JTC1/SC29/WG11 N2996, MPEG-7 Requirement Document (V1.0), Melbourne, Oct. 1999.

[41] MPEG, ISO/IEC JTC1/SC29/WG11 N2999, MPEG-7 Development Process, Melbourne, Oct. 1999.

[42] E. Selberg and O. Etsioni, Multi-Service Search and Comparison Using the MetaCrawler, *Proceedings of the 4th Int. WWW Conference,* Dec. 1995.

[43] A.B. Benitez, M. Beigi, and S.F. Chang, Using Relevance Feedback in Content Based Image Meta-Search Engine, *IEEE Internet Computing,* vol. 2, no. 4, July-Aug. 1998.

[44] M.H. Chignell, J. Gwizdka, and R.C. Bodner, Discriminating Meta-Search: A Framework for Evaluation, *Information Processing and Management,* vol. 35, no. 3, May 1999, pp. 337–362.

Chapter 15

Rapid Similarity Retrieval from Image and Video

Kim Shearer, Svetha Venkatesh, and Horst Bunke

15.1 Introduction

The rapidly increasing power of computers has led to the development of novel applications such as multimedia, image, and video databases. These applications take advantage of the increasing processing power and storage of computers to rapidly process large amounts of data. The challenge has now become one of developing suitable tools and methods for manipulation of the data that has become available. Given the enormous amount of information contained in a multimedia data stream, it is sensible that a deeper comprehension of the data stream may need to be achieved through the integration of many independent analyses on different aspects of it.

One method of indexing and retrieval that has been developed for image databases is based on the qualitative spatial relationships between key objects present in the images. The representation of indices, and retrieval — both exact and inexact — from these indices, has been the subject of a large volume of work [1]–[9]. By using qualitative relationships it is possible to provide a simple-to-use and intuitive interface for query by iconic example. While this work has produced successful image database systems, little work has been done to extend such systems further [10]. The most significant disincentive for extending this work has been the difficulty in the automation of the initial key object annotation. In many cases it has been necessary to produce the required annotations by hand.

Recent developments have reduced the annotation problem, and further developments seem likely to make this difficulty manageable. Motion tracking has been applied in a number of areas and has received much attention in recent years [11]–[16]. Work has already begun using the motion blocks of current MPEG representation as objects for the purpose of video data indexing [17]–[19]. The ability to track a body in motion through a video shot has reduced the laborious process of hand annotation to a process of labeling the detected objects. In addition, as the motion picture representation standard moves toward content-based coding (MPEG–4), more detailed object indexing will become possible.

With the production of suitable object annotation becoming less difficult for video processing, it has become possible to apply indices based on qualitative spatial relationships to video. In recent work, Shearer et al. [20, 21] have extended a qualitative spatial representation to the indices of video data. This form of index provides an efficient encoding of spatial relationships between key objects in a video stream and an efficient method for exact query resolution. In general, encodings that use spatial information to index images or video allow searches for exact pictorial or subpicture matches using a compact encoding such as 2D strings [3, 6, 20, 22].

However, for the case of inexact matching between two images, referred to as similarity retrieval, the best expression is given by inexact isomorphism detection between the graphs representing the two images [23, 24].

In inexact isomorphism detection, an individual image can be encoded as a graph by taking a vertex to represent each object in an image, and joining each pair of vertices with an edge labeled with the spatial relationship between the two corresponding objects. Although this is a plausible representation of images using graphs, other representations are possible, and in some cases are preferable. When this representation is used for video, temporal redundancy can be used to reduce the representation and matching complexity by recording and examining only the initial spatial configuration and changes to the spatial relationships. In this manner each graph instance expresses the state for a block of the video stream, rather than for a single frame.

There are two distance measures that may be used as the similarity measure for inexact isomorphism detection: edit distance and largest common subgraph. However, algorithms using either of these similarity measures for inexact isomorphism detection based on spatial relationship information are computationally complex. Given that the retrieval process for multimedia databases is usually one of browsing with progressive refinement, it is not advantageous to use algorithms that have complexity exponential in the size of the database elements.

A new approach to graph isomorphism algorithms has been developed by Messmer and Bunke [25]–[27]. This approach takes advantage of a priori knowledge of a database of model graphs to build an efficient index structure offline. By taking advantage of common structure between models, this new approach is able to reduce online isomorphism detection time significantly. The algorithms proposed by Messmer and Bunke are capable of solving the exact subgraph isomorphism problem and can detect inexact subgraph isomorphisms using a flexible edit distance measure. However, the edit distance measure is not suitable for similarity retrieval for image and video databases. The most significant problem in the application of the edit distance measure to image and video similarity is that the graph edit operations may have no suitable physical interpretation. While vertex deletion implies exclusion of an object from the common part of two graphs, and a constant cost could be applied, altering the label of a vertex, which would represent substitution of one object for another, does not carry an obvious cost assignment. Moreover, changing an edge label presents another problem: What is a suitable measure for the distance between different spatial relationships of two objects? Furthermore, there is no meaningful comparison between vertex substitution and edge label changes. Edit distance-based isomorphism algorithms also suffer from bias when used to compare an input against multiple model graphs. By the nature of edit distance algorithms, if the input graph is smaller than the models, smaller model graphs are more likely to be chosen as similar, and vice versa.

A more appropriate measure for image and video similarity is the largest common subgraph (LCSG) (or maximal common subgraph) between graphs that represent images. When two images, I_1 and I_2, are represented as two graphs, G_1 and G_2, which encode spatial relationships between key objects, the largest common subgraph between G_1 and G_2 represents the largest collection of objects found in I_1 and I_2 that exhibit the same relationships to each other (under the chosen matching type) in both images. The LCSG is thus an intuitive and easily interpreted measure for similarity retrieval in image and video databases.

The algorithm usually used in determining the LCSG is the maximal clique detection method proposed by Levi [28]. Unfortunately, the maximal clique detection algorithm can be a computationally complex method, with the worst-case performance requiring $O((nm)^n)$ time, where n is the number of vertices in the input graph and m is the number of vertices in the model. This worst-case performance is produced by a complete graph with all vertices having the same label. It is this high computational complexity that makes indices based on spatial relationships unsuitable for application to large databases.

This chapter describes two algorithms for detection of the LCSGs between an input graph and a database of model graphs. These algorithms extend the algorithms proposed by Messmer and Bunke [29] to detect LCSGs. Two major advantages of the new LCSG algorithms are that they use a preprocessing step over the database to provide rapid online classification of queries and that the database may be extended incrementally.

Although this chapter deals with LCSG detection as an application for image and video similarity retrieval, the new algorithms can be applied to other representations, provided they have the high degree of common structure expected in video databases. In this chapter a video is treated as a simple sequence of images. With this treatment there is a high degree of common structure between frames, which is subsumed by the approach of representing the state of a block of video by a single graph. This will lead to the classification of multiple frames by one element of the classification structure.

The algorithms presented may be applied to labeled and attributed, directed or undirected graphs. Thus, other encodings of image and video information, or other datasets of relational structures, may also find advantage in these algorithms.

The remainder of the introductory section will provide definitions required for the key graph-related concepts used in this work. The video encoding used for testing the algorithms will then be briefly outlined. The algorithms by Messmer and Bunke, upon which the LCSG algorithms are based, will then be explained, followed by a detailed description of the new algorithms. The final section of this chapter will discuss results of experiments with the new algorithms and comparisons with other available algorithms.

15.1.1 Definitions

DEFINITION 15.1 *A graph is a 4-tuple $G = (V, E, \mu, \nu)$, where*

- *V is a set of vertices*
- *$E \subseteq V \times V$ is the set of edges*
- *$\mu : V \to L_V$ is a function assigning labels to vertices*
- *$\nu : E \to L_E$ is a function assigning labels to the edges.*

DEFINITION 15.2 *Given a graph $G = (V, E, \mu, \nu)$, a subgraph of G is a graph $S = (V_S, E_S, \mu_S, \nu_S)$ such that*

- *$V_S \subseteq V$*
- *$E_S = E \cap (V_S \times V_S)$*
- *μ_s and ν_S are the restrictions of μ and ν to V_S and E_S, respectively; i.e.,*

$$\mu_s(v) = \begin{cases} \mu(v) & \text{if } v \in V_S \\ \text{undefined} & \text{otherwise} \end{cases} \qquad \nu_s(e) = \begin{cases} \nu(e) & \text{if } e \in E_S \\ \text{undefined} & \text{otherwise .} \end{cases}$$

The notation $S \subseteq G$ is used to indicate that S is a subgraph of G.

DEFINITION 15.3 *A bijective function $f : V \to V'$ is a graph isomorphism from a graph $G = (V, E, \mu, \nu)$ to a graph $G' = (V', E', \mu', \nu')$ if*

1. $\mu(v) = \mu'(f(v)) \; \forall v \in V$

2. *For any edge* $e = (v_1, v_2) \in E$ *there exists an edge* $e' = (f(v_1), f(v_2)) \in E'$ *such that* $\nu(e) = \nu(e')$, *and for any* $e' = (v_1', v_2') \in E'$ *there exists an edge* $e = (f^{-1}(v_1'), f^{-1}(v_2')) \in E$ *such that* $\nu(e') = \nu(e)$.

DEFINITION 15.4 *An injective function* $f : V \rightarrow V'$ *is a subgraph isomorphism from* G *to* G' *if there exists a subgraph* $S \subseteq G'$ *such that* f *is a graph isomorphism from* G *to* S.

Note that finding a subgraph isomorphism from G to G' implies finding a subgraph of G' isomorphic to the whole of G. The injective nature of this mapping becomes important in later discussion.

DEFINITION 15.5 S *is an LCSG of two graphs* G *and* G', *where* $S \subseteq G$ *and* $S \subseteq G'$, *iff* $\forall S' : S' \subseteq G \wedge S' \subseteq G' \implies |S'| \leq |S|$.

There may or may not be a unique LCSG for any two graphs G and G'.

15.2 Image Indexing and Retrieval

In order to discuss algorithms for similarity retrieval of video based on spatial relationships, it is necessary to present the encoding and matching scheme used. The encoding in this chapter is based on the 2D B string notation [24] for pictorial databases.

Numerous descriptions are possible for qualitative spatial relationships between objects. The most common basis for such descriptions is the interval relations proposed by Allen [30] for qualitative temporal reasoning. Allen specifies 13 possible relationships that may occur between two intervals in one dimension (see Table 15.1). For two objects in a two-dimensional picture, the qualitative relationship between the two may be described by a one-dimensional relationship along each of two axes. This gives a total of 169 possible relationships between two objects in two dimensions. The two axes, one horizontal and one vertical, are named either the u and v axes or the x and y axes. The usual convention is for u or x to refer to the horizontal axis, with v or y referring to the vertical axis.

This description of the qualitative spatial relationships between two objects is generally used as the basis for matching in image indexing and retrieval. Using the relationships between two objects, a less precise description has been derived by Lee et al. [24], called *relationship categories*. Lee et al. partitioned the 169 possible spatial relationship pairs in two dimensions into five categories. These categories group together relationship pairs that have similar appearance characteristics. The five categories of relationship are as follows.

DEFINITION 15.6

1. *Disjoint* — The two objects a and b do not touch or overlap; that is, there is a *less than* $<$ operator along at least one axis.

2. *Meets* — The two objects a and b touch but do not overlap; thus, they have a *meets* $|$ relationship along one axis and a nonoverlapping relationship along the other.

Table 15.1 Possible Interval Relationships (Allen [30])

<table>
<tr><th>Relation</th><th>Symbol</th><th>Example</th><th>Relation</th><th>Symbol</th><th>Example</th></tr>
<tr><td>less than</td><td>a<b</td><td></td><td>meets</td><td>a|b</td><td></td></tr>
<tr><td>overlaps</td><td>a/b</td><td></td><td>ends</td><td>a]b</td><td></td></tr>
<tr><td>contains</td><td>a%b</td><td></td><td>begins</td><td>a[b</td><td></td></tr>
<tr><td>equals</td><td>a=b</td><td></td><td>begins inverse</td><td>a['b</td><td></td></tr>
<tr><td>contains inverse</td><td>a%'b</td><td></td><td>ends inverse</td><td>a]'b</td><td></td></tr>
<tr><td>overlaps inverse</td><td>a/'b</td><td></td><td>meets inverse</td><td>a|'b</td><td></td></tr>
<tr><td>less than inverse</td><td>a<'b</td><td></td><td></td><td></td><td></td></tr>
</table>

3. *Contains* — Object a contains object b; that is, object a *contains* %, *begins* [or *ends*] object b along both axes.

4. *Belongs to* — Object b contains object a; that is, object b *contains* %, *begins* [or *ends*] object a along both axes.

5. *Overlaps* — The two objects do not fall into any of the above categories.

These relationship categories are useful in approximate matching, as they exhibit the properties of rotation and scaling invariance.

In general, each qualitative spatial representation defines three types of matching, allowing differing levels of precision in the matching process. Each of the types of matching, called type 0, type 1, and type 2, defines the form of correspondence required in the relationships of two objects present in two pictures. Offering three types of matching with varying levels of precision facilitates the process of browsing with progressive refinement.

The form of matching adopted in this work is derived from the B string notation [24], which is the underlying notation used in the video index. Type 2 matching is the most exact form of matching for B strings. It specifies that the spatial relationship pair between two objects a and b in one picture P_1 must be duplicated in another picture P_2 for them to be considered a match. Type 0 matching is the least strict form, requiring only that the relationship pair between two objects a and b in a picture P_1 fall in the same relationship category as the relationship pair for a and b in picture P_2. Between these two types of match is type 1 matching, which requires that two objects a and b satisfy the type 0 match in pictures P_1 and P_2 and that the orthogonal relationships between a and b are the same in P_1 and P_2. Orthogonal relationships are determined by the extent of an object viewed relative to another object. The inclusion of orthogonal relationships in the matching scheme, in addition to type 0 matching, restricts type 1 matching to a similar rotational position, and so is no longer rotationally invariant.

Given these definitions of matching between two pictures for pairs of objects, general pictorial matches may be defined as follows.

DEFINITION 15.7 *Two pictures P and Q are said to be a type n match if:*

1. *For each object $o_i \in P$ there exists an $o_j \in Q$ such that $o_i \equiv o_j$, and for each object $o_j \in Q$ there exists an $o_i \in P$ such that $o_j \equiv o_i$*

2. *For all object pairs o_i, o_j where $o_i \in P$ and $o_j \in Q$, o_i and o_j are a type n match.*

If it is assumed that most image and video database retrieval will be performed using a combination of browsing and progressive refinement, it is unlikely that a complete match will be found in many cases. Most retrieval is likely to involve detecting a subpicture within the database elements for finding the most similar database element. This process of searching for partial matches is termed similarity retrieval. A *subpicture match* between two pictures is defined as follows.

DEFINITION 15.8 *A subpicture match between P and Q is defined as:*

1. *For all objects $o_i \in Q'$, where $Q' \subset Q$, there exists an object $o_j \in P'$, where $P' \subset P$, such that $o_i \equiv o_j$*

2. *For all objects $o_j \in P'$, where $P' \subset P$, there exists an object $o_i \in Q'$, where $Q' \subset Q$, such that $o_j \equiv o_i$*

3. *For all object pairs o_i, o_j where $o_i \in P'$ and $o_j \in' Q$, o_i and o_j are a type n match.*

A subpicture match is therefore a set of objects present in both P and Q that have a type n spatial relationship match between P and Q for all pairs of objects in the set.

15.3 Encoding Video Indices

The techniques described in the previous section for representing and retrieving pictorial information may be adapted to video databases. Several adaptations are reported in [10, 20]. The basic principle is to represent the initial frame of a video segment in full, then represent only the changes of the spatial relationships as the video progresses. The work in this chapter uses a 2D B string to represent the spatial relationships of objects in the initial frame of the video segment, then represents the changes to spatial relationships during the video as edits to this string. In this way, the edits to the 2D string represent changes in the state of the objects that are indexed, each state being valid for a block of frames in the video. The work performed by this group represents changing spatial relationships in such a way that exact matching may be performed directly from the edits, without expansion to the full 2D B strings. This is an advantage over earlier notations. Similarity retrieval is performed in the same manner as for simple pictorial information, that is, using subgraph isomorphism detection.

In order to perform similarity retrieval it is necessary to represent the video information and query input as graphs. For this work the user's query, which is given online, will be referred to as either the query or the input. The elements of the database will be referred to as the models. In video database retrieval the operands of the retrieval process are the objects in the input and video, and the spatial relationships between them. Each input, or state representing a block of frames in a model, is represented as a graph by creating a vertex for each object, labeled with the object's identifier, and joining the vertices with edges labeled with the spatial relationships between the appropriate objects. This leads to a complete labeled graph, with edge direction used to disambiguate the edge labels. In actual implementation the labels of the edges are the relationship category (definition 15.6) of the relationships between the two objects, since all three types of match require at least a type 0 match, so an efficient initial comparison is made possible. The actual spatial relationships of the two objects are used as attributes of the edge and are examined if the type 0 label forms a match.

An example of the representation used is given in Figure 15.1. This figure presents two pictures and the graphs that would be used to represent them. Inspection of the two pictures

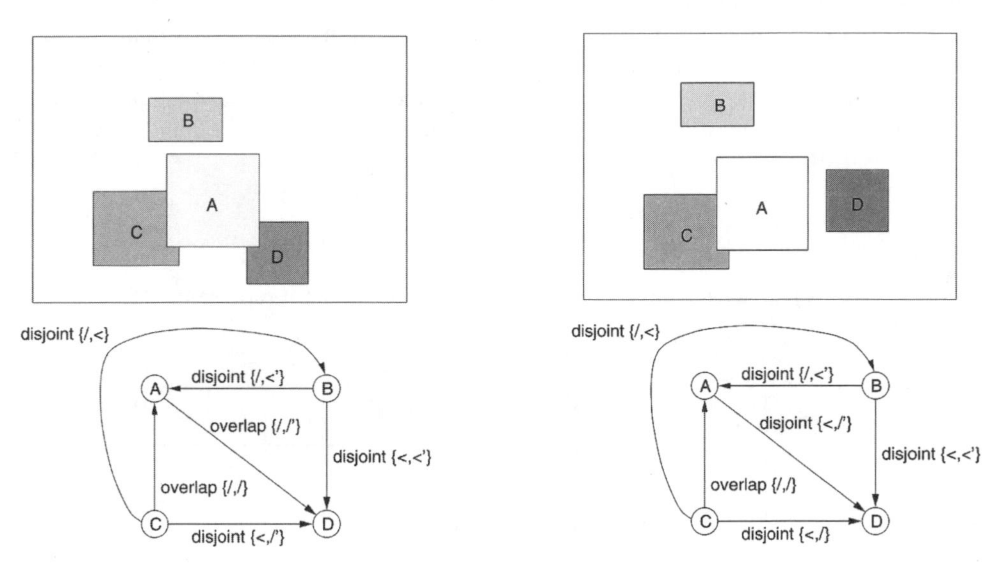

FIGURE 15.1
Two pictures with possible graph encodings.

shows there is a clear resemblance of one to the other, in that the objects A, B, and C display similar relationships in the pictures. Although the object labeled B does move in a quantitative sense, the qualitative relationships between objects A and B and objects B and C do not change. Therefore, if the object D and all edges incident on vertex D are deleted from the two pictures, the remaining parts are a qualitative match. Similarity retrieval is the process of detecting the largest such common parts between an input and the database. Thus, similarity retrieval is the search for the models in the database that contain the largest subgraph isomorphic to the query input. Such subgraphs are referred to as the LCSGs between two graphs.

If this encoding of images as graphs is used, it is possible to extract the largest common part of two images by comparing their representative graphs. When the object state that represents a block of frames from a video sequence is encoded as a graph, similarities may be detected between states, and between an input and a state. The algorithms presented in the following section may be applied to similarity retrieval for both image and video databases. Matching is performed between a single input query and a database of models, the models being known before matching begins. The compilation of multiple model images into a database explicitly takes advantage of the temporal redundancy between frames within a video. It is partially this similarity that leads to the efficiency of the proposed algorithms.

We shall now present the algorithms for efficient subgraph isomorphism detection and the results gained from applying these algorithms to data encoded as outlined in this section.

15.4 Decision Tree Algorithms

The decision tree algorithm [27] is important for image and video databases because it is intended to detect subgraph isomorphisms from the input graph to the model graphs. From Definition 15.4 it can be seen that this implies the algorithm detects subgraphs of the model graphs that are isomorphic to the entire input graph. The most common form of query in the image and video database area is query by pictorial example, where the query is an iconic sketch depicting the required relationships for the objects of interest. Thus, when the query is encoded as a graph, query resolution requires detection of model graphs that contain a subgraph isomorphic to the input graph.

The decision tree used in this algorithm is constructed from the adjacency matrix representation of the model graphs. Each graph G can be represented by a matrix M which describes the vertex labels along the diagonal and the edge labels in the off–diagonal elements. The elements of the adjacency matrix M for a graph $G = (V, E, \mu, \nu)$ are defined as follows:

$$M_{ij} = \begin{cases} \mu(v_i) & \text{if } i = j \\ \nu((v_i, v_j)) & \text{if } i \neq j \, . \end{cases} \tag{15.1}$$

In Figure 15.2 an adjacency matrix is given alongside the graph it represents. Here the edges

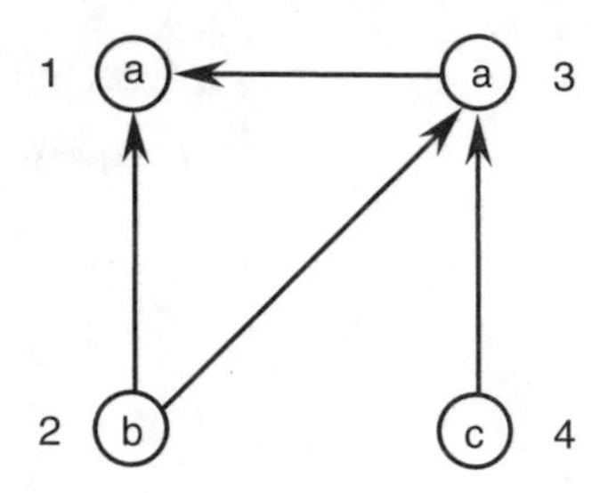

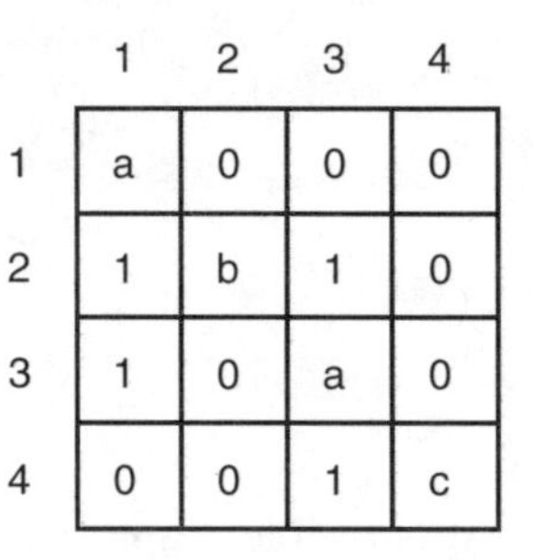

	1	2	3	4
1	a	0	0	0
2	1	b	1	0
3	1	0	a	0
4	0	0	1	c

FIGURE 15.2
Graph with adjacency matrix.

are unlabeled and so are represented in the adjacency matrix as either 1 for present or 0 for absent.

Adjacency matrices have the property that each permutation of an adjacency matrix M represents a graph isomorphic to the graph represented by M. Thus,

$$M' = PMP^T \tag{15.2}$$

where P is a permutation matrix with the sum of each row and column equal to 1. This equation gives the transformation of an adjacency matrix M to an isomorphic adjacency matrix M'. Isomorphism detection between two graphs can therefore be expressed as the task of finding a suitable permutation matrix that transforms the adjacency matrix of one graph to the adjacency matrix of the other.

Rapid graph isomorphism detection can be provided by producing a decision tree constructed from the adjacency matrices that represent all isomorphisms of a graph. The decision tree algorithm constructs the decision tree using the *row–column elements* of each adjacency matrix. A row–column element r_i of a matrix M consists of the following matrix elements: $M_{ij} : 0 \leq j \leq i$ and $M_{ji} : 0 \leq j \leq i$. As can be seen in Figure 15.3, this gives a set of partitions of the

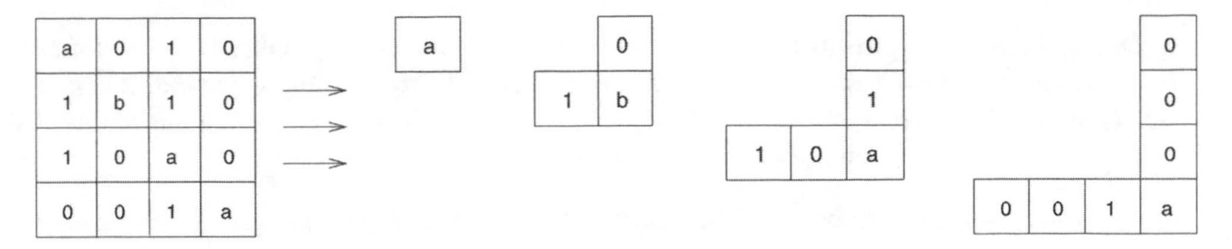

FIGURE 15.3
Row–column elements of a matrix.

matrix, each of which contains one vertex label and connectivity information for that vertex and all vertices above and left in the matrix.

The decision tree is constructed beginning with a single unlabeled root node. The first-level descendants of this root node are arc labeled with each unique, one-element, row–column element. In Figure 15.4 there are six possible permutations in total; however, there are only two unique one-element row–column elements. Therefore, there are two descendants from the root node, with arcs labeled a and b. From each of these nodes the tree is continued by adding an arc for each possible subsequent row–column element. Thus, from the arc labeled b there are three possible next row–column elements among the adjacency matrices, so there are three descendants. The process continues until the tree expresses all possible adjacency matrices, with the model and its permutations being the leaves of the tree, as shown in Figure 15.4. Additional graphs may be added to the decision tree using the same process, creating new nodes and arcs where necessary. The advantage for graphs sharing isomorphic subgraphs is immediately seen, in that each shared subgraph will be represented only once in the tree. A simple example of this is seen in Figure 15.4, with one node used as the ancestor of the nodes representing matrices $\{C\}$ and $\{F\}$.

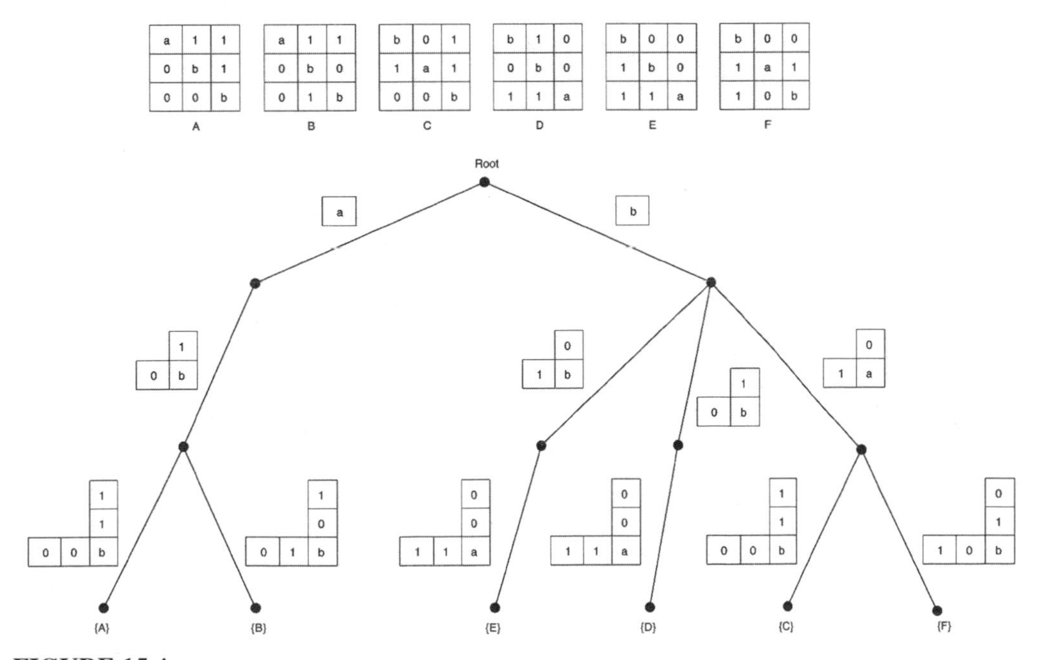

FIGURE 15.4
Decision tree for example graph.

Detection of isomorphisms from an input graph, G_I, to the model graphs, M_i, is performed using the adjacency matrix representing the input graph. The row–column elements that make up G_I are used to navigate the arcs of the decision tree until one of two termination conditions is encountered. The termination conditions are:

1. All row–column elements of the matrix G_I have been used to navigate an arc.
2. There is no arc descending from the current node labeled with a row–column element which matches the next row–column element from G_I.

If the first termination condition is encountered, then all models associated with the final node reached have a subgraph isomorphic to the input G_I. The second termination condition indicates that there is no subgraph isomorphism from the input G_I to any of the models represented in the decision tree.

Graph and subgraph isomorphism detection by decision tree provides an algorithm that is exceptionally rapid for online classification of input graphs. However, while the computational complexity is favorable, the space complexity is the major disadvantage. The best space complexity that can be achieved for a full decision tree is $O(L \mid l_v \mid (1 + \mid l_e \mid^2)^n)$. Here L is the number of models in the database, l_v is the number of distinct vertex labels, l_e is the number of distinct edge labels, and n is the number of vertices in the model graphs. There are a number of heuristics available for pruning the decision tree with restrictions on the applicability of the algorithm [31].

For applications in which the number of labels is limited, the decision tree algorithm detects graph and subgraph isomorphisms between the input and the database of models in time $O(n^2)$, where n is the number of vertices in the input. The computational complexity is therefore independent of both the number of models in the database and the number of vertices in the models. This compares well to the algorithm usually applied to exact isomorphism detection, Ullman's algorithm [32], which has computational complexity of $O(Lm^n n^2)$, with m being the number of vertices in the models.

Although the decision tree algorithm does provide a rapid solution for exact graph and subgraph isomorphism detection, as noted earlier, there are frequently no exact matches in a database. More often the task is to detect the best inexact matches, to allow progressive refinement of the query. The decision tree algorithm as proposed by Messmer and Bunke has no useful inexact isomorphism detection method. The next section presents an extension of the original algorithm to inexact isomorphism detection using the LCSG as the measure of similarity. This algorithm is well suited to the similarity retrieval task most often found in image and video databases.

15.4.1 Decision Tree-Based LCSG Algorithm

Detection of the largest common subgraph between two graphs by traditional algorithms has computational complexity that is exponential in the number of vertices of both graphs. Given a database of model graphs and an input, this is compounded by the linear factor of the number of models. The decision tree algorithm presented in the previous section offers fast detection of the LCSG between an input graph and a database of model graphs, which makes it possible to detect numerous isomorphisms while retaining rapid execution. The algorithm presented here for detection of the LCSG between an input graph and a database of model graphs takes this approach.

When the decision tree algorithm terminates with failure, there will be one or more row–column elements that have not been used to descend an arc in the decision tree. In the simplest case there will be only one such row–column element, which is the cause of failure. However, in many cases there will be other row–column elements below the one that causes failure, and it is possible that one or more of these row–column elements could be used to descend further

b	0	1
1	a	0
0	1	b

b	1	0
0	b	1
1	0	a

FIGURE 15.5
Input graph adjacency matrices.

through the decision tree. Figure 15.5a shows an adjacency matrix which, when classified against the decision tree in Figure 15.4, matches only the first row–column element. There is no match for the second row–column element; however, if the matrix is permuted as in Figure 15.5b, such that the second and third row–column elements are exchanged, then two row–column elements may be matched. Clearly the deepest such descent through the decision tree, representing the largest portion of the input adjacency matrix being used, gives a subgraph isomorphism using as many vertices as possible for the particular input graph and database of models. This is an LCSG between the input and the models in the database.

In order to determine such a deepest descent it is necessary to permute the input adjacency matrix such that the row–column elements that cannot be used are at the bottom right of the matrix. To achieve this the adjacency matrix is classified as deeply as possible, then the row–column element that has caused the failure is permuted to the final row–column element position in the matrix. In order to prevent repeated attempts to match the same row–column element, a temporary counter is kept of the number of available vertices in the input. This counter is decremented each time a row–column element is permuted out of the available set. At some point all possible matching row–column elements will have been used to descend the tree. At this point the input adjacency matrix is partitioned into two parts, as shown in Figure 15.6. In this figure the matrix is classified against the decision tree of Figure 15.4; the

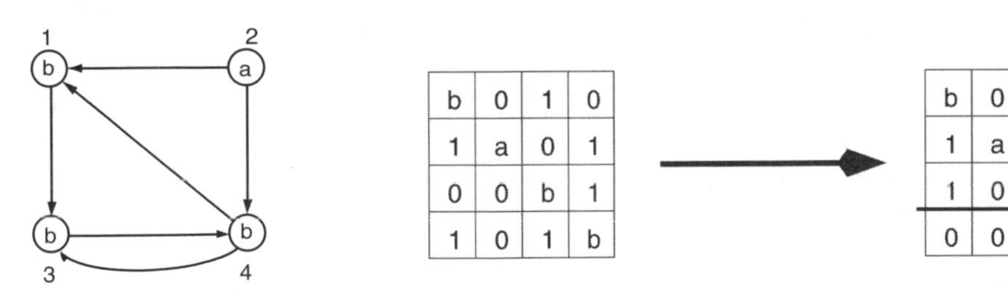

FIGURE 15.6
Partitions of the adjacency matrix.

initial descent uses the first two row–column elements. The third row–column element has no matching arc and so is permuted to the final row, allowing one more row–column element to be matched. This permuted matrix is shown in Figure 15.6 to the right of the arrow, the unused row–column element separated by a solid line. The upper left-hand portion of the matrix represents a potential LCSG between the input graph and the model graphs. It is only a potential LCSG because the inclusion of any one vertex in the matching part of the input graph can exclude the possibility of detecting a true LCSG. This possibility is illustrated in Figure 15.7.

Figure 15.7 shows an input graph and a number of adjacency matrices at different stages of classification. The matrix in Figure 15.7b is the original adjacency matrix used to represent the

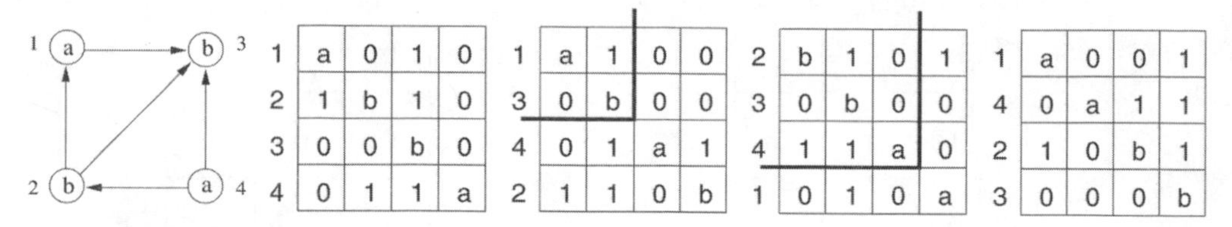

FIGURE 15.7
Permutation of matrices during classification.

graph. When this matrix is used to classify the graph against the decision tree in Figure 15.4, the initial descent cannot match the second row–column element. This row–column element is therefore permuted to the last position in the matrix. The third row–column element is used to descend further through the decision tree; however, the fourth row–column element also cannot be matched, thus terminating classification because there are no further row–column elements. The final permuted matrix from the initial descent is given in Figure 15.7c. Here the two matched row–column elements are separated from the unmatched row–column elements by a solid line. Figure 15.7d shows the resulting partitions of the matrix if the first row–column element is permuted to the final position. With this permutation the first three row–column elements can be used to descend the decision tree, giving a larger common subgraph than the initial descent. Thus, the inclusion of the vertex labeled 1 prevents the detection of the largest possible subgraph.

This shows the necessity for backtracking once a possible solution has been detected. Figures 15.7b, c, and d show the first three steps of the algorithm. First the tree is descended as far as possible without permutation. This gives a match for the first row–column element only (Figure 15.7b). Permutations are then performed to descend as far as possible (Figure 15.7c). When the maximum depth is reached, the best size is recorded before backtracking occurs. Backtracking is performed by taking the final row–column element used in the descent and permuting it to the final position, also decrementing the dimension of the matrix. Effectively this discards the row–column element from possible matches. Note that this differs from decrementing the counter of available vertices. The dimension of the matrix is decremented when a row–column element has been tested and is excluded from further descents; the counter of available vertices is decremented to indicate that, for this descent, the row–column element is not usable.

Figure 15.7e shows the matrix that results from backtracking from the best initial decent of Figure 15.7c. Here row–column element 3 has been permuted to the end, allowing row–column element 4 to be used, if possible. In this case, row–column element 4 does not provide a match. Discarding row–column element 4 would decrease the available vertex count to one, leaving only the first row–column element, indicating that all combinations beginning with row–column element 1 have been tested. The next permutation is formed by continuing the process of discarding the last used row–column element, in this case row–column element 1, giving the matrix in Figure 15.7d. For this matrix all available vertices are used in the descent. The next step is therefore to permute row–column element 2 to the final position, attempting to match row–column elements 3 and 4. In this case such a descent is clearly redundant because the set of row–column elements 2, 3, and 4 have already been matched and are a superset. The final piece needed for this algorithm is thus a method for pruning the search space.

Pruning is undertaken using the counter of the number of available vertices and the number of vertices in the best subgraph detected. Once the initial descent has been completed, there is a candidate LCSG with number of vertices N_c. This number is updated each time a better subgraph is found. Using this value, descent of the decision tree may be terminated at any time that the number of nodes permuted to the final position in this descent, plus N_c, is at least equal to the current matrix dimension. As an example, assume a candidate subgraph has been found with 4 vertices, from an input with 7 vertices. When classification reaches the third descendant of the root in the decision tree, the first two row–column elements will have been permuted to the final position and the dimension of the matrix reduced to 5. Therefore, only one permutation can be performed on any descent without reducing the largest possible number of vertices to less than the current best.

In the extreme case, if the first descent for an input matrix of dimension d reaches a depth of $d - 1$, then classification can be terminated for the second descendant from the root if any permutation is needed. Further descendants from the root will not be examined because the first two row–column elements have already been eliminated. This gives a computational complexity for the algorithm of $O(2^n n^3)$, where n is the number of vertices in the input graph. This compares favorably with the maximal clique algorithm, which has complexity $O(L(nm)^n)$, where m is the number of vertices in the models and L is the number of models in the database.

15.5 Decomposition Network Algorithm

The decomposition network algorithm detects graph and subgraph isomorphisms from the models to the input. It can therefore be used to find models that are subgraphs of the input.

The decomposition algorithm represents a database of graphs by first decomposing them into a network of subgraphs. A decomposition of a graph is formally defined as follows.

DEFINITION 15.9 *Let $B = \{G_1, \ldots, G_L\}$ be a set of model graphs. A* decomposition *of B, $D(B)$, is a finite set of 4–tuples (G, G', G'', E), where*

1. *G, G', and G'' are graphs with $G' \subset G$ and $G'' \subset G$*
2. *E is a set of edges such that $G = G' \cup_E G''$*
3. $\forall\, G_i \,\exists\, (G, G', G'', E) \in D(B) : G = G_i, i = 1, \ldots, L$
4. $\forall\, (G, G', G'', E) \in D(B) \,\nexists (G_1, G_1', G_1'', E_1) \in D(B) : G = G_1$
5. *For each 4–tuple (G, G', G'', E)*
   (a) *if G' consists of more than one vertex, then there exists a 4–tuple $(G_1, G_1', G_1'', E_1) \in D(B)$ such that $G' = G_1$*
   (b) *if G'' consists of more than one vertex, then there exists a 4–tuple $(G_2, G_2', G_2'', E_2) \in D(B)$ such that $G'' = G_2$*
   (c) *if G' consists of one vertex, then there exists* no *4–tuple $(G_3, G_3', G_3'', E_3) \in D(B)$ such that $G' = G_3$*
   (d) *if G'' consists of one vertex, then there exists* no *4–tuple $(G_4, G_4', G_4'', E_4) \in D(B)$ such that $G'' = G_4$.*

In part 2 of the definition, there is an additional symbol $\cup_E$, which is defined as follows.

DEFINITION 15.10 *Given two graphs $G_1 = (V_1, E_1, \mu_1, \nu_1)$ and $G_2 = (V_2, E_2, \mu_2, \nu_2)$, where $V_1 \cap V_2 = \emptyset$, and a set of edges $E' = (V_1 \times V_2) \cup (V_2 \times V_1)$ with a labeling function*

$\nu : E' \rightarrow L_E$, the union of G_1 and G_2 with respect to E' ($G_1 \cup_{E'} G_2$) is the graph $G = (V, E, \mu, \nu)$ such that

1. $V = V_1 \cup V_2$

2. $E = E_1 \cup E_2 \cup E'$

3. $\mu(v) = \begin{cases} \mu_1(v) & \text{if } v \in V_1 \\ \mu_2(v) & \text{if } v \in V_2 \end{cases}$

4. $\nu(v) = \begin{cases} \nu_1(v) & \text{if } e \in E_1 \\ \nu_2(v) & \text{if } e \in E_2 \\ \nu(v) & \text{if } e \in E'. \end{cases}$

From the conditions in Definition 15.9, we see that G' and G'' are two subgraphs of G that may be joined by the edges in the set E to form the graph G.

A decomposition of a graph G gives a set of 4–tuples that progressively break down G into smaller and smaller subgraphs, ending with the individual vertices. Each 4–tuple consists of an initial graph G, two subgraphs G' and G'', and the set of edges E, which join G' and G'' to form G.

The advantages of the decomposition network algorithm come from the observation that if we take two similar graphs that contain common subgraphs, there may be common 4–tuples in their decomposition. Merging the two decompositions, but representing each common subgraph only once, gives a network that represents two graphs but is more compact than the two decompositions in isolation. Figure 15.8 shows two model graphs and a decomposition that may be constructed to represent them. This shows the advantages of the representation, with two model graphs requiring only two internal nodes in the decomposition network. In practice, decompositions are not compiled in isolation; rather, an attempt is made to optimize the network produced for the initial set of models $M_1, M_2, \ldots, M_L$. Once a network has been constructed, additional graphs are compiled into it in an incremental fashion using heuristics to achieve a good decomposition.

There are numerous possible decompositions of a graph, some of which will be more suitable for combination with an existing decomposition network. The size of the decomposition network that represents a database of graphs will depend not only on the proportion of common structure but also on the strategies used in the construction of the decomposition. Various factors can be used to guide the construction, such as choosing the LCSG that may be found or minimizing the number of elements of the edge set E at each point. A more complete discussion of this can be found in Messmer's thesis [31].

Once a database of models is compiled into a single decomposition network, the search for subgraph isomorphisms from the models to the input can be performed using this network. The initial step is to map each vertex of the input graph to all possible single vertex nodes in the network. The nodes that are activated by this mapping are then merged, where possible, to give subgraphs of two vertices. At each merge step, given two activated nodes $(G_1, {G_1}', {G_1}'', E_1)$ and $(G_2, {G_2}', {G_2}'', E_2)$, the task is to determine the existence of a further node (G, G', G'', E) where

$$(G' = G_1 \wedge G'' = G_2) \vee (G' = G_2 \wedge G'' = G_1),$$

$$(E - (E_1 \cup E_2) = E') \wedge (G = G_1 \cup_{E'} G_2).$$

Any node that successfully merges two of its ancestors is also considered to be activated and passes the mapping it has discovered to its descendants. This process continues until it is no

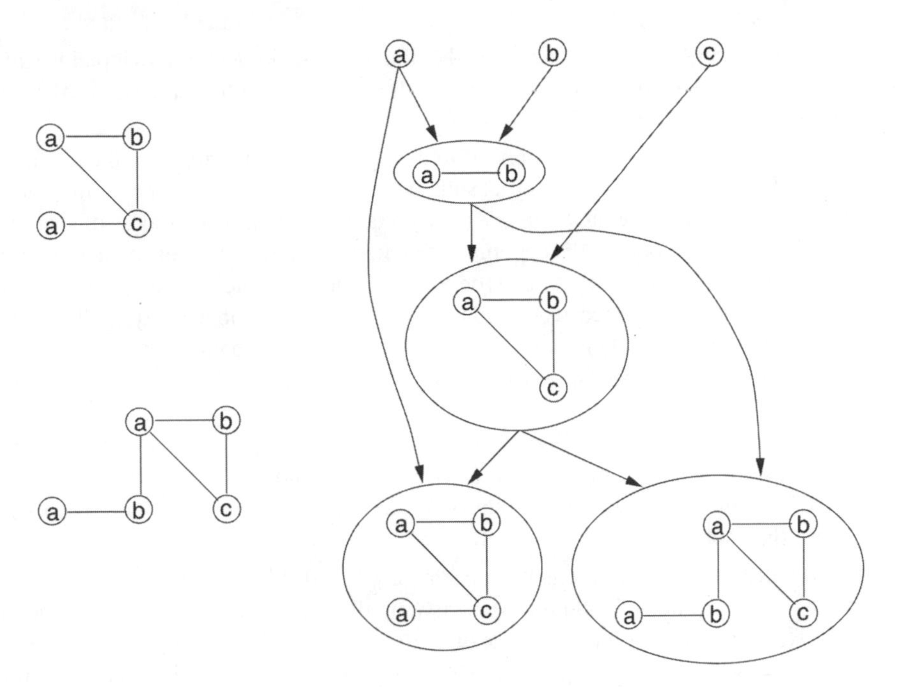

FIGURE 15.8
Two example graphs and their decomposition network.

longer possible to merge any further nodes, or a node (G, G', G'', E) representing a model graph from the database is reached.

Figure 15.9 shows an example graph and decomposition network, with the sets of vertices that activate nodes indicated on the arcs. There are two instances of vertex c in the input, both

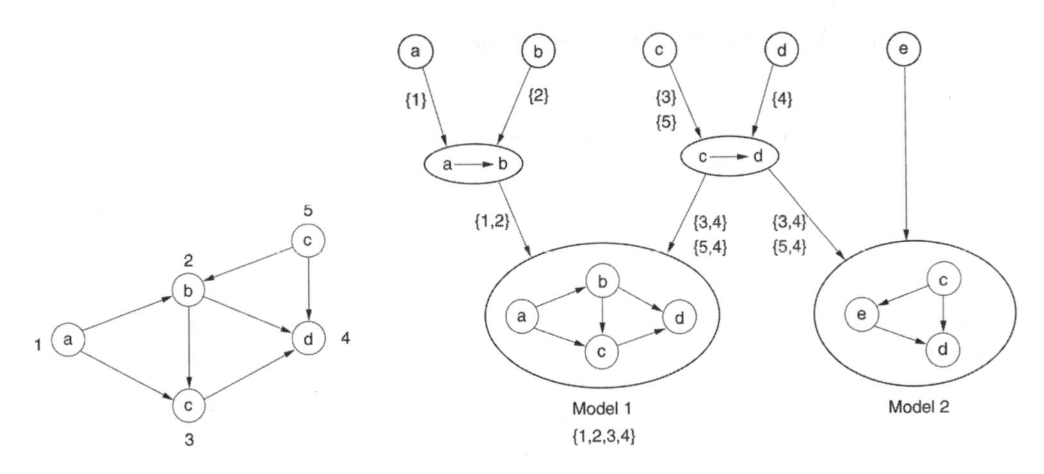

FIGURE 15.9
Node activation in a decomposition network. (a) Query input; (b) Network for two graphs.

of which contribute to a subgraph $c \to d$. This gives two mappings for the subgraph $c \to d$ in the decomposition network. Only one of these mappings is able to contribute to model 1. The other mapping $(5, 4)$ does not have edges in the input that conform to the edge set of the node representing model 1. Each network node $N_i = (G_i, G', G'', E_i)$ tests the mappings passed by its ancestors by examining each edge in the input between the mapped vertices and ensuring that these edges match the edge set E_i. If the edges are found to be correct, the new vertex mapping is passed to all children of N_i.

The network algorithm may be extended in a straightforward manner to inexact matching using an edit distance measure. The initial step for inexact matching is to perform matching as described above to determine the existence of any exact matches. The process of searching for exact matches activates nodes in the decomposition network with all the exact matches possible for the input. If there are no exact matches detected, or inexact matches are required regardless, the edit distance algorithm proceeds to find the least cost edit sequence that modifies the input to be isomorphic to a model. This can be achieved by introducing a null vertex mapping at each initial node in the decomposition network. Mapping a vertex to the null vertex δ represents deleting the vertex, with an associated cost. Each of the initial nodes is then inserted into the list of active nodes, sorted by increasing cost. The first element of this list is then taken from the list as the least cost inexact match to this point, and this least cost inexact match is propagated through the decomposition network. All nodes that are activated by this inexact mapping are then sorted into the list of active nodes.

Each time all possible activations have been sorted into the active node list, the least cost active node is again removed and processed. Eventually this leads to an active inexact match, which represents a model reaching the head of the list. At this point the cost of the inexact isomorphism between the input and the model is calculated and inserted into the active node list as the last active node with that cost, providing a natural cut point. Once a node representing a model graph that is marked as having been seen reaches the first element in the active node list, there can be no lower cost inexact isomorphism to a different model.

The algorithm used to determine the cost of each active node is clearly important to the computational complexity of this method. The experimental results for the decomposition method refer to an algorithm that uses the A* algorithm with look-ahead to determine the next active node to process.

15.5.1 Decomposition-Based LCSG Algorithm

The decomposition-based isomorphism detection algorithm may be separated into two parts. One part is the representation of multiple model graphs in a decomposition network, the other is the algorithm that uses the network to find graph or subgraph isomorphisms. Messmer and Bunke provide two algorithms that operate over the decomposition network, one to find exact isomorphisms from the models to the input and one to provide inexact isomorphism detection using an edit distance measure. These algorithms give examples of how the decomposition network representation may be employed for efficient isomorphism detection. Given the efficiency of the algorithms, it is reasonable to seek to use this representation to solve other problems for graph representations. As noted earlier, the edit distance formulation of inexact graph isomorphism detection is often inappropriate for a task; we therefore present an algorithm that uses a decomposition network to detect the LCSG between an input graph and a database of model graphs. This is not a simple modification of the edit distance algorithm, as the two measures require a fundamentally different approach.

The new algorithm for the detection of the LCSG introduces one additional symbol, a wild-card vertex label. This label, indicated by the symbol ?, is used to map vertices from the input graph for which there is no correct mapping in the models of the decomposition network. The initial step of the LCSG algorithm, as for the edit distance algorithm, is to detect exact isomorphisms between the input and the models. This has the effect of instantiating all possible mappings in the network. If there are no exact isomorphisms, a wild-card label is introduced at each initial node of the network. These wild-card mappings are then combined with other nodes to complete mappings for which there is no exact isomorphism. Figure 15.10 shows a decomposition network, annotated with the mappings for an example input. The absence of

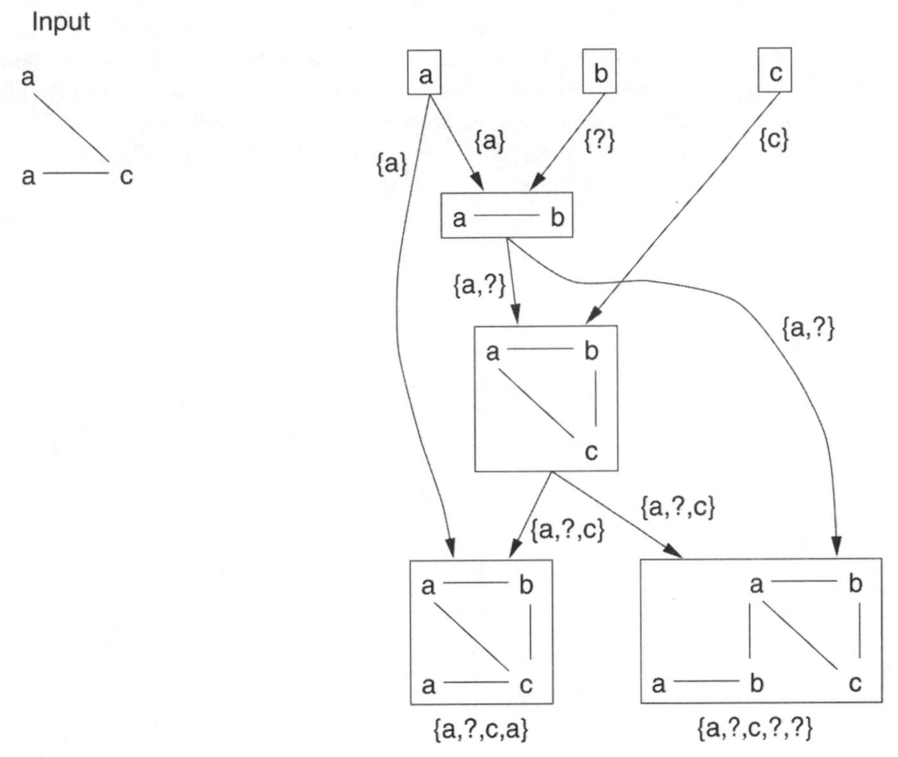

FIGURE 15.10
Example using wild-cards to complete mappings.

a vertex labeled b in the input is compensated for by mapping it to a wild-card. This allows detection of a subgraph of order three in the left-hand model. This mechanism is similar to the mapping of unmatched vertices in edit distance measures. The differences between LCSG and edit distance are:

- The LCSG algorithm is not concerned with possible edge mappings.
- A different control algorithm is required for detection of the best solution.

It remains to provide a control algorithm that uses the wild-card mappings to ensure all best solutions are detected, as efficiently as possible.

If all possible mappings of vertices in the decomposition network to the vertices in the union of input and the wild-card vertex label are examined, then the LCSG will be found. Since this will lead to an inefficient algorithm, a method of pruning the search space is required, as it was for the decision tree. When a permutation of the input adjacency matrix is performed in the decision tree LCSG algorithm, an input vertex is explicitly excluded from the isomorphism; thus, the size of the largest LCSG is decreased. This allows the implementation of an effective pruning process, as we know at each point the size of the largest possible common subgraph that a descent can yield. However, when a wild-card is included in a vertex mapping at a node in the decomposition network, one node of a model is excluded from the isomorphism. Thus, it is unknown whether the possible number of vertices in the LCSG has been reduced. This is because, although a vertex label is mapped to ? at one node, there may be another instance of the vertex label in another section of the network that is correctly mapped. The correct mapping may belong to a branch of the network that is merged with the branch of the wild-card mapping further down the network. This is illustrated in Figure 15.11. In Figure 15.11, the node containing a, b, and c (N_1) can only be activated by mapping b to ?. However, the vertex labeled b is not omitted from the final mapping because it is included in the node containing c, b, and e (N_2). The subgraphs of N_1 and N_2 are merged, combining their mappings to produce the best outcome. In fact, the vertex mapping at node N_1 that leads to the LCSG is a, ?, ?. This mapping at N_1 is merged with the mapping b, c, e at node N_2. Due to the lack of knowledge of mappings on other branches, it is not possible to determine the number of input vertices that may be mapped correctly at each node. It is hence necessary to find an alternative measure of fitness.

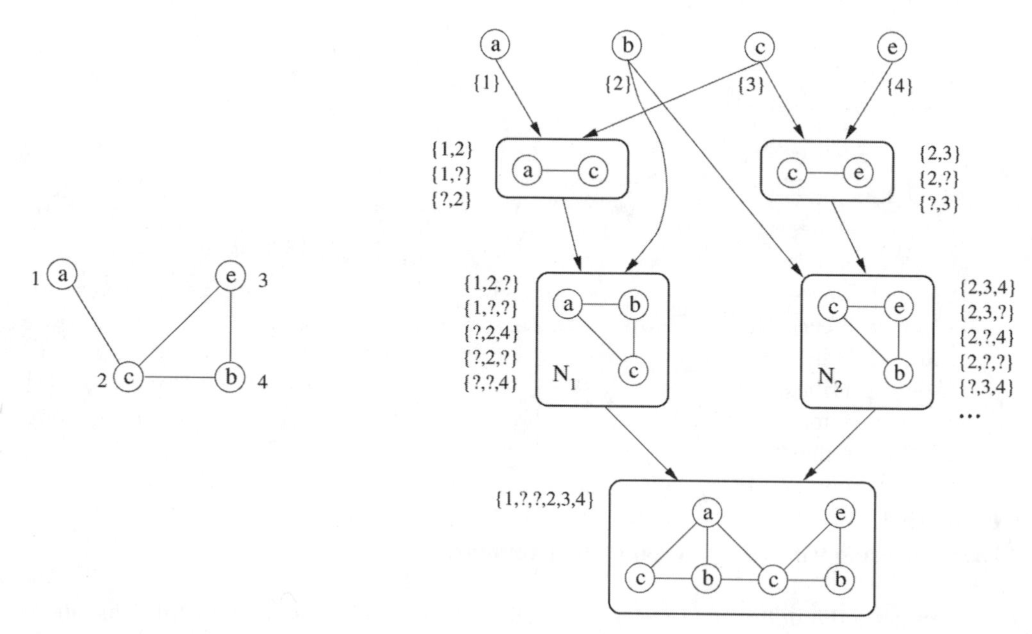

FIGURE 15.11
Merge example network.

The measure of fitness used in the edit distance algorithm progressively increases in a monotonic fashion during the descent of the network. At each node, the least error cost isomorphism is calculated as the fitness value to allow the use of an A* search strategy. A similar property is desired for the measure used for the LCSG algorithm.

The measure used by the LCSG algorithm to determine fitness of an activated node is the number of vertices mapped to a wild-card subtracted from the order of the largest model to which the node contributes. To enable this calculation to be made, additional information must be included in the network. During initial construction of the decomposition network, each new node has the order of that largest model to which the node contributes, set to the order of the initial model. As more models are added, each node that is visited in compiling a new model into the network compares the recorded maximum order of models contributed to, against the order of the new model, updating if the new model is larger. In Figure 15.12, each node of the decomposition network has been annotated with the order of the largest model to which the node contributes, and the fitness measure for the example input. Note that the node representing the subgraph $\{c, d\}$ has a second possible node mapping indicated, with one node mapped to ?. Such additional mappings are possible at each network node; however, this particular mapping is the only alternative mapping required for classification of the input graph provided, and so

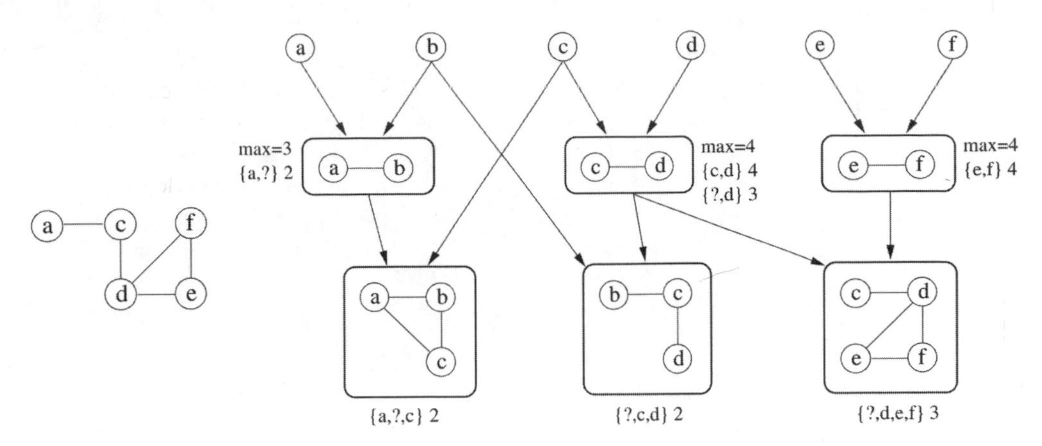

FIGURE 15.12
Fitness measure example.

it is the only additional mapping indicated in the figure. This fitness measure gives an upper limit to the possible size of the LCSG to which each network node can contribute, and this measure is monotonically decreasing over the network descent. Unfortunately, the measure provides only an estimate of the best order, which may be a large overestimate in early parts of the network, becoming more tightly constrained during the descent. The measure could be limited initially by the order of the input; however, the original measure guides the search to nodes that contribute to the largest models first, which is a reasonable initial guideline. This fitness measure may be refined and greatly improved by introducing a vertex frequency table at each node of the decomposition network.

The main source of inaccuracy when calculating the fitness measure at a node is that we do not know whether a vertex that is mapped to a wild-card may be correctly mapped in a separate branch of the network. For example, given the input graph of Figure 15.13a and the decomposition network from Figure 15.10, the node containing the vertices a, b, and c should have the mapping ?, b, c. At this point we cannot say that the maximum possible LCSG has been reduced to two, because there is another branch of the network that may produce a mapping for the vertex labeled a. In fact, in that example, another branch does produce a mapping for the vertex labeled a, which leads to a common subgraph of order three. This illustrates why an overestimate must be used for the fitness measure. However, in some situations it is possible

FIGURE 15.13
Example graphs. (Left) Example input graph. (Right) Example frequency table.

to improve the fitness measure by recording, at each node, the maximum number of instances of each vertex label that occur in any models to which the node contributes. This information is stored in the *vertex frequency* table for each node. In Figure 15.14 the nodes labeled N_1 and N_2 have their vertex frequency tables shown in Table 15.2.

If a vertex contained in the input graph is mapped to ?, the vertex frequency table can be examined to see whether there is any other instance of the vertex that may be correctly mapped. In Figure 15.14 there is no more than one vertex labeled a in any of the models to which a node

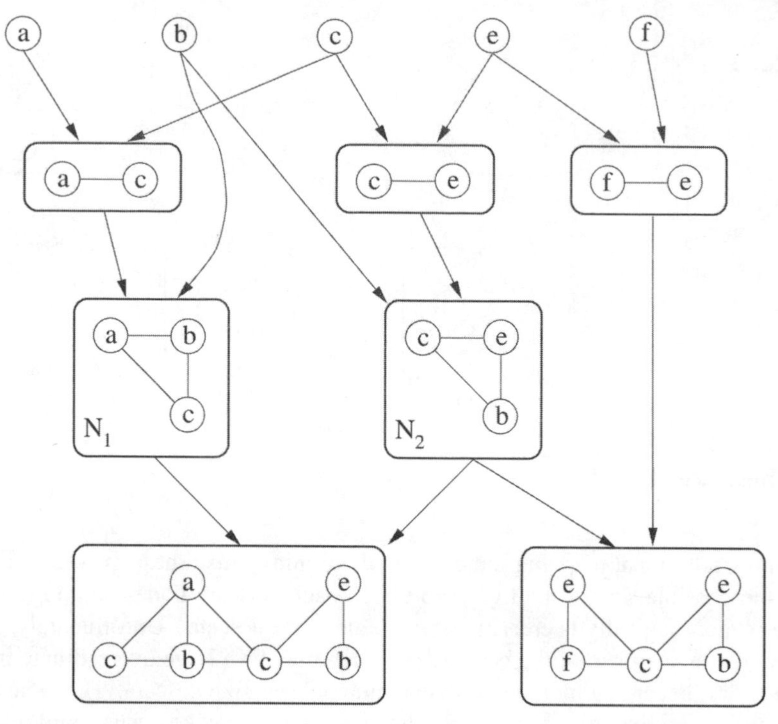

FIGURE 15.14
Graph for vertex frequency tables.

Table 15.2 Vertex Frequency Tables for Two Nodes

Node N_1		Node N_2	
Label	Number	Label	Number
a	1	a	1
b	2	b	2
c	2	c	2
e	1	e	2
		f	1

N_1 contributes, so that if node N_1 has mapped a vertex labeled a to ?, the best possible order for the LCSG is the order of the input minus one. The graph in Figure 15.13b illustrates this case. When this input is classified using the network of Figure 15.14, node N_1, which contains the vertices a, b, and c, is reached with the mapping ?, b, c, with a mapped to ? because there is no edge between a and c. The vertex frequency table for node N_1 is shown in Table 15.2. Thus, the a that has been mapped to a wild-card is the only vertex labeled a in any of the models that may be reached from this node. Therefore, the LCSG order that may be achieved from this node is three, which is the order of the input minus one.

In addition to this, there is an extremely useful additional effect that can be implemented using the vertex frequency table and which can greatly reduce the search space. At each initial node of the decomposition network, the vertices of the input can be compared to the vertex frequency table of the node to determine the number of common vertices. Using the number of common vertices, rather than the order of the largest model to which a node contributes, gives a much more accurate measure of the possible order for the LCSG. This measure can be computed at each node during matching, and the number of vertices known to be mapped out can be subtracted from this new estimate. In practice, this reduces computation time by a factor proportional to the number of nodes in the input.

This addition of the vertex frequency table gives an efficient algorithm for detection of the LCSG between an input graph and a database of models that has a reasonable space requirement such that it may be used for general applications. Video and image databases generally provide a high level of similar structure, making the decomposition network representation an attractive option. The theoretical computational complexity of the LCSG algorithm is lower than that for a general edit distance measure, although, as has been noted, the computational complexity of an edit distance algorithm is dependent on the costs applied to the edit operations.

15.6 Results of Tests Over a Video Database

The video database used in the experiments was drawn mainly from our Campus Guide database [20, 21]. These video clips depict a guide walking between various locations on the Curtin University of Technology campus. In addition to the clips from the Campus Guide, there are a number of other clips of park and city scenes and a small number of disparate clips of completely different types of scenes.

Retrieval examples can be seen in Figures 15.15 and 15.16. In Figure 15.15, a single iconic sketch, shown in Figure 15.17a, has been presented to the query system. This query requests

FIGURE 15.15
Retrieval of video segments using spatial relationships.

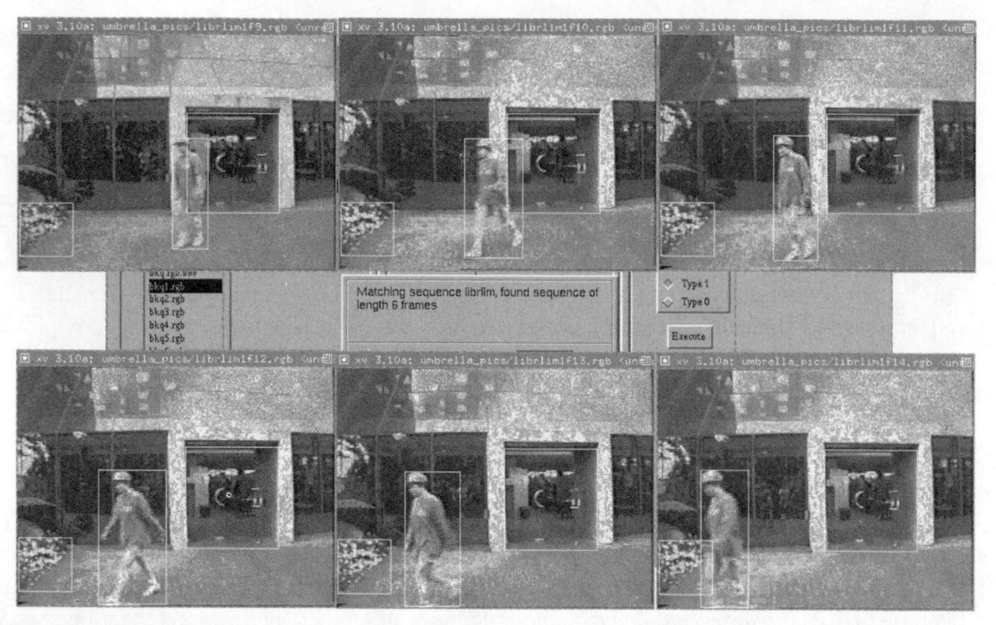

FIGURE 15.16
Retrieval of video segments using a sequence of sketched spatial relationships.

an image or video frame with objects Guide, Exit, and Garden, where the Guide overlaps with the Exit. The retrieval results in the detection of two matching frames in two sequences. Figure 15.15 shows the interface along with the image display tool for each of the sequences, as well as the control window for each of the image display tools giving a choice of two frames for the sequence displayed.

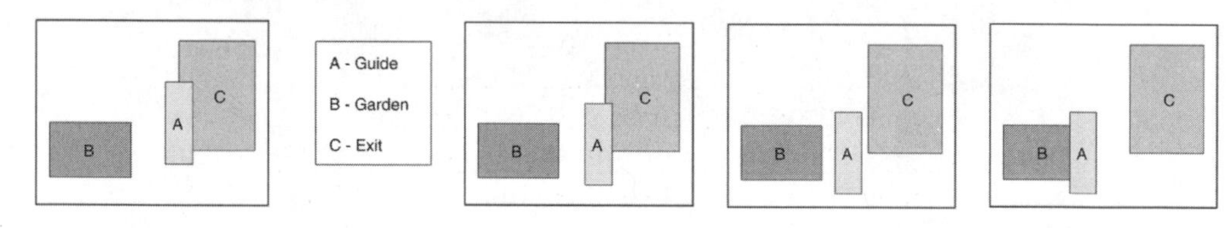

FIGURE 15.17
Iconic sketches used as queries for video retrieval figures.

Figure 15.16 shows the results of retrieval for the sequence of iconic sketches in Figure 15.17. The initial sketch in this sequence is the sketch from the retrieval of Figure 15.15. In Figures 15.17b, c, and d we see the object Guide moving left away from the object Exit to overlap the object Garden. Figure 15.16 shows the interface along with the image display tool showing each of the frames from the video sequence that constitute a match for the series of iconic sketches.

The database used in the experiments consists of clips between 4 and 20 s. The number of objects contained in each clip ranges from 12 to 19 objects. The number of changes in object relationships is important for the representation used and varies from 71 changes in the shortest clip to 402 in the longest clip. This number of changes per second is typical, although an increased number of objects in the video will lead to a greater number of changes per second.

All times given in the tables are in milliseconds and are averaged over a number of executions. Results from two types of tests are reported:

- Results from a number of different queries, averaged to give an indication of general performance
- Tests of a single query, intended to test a specific characteristic.

Table 15.3 is an example of the first type of test. In this table we report the results of a variety of queries over the database, in order to give a general measure for the performance of the algorithms. The queries in this table are a selection of frames from the videos included

Table 15.3 General Performance of Exact Decomposition vs. Ullman's Algorithm

Algorithm	Mean	Min	Max	σ
Ullman	624.4 μs	403 μs	982 μs	186.1 μs
Network exact	45.2 μs	27 μs	86 μs	19.5 μs
Network inexact	176.1 μs	145 μs	228 μs	23.5 μs

in the database, each one of which has at least one exact match. The results are given as minimum, maximum, mean, and standard deviation (σ) and are intended to assess the general performance of the exact algorithms.

In Table 15.4, the second type of test is presented. Here we are interested in the results of a single query with a particular characteristic. In this table the effect of increases in the resultant error measure is being assessed, so each query generates a different error measure value and is a specific individual test. The results of each query are reported as a separate measure in order to assess specific facets of the performance of the algorithms. Results in both of these tables are discussed in detail in the following sections.

Table 15.4 Effect of Increasing Error Distance on Isomorphism Detection

A* With Look-Ahead			Inexact Decomposition		
Query	Error	Time	Query	Error	Time
liblr.10	0	10737 μs	liblr.10	0	172 μs
liblr.0	0	9073 μs	liblr.0	0	164 μs
wayq	6	9122 μs	wayq	6	223 μs
libq	12	35674 μs	libq	12	2851 μs

15.6.1 Decomposition Network Algorithm

The decomposition network algorithm finds models isomorphic to a subgraph of the input graph, that is, model graphs that are the same as some part of the input graph. This is not the form of isomorphism that is generally of interest in video database retrieval. The most common query operations encountered in image and video database work are similarity retrieval and subpictures of the models isomorphic to the input. The user will typically create a pictorial query by arranging icons in an example frame; retrieval is then the task of finding a subpicture in a model that is similar to the input picture.

Although the exact decomposition algorithm is not suitable for either of these tasks, it may be used to find components that make up a picture, and would be useful in detecting compound objects within a picture. Such a technique may be useful in the automated classification of images and video, by registering components of an image, and also in object tracking. However, there are a number of facets in applying the decomposition algorithm to a real problem, which will need to be studied.

The initial test was performed by loading all clips from the guide database and detecting isomorphisms to a single example frame chosen from each clip. Thus, for every clip in the database, a random frame was chosen, and each of these frames was presented to the system as a query. This is a simple test of the performance of the decomposition algorithm in general, over a variety of differing queries. Table 15.3 gives the average, maximum, minimum, and standard deviation for the decomposition exact algorithm and Ullman's algorithm. This table clearly illustrates the advantage of the decomposition algorithm. The times taken are an order of magnitude less than those for the best standard algorithm. The relative variation between the minimum and maximum for the decomposition algorithm was much higher than for the standard algorithm; however, because the worst case for the decomposition algorithm was approximately one fifth the best time for Ullman's algorithm, this cannot be a major criticism. The relative variations in performance between queries was similar for Ullman's algorithm and the decomposition exact algorithm; that is to say, they produce better or worse performances on the same queries.

It should be noted that all the clips in the guide database draw their key objects from a similar domain, which could contribute to a bias in the test. In order to examine this effect we used a database of clips of fish tanks to form a base decomposition network, then added clips of children playing in a park. These clips contain mostly disjoint object sets; thus, any bias should become apparent. Initially, half of the fish tank clips were compiled into the database, and a representative frame from each clip was used as a query to assess the performance of both algorithms. The remainder of the fish tank clips was then added and the algorithms were tested again. Finally, a number of clips of park scenes equal to half the number of fish tank scenes were added, and a final series of tests was performed. The results are presented in Table 15.5.

Table 15.5 Effect of Dissimilar Clips

	Ullman	Network
Half fish	140 μs	114 μs
All fish	277 μs	137 μs
Fish and park	365 μs	149 μs

As shown in Table 15.5, the performance of the decomposition algorithm does not degrade significantly with the introduction of dissimilar objects. The working set size of the decomposition exact algorithm was comparable to that of Ullman's algorithm, differing by less than 10%. This evidence supports the statement that for the purpose of detecting database pictures, which are the same as subpictures of an input picture, the decomposition model offers a highly efficient and well-conditioned solution.

15.6.2 Inexact Decomposition Algorithm

Error-correcting algorithms are now considered over the guide database of clips, comparing the A* (with look-ahead) algorithm with the decomposition inexact algorithm. The initial test performed for Ullman's algorithm and the exact decomposition algorithm was also undertaken using the inexact decomposition algorithm. The results, as seen in Table 15.3, indicate a performance improvement similar to that of the exact decomposition algorithm. Although the inexact decomposition algorithm is, as expected, somewhat slower than the exact algorithm, it is still considerably quicker than the best standard algorithm for exact matching. This gives a clear indication of the performance possible using preprocessing methods.

Table 15.3 gives performance for only those queries for which there exists an exact graph isomorphism between the input graph and at least one model. More important for inexact algorithms is their performance when no exact match exists. Both the A* and decomposition inexact algorithms display response times that increase greatly as the error measure of the closest match increases. This is due to the exponential increase in equal cost alternatives as the nest error value is increased.

The GUB toolkit system [31] used provides the following operations which contribute to the error distance measure: substitution of vertex/edge labels, deletion of vertices/edges,and insertion of vertices/edges. The cost assigned to each of these components for the video database tests is given in Table 15.6. Insertion and deletion of edges is assigned an infinite cost, since the representation used yields a complete graph for each picture or frame. The range of costs for substitution of edge labels is [0–17], with the cost of substitution being zero for the same relationship and increasing with the number of relationships separating the two labels (see Figure 15.1).

Table 15.6 Costs Applied for the Edit Distance Algorithm

	Vertex	Edge
Deletion	1	∞
Insertion	1	∞
Substitution	5	[0–17]

A major factor in the time required to satisfy queries is the error measure value. Table 15.4 shows the execution time required by the A* algorithm and the inexact decomposition algorithm for four different queries against the database. The queries used in the table have increasing error values and show the rate of increase for the two algorithms. Even in the worst case, the inexact decomposition algorithm has a sufficient performance edge over the traditional algorithm, to produce a far more rapid response, although a comparison of queries with error measure 6 and 12 shows a fourfold increase in execution time for the A* algorithm and a factor of 12 increase for the inexact decomposition. Table 15.7 shows a similar trend, with greater detail because it was produced by beginning with a frame from one of the sequences as the query, then altering it to gradually increase the error returned.

The results in Table 15.4 indicate that although the network inexact method is considerably faster than the A* algorithm, it may not be sufficiently rapid to be used when a large error is expected. For a similarity measure that may be limited with respect to the maximum error allowed, the network inexact algorithm would be a useful tool. Given an error limit, the search could be pruned significantly when processing the database. This would indicate that when a well-directed search is being performed, we could successfully use the network inexact algorithm. However, for an exhaustive search of a database, we would expect poor performance from the network inexact algorithm, with response time possibly exceeding reasonable limits. However, it is important to remember that the performance returned will be an improvement over current algorithms.

The space requirement of the network algorithm, in both exact and inexact forms, is only marginally greater than that of the traditional algorithms.

Table 15.7 Effect of Increasing Error Distance on the Inexact Decomposition Algorithm

Inexact Network			
Error	Time	Error	Time
1	172 μs	7	558 μs
2	184 μs	8	838 μs
3	207 μs	9	883 μs
4	227 μs	10	1772 μs
5	328 μs	11	2474 μs
6	350 μs	12	3273 μs

15.6.3 Decision Tree

The decision tree algorithm provides easily the most rapid classification. In Table 15.8 the decision tree algorithm is compared to the fastest of the previously discussed algorithms, the exact network algorithm. The decision tree algorithm is clearly faster, with mean added to standard deviation for the decision tree algorithm being less than the mean for the exact network algorithm. The results in this table were taken over two thirds of the guide database. The final third could not be compiled into the decision tree because the number of objects in the clips caused too great a space requirement. The clips that caused this difficulty have 19 objects each. Thus, the decision tree network provides extremely rapid classification for applications where the number of vertex and edge labels is limited. The limit depends on various factors but can in general be thought of as approximately 15 labels of each type. As the number of labels increases beyond this, the decision tree quickly increases to greater than practical size.

Table 15.8 General Performance of the Decision Tree Algorithm

	Mean	Min	Max	σ
Exact network	27.4 μs	15 μs	39 μs	7.1 μs
Decision tree	14.1 μs	3 μs	36 μs	10.9 μs

15.6.4 Results of the LCSG Algorithms

The results in Table 15.9 parallel those in Table 15.4 for edit distance algorithms. Both tables present classification time for a sequence of queries with increasing error. Comparing the results in Table 15.9 against the results in Table 15.4 shows that both of the LCSG algorithms are faster than any of the edit distance algorithms. The decision tree (DT) LCSG algorithm is, as expected, far faster than any other algorithm. The decomposition network (DN) LCSG algorithm is slower than the decision tree LCSG algorithm, but it is still significantly faster than the decomposition algorithm based on edit distance.

In Table 15.10, results are presented for classification between an input and a single model graph. Each of the models was constructed to provide worst-case performance from the LCSG algorithms. Thus, the LCSG algorithms are disadvantaged by the construction of the graphs and also by the lack of multiple graphs to share common structure. Examination of the results in Table 15.10 shows that even in this worst-case scenario the LCSG algorithms

Table 15.9 Performance of LCSG Algorithm

Query	DT LCSG	DN LCSG
liblr.10	26 μs	78 μs
liblr.0	22 μs	65 μs
wayq	7 μs	46 μs
libq	14 μs	72 μs

both outperform the A* algorithm by a large margin. The decision tree LCSG algorithm and the network algorithm display similar performance, with the decomposition network LCSG algorithm slightly slower, as expected given the construction of the models.

Table 15.10 Approximate Match Between Two Graphs

Query	A*	Network	DT LCSG	DN LCSG
15.1	158 μs	23 μs	22 μs	34 μs
15.2	160 μs	22 μs	23 μs	35 μs
15.3	164 μs	22 μs	22 μs	34 μs

Table 15.11 Approximate Match Against 11 Graphs

Query	A*	DN Inexact	DT LCSG	DN LCSG
11.6	282 μs	30 μs	17 μs	38 μs
12.2	53 μs	15 μs	5 μs	14 μs
14.1	84 μs	18 μs	7 μs	16 μs
15.2	642 μs	34 μs	23 μs	45 μs
15.3	692 μs	36 μs	22 μs	46 μs
16.1	136 μs	20 μs	8 μs	20 μs

Table 15.11 extends the results from Table 15.10 to multiple model graphs. Each of the 11 model graphs used in this experiment were constructed to deliberately produce poor performance from the LCSG algorithms. These results once again show the advantage of the decision tree algorithm, with the decision tree LCSG being by far the most rapid. With the extension to multiple graphs, the margin between the two decomposition network algorithms is decreased. The edit distance-based decomposition network algorithm is still faster in most cases, but by a smaller margin than for single model classification. The A* algorithm is far slower than any of the other algorithms in all cases, even for so small a number of model graphs.

Discussion of the results presented in this section must include the consideration that the edit distance algorithms (A* and DN inexact) do perform different computations from the LCSG algorithms. However, the algorithms are used for the same application, so comparison is useful. The nature of the different computations will mean that for the majority of applications the LCSG version of the decomposition network algorithm will perform better than the edit distance-based version. The decision tree LCSG algorithm provides the most rapid performance in every case, even with models constructed to provide worst-case performance. The choice between the two LCSG algorithms will be based on the characteristics of the problem, which will determine the space requirement for the decision tree LCSG algorithm.

The decision as to whether an edit distance or LCSG algorithm is most suitable for a task will usually be determined by the problem itself. In most cases a task will clearly favor either edit distance or LCSG as a measure of similarity, with one providing a more appropriate correspondence to the problem. The final consideration is the amount of common structure between the model graphs, with the new algorithm's performance improving as the amount of common structure increases.

15.7 Conclusion

This chapter has presented new algorithms in the field of graph and subgraph isomorphism detection. Three of these algorithms — the decomposition, inexact decomposition, and exact decision tree algorithms — were developed in earlier work. The contributions of this chapter are to examine the performance of these algorithms over actual video database data and the presentation of two new algorithms for largest common subgraph detection between an input graph and a database of model graphs.

The existing algorithms are suitable for certain problems in image and video database retrieval; however, each is limited in this application. The two largest common subgraph algorithms provide suitable solutions to the task of similarity retrieval for images and video in large databases. The strength of the two new algorithms is shown in the results over video database data. Each of the new algorithms provides considerably better online classification time than previous algorithms. Experiments performed show that even for worst-case situations the new algorithms perform very well.

Either of the new algorithms presented in this chapter provides a tractable solution to similarity retrieval for large databases of relational structures. As the capability of desktop computing equipment continues to increase, and the prices continue to decrease, a greater variety of complex information types will become available. Given a large database of such information known a priori, which can be indexed and retrieved using a graph representation, the new largest common subgraph algorithms may prove highly useful.

References

[1] S. Chang and S. Liu, "Picture indexing and abstraction techniques for pictorial databases," *IEEE Transactions on Pattern Analysis and Machine Intelligence,* vol. 6, pp. 475–484, July 1984.

[2] H. Tamura and N. Yokoya, "Image database systems: A survey," *Pattern Recognition,* vol. 17, no. 1, pp. 29–43, 1984.

[3] S. Chang, Q. Shi, and C. Yan, "Iconic indexing by 2D strings," in *Proceedings of the IEEE Workshop on Visual Languages,* Dallas, TX, June 1986. Also in *IEEE Transactions on Pattern Analysis and Machine Intelligence,* vol. 9, pp. 413–428, May 1987.

[4] S.K. Chang, C.W. Yan, D.C. Dimitroff, and T. Arndt, "An intelligent image database system," *IEEE Transactions on Software Engineering,* vol. 14, pp. 681–688, May 1988.

[5] S.K. Chang and Y.Li, "Representation of multi-resolution symbolic and binary pictures using 2D H-strings," in *IEEE Workshop on Language for Automation,* pp. 190–195, 1988.

[6] S. Chang, E. Jungert, and T. Li, "Representation and retrieval of symbolic pictures using generalized 2D strings," *SPIE Proceedings of Visual Communications and Image Processing IV,* vol. 1199, pp. 1360–1372, 1989.

[7] M.-C. Yang, "2D B-string representation and access methods of image database," Master's thesis, Department of Computer Science and Information Engineering, National Chiao Tung University, Hsinchu, Taiwan, July 1990.

[8] T. Arndt and S. Chang, "An intelligent image database system," *Proceedings of the IEEE Workshop on Visual Languages,* pp. 177–182, 1989.

[9] C.C. Chang and S.Y. Lee, "Retrieval of similar pictures on pictorial databases," *Pattern Recognition,* vol. 24, no. 7, pp. 675–680, 1991.

[10] T. Arndt and S.-K. Chang, "Image sequence compression by iconic indexing," in *1989 IEEE Workshop on Visual Languages,* pp. 177–182, IEEE Computer Society, October 1989.

[11] D. Daneels, D. Van Campenhout, W. Niblack, W. Equitz, R. Braber, E. Bellon, and F. Firens, "Interactive outliner: An improved approach using active geometry features," in *IS&T/SPIE Conference on Storage and Retrieval for Image and Video Databases II,* pp. 226–233, 1993.

[12] S. Ayer and H.S. Sawhney, "Layered representation of motion video using robust maximum-likelihood estimation of mixture models and MDL encoding," in *Proceedings of the IEEE International Conference on Computer Vision,* June 1995.

[13] J. Ashley, R. Barber, M. Flickner, J. Hafner, D. Lee, W. Niblack, and D. Petkovic, "Automatic and semi-automatic methods for image annotation and retrieval in QBIC," *SPIE Proceedings of Storage and Retrieval for Image Video Databases III,* pp. 24–35, 1995.

[14] A. Blake, R. Curwen, and A. Zisserman, "Affine-invariant contour tracking with automatic control of spatiotemporal scale," in *Proceedings of the 4th International Conference on Computer Vision,* pp. 66–75, 1993.

[15] D. Chetverikov and J. Verstóy, "Tracking feature points: A new algorithm," in *Proceedings of the International Conference on Pattern Recognition* (A.K. Jain, S. Venkatesh, and B.C. Lovell, eds.), pp. 1436–1438, IAPR, IEEE, August 1998.

[16] Y. Mae and Y. Shirai, "Tracking moving object in a 3–D space based on optical flow and edges," in *Proceedings of the International Conference on Pattern Recognition* (A.K. Jain, S. Venkatesh, and B.C. Lovell, eds.), pp. 1439–1441, IAPR, IEEE, August 1998.

[17] L. Gu, "Scene analysis of video sequences in the MPEG domain," in *Proceedings of the IASTED International Conference: Signal and Image Processing,* October 1997.

[18] M.A. Smith and T. Kanade, "Video skimming and characterization through the combination of image and understanding techniques," in *IEEE International Workshop on Content Based Access of Image and Video Databases,* pp. 61–70, 1998.

[19] N.O. Stoffler and Z. Schnepf, "An MPEG-processor-based robot vision system for real–time detection of moving objects by a moving observer," in *Proceedings of the International Conference on Pattern Recognition,* (A.K. Jain, S. Venkatesh, and B.C. Lovell, eds.), pp. 477–481, IAPR, IEEE, August 1998.

[20] K.R. Shearer, S. Venkatesh, and D. Kieronska, "Spatial indexing for video databases," *Journal of Visual Communication and Image Representation,* vol. 7, pp. 325–335, December 1997.

[21] K.R. Shearer, D. Kieronska, and S. Venkatesh, "Resequencing video using spatial indexing," *Journal of Visual Languages and Computing,* vol. 8, 1997.

[22] S. Lee and F. Hsu, "Spatial reasoning and similarity retrieval of images using 2D C-string knowledge representation," *Pattern Recognition,* vol. 25, no. 3, pp. 305–318, 1992.

[23] S. Lee, M. Shan, and W. Yang, "Similarity retrieval of iconic image database," *Pattern Recognition,* vol. 22, no. 6, pp. 675–682, 1989.

[24] S. Lee, M. Yang, and J. Chen, "Signature file as a spatial filter for iconic image database," *Journal of Visual Languages and Computing,* vol. 3, pp. 373–397, 1992.

[25] B.T. Messmer and H. Bunke, "A new algorithm for error-tolerant subgraph isomorphism detection," *IEEE Transactions on Pattern Analysis and Machine Intelligence,* vol. 20, pp. 493–504, May 1998.

[26] B.T. Messmer and H. Bunke, "Error-correcting graph isomorphism using decision trees," *International Journal of Pattern Recognition and Artificial Intelligence,* vol. 12, pp. 721–742, September 1998.

[27] H. Bunke and B. Messmer, "Recent advances in graph matching," *International Journal of Pattern Recognition and Artificial Intelligence,* vol. 11, no. 1, 1997.

[28] G. Levi, "A note on the derivation of maximal common subgraphs of two directed or undirected graphs," *Calcolo,* vol. 9, pp. 341–354, 1972.

[29] B.T. Messmer and H. Bunke, "Subgraph isomorphism detection in polynomial time on preprocessed model graphs," in *Second Asian Conference on Computer Vision,* pp. 151–155, 1995.

[30] J.F. Allen, "Maintaining knowledge about temporal intervals," *Communications of the ACM,* vol. 26, pp. 832–843, November 1983.

[31] B.T. Messmer, "Efficient graph matching algorithms for preprocessed model graphs," Ph.D. thesis, Institut fur Informatik und angewandte Mathematik, Universitat Bern, Switzerland, 1995.

[32] J.R. Ullman, "An algorithm for subgraph isomorphism," *Journal of the Association for Computing Machinery,* vol. 23, no. 1, pp. 31–42, 1976.

Chapter 16

Video Transcoding

Tzong-Der Wu, Jenq-Neng Hwang, and Ming-Ting Sun

16.1 Introduction

Video transcoding deals with converting a previously compressed video signal into another compressed signal with a different format or bit rate. As the number of different video compression standards (e.g., H.261, H.263, MPEG-1, MPEG-2, MPEG-4) increases and the variety of bit rates at which they are operated for different applications increases, there is a growing need for video transcoding. In this chapter, we focus on the specific problem of transcoding for bit rate reduction.

Why do we need bit rate reduction? In transmitting a video bitstream over a heterogeneous network, the diversity of channel capacities in different transmission media often gives rise to problems. More specifically, when connecting two transmission media, the channel capacity of the outgoing channel may be less than that of the incoming channel so that bit rate reduction is necessary before sending the video bitstream over the lower-bit-rate channel. In the application of video on demand (VOD), the server needs to distribute the same encoded video stream to several users through channels with different capacities, so the encoded video stream needs to be transcoded to specific bit rates for each outgoing channel. This problem also occurs in multipoint video conferencing, where the bundle of multiple video bitstreams may exceed the capacity of the transmission channel and require a bit rate reduction.

The simplest way to implement transcoding is to concatenate a decoder and an encoder. The decoder decompresses the bitstream, which was encoded at a bit rate R_1, and then the encoder encodes this reconstructed video at a lower bit rate R_2. This strategy is relatively inefficient because very high computational complexity is required. Because there is some reusable information in the incoming bitstream and there are common parts in both the decoder and the encoder, simplifying this cascaded architecture to reduce the complexity is possible.

Keesman et al. [1] introduced an architecture by assuming that the motion vectors decoded from the bitstreams can be reused by the encoder. Based on the principle that the discrete cosine transform (DCT) and the inverse DCT (IDCT) are linear operations, they reduced the number of DCT and IDCT modules from 3 to 2, and the number of frame memory from 2 to 1. Moreover, they merged the common motion compensation (MC) modules in the decoder and encoder. A similar transcoder was also presented in [2]. These architectures result in transcoders with much reduced complexity.

In [1] and [2], the whole transcoding process is performed in the DCT domain except for the MC module. To perform the motion compensation, the macroblocks in the DCT domain need to be transformed to the pixel domain. The motion-compensated macroblocks in the pixel domain also need to be transformed back to the DCT domain for further processing. In the simplified architecture, the most time-consuming parts are in the DCT and IDCT modules, so it is highly desirable if these two modules can be avoided in this architecture. Using the DCT domain interpolation algorithm developed for transform domain video composition [3], the transcoding can be performed without the DCT and IDCT modules.

As described above, to speed up the transcoding operation, a video transcoder usually reuses the decoded motion vectors when re-encoding the video sequences at a lower bit rate. In many situations, frame-skipping may occur in the transcoding process. When frame-skipping occurs in the transcoder, the motion vectors from the incoming bitstream are not directly applicable because the motion estimation of the current frame is no longer based on the immediately previous frame. The motion vectors referring the current frame to its previous nonskipped frame need to be obtained. To reduce the computational complexity of obtaining these motion vectors, a combination of bilinear interpolation and search range reduction methods can be used [4]. The bilinear interpolation roughly estimates the motion vectors from the incoming motion vectors. After the bilinear interpolation, motion estimation can be performed in a smaller search range to fine-tune the estimated motion vectors.

For video transport over a narrow-bandwidth channel, usually even a good bit rate control strategy cannot allocate sufficient bits to each frame, specifically when the frames are very complex or the objects in the frame are very active. In these cases, frame-skipping is a good strategy to both conform to the limitation of the channel bandwidth and maintain the image quality at an acceptable level. Actually, low-bit-rate video coding standards such as H.261 and H.263 allow frames to be skipped. If the frames are randomly skipped, some important frames may be discarded, resulting in an abrupt motion. To determine the best candidates for frame-skipping, a dynamic frame-skipping approach was introduced to tackle this problem [4, 5]. This approach can be applied to video transcoders for the bit rate reduction.

Video transcoding is also used for video combining in a multipoint control unit (MCU). The MCU is a central server that controls multiple incoming video streams to each conference participant in a multipoint conference. To combine the multiple incoming video streams, the MCU can perform the multiplexing function in the coded domain, but this simple scheme is not flexible and cannot take advantage of image characteristics to improve the quality of the combined images. With a transcoding approach for video combining, it is possible to achieve better video quality by allocating bits in a better way [6].

This chapter is organized as follows. Section 16.2 describes the simplified architecture of a pixel domain transcoder. Section 16.3 discusses the DCT domain transcoder. The motion vector interpolation scheme and the dynamic frame-skipping strategy will be introduced in Section 16.4. Section 16.5 compares the coded domain approach and transcoding approach for video combining. A brief summary is presented in Section 16.6.

16.2 Pixel-Domain Transcoders

16.2.1 Introduction

In this section, the pixel domain transcoders for dynamically adjusting the bit rate according to the new bandwidth constraints are discussed. The basic architecture of a pixel domain transcoder can be a decoder followed by an encoder. However, due to the high computational complexity, it is not very efficient. Since the incoming bitstream contains a lot of useful information, it need not be decoded completely before re-encoding. Some data such as motion vectors can be reused. The header information at each coding level can help the transcoder do a more efficient encoding. The following section describes the simplified architecture proposed in [1].

16.2.2 Cascaded Video Transcoder

The basic structure of the cascaded transcoder is as presented in Figure 16.1. This configuration is called a cascaded architecture, which consists of a decoder followed by an encoder. In Figure 16.1, the transcoder receives a video sequence coded at a bit rate R_1 and converts it into a lower bit rate R_2.

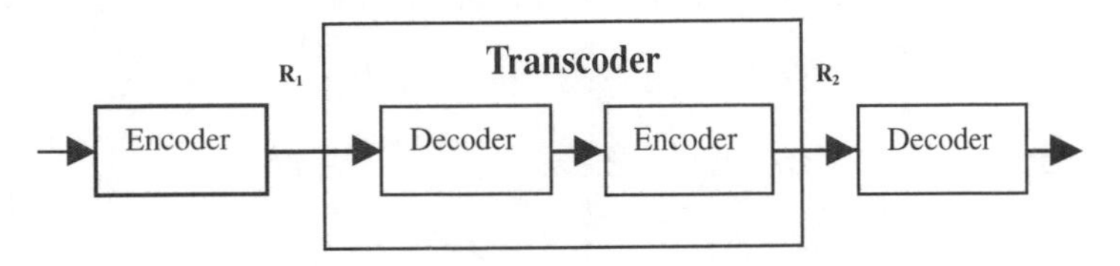

FIGURE 16.1
Basic configuration of a cascaded transcoder.

The detailed architecture of the cascaded transcoder is shown in Figure 16.2. At the decoder, the motion-compensated image from the previous frame is added to the decompressed prediction error, which was encoded at a bit rate R_1 to reconstruct the original image. This reconstructed image is sent to the encoder for compression at a lower bit rate R_2.

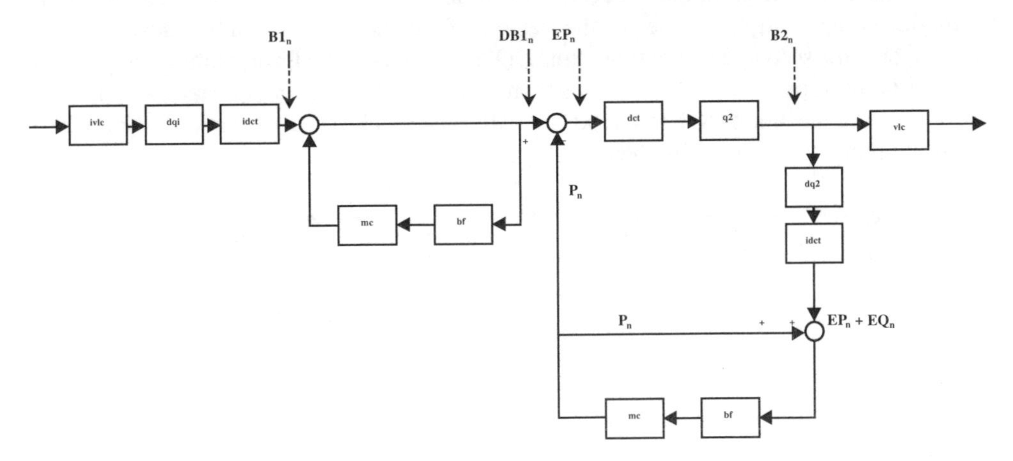

FIGURE 16.2
Cascaded video transcoder. $B1_n$, $DB1_n$, EP_n, $B2_n$, P_n, EP_n, and EQ_n are the video signals. Subscript n represents the frame number, $B1_n$ is the decompressed prediction error, $DB1_n$ is the reconstructed image, P_n is the predicted frame, EP_n is the prediction error, and EQ_n is the error produced by the second quantizer. $B2_n$ is the compressed DCT coefficient. The abbreviation ivlc represents inverse variable-length coding; dq1, dequantization using quantization level 1; idct, inverse discrete cosine transform; mc, motion compensation; bf, frame buffer; dct, discrete cosine transform; q2, quantization using quantization level 2; dq2, dequantization using quantization level 2; and vlc, variable-length coding. The left and right parts of this figure are the decoder and the encoder, respectively.

If the frames are not skipped at the encoder, the decoded motion vectors can be reused so that the motion estimation can be avoided. Since motion estimation is the most time-consuming module in the encoder, avoiding motion estimation will reduce the computational complexity of the transcoder significantly.

From this cascaded architecture, a significant simplification can be achieved if the picture types are the same before and after transcoding. The following section describes this simplification procedure, which was introduced in [1].

16.2.3 Removal of Frame Buffer and Motion Compensation Modules

Motion compensation is a linear operation. Assuming $MC(x)$ is the function of performing motion compensation to a variable x of a macroblock, then

$$MC(x+y) = MC(x) + MC(y) . \tag{16.1}$$

Referring to Figure 16.2, we can simplify the architecture as follows:

$$\begin{aligned} EP_{n+1} &= DB1_{n+1} - P_{n+1} \\ &= B1_{n+1} + MC\,(DB1_n) - MC\,(P_n + EP_n + EQ_n) \\ &= B1_{n+1} + MC\,(DB1_n) - MC\,(DB1_n - EP_n + EP_n + EQ_n) \\ &= B1_{n+1} + MC\,(DB1_n) - MC\,(DB1_n) - MC\,(EQ_n) \\ &= B1_{n+1} - MC\,(EQ_n) . \end{aligned} \tag{16.2}$$

In expression (16.2), n represents the current frame number and $n+1$ represents the following frame number. Referring to the deduced expression, only $B1$ remains at the decoder. Thus, the motion compensation module and the frame buffer at the decoder can be discarded. At the encoder, only the second quantization error, EQ, needs to be stored, and motion compensation needs to be performed for the following frame. Therefore, we can connect a line from the position before the block DCT to the position after the block IDCT to cancel EP. The result of this simplified architecture is shown in Figure 16.3.

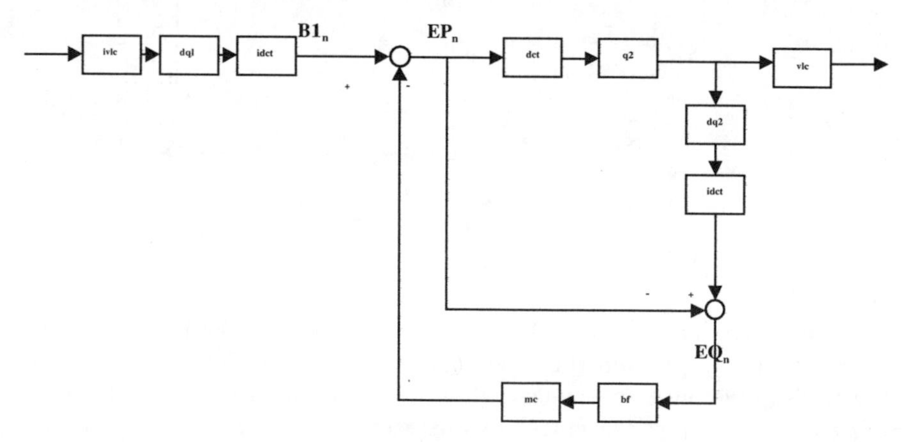

FIGURE 16.3
Architecture of the transcoder after removal of the motion compensation module from the decoder.

16.2.4 Removal of IDCT Module

Another simplification can be done in the transcoder of Figure 16.3. In the video transcoder, except for the motion estimation, the most time-consuming parts are the DCT and IDCT modules. From Figure 16.3, removal of the IDCT module from the decoder is possible. The DCT and IDCT are also linear operations. Therefore, we can perform the following operations to further simplify the architecture:

$$(DCT(IDCT)(A)) = A\ ,$$
$$DCT(A + B) = DCT(A) + DCT(B)\ . \tag{16.3}$$

Moving the DCT and IDCT modules in the architecture according to (16.3) results in the simplified block diagram shown in Figure 16.4. Now the decoder does not need to perform the IDCT operations. In Figure 16.4, the architecture is much simplified compared to the original architecture in Figure 16.2.

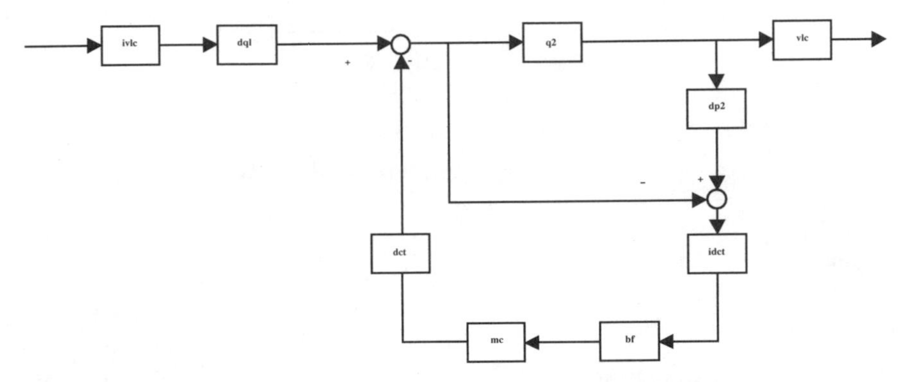

FIGURE 16.4
Simplified architecture of transcoder.

The procedure introduced in this section can be applied to either MPEG-2 [7], H.261 [8], or H.263 [9] bitstreams. However, if frames are skipped in the transcoding process, special care must be taken when composing motion vectors from the decoded incoming motion vectors. This subject will be addressed in Section 16.4.

16.3 DCT Domain Transcoder

16.3.1 Introduction

As can be seen in the simplified architecture derived in the last section, the whole transcoding process is performed in the DCT domain except the motion compensation loop. In this simplified architecture, the most time-consuming part is in the DCT and IDCT modules. It is highly desirable if these two modules can be removed from the architecture. Because the motion-compensated macroblocks usually do not exactly fall on the original macroblock boundaries, to perform the motion compensation in the DCT domain, an interpolation method is necessary to estimate the DCT coefficients of the shifted macroblock. This can be achieved using the DCT domain interpolation algorithm developed for transform domain video composition in [3]. This algorithm calculates the DCT coefficients of a shifted block from its four neighboring DCT blocks and the motion vector, directly in the DCT domain. Thus, the DCT and IDCT functions can be removed. In this section, we also introduce a half-pixel interpolation method with a similar principle.

16.3.2 Architecture of DCT Domain Transcoder

After performing the DCT coefficient interpolation, we do not need the DCT and IDCT modules in Figure 16.4. The resultant DCT domain transcoder is shown in Figure 16.5. In Figure 16.5, the whole transcoding process is performed in the DCT domain. The frames stored in memory for reference are the differences of the quantization errors in the DCT domain.

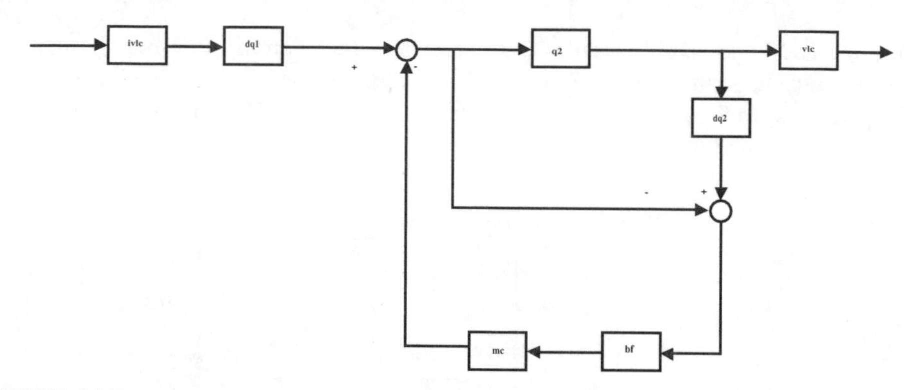

FIGURE 16.5
DCT domain transcoder.

16.3.3 Full-Pixel Interpolation

To exactly calculate the DCT coefficients of a shifted macroblock, we use the fact that the shift operation is equivalent to performing a convolution with a shifted delta function. In the pixel domain, the pixel values in the overlapped region can be obtained from four neighboring blocks by applying a series of windowing and shifting (convolution) operations. The matrices of convolution masks are determined by the motion vectors.

$$V_i\,(H_i) = \begin{bmatrix} 0 & 0 \\ I & 0 \end{bmatrix} \text{ or } \begin{bmatrix} 0 & I \\ 0 & 0 \end{bmatrix}, \tag{16.4}$$

where $V_i(H_i)$ are vertical (or horizontal) shift convolution masks, which are sparse matrices. The size of I is determined by the overlapped width between the shifted macroblock and each neighboring macroblock in the vertical or horizontal direction.

Because convolution in the pixel domain is equivalent to multiplication in the DCT domain, to directly perform this function in the DCT domain, the following formulation originally proposed for transform domain video composition [3] can be used:

$$DCT(\text{Full})_i = \sum_{i=1}^{4} DCT\,(V_i)\,DCT\,(NEB_i)\,DCT\,(H_i) \tag{16.5}$$

where NEB_i is one of the four neighboring blocks. The DCT coefficients V_i and H_i can be precomputed and stored in a table to accelerate the operation. Moreover, from the decoded run-length sequence of NEB_i, we can reduce the calculation complexity in the case of zero dequantized DCT coefficients. Except for the rounding error, the DCT domain interpolation gives exactly the same values as those from using the pixel domain approach.

16.3.4 Half-Pixel Interpolation

The above interpolation method can only perform full-pixel motion compensation. For half-pixel motion compensation, we can use the following two special sparse matrices:

$$H = \begin{bmatrix} 1/2 & 0 & \dots & \dots & 0 \\ 1/2 & 1/2 & \cdot & \vdots & \vdots \\ 0 & 1/2 & \cdot & 0 & \vdots \\ \vdots & 0 & \cdot & 1/2 & 0 \\ 0 & \dots & 0 & 1/2 & 1/2 \end{bmatrix}, \tag{16.6}$$

$$V = \begin{bmatrix} 1/2 & 1/2 & 0 & \dots & 0 \\ 0 & 1/2 & \cdot & \vdots & \vdots \\ \vdots & 0 & \cdot & 1/2 & 0 \\ \vdots & \vdots & \cdot & 1/2 & 1/2 \\ 0 & \dots & \dots & 0 & 1/2 \end{bmatrix}, \tag{16.7}$$

to average the coefficients in the horizontal and vertical directions.

Therefore,

$$DCT(\text{Half})_h = DCT(\text{Full})DCT(H)\,, \tag{16.8}$$

$$DCT(\text{Half})_v = DCT(\text{Full})DCT(V) \tag{16.9}$$

are the expressions used for interpolating horizontal and vertical half-pixel DCT coefficients, respectively. The DCT coefficients of the above matrices can also be precomputed and stored in a table to accelerate the operation.

A significant reduction in complexity can be achieved by using the above DCT coefficient interpolation method. This method will highly reduce the computational complexity of a transcoder, without any degradation in video quality.

16.4 Frame-Skipping in Video Transcoding

16.4.1 Introduction

For the low-bit-rate video compression standards, such as H.261 [8] and H.263 [9], the frames can be skipped to keep the generated bit rate from exceeding the channel bandwidth. In the process of transcoding, if the bandwidth of the outgoing channel is not enough to allocate bits for achieving good video quality, frame-skipping is a good strategy to control the bit rate and at the same time keep the image quality at an acceptable level.

It is possible to apply the frame-skipping strategy in a video transcoder to reduce the bit rate. As discussed in the first two sections, usually the motion vectors decoded from the bitstream can be reused to speed up the re-encoding process. However, when frames are skipped, the motion vectors cannot be reused because the motion vectors of each transcoded frame can no longer be estimated from its immediately previous frame, which has been skipped. If frame-skipping is necessary, the frames need to be decoded completely and the motion estimation needs to be performed again to search the motion vectors, which requires numerous computations.

To overcome this motion vector reuse problem, a bilinear interpolation method can be used to estimate the motion vectors from the current frame to its previous nonskipped frame if some frames between them are skipped. The interpolated motion vectors can serve as the search centers for the motion estimation. Since there is a maximum search range constraint in the motion estimation, and the interpolated motion vectors usually are not at the center of the search area, after the interpolation, we can use a smaller search range which reduces the computations in the motion estimation. The distance from the position of the interpolated motion vector to the original search-area boundary determines the size of the new search area.

The frame-rate control problem in the video transcoder is also considered in this section. Randomly skipping frames based on the consumed bits of the last encoded frame [10] may cause abrupt motion in the decoded video sequence. Taking advantage of known motion vectors in the video transcoder, a dynamic frame-skipping scheme can be used to dynamically estimate the motion activities of each frame so that the skipping of video frames can be done in an optimal way.

16.4.2 Interpolation of Motion Vectors

As described above, in video transcoding, usually the motion vectors decoded from the bitstream can be reused to speed up the re-encoding process. When frames are skipped, the motion estimation for the nonskipped frames requires the path of motion from the current frame to its previous nonskipped frame, which could be many frames away.

Since the motion vectors between every adjacent frame are known, the problem of tracing the motion from frame 4 to frame 1 in Figure 16.6 can be solved by repeated applications of bilinear interpolations [5].

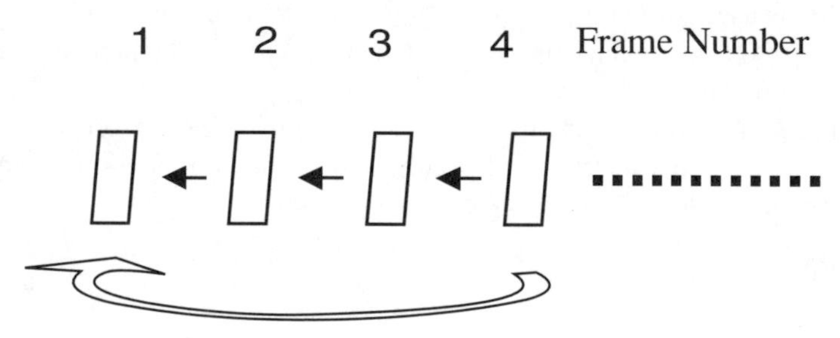

FIGURE 16.6
Motion tracing example.

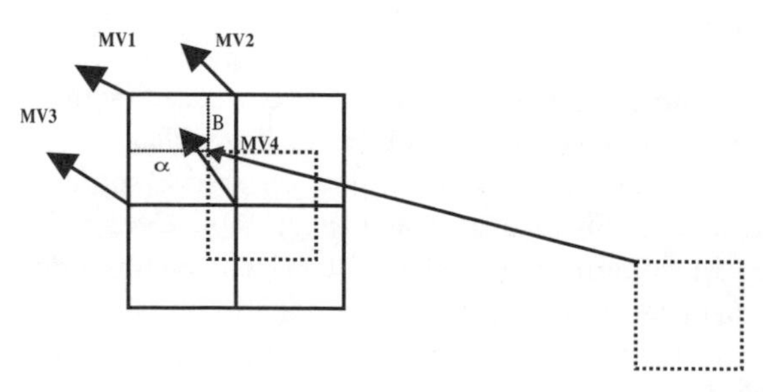

FIGURE 16.7
Interpolation of motion vectors.

Referring to Figure 16.7, a shifted macroblock is located at a position overlapped with four neighboring macroblocks. The bilinear interpolation is defined as

$$MV_{\text{int}} = (1-\alpha)(1-\beta)MV_1 + (\alpha)(1-\beta)MV_2 + (1-\alpha)(\beta)MV_3 + (\alpha)(\beta)MV_4 , \quad (16.10)$$

where $MV_1 \sim MV_4$ are the motion vectors of the four neighboring macroblocks; α and β are weighting constants determined by the pixel distances to MV_1 in the horizontal and vertical directions. The weighting constants of each neighboring block are inversely proportional to the pixel distances:

$$\text{Weighting} \propto \frac{1}{\text{Pixel Distance}} . \quad (16.11)$$

Repeatedly tracing the motion in this manner creates an interpolated motion vector for each macroblock in the current frame to its previous nonskipped frame.

16.4.3 Search Range Adjustment

The bilinear interpolation method derived interpolated motion vectors from the decoded incoming motion vectors. However, the derived motion vectors are not exactly equal to the optimal motion vectors. Therefore, further refinement can be performed. Taking advantage of the interpolated motion vectors, this time we can use a smaller search range to estimate the motion vectors. For each macroblock, we use the new position located by the composed interpolated motion vectors as the search center for the final motion estimation stage. We use the notation $MV_{\max}$ to define the maximum search range specified in most video coding standards. If the composed motion vectors for macroblock i in the horizontal and vertical directions are MV_X_i and MV_Y_i, then the search distance D_i to the new search center for each macroblock i can be described by the following:

$$\begin{aligned} &\text{For } i = 1 \text{ to Maximum Number of Macroblocks} \\ &D_i = MV_{\max} - MAX\left(|MV\_X_i| , |MV\_Y_i|\right) , \end{aligned}$$

where $MAX(|MV\_X_i|, |MV\_Y_i|)$ selects the maximum value between $|MV\_X_i|$ and $|MV\_Y_i|$, so that the allowable maximum motion vector magnitude will conform to the standards. This results in a smaller search range and lower computation load in the motion vector refinement. From the computer simulations, it is seen that this process does not introduce quality degradations.

16.4.4 Dynamic Frame-Skipping

The goal of the dynamic frame-skipping strategy is to make the motion of the decoded sequence smoother. In the video transcoder, since the motion vectors are known, they can serve as a good indication for determining which frame to skip. For example, we can set a threshold and if the accumulated magnitude of motion vectors in the frame exceeds this threshold, this frame will be encoded; otherwise, this frame will be skipped. The threshold is recursively reset after transcoding each frame because the number of encoded frames should be dynamically adjusted according to the variation of the generated bits produced by the unskipped frames. The overall algorithm can be described by the flowchart in Figure 16.8.

16.4.5 Simulation and Discussion

For motion estimation in transcoding with frame-skipping, Table 16.1 shows the percentage of the transcoding computations as well as the number of generated bits compared to using

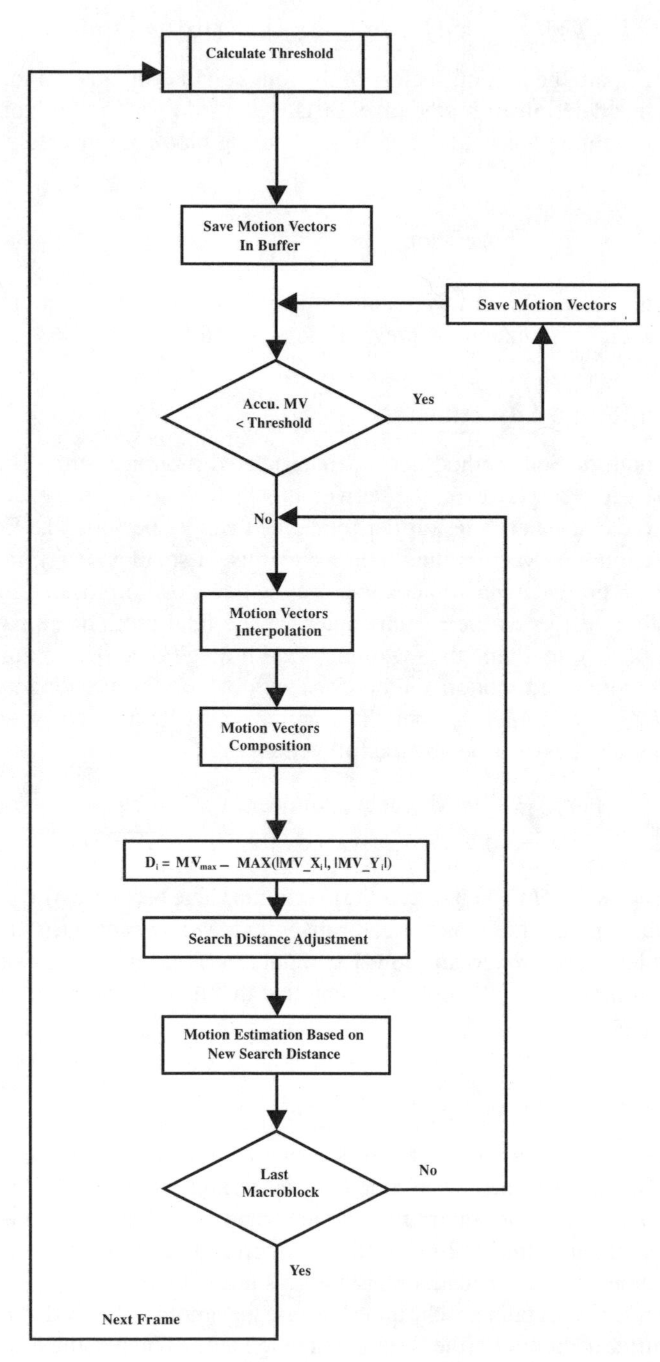

FIGURE 16.8
Flowchart of motion vector interpolation.

the standard full-search motion estimation approach. Because the search range is dynamically reduced, the average re-encoding time is only about half of the encoding time using the full-search motion estimation approach, while maintaining the same video quality. Table 16.2 shows the peak signal-to-noise ratio (PSNR) of using the full-search motion estimation and the approach described in this section. As shown in the table, the interpolated motion vectors followed by the smaller search-range motion estimation produce coded video sequences without any video quality degradations.

Table 16.1 Transcoding Time and Generated Bits

Sequences (400 frames)	Transcoding Time (% of full search approach)	Generated Bits (% of difference)
Foreman	58.97%	4.52%
Grandma	40.42%	4.09%
Mthr_dotr	46.71%	3.05%
Salesman	41.25%	1.23%

Table 16.2 PSNR Differences

Sequences (400 frames)	Average PSNR (original algorithm)	Average PSNR (simplified algorithm)	Difference
Foreman	33.75	33.73	−0.02
Grandma	33.31	33.27	−0.04
Mthr_dotr	33.52	33.49	−0.03
Salesman	31.58	31.60	+0.02

In Figure 16.9, four video sequences (400 frames each) are transcoded using the full-search motion estimation approach and the interpolation and dynamic search-range adjustment approach. The solid lines are the results of using the standard full-search motion re-estimation algorithm, and the "+" lines are the results of using the interpolation and dynamic search-range adjustment approach. The two lines almost completely overlap. Both simulations use the frame-skipping scheme in TMN5 [10], where frame-skipping is determined by how much the consumed bits of the last frame exceed the average target bits per frame.

The comparative results of using TMN5 frame-skipping and the dynamic frame-skipping scheme are shown in Figure 16.10. In the figure, the solid lines represent the motion activities of the Foreman sequence from frame 200 to frame 300 and the dotted lines represent the coded frame number when the sequence is transcoded from 128 to 64 kbps. The results shown in the upper plot indicate that the frames are skipped in a random way when using the TMN5 scheme. The results using the dynamic frame-skipping scheme are shown in the lower plot. In the dynamic frame-skipping approach, most of the frames are skipped during the low motion activity period; therefore, the video displayed at the receiver site will be smoother. Subjective video quality of the results is shown in Figures 16.11 and 16.12.

As shown in Figure 16.11, objects from frame 287 to frame 291 are active: the man moves out from the frame and the background shifts to the left side. When using the TMN5 skipping scheme, as shown in the upper line, only two frames are coded; therefore, the man disappears suddenly. However, if the dynamic skipping scheme is used, as shown in the lower line, five frames are coded; therefore, the man moves out smoothly and the background shifts to the left side gradually.

Figure 16.12 shows the encoded frames from frame 200 to frame 215 using two frame-skipping approaches. The objects in these frames are in low motion. The TMN5 scheme

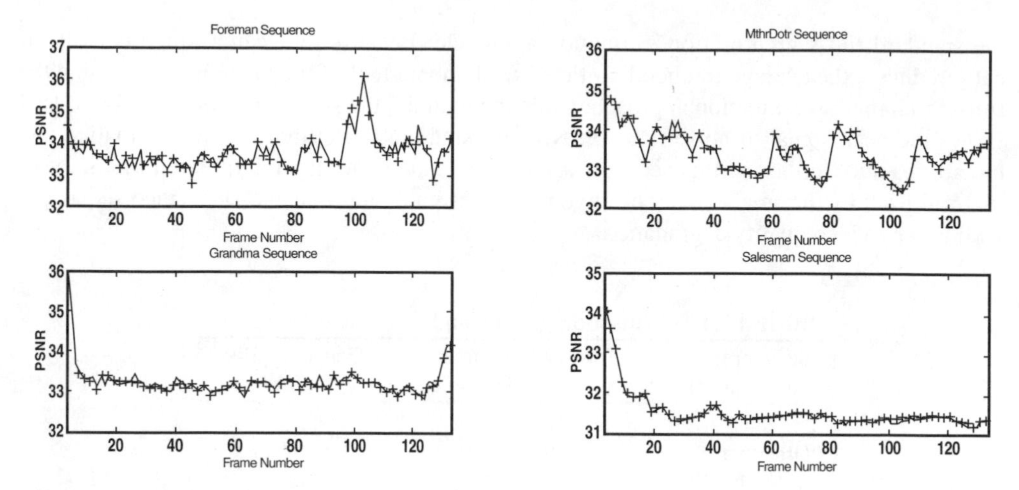

FIGURE 16.9
PSNR plots of four sequences using two motion vector search schemes. Solid lines: full-search scheme; "+" lines: motion vector interpolation and search-range adjustment scheme.

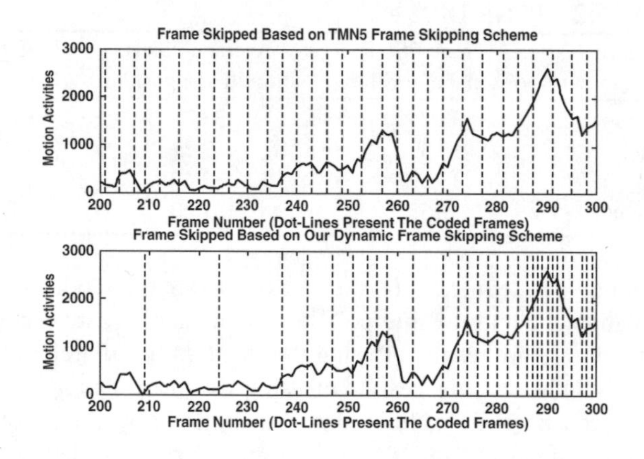

FIGURE 16.10
Coded frame numbers using the TMN5 frame-skipping scheme and the dynamic frame-skipping scheme.

encodes all five similar frames, but the dynamic frame-skipping scheme encodes just one frame to represent these five frames; thus, the consumed bits are much reduced. Because the saved bits from low-motion frames can be used for active frames, the picture quality of the active frames can be improved.

16.5 Multipoint Video Bridging

16.5.1 Introduction

An important application of video transcoders is for the MCU in multipoint video conferencing applications. Multipoint video conferencing is a natural extension of point-to-point video

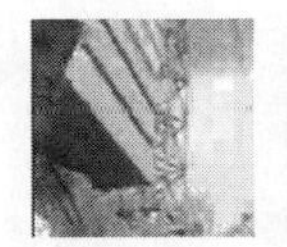 

Fm 287 Fm 291

(Two Frames Are Coded: Using Frame Skipping Scheme in TMN5)

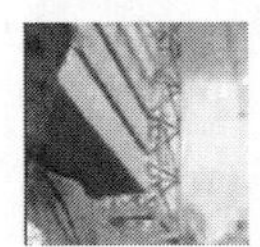    

Fm 287 Fm 288 Fm 289 Fm 290 Fm 291

(Five Frames Are Coded: Using Dynamic Frame-Skipping Scheme)

FIGURE 16.11
Two frame-skipping schemes in active frames.

Fm 201 Fm 204 Fm 207 Fm 209 Fm 212

(Five Frames Are Coded: Using Frame Skipping Scheme in TMN5)

Fm 209

(One Frame Is Coded: Using Dynamic Frame Skipping Scheme)

FIGURE 16.12
Two frame-skipping schemes in low-motion frames.

conferencing. With the rapid growth of video conferencing, the need for multipoint video conferencing is also growing [11]–[19]. Multipoint video conferencing technology involves networking, video combining, and presentation of multiple coded video signals. The same technology can also be used for distance learning, remote collaboration, and video surveillance involving multiple sites. For a multipoint video conference over a wide-area network, the conference participants are connected to an MCU in a central office [11]–[17]. A video combiner in the MCU combines the multiple coded digital video streams from the conference participants into a coded video bitstream that conforms to the syntax of the video coding standard and sends it back to the conference participants for decoding and presentation. The application scenario of four persons participating in a four-point video conference is shown in Figure 16.13. Although in the following we will use the four-point video conference as an example, the discussion can be easily generalized to multipoint video conferencing involving more sites.

One approach used in the video combiner to combine the multiple coded digital video streams is the coded domain combining approach. In this approach, the video combiner

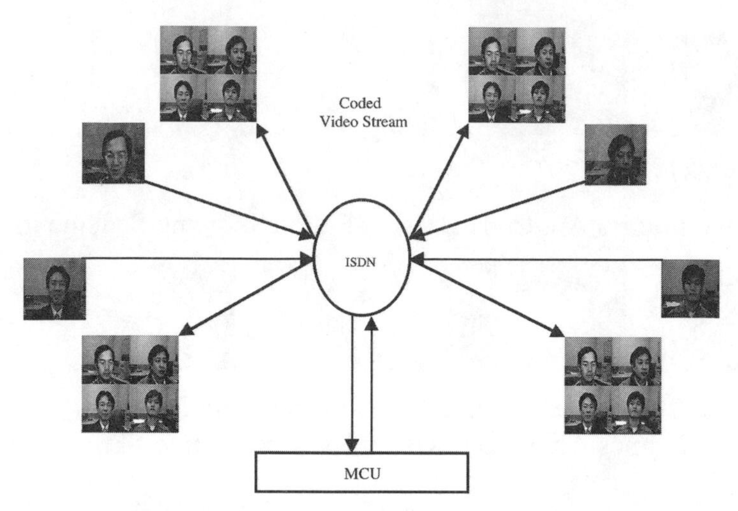

FIGURE 16.13
An example of a four-point video conference over a wide-area network.

modifies the headers of the individual coded video bitstreams from the conference participants, multiplexes the bitstreams, and generates new headers to produce a legal coded and combined video bitstream conforming to the video coding standard. For example, a coded domain combiner can modify the headers of four coded QCIF (quarter common intermediate format, 176×144 pixels) video sequences and concatenate them into a CIF (common intermediate format, 352×288 pixels) coded video sequence [12]. Using the coded domain combining approach, since the combiner only needs to perform the multiplexing and header modification functions in concatenating the video bitstreams, the required processing is minimal. Also, because it does not need to decode and re-encode the video streams, it does not introduce quality degradation. However, the coded domain combining approach offers limited flexibility for users to manipulate the video bitstreams (e.g., for resize, reposition, chroma-keying, change of bit rates, etc.). Another solution is to use the transcoding approach. When using the transcoding approach for video combining, the video combiner decodes each coded video stream, combines the decoded video streams, and re-encodes the combined video at the transmission channel rate. Figure 16.14 shows a block diagram of video combining for multipoint video conferencing using the transcoding approach. The transcoding approach requires more computations; it must decode the individual video stream and encode the combined video signal. The video quality using the transcoding approach will also potentially suffer from the double-encoding process because the video from each user will need to be decoded, combined, and then re-encoded, which introduces additional degradation.

Since the coded domain combiner simply concatenates bitstreams, the combined bit rate will be about the sum of the bit rates of all the individual bitstreams. For example, in a four-point video conference where the bit rate of each participant's encoded video is 32 kbps, the combined video bit rate will be about 128 kbps. Thus, for each conference participant, the input and output video bit rates are highly asymmetrical. For a four-point video conference over a wide-area network such as an integrated service digital network (ISDN), where about 128 kbps is available in the 2-B channels, each user can encode their video at only about 32 kbps so that the combined video can fit in the 2-B channels. Also, with the coded domain combining approach discussed in [12], it is difficult to utilize the video characteristics in a multipoint video conference, where only one or two persons are active at one time while other persons are just listening and have little motion. The active persons need higher bit rates to

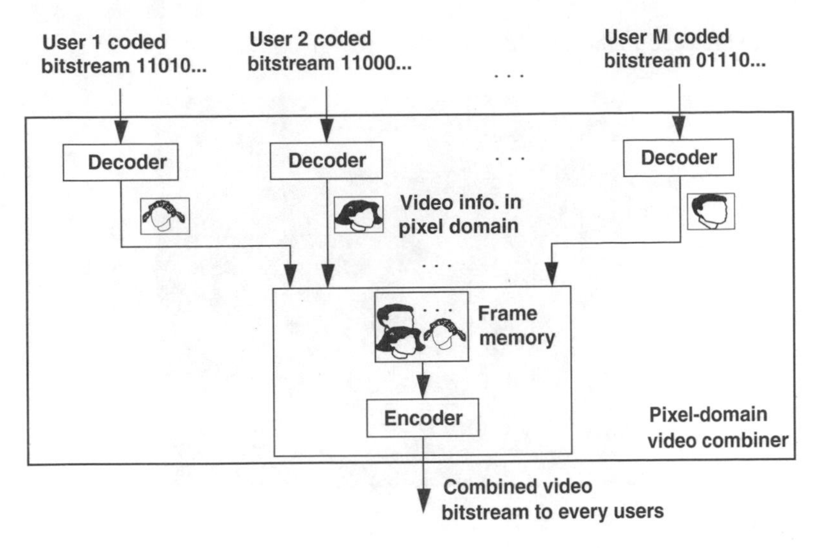

FIGURE 16.14
Video combining using the transcoding approach.

produce good quality video while the inactive persons only need lower bit rates to produce good quality video. With coded domain video combining, which allocates equal bandwidth to all participants, the bandwidth usually is too limited for an active person but is more than enough for an inactive person. It is difficult to dynamically allocate extra bandwidth to active persons using the coded domain combining approach.

The transcoding approach allows each conference participant to encode the video using the full bandwidth of the transmission channel (using the previous example, each participant will encode their video at 128 kbps instead of 32 kbps). Since each user can use the full bandwidth, and in the re-encoding process the rate-control strategies in video coding algorithms will dynamically allocate more bits to the active frames or areas, the overall quality can be more uniform compared to the coded domain combining approach where the quality of the active person may be much worse than the inactive persons. The result is that the transcoding approach can provide a much more uniform video quality among conference participants. With complexity reduction techniques such as those discussed in the previous sections, more flexibility in video manipulation [13, 14], and better video quality, transcoding is a very practical approach for video combining in multipoint video conferencing over a symmetrical wide-area network.

16.5.2 Video Characteristics in Multipoint Video Conferencing

In a multipoint video conference, most of the time only one or two persons are active at one time. As stated in the last subsection, the active persons need higher bit rates to produce good quality video while the inactive persons only need lower bit rates to produce good quality video. The following example describes a typical situation. In a four-point video conferencing session, each conference participant's video was recorded in QCIF. The four video sequences were combined into a CIF sequence. To describe the effects of using the coded domain and the transcoding approaches, we generated a 400-frame combined video sequence in which the first person is most active in the first 100 frames, the second person is most active in the second 100 frames, and so on. A frame in the combined video sequence is shown in Figure 16.15. The four individual video sequences and the combined video sequence (with 400 frames in each sequence) are used in the simulation.

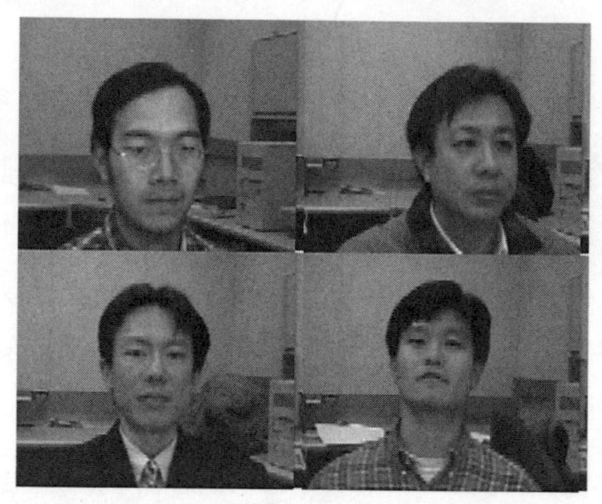

FIGURE 16.15
A frame in the combined four-point video conferencing sequence.

The bit rate required to encode a video sequence is highly related to the motion activity in the sequence. The accumulated magnitudes of all the motion vectors estimated for the macroblocks in the current frame will be used as the quantitative measure for the motion activity (MA); that is,

$$MA = \sum_{i=1}^{N} |(MVi(X))| + |(MVi(Y))| \tag{16.12}$$

where N is the total number of macroblocks in the current frame and $MVi(X)$ and $MVi(Y)$ are the horizontal and vertical components of the motion vector of the ith macroblock, respectively. The MAs for the four persons and the combined video sequence are shown in Figure 16.16.

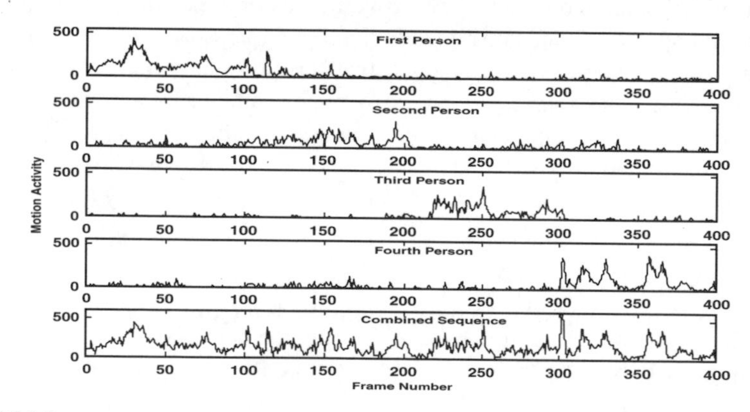

FIGURE 16.16
Motion activity of a multipoint video conference.

In Figure 16.16, the top four curves correspond to the four persons in the upper left, upper right, lower left, and lower right of Figure 16.15, respectively. The bottom curve represents the MA of the combined video. From the figure, although there are short periods of time when multiple persons are active, most of the time when a person is active, other persons are relatively inactive. The overall MA of the combined video is relatively random.

16.5.3 Results of Using the Coded Domain and Transcoding Approaches

The simulations were performed using the H.263 compression algorithm. Compared to the coded domain approach, the transcoding approach loses from the double-encoding aspect but gains from the dynamic bit-allocation aspect. The video combiner using the transcoding approach will dynamically reallocate suitable bit rates to individual video streams based on the motion activity in the re-encoding process. This is because the rate-control algorithm in the H.263 software will assign more bits to those parts that have more motion and fewer bits to those parts that have little motion. This results in a more uniform overall video quality.

In the following simulations, four video sequences of the conference participants are encoded using the H.263 software and then decoded, combined, and re-encoded using the coded domain and transcoding approaches. For the coded domain approach, the individual video is coded at 32 and 64 kbps so that the combined video can be fit in a transmission channel of 128 and 256 kbps, respectively. For the transcoding approach, the individual video is encoded at 128 and 256 kbps. The coded video streams are then decoded and combined. The combined video is then encoded at 128 and 256 kbps, respectively. The resulting video quality is compared using the PSNRs and subjective evaluation.

Figure 16.17 shows the PSNRs of a conference participant (the one at the lower right corner in Figure 16.15) in the combined video sequence for the coded domain and transcoding approaches. In the sequence, this participant is most active from frame 300 to frame 400 (as shown in the MA plot in Figure 16.16). It can be seen from Figure 16.17, for the coded domain approach, that the PSNR drops drastically when the person becomes active, because it is more difficult to encode the video with higher motion. The transcoding approach offers a much better quality compared to the coded domain approach when the person is active, due to the inherent dynamic bit allocation based on motion activities in the rate-control algorithm of the H.263 software. The overall quality of the transcoding approach is also much more uniform. This improvement is significant since in practical situations, most people will pay the most attention to the active persons. Thus, the quality improvement of the active persons will directly translate into improvement in the overall perceived quality for multipoint video conferencing. A frame (no. 385) in the combined sequence is shown in Figure 16.18. It can be seen that the video quality of the active person (at the lower right corner) with the transcoding approach is much higher than that with the coded domain approach.

Due to the dynamic bit allocation based on the motion activities, there will be fewer bits assigned to the conference participants who are not active. However, the bit rates allocated to the inactive persons usually are sufficient to produce a good quality video. The rate-control algorithm of the encoding software will usually provide a good balance of the bit rates assigned to all the participants. From the subjective evaluation, the degradation to the video quality of the inactive persons is not very visible. This can also be seen from Figure 16.18.

Figure 16.19 shows the PSNR of each frame in the overall combined video sequence. Although in every 100-frame period there is a most active person, during frames 301 to 400 the motion activities are higher than at other times. Thus, there is a significant PSNR drop using the coded domain approach during frames 301 to 400 since the bandwidths are not enough to encode the active persons with good quality. On the other hand, using the transcoding approach, the overall PSNR drops only slightly. Again, the transcoding approach offers a more uniform overall quality due to the dynamic bit allocation in the re-encoding of the combined video in the multipoint video conferencing sequence.

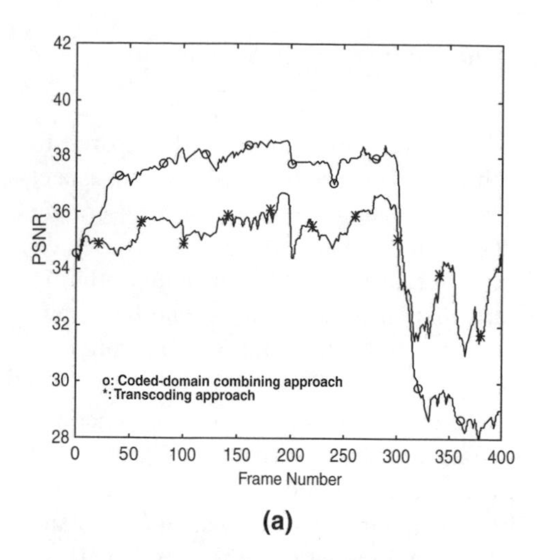

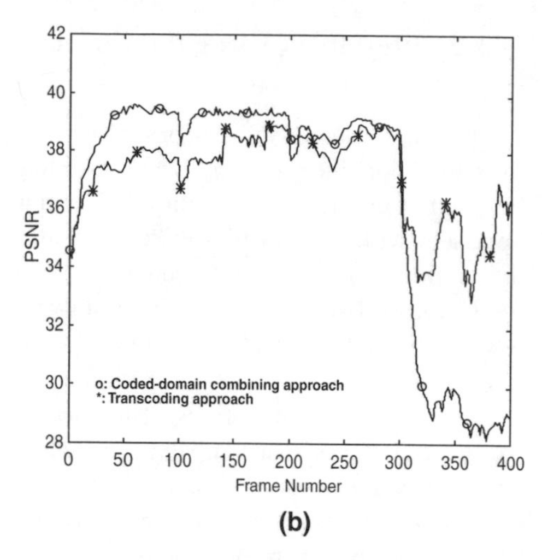

FIGURE 16.17
PSNRs of a conference participant who is most active between frame 300 and frame 400 (a) with a channel rate of 128 kbps and (b) with a channel rate of 256 kbps.

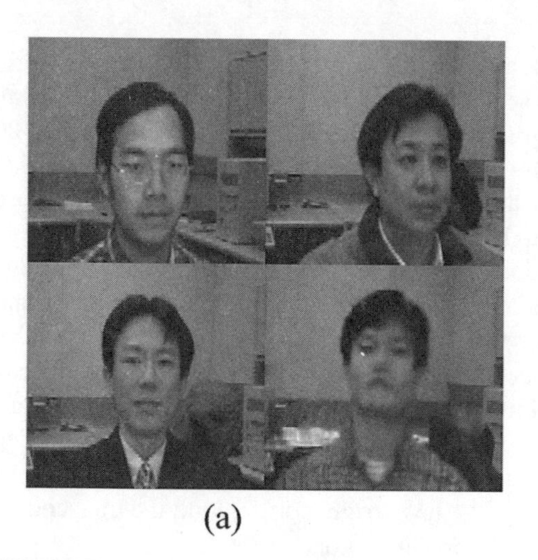
(a)

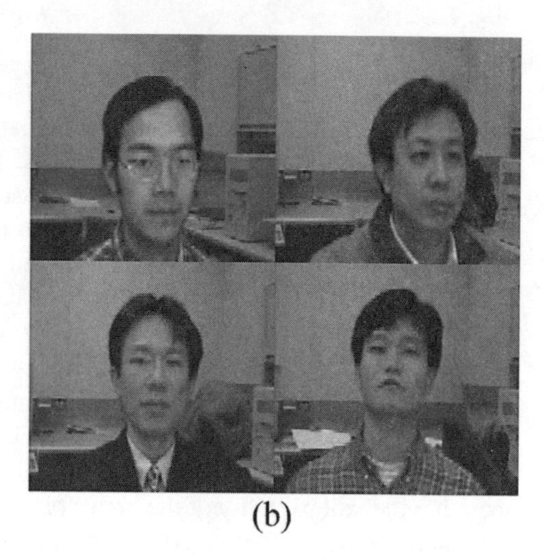
(b)

FIGURE 16.18
Frame 385 of the combined video sequence using (a) the coded domain approach and (b) the transcoding approach. The active person is at the lower right corner.

16.6 Summary

This chapter discussed various issues concerning video transcoding. It has been shown that a DCT domain transcoder can significantly reduce computational complexity. This approach uses an interpolation method to estimate the DCT coefficients of a shifted macroblock. With this estimation, a much-simplified transcoder architecture is derived.

This chapter also demonstrated that a motion vector interpolation method could be used to compose motion vectors when frames are skipped in the transcoding process. Based on these estimated motion vectors, a dynamic frame-skipping scheme can decide which frames to skip.

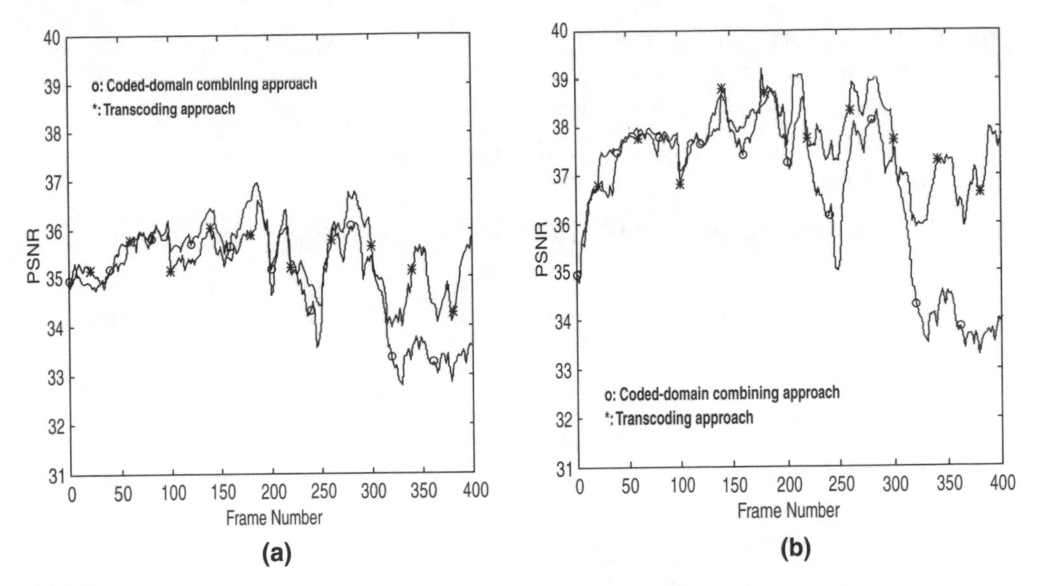

FIGURE 16.19
PSNR of the combined video sequence with a transmission rate of (a) 128 kbps and (b) 256 kbps.

The results show that the decoded sequence presents smoother motion.

This chapter also demonstrated that the transcoding approach can be applied to video combining. The overall results of using transcoding are better than those for the coded domain approach, which merely performs the multiplexing function. The dynamic bit allocation in the transcoder can improve the overall picture quality.

References

[1] G. Keesman, R. Hellinghuizen, F. Hoeksema, and G. Heideman, "Transcoding of MPEG Bitstreams," *Signal Processing: Image Communication,* pp. 481–500, 1996.

[2] P. Assuncao and M. Ghanbari, "Post-Processing of MPEG-2 Coded Video for Transmission at Lower Bit Rates," *IEEE International Conference on Acoustics, Speech, and Signal Processing, ICASSP'96,* vol. 4, pp. 1998–2001, May 1996.

[3] S.-F. Chang and D.G. Messerschmitt, "A New Approach to Decoding and Compositing Motion Compensated DCT-Based Images," *IEEE International Conference on Acoustics, Speech, and Signal Processing, ICASSP'93,* vol. 5, pp. 421–424, April 1993.

[4] J.N. Hwang and T.D. Wu, "Dynamic Frame Rate Control in Video Transcoding," *Asilomar'98 Conference,* Monterey, CA, November 1998.

[5] A. Lan and J.N. Hwang, "Context Dependent Reference Frame Placement for MPEG Video Coding," *IEEE ICASSP'97,* vol. 4, pp. 2997–3000, April 1997.

[6] M.-T. Sun, T.-D. Wu, and J.-N. Hwang, "Dynamic Bit-Allocation in Video Combining for Multipoint Conferencing," *International Symposium on Multimedia Information Processing, ISMIP'97,* Taipei, Taiwan, pp. 350–355, December 11–13, 1997.

[7] ISO/IEC (MPEG-2), "Generic Coding of Moving Pictures and Associated Audio," March 1994.

[8] ITU-U Recommendation H.261, "Video Codecs for Audiovisual Services at p $\times$ 64 kb/s," May 1992.

[9] ITU-U Recommendation H.263, "Video Coding for Low Bit Rate Communication," May 1997.

[10] ITU Study Group 15, Working Party 15/1, "Video Codec Test Model (TMN5)," 1995.

[11] S. Oka and Y. Misawa, "Multipoint Teleconferencing Architecture for CCITT Standard Videoconference Terminals," *Visual Communications and Image Processing '92,* Boston, MA, pp. 1502–1511, November 18–20, 1992.

[12] M.S.M. Lei, T.C. Chen, and M.T. Sun, "Video Bridging Based on H.261 Standard," *IEEE Trans. on Circuits & Systems for Video Technology,* vol. 4, no. 4, pp. 425–437, August 1994.

[13] M.E. Lukacs and D.G. Boyer, "A Universal Broadband Multipoint Teleconferencing Service for the 21st Century," *IEEE Communications Magazine,* vol. 33, no. 11, pp. 36–43, November 1995.

[14] D.G. Boyer, M.E. Lukacs, and M. Mills, "The Personal Presence System Experimental Research Prototype," *IEEE Conference on Communications,* vol. 2, pp. 1112–1116, June 23–27, 1996.

[15] T. Arakaki, E. Kenmouku, and T. Ishida, "Development of Multipoint Teleconference System Using Multipoint Control Unit (MCU)," *Pacific Telecommunications Council Fifteenth Annual Conference Proceedings,* Honolulu, HI, vol. 1, pp. 132–137, January 17–20, 1993.

[16] ITU-T Recommendation H.231, "Multipoint Control Units for Audiovisual Systems Using Digital Channels up to 1920 kb/s."

[17] M.H. Willebeek-LeMair, D.D. Kandlur, and Z.Y. Shae, "On Multipoint Control Units for Videoconferencing," *Proceedings 19th Conference on Local Computer Networks,* Minneapolis, MN, pp. 356–364, October 2–5, 1994.

[18] R. Gaglianello and G. Cash, "Montage: Continuous Presence Teleconferencing Utilizing Compressed Domain Video Bridging," *IEEE International Conference on Communications,* Seattle, WA, vol. 1, pp. 573–581, June 18–22, 1995.

[19] M.H. Willebeek-LeMair and Z.Y. Shae, "Videoconferencing Over Packet-Based Networks," *IEEE Journal on Selected Areas in Communications,* vol. 15, no. 6, pp. 1101–1114, August 1997.

Chapter 17

Multimedia Distance Learning

Sachin G. Deshpande, Jenq-Neng Hwang, and Ming-Ting Sun

17.1 Introduction

The ultimate goal of distance learning systems is to provide the remote participant most of the capabilities and the experience close to that enjoyed by an in-class participant. In fact, a non-real-time distance learning system can provide features that produce a better environment than a live class. There has been rapid progress in digital media compression research. Starting with MPEG-1 [1], MPEG-2 [2] at the higher bit rate range (Mbps) to H.261 [3] ($p \times 64$ Kbps) and H.263(+) [4] ($\geq$ 8 Kbps) at the low-bit-rate range, digital video coding standards can achieve a bit rate and quality suitable for the various network bandwidths in the current heterogeneous networking environments. The MPEG-4 [5] video coding standard is currently being formalized and aims at providing content-based access along with compression. In digital audio coding, MPEG, GSM [6], and G.723.1 [7] are among the state-of-the-art coding standards.

Because of the Internet and the heterogeneous network structure typically in use currently, end users have various network bandwidths. A T1 line can support 1.5 Mbps, whereas a modem user may have only a 28.8 or 56 Kbps connection. Corporate intranets and campus local area networks (LANs) can support bandwidths of several Mbps. Cable modem [9] and asymmetric digital subscriber line (ADSL) [10] technologies promise to bring high-speed connections to home users, but they are not ubiquitous yet. Because of this, a scalable encoder and layered coding is preferred because a low-bandwidth user can decode only the base layer and clients with higher bandwidth capability can decode enhancement layer(s) in addition to the base layer. H.263+ offers a negotiable optional coding mode (Annex O) that supports temporal, signal-to-noise ratio (SNR), and spatial scalability.

There are several distance learning systems available, notably Stanford-online [8, 11], and various industry products including those from RealNetworks [12], Microsoft Vxtreme [13], and Microsoft Netshow [14]. The majority of the services and products in this category allow a real-time streaming broadcast of live contents (classes or lecture talks for the distance learning systems) or on-demand streaming of stored contents. On the other hand, there are various videophones, video conferencing, and net meeting software products on the market, including Intel Video Phone [15], White Pine's Enhanced CU-SEEME [16], Meeting Point, and Microsoft NetMeeting [17], which support real-time interactive two-way communications. With affordable audio–video capture cards and cameras, the progress in low-bit-rate coding, and the so-called x2 technology [18], home users can participate in a video (and audio) conference at 56 Kbps or lower. It is thus the next step to combine these two types of services and provide a virtual distance learning classroom that is *interactive* [24, 27]. The live class interaction or talk will be streamed (unicast/multicast) to remote sites where students can participate by asking questions in real time, under the control of a central multipoint controller unit (MCU). Another mode allows a student to view on-demand video courses and ask questions during the office hours of the instructor or a teaching assistant.

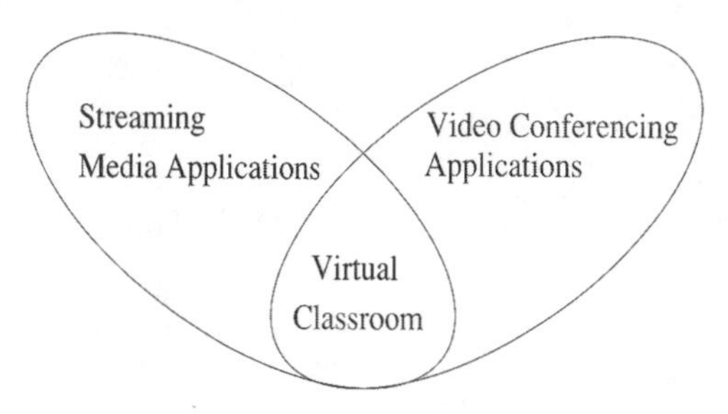

FIGURE 17.1
Virtual classroom application characteristics.

The majority of the current products mentioned above need a plug-in to be able to access the synchronized media on the Internet using a Web browser. Sun Microsystems' Java Media FrameWork (JMF) [19] is aimed at providing a framework that will make synchronized multimedia possible in the future without special plug-ins. However, Java has typically been slow for real-time multimedia decoding. The Just In Time (JIT) compiler and native methods offer a somewhat acceptable solution to this. However, JMF 1.0 does not support capturing and recording of media. This is planned to be a feature of JMF 2.0. Real media architecture [20] is also similar to JMF. For networking, real-time transport protocol (RTP) [21] (and real-time transport control protocol — RTCP) on top of user datagram protocol (UDP) is preferred for low-overhead media delivery of real-time data. Real-time streaming protocol (RTSP) [22] is the application-level protocol that allows control over the delivery of real-time media and is often described as "network remote control." The International Telecommunication Union (ITU-T) recommendation H.323 [23] is a standard for packet-based (including Internet) multimedia communication systems. A majority of the videophone and video conferencing products are H.323 compliant. This standardization helps in their interoperability.

For non-real-time (on-demand) distance learning applications, it is important to add multimedia features that will enrich the user experience. Multimedia features such as hypervideo links cross-referencing the course materials and video, and synchronized and automatic database queries listing the additional references or course notes in addition to the synchronized video, audio, text captions, and slides are useful and effective [25, 26]. Non-real-time multimedia distance learning applications need special tools that help the instructor embed multimedia features and synchronize various media. Automating this content creation as much as possible is important. As an example, a speech recognition system can automatically create the text captions for a talk. Thus, manual typing or text extraction (from electronic notes) of the captions is avoided.

In Section 17.2 we present the design and the development of a real-time virtual classroom distance learning system. A crucial component of the virtual classroom is the way electronic and handwritten slides are handled. This is discussed in Subsections 17.2.1 and 17.2.2, respectively. Section 17.3 describes various multimedia features useful for a non-real-time distance learning system. Section 17.4 various issues are raised. Section 17.5 provides a summary and a conclusion.

17.2 Interactive Virtual Classroom Distance Learning Environment

A virtual classroom environment aims at simulating a real classroom for remote participants. Thus, the remote participant can receive a live class feed and is also able to interact and participate in the class by asking questions. Since there are possibly multiple remote participants, a centralized MCU handles the question requests from the remote sites. Figure 17.2 shows an example of a virtual classroom distance learning environment.

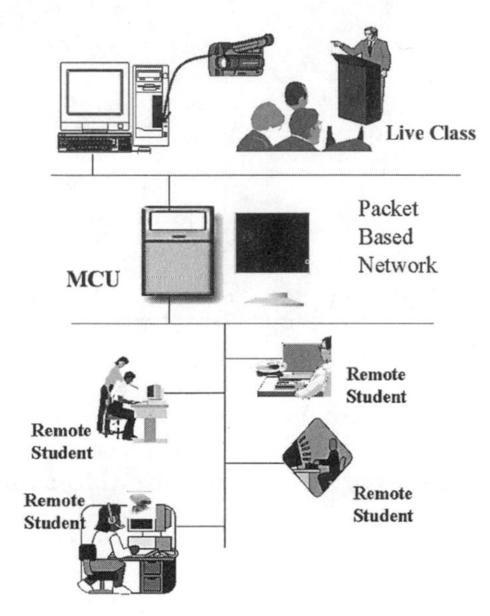

FIGURE 17.2
Virtual classroom *interactive* distance learning environment.

All the participating terminals can passively receive the live class broadcast and actively communicate in a point-to-point fashion with the MCU. Remote users send a request for a question they want to ask to the instructor. The MCU puts this request on a waiting queue of question requests. Under the control of the instructor and based on a fairness policy, the MCU responds to let the remote participant start streaming his video and audio and ask a question. The participants transmit their audio, video, and/or datastreams to the MCU. The MCU consists of [23] a multipoint controller (MC), which controls the remote participants' question requests, and a multipoint processor (MP), which switches between the media streams from the class and the remote terminals under the control of the MC. The MC is also responsible for streaming this media to all the remote terminals. After finishing the question the particular remote user relinquishes control (or the MC timeout for a user occurs) and the instructor responds to the question or is ready to take further questions and to continue the talk. Figures 17.3 and 17.4 show a typical virtual classroom session with the live class server and the remote participant client during the instructor's video and during the question time.

The remote participants are classified, based on their transmission capabilities, into the following categories:

- Users with both audio and video transmission capability
- Users with audio-only transmission capability
- Users with no transmission capability.

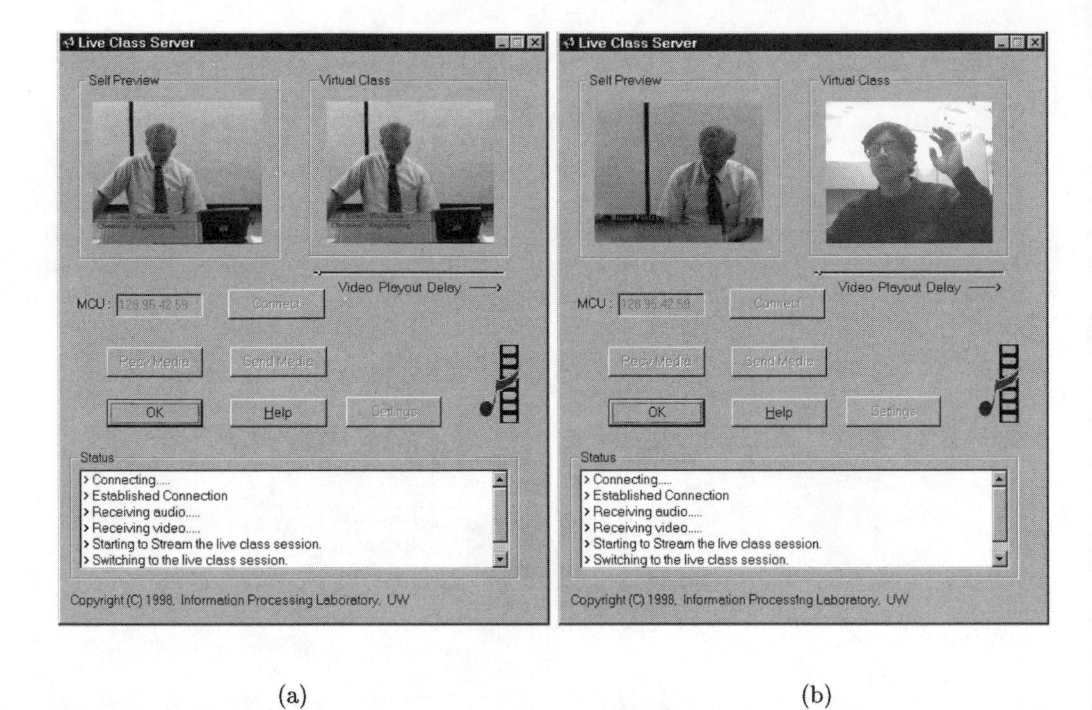

FIGURE 17.3
Virtual classroom live class server. (a) Instructor's video; (b) remote student's video during question time.

This selection is done during the initial capabilities exchange between the remote participants and the MC. The users who do not have any transmission capability can join the virtual classroom as passive recipients. Similarly, it is also possible to allow a remote participant with only audio transmission capability to use it to transmit his or her question. In this case the MC will transmit the audio for their question along with the video of the live class during their question time. It is also possible to maintain a central database of currently registered remote participant students for the class and display their picture during the time their audio is being transmitted. For remote participants who have frequently asked questions (FAQs) and are not willing to ask them in public, an interface allows them to send these queries to the MCU, which will automatically handle them by querying a precompiled database or handbook. Only the student who initiated this query will get the results. The MCU also handles the job of recording the live interactive session, so that it is immediately available after the live class, as a stored on-demand system.

In a situation where participants have the capability and network bandwidth to receive two live feeds simultaneously, the MCU can transmit a separate class feed and remote participants feed. In this case the remote participants feed consists of the question requests and during the idle time when there are no questions in the queue, the MCU polls the remote sites according to some fair algorithm and streams the media from all the remote participants (one at a time) with a transmit capability. In this scenario it is also possible to do segmentation and create a composite video at the MCU. The composed video will consist of video object planes (VOPs) [5] where only one VOP has the currently active remote participant and all the other VOPs are static and have the previously segmented frames of the other remote participants. The MCU can also handle the job of transcoding the compressed audio–video data. The transcoding could be done either in the pixel domain or in the compressed domain. The transcoding can be useful

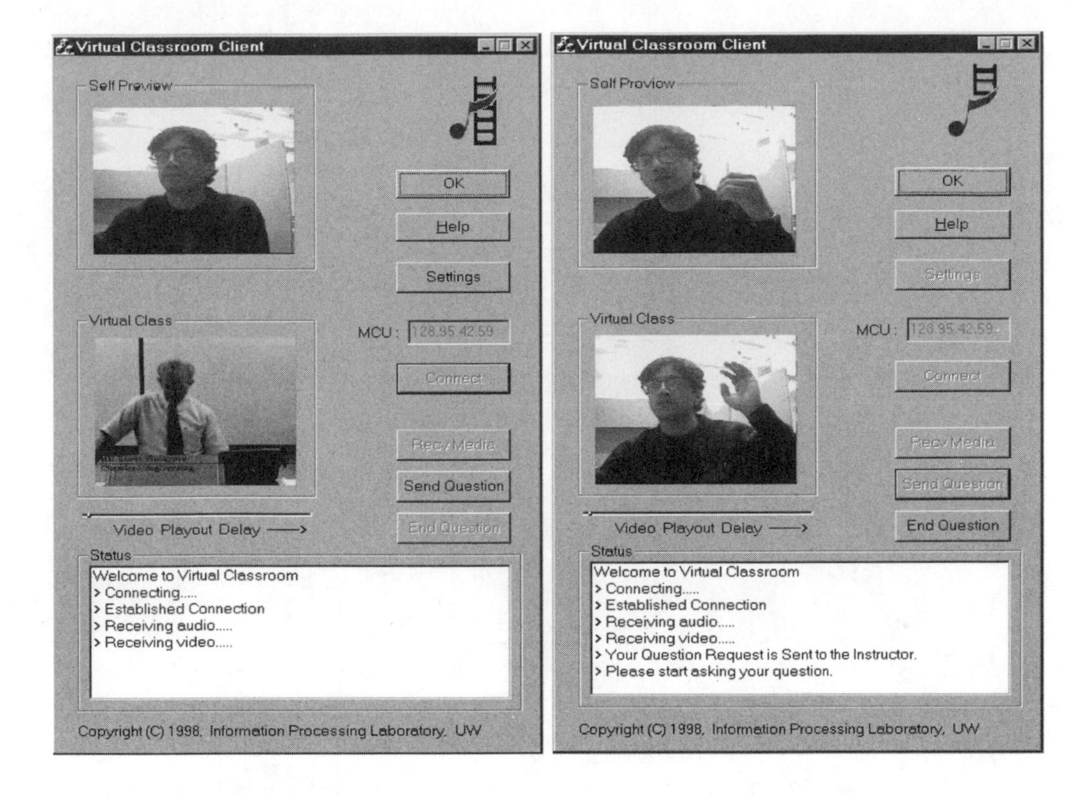

FIGURE 17.4
Virtual classroom client. (a) Instructor's video; (b) remote student's video during question time.

in a variety of scenarios (e.g., for bit rate conversion, compression formats, inter conversion, etc.).

Figure 17.5 shows the protocol used between the MCU and the live class server. Table 17.1 gives a brief description of each message exchanged between the two. A reliable communication mechanism (e.g., TCP/IP) is used for all the control (CTRL) signaling. UDP is used for all the real-time media (video, audio) data (DATA) because of its low overhead. Figure 17.6 and Table 17.2 similarly show the protocol used between the MCU and a client with both audio and video transmission capability. The MCU is connected to one live class server and multiple clients at any given time.

We now discuss various methods to improve a remote participant's experience in a virtual classroom.

17.2.1 Handling the Electronic Slide Presentation

Many instructors use electronic presentation material in their classes. During the course of the lecture the instructor flips these slides according to the flow of his or her narrative. The flipping times are not known a priori. The low-bit-rate video encoding standards are not effective in encoding the slides used in a typical classroom.

Figure 17.7 shows a display at the remote client side of the slide data which was encoded using the conventional video coding standard. This video was taken from the University of

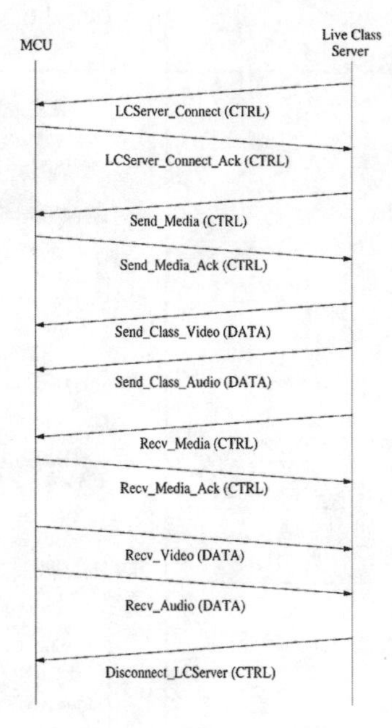

FIGURE 17.5
Interaction between the MCU and the live class server.

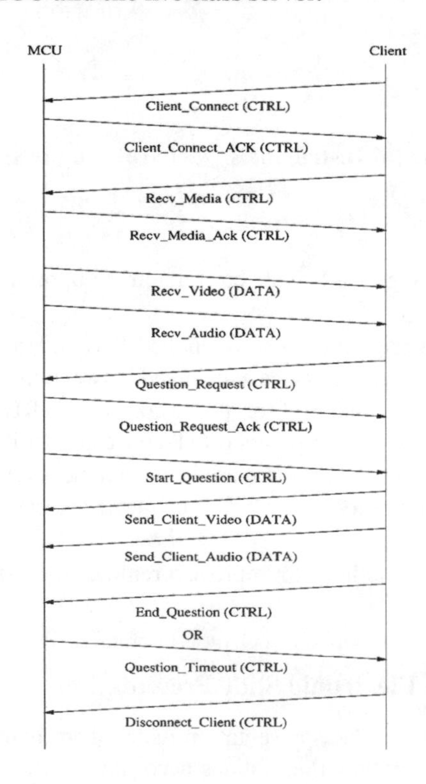

FIGURE 17.6
Interaction between the MCU and a remote participant client with audio and video transmission capability.

Table 17.1 Messages Exchanged Between the MCU and the Live Class Server

Message	Type	Description
LCServer_Connect	Control	Live class server initiates a connection and registers with the MCU; this is the start of a virtual classroom session
LCServer_Connect_Ack	Control	MCU acknowledges live class server connection
Send_Media	Control	Live class server notifies MCU that it is ready to send media data
Send_Media_Ack	Control	MCU acknowledges and asks the live class server to start streaming the media (video, audio) data
Send_Class_Video	Data	Live class server begins streaming the compressed video data to the MCU
Send_Class_Audio	Data	Live class server begins streaming the compressed audio data to the MCU
Recv_Media	Control	Live class server request to the MCU to receive the virtual classroom media data
Recv_Media_Ack	Control	MCU acknowledges the receive media request
Recv_Video	Data	MCU starts streaming the virtual classroom video data to the live class server; this allows the instructor to receive remote student question video during the question time
Recv_Audio	Data	MCU starts streaming the virtual classroom audio data to the live class server; this allows the instructor to receive remote student question audio during the question time
Disconnect_LCServer	Control	The live class server disconnects from the MCU; this is the end of the virtual classroom session

Washington's Televised Instructions in Engineering (TIE) class. As is obvious from the figure, the display is too small and the resolution is low. This impairs the effectiveness of the virtual classroom. Also, the instructor typically pans or zooms on the slide to give the students a better view. This results in a lot of wasted bits when the video sequence is encoded. The solution to this is to send the original electronic slide to the client machine only once.

We have designed and developed a real-time interactive Web-based presentation system to overcome the above drawback. This Web-based client–server system is called "slidecast." Currently, a few Web presentation systems (e.g., Contigo's Itinerary Web Presenter 2.1 [28]) exist. However, they are restrictive and lack the features required for a distance learning environment. In our system the instructor selects an appropriate slide URL at the server when he wants to flip the next slide. The remote participant's client (a Java applet in a Web browser) automatically flips the slide to the new one every time it is flipped by the instructor. The instructor can also draw, mark, or point on the slides in real time, and the same drawings, markings, or pointers will appear on the client's slides. The markings are retained so that the

Table 17.2 Messages Exchanged Between the MCU and a Remote Participant Client

Message	Type	Description
Client_Connect	Control	A remote client initiates a connection and registers with the MCU; this is the start of a virtual classroom session for the client
Client_Connect_Ack	Control	MCU acknowledges client connection
Recv_Media	Control	Client request to the MCU to receive the virtual classroom media data
Recv_Media_Ack	Control	MCU acknowledges the receive media request
Recv_Video	Data	MCU starts streaming the virtual classroom video data to the client; the client receives the instructor's video or a remote student video (during the question time)
Recv_Audio	Data	MCU starts streaming the virtual classroom audio data to the live class server; the client receives the instructor's audio or the remote student audio (during the question time)
Question_Request	Control	This client notifies the MCU that it is interested in asking the instructor a question; this is similar to *raising the hand* by a student in a real classroom
Question_Request_Ack	Control	The MCU acknowledges the question request and puts the request in a queue of possibly pending questions; this allows the instructor to answer the questions at a suitable time
Start_Question	Control	The MCU allows and asks the remote client to start its question; this happens after the instructor permits the question request
Send_Client_Video	Data	Client starts streaming the compressed video data to the MCU
Send_Client_Audio	Data	Client starts streaming the compressed audio data to the MCU
End_Question	Control	Client notifies the MCU that it has finished its question
Question_Timeout	Control	The MCU can end the remote participant's question after a fixed timeout interval; this is provided so that the MCU has the total control of the session
Disconnect_Client	Control	The client disconnects from the MCU; this is the end of the virtual classroom session for this client

instructor can go back and forth between the slides. Markings can also be cleared at any time. The system also allows the instructor to send text instructions to the remote student's clients. Similarly, students can ask questions in real time but do not bother the instructor, as would happen in a real classroom setting, by sending text to the instructor in real time. A private chat between the students is also possible. The system is also be capable of handling late arrivals (i.e., those remote participants who join the session in the middle will get all the slides along with the markings and text that has already been sent in the session). It is also possible to use the slide presenter server as a whiteboard controlled by the instructor.

Figure 17.8 shows the slide presenter server which the instructor can use for presentations. The figure shows a list of slide URLs with navigation controls, connected remote participants, the slide presentation area, the drawing and marking tools, and a text chat area. Figure 17.9 shows the client's view of the presentation. This includes the slides along with the markings, text chat, a list of URLs, and a list of remote participants.

Security and privacy of the remote participants is an important issue in the design of the Web-based presentation system. In general, (untrusted) applets loaded over the net are prevented from reading and writing files on the client file system and from making network connections

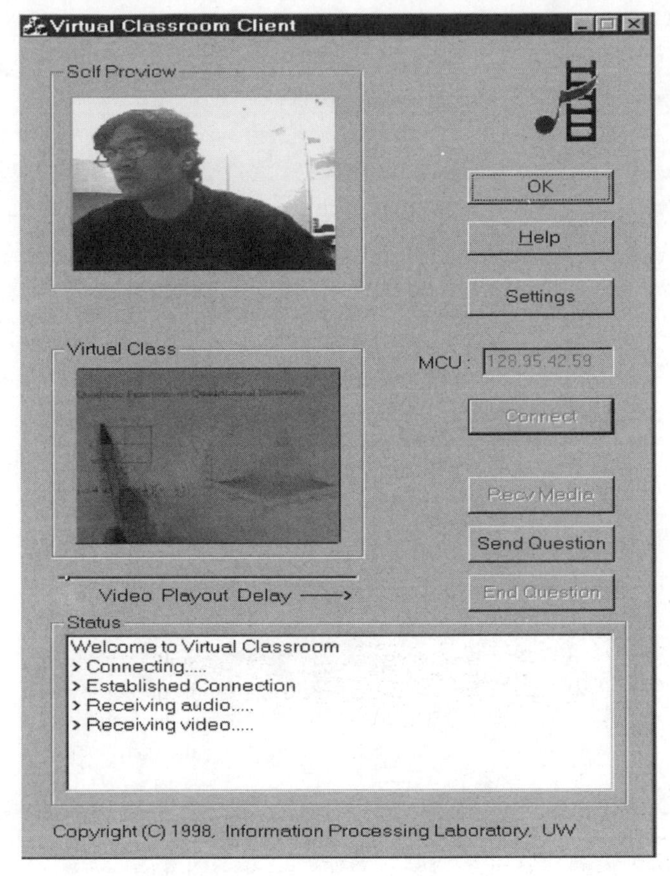

FIGURE 17.7
Display of the video-encoded slides at the remote client site.

except to the originating host. This is a very useful security feature. However, this leads to two possible architectures for our slide presentation system. Figures 17.10a and b show the possible architectures.

In the first architecture (Figure 17.10a), the slide server and the Web server run on the same host. Thus, the remote participant (using the applet downloaded from this Web server) can open a network connection with the slide server. In the second architecture (Figure 17.10b), the Web server and slide MCU are on the same host. The slide presentation server and all the remote slide clients carry out their data and control traffic through the slide MCU.

17.2.2 Handling Handwritten Text

Another popular form of presentation is for the instructor to write notes during the class either on a whiteboard or on a piece of paper. In the University of Washington's TIE classes, an instructor typically writes on a piece of paper. A video of this writing, compressed at low bit rates using the current low-bit-rate video coding techniques, results in a decoded video in which the written material is not very legible. This can be seen in Figure 17.11. We have proposed a new video coding scheme suitable for coding the handwritten text in a much more efficient and robust way utilizing and extending the well-known still image coding algorithms. We use the JBIG [29, 30] standard for bilevel image encoding to compress the handwritten text. The video of the handwritten text typically need not be sent at a very high frame rate.

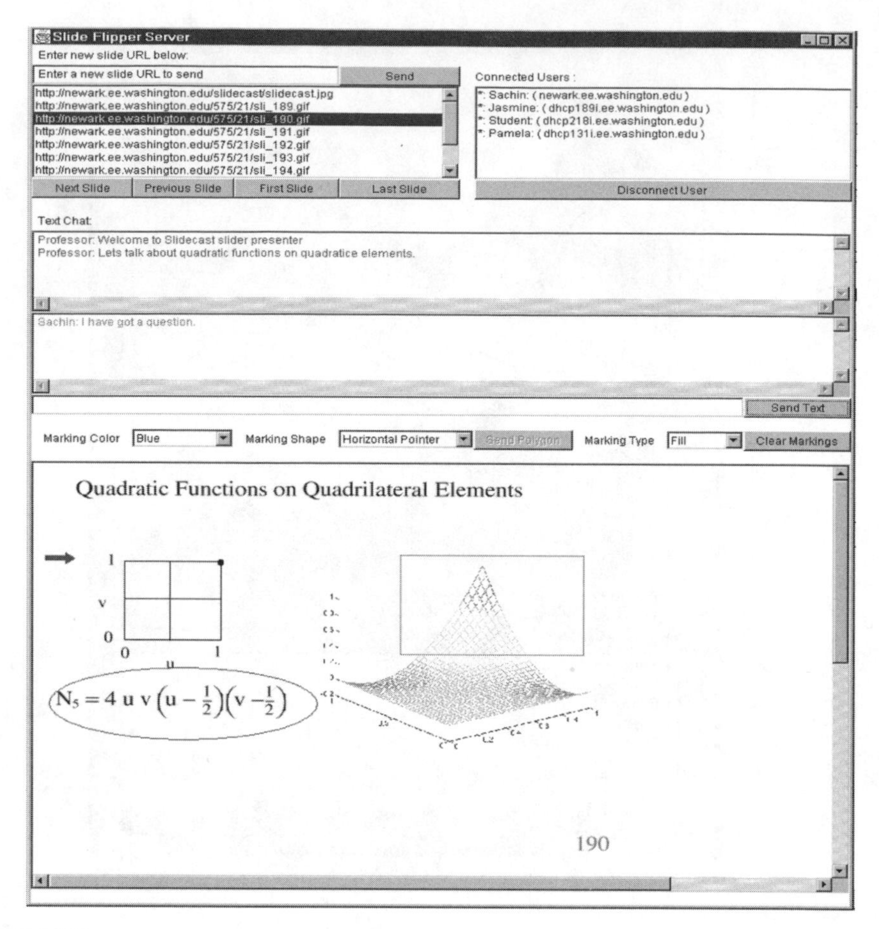

FIGURE 17.8
Slide presenter server.

This is because the speed at which an instructor writes is restricted. This will also result in a lower required bit rate. This also allows us to use a higher spatial resolution (CIF) for the text video. The overall scheme also results in a bit rate savings. It is also possible for those remote participants with a very low connection speed to receive only the audio and the slides (electronic and handwritten).

Figure 17.12a shows the original video frame with handwritten text material. Figure 17.12b shows the output after using the JBIG compression. The difficulties involved in this scheme include the threshold determination and avoiding and/or tracking the hand movements.

17.3 Multimedia Features for On-Demand Distance Learning Environment

Course-on-demand distance learning systems use non-real-time encoding. Thus, it is possible to add multimedia features and annotations to these media. This is typically done by the instructor, teaching assistant, or trained staff member. The challenge in this type of system

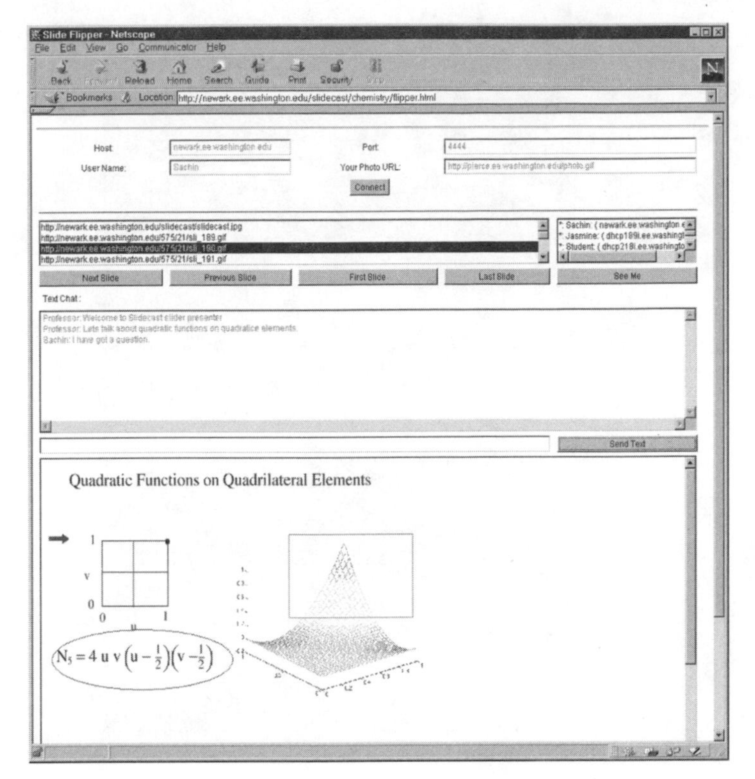

FIGURE 17.9
Slide flipper client Java applet in a browser.

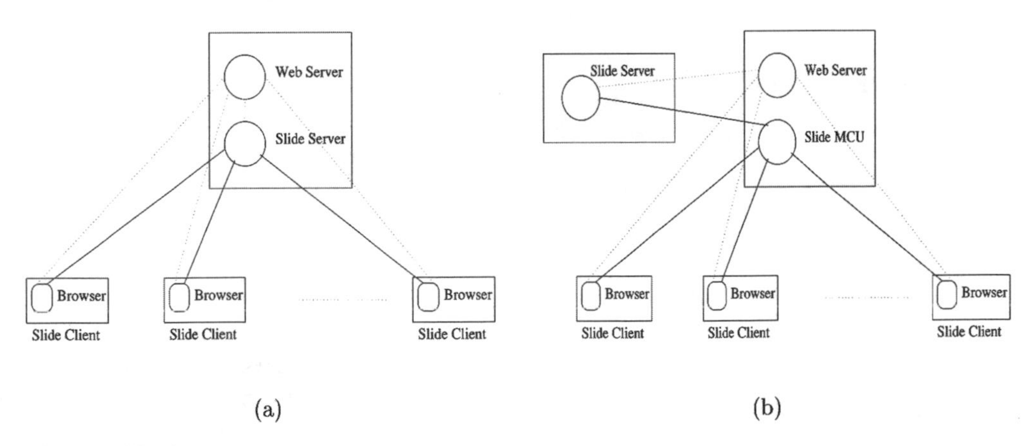

FIGURE 17.10
Slide presentation system, possible architectures.

is to automate most of the feature creation. Also, user-friendly content creation and editing tools help the instructor easily add such features. The created contents are then uploaded to a media server which can serve multiple clients by streaming the stored contents. There are many advantages to using a separate media server as opposed to the Web server to stream the contents. HTTP used by the Web server is not designed with the goal of streaming media to clients with heterogeneous network bandwidths. A media server can use various application-

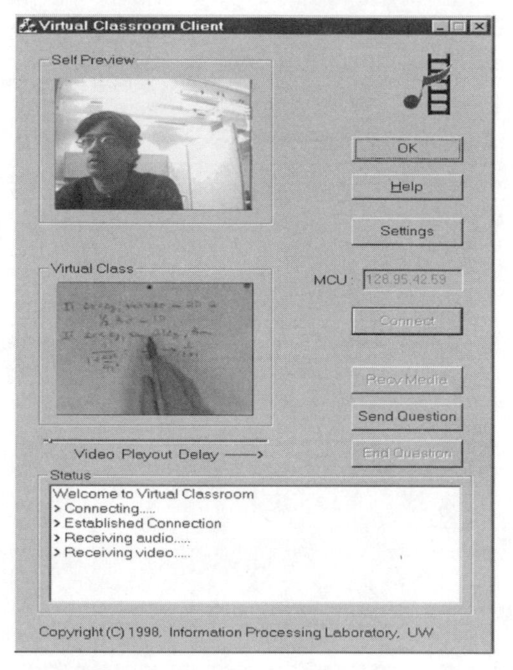

FIGURE 17.11
Client-side display of handwritten material in a live class.

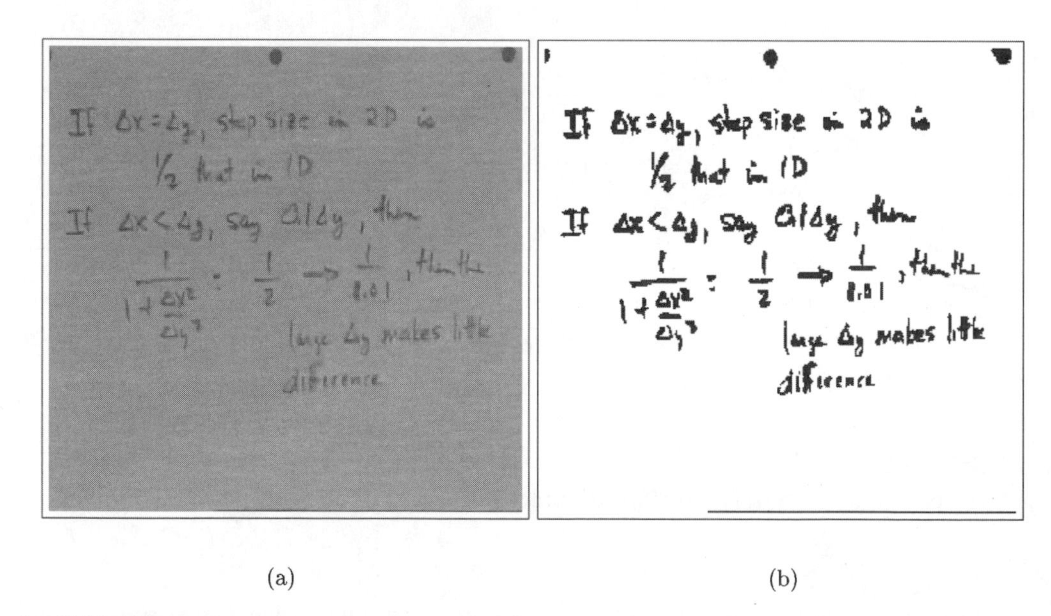

(a) (b)

FIGURE 17.12
(a) Uncompressed video frame of handwritten material in a live class; (b) JBIG compressed video frame.

level protocols (e.g., RTSP [22]) to control and initiate the media streaming. Similarly, RTP and RTCP [21] have been designed to handle data with real-time properties.

It is important to add multimedia features to the video and audio streams. For distance learning systems, we have developed and found useful the following features [25, 26]:

- Hypervideo links: These are similar in concept to hypertext. Many instructors make use of a whiteboard during their classes. Thus, there could be various hyperregions on a video frame. These regions would be bounding boxes for something important written on the whiteboard. Clicking on this hyperregion would allow the user to jump to another class (perhaps a previous sequence class) that gives details about the material in the hyperregion. As an example, in a graduate-level class on digital video coding, the instructor mentions DCT as a popular transform coding method. Then the video frame region where the DCT equation is written on the whiteboard could be hyperlinked with a video frame of an undergraduate-level class on transform coding. The transform coding class similarly could have cross-linking with, say, a basic class on Fourier theory.

- Database queries: Associated with each hyperregion will also be an automatic database query. Thus, students interested in more reference material or papers could get such material either from precompiled course notes or other databases. It is also possible to link this to the university library search, which can list relevant papers or texts available on the subject. Continuing the same example as above, the DCT hyperregion will have an automatic query for the keyword *dct*.

- Text captions: The instructor's talk will appear as text captions which are synchronized with the video and audio. It is thus possible for students with various needs including hearing problems to read the text if they find it difficult to follow the instructor's accent or speed. These text captions can also be downloaded and saved during the progression of the class.

- Slides: Apart from using the whiteboard, instructors also make use of electronic slides during their lectures. These slides are usually available in an electronic form. It is thus possible to have a synchronized presentation with the video, audio, text captions, and slides. All these media together can give the student a feeling similar to that of a live class.

- Random access and keyword jumps: The standard VCR trick modes are provided for the class media. Also, there are various keywords that correspond to the topics covered in the class. Clicking on these keywords would allow students to jump to the appropriate position in the media. Thus, advanced-level students can skip the material they are familiar with and only go through the new material.

We have developed a media browser that supports these features. A screen shot of the browser is shown in Figure 17.13. Figure 17.14 shows the concept of hypervideo links and automatic database queries between two different class videos.

17.3.1 Hypervideo Editor Tool

Creating the above-mentioned features and synchronizing various media require a multimedia authoring tool [25, 26]. The hypervideo editor tool allows marking of various hyperregions on the video frames. We have developed this tool to assist the instructor in the creation of the hypervideo links. The user interface and a typical session of this tool is shown in Figure 17.15.

The instructor can mark rectangular regions on the frames as shown in the figure. The hypervideo entries for each region can be edited by using a form associated with the marked

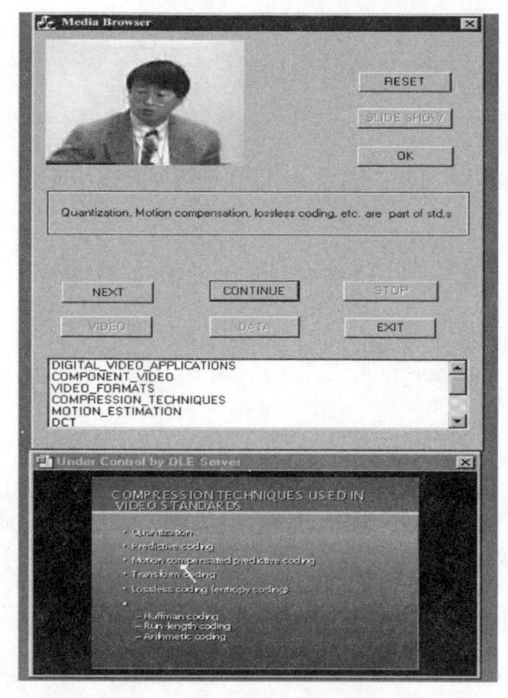

FIGURE 17.13
Screen shot of distance learning media browser.

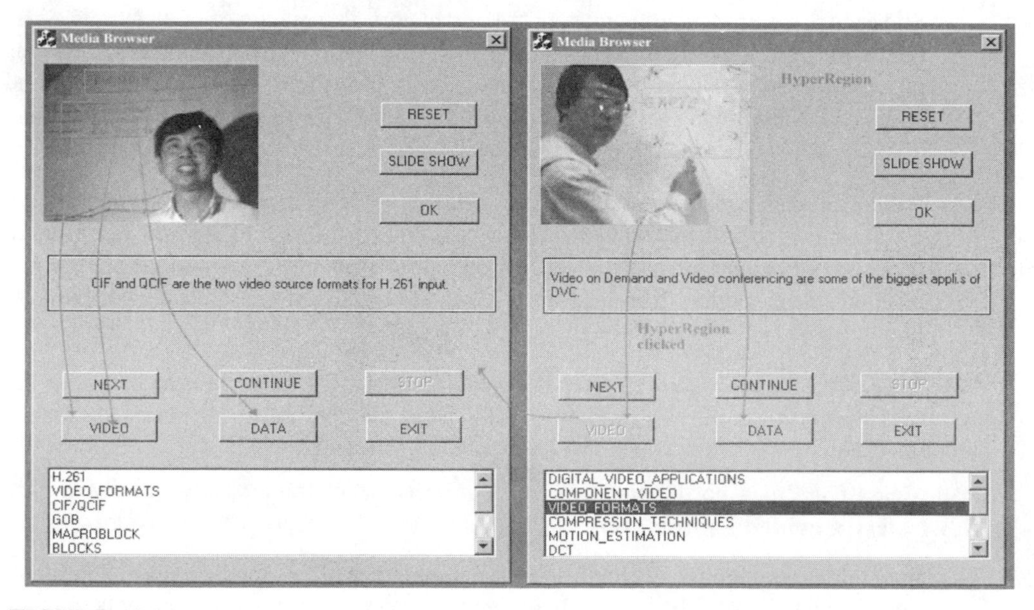

FIGURE 17.14
Hyperlinking between different class videos.

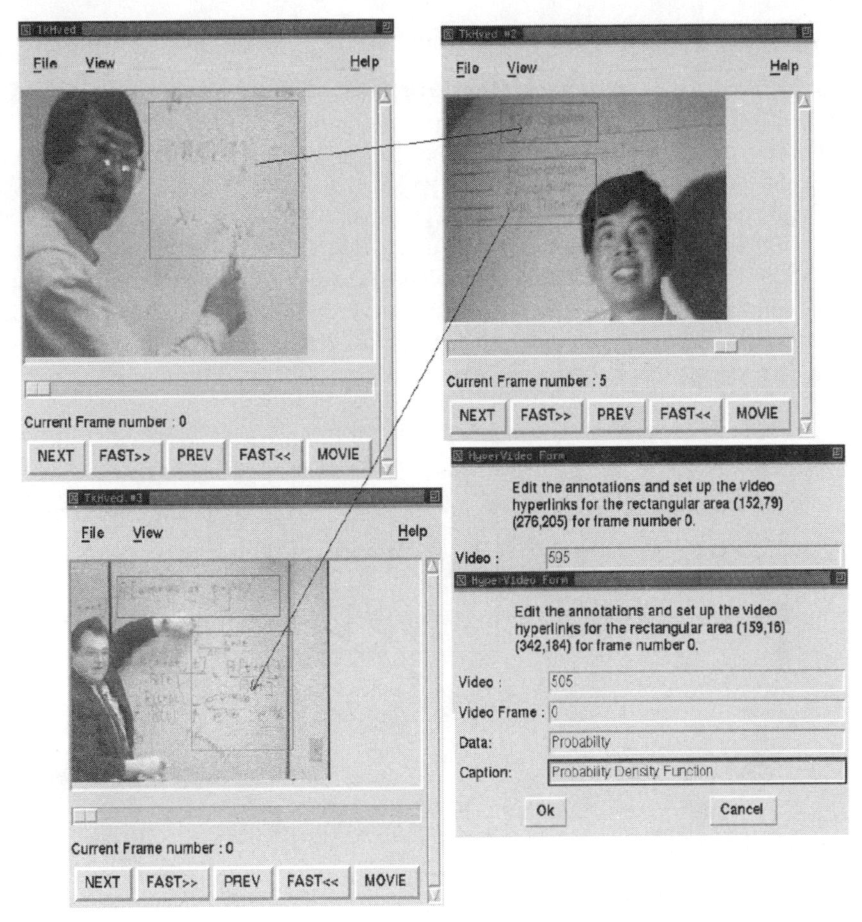

FIGURE 17.15
Display of a typical session of interactive hyperlink creation using the hypervideo editor tool.

regions. The figure shows arrows which indicate a typical scenario in which the hyperlinks for different courses are created, cross-referencing each other. The arrows indicate that the instructor wants the course material for the top-left video associated by a hyperlink with the marked course material in the top-right video. Similarly, the instructor is editing another hyperlink between the appropriate frames of the top-right video and the bottom-left video. The instructor can navigate the video using `Next`, `Prev` buttons, which allow traveling one frame at a time in the forward and backward directions. In the distance learning application, typically the video data frames are available at rates ranging from 10 to 30 frames per second. In this case there may not be significant change in adjacent frames and typically the user will set up the hyperlinks after every several frames. In such a case a fast-forward and fast-backward random access is useful, which allows the user to edit only those frames. `Fast>>` and `Fast<<` buttons allow this functionality. In addition, a slider bar is provided which allows users to randomly jump to any frame of the sequence for editing. The `Movie` option allows users to watch all the frames of the sequence as a movie.

We are currently working on a tool that automates the hyperlink content creation. This is explained in the next subsection.

17.3.2 Automating the Multimedia Features Creation for On-Demand System

It was observed that a considerable effort is necessary to cross-link between the various class video sequences. Thus, automation of this process is highly desirable. We are developing a tool that allows the instructor to mark hyperregions on the slides and/or video frames and type in a keyword for the region. The authoring tool will automatically search a database for this keyword and assist the instructor with a list of possible hyperlinks to videos and slides of similar type. The authored material will then be used in a video-on-demand system. The architecture is shown in Figure 17.16. The Web server will host the electronic slides. The client applet will also be downloaded from this host. The media server will host the compressed media (video and audio). The database server will have a database of all the slides and associated links. The presentation server will handle the client requests for additional information.

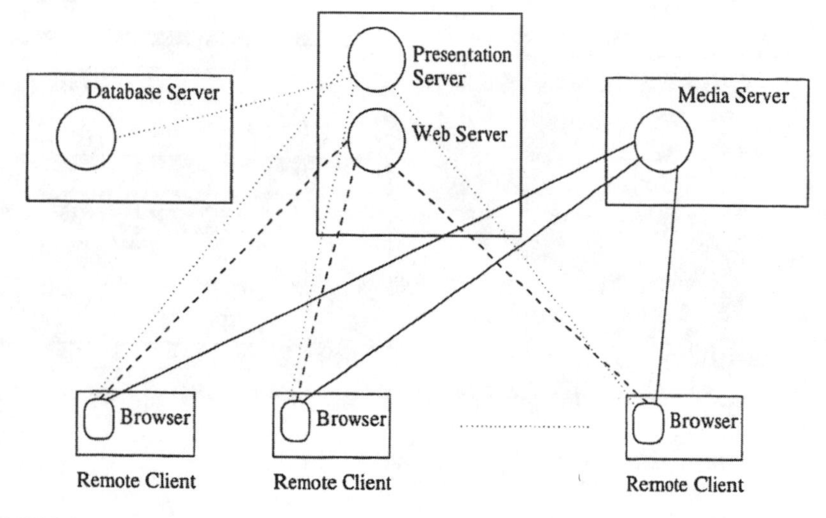

FIGURE 17.16
On-demand system architecture.

17.4 Issues in the Development of Multimedia Distance Learning

The issues involved in the design of multimedia distance learning systems can be classified at two levels. At one level we have to look at technology and at the other level the effectiveness of the system for students and for the instructors. It is important that a state-of-the-art distance learning system is ultimately useful and allows a natural flow of the class sequence without feeling awkward or requiring extra effort. It is also important to remember that a live class has a different set of characteristics than a video conference or a video broadcast. In the real-time virtual classroom, the MCU should be able to totally control the session flow and the instructor should be able to override the MCU in case of any unforeseen situation. The security of the virtual classroom session and authentication of the remote users is also an important aspect. The MCU is responsible for handling these issues.

The major research aspects of the distance learning system are related to media compression and media delivery. Network delivery of the compressed media, synchronization of different media streams, error mitigation schemes to handle lost packets, handling packet delay, and multicast routing of the media data are active research aspects. For the encoder, developing scalable, fast, high-quality software-only compression is the main research area. We will very briefly look at some of the issues that concern a multimedia distance learning system.

17.4.1 Error Recovery, Synchronization, and Delay Handling

The recovery of lost packets, synchronization of different media streams, and end-to-end delay minimization are three interrelated aspects. For an interactive system, the end-to-end delay tolerable is about 250 to 400 ms. Because of the use of RTP/UDP for real-time transport of media streams, the packets could be lost, get duplicated, or arrive out of order. Because of high delay sensitivity of the interactive virtual classroom, a very late arriving packet is in fact treated as a lost packet. The lost packets are typically detected using missing RTP sequence numbers. There exist various approaches to handle lost packets. The schemes could be classified based on the amount of extra processing requirements at the server or client side [31]. For an interactive application, the delay induced by forward error correction (FEC) [32] or interleaving-based schemes [33] may not be acceptable. Similarly, the retransmission-based schemes [33] are also infeasible because of the delay considerations. Thus, the most popular method is for the encoder to add some media-dependent redundant information in the bitstream, and for the decoder to use this information along with error concealment methods. The delay jitter is handled by buffering at the receiver end. The RTP time stamp is used for determining the playout times for the packets. The synchronization between various media streams is handled by the time stamp information for the individual streams. Typically, one stream is used as a so-called master stream.

17.4.2 Fast Encoding and Rate Control

The single most computationally intensive component of a typical video encoder is motion estimation. Thus, fast and robust motion estimation methods [34] are used for improving the speed of encoding. However, with the currently available computing power, even on a personal computer, the main bottleneck for high-quality encoding is largely due to the low bandwidth connections at the client sites. Thus, a layered encoding approach with a good rate control mechanism is very important.

17.4.3 Multicasting

With a large number of concurrent connections to the server or MCU, the bandwidth requirement at the server end for unicast sessions would be huge and would not scale well. Unicast transmission requires sending the same datagram packet to each connected user separately. This results in an inefficient use of network bandwidth. Multicasting [35] attempts to solve this problem and avoids the delivery of the same media data multiple times on the same links. In multicasting, a packet is transmitted just one time to multiple users. The packet will be duplicated only at a router when the paths diverge.

17.4.4 Human Factors

Human factors play a significant role in the effectiveness and success of a multimedia distance learning system. From the instructor's perspective, he or she should not be distracted from the real classroom and should not have to make any clumsy effort. In addition, the instructor should have ultimate control over the virtual classroom session. From the remote participating client's point of view, he or she should get an experience close to or even better than that in a live classroom.

17.5 Summary and Conclusion

In this chapter we discussed two forms of distance learning systems:

- Live, interactive
- On demand.

The design of the interactive virtual classroom is aimed at giving the instructor total control of the entire session. We also proposed useful multimedia features aimed at enriching the user experience. A majority of the research topics related to media compression and multimedia networking are directly related to the multimedia distance learning system. Human factors also play a significant role in the widespread and successful use of multimedia distance learning.

References

[1] ISO/IEC 11172-2: Information technology — Coding of moving pictures and associated audio for digital storage media at up to about 1.5 Mbit/s — Part 2: Video, 1993.

[2] ISO/IEC 13818-2: Information technology — Generic coding of moving pictures and associated audio information: Video, 1996.

[3] ITU-T Video codec for audiovisual services at p x 64 kbit/s, Recommendation H.261, March 1993.

[4] ITU-T Video coding for low bit rate communication, Draft H.263, January 1998.

[5] ISO/IEC 14496-2: Information technology — Coding of audio visual objects: Visual, Committee Draft, November 1997.

[6] S.M. Redl, M.K. Weber, and M.W. Oliphant, *An Introduction to GSM,* Artech House, Boston, 1995.

[7] ITU-T Recommendation G.723.1, Speech coders: Dual rate speech coder for multimedia communications transmitting at 5.3 and 6.3 kbit/s, 1996.

[8] Stanford-online Web site, `http://stanford-online.stanford.edu`.

[9] C.R. Lewart, *The Ultimate Modem Handbook: Your Guide to Selection, Installation, Troubleshooting, and Optimization,* Prentice-Hall, Englewood Cliffs, NJ, 1997.

[10] W.J.Goralski, *ADSL (Computer Communications),* McGraw-Hill, New York, 1998.

[11] D. Harris and A. DiPaolo, "Advancing Asynchronous Distance Education Using High-Speed Networks," *IEEE Trans. on Education,* vol. 39, no. 3, pp. 444–449, August 1996.

[12] Real Networks Web site, `http://www.real.com`.

[13] Microsoft's Vxtreme Web site, `http://www.microsoft.com/netshow/vxtreme`.

[14] Microsoft's NetShow Web site, `http://www.microsoft.com/netshow`.

[15] Intel Video Phone Web site, `http://www.intel.com/proshare/videophone/`.

[16] White Pine's CU-SeeMe Web site, `http://www.cuseeme.com`.

[17] Microsoft's NetMeeting Web site, `http://www.microsoft.com/netmeeting`.

[18] 3com and US Robotics x2 technology Web site, `http://x2.usr.com`.

[19] Sun Microsystem's Java Media Framework Web site, `http://java.sun.com/products/java-media/jmf/`.

[20] Real Media Architecture Web site, `http://www.real.com/realmedia`.

[21] H. Schulzrinne, S. Casner, R. Frederick, and V. Jacobson, RTP: A Transport Protocol for Real-Time Applications, Internet Draft, Internet Engineering Task Force (IETF), August 1998.

[22] H. Schulzrinne, A. Rao, and R. Lanphier, Real Time Streaming Protocol (RTSP), Internet Draft, Internet Engineering Task Force (IETF), January 1998.

[23] ITU-T Recommendation H.323, Packet based multimedia communications systems, 1998.

[24] J.-N. Hwang, S.G. Deshpande, and M.-T. Sun, "A Virtual Classroom for Real-Time Interactive Distance Learning," *ISCAS,* vol. 3, pp. 611–614, June 1998.

[25] J.-N. Hwang, S.G. Deshpande, and M.-T. Sun, "Multimedia Features for Course-on-Demand in Distance Learning," *IEEE Signal Processing Society, 1st Workshop on Multimedia Signal Processing,* pp. 513–518, Princeton, NJ, June 1997.

[26] J.-N. Hwang, J. Youn, S. Deshpande, and M.-T. Sun, "Video Browsing for Course-on-Demand in Distance Learning," *International Conference on Image Processing,* vol. II, pp. 530–533, Santa Barbara, CA, October 1997.

[27] S.G. Deshpande and J.-N. Hwang, "A Real-Time Interactive Multimedia Distance Learning System," *ISMIP* 98, (to appear).

[28] Contigo's Itinerary Web Presenter 2.1, `http://www.contigo.com`.

[29] ITU-T T.82, Information technology — Coded representation of picture and audio information — Progressive bi-level image compression, 1993.

[30] ISO/IEC JTC1 SC29 Working Group 1, WD14492, JBIG-2, Working Draft, August 1998, Johns Hopkins University Press, Baltimore, 1989.

[31] Y. Wang and Q.-F. Zhu, "Error Control and Concealment for Video Communication: A Review," *Proc. of IEEE,* vol. 86, no. 5, pp. 974–997, May 1998.

[32] J. Rosenberg and H. Schulzrinne, An RTP Payload Format for Generic Forward Error Correction, Internet Engineering Task Force, Internet Draft, 30 July 1998.

[33] C. Perkins, Options for Repair of Streaming Media, Internet Engineering Task Force, Internet Draft, 13 March 1998, (expired).

[34] F. Dufaux and F. Moscheni, "Motion Estimation Techniques for Digital TV: A Review and a New Contribution," *Proc. IEEE,* vol. 83, no. 6, pp. 858–876, June 1995.

[35] V. Kumar, *MBone: Interactive Multimedia on the Internet,* Macmillan, New York, 1995.

Chapter 18

A New Watermarking Technique for Multimedia Protection

Chun-Shien Lu, Shih-Kun Huang, Chwen-Jye Sze, and Hong-Yuan Mark Liao

18.1 Introduction

18.1.1 Watermarking

Owing to the popularity of the Internet, the use and transfer of digitized media are increasing. However, this frequent use of the Internet has created the need for security. It is imperative to protect information to prevent intentional or unwitting use of information by someone other than the rightful owner. A commonly used method is to insert watermarks into original information to declare rightful ownership. This is the so-called watermarking technique. A watermark can be a visible or invisible text, binary stream, audio, image, or video. It is embedded in an original source and is expected to tolerate attacks of any kind. A valid watermarking procedure enables one to judge the owner of media contents via a retrieved watermark even if it is attacked and is, thus, fragmentary.

An effective watermarking procedure should satisfy or consider the following requirements:

1. **Transparency:** The inserted watermark should be perceptually invisible. This demand is most challenging for images with large homogeneous areas.

2. **Robustness:** A secure watermark should be difficult to remove or destroy, or at least the watermarked image must be severely degraded before the watermark is lost. Typical intentional or unwitting attacks include:

   - Common digital processing: A watermark should survive after image blurring, compression, dithering, printing and scanning, etc.
   - Subterfuge attacks (collusion and forgery) [4]: A watermark should be resistant to combinations of the same image watermarked with different watermarks (collusion). In addition, a watermark should be robust to repeated watermarking (forgery).
   - Geometric distortions: A watermark should be able to survive attacks which use general geometric transformation, such as cropping, rotation, translation, and scaling.

3. **Capacity:** Capacity [23, 27] is the issue allowing embedding of the maximum number of distinguishable watermarks. Cox et al. [4] discovered that the significant components of

an image have a perceptual capacity that allows watermark insertion without perceptual degradation. In other words, any attempt to remove or destroy embedded watermarks will influence the significant components of an image and thus lead to fidelity degradation.

4. **Public watermarking:** Authentication without using original sources is necessary for two reasons [35]: (1) searching for the original image in large digital libraries is time consuming, and (2) application of "Web-crawling" detection. Source-based and destination-based approaches are two major watermarking schemes [23]. The source-based approach focuses on ownership authentication/identification. A unique watermark is detected or extracted to determine the owner of data. It is desirable to confirm ownership by retrieving the watermark without the original image. On the other hand, the destination-based method can be used to trace the end user when illegal use such as reselling occurs. Under these circumstances, the existence of original images is allowed.

5. **Resolving rightful ownership deadlock:** A watermark should unambiguously certify the true occupant. Craver et al. [5] took the initiative in presenting and solving this problem. It is in fact very important and is usually ignored in most watermarking schemes. Qiao and Nahrstedt also solved the problem of rightful ownership and were the first to provide protection of customers' rights [24].

18.1.2 Overview

In the literature, Koch and Zhao [13] transformed an image by using a block DCT transform and then a pseudo-random number generator to select a subset of blocks. A triplet of blocks with midrange frequencies was slightly revised to yield a binary sequence watermark. This seems reasonable because low-frequency components are perceptually important but easy to sense after modification, and high-frequency components are easy to tamper with. Macq and Quisquater [18] suggested hiding data in the least significant bits such that the embedded data is imperceptible. Their watermark is easy to destroy using attacks such as low-pass filtering. Cox et al. [4] proposed a global DCT-based spread spectrum approach to hide watermarks. They believe that signal energy present in any frequency is undetectable if a narrowband signal is transmitted over a much broader bandwidth. Ideally, this will lead to a watermark that spreads over all frequencies so that the energy in any single frequency is very small and, thus, undetectable. Their watermark is of fixed length and is produced using a Gaussian distribution with zero mean and unit variance. They distributed as fairly as possible the watermark to the first 1000 largest AC coefficients. An objective measurement was proposed to evaluate the similarity between the original and extracted watermarks. Hsu and Wu [11] used multi-resolution representations for the host image and the binary watermark. The middle frequencies in the transformed wavelet domain were selected for modification using a residual mask. Their method has been shown to be effective for JPEG-based compression at higher bit rates. A similar work was proposed by Hsu and Wu using the discrete cosine transform (DCT) [12]. Some commercially available watermarking software programs, such as SysCoP and EikonaMark [20], also embed an identification word of finite length into specified positions (determined by a secret key) in an image. However, there are limitations in the above-mentioned methods: (1) the length of a watermark is short and bounded, and (2) it is unclear where the watermark can be hidden and to what extent modifications can be done to meet the transparency and robustness requirements.

To counter the above-mentioned drawbacks, characteristics of the human visual system (HVS) have been incorporated into some schemes [6, 23, 32]. It is very meaningful and reasonable to take the HVS into consideration because of its inherent features. If one can modify an image based on rules taken from the HVS, then it will be easier to generate an imperceptible watermark with maximum capacity, and the length and strength of a watermark can be adaptive to the host image. Many existing watermarking techniques generate binary sequences or small texts as watermarks. The former are visually meaningless while the latter might be easily removed or destroyed. For example, the SysCoP watermarking technique hides an eight-character watermark and uses an eight-digit secret key. Cox et al.'s watermark [4] is composed of a binary sequence that is statistically undetectable but visually meaningless. Hsu and Wu's watermark [11] is a recognizable binary text, but it is simple and easy to destroy. Basically, a watermarking scheme that does not sufficiently utilize the capacity of a host image may cause the potential length and strength of a watermark to be bounded. Podilchuk and Zeng [23] proposed a perceptual model-based watermarking scheme, but their watermark is image dependent. In other words, their watermarks cannot be specified in advance.

One of the important issues in watermarking is the need to access the original image to reliably extract the embedded watermarks [4, 11, 12, 23]. Without the original image, the extracted watermarks may be degraded [2, 14]. Some researchers [25, 33] have pointed out that the original image may help to overcome geometric transformation attacks. However, for security reasons, a watermarking technique that does not use the original source is always preferable.

Another important issue in watermarking is the need to design an effective watermarking technique such that embedded watermarks will survive attacks. StirMark [20] and unZign [34] are two software programs available on the WWW that can be used to verify the robustness of watermarking. StirMark and unZign are powerful because watermarks generated by most existing watermarking techniques cannot resist their attacks. To the best of our knowledge, there is still no research reported in the watermarking literature that provides any results of robustness evaluation under attacks from StirMark and unZign.

Some surveys regarding watermarking methods can be found in [4, 8, 10, 22, 32, 38]. In this chapter, we propose an HVS-based watermarking algorithm with both gray-scale and binary watermarks taken into consideration. Our method will attain the following goals:

- The watermark is as large and as strong as possible.
- Embedded watermarks are meaningful and recognizable.
- Watermarks are resistant to common attacks, in particular from StirMark and unZign.
- The watermark recovery process does not use the original source.

18.2 Human Visual System-Based Modulation

To satisfy the demand for maximum perceptual capacity, a model based on the human visual system is introduced here. Some previously proposed systems [1, 36] that base their designs on the human perceptual model have played an important role in the field of image compression. Basically, these systems take into account the structures of complex natural images. More specifically, masking, the effect of a visual model, refers to the fact that a component in a given visual signal may become imperceptible in the presence of another signal called a masker. In this situation a signal raises the visual *threshold* for other signals around it. Three commonly encountered masking types are frequency masking, luminance masking, and contrast masking. The first one is image independent but depends on the visual environment. The other two are image dependent. Frequency masking specifies the sensitivity of human eyes to sine wave gratings at various frequencies. For a given visual distance and display resolution, it can determine the just noticeable distortion (JND) for each spatial frequency from specified wave functions. Psychologists have experimented with several contrast sensitivity functions from some specific wave functions, such as the DCT basis function [19] and wavelet [37]. Since wavelet transform is very powerful in image representation, we will use the wavelet-based frequency masking model [37] for watermarking. The frequency masking map with a four-level wavelet transform and display visual resolution (DVR) 32 [37] is illustrated in Figure 18.1. Whiter gray values imply higher JND values.

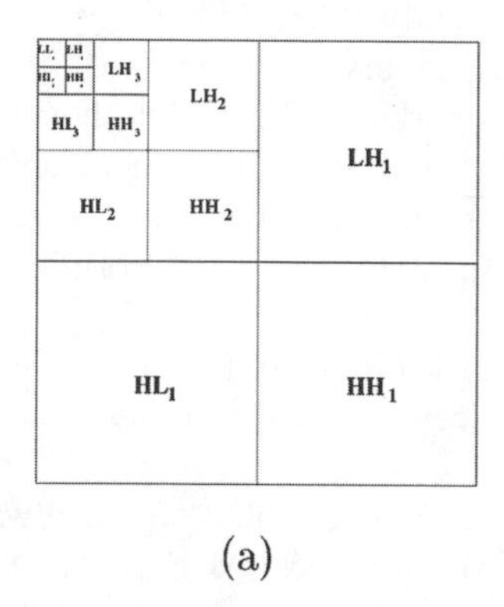

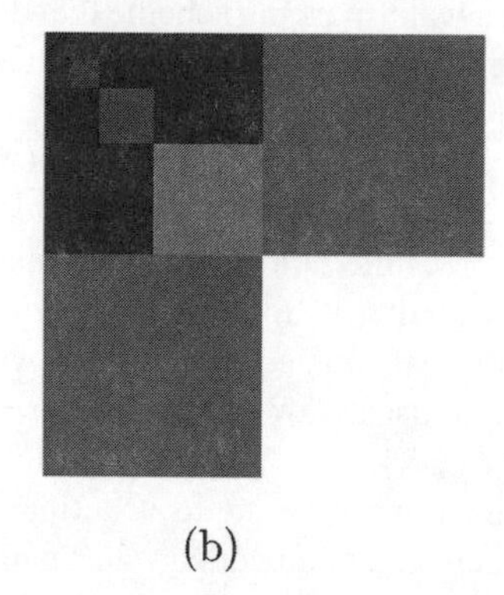

FIGURE 18.1
(a) Four-level wavelet transform; (b) frequency masking map corresponding to four-level wavelet transform [37].

Two very popular watermarking techniques that employed perceptual significance were presented in [4, 23]. Cox et al. [4] used spread spectrum embedding to hide a watermark:

$$I_i^* = I_i \left(1 + \alpha \cdot n_i\right) , \tag{18.1}$$

where I_i and I_i^* are DCT coefficients before and after modulation, respectively, and n_i is the watermark sequence; α is a weight that controls the trade-off between transparency and robustness. In [23], Podilchuk and Zeng presented two watermarking schemes based on the human visual model: image adaptive-DCT (IA-DCT) and image adaptive wavelet (IA-W) schemes. The watermark embedder for both the IA-DCT and IA-W approaches can be described in general as

$$I_{u,v}^* = \begin{cases} I_{u,v} + J_{u,v} \cdot n_{u,v}, & \text{if } I_{u,v} > J_{u,v}, \\ I_{u,v}, & \text{otherwise} \end{cases} \tag{18.2}$$

where $J_{u,v}$ is the masking value of a DCT- or wavelet-based visual model, and $n_{u,v}$ is the sequence of watermark values. It is found from both in embedding schemes that modulations take place in the perceptually significant coefficients with the modification quantity specified by a weight. The weight is heuristic [4] or depends on a visual model [23]. Cox et al. [4] and Podilchuk and Zeng [23] both adopted a similar detector response measurement, described by

$$Sim\left(n, n^*\right) = \frac{n \cdot n^*}{\sqrt{n^* \cdot n^*}} , \tag{18.3}$$

where n^* is the extracted watermark sequence. If the signs of a pair of elements in n and n^* are the same, then they contribute to the detector response. A higher value of $Sim(n, n^*)$ means a higher probability that n^* is a genuine watermark. High correlation values can only be achieved if the transformed coefficients are modulated and distorted along the same direction during the embedding and attacking processes, respectively. This is very important if a watermark detector is to get a higher similarity value. However, we find from (18.1) and (18.2) that the directions of modulation are random. A positive coefficient can be updated with a positive or negative quantity, and a negative coefficient can be changed with a positive or negative quantity. Furthermore, the works of [4, 23] did not consider the relationship between the signs of correlation pairs (the transformed coefficients and the watermark values). This is why many attacks can successfully defeat the above-mentioned watermarking schemes.

In this chapter, we shall seriously treat the modulation problem. The modulation strategies described in [4, 23] are called random modulation here, in contrast to our attack-adaptive modulation mechanism. In [16, 17], we noted that if a modulation strategy operates by adding a negative quantity to a positive coefficient or by adding a positive quantity to a negative coefficient, then we call it negative modulation. Otherwise, it is called positive modulation if the sign of the added quantity is the same as that of the transformed coefficient.

18.3 Proposed Watermarking Algorithms

We shall develop two watermarking algorithms based on the assumption that the original image (host image) is gray scale. Our watermark can be either a binary watermark or a gray-scale watermark, and its maximum size can be as large as that of the host image. The wavelet transform adopted in this work is constrained such that the size of the lowest band is 16×16. A four-level decomposition via wavelet transform is shown in Figure 18.1a.

18.3.1 Watermark Structures

A gray-scale watermark with "ACADEMIA SINICA" and its corresponding Chinese text in a pictorial background is shown in the middle-left part of Figure 18.2. The bottom-left part of Figure 18.2 shows a binary watermark with Chinese text that means "ACADEMIA SINICA." For embedding binary watermarks, we do not take the irrelevant background [11, 12, 35] into account; rather, we hide the watermark pixels (on the foreground) only. That is, the watermark pixels or the foreground pixels constituting the text are embedded. For gray-scale watermarks, the characteristic we adopt is that a recognizable watermark is extracted for subjective judgment instead of an objective decision determined by some similarity measurements. Another benefit of using a gray-scale watermark is that it can avoid checking the existence of a single watermark pixel. Under different attacks, a gray-scale watermark has a greater chance to survive. This is because a gray-scale watermark can always preserve a certain degree of *contextual information* even after attacks. It is well known that to embed a gray-scale watermark with its original intensities is extremely difficult since the transparency requirement is easily violated. Therefore, we shall apply the human visual model to make the embedded gray-scale watermark look like the original, but it is in fact "compressed."

18.3.2 The Hiding Process

Our watermarking technique is detailed in this section. First, the host image and a visually recognizable gray-scale watermark are transformed by discrete wavelet transform. It is noted that the binary watermark does not have to be transformed.

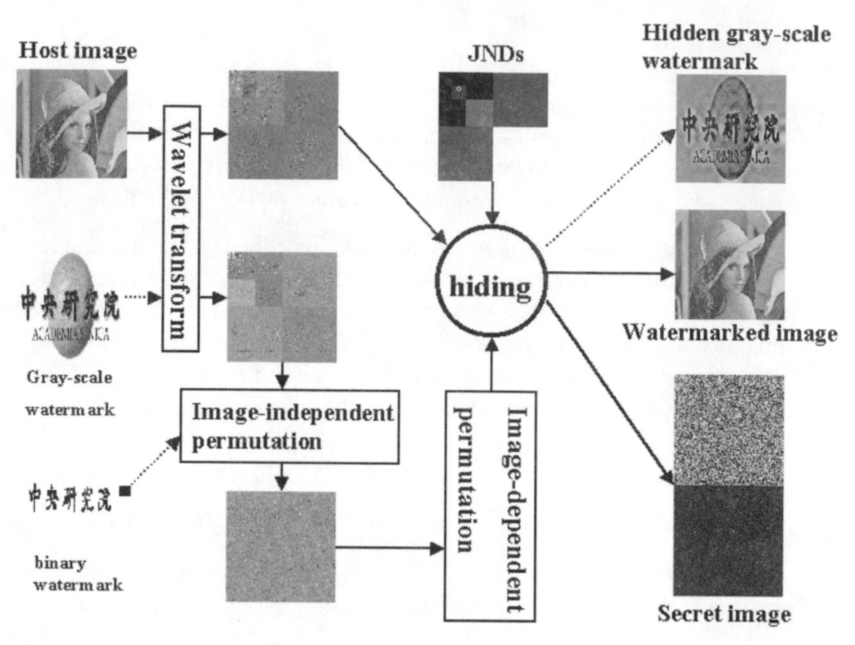

FIGURE 18.2
The flow chart of our watermark hiding process.

Image-Independent and Image-Dependent Permutations

In order to make the transformed gray-scale watermark and the nontransformed binary watermark statistically undetectable, they are spatially transformed using a chaotic system called "toral automorphism" [35]. After the chaotic mixing process, the watermark is converted into a noise-like form, which can guarantee undetectability. Basically, the toral automorphism is a kind of image-independent permutation. Next, an image-dependent permutation is executed to increase the security level and to select places for hiding the two types of watermarks. The image-dependent permutation used in our approach is in the form of a mapping function. The host image and the gray-scale watermark in the wavelet domain are a one-to-one mapping. We design such a mapping function based on the significance of the wavelet coefficients. The security level is raised because the watermarks are embedded into those components with larger coefficients. These significant coefficients are more secure than the insignificant coefficients, especially when compression is performed. The reason is that under compression attack, the significant coefficients are more likely to survive. That is, the amount of modulation is proportional to the scalogram of the wavelet transform. The concept of significance is inspired by Shapiro's EZW compression scheme [28]. The larger the magnitude of a wavelet coefficient, the more significant it is. After the wavelet coefficients of the host image and the gray-scale watermark are sorted, the mapping function $m(.,.)$ is defined as

$$m\left(x_h, y_h\right) = \left(x_m, y_m\right) , \tag{18.4}$$

where (x_h, y_h) represents the position of a wavelet coefficient, $H_{s,o}^{(k)}(x_h, y_h)$, in the host image; and (x_m, y_m) represents the position of a wavelet coefficient, $M_{s,o}^{(k)}(x_m, y_m)$, obtained from the watermark. Both (x_h, y_h) and (x_m, y_m) correspond to the kth largest wavelet coefficients (where $1 \le k \le N \times M$ and $N \times M$ is the image size) with the same scale (s) and orientation (o), respectively.

In order to obtain better security for a binary watermark, we propose to doubly insert the active pixels of the binary watermark into the wavelet domain of the host image. Double insertion has the merit of cross-supporting the existence of a watermark pixel, in particular when watermarked images are attacked. Therefore, one can consider the mapping function for hiding binary watermarks a two-to-one function:

$$m\left(x_p, y_p\right) = (x_m, y_m) \ ,$$
$$m\left(x_n, y_n\right) = (x_m, y_m) \ , \tag{18.5}$$

where (x_p, y_p) and (x_n, y_n) are the positions of the host image's positive and negative wavelet coefficients, respectively; (x_m, y_m) is the position of a foreground pixel in a binary watermark.

Without the mapping function(s), it is hard to guess the location of the embedded watermark. Embedding watermarks into the significant frequency components implies automatic adaptation to the local scalogram of wavelet transformed host images.

Compression-Adaptive Hiding

After completing the secret mapping process, we then consider how to modify the host image's wavelet coefficients and consider the extent of modification. The most important problem in watermarking is that the watermarked image should have no visual artifact. Previously, Bender et al. [3] altered the intensities of a host image within a small range and hoped the update would be perceptually unnoticed. However, they did not address clearly what the range of modification should be in order to obtain perceptual invisibility. Here, we shall take the HVS into account.

Since images (or other media) need to be transmitted via networks, the compression procedure has to be used so that the traffic jam problem can be avoided. Therefore, any watermarking technique has to take the compression-style "attack" into consideration. In this chapter, we shall embed the watermark by modulating the coefficients in the wavelet transformed domain; therefore, the modulation quantity of wavelet coefficients should be able to adapt to a general compression process. Usually, when an image is compressed at a specific ratio, the absolute values of its transformed insignificant coefficients are reduced to small values or zero. On the other hand, the absolute values of the significant coefficients are also decreased by a certain amount. We have observed that if the sign of the modulated quantity is the same as that of a wavelet coefficient to be updated, then the corresponding watermarked image will preserve better image quality after compression. However, the robustness of the compressed watermarked image will be degraded under the same conditions. This is because the robustness issue, which is closely related to detector response, has been violated. Hence, we can say that there is a trade-off between image quality and robustness. Figure 18.3 illustrates this phenomenon using EZW-based compression at a ratio of 80:1. Figures 18.3a and b show compressed Lena images that were positively and negatively modulated, respectively. It is obvious that the face portion (including the eyes, mouth, and nose) shown in Figure 18.3a is clearer than that in Figure 18.3b. Although the image quality of Figure 18.3a is better than that of Figure 18.3b, we also find some apparent edges around the hat areas. This is because the number of positive and negative wavelet coefficients is not the same.

To embed a watermark safely, the negative modulation strategy is adopted. That is, the sign of a wavelet coefficient and that of its corresponding modulation quantity should be different. The positive modulation strategy, which makes both the wavelet coefficient and its corresponding modulation quantity the same sign, is advantageous for image quality but sacrifices robustness. Another study regarding the robustness issue can be found in [16, 17].

In what follows, we demonstrate how the concept of JNDs [37] can be realized and used in developing a watermarking technique. The JND-based watermarking techniques are discussed in the subsequent sections, gray-scale watermarking first and then binary watermarking. We

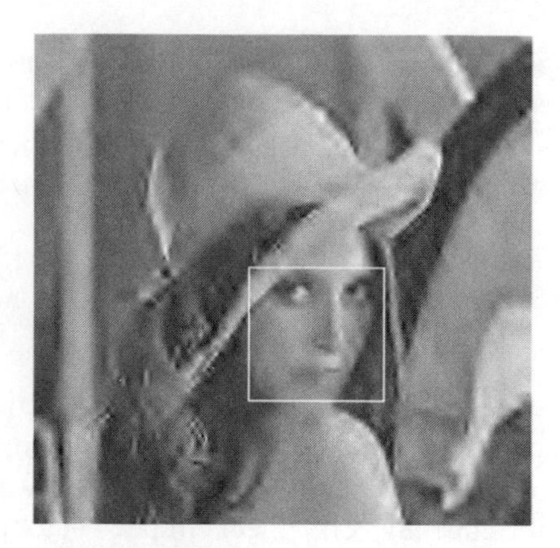

FIGURE 18.3
The trade-off between image quality (at a compression ratio of 80:1) and robustness. (a) Positively modulated image after compression; (b) negatively modulated image after compression.

divide the hiding process into two parts, the lowest frequency (LL_4) part and the part for the remaining frequencies. It is noted that the lowest frequency wavelet coefficient corresponds to the largest portion of decomposition.

JND-Based Modifications on a Gray-Scale Watermark

The purpose of using a gray-scale watermark instead of a binary one is to make the watermark much more easily "understood" due to its stronger contextual correlation. The watermarking process for a gray-scale watermark is as follows:

$$H^m_{s,o}(x_h, y_h) = \begin{cases} H_{s,o}(x_h, y_h) + sgn\left(H_{s,o}(x_h, y_h)\right) \times J_{s,o}(x_h, y_h) \\ \qquad \times \frac{|M_{s,o}(x_m, y_m)|}{\max(M_{s,o}(.,.))} \times w, \\ \qquad \text{if } \left|H_{s,o}(x_h, y_h)\right| > J_{s,o}(x_h, y_h) \\ H_{s,o}(x_h, y_h), \\ \qquad \text{otherwise}, \end{cases} \tag{18.6}$$

where

$$sgn\left(H_{s,o}(x_h, y_h)\right) = \begin{cases} -1, \text{ if } H_{s,o}(x_h, y_h) \geq 0.0 \\ 1, \text{ if } H_{s,o}(x_h, y_h) < 0.0 . \end{cases} \tag{18.7}$$

The function $sgn(\cdot)$ is designed for negative modulation; $H^m_{s,o}(x_h, y_h)$ is the wavelet coefficient of the watermarked image to be determined; $H_{s,o}(x_h, y_h)$ and $J_{s,o}(x_h, y_h)$ represent the host image's wavelet coefficient and the JND value at position (x_h, y_h), scale s, and orientation o, respectively; $M_{s,o}(x_m, y_m)$ represents the wavelet coefficient of the gray-scale watermark; $\max(M_{s,o}(.,.))$ represents the maximum $M_{s,o}(.,.)$ value obtained among different locations; the relationship between (x_m, y_m) and (x_h, y_h) has been defined in (18.4); and w is a weight used to control the maximum possible modification that will lead to the least image quality degradation, and is defined as

$$w = \begin{cases} w_L, \text{ if } H_{s,o}(x_h, y_h) \in LL_4 \\ w_H, \text{ if } H_{s,o}(x_h, y_h) \notin LL_4 . \end{cases} \tag{18.8}$$

When hiding a gray-scale watermark, the embedded watermark, M^e, is expressed as [according to (18.6)]

$$M^e_{s,o}(x_h, y_h) = \begin{cases} sgn\left(H_{s,o}(x_h, y_h)\right) \times J_{s,o}(x_h, y_h) \times \frac{|M_{s,o}(x_m, y_m)|}{\max(M_{s,o}(.,.))} \times w, & \text{if } \left|H_{s,o}(x_h, y_h)\right| > J_{s,o}(x_h, y_h) \\ 0, & \text{otherwise}. \end{cases} \quad (18.9)$$

JND-Based Modification of a Binary Watermark

Each foreground pixel $s(x, y) \in \mathcal{S}$ is doubly embedded by modification of the corresponding positive and negative wavelet coefficients of the host image according to the mapping function depicted in (18.5). For a positive wavelet coefficient, $H_{s,o}(x_p, y_p)$, the subtraction operation is triggered while the addition operation is inhibited:

$$\text{Subtraction: } H^m_{s,o}\left(x_p, y_p\right) = H_{s,o}\left(x_p, y_p\right) + sgn\left(H_{s,o}\left(x_p, y_p\right)\right) J_{s,o}\left(x_p, y_p\right) \times |w| \,. \quad (18.10)$$

For a negative wavelet coefficient, $H_{s,o}(x_n, y_n)$, the addition operation is triggered while the subtraction operation is inhibited:

$$\text{Addition: } H^m_{s,o}(x_n, y_n) = H_{s,o}(x_n, y_n) + sgn\left(H_{s,o}(x_n, y_n)\right) J_{s,o}(x_n, y_n) \times |w| \,. \quad (18.11)$$

$sgn(\cdot)$ is defined as in (18.7); $H^m_{s,o}(x_i, y_i)(i = p, n)$ is the wavelet coefficient of the watermarked image; $H_{s,o}(x_i, y_i)$ and $J_{s,o}(x_i, y_i)$ represent the host image's wavelet coefficient and the JND value at position (x_h, y_h), scale s, and orientation o, respectively; the relationship between (x, y) and $(x_i, y_i)(i = p, n)$ has been defined in (18.5); and w is a weight used to control the maximum possible modification without degrading the image quality, and can be defined as in (18.8) or generated as a sequence of random numbers for the reason discussed in Section 18.5.

After the negative modulation stage, the absolute magnitudes of the modified wavelet coefficients become smaller. The absolute value of an attacked (or modified) wavelet coefficient should retain the above relation if it corresponds to a valid binary watermark pixel. If this absolute value increases after an attack, then we conclude that its corresponding pixel does not belong to a binary watermark.

Finally, the inverse wavelet transform is applied to the modulated wavelet coefficients (in gray-scale or binary watermark hiding) to generate the watermarked image.

18.3.3 Semipublic Authentication

Original source protection is considered extremely important in watermarking. In some watermarking schemes, if one wishes to extract watermarks from a watermarked image, the original source is required. However, it is always preferable to detect/extract watermarks without accessing the original source since it is dangerous to retrieve them through the Internet. In the literature, if a watermarking technique does not need the original source to extract the watermark, then the extracted watermark will be somewhat degraded [2, 14]. Here, we will present a technique to extract watermarks without using the original images. In the proposed method, an extra set of secret parameters (in fact, it is a secret image) instead of a secret key only is required. Our assertion is that it is more secure to use a secret image rather than the original source. This is because even if the secret image is intercepted, there will be no *computationally feasible* way to figure out its contents.

In the watermark recovery process, image-dependent permutation mapping is also required to locate the watermark. Hsu and Wu [11, 12] obtained this information by either saving it as a file during the embedding step or recomputing it from the original image and the watermark. In our scheme, the original image is not directly used. What we do is combine the wavelet coefficients of the original image and the results of image-dependent permutation mapping to obtain the secret image. Under these circumstances, it is difficult for one to guess the contents correctly, and the watermark can be reconstructed without any degradation, even if the original source is absent.

Our fusion process is depicted in Figure 18.4. Each pixel of these images consists of 4 bytes (32 bits). In order to get a secure combination, the least significant 16 bits of the image-dependent permutation are replaced by the least significant 16 bits of the wavelet transformed image to form the first part of the secret image. Similarly, the least significant 16 bits of the

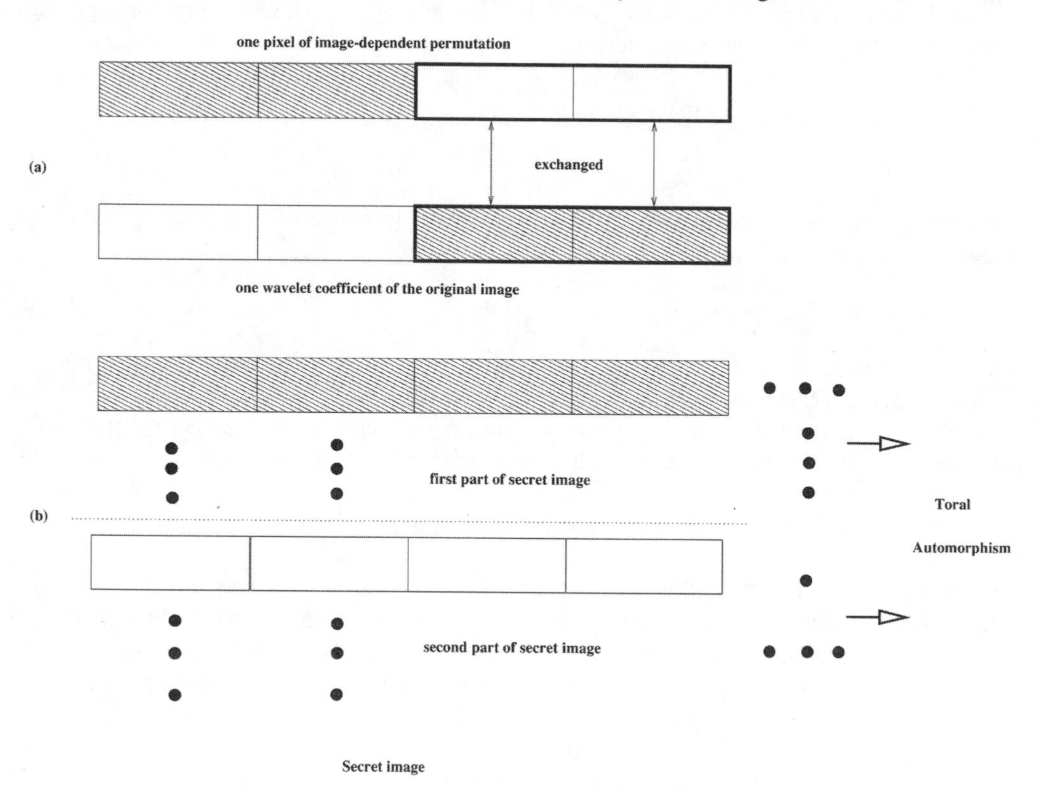

FIGURE 18.4
Secret image: integrating the mapping image of the image-dependent permutation and the wavelet coefficients of the original image in the watermark embedding process.

wavelet transformed original image are replaced by the least significant 16 bits of the image-dependent permutation to form the second part of the secret image. Certainly, other interleaved combinations may be used too. After the replacement process, the shaded line areas shown in Figure 18.4a form the first part of the secret image, and the white areas form the second part of the secret image. Finally, the toral automorphism is imposed on the individual parts of the fused image. There are two reasons why an unauthorized person would have difficulty interpreting our secret image. First, one has to know the number of iterations as well as the parameter used in the toral automorphism. Second, one has to know how the two parts of the secret image have been combined. To reconstruct the hidden watermarks, the secret image is "decompressed" such that the actual wavelet coefficients can be used for the purpose of comparison, and the results can be understood in the real world. There exists a relation between the quality of the extracted watermark and whether or not the original image is needed for authentication. It is well known that if a watermarking algorithm does not use the original image, then the quality of the extracted watermark will be somewhat degraded. When a secret image is adopted, use of the original image is no longer necessary; thus, the quality and authentication problems are simultaneously solved. Figure 18.2 illustrates the whole process of our watermark hiding technique.

18.4 Watermark Detection/Extraction

In this section, we describe in detail how watermarks can be extracted from a watermarked image. Usually, the watermarked image (possibly distorted) and a secret parameter or a set of parameters are the necessary components for watermark extraction. In our method, the original image is not required, but a secret image is needed. The first step in watermark extraction is to perform wavelet transform on the watermarked image. The corresponding wavelet coefficients of the host image and the positions where the extracted watermark (M^e) should be located are both retrieved by decompressing the secret image, SI. Next, the wavelet coefficients of the host image and of the distorted watermarked image are subtracted to obtain the information of the embedded watermark.

18.4.1 Gray-Scale Watermark Extraction

In gray-scale watermark extraction, the subtracted results are simply regarded as the wavelet coefficients of the extracted watermark. They are expressed as

$$M^e_{s,o}\left(m\left(x_h, y_h\right)\right) = H^m_{s,o}\left(x_h, y_h\right) - H^{SI}\left(x_h, y_h\right) , \tag{18.12}$$

where $H^m_{s,o}(x_h, y_h)$ represents the wavelet coefficient of the watermarked image at position (x_h, y_h) with scale s and orientation o; $m(x_h, y_h)$ is the retrieved location of the gray-scale watermark; and $H^{SI}(x_h, y_h)$ is the retrieved wavelet coefficient of the host image. The relationship between (x, y) and (x_h, y_h) is a one-to-one mapping function which has been defined in (18.4). Finally, the inverse image-independent permutation and inverse wavelet transform are executed to obtain the reconstructed watermark.

18.4.2 Binary Watermark Extraction

The binary watermark detection process has two major steps — determining the subtracted results and designing the decision-making mechanism. We describe these two steps in detail.

Subtracted results:

$$\begin{aligned} Diff_{\text{Subtraction}}\left(x_p, y_p\right) &= H^m_{s,o}\left(x_p, y_p\right) - H^{SI}\left(x_p, y_p\right), \\ Diff_{\text{Addition}}\left(x_n, y_n\right) &= H^m_{s,o}\left(x_n, y_n\right) - H^{SI}\left(x_n, y_n\right) , \end{aligned} \tag{18.13}$$

where $Diff_{\text{Subtraction}}(x_p, y_p)$ and $Diff_{\text{Addition}}(x_n, y_n)$ denote the subtracted results corresponding, respectively, to the positive and negative wavelet coefficient sequences; $H^m_{s,o}(x_i, y_i)$ $(i = p, n)$ denotes the watermarked image's positive and negative coefficients; and $H^{SI}(x_p, y_p)$ and $H^{SI}(x_n, y_n)$ are the retrieved positive and negative wavelet coefficients of the host image derived from the secret image.

Decision:

$$M^e\left(m\left(x_p, y_p\right)\right)=\begin{cases} s\left(m\left(x_p, y_p\right)\right), & \text{if } Diff_{\text{Addition}}\left(x_n, y_n\right)>0 \\ & \text{or } Diff_{\text{Subtraction}}\left(x_p, y_p\right)<0 \\ None, & \text{otherwise}, \end{cases} \tag{18.14}$$

where $m(x_p, y_p)$ is the retrieved location of the binary watermark, and $s(m(x_p, y_p))$ is the value of a foreground pixel in the original binary watermark. Notice that $m(x_p, y_p)$ is the same as $m(x_n, y_n)$. The relationship between (x, y), (x_p, y_p), and (x_n, y_n) is a two-to-one mapping as indicated in (18.5). The decision operation shown in (18.14) shows the merit of alternating support of double hiding. That is, the "*OR*" operation is adopted to determine the final result from the two detected watermarks. Finally, the inverse image-independent permutation is executed to obtain the reconstructed watermark.

18.4.3 Dealing with Attacks Including Geometric Distortion

In this section, we present a relocation strategy to deal with attacks generating asynchronous phenomena. StirMark [20] and unZign [34] are two very strong attackers against many watermarking techniques. From analysis of StirMark [20], it is known that StirMark introduces an unnoticeable quality loss in the image with some simple geometric distortions. In addition, a jitter attack [21] is another type of attacker, which leads to spatial errors in images that are perceptually invisible. Basically, these attackers cause asynchronous problems. Experience tells us that an embedded watermark that encounters these attacks is often more severely degraded than those encountering other attacks. Moreover, the behaviors of other unknown attacks are also not predictable. It is important to deal with the encountering attack in a clever way so that damage caused by a StirMark attack can be minimized. This is because the orders of wavelet coefficients are different before and after an attack and might be varied extremely for attacks with an inherent asynchronous property. Consequently, in order to recover a "correct" watermark, the attacked watermarked image's wavelet coefficients should be relocated to proper positions before watermark detection. The relocation operation is described in the following. First, the wavelet coefficients of the watermarked image (before attacks) and those of the attacked watermarked image are sorted, respectively. The wavelet coefficients of the watermarked image after attacks are rearranged into the same order as those of the watermarked image before attacks. Generally speaking, by preserving the orders, damage to the extracted watermark can always be reduced. Owing to the similarity between StirMark and unZign, it is expected that unZign can be dealt with the same way. The improved results are especially remarkable for attacks including geometric distortions, such as flip, rotation, jitter, StirMark, and unZign. Figure 18.5 shows the whole process of our watermark detection/extraction technique.

18.5 Analysis of Attacks Designed to Defeat HVS-Based Watermarking

Although the human visual model helps to maximize the capacity of an embedded watermark, it is not entirely understood whether it provides a higher degree of security, because the watermarked image carries a clue about the strength and location of the watermark [7]. For example, Cox et al. [4] hid their watermarks in the first 1000 AC coefficients in the DCT domain. Podilchuk and Zeng's watermark [23] was embedded according to the masking effects of human perception. In order to prevent the embedded watermark from being successfully attacked, we adopt the same visual model but use a different modulation function. It is noted

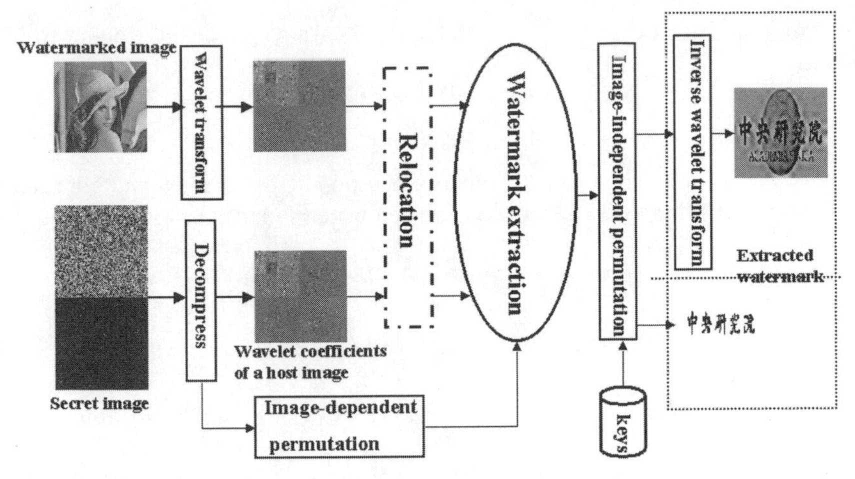

FIGURE 18.5
The flow chart of our watermark detection/extraction process.

that the positions where the watermark is hidden may be blabbed in the above-mentioned schemes [4, 23]. Fortunately, the amount of modification is a random number and is difficult to predict. Furthermore, the length of the hidden recognizable watermark is not fixed and is also difficult to predict. In this section, we start by examining the effect of a proposed reverse operation which is imposed on the HVS-based watermarked image. This operation is similar to scenario 1 in Carver et al.'s interpretation attack [5]. The pirate may not have to contest with the actual owner for the watermarked resources, but he/she can destroy the watermark instead. Therefore, the pirates still accomplish the goal of peculating the watermarked resources. The above-mentioned attack is the so-called removal attack [10]. We also introduce a way to prevent a pirate from using the HVS-based concept to execute a valid attack.

Let H be the host image and M the gray-scale watermark to be hidden; the watermarked image denoted as H^m is expressed as in (18.6). Suppose now that a pirate constructs a counterfeit watermark $\bar{M}$ and obtains a faked host image $\bar{H}$; they are related to the watermarked image H^m by means of a removal operation $\ominus$:

$$H^m \ominus \bar{M} = \bar{H} \, . \tag{18.15}$$

Usually, pirates will seek to fulfill the following conditions; that is,

$$M \subset \bar{M}$$

and

$$Sim\left(H, \bar{H}\right) \approx 1 \, .$$

Assume that the embedding process is known to the pirates except for the secret keys. If a forged watermark is made based on knowledge of the human visual model, then we can check whether the watermarked image is vulnerable to an HVS-based removal attack. According to our watermark hiding procedure [Eq. (18.6)], the positive wavelet coefficients become smaller and the negative ones larger after modulation. Therefore, a forged watermark can be made based on the concept of JNDs of the human visual model [37]:

$$\bar{M}_{s,o}(x, y) = sgn\left(H^m_{s,o}(x, y)\right) J_{s,o}(x, y) \times \ ratio \ \times \bar{w} \, , \tag{18.16}$$

where $sgn(\cdot)$ has been defined in (18.7) and $\bar{w}$ represents the predicted weight, which can be defined as

$$\bar{w} = \begin{cases} \bar{w}_L, & \text{if } H^m_{s,o}(x,y) \in LL_4\,, \\ \bar{w}_H, & \text{if } H^m_{s,o}(x,y) \notin LL_4\,, \end{cases}$$

where $\bar{w}_L$ and $\bar{w}_H$ are the predicted weights corresponding to the low and high frequencies, respectively. According to (18.15), the counterfeit watermark can be expressed as

$$\bar{H}_{s,o}(x,y) = H^m_{s,o}(x,y) - \bar{M}_{s,o}(x,y)$$

$$= \begin{cases} H_{s,o}(x,y) + sgn\left(H_{s,o}(x,y)\right) J_{s,o}(x,y) \times \frac{|M_{s,o}(x,y)|}{\max\left(M_{s,o}(.,.)\right)} \\ \qquad \times w - sgn\left(H^m_{s,o}(x,y)\right) J_{s,o}(x,y) \times \ ratio \ \times \bar{w}, \\ \qquad \text{if } \left|H_{s,o}(x,y)\right| > J_{s,o}(x,y) \\ H_{s,o}(x,y), \qquad \text{otherwise}\,. \end{cases} \tag{18.17}$$

In what follows, we shall analyze the positive wavelet coefficients of (18.17) to check the similarity between the host image and the counterfeit image. The analysis on the negative wavelet coefficients is the same and is, thus, omitted. The positive wavelet coefficients of $\bar{H}$ can be expressed as

$$\bar{H}_{s,o}(x,y) = H_{s,o}(x,y) + \left(\bar{w} \times \ ratio \ - \frac{\left|M_{s,o}(x,y)\right|}{\max\left(M_{s,o}(.,.)\right)} \times w\right) \times J_{s,o}(x,y)\,. \tag{18.18}$$

Note that $sgn(H_{s,o}(x,y))$ is the same as $sgn(H^m_{s,o}(x,y))$. If $\bar{w} \times ratio$ can be precisely predicted and its value happens to be $\frac{|M_{s,o}(x,y)|}{\max(M_{s,o}(.,.))} \times w$, then the faked image is equivalent to the original image and the watermark may be entirely removed. But the above situation is only an ideal case. It is noted that the term $\frac{|M_{s,o}(x,y)|}{\max(M_{s,o}(.,.))}$ originates in the gray-scale watermark, which is designed to be hidden. As described in Section 18.3.1, only the compressed version of the designed watermark is embedded. Hence, pirates will not have any a priori knowledge they can use to predict the originally designed watermark. Furthermore, the number of modified coefficients (watermark length) should also be guessed approximately. Our experiments demonstrate that it is very difficult to predict a *ratio* (which is random) and then use it to remove or degrade the embedded watermark even when $\bar{w}$ is very close to w. For a binary watermark, the random term is the weight, w, defined in (18.10) and (18.11).

18.6 Experimental Results

We have conducted a series of experiments to corroborate the effectiveness of the proposed method. Different kinds of attacks, including some digital processing and two attackers — StirMark and unZign — were used to check whether the embedded watermark was transparent and robust. StirMark and unZign were considered to be two very powerful watermark attackers because they have successfully destroyed many watermarks made by some commercially available watermarking software programs such as Digimac, SysCoP, JK_PGS, EikonaMark, and Signnum Technologies [20]. A basic requirement for a watermark attacker is that it should "destroy" the watermark while preserving the watermarked image to some extent. A watermark attacker that destroys both the watermark and the watermarked image is, in fact, useless. The two watermarks (one binary and one gray scale) used in our experiments are shown on the left side of Figure 18.2. The watermarked image shown in the middle right part of Figure 18.2 was watermarked using the gray-scale watermark. The experimental results are reported in the following.

18.6.1 Results of Hiding a Gray-Scale Watermark

The popular Lena image 256 × 256 in size was used in the experiments and the watermarked image had 35 dB peak signal-to-noise ratio (PSNR) with no visual artifacts.

Blurring attack: The watermarked image was strongly blurred using a low-pass filter with a Gaussian variance 7 (window size of 15 × 15). Basically, the high-frequency components of the watermarked image were removed after this processing. The result is shown in Figure 18.6, in which the retrieved watermark is very recognizable.

(a) Blurred watermarked image

(b) Retrieved watermark

FIGURE 18.6
Blurring attack.

Median filtering attack: Similarly, the watermarked image was median filtered using a 7 × 7 mask. The results after median filtering attack are shown in Figure 18.7, in which the retrieved watermark is again very recognizable.

Image rescaling: The watermarked image was scaled to one quarter of its original size and upsampled to its original dimensions. We can see from Figure 18.8a that many details were lost. Figure 18.8b shows the retrieved watermark.

JPEG compression attack: The JPEG compression algorithm is one of the attacks that the watermark should be most resistant to. The effect of JPEG compression was examined using a very low quality factor of 5%. Figure 18.9a shows the watermarked image after JPEG compression. Visually, the watermarked image is severely damaged because it contains apparent blocky effects. The watermark shown in Figure 18.9b is the extracted result. It is apparent that the retrieved watermark is very clear.

EZW compression attack [26]: The embedded zero tree (EZW) compression algorithm is another one of the attacks that the watermark should be most resistant to. The watermarked image was attacked by SPIHT compression (a member of the EZW compression family). The compressed result is shown in Figure 18.10a (compression ratio 128:1). It is obvious that the degradation is quite significant. However, the reconstructed watermark shown in Figure 18.10b is again recognizable.

(a) Median filtered watermarked image

(b) Retrieved watermark

FIGURE 18.7
Median filtering attack.

(a) Watermarked image after rescaling

(b) Retrieved watermark

FIGURE 18.8
Rescaling attack.

Jitter attack: Jitter attack [20] is a kind of attack that introduces geometric distortions into a watermarked image. Figure 18.11a shows the result after jitter attack with four pairs of columns deleted and duplicated. It is obvious that the damage caused by this attack is invisible. Figure 18.11b shows the retrieved watermark without introducing any synchronization processing. Again, the retrieved result is still recognizable.

StirMark attack: The StirMark-attacked watermarked image is shown in Figure 18.12a. All the default parameters of StirMark software were used. It is perceived that the watermarked image before and after the attack is very much the same, but it is, in fact, geometrically distorted. These distortions are illustrated in Figure 18.12b using meshes. From the distorted meshes, it is easy to see the power of StirMark. Asynchronization is the reason why many commercially available watermarking algorithms [20] fail under its attack. Figure 18.12c shows the extracted watermark. Figure 18.12d shows the watermarked image attacked by applying StirMark five times. The extracted watermark shown in Figure 18.12e is surprisingly good. Apparently, the proposed relocation technique broke down the effects caused by StirMark attack.

unZign attack: unZign acts like StirMark, and a report comparing them can be found in [20]. Many commercial watermarking algorithms [20] have also failed under this attack.

(a)

(b)

FIGURE 18.9
JPEG attack. (a) Watermarked image attacked by JPEG compression with a quality factor of 5% (without smoothing); (b) watermark retrieved from (a).

(a)

(b)

FIGURE 18.10
EZW attack. (a) Watermarked image degraded by SPIHT with a compression ratio of 128:1 (without smoothing); (b) retrieved watermark.

However, the retrieved results illustrated in Figure 18.13b show that both the Chinese and English texts were well recovered.

Combination attacks: Combination attacks were also conducted to test our approach. The watermarked image after StirMark attack, JPEG compression (5%), and blurring (7×7) is shown in Figure 18.14a, and the retrieved watermark is shown in Figure 18.14b. Figure 18.14a rotated by 180° is shown in Figure 18.14c, and the corresponding retrieved watermark is shown in Figure 18.14d. It is apparent that the extracted watermark is still in good condition.

Collusion: Several watermarks could be inserted, aiming at making the original mark unreadable. Here, five different watermarks were embedded into the same host image, separately, and averaged. The watermarked image after collusion attack is shown in Figure 18.15a. Figure 18.15b shows the retrieved watermark.

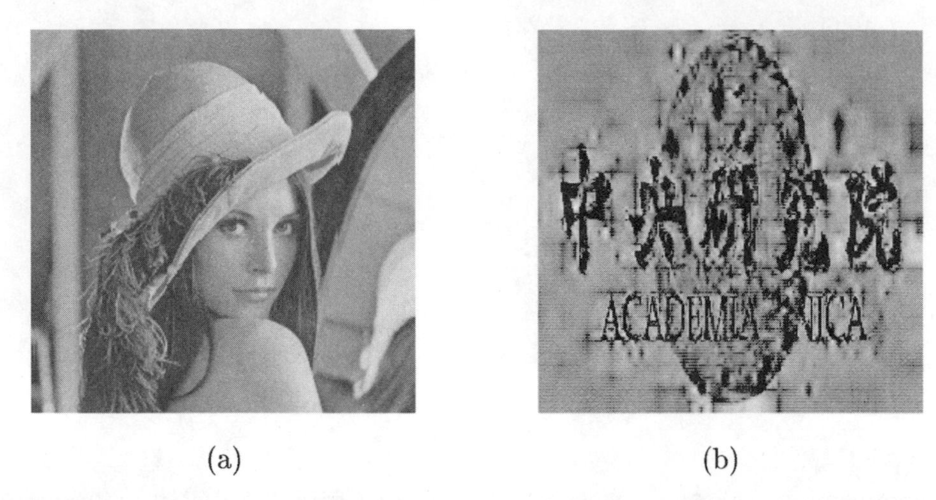

FIGURE 18.11
Jitter attack [four pairs of columns, (90, 200), (50, 150), (120, 110), and (1, 3), were deleted/duplicated]. (a) Jitter-attacked watermarked image; (b) retrieved watermark.

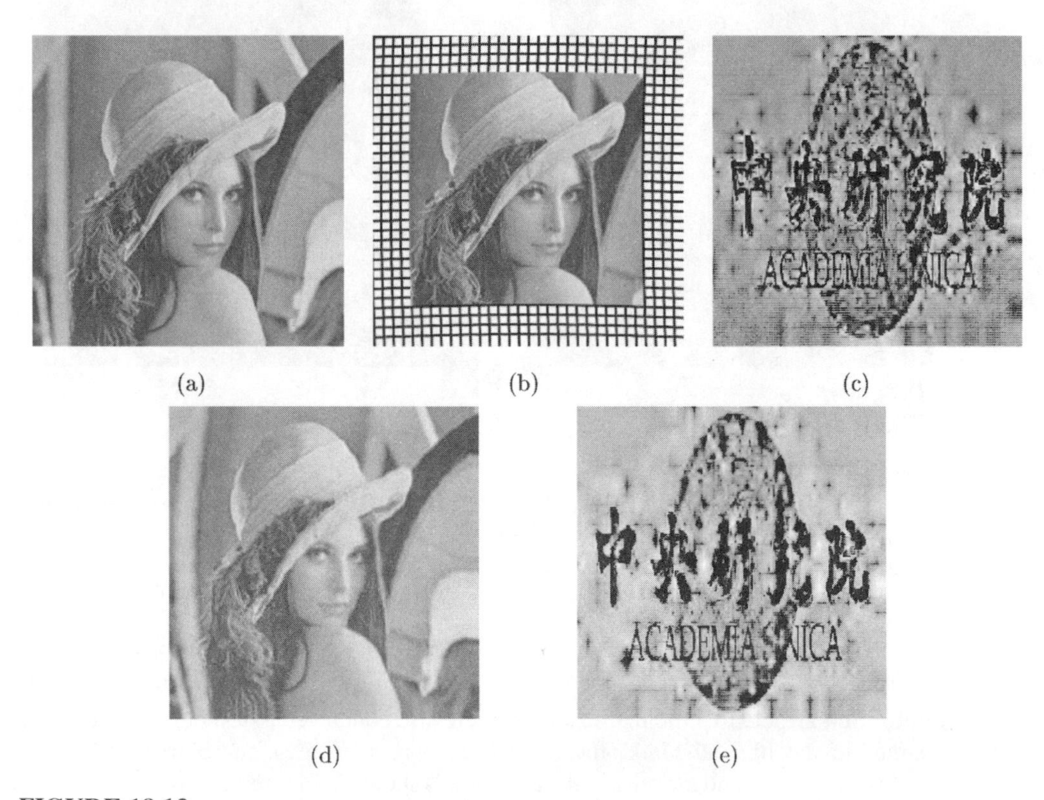

FIGURE 18.12
StirMark attack (all default parameters). (a) StirMark-attacked watermarked image (one time); (b) distorted mesh caused by StirMark attack; (c) watermark retrieved from (a); (d) watermarked image attacked by applying StirMark five times; (e) watermark retrieved from (d).

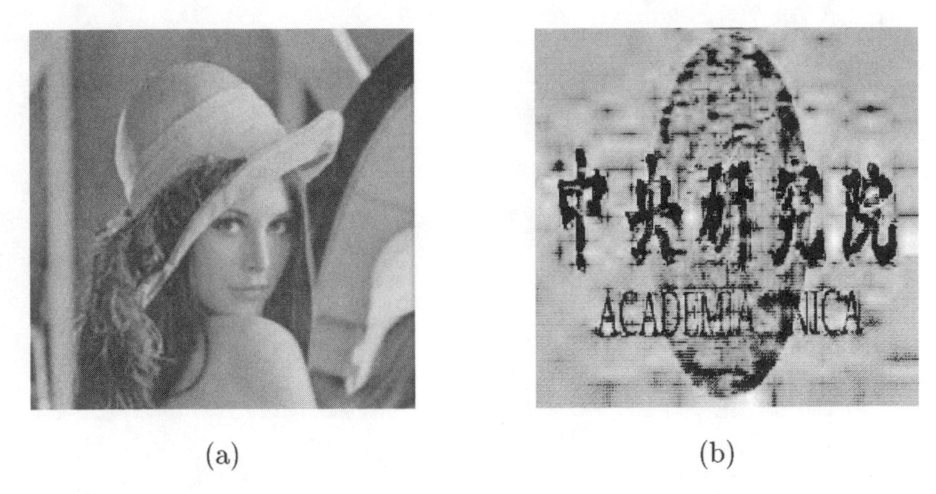

(a) (b)

FIGURE 18.13
unZign attack. (a) unZign-attacked watermarked image; (b) retrieved watermark.

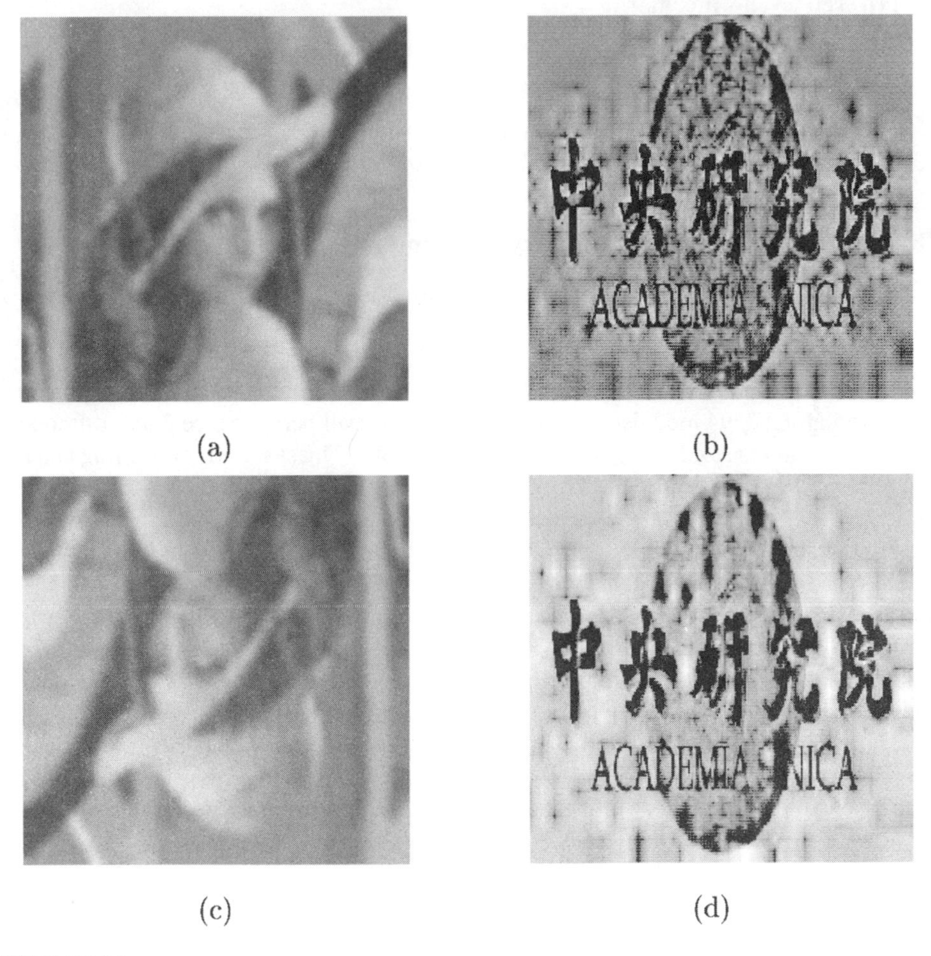

(a) (b)

(c) (d)

FIGURE 18.14
Combination attacks [StirMark + JPEG (5%) + blurring (7×7)]. (a) Attacked watermarked image; (b) watermark retrieved from (a); (c) image in (a) rotated by 180°; (d) watermark retrieved from (c).

(a)

(b)

FIGURE 18.15
Collusion attack (five watermarked images were averaged). (a) Collusion-attacked image; (b) retrieved watermark.

In this experiment, the strength of an attack was always set to be strong enough if the parameters of the attack are available to change. Although the image degradation may be so heavy that it is not accepted in practical applications, the mark is still easily recovered.

18.6.2 Results of Hiding a Binary Watermark

The Rainbow image 256 $\times$ 256 in size was used in this experiment. The contents of the binary watermark were the Chinese text of "ACADEMIA SINICA" with 1497 foreground pixels. The generated watermarked image has 27.23 dB PSNR, but again there are no visual artifacts even when the PSNR value is relatively low. Owing to our double hiding strategy, there are in total 2994 modulated coefficients. The overall performance under different kinds of attacks is summarized in Figure 18.16. The sizes of the masks used for blurring and median filtering were 15 $\times$ 15 and 11 $\times$ 11, respectively. The rescaling attack was carried out in the same way as in the gray-scale watermark case. The variance of the Gaussian noise was 16, and the jitter attack delete/update 5 pairs of columns. The numbers 5% and 128:1 below the JPEG and SPIHT compressions denote the quality factor and the compression ratio, respectively. The numbers 1 and 5 below the StirMark attacks represent the number of times StirMark attack was executed. Figure 18.16 also shows a combination of attacks which include StirMark, flip, and jitter attacks. The normalized correlation value (between 0 and 1) was calculated using the similarity measurement [11]. Higher values represent better matching between the retrieved watermark and the original watermark. The highest correlation value, 1, was obtained for all the attacks described in Figure 18.16. This series of experiments demonstrates that the double hiding scheme is indeed a super mechanism for embedding a binary watermark.

18.7 Conclusion

A robust watermarking scheme has been developed in this chapter to protect digital images. The JND threshold values provided by the human visual model have been introduced to decide

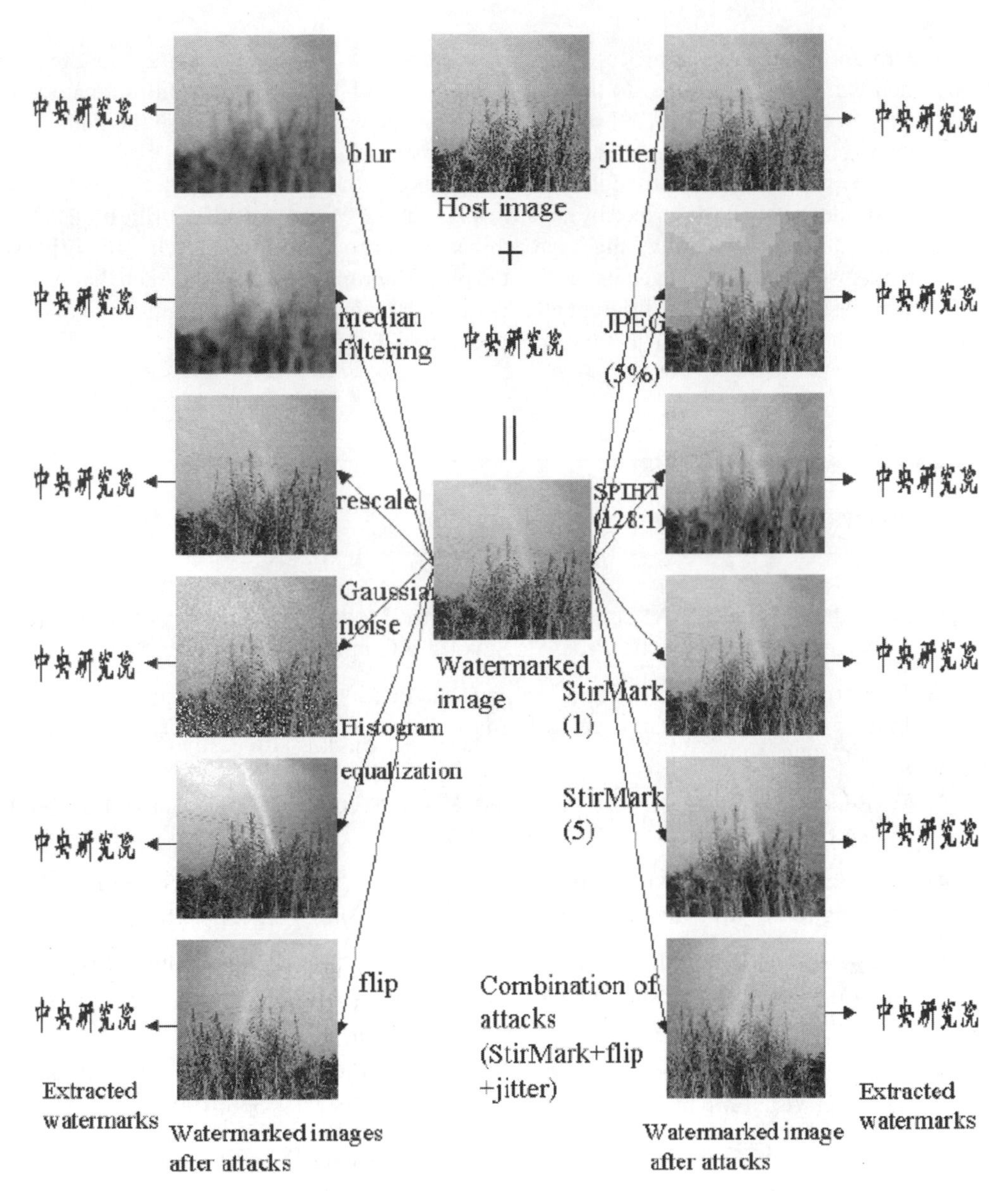

FIGURE 18.16
Performance of our binary watermark hiding/detection under various attacks.

the maximum perceptual capacity allowed to embed a watermark. The proposed scheme has the following characteristics: (1) it uses the contextual information of a gray-scale watermark; (2) it applies multiple modulations to enhance security for a binary watermark hiding; (3) it adapts to the local scalogram of a wavelet transformed host image; (4) it defends against attacks with geometric distortions using a relocation technique; and (5) it authenticates without (indirectly) using the host image. Experiments have demonstrated that our watermarking scheme can achieve not only transparency, but also robustness, and that authentication can be done without the original source.

For a binary watermark considered in this chapter, the foreground pixels in the spatial domain are directly embedded in the space/frequency domain of a host image. If the number of watermark pixels is large or if the number of embedded binary watermarks is more than one, then the capacity required to hide these data should be considered. One solution is to use binary wavelet transform [29] to transform a binary watermark, because the number of foreground pixels will be reduced in the wavelet domain.

In this chapter, the negative modulation strategy has been adopted. In practice, some attacks still cannot adequately be described by negative modulation. We have dealt with the modulation problem, and a novel watermarking approach has been proposed [16, 17]. In [16, 17], two watermarks were embedded in a host image and played complementary roles such that at least one watermark survived after different attacks. This explains why either positive modulation or negative modulation alone is not enough for multimedia protection.

References

[1] A.J. Ahumada and H.A. Peterson, "Luminance-Model-Based DCT Quantization for Color Image Compression," *Proc. SPIE,* vol. 1666, pp. 365–374, 1992.

[2] M. Barni, F. Bartolini, V. Cappellini, and A. Piva, "Copyright Protection of Digital Images by Embedded Unperceivable Marks," *Image and Vision Computing,* vol. 16, pp. 897–906, 1998.

[3] W. Bender, D. Gruhl, N. Morimoto, and A. Lu, "Techniques for Data Hiding," *IBM Systems Journal,* vol. 25, pp. 313–335, 1996.

[4] I.J. Cox, J. Kilian, F.T. Leighton, and T. Shamoon, "Secure Spread Spectrum Water-Marking for Multimedia," *IEEE Trans. Image Processing,* vol. 6, pp. 1673–1687, 1997.

[5] S. Craver, N. Memon, B.-L. Yeo, and M.M. Yeung, "Resolving Rightful Ownerships with Invisible Watermarking Techniques: Limitations, Attacks, and Implications," *IEEE Journal on Selected Areas in Communications,* vol. 16, pp. 573–586, 1998.

[6] J.F. Delaigle, C. De Vleeschouwer, and B. Macq, "Watermarking Algorithms Based on a Human Visual Model," *Signal Processing,* vol. 66, pp. 319–336, 1998.

[7] J. Fridrich, A.C. Baldoza, and R.J. Simard, "Robust Digital Watermarking Based on Key-Dependent Basis Functions," *Second Int. Workshop on Information Hiding,* pp. 143–157, 1998.

[8] J. Fridrich, "Applications of Data Hiding in Digital Images," *Tutorial for the ISPACS Conference,* 1998.

[9] F. Hartung and B. Girod, "Watermarking of Uncompressed and Compressed Video," *Signal Processing,* vol. 66, pp. 283–302, 1998.

[10] F. Hartung, J.K. Su, and B. Girod, "Spread Spectrum Watermarking: Malicious Attacks and Counterattacks," *Proc. SPIE: Security and Watermarking of Multimedia Contents,* vol. 3657, 1999.

[11] C.T. Hsu and J.L. Wu, "Multiresolution Watermarking for Digital Images," *IEEE Trans. CAS II: Analog and Digital Signal Processing,* vol. 45, pp. 1097–1101, 1998.

[12] C.T. Hsu and J.L. Wu, "Hidden Digital Watermarks in Images," *IEEE Trans. Image Processing,* vol. 8, pp. 58–68, 1999.

[13] E. Koch and J. Zhao, "Toward Robust and Hidden Image Copyright Labeling," *Proc. Nonlinear Signal and Image Processing Workshop,* Greece, 1995.

[14] M. Kutter, F. Jordan, and F. Bossen, "Digital Signature of Color Images Using Amplitude Modulation," *Journal of Electronic Imaging,* vol. 7, pp. 326–332, 1998.

[15] S.H. Low and N.F. Maxemchuk, "Performance Comparison of Two Text Marking Methods," *IEEE Journal on Selected Areas in Communications,* vol. 16, pp. 561–572, 1998.

[16] C.S. Lu, H.Y. Mark Liao, S.K. Huang, and C.J. Sze, "Cocktail Watermarking on Images," to appear in *3rd Int. Workshop on Information Hiding,* Dresden, Germany, September 29–October 1, 1999.

[17] C.S. Lu, Y.V. Chen, H.Y. Mark Liao, and C.S. Fuh, "Complementary Watermarks Hiding for Robust Protection of Images Using DCT," to appear in *Int. Symposium on Signal Processing and Intelligent Systems, A Special Session on Computer Vision,* China, 1999. (invited paper)

[18] B.M. Macq and J.J. Quisquater, "Cryptology for Digital TV Broadcasting," *Proceedings of the IEEE,* vol. 83, pp. 944–957, 1995.

[19] H.A. Peterson, "DCT Basis Function Visibility Threshold in RGB Space," *SID Int. Symposium Digest of Technical Papers, Society of Information Display,* pp. 677–680, 1992.

[20] F. Petitcolas and M.G. Kuhn, "StirMark 2.3 Watermark Robustness Testing Software," `http://www.cl.cam.ac.uk/~fapp2/watermarking/stirmark/`, 1998.

[21] F. Petitcolas, R.J. Anderson, and M.G. Kuhn, "Attacks on Copyright Marking Systems," *Second Workshop on Information Hiding,* pp. 218–238, 1998.

[22] F. Petitcolas, R.J. Anderson, and M.G. Kuhn, "Information Hiding: A Survey," to appear in *Proc. IEEE Special Issue on Protection of Multimedia Content,* 1999.

[23] C.I. Podilchuk and W. Zeng, "Image-Adaptive Watermarking Using Visual Models," *IEEE Journal on Selected Areas in Communications,* vol. 16, pp. 525–539, 1998.

[24] L. Qiao and K. Nahrstedt, "Watermarking Schemes and Protocols for Protecting Rightful Ownership and Customer's Rights," *J. Image Comm. and Image Representation,* vol. 9, pp. 194–210, 1998.

[25] J.J.K. Ruanaidh and T. Pun, "Rotation, Scale, and Translation Invariant Spread Spectrum Digital Image Watermarking," *Signal Processing,* vol. 66, pp. 303–318, 1998.

[26] A. Said and W.A. Pearlman, "A New, Fast, and Efficient Image Codec Based on Set Partitioning in Hierarchical Trees," *IEEE Trans. Circuits and Systems for Video Technology,* vol. 6, pp. 243–250, 1996.

[27] S.D. Servetto, C.I. Podilchuk, and K. Ramchandran, "Capacity Issues in Digital Image Watermarking," *5th IEEE Conf. Image Processing,* 1998.

[28] J.M. Shapiro, "Embedded Image Coding Using Zerotrees of Wavelet Coefficients," *IEEE Trans. Signal Processing,* vol. 41, pp. 3445–3462, 1993.

[29] M.D. Swanson and A.H. Tewfik, "A Binary Wavelet Decomposition of Binary Images," *IEEE Trans. Image Processing,* vol. 5, 1996.

[30] M.D. Swanson, B. Zhu, and A.H. Tewfik, "Multiresolution Scene-Based Video Watermarking Using Perceptual Models," *IEEE Journal on Selected Areas in Communications,* vol. 16, pp. 540–550, 1998.

[31] M.D. Swanson, B. Zhu, A.H. Tewfik, and L. Boney, "Robust Audio Watermarking Using Perceptual Masking," *Signal Processing,* vol. 66, pp. 337–356, 1998.

[32] M.D. Swanson, M. Kobayashi, and A.H. Tewfik, "Multimedia Data-Embedding and Watermarking Technologies," *Proc. of the IEEE,* vol. 86, pp. 1064–1087, 1998.

[33] A.Z. Tirkel, C.F. Osborne, and T.E. Hall, "Image and Watermark Registration," *Signal Processing,* vol. 66, pp. 373–384, 1998.

[34] unZign Watermark Removal Software, `http://altern.org/watermark/`, 1997.

[35] G. Voyatzis and I. Pitas, "Digital Image Watermarking Using Mixing Systems," *Computers & Graphics,* vol. 22, pp. 405–416, 1998.

[36] A.B. Watson, "DCT Quantization Matrics Visually Optimized for Individual Images," *Proc. SPIE Conf. Human Vision, Visual Processing, and Digital Display IV,* vol. 1913, pp. 202–216, 1993.

[37] A.B. Watson, G.Y. Yang, J.A. Solomon, and J. Villasenor, "Visibility of Wavelet Quantization Noise," *IEEE Trans. Image Processing,* vol. 6, pp. 1164–1175, 1997.

[38] J. Zhao and E. Koch, "A General Digital Watermarking Model," *Computers & Graphics,* vol. 22, pp. 397–403, 1998.

Chapter 19

Telemedicine: A Multimedia Communication Perspective

Chang Wen Chen and Li Fan

19.1 Introduction

With the rapid advances in computer and information technologies, multimedia communication has brought a new era for health care through the implementation of state-of-the-art telemedicine systems. According to a formal definition recently adopted by the Institute of Medicine, telemedicine can be defined as "the use of electronic information and communication technologies to provide and support health care when distance separates the participants" [1].

The envisioning of health care service performed over a distance first appeared in 1924 in an imaginative cover for the magazine *Radio News* which described a "radio doctor" who could talk with the patient by a live picture through radio links [1]. However, the technology to support such a visionary description, namely television transmission, was not developed until 3 years later, in 1927. According to a recent review, the first reference to telemedicine in the medical literature appeared in 1950 [2], with a description of the transmission of radiological images by telephone over a distance of 24 miles. An interactive practice of telemedicine, as a signature mode currently perceived by many people, began in the 1960s when two-way, closed-circuit, microwave televisions were used for psychiatric consultation by the clinicians at the Nebraska Psychiatric Institute [3]. Although these pioneering efforts have demonstrated both technical and medical feasibilities and received enthusiastic appraisal from the health care recipients of telemedicine [4], the issue of cost-effectiveness was debated at a premature stage by the telemedicine authorities, especially the major funding agencies. The prevailing fear was that as the technologies for telemedicine became more sophisticated, the cost for telemedicine would only increase [3]. Such fear has been proven to the contrary, as many applications of telemedicine are now considered to have potential in reducing health care costs or reducing rates of cost escalation [1]. The rapid advances of modern communication and information technologies, especially multimedia communication technologies in the 1990s, have been a major driving force for the strong revival of telemedicine today.

In this chapter, we will examine recent advances in telemedicine from a multimedia communication perspective. We will first demonstrate the needs for multimedia communication to improve the quality of health care in various telemedicine applications. We will then present examples of telemedicine systems that have adopted multimedia communication over different communication links. These diverse applications illustrate that multimedia systems can be designed to suit a wide variety of health care needs, ranging from teleconsultation to emergency medicine.

19.2 Telemedicine: Need for Multimedia Communication

Notice that there are three essential components in the definition of telemedicine adopted by the Institute of Medicine: (1) information and communication technologies, (2) distance between the participants, and (3) health or medical uses. With a well-designed telemedicine system, improved access to care and cost savings can be achieved by enabling doctors to remotely examine patients. The distance separating the participants prevents doctors from engaging in traditional face-to-face medical practice. However, this also creates the opportunity for information and communication technologies to be integrated with health care services in terms of patient care, medical education, and research. In general, patient care focuses on quality care with minimum cost; education focuses on training future health care professionals and promoting patient and community health awareness; research focuses on new discoveries in diagnostic and therapeutic methods and procedures. In each of these service categories, multimedia communication can facilitate an enabling environment to take full advantage of what the state-of-the-art computing and information technologies are able to offer to overcome the barriers created by distance separating the participants.

In the case of patient care, a telemedicine system should be able to integrate multiple sources of patient data, diagnostic images, and other information to create a virtual environment similar to where the traditional patient care is undertaken. During the last decade, we have witnessed major advances in the development of hospital information systems (HIS) and picture archiving and communication systems (PACS). The integration of both HIS and PACS can provide a full array of information relevant to patient care, including demographics, billing, scheduling, patient history, laboratory reports, related statistics, as well as various diagnostic medical images. With a remote access mode for participants, the interaction between the participants as well as between the participants and the systems of HIS and PACS will undoubtedly need to operate with multiple media of information transmission. To seamlessly integrate multiple media into a coherent stream of operations for a telemedicine system, the development of multimedia communication technology suitable for health care application will become increasingly important. For example, Professor H.K. Huang and his group at the University of California at San Francisco have shown that a networked multimedia system can play an important role for managing medical imaging information suitable for remote access [5]. The information types considered in this system include still 2D or 3D images, video or cine images, image headers, diagnostic reports (text), physicians' dictation (sounds), and graphics. The networked multimedia system capable of integrating multimedia information meets the need to organize and store large amounts of multimodel image data and to complement them with relevant clinical data. Such a system enhances the ability of the participating doctor to simultaneously extract rich information embedded in the multimedia data.

There are two major modes of telemedicine involving patient care: teleconsultation and telediagnosis [6]. Teleconsultation is the interactive sharing of medical images and patient information between the doctor at the location of the patient and one or more medical specialists at remote locations. Figure 19.1 illustrates a typical teleconsultation system. In this case, the primary diagnosis is made by the doctor at the location of the patient, while remote specialists are consulted for a second opinion to help the local doctor arrive at an accurate diagnosis. In addition to video conferencing transmitting synchronized two-way audio and video, networked multimedia communication is very much desired to access HIS and PACS, to share relevant medical images and patient information. The verbal and nonverbal cues are supported through the video conferencing system to mimic a face-to-face conversation. An integration of the networked multimedia system and the video conferencing system is needed to enable a clear and uninterrupted communication among the participants. However, some loss of the

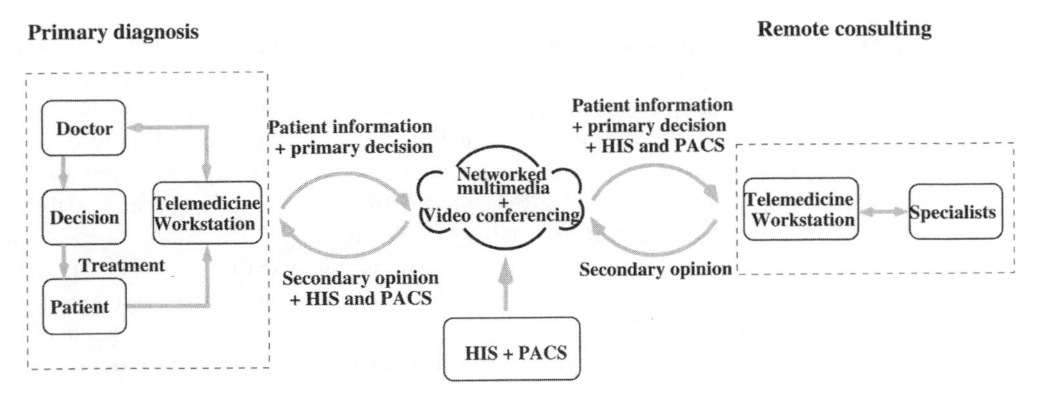

FIGURE 19.1
Illustration of a typical teleconsultation system.

image quality in the case of video conferencing may be acceptable. One successful example of such a telemedicine system is the WAMI (Washington, Alaska, Montana, and Idaho) Rural Telemedicine Network [6]. Telediagnosis, on the other hand, refers to the interactive sharing of medical images and patient information through a telemedicine system, while the primary diagnosis decision is made by the specialists at a remote location. Figure 19.2 illustrates a typical telediagnosis system. To ensure diagnosis accuracy, no significant loss of the image quality is allowed in the process of acquisition, processing, transmission, and display. For synchronous telediagnosis, high communication bandwidth is required to support interactive multimedia data transfer and diagnosis-quality video transmission. For asynchronous telediagnosis, lower communication bandwidth is acceptable because the relevant images, video, audio, text, and graphics are assembled to form an integrated multimedia file to be delivered to the referring physician for off-line diagnosis. In the case of emergency medicine involving a trauma patient, telediagnosis can be employed to reach a time-critical decision on whether or not to evacuate the patient to a central hospital. Such a mode of telediagnosis operation was successfully implemented during the Gulf War by transmitting X-ray computed tomography (CT) images over a satellite teleradiology system to determine whether a wounded soldier could be treated at the battlefield or should be evacuated [7]. In this case, high communication bandwidth was available for telediagnosis operation.

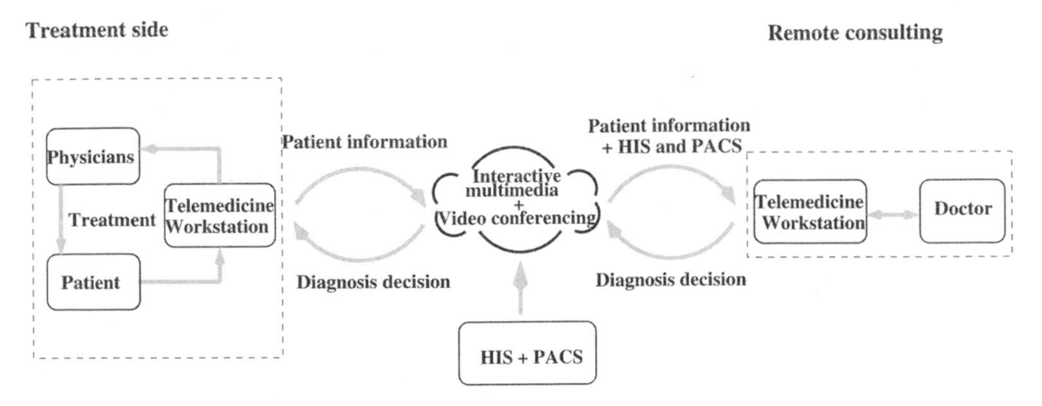

FIGURE 19.2
Illustration of a typical telediagnosis system.

Examples of clinical applications of telemedicine in different medical specialties include teleradiology, telepathology, teledermatology, teleoncology, and telepsychiatry. Among them, teleradiology is a primary image-related application. Teleradiology has been considered a practical cost-effective method of providing professional radiology services to underserved areas for more than 30 years. It has now been widely adopted to provide radiology consultations from a distance. Teleradiology uses medical images acquired from various radiological modalities including X-ray, CT, magnetic resonance imaging (MRI), ultrasound (US), positron emission tomography (PET), single-photon emission-computed tomography (SPECT), and others. Associated with these medical images, relevant patient information in the form of text, graphics, and even voice should also be transmitted for a complete evaluation to reach an accurate clinical decision. Figure 19.3 illustrates a typical teleradiology system. The need for multimedia communication is evident when such a teleradiology system is implemented.

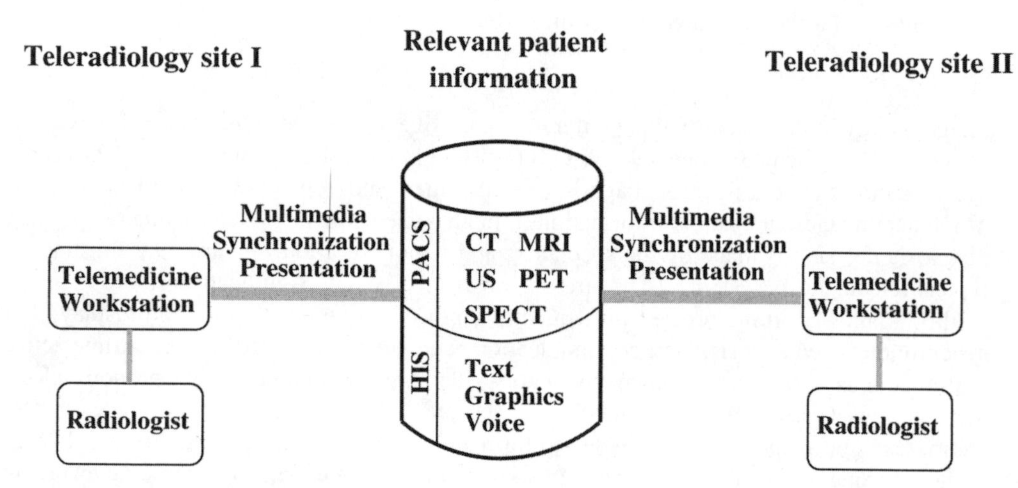

FIGURE 19.3
Illustration of a typical teleradiology system.

In the case of medical education, a telemedicine system generally includes video conferencing with document and image sharing capabilities. The modes of operation for the telemedicine system used for remote medical education include one-to-one mentoring, online lecturing, and off-line medical education. Depending on the mode of operation, such a telemedicine system may use either point-to-point or point-to-multipoint communication. In general, multimedia presentation is desired because the education may involve clinical case study using medical images, video, and patient history data. Similar telemedicine systems can also be designed for public access to community health care resources. With Internet and World Wide Web resources, health care information can be readily obtained for the formal and informal provision of medical advice, and continuing medical education can be implemented at multiple sites with effective multimedia presentations.

In the case of medical research, the telemedicine system can be used to collect patients' data from distinct physical locations and distribute them to multiple sites in order to maximize the utilization of all available data. One prominent application of such a telemedicine system is the research on medical informatics in which distributed processing of multimedia medical information at separate physical sites can be simultaneously executed. Such a mode of operation is also very useful when research on public health is conducted. In general, public health research involves massive and timely information transfer, such as for disease monitoring. For public health research, we expect that a telemedicine system with advanced multimedia communication capability will be able to provide the connectivity needed for mass education on disease prevention and the global network needed for disease monitoring.

In summary, the required multimedia communication infrastructure for telemedicine depends on the type of telemedicine applications. However, the need for advanced multimedia technology is clear. It is the enhanced multimedia communication capability that distinguishes the present state-of-the-art telemedicine systems from the early vision of "radio doctor" consisting of only live pictures of the doctor and the patient.

19.3 Telemedicine over Various Multimedia Communication Links

There have been numerous applications in telemedicine, both clinical and nonclinical, as we have discussed in the previous section. Although the capability of multimedia communication is desired in nearly all telemedicine applications, the required communication capacity in terms of bandwidth, power, mobility, and network management can be quite different from one application to another. Traditionally, plain old telephone service (POTS) has been the primary network for telecommunications applications. Early telemedicine applications started with the POTS in which the transmission of radiological images by telephone over a distance of 24 miles was reported in 1950 [2]. However, modern telemedicine applications have recently moved quickly toward making use of advanced high-performance communication links, such as integrated service digital network (ISDN), asynchronous transfer mode (ATM), the Internet, and wireless mobile systems. In this section, we discuss how different communication links can be used in various telemedicine applications to enhance their multimedia communication capabilities.

19.3.1 Telemedicine via ISDN

ISDN is essentially a high-speed digital telephony service that carries simultaneous transmission of voice, data, video, image, text, and graphics information over an existing telephone system. It originally emerged as a viable digital communication technology in the early 1980s. However, its limited coverage, high tariff structure, and lack of standards stunted its growth early on [8]. This situation changed in the 1990s with the Internet revolution, which increased demands for more bandwidth, decreasing hardware adapter costs, and multiple services. In North America, efforts were made in 1992 to establish nationwide ISDN systems to interconnect the major ISDN switches around the United States and Canada. In 1996, ISDN installations almost doubled from 450,000 to 800,000, and they were expected to reach 2,000,000 lines by the year 1999.

ISDN provides a wide range of services using a limited set of connection types and multipurpose user–network interface arrangements. It is intended to be a single worldwide public telecommunication network to replace the existing public telecommunication networks which are currently not totally compatible among various countries. There are two major types of ISDNs, categorized by capacity: narrowband ISDN and broadband ISDN (B-ISDN). Narrowband ISDN is based on the digital 64-Kbps telephone channel and is therefore primarily a circuit-switching network supported by frame relay protocols. A transmission rate ranging from 64 Kbps to 1.544 Mbps can be provided by the narrowband ISDN. Services offered by narrowband ISDN include (1) speech; (2) 3.1-KHz audio; (3) 3-KHz audio; (4) high-speed end-to-end digital channels at a rate between the basic rate of 64 Kbps and the super-rate of 384 Kbps; and (5) packet-mode transmission. B-ISDN provides very high data transmission rates on the order of 100s Mbps with primarily a packet-switching model [9]. In 1988, the International Telecommunication Union (ITU) defined the ATM as the technology for B-ISDN to support the packet-switching mode. The transmission rate of B-ISDN ranges from 44.736 Mbps, or DS3 in the digital signal hierarchy, to 2.48832 Gbps, or OC-48 in the optical carrier hierarchy in synchronous optical networks (SONETs). A variety of interactive and distribution services can be offered by B-ISDN. Such services include (1) broadband video telephony and video conferencing; (2) video surveillance; (3) high-speed file transfer; (4) video and document retrieval service; (5) television distribution; and potentially many other services.

The characteristics of ISDN to provide multimedia and interactive services naturally led to its application in telemedicine. Figure 19.4 illustrates an ISDN-based telemedicine system in which transmission of multiple media data is desired and interactivity of the communication is required. In addition, the current ISDN systems are fundamentally switch-based wide area networking services. Such switch-based operations are more suitable for telemedicine because many of its applications need network resources with guaranteed network bandwidth and quality of service (QoS). In general, switch-based ISDNs, especially the B-ISDN, are able to meet the requirements of bandwidth, latency, and jitter for multimedia communications in many telemedicine applications. From a practical point of view, the advantages of ISDN are immediately ready in many areas, the telecommunications equipment and line rates are inexpensive, and there are protocol supports among existing computing hardware and software [10]. Another characteristic of ISDN is its fast establishment of bandwidth for multimedia communication within a very short call setup time. This matches well with the nature of the telemedicine applications in which the need is immediate and the connection lasts for a relatively short period of time. In addition, the end-to-end digital dial-up circuit can transcend geographical or national boundaries. Therefore, an ISDN connection can offer automatic translation between European and U.S. standards [11].

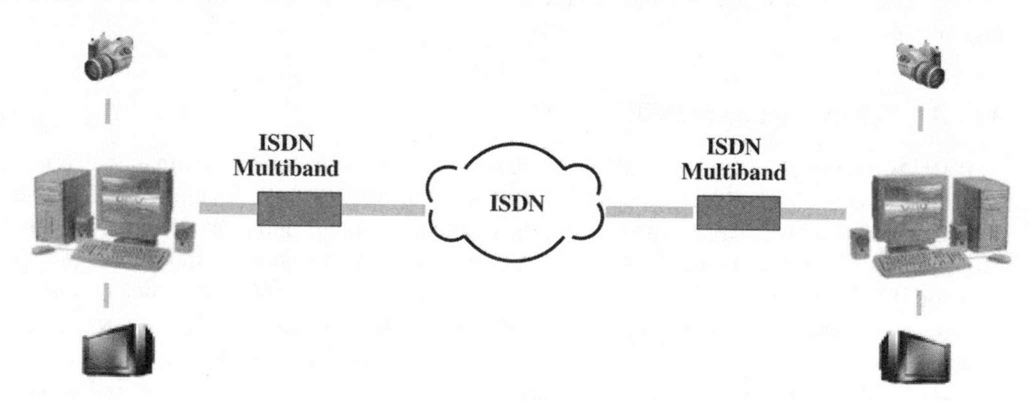

FIGURE 19.4
Illustration of a typical ISDN-based telemedicine system.

Because of its worldwide deployment, the ISDN has also been used to implement telemedicine applications outside Europe and North America. A telemedicine project via ISDN has been successfully implemented in Taiwan, China [12]. In this project, a telemedicine link was established between the Tri-Service General Hospital (TSGH) and the Lian-Jian County Hospital (LJCH), Taiwan. Lian-Jian County consists of several islands located 140 miles northwest of Taiwan island with a population of 4000. However, the LJCH has only five physicians, without residence training. Therefore, the telemedicine system was expected to provide better health care services to the residents in Lian-Jian County while reducing unnecessary patient transfer. On-the-job training of county hospital physicians has also been provided.

The telemedicine system consists of two teleconsultation stations located at TSGH and LJCH, respectively, and a multimedia electronic medical record system at TSGH for storing the multimedia medical records of patients. Each teleconsultation station is equipped with a video conferencing system, a high-resolution teleradiology workstation for displaying multimedia electronic medical records, a film digitizer for capturing medical images, and document cameras for online hard-copy documents capture. The two sites are linked by six basic rate interface (BRI) ISDNs with a total bandwidth of 768 Kbps to transfer images and real-time audio–video data.

The TSGH–LJCH telecommunication system was in operation in May 1997. Between May and October 1997, 124 cases were successfully teleconsulted. Assessments show that the telemedicine system achieved the previously set goals. Surveys were also conducted to investigate how people would accept this new health care technology. The results show that 81% of doctors at TSGH and 100% of doctors and 85% of patients at LJCH think the teleconsultation services are valuable and should be continued.

It is evident that current ISDN systems offering integrated multimedia communication are suitable for many telemedicine applications. However, current bandwidth limitations confine the applications to mainly teleconsultation over video conferencing format. With large-scale deployment of the B-ISDN systems worldwide in the future [13], we expect a much improved multimedia communication quality in telemedicine applications that are based on ISDN systems.

19.3.2 Medical Image Transmission via ATM

The bandwidth limitation of ISDNs prohibits the transmission of larger size medical images. Even at a primary rate of 1.92 Mbps, transfer of medical images of 250 Mb over ISDN would require 130 s without compression and 6.5 s with 20:1 compression. Such applications of medical image transfer would call for another switch-based networking technology, the ATM.

In general, ATM is a fast-packet switching mode that allows asynchronous operation between the sender clock and the receiver clock. It takes advantage of the ultra-high-speed fibers that provide low bit error rates (BERs) and high switching rates. ATM has been selected by the ITU as the switching technology, or the transfer mode, for the future B-ISDN, which is intended to become the universal network to transport multimedia information at a very high data rate. ATM is regarded as the technology of the 21st century because of it ability to handle future expanded multimedia services.

The advantages of ATM include higher bandwidth, statistical multiplexing, guaranteed QoS with minimal latency and jitter, flexible channel bandwidth allocation, and seamless integration of local area networks (LANs) and global wide area networks (WANs) [14]. The higher bandwidth of ATM is sufficient to support the entire range of telemedicine applications, including the transfer of large medical images. Figure 19.5 shows a typical ATM-based telemedicine system used to transfer massive medical images. For the transfer of the same size (250 Mb) medical image over ATM at the transmission rate of 155 Mbps, only 1.6 s without compression and 0.08 s with 20:1 compression are required. Statistical multiplexing can integrate various types of service data, such as video, audio, image, and patient data, so that the transport cost can be reduced and the bandwidth can be dynamically allocated according to the statistical measures of the network traffic. Such statistical multiplexing offers the capability to allow a connection to deliver a higher bandwidth only when it is needed and is very much suitable for the bursty nature of transferring medical images. The ATM's guaranteed QoS and minimal latency and jitter are significant parameters when establishing a telemedicine system, especially when interactive services such as teleconsultation and remote monitoring are desired. However, the disadvantages of ATM-based telemedicine systems are the current high cost and scarcity of ATM equipment and deployment, especially in rural areas. We expect these costs to decrease steadily as the ATM gains more user acceptance and the ATM market increases.

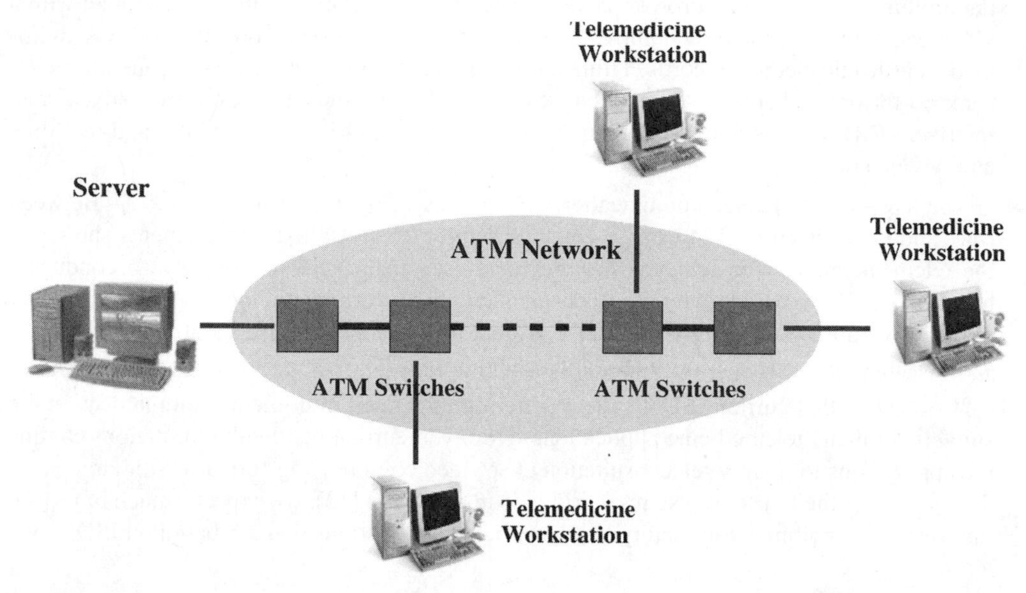

FIGURE 19.5
Illustration of a typical ATM-based telemedicine system.

One successful example of medical image transmission via ATM is the European High-Performance Information Infrastructure in Medicine no. B3014 (HIM3) project started in March 1996 and completed in July 1997 [15]. This work aimed at testing the medical usability of the European ATM network for DICOM image transmission and telediagnosis. This cooperative project was carried out by the Department of Radiology, University of Pisa, Italy, and St-Luc University Hospital, Brussels, Belgium. The Pisa site was connected to the Italian ATM pilot and the St-Luc University Hospital was connected to the Belgium ATM network. A link between the two sites was established via the international connections provided by the European JAMES project.

DICOM refers to the digital imaging and communication in medicine standard developed mainly by the American College of Radiology (ACR) and the National Electrical Manufacturers Association (NEMA) in the U.S., with contributions from standardization organizations of Europe and Asia. The standard allows the exchange of medical images and related information between systems from different manufacturers. In the project reported in [15], the use of DICOM was limited to remote file transfer from image servers accessed via an ATM backbone. Users could select and transfer medical images to their own DICOM-compatible viewing stations for study. The project also included interactive telediagnosis using a multi-platform telemedicine package with participation by radiologists in both hospitals. It was concluded that such an ATM-based telemedicine project was successful from both a technical and a medical point of view. This project also illustrated that simultaneous multimedia interaction with huge amounts of data transmission can be implemented with ATM technology.

19.3.3 Telemedicine via the Internet

The communication links through ISDN and ATM offer switch-based networking for telemedicine applications. For many telemedicine applications, the guaranteed network bandwidth and QoS are critical. However, switches are fundamentally exclusive, connecting operations that are efficient in terms of network resource sharing. For some telemedicine applications in which exclusive connection between the participants can be compromised, the communication links can be established via the routed networks. In fact, most wide-area data networks today are routed networks. One important characteristic of the routed network is its capability to work at a high level in the protocol hierarchy and efficiently exchange packets of information between networks of similar or different architecture. Such capability enables efficient sharing of the network resources.

One giant routed network today is the Internet, a worldwide system of computer networks, or a global network of networks. The Internet began as a project of the Advanced Research Projects Agency (ARPA) of the U.S. Department of Defense in 1969 and was therefore first known as ARPANet. The original aim was to link scientists working on defense research projects around the country. During the 1980s, the National Science Foundation (NSF) took over responsibility for the project and extended the network to include major universities and research sites. Today, the Internet is a public, cooperative, and self-sustaining facility accessible to hundreds of millions of people worldwide — the majority of countries in the world are linked in some way to the Internet.

The basic communication protocol of the Internet is the transmission control protocol/internet protocol (TCP/IP), a two-layered program. The higher layer of TCP/IP is the transmission control protocol. It manages the assembling of a message into small packets that are transmitted over the Internet and the reassembling of the received packets into the original message. The lower level is the Internet protocol, which handles the address part of each packet so that it can be transmitted to the right destination. Therefore, a message can be reassembled correctly even if the packets are routed differently. Some higher protocols based on TCP/IP are (1) the World Wide Web (HTTP) for multimedia information; (2) Gopher (GOPHER) for hierarchical menu display; (3) file transfer protocol (FTP) for downloading files; (4) remote login (TELNET) to access existing databases; (5) usenet newsgroups (NNTP) for public discussions; and (6) electronic mail (SMTP) for personal mail correspondence.

With its worldwide connection and shared network resources, the Internet is having a tremendous impact on the development of telemedicine systems. There are several advantages in implementing telemedicine applications via the Internet. First, the cost of implementing a telemedicine application via the Internet can be minimal because communication links can make use of existing public telecommunication networks. Second, the capability for universal user interface through any Internet service provider enables access from all over the world. Third, because WWW browsers are supported by nearly all types of computer systems, including PCs, Macintoshes, and workstations, information can be accessed independent of the platform of the users. Moreover, the WWW supports multimedia information exchange, including audio, video, images, and text, which can be easily integrated with HIS and PACS for various telemedicine applications. Figure 19.6 illustrates a typical telemedicine system based on the Internet.

A successful project using the Internet and the WWW to support telemedicine with interactive medical information exchange is reported in [16]. The system, based on Java, was developed by a group of Chinese researchers and is able to provide several basic Java tools to meet the requirements of desired medical applications. It consists of a file manager, an image tool, a bulletin board, and a point-to-point digital audio tool. The file manager manages all medical images stored on the WWW information server. The image tool displays the medical image downloaded from the WWW server and establishes multipoint network connections with other clients to provide interactive functionality. The drawing action of one physician on the image can be displayed on all connected clients' screens immediately. The bulletin board is a multipoint board on which a physician can consult with other physicians and send back

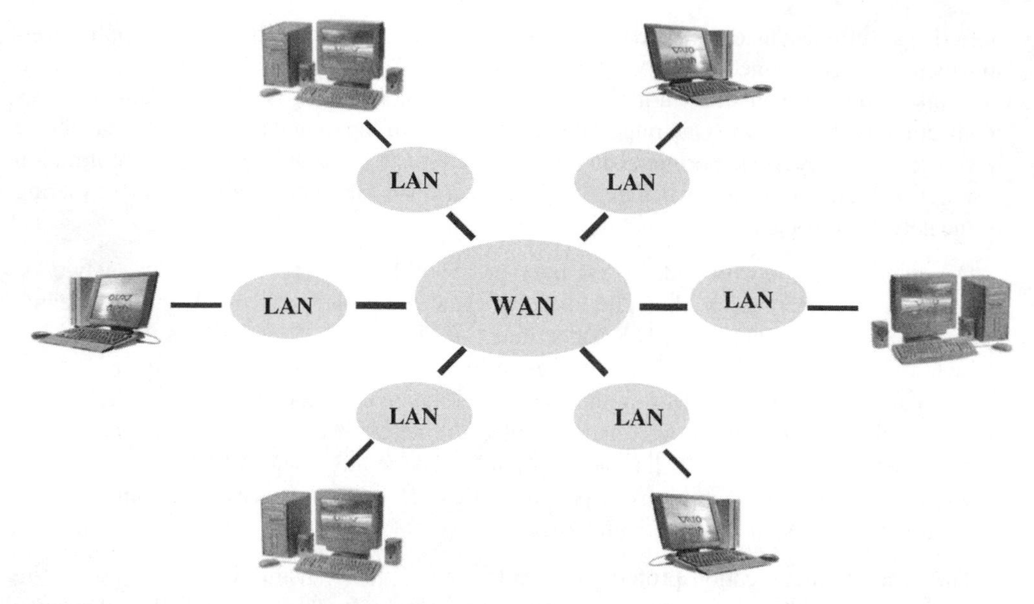

FIGURE 19.6
Illustration of a typical Internet-based telemedicine system.

the diagnosis in plain text format. The point-to-point digital audio tool enables two physicians to communicate directly by voice.

The designed telemedicine system was implemented on a LAN connected to the campus network of Tsinghua University, China. The backbone of the LAN is a 10-Mbps Ethernet thin cable. PCs using Windows NT and Windows 95, as well as a Sun workstation using a Unix operating system, were linked together as clients. Unlike many other systems designed for teleconsultation using specific protocols, this system provides a hardware-independent platform for physicians to interact with one another and access medical information over the WWW. With the explosive growth of the Internet, we expect to witness a entirely new array of telemedicine applications that make full use of the continuously improving capacity of the Internet in terms of backbone hardware, communication links, protocol, and new software.

19.3.4 Telemedicine via Mobile Wireless Communication

Access to communication and computer networks has largely been limited to wired links. As a result, most telemedicine applications discussed in the previous sections have been implemented through wired communication links. However, the wire link infrastructure may not be possible in some medical emergency situations or on the battlefield. A natural extension of the desired telemedicine services to these applications would be to make use of wireless communication links encompassing mobile or portable radio systems. Historically, mobile wireless systems were largely dominated by military and paramilitary users. Recently, with the rapid development in VLSI, computer and information technologies, mobile wireless communication systems have become increasingly popular in civil applications. Two ready examples are the cordless telephone and the cellular phone.

In contrast to wired communications that rely on the existing link infrastructure, wireless communication is able to provide universal and ubiquitous *anywhere, anytime* access to remote locations. Such telecommunication technology is especially favorable when users are in moving vehicles or in disaster situations. Various technologies are used to support this wireless communication. One widely adopted technology is the code division multiple access (CDMA) technique, which uses frequency spreading [17]. After digitizing the data, CDMA spreads it out over the entire available bandwidth. Multiple calls are overlaid on the channel, with each assigned a unique sequence code. One prominent characteristic of CDMA is the privacy ensured by code assignment. It is also robust against impulse noise and other electromagnetic interference. CDMA has been successfully adopted in wireless LANs, cellular telephone systems, and mobile satellite communications. Cellular technology has been used in mobile telephone systems. It uses a microwave frequency with the concept of frequency reuse, which allows a radio frequency to be reused outside the current coverage area. By optimizing the transmit power and the frequency reuse assignment, the limited frequency can be used to cover broader areas and serve numerous customers. Mobile satellite communication has also been making rapid progress to serve remote areas where neither wired links nor cellular telephones can be deployed. It provides low- or medium-speed data transmission rates to a large area covered by the satellite. At the mobile receiver end, directional antennas are generally equipped for intended communication.

Although mobile wireless communication, compared with wireline networks, has some limitations, such as lower transmission speed due to the limited spectrum, the universal and ubiquitous access capability makes it extremely valuable for many telemedicine applications that need immediate connection to central hospitals and mobile access to medical databases. The most attractive characteristic of wireless communication is its inherent ability to establish communication links in moving vehicles, disaster situations, and battlefield environments.

An early success of a telemedicine system via mobile satellite communication (MSC) was reported in [18]. Figure 19.7 illustrates such a wireless telemedicine system. The system was established in Japan through cooperation among the Communication Research Laboratory of the Ministry of Post and Telecommunications, the Electronic Navigation Research Institute of the Ministry of Transport, and the National Space Development Agency of Japan. The telemedicine system includes the three-axis geostationary satellite, ETS-V, a fixed station providing basic health care services located in the Kashima ground station of the Communication Research Laboratory, and a moving station with patients either on a fishery training ship or on a Boeing 747 jet cargo plane.

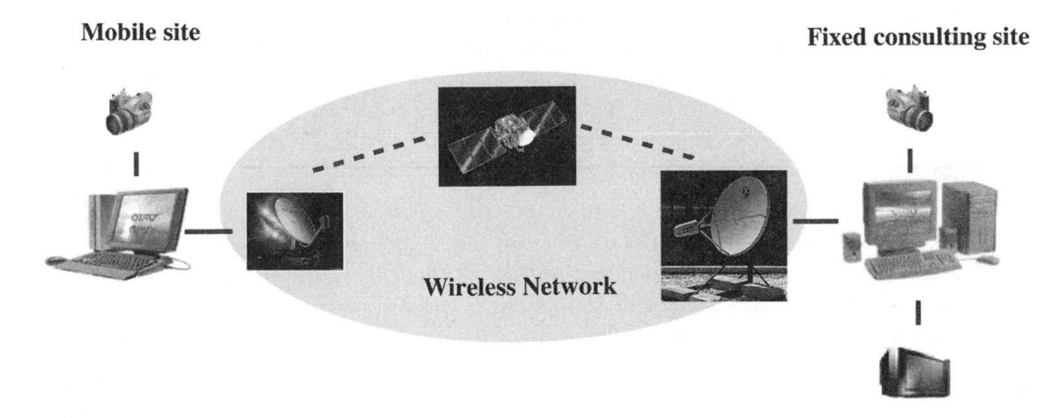

FIGURE 19.7
Illustration of a satellite-based wireless telemedicine system.

The system was capable of multimedia communication, transmitting color video images, audio signals, ecocardiograms (ECGs), and blood pressures simultaneously from the mobile station through the satellite to the ground station. The system was also able to transmit audio signals and error control signals to the mobile station in full duplex mode. To ensure a reliable transmission of vital medical information in the inherent error-prone wireless environment, the system adopted error control techniques to protect the ECG and blood pressure signals. In particular, an automatic repeat request (ARQ) has been applied to ECG signals and forward error correction (FEC) has been applied to blood pressure signals. Experimental results show that telemedicine via mobile satellite communication is feasible and may have a significant implication on health care services through mobile and remote access.

Another fine example of mobile wireless communication in telemedicine is the mobile medical database approach for battlefield environments proposed in [19]. The proposed mobile system enables medical personnel to treat a soldier in the field with the capability of real-time, online access to medical information databases that support the care of the individual injured soldiers. With mobile wireless access, the amount of evacuation or patient movement can be reduced.

Many telemedicine applications based on mobile wireless communication can be envisioned. One example of such applications is emergency medicine in a moving vehicle, such as an aircraft, ship, or ambulance. The treatment of a stroke or other severe injury by the onboard medical personnel may be greatly enhanced with a live telemedicine system that connects the vehicle with a medical specialist. Another example is emergency medicine in a disaster area. In the case of an earthquake or flood, ground communication links may well be in disorder. In these cases, emergency medicine may have to rely on mobile wireless communication for the rescue members to receive pertinent instructions from medical specialists to effectively select the most serious cases for treatment. In summary, mobile wireless communication certainly is able to provide another importance dimension to expand telemedicine services to situations where wired links are beyond reach.

19.4 Conclusion

We have discussed various telemedicine applications from the multimedia communication perspective. Rapid advances in computer, information, and communication technologies have enabled the development of high-performance multimedia communication systems. With enhanced multimedia communication capability, telemedicine systems are able to offer many health care services that could only be dreamed about just a few years ago. With mobile wireless communication booming over the entire world, universal and ubiquitous access to a global telemedicine system will soon become a reality.

Although the great potential of telemedicine will undoubtedly be realized with continued advances in computer, information, and communication technologies, great challenges remain. Many of these challenges are dependent on factors other than the technologies supporting telemedicine. They include the lack of a comprehensive study on cost-effectiveness, the lack of standards for telemedicine practice, and the obstacles presented by the human factor and public policies. Only after these nontechnological issues are also duly resolved can telemedicine achieve its maximum potential.

References

[1] M.J. Field, *Telemedicine: A Guide to Assessing Telecommunications in Health Care.* Washington, D.C.: National Academy Press, 1996.

[2] K.M. Zundel, "Telemedicine: History, applications, and impact on librarianship," *Bulletin of the Medical Library Association,* vol. 84, no. 1, pp. 71–79, 1996.

[3] R.L. Bashshur, P.A. Armstrong, and Z.I. Youssef, *Telemedicine: Explorations in the Use of Telecommunications in Health Care.* Springfield, IL: Charles C. Thomas, 1975.

[4] R. Allan, "Coming: The era of telemedicine," *IEEE Spectrum*, vol. 7, pp. 30–35, December 1976.

[5] S.T. Wong and H. Huang, "Networked multimedia for medical imaging," *Multimedia in Medicine,* pp. 24–35, April–June 1997.

[6] J.E. Cabral, Jr. and Y. Kim, "Multimedia system for telemedicine and their communications requirements," *IEEE Communications Magazine,* pp. 20–27, July 1996.

[7] M.A. Cawthon et al., "Preliminary assessment of computed tomography and satellite teleradiology from Operation Desert Storm," *Invent. Radiol.*, vol. 26, pp. 854–857, 1991.

[8] C. Dhawan, *Remote Access Networks.* McGraw-Hill, New York, 1998.

[9] S.V. Ahamed and V.B. Lawrence, *Intelligent Broadband Multimedia Networks.* Kluwer Academic Publishers, 1997.

[10] S. Akselsen, A. Eidsvik, and T. Fokow, "Telemedicine and ISDN," *IEEE Communication Magazine,* vol. 31, pp. 46–51, 1993.

[11] I. McClelland, K. Adamson, and N. Black, "Telemedicine: ISDN & ATM — the future?," *Annual International Conference of the IEEE Engineering in Medicine and Biology — Proceedings,* vol. 17, pp. 763–764, 1995.

[12] T.-K. Wu, J.-L. Liu, H.-J. Tschai, Y.-H. Lee, and H.-T. Leu, "An ISDN-based telemedicine system," *Journal of Digital Imaging,* vol. 11, pp. 93–95, 1998.

[13] G. Pereira, "Singapore pushes ISDN," *The Institute, IEEE,* December 1990.

[14] P. Handel, M. Huber, and S. Schroder, *ATM Networks: Concepts, Protocols, Applications.* Addison-Wesley, Reading, MA, 1993.

[15] E. Neri, J.-P. Thiran, et al., "Interactive DICOM image transmission and telediagnosis over the European ATM network," *IEEE Trans. on Information Technology in Biomedicine,* vol. 2, no. 1, pp. 35–38, 1998.

[16] J. Bai, Y. Zhang, and B. Dai, "Design and development of an interactive medical teleconsultation system over the World Wide Web," *IEEE Trans. on Information Technology in Biomedicine,* vol. 2, no. 2, pp. 74–79, 1998.

[17] M.D. Yacoub, *Foundations of Mobile Radio Engineering.* CRC Press, Boca Raton, FL, 1993.

[18] H. Murakami, K. Shimizu, K. Yamamoto, T. Mikami, N. Hoshimiya, and K. Konodo, "Telemedicine using mobile satellite communication," *IEEE Trans. on Biomedical Engineering,* vol. 41, no. 5, pp. 488–497, 1994.

[19] O. Bukhres, M. Mossman, and S. Morton, "Mobile medical database approach for battlefield environments," *Australian Computer Journal,* vol. 30, pp. 87–95, 1994.

Index